Molecular Driving Forces

Statistical Thermodynamics in Chemistry and Biology

Molecular Driving Forces

Statistical Thermodynamics in Chemistry and Biology

Ken A. Dill   **Sarina Bromberg**

With the assistance of Dirk Stigter on the Electrostatics chapters

GS Garland Science
Taylor & Francis Group

NEW YORK AND LONDON

Figure Acknowledgements

The following figures are gratefully used with permission: 22.6, 22.8a, 28.9, 29.3, 29.4, 29.6a, 29.6b, 29.9, 29.10, 29.16 and 32.4 American Association for the Advancement of Science. 10.5, 11.14, 19.6, 22.12, 22.13, 25.10, 25.26, 26.5, 28.18a, 28.18b, 32.7 and 32.9 Academic Press. 9.1, 9.2, 9.4, 19.7, 28.1, 30.1b, 25.20, 25.27, 26.14, 29.11, 30.5, 30.10, 31.11, 33.7, 33.10 and 33.12 American Chemical Society. 25.17 and 11.1 WA Benjamin. 30.6 American Institute of Physics. 14.8 Butterworth Heinemann. 27.4b and 11.13 Cambridge University Press. 8.12, 18.8, 25.13, 26.7, 32.3 and 32.12 Oxford University Press, Inc. 11.16 Cold Spring Harbor University Press. 1.16 Columbia University Press. 4.1 and 31.3 Cornell University Press. 30.13 Chemical Rubber Company. 9.5 Deutsche Bunsen-Gesellschaft für Physikalische Chemie. 13.3, 16.7, 23.3 and 29.13 Dover Publications. 26.17a and 32.10 Elsevier. 18.9, 18.10, 25.28a, 26.15, 27.6 and 27.16 Springer-Verlag. 10.10 and 13.5 McGraw Hill. 18.5, 22.8a, 27.11 and 27.12 MIT Press. 26.17b American Physical Society. 19.9 Portland Press. 10.6 and 25.9 Pearson Education, Inc. 29.18, 18.9 and 18.13 Princeton University Press. 25.28b and 29.8 The Royal Society of Chemistry. 28.2 and 28.13 University Science Books. 28.23 and 32.6 WH Freeman. 9.8, 11.2, 11.12, 16.11, 18.14, 19.8, 21.7, 23.6, 24.3, 25.2, 25.15, 25.21, 26.3, 27.4a, 27.18, 30.2, 31.2 and 31.8 John Wiley & Sons. 19.2 and 19.5 Reinhold Publishing Corp. 9.6 Munks Gaard. 25.4, 26.9, 27.5 and 30.7 National Academy of Sciences USA. 16.5 and 29.14 Plenum Press. 28.7 and 28.19a American Society for Biochemistry & Molecular Biology.

Table Acknowledgements

The following tables are gratefully used with permission: 18.2, 18.3 American Chemical Society. 20.2 and 30.3 Dover Publications. 26.1 Oxford University Press. 11.3, 13.1 and 16.1 Pearson Education. 8.2 Prentice-Hall. 29.1 The Royal Society of Chemistry. 9.1, 11.2, 11.4, 14.2, 18.1 and 25.6 John Wiley & Sons. 11.1 W. W. Norton & Company, Inc. 30.1 and 30.7 National Academy of Sciences USA. 22.3 Sinauer.

10-digit ISBN 0 8153 2051 5
13-digit ISBN 978 0 8153 2051 7

Library of Congress Cataloging-in-Publication Data

Dill, Ken A.

  Molecular driving forces: statistical thermodynamics in chemistry and biology / Ken A. Dill, Sarina Bromberg
    p. cm.
  Includes bibliographical references and index.
  ISBN 0-8153-2051-5
  1. Statistical thermodynamics. I. Bromberg, Sarina. II. Title.

QC311.5 .D55 2002
536'.7–dc21   2001053202

Published by Garland Science, a member of the Taylor & Francis Group, an informa business
270 Madison Avenue, New York, NY 10016, USA and
2 Park Square, Milton Park, Abingdon, Oxon, OX14 4RN, UK

Printed in the United States of America
15 14 13 12 11 10 9 8 7 6 5 4 3

About the Authors

Ken A. Dill is Professor of Pharmaceutical Chemistry and Biophysics at the University of California, San Francisco. He received his undergraduate training at the Massachusetts Institute of Technology, his PhD from the University of California, San Diego, and did postdoctoral work at Stanford. A researcher in biopolymer statistical mechanics and protein folding, he has been the President of the Biophysical Society and received the Hans Neurath Award from the Protein Society in 1998.

Sarina Bromberg received her BFA at the Cooper Union for the Advancement of Science and Art, her PhD in molecular biophysics from Wesleyan University, and her postdoctoral training at the University of California, San Francisco. She writes, edits and illustrates scientific textbooks.

Dirk Stigter received his PhD from the State University of Utrecht, The Netherlands, and did postdoctoral work at the University of Oregon, Eugene. His research is in colloid chemistry, detergent micelles, and DNA.

Dedicated to Austin, Peggy, Jim, Jolanda, Tyler, and Ryan

Contents

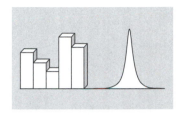

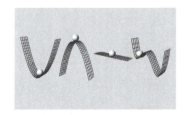

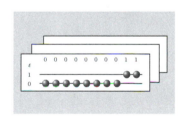

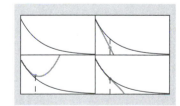

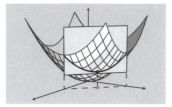

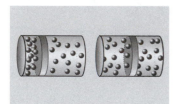
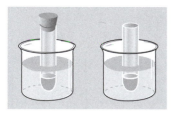

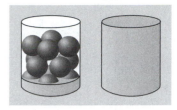

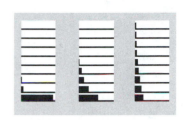

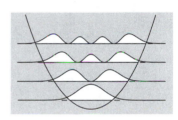

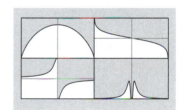

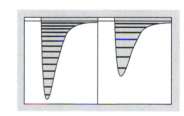

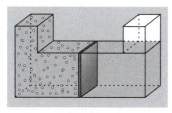

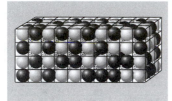

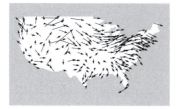

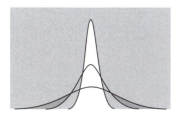

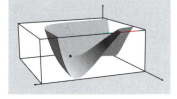

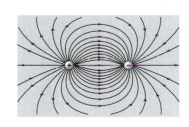

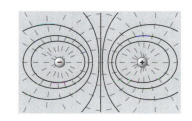

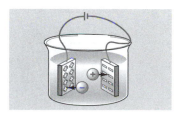

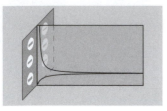

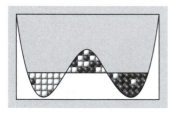

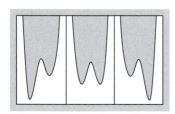

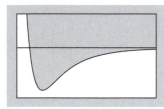

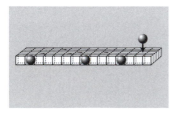

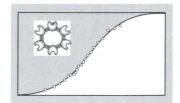

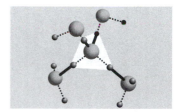

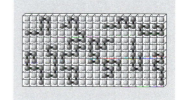

Preface

What forces drive atoms and molecules to bind, to adsorb, to dissolve, to permeate membranes, to undergo chemical reactions, and to undergo conformational changes? This is a textbook on statistical thermodynamics. It describes the forces that govern molecular behavior. Statistical thermodynamics uses physical models, mathematical approximations, and empirical laws that are rooted in the language of *entropy, distribution function, energy, heat capacity, free energy, and partition function*, to predict the behaviors of molecules in physical, chemical, and biological systems.

This text is intended for graduate students and advanced undergraduates in physical chemistry, biochemistry, bioengineering, polymer and materials science, pharmaceutical chemistry, chemical engineering, and environmental science.

We had three goals in mind as we wrote this book. First, we tried to make extensive connections with experiments and familiar contexts, to show the practical importance of this subject. We have included many applications in biology and polymer science, in addition to applications in more traditional areas of chemistry and physics. Second, we tried to make this book accessible to students with a variety of backgrounds. So, for example, we have included material on probabilities, approximations, partial derivatives, vector calculus, and on the historical basis of thermodynamics. Third, we strove to find a vantage point from which the concepts are revealed in their simplest and most comprehensible forms. For this reason, we follow the axiomatic approach to thermodynamics developed by HB Callen, rather than the more traditional inductive approach; and the Maximum Entropy approach of Jaynes, Skilling and Livesay, in preference to the Gibbs ensemble method. We have drawn from many excellent texts, particularly those by Callen, Hill, Atkins, Chandler, Kubo, Kittel and Kroemer, Carrington, Adkins, Weiss, Doi, Flory, and Berry, Rice and Ross.

Our focus here is on molecular driving forces, which overlaps with—but is not identical to—the subject of thermodynamics. While the power of thermodynamics is its generality, the power of statistical thermodynamics is the insights it gives into microscopic interactions through the enterprise of model-making. A central theme of this book is that making models, even very simple ones, is a route to insight and to understanding how molecules work. A good theory, no matter how complex its mathematics, is usually rooted in some very simple physical idea.

Models are mental toys to guide our thinking. The most important ingredients in a good model are predictive power and insight into the causes of the predicted behavior. The more rigorous a model, the less room for ambiguity. But models don't need to be complicated to be useful. Many of the key insights in statistical mechanics have come from simplifications that may seem unrealistic at first glance: particles represented as perfect spheres with atomic detail left out, neglecting the presence of other particles, using crystal-like lattices of particles in liquids and polymers, and modelling polymer chains as random flights, etc. To borrow a quote, statistical thermodynamics has a history of what might be called the *unreasonable effectiveness of unrealistic simplifications*. Perhaps the classic example is the two-dimensional Ising model of magnets as two types of arrows, up spins or down spins, on square lattices. Lars Onsager's famous solution to this highly simplified model was a major contribution to the modern revolution in our understanding of phase transitions and critical phenomena.

We begin with entropy. Chapter 1 gives the underpinnings in terms of probabilities and combinatorics. Simple models are used in chapters 2 and 3 to show how entropy is a driving force. This motivates more detailed treatments throughout the text illustrating the Second Law of thermodynamics and the concept of equilibrium. Chapters 1, 4, and 5 lay out the mathematical foundations—probability, approximations, multivariate calculus—that are needed for the following chapters.

These threads culminate in chapter 6, which defines the entropy and gives the Boltzmann distribution law, the lynch-pin of statistical thermodynamics. The key expressions, $S = k \ln W$ and $S = -k \sum p_i \ln p_i$, are often regarded in physical chemistry texts as given, but here we provide optional material in which we derive these expressions from a principle of fair apportionment, based on treatments by Jaynes, Skilling, Livesay, and others.

The principles of thermodynamics are described in chapters 7—9. The statistical mechanics of simple systems follows in chapters 10 and 11. While temperature and heat capacity are often regarded as needing no explanation (perhaps because they are so readily measured), our chapter 12 uses simple models to shed light on the physical basis of those properties. Chapter 13 applies the principles of statistical thermodynamics to chemical equilibria.

Chapters 14—16 develop simple models of liquids and solutions. We use lattice models here, rather than ideal solution theories, because such models give more microscopic insight into real molecules and into the solvation processes that are central to computational chemistry, biology, and materials science. For example, theories of mixtures often begin from the premise that Raoult's and Henry's laws are experimental facts. Our approach, instead, is to show why molecules are driven to obey these laws. An equally important reason for introducing lattice models here is as background. Lattice models are standard tools for treating complex systems: phase transitions and critical phenomena in chapters 25 and 26, and polymer conformations in chapters 30—33.

We explore the dynamic processes of diffusion, transport, and physical and chemical kinetics in chapters 18 and 19 through the random-flight model, the Langevin model, Onsager relations, time correlation functions and transition-state theory.

We treat electrostatics in chapters 20—23. Our treatment is more extensive than in other physical chemistry texts because of the importance, in our view, of electrostatics in understanding the structures of proteins, nucleic acids, micelles and membranes; for predicting protein- and nucleic acid–ligand interactions and the behaviors of ion channels; as well as for the classical areas of electrochemistry and colloid science. We develop the Nernst and Poisson-Boltzmann equations and the Born model, modern workhorses of quantitative biology. Chapter 24 describes intermolecular forces.

We describe simple models of complex systems, including polymers, colloids, surfaces, and catalysts. Chapters 25 and 26 focus on cooperativity: phase equilibria, solubilities, critical phenomena, and conformational transitions, described through mean-field theories, the Ising model, helix-coil model, and Landau theory. Chapters 27 and 28 describe binding polynomials, essential to modern pharmaceutical science. Chapters 29 and 30 describe water, the hydrophobic effect, and ion solvation. And chapters 31—33 focus on the conformations of polymers and biomolecules that give rise to the elasticity of rubber, the viscoelasticities of solutions, the immiscibilities of polymers, reptational motion, and the folding of proteins and RNA molecules.

Acknowledgements

We owe a great debt to Brad Anderson, Victor Bloomfield, Robert Cantor, Hue Sun Chan, John Chodera, Margaret Daugherty, John von Drie, Roland Dunbrack, Burak Erman, Tony Haymet, Peter Kollman, ID Kuntz, Michael Laskowski, Alenka Luzar, Andy McCammon, Chris Miller, Terry Oas and Eric Toone and their students, Rob Phillips, Miranda Robertson, Kim Sharp, Krista Shipley, Noel Southall, Vojko Vlachy, Peter von Hippel, Hong Qian, Ben Widom, and Bruno Zimm for very helpful comments on this text. We owe special thanks to Jan WH Schreurs, Eugene Stanley, and John Schellman for careful reading and detailed criticism of large parts of various drafts of it. We are grateful to Richard Shafer and Ron Siegel, who in co-teaching this course have contributed considerable improvements, and to the UCSF graduate students of this course over the past several years, who have helped form this material and correct mistakes. We thank Claudia Johnson who created the original course notes manuscript—with the good spirits and patience of a saint. We are deeply grateful to the talented people who worked so hard on book production: Danny Heap and Patricia Monohon who keyboarded, coded, extended and converted the manuscript into this book format; to Jolanda Schreurs who worked tirelessly on the graphics; and to Emma Hunt, Matthew Day and Denise Schanck, who patiently and professionally saw us through to bound books. The preparation of this text was partially funded by a grant from the Polymer Education Committee of the Divisions of Polymer Chemistry and Polymeric Materials of the American Chemical Society.

1 Principles of Probability

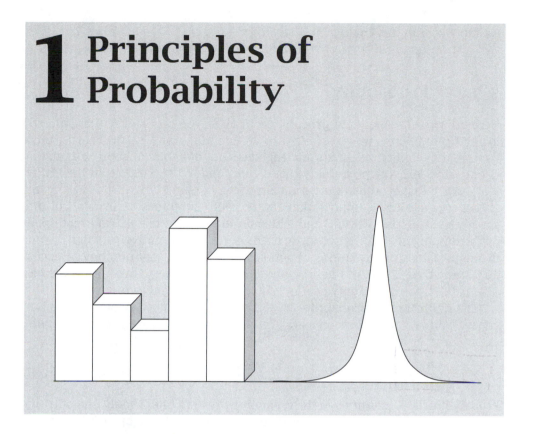

The Principles of Probability Are the Foundations of Entropy

Fluids flow, boil, freeze, and evaporate. Solids melt and deform. Oil and water don't mix. Metals and semiconductors conduct electricity. Crystals grow. Chemicals react and rearrange, take up heat and give it off. Rubber stretches and retracts. Proteins catalyze biological reactions. What forces drive these processes? This question is addressed by statistical thermodynamics, a set of tools for modeling molecular forces and behavior, and a language for interpreting experiments.

The challenge in understanding these behaviors is that the properties that can be measured and controlled, such as density, temperature, pressure, heat capacity, molecular radius, or equilibrium constants, do not predict the tendencies and equilibria of systems in a simple and direct way. To predict equilibria, we must step into a different world, where we use the language of *energy, entropy, enthalpy*, and *free energy*. Measuring the density of liquid water just below its boiling temperature does not hint at the surprise that just a few degrees higher, above the boiling temperature, the density suddenly drops more than a thousandfold. To predict density changes and other measurable properties, you need to know about the driving forces, the entropies and energies. We begin with entropy.

Entropy is one of the most fundamental concepts in statistical thermodynamics. It describes the tendency of matter toward disorder. The concepts that we introduce in this chapter, *probability, multiplicity, combinatorics, averages,* and *distribution functions,* provide a foundation for describing entropy.

What Is Probability?

Here are two statements of probability. In 1990, the probability that a person in the United States was a scientist or an engineer was 1/250. That is, there were about a million scientists and engineers out of a total of about 250 million people. In 1992, the probability that a child under 13 years old in the United States ate a fast-food hamburger on any given day was 1/30 [1].

Let's generalize. Suppose that the possible outcomes or events fall into categories A, B, or C. 'Event' and 'outcome' are generic terms. An event might be the flipping of a coin, resulting in heads or tails. Alternatively it might be one of the possible conformations of a molecule. Suppose that outcome A occurs 20% of the time, B 50% of the time, and C 30% of the time. Then the probability of A is 0.20, the probability of B is 0.50, and the probability of C is 0.30.

The **definition of probability** is: If N is the total number of possible outcomes, and n_A of the outcomes fall into category A, then p_A, the probability of outcome A, is

$$p_A = \left(\frac{n_A}{N} \right).$$ (1.1)

Probabilities are quantities in the range from zero to one. If only one outcome is possible, the process is *deterministic*—the outcome has a probability of one. An outcome that never occurs has a probability of zero.

Probabilities can be computed for different combinations of events. Consider one roll of a six-sided die, for example (die, unfortunately, is the singular of dice). The probability that a **4** appears face up is 1/6 because there are $N = 6$ possible outcomes and only $n_4 = 1$ of them is a **4**. But suppose you roll a six-sided die three times. You may ask for the probability that you will observe the sequence of two **3**'s followed by one **4**. Or you may ask for the probability of rolling two **2**'s and one **6** in any order. The rules of probability and combinatorics provide the machinery for calculating such probabilities. Here we define the relationships among events that we need to formulate the rules.

Definitions: Relationships Among Events

MUTUALLY EXCLUSIVE.   Outcomes $A_1, A_2, \ldots, A_t$ are mutually exclusive if the occurrence of each one of them precludes the occurrence of all the others. If A and B are mutually exclusive, then if A occurs, B does not. If B occurs, A does not. For example, on a single die roll, **1** and **3** are mutually exclusive because only one number can appear face up each time the die is rolled.

COLLECTIVELY EXHAUSTIVE.   Outcomes $A_1, A_2, \ldots, A_t$ are collectively exhaustive if they constitute the entire set of possibilities, and no other outcomes are possible. For example, [heads, tails] is a collectively exhaustive set of outcomes for a coin toss, provided that you don't count the occasions when the coin lands on its edge.

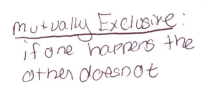

mutually Exclusive:
if one happens the
other does not

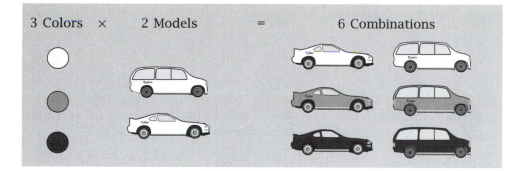

Figure 1.1 If there are three car colors for each of two car models, there are six different combinations of color and model, so the multiplicity is six.

INDEPENDENT. Events $A_1, A_2, \ldots, A_t$ are independent if the outcome of each one is unrelated to (or not *correlated* with) the outcome of any other. The score on one die roll is independent of the score on the next, unless there is trickery.

→ outcome of one doesn't Depend on outcome of another.

MULTIPLICITY. The multiplicity of events is the total number of ways in which different outcomes can possibly occur. If the number of outcomes of type A is n_A, the number of outcomes of type B is n_B, and the number of outcomes of type C is n_C, the total number of possible combinations of outcomes is the multiplicity W:

→ total # of Possible Combinations

$$W = n_A n_B n_C. \tag{1.2}$$

Figure 1.1 shows an example of multiplicity.

The Rules of Probability Are Recipes for Drawing Consistent Inferences

The addition and multiplication rules permit you to calculate the probabilities of certain combinations of events.

ADDITION RULE. If outcomes $A, B, \ldots, E$ are mutually exclusive, and occur with probabilities $p_A = n_A/N, p_B = n_B/N, \ldots, p_E = n_E/N$, then the probability of observing either A OR B OR ..., OR E (the union of outcomes expressed as $A \cup B \cup \cdots \cup E$) is the sum of the probabilities:

$$p(A \text{ OR } B \text{ OR} \ldots, \text{ OR } E) = \frac{n_A + n_B + \cdots + n_E}{N}$$

$$= p_A + p_B + \cdots + p_E. \tag{1.3}$$

The addition rule holds only if two criteria are met: the outcomes are *mutually exclusive,* and we seek the probability of one outcome OR another outcome.

When they are not divided by N, the broader term for the quantities n_i, ($i = A, B, \ldots, E$) is *statistical weights.* If outcomes $A, B, \ldots, E$ are both collectively exhaustive and mutually exclusive, then

$$n_A + n_B + \cdots + n_E = N, \tag{1.4}$$

and dividing both sides of Equation (1.4) by N, the total number of trials, gives

$$p_A + p_B + \cdots + p_E = 1. \tag{1.5}$$

MULTIPLICATION RULE. If outcomes $A, B, \ldots, E$ are independent, then the probability of observing A AND B AND $\ldots$ AND E (the intersection of outcomes, expressed as $A \cap B \cap \cdots \cap E$) is the product of the probabilities,

$$p(A \text{ AND } B \text{ AND } \ldots \text{ AND } E) = \left(\frac{n_A}{N}\right)\left(\frac{n_B}{N}\right)\cdots\left(\frac{n_E}{N}\right)$$
$$= p_A p_B \cdots p_E. \tag{1.6}$$

The multiplication rule applies when the outcomes are *independent* and we seek the probability of one outcome AND another outcome AND possibly other outcomes. A more general multiplication rule, described on page 7, applies even when outcomes are not independent.

Here are a few examples using the addition and multiplication rules.

EXAMPLE 1.1 Rolling a die. What is the probability that either a **1** or a **4** appears on a single roll of a die? The probability of a **1** is $1/6$. The probability of a **4** is also $1/6$. The probability of either a **1** OR a **4** is $1/6 + 1/6 = 1/3$, because the outcomes are mutually exclusive (**1** and **4** can't occur on the same roll) and the question is of the OR type.

EXAMPLE 1.2 Rolling twice. What is the probability of a **1** on the first roll of a die and a **4** on the second? It is $(1/6)(1/6) = 1/36$, because this is an AND question, and the two events are independent. This probability can also be computed in terms of the multiplicity. There are six possible outcomes on each of the two rolls of the die, giving a product of $W = 36$ possible combinations, one of which is **1** on the first roll and **4** on the second.

EXAMPLE 1.3 A sequence of coin flips. What is the probability of getting five heads on five successive flips of an unbiased coin? It is $(1/2)^5 = 1/32$, because the coin flips are independent of each other, this is an AND question, and the probability of heads on each flip is $1/2$. In terms of the multiplicity of outcomes, there are two possible outcomes on each flip, giving a product of $W = 32$ total outcomes, and only one of them is five successive heads.

EXAMPLE 1.4 Another sequence of coin flips. What is the probability of two heads, then one tail, then two more heads on five successive coin flips? It is $p_H^2 p_T p_H^2 = (1/2)^5 = 1/32$. You get the same result as in Example 1.3 because p_H, the probability of heads, and p_T, the probability of tails, are both $1/2$. There are a total of $W = 32$ possible outcomes and only one is the given sequence. The probability $p(n_H, N)$ of observing one particular sequence of N coin flips having exactly n_H heads is

$$p(n_H, N) = p_H^{n_H} p_T^{N - n_H}. \tag{1.7}$$

If $p_H = p_T = 1/2$, then $p(n_H, N) = (1/2)^N$.

EXAMPLE 1.5 Combining events—both, either/or, or neither.   If independent events A and B have probabilities p_A and p_B, the probability that *both* events happen is $p_A p_B$. What is the probability that A happens AND B does not? The probability that B does not happen is $(1 - p_B)$. If A and B are independent events, then the probability that A happens and B does not is $p_A(1 - p_B) = p_A - p_A p_B$. What is the probability that *neither* event happens? It is

$$p(\text{not } A \text{ AND not } B) = (1 - p_A)(1 - p_B), \tag{1.8}$$

where p (not A AND not B) is the probability that A does not happen AND B does not happen.

EXAMPLE 1.6 Combining events—something happens.   What is the probability that *something* happens, that is, A OR B OR both happen? This is an OR question but the events are independent and not mutually exclusive, so you cannot use either the addition or multiplication rules. You can use a simple trick instead. The trick is to consider the probabilities that events *do not* happen, rather than that events *do* happen. The probability that something happens is $1 - p(\text{nothing happens})$:

$$1 - p(\text{not } A \text{ AND not } B) = 1 - (1 - p_A)(1 - p_B) = p_A + p_B - p_A p_B. \tag{1.9}$$

Multiple events can occur as ordered sequences in *time*, such as die rolls, or as ordered sequences in *space*, such as the strings of characters in words. Sometimes it is more useful to focus on collections of events rather than the individual events themselves.

Elementary and Composite Events

Some problems in probability cannot be solved directly by applying the addition or multiplication rules. Such questions can usually be reformulated in terms of *composite events* to which the rules of probability can be applied. Example 1.7 shows how to do this. Then on page 13 we'll use reformulation to construct probability distribution functions.

EXAMPLE 1.7 Elementary and composite events.   What is the probability of a **1** on the first roll of a die OR a **4** on the second roll? If this were an AND question, the probability would be $(1/6)(1/6) = 1/36$, since the two rolls are independent, but the question is of the OR type, so it cannot be answered by direct application of either the addition or multiplication rules. But by redefining the problem in terms of composite events, you can use those rules. An individual coin toss, a single die roll, etc. could be called an elementary event. A composite event is just some set of elementary events, collected together in a convenient way. In this example it's convenient to define each composite event to be a pair of first and second rolls of the die. The advantage is that the complete list of composite events is mutually exclusive. That allows us to frame the problem in terms of an OR question and use the multiplication and addition rules. The composite events are:

[1, 1]*	[1, 2]*	[1, 3]*	[1, 4]*	[1, 5]*	[1, 6]*
[2, 1]	[2, 2]	[2, 3]	[2, 4]*	[2, 5]	[2, 6]
[3, 1]	[3, 2]	[3, 3]	[3, 4]*	[3, 5]	[3, 6]
[4, 1]	[4, 2]	[4, 3]	[4, 4]*	[4, 5]	[4, 6]
[5, 1]	[5, 2]	[5, 3]	[5, 4]*	[5, 5]	[5, 6]
[6, 1]	[6, 2]	[6, 3]	[6, 4]*	[6, 5]	[6, 6]

The first and second numbers in the brackets indicate the outcome of the first and second rolls respectively, and * indicates a composite event that satisfies the criterion for 'success' (**1** on the first roll OR **4** on the second roll). There are 36 composite events, of which 11 are successful, so the probability we seek is 11/36.

Since many of the problems of interest in statistical thermodynamics involve huge systems ($\sim 10^{23}$), we need a more systematic way to compute composite probabilities than enumerating them all.

To compute this probability systematically, collect the composite events into three mutually exclusive classes, A, B, and C, about which you can ask an OR question. Class A includes all composite events with a **1** on the first roll AND anything but a **4** on the second. Class B includes all events with anything but a **1** on the first roll AND a **4** on the second. Class C includes the one event in which we get a **1** on the first roll AND a **4** on the second. A, B, and C are mutually exclusive categories. This is an OR question, so add p_A, p_B, and p_C to find the answer:

$$p(\mathbf{1}\ \text{first OR } \mathbf{4}\ \text{second}) = p_A(\mathbf{1}\ \text{first AND anything but } \mathbf{4}\ \text{second})$$
$$+ p_B(\text{anything but } \mathbf{1}\ \text{first AND } \mathbf{4}\ \text{second})$$
$$+ p_C(\mathbf{1}\ \text{first AND } \mathbf{4}\ \text{second}). \quad (1.10)$$

The same probability rules that apply to elementary events also apply to composite events. Moreover, p_A, p_B, and p_C are each products of elementary event probabilities because the first and second rolls of the die are independent:

$$p_A = \left(\frac{1}{6}\right)\left(\frac{5}{6}\right),$$

$$p_B = \left(\frac{5}{6}\right)\left(\frac{1}{6}\right),$$

$$p_C = \left(\frac{1}{6}\right)\left(\frac{1}{6}\right).$$

Add p_A, p_B and p_C: $p(\mathbf{1}\ \text{first OR } \mathbf{4}\ \text{second}) = 5/36 + 5/36 + 1/36 = 11/36$. This example shows how elementary events can be grouped together into composite events so as to take advantage of the addition and multiplication rules. Reformulation is powerful because virtually any question can be framed in terms of combinations of AND and OR operations. With these two rules of probability, you can draw inferences about a wide range of probabilistic events.

Two events can have a more complex relationship than we have considered so far. They are not restricted to being either independent or mutually exclusive. More broadly, events can be *correlated*.

Correlated Events Are Described by Conditional Probabilities

Events are correlated if the outcome of one depends on the outcome of the other. For example, if it rains on 36 days a year, the probability of rain is $36/365 \approx 0.1$. But if it rains on 50% of the days when you see dark clouds, then the probability of observing rain (event B) depends upon, or is conditional upon, the appearance of dark clouds (event A). Example 1.8 and Table 1.1 demonstrate the correlation of events when balls are taken out of a barrel.

EXAMPLE 1.8 Balls taken from a barrel with replacement. Suppose a barrel contains one red ball, R, and two green balls, G. The probability of drawing a green ball on the first try is 2/3, and the probability of drawing a red ball on the first try is 1/3. What is the probability of drawing a green ball on the second draw? That depends on whether or not you put the first ball back into the barrel before the second draw. If you replace each ball before drawing another, then the probabilities of different draws are uncorrelated with each other. Each draw is an independent event.

However, if you draw a green ball first, and don't put it back in the barrel, then 1 R and 1 G remain after the first draw, and the probability of getting a green ball on the second draw is now 1/2. The probability of drawing a green ball on the second try is different from the probability of drawing a green ball on the first try. It is *conditional* on the outcome of the first draw.

Here are some definitions and examples describing the conditional probabilities of correlated events.

CONDITIONAL PROBABILITY. The conditional probability $p(B \mid A)$ is the probability of event B, *given that* some other event A has occurred. Event A is the *condition* upon which we evaluate the probability of event B. In Example 1.8, event B is getting a green ball on the second draw, event A is getting a green ball on the first draw, and $p(G_2 \mid G_1)$ is the probability of getting a green ball on the second draw, given a green ball on the first draw.

JOINT PROBABILITY. The joint probability of events A and B is the probability that both events A AND B occur. The joint probability is expressed by the notation $p(A \text{ AND } B)$, or more concisely by $p(AB)$.

GENERAL MULTIPLICATION RULE (BAYES RULE). If outcomes A and B occur with probabilities $p(A)$ and $p(B)$, the joint probability of events A AND B is

$$p(AB) = p(B \mid A)p(A) = p(A \mid B)p(B). \tag{1.11}$$

If events A and B happen to be independent, the pre-condition A has no influence on the probability of B. Then $p(B \mid A) = p(B)$, and Equation (1.11) reduces to $p(AB) = p(B)p(A)$, the multiplication rule for independent events. A probability $p(B)$ that is not conditional is called an *a priori* probability. The conditional quantity $p(B \mid A)$ is called an *a posteriori* probability. The general multiplication rule is general because independence is not required. It defines the probability of the *intersection* of events, $p(AB) = p(A \cap B)$.

Table 1.1 All of the probabilities for the three draws without replacement described in Examples 1.8 and 1.9.

1st Draw	2nd Draw	3rd Draw
$p(R_1) = 1/3 \longrightarrow$	$\begin{array}{l} p(R_2 \mid R_1)p(R_1) \\ 0 \cdot (1/3) = 0 \\[6pt] p(G_2 \mid R_1)p(R_1) \\ 1 \cdot (1/3) = 1/3 \end{array}$	$\longrightarrow \begin{array}{l} p(G_3 \mid G_2R_1)p(G_2R_1) \\ 1 \cdot (1/3) = 1/3 \end{array}$
$p(G_1) = 2/3 \longrightarrow$	$\begin{array}{l} p(R_2 \mid G_1)p(G_1) \\ (1/2) \cdot (2/3) = 1/3 \\[6pt] p(G_2 \mid G_1)p(G_1) \\ (1/2) \cdot (2/3) = 1/3 \end{array}$	$\begin{array}{l} \longrightarrow p(G_3 \mid R_2G_1)p(R_2G_1) \\ 1 \cdot (1/3) = 1/3 \\[6pt] \longrightarrow p(R_3 \mid G_2G_1)p(G_2G_1) \\ 1 \cdot (1/3) = 1/3 \end{array}$

GENERAL ADDITION RULE. A general rule can also be formulated for the union of events $p(A \cup B) = p(A) + p(B) - p(A \cap B)$ when we seek the probabilitiy of A OR B for events that are not mutually exclusive. When A and B *are* mutually exclusive, $p(A \cap B) = 0$ and the general addition rule reduces to the simpler addition rule on page 3. When A and B are independent, $p(A \cap B) = p(A)p(B)$ and the general addition rule gives the result in Example 1.6.

DEGREE OF CORRELATION. The degree of correlation g between events A and B can be expressed as the ratio of the conditional probability of B, given A, to the unconditional probability of B alone. This indicates the degree to which A influences B:

$$g = \frac{p(B \mid A)}{p(B)} = \frac{p(AB)}{p(A)p(B)}. \tag{1.12}$$

The second equality in Equation (1.12) follows from the general multiplication rule, Equation (1.11). If $g = 1$, events A and B are independent and not correlated. If $g > 1$, events A and B are positively correlated. If $g < 1$, events A and B are negatively correlated. If $g = 0$ and A occurs then B will not. If the *a priori* probability of rain is $p(B) = 0.1$, and if the conditional probability of rain, given that there are dark clouds, A, is $p(B \mid A) = 0.5$, then the degree of correlation of rain with dark clouds is $g = 5$. Correlations are important in statistical thermodynamics. For example, attractions and repulsions among molecules in liquids can cause correlations among their positions and orientations.

EXAMPLE 1.9 Balls taken from that barrel again. As before, start with three balls in a barrel, one red and two green. The probability of getting a red ball on the first draw is $p(R_1) = 1/3$, where the notation R_1 refers to a red ball on the first draw. The probability of getting a green ball on the first draw is $p(G_1) = 2/3$. If balls are not replaced after each draw, the joint probability for getting a red ball first and a green ball second is $p(R_1G_2)$:

$$p(R_1G_2) = p(G_2 \mid R_1)p(R_1) = (1)(1/3) = 1/3. \tag{1.13}$$

Conditional probabilities are useful in a variety of situations including card games and horse races, as the following example shows.

EXAMPLE 1.10 A gambling equation. Suppose you have a collection of mutually exclusive and collectively exhaustive events $A, B, \ldots, E$, with probabilities $p_A, p_B, \ldots, p_E$. These could be the probabilities that horses $A, B, \ldots, E$ will win a race (based on some theory, model, or prediction scheme), or that card types A to E will appear on a given play in a card game. Let's look at a horse race [2].

Suppose you have some information, such as the track records of the horses, that predicts the *a priori* probabilities that each horse will win. Figure 1.2 gives an example. Now as the race proceeds, the events occur in order, one at a time: one horse wins, then another comes in second, and another comes in third. Our aim is to compute the conditional probability that a particular horse will come in second, given that some other horse has won. The *a priori* probability that horse C will win is $p(C)$. Now *assume* that horse C has won, and you want to know the probability that horse A will be second, $p(A$ is second $\mid C$ is first). From Figure 1.2, you can see that this conditional probability can be determined by eliminating region C, and finding the fraction of the remaining area occupied by region A:

$$p(A \text{ is second} \mid C \text{ is first}) = \frac{p(A)}{p(A) + p(B) + p(D) + p(E)}$$

$$= \frac{p(A)}{1 - p(C)}. \tag{1.14}$$

$1 - p(C) = p(A) + p(B) + p(D) + p(E)$ follows from the mutually exclusive addition rule.

The probability that event i is first is $p(i)$. Then the conditional probability that event j is second is $p(j)/(1 - p(i))$. The joint probability that i is first, j is second, and k is third is

$$p(i \text{ is first}, j \text{ is second}, k \text{ is third}) = \frac{p(i)p(j)p(k)}{[1 - p(i)][1 - p(i) - p(j)]}. \tag{1.15}$$

Equations (1.14) and (1.15) are useful for computing the probability of drawing the queen of hearts in a card game, once you have seen the seven of clubs and the ace of spades. It is also useful for describing the statistical thermodynamics of liquid crystals, and ligand binding to DNA (see page 552).

Combinatorics Describes How to Count Events

Combinatorics, or counting events, is central to statistical thermodynamics. It is the basis for entropy, and the concepts of *order* and *disorder*, which are defined by the numbers of ways in which a system can be configured. Combinatorics is concerned with the *composition* of events rather than the *sequence*

(a) Who will win?

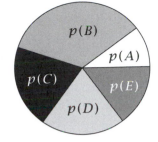

(b) Given that C won...

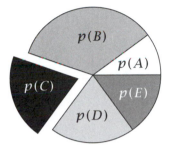

(c) Who will place second?

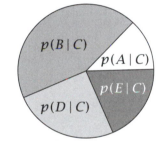

Figure 1.2 (a) *A priori* probabilities of outcomes A to E, such as a horse race. (b) To determine the *a posteriori* probabilities of events A, B, D and E, given that C has occurred, remove C and keep the relative proportions of the rest the same. (c) *A posteriori* probabilities that horses A, B, D and E will come in second, given that C won.

of events. For example, compare the following two questions. The first is a question of sequence: What is the probability of the specific sequence of four coin flips, $HTHH$? The second is a question of composition: What is the probability of observing three H's and one T in *any* order? The sequence question is answered by using Equation (1.7): this probability is 1/16. However, to answer the composition question you must count the number of different possible sequences with the specified composition: $HHHT$, $HHTH$, $HTHH$, and $THHH$. The probability of getting three H's and one T in any order is 4/16 = 1/4. When you seek the probability of a certain *composition* of events, you count the possible sequences that have the correct composition.

EXAMPLE 1.11 Permutations of ordered sequences. How many permutations, or different sequences, of the letters **w**, **x**, **y**, and **z** are possible? There are twenty-four:

wxyz	wxzy	wyxz	wyzx	wzxy	wzyx
xwyz	xwzy	xywz	xyzw	xzwy	xzyw
ywxz	ywzx	yxwz	yxzw	yzwx	yzxw
zwyx	zwxy	zxwy	zxyw	zywx	zyxw

How can you compute the number of different sequences without having to list them all? You can use the strategy developed for drawing letters from a barrel without replacement. The first letter of a sequence can be any one of the four. After drawing one, the second letter of the sequence can be any of the remaining three letters. The third letter can be any of the remaining two letters, and the fourth must be the one remaining letter. Use the definition of multiplicity W (Equation (1.2)) to combine the numbers of outcomes n_i, where i represents the position 1, 2, 3, or 4 in the sequence of draws. We have $n_1 = 4$, $n_2 = 3$, $n_3 = 2$, and $n_4 = 1$, so the number of permutations is $W = n_1 n_2 n_3 n_4 = 4 \cdot 3 \cdot 2 \cdot 1 = 24$.

In general, for a sequence of N *distinguishable* objects, the number of different permutations W can be expressed in factorial notation

$$W = N(N-1)(N-2)\cdots 3 \cdot 2 \cdot 1 = N!$$
$$= 4 \cdot 3 \cdot 2 \cdot 1 = 24. \tag{1.16}$$

EXAMPLE 1.12 Letters of the alphabet. Consider a barrel containing one each of the twenty-six letters of the alphabet. What is the probability of drawing the letters out in exactly the order of the alphabet, **A** to **Z**? The probability of drawing the **A** first is 1/26. If you replace each letter after it is drawn, the probability of drawing the **B** on the second try would be 1/26, and the probability of drawing the alphabet in order would be $(1/26)^{26}$. But if each letter were *not* replaced in the barrel, the probability of drawing the **B** on the second trial would be 1/25. The probability of drawing the **C** on the third trial would be 1/24. Without replacement, the probability of drawing the exact sequence of the alphabet is

$$p(\mathbf{ABC\ldots XYZ}) = \frac{1}{26 \cdot 25 \cdot 24 \cdots 2 \cdot 1} = \frac{1}{N!}, \tag{1.17}$$

where $N = 26$ is the number of letters in the alphabet. $N!$ is the number of permutations, or different sequences in which the letters could be drawn. $1/(N!)$ is the probability of drawing one particular sequence.

The Factorial Notation

The notation $N!$, called N factorial, denotes the product of the integers from one to N:

$$N! = 1 \cdot 2 \cdot 3 \cdots (N - 2)(N - 1)N.$$

0! is defined to equal one.

In Examples 1.11 and 1.12, all the letters are distinguishable from each other: **w**, **x**, **y**, and **z** are all different. But what happens if some of the objects are indistinguishable from each other?

EXAMPLE 1.13 Counting sequences of distinguishable and indistinguishable objects. How many different arrangements are there of the letters **A**, **H**, and **A**? That depends on whether or not you can tell the **A**'s apart. Suppose first that one **A** has a subscript 1 and the other has a subscript 2: A_1, **H**, and A_2. Then all the characters are distinguishable, as in Examples 1.11 and 1.12, and there are $W = N! = 3! = 6$ different arrangements of these three distinguishable characters:

$$\mathbf{HA_1A_2} \qquad \mathbf{A_1HA_2} \qquad \mathbf{A_1A_2H} \qquad \mathbf{HA_2A_1} \qquad \mathbf{A_2HA_1} \qquad \mathbf{A_2A_1H.}$$

However, now suppose that the two **A**'s are indistinguishable from each other: they have no subscripts. There are now only $W = 3$ distinguishable sequences of letters: **HAA**, **AHA**, and **AAH**. (Distinguishable is a term that applies either to the letters or to the sequences. We have three distinguishable sequences, each containing two distinguishable letters, **A** and **H**.) The previous expression $W = N!$ overcounts by a factor of two when two **A**'s are indistinguishable. This is because we have counted each sequence of letters, say **AAH**, twice—A_1A_2H and A_2A_1H. Written in a more general way, the number of distinguishable sequences is $W = N!/N_A! = 3!/2! = 3$. The $N!$ in the numerator comes from the number of permutations as if all the characters were distinguishable from each other, and the $N_A!$ in the denominator corrects for overcounting. The overcounting correction 2! is simply the count of all the permutations of the indistinguishable characters, the number of ways in which the **A**'s can be arranged among themselves.

EXAMPLE 1.14 Permutations of mixed sequences. Consider the word **cheese** as $\mathbf{che_1e_2se_3}$, in which the **e**'s are distinguished from each other by a subscript. Then $N = 6$ and there are $6! = 720$ distinguishable ways of arranging the characters. By counting in this way, we have reckoned that $\mathbf{che_1e_2se_3}$ is different from $\mathbf{che_2e_1se_3}$. This correct spelling is counted exactly six times because there are six permutations of the subscripted **e**'s. There are also exactly six permutations of the **e**'s in every other specific sequence. For example:

For the word **freezer**, you have three indistinguishable **e**'s and two indistinguishable **r**'s. There are $7!/(3!2!)$ permutations of the letters that spell freezer. In general, for a collection of N objects with t categories, of which n_i objects in each category are *indistinguishable* from one another, but distinguishable from the objects in the other $t - 1$ categories, the number of permutations W is

$$W = \frac{N!}{n_1!n_2!\cdots n_t!}. \tag{1.18}$$

When there are only two categories (success/failure, or heads/tails, ...), $t = 2$ so $W(n, N)$, the number of sequences with n successes out of N trials, is

$$W(n, N) = \binom{N}{n} = \frac{N!}{n!(N-n)!}, \tag{1.19}$$

where the shorthand notation $\binom{N}{n}$ for combinations is pronounced 'N choose n.' Example 1.15 applies Equation (1.19) to coin flips and die rolls.

EXAMPLE 1.15 Counting sequences of coin flips and die rolls. You flip a coin $N = 4$ times. How many different sequences have three heads? According to Equation (1.19),

$$W(n_H, N) = \frac{N!}{n_H!n_T!} = \frac{4!}{3!1!} = 4.$$

They are $THHH$, $HTHH$, $HHTH$, and $HHHT$. How many different sequences have two heads?

$$W(2, 4) = \frac{4!}{2!2!} = 6.$$

They are $TTHH$, $HHTT$, $THTH$, $HTHT$, $THHT$, and $HTTH$.

You flip a coin one hundred and seventeen times. How many different sequences have thirty-six heads?

$$W(36, 117) = \frac{117!}{36!81!} \approx 1.84 \times 10^{30}.$$

We won't write the sequences out.

You roll a die fifteen times. How many different sequences have three **1**'s, one **2**, one **3**, five **4**'s, two **5**'s, and three **6**'s? According to Equation (1.18),

$$W = \frac{15!}{3!1!1!5!2!3!} = 151,351,200.$$

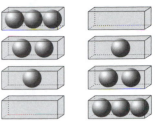

(a) Balls in Boxes

EXAMPLE 1.16 What is the probability of drawing a royal flush in poker?
There are four different ways to draw a royal flush in poker—an ace, king, jack, queen, and ten, all from any one of the four suits. To compute the probability, you need to know how many five-card hands there are in a deck of 52 cards. Use the barrel metaphor: put the 52 cards in the barrel. On the first draw, there are 52 possibilities. On the second draw, there are 51 possibilities, etc. In five draws, there are

$$\frac{52 \cdot 51 \cdot 50 \cdot 49 \cdot 48}{5!} = \frac{52!}{5!(52-5)!} = 2,598,960$$

possible poker hands. The 5! in the denominator corrects for all the possible permutations of each sequence (you don't care whether you draw the king or the ace first, for example). The probability is $4/(2,598,960) = 1.5 \times 10^{-6}$ that you will draw a royal flush.

Here's an example of a type of counting problem in statistical thermodynamics.

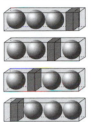

(b) Moveable Walls

EXAMPLE 1.17 Bose–Einstein statistics. How many ways can n indistinguishable particles be put into M boxes, with any number of particles per box? This type of counting is needed to predict the properties of particles called *bosons*, such as photons and He^4 atoms. Bose–Einstein statistics counts the ways that n particles can be distributed in M different energy levels, when several particles can occupy the same quantum mechanical energy levels. For now, our interest is not in the physics, but just in the counting problem. Figure 1.3 shows that one way to count the number of arrangements is to think of the system as a linear array of n particles interspersed with $M-1$ movable walls that partition the system into M boxes (spaces between walls). There are $M+n-1$ objects, counting walls plus particles. The n particles are indistinguishable from each other. The $M-1$ walls are indistinguishable from the other walls. Because the walls are distinguishable from the particles, the number of arrangements is

$$W(n,M) = \frac{(M+n-1)!}{(M-1)!n!}. \tag{1.20}$$

Figure 1.3 Three Bose–Einstein particles in two boxes for Example 1.17: (a) There are four ways to partition $n = 3$ balls into $M = 2$ boxes when each box can hold any number of balls. (b) There are also four ways to partition three balls and one movable wall.

Collections of Probabilities Are Described by Distribution Functions

The probabilities of events can be described by *probability distribution functions*. For t mutually exclusive outcomes, $i = 1,2,3,\ldots,t$, the distribution function is $p(i)$, the set of probabilities of all the outcomes. Figure 1.4 shows a probability distribution function for a system with $t = 5$ outcomes.

A property of probability distribution functions is that the sum of the probabilities equals one. Because the outcomes are mutually exclusive and collectively exhaustive, Equations (1.3) and (1.5) apply and

$$\sum_{i=1}^{t} p(i) = 1. \tag{1.21}$$

Figure 1.4 A probability distribution function. The possible outcomes are indexed on the x-axis. The probability of each outcome is shown on the y-axis. In this example, outcome **4** is the most probable and outcome **3** is the least probable.

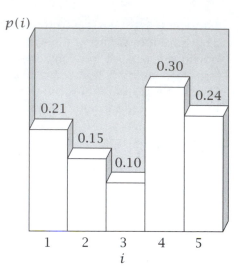

For some types of events the order of the outcomes $i = 1, 2, 3, \ldots, t$ has meaning. For others it does not. For statistical thermodynamics the order usually has meaning and i represents the value of some physical quantity. On the other hand, the index i may be just a label. The index $i = 1, 2, 3$ can represent the colors of socks, [red, green, blue], or [green, red, blue], where the order is irrelevant. Probability distributions can describe either case.

Summations

The sigma notation means to sum terms. For example,

$$\sum_{i=1}^{6} i p_i = p_1 + 2p_2 + 3p_3 + 4p_4 + 5p_5 + 6p_6 \qquad (1.22)$$

means 'sum the quantity $i p_i$ from $i = 1$ up to $i = 6$.' Sometimes the index i above and/or below the sigma is omitted in concise shorthand expressions.

Continuous Probability Distribution Functions

In some situations, the outcomes of an event are best represented by a continuous variable x rather than by a discrete variable. For example, a particle might have some probability $p(x)dx$ of being between position $x = 1.62$ and $x + dx = 1.63$ centimeters or $p(\theta)d\theta$ of having an orientation angle between $\theta = 25.6$ and $\theta + d\theta = 25.8$ degrees. If x is continuous, $p(x)$ is called a *probability density*, because it is a probability per unit interval of x. If x ranges from $x = a$ to $x = b$, Equation (1.21) becomes

$$\int_{a}^{b} p(x)\, dx = 1. \qquad (1.23)$$

Some distribution functions aren't *normalized*: the statistical weights do not sum to one. For a continuous distribution function $\psi(x)$ where x ranges from a to b, you can normalize to form a proper probability distribution function. Find the normalization constant ψ_0 by integrating over x:

$$\psi_0 = \int_a^b \psi(x)\, dx. \tag{1.24}$$

The normalized probability density is

$$p(x) = \frac{\psi(x)}{\psi_0} = \frac{\psi(x)}{\displaystyle\int_a^b \psi(x)\, dx}. \tag{1.25}$$

The Binomial and Multinomial Distribution Functions

Some probability distribution functions occur frequently in nature, and have simple mathematical expressions. Two of the most useful ones are the binomial and multinomial distribution functions. These will be the basis for our development of the concept of entropy in Chapter 2. The binomial distribution describes processes in which each independent elementary event has two mutually exclusive outcomes such as heads/tails, yes/no, up/down, or occupied/vacant. Independent trials with two such possible outcomes are called *Bernoulli trials*. Let's label the two possible outcomes ● and ◗. Let the probability of ● be p. Then the probability of ◗ is $1 - p$. We choose composite events that are pairs of Bernoulli trials. The probability of ● followed by ◗ is $P_{●◗} = p(1 - p)$. The probabilities of the four possible composite events are

$$P_{●●} = p^2, \qquad P_{●◗} = p(1 - p),$$
$$P_{◗●} = (1 - p)p, \qquad \text{and} \qquad P_{◗◗} = (1 - p)^2. \tag{1.26}$$

This set of composite events is mutually exclusive and collectively exhaustive.

The same probability rules apply to the composite events that apply to elementary events. For example, Equation (1.21) for the normalization of discrete distributions requires that the probabilities must sum to one:

$$P_{●●} + P_{●◗} + P_{◗●} + P_{◗◗} = p^2 + 2p(1 - p) + (1 - p)^2$$
$$= (p + (1 - p))^2 = 1. \tag{1.27}$$

In Example 1.7 we defined composite events as pairs of elementary events. More generally, a composite event is a sequence of N repetitions of independent elementary events. The probability of a *specific sequence* of n ●'s and $N - n$ ◗'s is given by Equation (1.7) (page 4). What is the probability that a series of N trials has n ●'s and $N - n$ ◗'s *in any order*? Equation (1.19) (page 12) gives the total number of sequences that have n ●'s and $N - n$ ◗'s. The product of Equations (1.7) and (1.19) gives the probability of n ●'s and $N - n$ ◗'s irrespective of their sequence. This is the **binomial distribution**:

$$P(n, N) = p^n (1 - p)^{N-n} \frac{N!}{n!(N - n)!}. \tag{1.28}$$

Because the set of all possible sequences of N trials is mutually exclusive and collectively exhaustive, Equations (1.3) and (1.5) apply, and the composite probabilities sum to one:

$$\sum_{n=0}^{N} P(n, N) = \sum_{n=0}^{N} p^n (1 - p)^{N-n} \frac{N!}{n!(N-n)!}$$

$$= (1 - p)^N + Np(1 - p)^{N-1}$$

$$+ \frac{N(N-1)}{2} p^2 (1 - p)^{N-2}$$

$$+ \ldots + Np^{N-1}(1 - p) + p^N$$

$$= (p + (1 - p))^N = 1. \tag{1.29}$$

A simple way to write all of the combinatoric terms in the binomial distribution is *Pascal's triangle*. Make a triangle in which each line is numbered $N = 0, 1, 2, \ldots$ and columns are numbered $n = 0, 1, 2, \ldots$. Compute $N!/(n!(N-n)!)$ at each position:

```
N = 0                    1
N = 1                 1     1
N = 2              1     2     1
N = 3           1     3     3     1
N = 4        1     4     6     4     1
N = 5     1     5    10    10     5     1
```

Each term in Pascal's triangle is the sum of the two terms to the left and right from the line above it. Pascal's triangle gives the coefficients in the expansion of $(x + y)^N$. For example, for $N = 4$, Equation (1.28) is

$$[p + (1 - p)]^4 = p^4 + 4p^3(1 - p) + 6p^2(1 - p)^2$$

$$+ 4p(1 - p)^3 + (1 - p)^4. \tag{1.30}$$

EXAMPLE 1.18 Distribution of coin flips. Figure 1.5 shows a distribution function, the probability $p(n_H, N)$ of observing n_H heads in $N = 4$ coin flips, given by Equation (1.28) with $p = 0.5$. This shows that in four coin flips, the most probable number of heads is two. It is least probable that all four will be heads or all four will be tails.

The multinomial probability distribution is a generalization of the binomial probability distribution. A binomial distribution describes two-outcome events such as coin flips. A multinomial probability distribution applies to t-outcome events where n_i is the number of times that outcome $i = 1, 2, 3, \ldots, t$ appears. For example, $t = 6$ for die rolls. For the multinomial distribution, the number of distinguishable outcomes is given by Equation (1.18): $W = N!/(n_1!n_2!n_3! \cdots n_t!)$. The **multinomial probability distribution** is

$$P(n_1, n_2, \ldots, n_t, N) = p_1^{n_1} p_2^{n_2} p_3^{n_3} \cdots p_t^{n_t} \left(\frac{N!}{n_1!n_2! \cdots n_t!} \right), \tag{1.31}$$

where each factor $n_i!$ accounts for the indistinguishability of objects in category i. The n_i's are constrained by the condition $\sum_{i=1}^{t} n_i = N$.

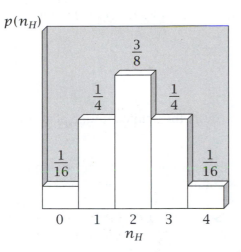

$p(n_H)$

Figure 1.5 The probability distribution for the numbers of heads in four coin flips in Example 1.18.

Distribution Functions Have Average Values and Standard Deviations

Averages

A probability distribution function contains all the information that can be known about a probabilistic system. A full distribution function, however, is rarely accessible from experiments. Generally, experiments can measure only certain averages or *moments* of the distribution. The **nth moment of a probability distribution function** $\langle x^n \rangle$ is

$$\langle x^n \rangle = \int_a^b x^n p(x)\, dx = \frac{\int_a^b x^n \psi(x)\, dx}{\int_a^b \psi(x)\, dx}. \tag{1.32}$$

Angle brackets $\langle\ \rangle$ are used to indicate the moments, also called the expectation values or averages, of a distribution function. For a *probability* distribution the zeroth moment always equals one, because the sum of the probabilities equals one. The first moment of a distribution function ($n = 1$ in Equation (1.32)) is called the mean, average, or expected value. For discrete functions

$$\langle i \rangle = \sum_{i=1}^{t} i p(i), \tag{1.33}$$

and for continuous functions

$$\langle x \rangle = \int_a^b x p(x)\, dx. \tag{1.34}$$

For distributions over t discrete values, the mean of a function $f(i)$ is

$$\langle f(i) \rangle = \sum_{i=1}^{t} f(i) p(i). \tag{1.35}$$

For distributions over continuous values, the mean of a function $f(x)$ is

$$\langle f(x) \rangle = \int_a^b f(x)p(x)\,dx = \frac{\int_a^b f(x)\psi(x)\,dx}{\int_a^b \psi(x)\,dx}. \tag{1.36}$$

Equations (1.33), (1.34), (1.35) and (1.36) quantify the familiar notion of average, as Example 1.19 shows.

EXAMPLE 1.19 Taking an average. The average of the set of numbers $[3, 3, 2, 2, 2, 1, 1]$ is 2. The average may be computed by the usual procedure of summing the numbers and dividing by the number of entries. Let's compute the average using Equation (1.33) instead. Since two of the seven outcomes are **3**'s, the probability of a **3** is $p(3) = 2/7$. Similarly, three of the seven outcomes are **2**'s, so $p(2) = 3/7$, and two of the seven outcomes are **1**'s, so $p(1) = 2/7$. The average $\langle i \rangle$ is

$$\langle i \rangle = \sum_{i=1}^{3} ip(i) = 1p(1) + 2p(2) + 3p(3)$$

$$= \left(\frac{2}{7}\right) + 2\left(\frac{3}{7}\right) + 3\left(\frac{2}{7}\right) = 2. \tag{1.37}$$

Here are two useful and general properties of averages, derived from the definition given in Equation (1.36):

$$\langle af(x) \rangle = \int af(x)p(x)\,dx = a\int f(x)p(x)\,dx$$

$$= a\langle f(x) \rangle, \quad \text{where } a \text{ is a constant.} \tag{1.38}$$

$$\langle f(x) + g(x) \rangle = \int (f(x) + g(x))\,p(x)\,dx$$

$$= \int f(x)p(x)\,dx + \int g(x)p(x)\,dx$$

$$= \langle f(x) \rangle + \langle g(x) \rangle. \tag{1.39}$$

Variance

The *variance* σ^2 is a measure of the width of a distribution. A broad flat distribution has a large variance, while a narrow peaked distribution has a small variance. The variance σ^2 is defined as the average square deviation from the mean,

$$\sigma^2 = \langle (x - a)^2 \rangle = \langle x^2 - 2ax + a^2 \rangle, \tag{1.40}$$

where $a = \langle x \rangle$ is the mean value, or first moment. We use a instead of $\langle x \rangle$ as a reminder here that this quantity is just a constant, not a variable. Using

Equation (1.39), Equation (1.40) becomes

$$\sigma^2 = \langle x^2 \rangle - \langle 2ax \rangle + \langle a^2 \rangle.$$

Using Equation (1.38),

$$\sigma^2 = \langle x^2 \rangle - 2a\langle x \rangle + a^2 = \langle x^2 \rangle - \langle x \rangle^2. \tag{1.41}$$

Second moments such as the variance are important for understanding heat capacities (Chapter 12), random walks (Chapters 4 and 18), diffusion (Chapter 18), and polymer chain conformations (Chapters 31–33). Moments higher than the second describe asymmetries in the shape of the distribution. Examples 1.20, 1.21, and 1.22 show calculations of means and variances for discrete and continuous probability distributions.

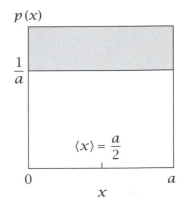

$p(x)$

$\dfrac{1}{a}$

$\langle x \rangle = \dfrac{a}{2}$

0 a

x

Figure 1.6 Flat distribution function, $0 \le x \le a$. The average value is $\langle x \rangle = a/2$ (see Example 1.21).

EXAMPLE 1.20 Coin flips: mean and variance. Compute the average number of heads $\langle n_H \rangle$ in $N = 4$ coin flips by using the distribution in Example 1.18 (page 16):

$$\langle n_H \rangle = \sum_{n_H=0}^{4} n_H p(n_H, N)$$

$$= 0\left(\frac{1}{16}\right) + 1\left(\frac{4}{16}\right) + 2\left(\frac{6}{16}\right) + 3\left(\frac{4}{16}\right) + 4\left(\frac{1}{16}\right) = 2, \quad \text{and}$$

$$\langle n_H^2 \rangle = \sum_{n_H=0}^{4} n_H^2 p(n_H, N)$$

$$= 0\left(\frac{1}{16}\right) + 1\left(\frac{4}{16}\right) + 4\left(\frac{6}{16}\right) + 9\left(\frac{4}{16}\right) + 16\left(\frac{1}{16}\right) = 5.$$

According to Equation (1.41), the variance σ^2 is

$$\sigma^2 = \langle n_H^2 \rangle - \langle n_H \rangle^2 = 5 - 2^2 = 1.$$

EXAMPLE 1.21 The average and variance of a continuous function. Suppose you have a flat probability distribution, $p(x) = 1/a$ (shown in Figure 1.6) for a variable $0 \le x \le a$. To compute $\langle x \rangle$, use Equation (1.34) (page 17):

$$\langle x \rangle = \int_0^a x p(x)\,dx = \frac{1}{a}\int_0^a x\,dx = \left(\frac{1}{a}\right)\frac{x^2}{2}\Big|_0^a = \frac{a}{2}.$$

Equation (1.32) gives the second moment $\langle x^2 \rangle$:

$$\langle x^2 \rangle = \int_0^a x^2 p(x)\,dx = \frac{1}{a}\int_0^a x^2\,dx = \left(\frac{1}{a}\right)\frac{x^3}{3}\Big|_0^a = \frac{a^2}{3}.$$

The variance is $\langle x^2 \rangle - \langle x \rangle^2 = a^2/3 - a^2/4 = a^2/12$.

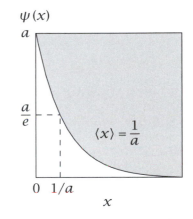

$\psi(x)$

a

$\dfrac{a}{e}$

$\langle x \rangle = \dfrac{1}{a}$

0 $1/a$

x

Figure 1.7 Exponential distribution function, $0 \le x \le \infty$. The average of $p(x) = ae^{-ax}$ is $\langle x \rangle = 1/a$ (see Example 1.22).

EXAMPLE 1.22 The average of an exponential distribution. Figure 1.7 shows a distribution function $\psi(x) = e^{-ax}$ over the range $0 \le x \le \infty$. First normalize $\psi(x)$ to make it a probability distribution. According to Equation (1.25), $p(x) = \psi(x)/\psi_0$. Integrate $\psi(x)$ to determine ψ_0:

$$\psi_0 = \int_0^\infty e^{-ax}\,dx = -\left(\frac{1}{a}\right)e^{-ax}\bigg|_0^\infty = \frac{1}{a} \qquad \text{for } a > 0.$$

The normalized distribution function is $p(x) = \psi(x)/\psi_0 = ae^{-ax}$. Now, to compute $\langle x \rangle$ for this distribution, use Equation (1.32):

$$\langle x \rangle = \int_0^\infty xp(x)\,dx = a\int_0^\infty xe^{-ax}\,dx$$

$$= -\left[e^{-ax}\left(x + \frac{1}{a}\right)\right]\bigg|_0^\infty = \frac{1}{a}.$$

EXAMPLE 1.23 Averaging the orientations of a vector. For predicting the conformations of a polymer or spectroscopic properties you might have a vector that is free to orient uniformly over all possible angles θ. If you want to compute its average projection on an axis, using quantities such as $\langle \cos\theta \rangle$ or $\langle \cos^2\theta \rangle$, put the beginning of the vector at the center of a sphere. If the vector orients uniformly, it points to any given patch on the surface of the sphere in proportion to the area of that patch.

The strip of area shown in Figure 1.8 has an angle θ with respect to the z-axis. The area of the strip is $(r\,d\theta)(2\pi\ell)$. Since $\ell = r\sin\theta$, the area of the strip is $2\pi r^2 \sin\theta\,d\theta$. A strip has less area if θ is small than if θ approaches $90°$. The fraction of vectors $p(\theta)$ that point to, or end in, this strip is

$$p(\theta) = \frac{2\pi r^2 \sin\theta\,d\theta}{\int_0^\pi 2\pi r^2 \sin\theta\,d\theta} = \frac{\sin\theta\,d\theta}{\int_0^\pi \sin\theta\,d\theta}. \qquad (1.42)$$

The average $\langle \cos\theta \rangle$ over all vectors is

$$\langle \cos\theta \rangle = \int_0^\pi \cos\theta\, p(\theta)\,d\theta = \frac{\int_0^\pi \cos\theta \sin\theta\,d\theta}{\int_0^\pi \sin\theta\,d\theta}. \qquad (1.43)$$

This integration is simplified by noticing that $\sin\theta\,d\theta = -d\cos\theta$, by letting $x = \cos\theta$, and by replacing the limits 0 and π by 1 and -1. Then Equation (1.43) becomes

$$\langle \cos\theta \rangle = \frac{\int_1^{-1} x\,dx}{\int_1^{-1} dx} = \frac{\dfrac{x^2}{2}\bigg|_1^{-1}}{x\bigg|_1^{-1}} = 0. \qquad (1.44)$$

Physically, this says that the average projection on the z-axis of uniformly distributed vectors is zero. You can also see this by symmetry: just as many vectors point forward $(0 < \theta \le 90°)$ as backward $(90° < \theta \le 180°)$, so the average is zero.

Later we will find the quantity $\langle \cos^2\theta \rangle$ to be useful. Following the same logic, you have

Figure 1.8 A vector that can orient in all directions can be represented as starting at the origin and ending on the surface of a sphere. The area $2\pi r \ell d\theta$ represents the relative proportion of all the vectors that land in the strip at an angle between θ and $\theta + d\theta$ relative to the z-axis.

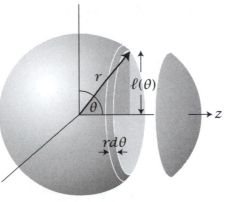

$$\langle \cos^2 \theta \rangle = \frac{\displaystyle\int_0^\pi \cos^2 \theta \sin \theta \, d\theta}{\displaystyle\int_0^\pi \sin \theta \, d\theta} = \frac{\displaystyle\int_1^{-1} x^2 \, dx}{\displaystyle\int_1^{-1} dx} = \frac{\left.\dfrac{x^3}{3}\right|_1^{-1}}{\left. x \right|_1^{-1}} = \frac{1}{3}. \tag{1.45}$$

Summary

Probabilities describe incomplete knowledge. The addition and multiplication rules allow you to draw consistent inferences about probabilities of multiple events. Distribution functions describe collections of probabilities. Such functions have mean values and variances. Combined with combinatorics—the counting of arrangements of systems—probabilities provide the basis for reasoning about entropy, and about driving forces among molecules, described in the next chapter.

Examples of Distributions

Here are some probability distribution functions that commonly appear in statistical mechanics.

Figure 1.9   Bernoulli

$$\psi(n) = p^n(1-p)^{N-n},$$

$$n = 0, 1, 2, \ldots, N. \qquad (1.46)$$

The Bernoulli distribution describes independent trials with two possible outcomes (see page 15). $\psi(n)$ is a distribution function, not a probability, because it is not normalized to sum to one.

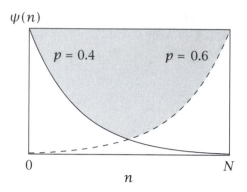

Figure 1.10   Poisson

$$p(n) = \frac{a^n e^{-a}}{n!},$$

$$n = 0, 1, 2, \ldots, N. \qquad (1.47)$$

The Poisson distribution approximates the binomial distribution when the number of trials is large and the probability of each one is small [3]. It is useful for describing radioactive decay, the number of vacancies in the Supreme court each year [4], the numbers of dye molecules taken up by small particles, or the sizes of colloidal particles.

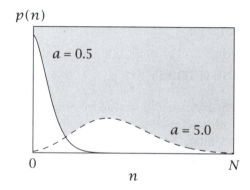

Figure 1.11   Gaussian

$$p(x) = \frac{1}{\sigma\sqrt{2\pi}}\left(e^{-x^2/2\sigma^2}\right),$$

$$-\infty \leq x \leq \infty. \qquad (1.48)$$

The Gaussian distribution is derived from the binomial distribution for large N [5]. It is important for statistics, error analysis, diffusion, conformations of polymer chains, and the Maxwell Boltzmann distribution law of gas velocities.

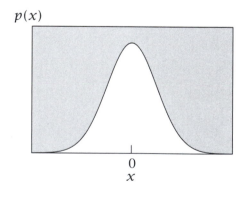

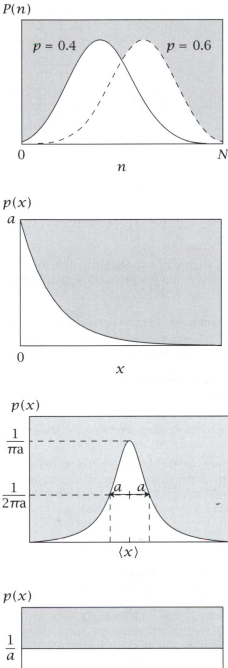

$P(n)$

$p = 0.4$ $p = 0.6$

0 n N

$p(x)$

a

0 x

$p(x)$

$\dfrac{1}{\pi a}$

$\dfrac{1}{2\pi a}$

a a

$\langle x \rangle$

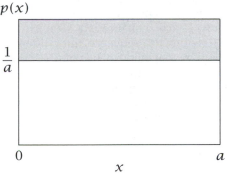

$p(x)$

$\dfrac{1}{a}$

0 x a

Figure 1.12 Binomial

$$P(n) = p^n (1-p)^{N-n}$$
$$\times \left(\frac{N!}{n!(N-n)!} \right),$$
$$n = 0, 1, 2, \ldots, N. \qquad (1.49)$$

The Binomial distribution for collections of Bernoulli trials is derived on page 15.

Figure 1.13 Exponential (Boltzmann)

$$p(x) = ae^{-ax},$$
$$0 \leq x \leq \infty. \qquad (1.50)$$

The exponential, or Boltzmann distribution, is central to statistical thermodynamics (see Chapters 6 and 10).

Figure 1.14 Lorentzian

$$p(x) = \frac{1}{\pi} \frac{a}{(x - \langle x \rangle)^2 + a^2},$$
$$-\infty \leq x \leq \infty. \qquad (1.51)$$

$2a$ is the width of the Lorentzian curve at the level of half the maximum probability. Lorentzian distributions are useful in spectroscopy [3].

Figure 1.15 Flat

$$p(x) = 1/a, \qquad (1.52)$$

where a is a constant independent of x (see Example 1.21, page 19).

Problems

1. Combining independent probabilities. You have applied to three medical schools: University of California at San Francisco (UCSF), Duluth School of Mines (DSM), and Harvard (H). You guess that the probabilities you'll be accepted are: $p(\text{UCSF}) = 0.10$, $p(\text{DSM}) = 0.30$, and $p(\text{H}) = 0.50$. Assume that the acceptance events are independent.

(a) What is the probability that you get in somewhere (at least one acceptance)?

(b) What is the probability that you will be accepted by both Harvard and Duluth?

2. Probabilities of sequences. Assume that the four bases A, C, T, and G occur with equal likelihood in a DNA sequence of nine monomers.

(a) What is the probability of finding the sequence AAATCGAGT through random chance?

(b) What is the probability of finding the sequence AAAAAAAAA through random chance?

(c) What is the probability of finding any sequence that has four A's, two T's, two G's, and one C, such as that in (a)?

3. The probability of a sequence (given a composition). A scientist has constructed a secret peptide to carry a message. You know only the composition of the peptide, which is six amino acids long. It contains one serine **S**, one threonine **T**, one cysteine **C**, one arginine **R**, and two glutamates **E**. What is the probability that the sequence **SECRET** will occur by chance?

4. Combining independent probabilities. You have a fair six-sided die. You want to roll it enough times to ensure that a **2** occurs at least once. What number of rolls k is required to ensure that the probability is at least 2/3 that at least one **2** will appear?

5. Predicting compositions of independent events. Suppose you roll a die three times.

(a) What is the probability of getting a total of two **5**'s from all three rolls of the dice?

(b) What is the probability of getting a total of *at least* two **5**'s from all three rolls of the die?

6. Computing a mean and variance. Consider the probability distribution $p(x) = ax^n$, $0 \le x \le 1$, for a positive integer n.

(a) Derive an expression for the constant a, to normalize $p(x)$.

(b) Compute the average $\langle x \rangle$ as a function of n.

(c) Compute $\sigma^2 = \langle x^2 \rangle - \langle x \rangle^2$ as a function of n.

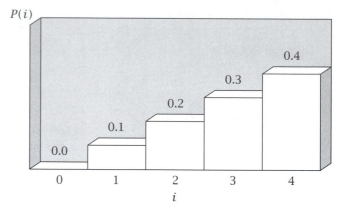

Figure 1.16 A simple probability distribution.

7. Computing the average of a probability distribution. Compute the average $\langle i \rangle$ for the probability distribution function shown in Figure 1.16.

8. Predicting coincidence. Your statistical mechanics class has twenty-five students. What is the probability that at least two classmates have the same birthday?

9. The distribution of scores on dice. Suppose that you have n dice, each a different color, all unbiased and six-sided.

(a) If you roll them all at once, how many distinguishable outcomes are there?

(b) Given two distinguishable dice, what is the most probable sum of their face values on a given throw of the pair? (That is, which sum between two and twelve has the greatest number of different ways of occurring?)

(c) What is the probability of the most probable sum?

10. The probabilities of identical sequences of amino acids. You are comparing protein amino acid sequences for homology. You have a twenty-letter alphabet (twenty different amino acids). Each sequence is a string n letters in length. You have one test sequence and s different data base sequences. You may find any one of the twenty different amino acids at any position in the sequence, independent of what you find at any other position. Let p represent the probability that there will be a 'match' at a given position in the two sequences.

(a) In terms of s, p, and n, how many of the s sequences will be perfect matches (identical residues at every position)?

(b) How many of the s comparisons (of the test sequence against each database sequence) will have exactly one mismatch at any position in the sequences?

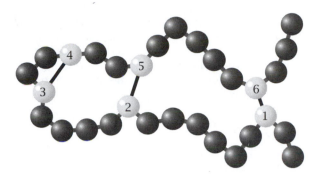

Figure 1.17 This disulfide bonding configuration with pairs 1-6, 2-5, and 3-4 is one of the many possible pairings. Count all the possible pairing arrangements.

11. The combinatorics of disulfide bond formation. A protein may contain several cysteines, which may pair together to form disulfide bonds as shown in Figure 1.17. If there is an even number n of cysteines, $n/2$ disulfide bonds can form. How many different disulfide pairing arrangements are possible?

12. Predicting combinations of independent events. If you flip an unbiased green coin and an unbiased red coin five times each, what is the probability of getting four red heads and two green tails?

13. A pair of aces. What is the probability of drawing two aces in two random draws without replacement from a full deck of cards?

14. Average of a linear function. What is the average value of x, given a distribution function $q(x) = cx$, where x ranges from zero to one, and $q(x)$ is normalized?

15. The Maxwell–Boltzmann probability distribution function. According to the kinetic theory of gases, the energies of molecules moving along the x-direction are given by $\varepsilon_x = (1/2)mv_x^2$, where m = mass and v_x is the velocity in the x-direction. The distribution of particles over velocities is given by the Boltzmann law, $p(v_x) = e^{-mv_x^2/2kT}$. This is the Maxwell–Boltzmann distribution (velocities may range from $-\infty$ to $+\infty$).

(a) Write the probability distribution $p(v_x)$, so that the Maxwell–Boltzmann distribution is correctly normalized.

(b) Compute the average energy $\langle (1/2)mv_x^2 \rangle$.

(c) What is the average velocity $\langle v_x \rangle$?

(d) What is the average momentum $\langle mv_x \rangle$?

16. Predicting the rate of mutation based on the Poisson probability distribution function. The evolutionary process of amino acid substitutions in proteins is sometimes described by the Poisson probability distribution function. The probability $p_s(t)$ that exactly s substitutions at a given amino acid position occur over an evolutionary time t is

$$p_s(t) = \frac{e^{-\lambda t}(\lambda t)^s}{s!},$$

where λ is the rate of amino acid substitutions per site per unit time. Fibrinopeptides evolve rapidly: $\lambda_F = 9.0$ substitutions per site per 10^9 years. Lysozyme is intermediate: $\lambda_L \approx 1.0$. Histones evolve slowly: $\lambda_H = 0.010$ substitutions per site per 10^9 years.

(a) What is the probability that a fibrinopeptide has no mutations at a given site in $t = 1$ billion years?

(b) What is the probability that lysozyme has three mutations per site in 100 million years?

(c) We want to determine the expected number of mutations $\langle s \rangle$ that will occur in time t. We will do this in two steps. First, using the fact that probabilities must sum to one, write $\alpha = \sum_{s=0}^{\infty} (\lambda t)^s / s!$ in a simpler form.

(d) Now write an expression for $\langle s \rangle$. Note that

$$\sum_{s=0}^{\infty} \frac{s(\lambda t)^s}{s!} = (\lambda t) \sum_{s=1}^{\infty} \frac{(\lambda t)^{s-1}}{(s-1)!} = \lambda t \alpha.$$

(e) Using your answer to part (d), determine the ratio of the expected number of mutations in a fibrinopeptide to the expected number of mutations in histone protein, $\langle s \rangle_{\text{fib}} / \langle s \rangle_{\text{his}}$ [6].

17. Probability in court. In forensic science, DNA fragments found at the scene of a crime can be compared with DNA fragments from a suspected criminal to determine the probability that a match occurs by chance. Suppose that DNA fragment A is found in 1% of the population, fragment B is found in 4% of the population, and fragment C is found in 2.5% of the population.

(a) If the three fragments contain independent information, what is the probability that a suspect's DNA will match all three of these fragment characteristics by chance?

(b) Some people believe such a fragment analysis is flawed because different DNA fragments do not represent independent properties. As before, suppose that fragment A occurs in 1% of the population. But now suppose that the conditional probability of B, given that A is $p(B|A) = 0.40$ rather than 0.040, and $p(C|A) = 0.25$ rather than 0.025. There is no additional information about any relationship between B and C. What is the probability of a match now?

18. Flat distribution. Given a flat distribution, from $x = -a$ to $x = a$, with probability distribution $p(x) = 1/(2a)$:

(a) Compute $\langle x \rangle$.

(b) Compute $\langle x^2 \rangle$.

(c) Compute $\langle x^3 \rangle$.

(d) Compute $\langle x^4 \rangle$.

19. Family probabilities. Given that there are three children in your family, what is the probability that:

(a) two are boys and one is a girl?

(b) all three are girls?

20. Evolutionary fitness. Suppose that the probability of having the dominant allele (D) in a gene is p and the probability of the recessive allele (R) is $q = 1 - p$. You have two alleles, one from each parent.

(a) Write the probabilites of all the possibilities: DD, DR, and RR.

(b) If the fitness of DD is f_{DD}, the fitness of DR is f_{DR}, and the fitness of RR is f_{RR}, write the average fitness in terms of p.

21. Ion-channel events. A biological membrane contains N ion-channel proteins. The fraction of time that any one protein is open to allow ions to flow through is q. Express the probability $P(m, N)$ that m of the channels will be open at any given time.

22. Joint probabilities: balls in a barrel. For Example 1.9, two green balls and one red ball drawn from a barrel without replacement:

(a) Compute the probability $p(RG)$ of drawing one red and one green ball in either order.

(b) Compute the probability $p(GG)$ of drawing two green balls.

23. Sports and weather. The San Francisco football team plays better in fair weather. They have a 70% chance of winning in good weather, but only a 20% chance of winning in bad weather.

(a) If they play in the Super Bowl in Wisconsin and the weatherman predicts a 60% chance of snow that day, what is the probability that San Francisco will win?

(b) Given that San Francisco lost, what is the probability that the weather was bad?

24. Monty Hall's dilemma: a game show problem. You are a contestant on a game show. There are three closed doors: one hides a car, and two hide goats. You point to one door, call it C. The gameshow host, knowing what's behind each door, now opens either door A or B, to show you a goat; say it's door A. To win a car, you now get to make your final choice: should you stick with your original choice C, or should you now switch and choose door B? (*New York Times*, July 21, 1991; *Sci Amer*, August 1998.)

References

[1] L Krantz. *What the Odds Are*. Harper Perennial, New York, 1992.

[2] WT Ziemba and DB Hansch. *Dr Z's Beat the Racetrack*. W Morrow & Co., New York, 1987.

[3] P Bevington and D Robinson. *Reduction and Error Analysis for the Physical Sciences*. McGraw-Hill, New York, 1992.

[4] S Ross. *A First Course in Probability*, 3rd edition. Macmillan, New York, 1988.

[5] F Reif. *Fundamentals of Statistical and Thermal Physics*. McGraw-Hill, New York, 1965.

[6] M Nei. *Molecular Evolutionary Genetics*. Columbia Press, New York, 1987, page 50.

Suggested Reading

W Feller, *An Introduction to Probability Theory and Its Applications*, Wiley, New York, 1968. Advanced treatise on principles and applications of probability theory.

PG Hoel, *Introduction to Mathematical Statistics*, Wiley, New York, 1984. Excellent review of uses of probabilities in statistics: statistical inference, hypothesis testing, and estimation.

E Nagel, *Principles of Probability Theory*, University of Chicago, Chicago, 1939. Philosophical foundations of probability theory.

S Ross, *A First Course in Probability*, 3rd edition, Macmillan, New York, 1988. An excellent introduction to probability theory.

RA Rozanov, *Probability Theory: Basic Concepts*, Dover, New York, 1969. A concise summary of probabilities and distribution functions.

2 Extremum Principles Predict Equilibria

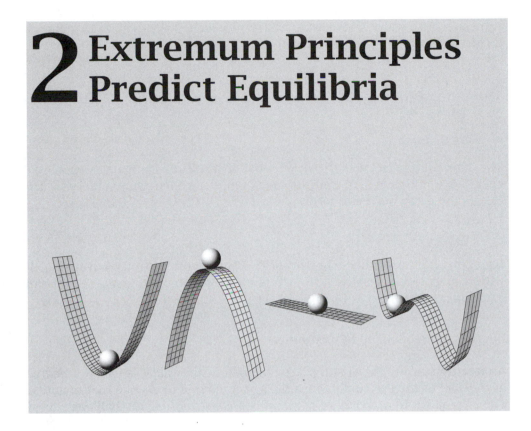

What Are Extremum Principles?

The forces on atoms and molecules can be described in terms of two tendencies. Molecules react, change conformations, bind, and undergo other chemical or physical changes in ways that cause a quantity called the energy to reach its minimum possible value, or a quantity called the entropy to reach its maximum possible value. We can predict the tendencies of matter by computing the minima or maxima of certain mathematical functions. These are called extremum (or variational) principles.

To illustrate, we first show that balls rolling downhill can be predicted by the minimization of energy. We describe the various types of extrema, called *states of equilibrium*—stable, unstable, neutral, and metastable states. Then, as a first step toward introducing the entropy and the Second Law of thermodynamics, we use the multiplicity concept from Chapter 1, and illustrate how gases exerting pressure, the mixing and diffusion of molecules, and rubber elasticity can be explained by a maximization principle.

EXAMPLE 2.1 Mechanical equilibrium is the state of minimum potential energy. For a ball of mass m on a hillside at height z, the gravitational *potential energy* $V(z)$ is

$$V(z) = mgz,$$

where g is the gravitational acceleration constant (see Table of Constants,

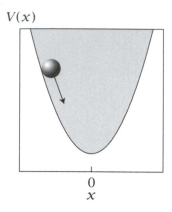

$V(x)$

0
x

Figure 2.1 The equilibrium position $x^* = 0$ for a ball in a quadratic valley has the minimum potential energy $V(x^*)$.

Appendix A). Figure 2.1 shows a valley in which the height $z(x)$ depends quadratically on the lateral position x:

$$z = x^2.$$

Substitute $z = x^2$ into $V(z) = mgz$ to find the potential energy of a ball in this quadratic valley:

$$V(x) = mgx^2.$$

The extremum principle says that the ball will roll to wherever the potential energy is a minimum. To find the minimum, determine what value, $x = x^*$, causes the derivative to be zero:

$$\left. \frac{dV(x)}{dx} \right|_{x^*} = 0 \quad \Longrightarrow \quad 2mgx^* = 0. \tag{2.1}$$

Equation (2.1) is satisfied when $x^* = 0$. Because the second derivative of V with respect to x is positive, this extremum is a minimum (the mathematics of extrema are discussed in Chapter 5, pages 65–72).

Degrees of Freedom and Constraints

A quantity such as the position x of the ball in the valley is called a *degree of freedom* of the system, because the system is free to change that quantity. The alternative to a *degree of freedom* is a *constraint*. If the ball were fixed at position $x = x_c$, it would be *constrained*. A system is not able to change a constraint. Constraints are imposed on the system from outside. A system governed by an extremum principle will change its degree of freedom until it reaches the maximum or minimum of some function allowed by the constraints. Systems can have multiple degrees of freedom and constraints.

What Is a State of Equilibrium?

A ball rolls downhill until it reaches the bottom, and then stays there. The bottom is its *equilibrium* position. Equilibrium defines where a system tends to go and stay. The *force f* on the ball defines the strength of its tendency toward equilibrium:

$$f = -\left(\frac{dV}{dx} \right). \tag{2.2}$$

The point of equilibrium is the point of zero net force. Therefore the state of equilibrium can either be defined as an extremum in the energy or as the point where the force is $f = 0$.

Stable, Unstable, Neutral, and Metastable States

Zero net force can be achieved in different ways. The potential energy is a function of the degrees of freedom, $V(x)$ in Example 2.1. Potential energy functions can take any shape. They may be perfectly flat, or smooth and curved, or rough and bumpy. An equilibrium state is defined as *stable* where a potential is at a minimum, *neutral* where a potential is flat, or *metastable* where a potential energy is at a minimum but other minima are deeper. Where a potential energy is at a maximum, a state is *unstable*. The stability of a system depends not just on whether there is zero force, but also on how the system responds to perturbations. Think about a ping-pong ball in a slight breeze. The breeze simply represents a small random fluctuating force.

The force on the ball due to gravity is $f(x) = -dV/dx$. Our sign convention specifies that as we move the ball in a direction to increase its potential energy (up), the force acts in the opposite direction (down) to reduce the energy.

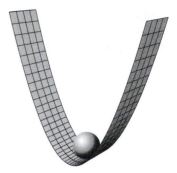

Figure 2.2 Stable.

STABLE EQUILIBRIUM.   For a system in a *stable* equilibrium at $x = x^*$, the energy is at a minimum as shown in Figure 2.2. The potential energy increases as the system moves away from a stable equilibrium:

$$V(x) - V(x^*) > 0 \qquad \text{for all } x \neq x^*.$$

A stable equilibrium can be identified by tests of the first and second derivatives:

$$\frac{dV}{dx} = 0 \quad \text{and} \quad \frac{d^2V}{dx^2} > 0 \quad \text{at } x = x^*.$$

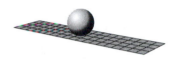

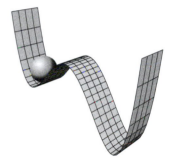

Figure 2.3 Neutral.

To move a system incrementally *toward* a stable equilibrium requires a downward step, $dV \leq 0$. If a small fluctuation displaces the system away from $x = x^*$, a stable system tends to return to $x = x^*$. If you put a ping-pong ball in a hole in your backyard, it will still be there later, even with the breeze blowing.

NEUTRAL EQUILIBRIUM.   In a *neutral* equilibrium, the potential energy surface is flat as shown in Figure 2.3:

$$\frac{dV}{dx} = 0 \qquad \text{for all } x.$$

If you put a ping-pong ball on a flat surface, when you return the breeze may have blown it elsewhere on the surface.

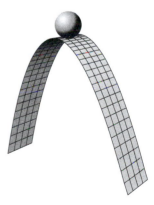

Figure 2.4 Metastable.

METASTABLE EQUILIBRIUM.   A system in *metastable* equilibrium is stable to small displacements, but unstable to larger displacements as shown in Figure 2.4.

$$V(x) - V(x^*) > 0 \qquad\qquad \text{for } |x - x^*| = \text{small}$$

$$\frac{dV}{dx} = 0 \quad \text{and} \quad \frac{d^2V}{dx^2} > 0$$

$$V(x) - V(x^*) < 0 \qquad\qquad \text{for } |x - x^*| = \text{large}.$$

In metastable equilibrium, a system is locally but not globally stable. If you put a ping-pong ball in a small hole on the side of a deeper ditch, and if the breeze is small, it will stay in the small hole. However, in a stiffer wind, you might later find the ball in the deeper ditch.

UNSTABLE STATES.   An *unstable* state is the top of a hill as shown in Figure 2.5. The system can reduce its energy by moving away from the point of

Figure 2.5 Unstable.

Table 2.1 $N = 4$

n	W	$\ln W$
4	$\dfrac{4!}{0!4!} = 1$	0
3	$\dfrac{4!}{1!3!} = 4$	1.386
2	$\dfrac{4!}{2!2!} = 6$	1.792
1	$\dfrac{4!}{3!1!} = 4$	1.386
0	$\dfrac{4!}{4!0!} = 1$	0
	Total $W = 16$	

Table 2.2 $N = 10$

n	W	$\ln W$
10	1	0.0
9	10	2.303
8	45	3.807
7	120	4.787
6	210	5.347
5	252	5.529
4	210	5.347
3	120	4.787
2	45	3.807
1	10	2.303
0	1	0.0
	Total $W = 1024$	

zero force x^* in either the $+x$ or $-x$ direction:

$$V(x) - V(x^*) < 0 \qquad \text{for x} \neq x^*.$$

To determine whether a state is stable or unstable requires a second-derivative test to see whether the curvature of $V(x)$ is up or down. For an unstable state

$$\frac{dV}{dx} = 0 \quad \text{and} \quad \frac{d^2V}{dx^2} < 0 \qquad \text{at } x = x^*.$$

Moving toward an unstable state is described by $dV \geq 0$. A ping-pong ball placed on top of a hill won't be there long. If the slightest breeze is blowing, the ball will roll off the hill.

The cover of this book shows an energy surface involving multiple degrees of freedom, and various stable, metastable and unstable states.

The Second Extremum Principle: Maximizing Multiplicity Predicts the Highest Probability Outcomes

Now we introduce a different extremum principle, one that predicts the distributions of outcomes in statistical systems, such as coin flips or die rolls. This will lead to the concept of entropy and the Second Law of thermodynamics.

Let's consider the difference between two questions of probability, one based on the *sequence* and the other on the *composition* of outcomes. First, in a series of 4 coin tosses, which *sequence*, $HHHH$ or $HTHH$, is more likely? The answer, given in Examples 1.3 and 1.4, is that all sequences are equally probable. Each sequence occurs with a probability of $(1/2)^4$. Contrast this with a question of *composition*. Which *composition* is more probable: 4 H's and 0 T's, or 3 H's and 1 T? Although each *sequence* is equally probable, each *composition* is not. 3 H's and 1 T is a more probable composition than 4 H's and 0 T's. Only 1 of the 16 possible sequences is composed of 4 H's and 0 T's, but 4 of the 16 possible sequences are composed of 3 H's and 1 T ($HHHT, HHTH, HTHH,$ and $THHH$).

Predicting Heads and Tails by a Principle of Maximum Multiplicity

A simple maximization principle predicts the most probable composition, 2 heads and 2 tails in this case. Table 2.1 shows the number of sequences $W(n, N) = N!/(n!(N - n)!)$, as a function of n, the number of heads in $N = 4$ coin flips. W is a maximum (six sequences) when there are 2 heads, $n = 2$. The most probable composition of 4 coin flips is 2H's and 2T's because more sequences (6 out of 16) have this composition than any other. Table 2.1 shows that the value of n ($n^* = 2$) that maximizes W also maximizes $\ln W$. Later, we will use $\ln W$ to define the entropy.

What happens if you have $N = 10$ coin flips instead of $N = 4$? Table 2.2 shows that the maximum multiplicity occurs again when the number of heads equals the number of tails. You would observe the 5H, 5T composition 24.6% of the time (252 arrangements out of 1024 possible). Any other composition is less probable. Nearly 90% of the time the composition will be within ±2 of 5H, 5T.

The multiplicity function $W(n)$ becomes increasingly peaked as you increase the number of coin flips N (see Figure 2.6). Suppose you have $N = 100$ coin flips. The number of sequences of coin-flips that have 50 heads and 50 tails is nearly a millionfold greater than the number of sequences that have 25 heads and 75 tails:

$$W(50, 100) = \frac{100!}{50!50!} = 1.01 \times 10^{29}, \quad \text{and}$$

$$W(25, 100) = \frac{100!}{25!75!} = 2.43 \times 10^{23}.$$

If you observe 25 heads and 75 tails more often than 50 heads and 50 tails in many sets of $N = 100$ coin flips, you would have evidence of bias. Furthermore, because there is only one sequence that is all heads, the probability of observing 100 heads in 100 flips of an unbiased coin is 9.9×10^{-30}, virtually zero.

The implication of these results is remarkable. While the random process of flipping coins can result in any composition of heads and tails, the composition has a strong tendency toward 50% H and 50% T. Indeed, even though this is a random process, if the number of trials is large enough, the composition of heads and tails becomes predictable with great precision.

Like the energy minimization principle, here we have a function $W(n)$, the maximum of which identifies the value of n that we are most likely to observe. To maximize W or $\ln W$ for a series of coin flips, we compute $dW(n)/dn$ for fixed N, and find the value $n = n^*$ that causes the derivative to equal zero. The result is that $n^* = N/2$. In this case the number of heads n acts like a degree of freedom of the system. n can take on any value from zero to N. Any individual sequence is just as likely as any other. The reason that the 50% heads, 50% tails composition is so strongly favored is that there are more distinguishable sequences of coin flips with that composition than any other. No individual coin flip has any particular tendency, or 'feels' any force. As we'll see, this is the nature of entropy as a driving force. Systems tend to be found in the states with the highest multiplicity. The maximization of W or $\ln W$ serves the same predictive role for the composition of coin flips as the minimization of energy serves for the position of the ball in the valley.

Now we will use very simple models to illustrate how this principle of maximum multiplicity also predicts the tendencies of gases to expand and exert pressure, the tendencies of molecules to mix and diffuse, and the tendency of rubber to retract when it is stretched.

Simple Models Show
How Systems Tend Toward Disorder

In the examples below and throughout the text, we will rely on one of the major tools of physics, simplified models. Simple models are caricatures. They seek to keep what is essential, and omit less important details. Model-makers are storytellers. Stories can be involved and detailed, or, like simple models, they can describe just the essentials. The art of model building is in recognizing

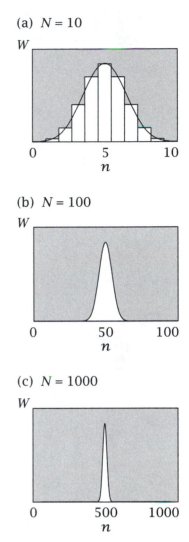

(a) $N = 10$

(b) $N = 100$

(c) $N = 1000$

Figure 2.6 The multiplicity function W of the number of heads n narrows as the total number of trials N increases.

Figure 2.7 For Example 2.2, Case A, three particles are distributed throughout $M_A = 5$ sites. In Case B, three particles are in $M_B = 4$ sites. In Case C the three particles fill $M_C = 3$ sites.

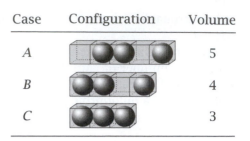

Case	Configuration	Volume
A		5
B		4
C		3

what is essential. Sometimes a given phenomenon can be described by different models. For example, we will find that both lattice models and quantum mechanics can give the ideal gas law. The best models are the least arbitrary—they have the fewest assumptions or adjustable parameters. A model is useful if it makes testable predictions (right or wrong), and if it helps encapsulate or unify our knowledge.

Here and in later chapters we use lattice models. In lattice models, atoms or parts of molecules are represented as hard spherical beads. Space is divided up into bead-sized boxes, called lattice sites, which are artificial, mutually exclusive, and collectively exhaustive units of space. Each lattice site holds either zero or one bead. Two or more beads cannot occupy the same site. The lattice model just captures the idea that particles can be located at different positions in space, and that no two particles can be in the same place at the same time. This is really all we need for some problems, such as the following.

EXAMPLE 2.2 Why do gases exert pressure? A lattice model. Imagine a gas of N spherical particles that are free to distribute throughout either a large volume or a small one. Why does the gas spread out into a large volume?

The tendency to spread out, called pressure, can be explained either in terms of mechanics or maximum multiplicity. These are two different perspectives of the same process. According to the mechanical interpretation, pressure results from particles banging against the container walls. In general, few problems can be solved by the mechanical approach. The multiplicity perspective is more powerful, particularly for complex problems. We illustrate the multiplicity perspective here.

Let's miniaturize the problem. Our aim is to devise a model that we can visualize easily, solve exactly, and that will capture the essence of the problem without mathematical complexity. Figure 2.7 shows $N = 3$ spherical particles spread out in five boxes (in A), contained in a smaller volume (in B), and bunched up against the left wall of a one-dimensional lattice container (in C). We take the volume M (the number of lattice sites in which the particles are distributed) as the degree of freedom: $M_A = 5$, $M_B = 4$, and $M_C = 3$.

What value of M maximizes the multiplicity? You can compute the multiplicity $W(N,M)$ of N particles in M lattice sites the same way that you count the number of distinguishable sequences of N heads in M coin flips. That is, the sequence [vacant, occupied, occupied, vacant, occupied] is just a set of binary outcomes like the coin flip sequence $[THHTH]$. Our premise is that every sequence is equally likely, no matter how its vacancies and occupancies are arranged. The number of distinguishable arrangements of vacancies and

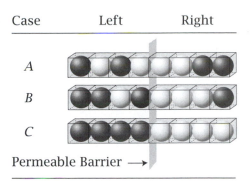

Case	Left	Right
A		
B		
C		

Permeable Barrier $\longrightarrow$

Figure 2.8 For Example 2.3, Case A, the composition is two ● and two ○ particles on the left, and two ● and two ○ on the right. In Case B, the composition is three ● and one ○ on the left and one ● and three ○ on the right. In Case C the composition is four ● and zero ○ on the left, and zero ● and four ○ on the right.

occupancies of N particles in M sites is

$$W(N, M) = \frac{M!}{N!(M-N)!},$$ (2.3)

so there are $W_A(3, 5) = 5!/(3!2!) = 10$ possible arrangements for $M_A = 5$, $W_B(3, 4) = 4!/(3!1!) = 4$ arrangements for $M_B = 4$, and $W_C(3, 3) = 3!/(3!0!) = 1$ arrangement for $M_C = 3$. The multiplicity increases as the volume increases. If the system has only these three possible volumes available, then the probability is $p_C = 1/15$ (that is $1/(1+4+10)$) that the particles will be bunched up on the left side of the container, $p_B = 4/15$ that the gas will expand to an intermediate degree, and $p_A = 2/3$ that the particles will be fully spread out.

If the degree of freedom of the system is its volume, the particles will spread out into the largest possible volume to maximize the multiplicity of the system. This is the basis for the force called *pressure*. In Chapters 7 and 24, after we have developed the thermodynamics that we need to define the pressure, we will show that despite the simplicity of this model, it accurately gives the ideal and van der Waals gas laws.

While the *pressure* is a force for expansion, the *chemical potential* is a force for mixing, as when a dye diffuses in water. The chemical potential describes the tendencies of substances to dissolve or to partition between different phases of matter or to transform into different chemical species. Here we use the lattice model to show why particles tend to mix.

EXAMPLE 2.3 Why do materials diffuse? A lattice model for the chemical potential. Suppose that you have four black particles ● and four white particles ○ in eight lattice sites. There are four lattice sites on the left and four lattice sites on the right, separated by a permeable wall. The total volume is fixed. All the lattice sites are occupied by either a ○ or a ● particle. Now the degrees of freedom are the numbers of ○ and ● particles on each side of the permeable wall, not the volume. Again we consider three cases. As in Example 2.2, Figure 2.8 shows one particular spatial configuration for each case, but several different configurations might be possible for a given composition. What is the most probable composition of ○ and ● particles on each side of the wall? (We predict compositions rather than particle sequences because this is what experiments can measure.)

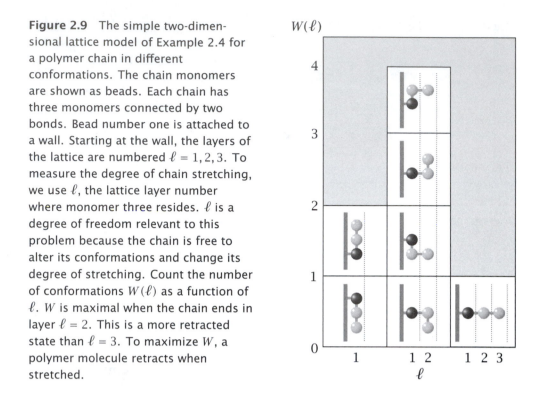

Figure 2.9 The simple two-dimensional lattice model of Example 2.4 for a polymer chain in different conformations. The chain monomers are shown as beads. Each chain has three monomers connected by two bonds. Bead number one is attached to a wall. Starting at the wall, the layers of the lattice are numbered $\ell = 1, 2, 3$. To measure the degree of chain stretching, we use ℓ, the lattice layer number where monomer three resides. ℓ is a degree of freedom relevant to this problem because the chain is free to alter its conformations and change its degree of stretching. Count the number of conformations $W(\ell)$ as a function of ℓ. W is maximal when the chain ends in layer $\ell = 2$. This is a more retracted state than $\ell = 3$. To maximize W, a polymer molecule retracts when stretched.

Once again, the statistical mechanical approach is to assume that each spatial configuration (sequence) is equally probable. Find the most probable mixture by maximizing the multiplicity of arrangements. For each given value of left and right compositions, the total multiplicity is the product of the multiplicities for the left and the right sides:

$$\text{Case } A: \quad W = W(\text{left}) \cdot W(\text{right}) = \frac{4!}{2!2!} \frac{4!}{2!2!} = 36.$$

$$\text{Case } B: \quad W = W(\text{left}) \cdot W(\text{right}) = \frac{4!}{1!3!} \frac{4!}{3!1!} = 16.$$

$$\text{Case } C: \quad W = W(\text{left}) \cdot W(\text{right}) = \frac{4!}{0!4!} \frac{4!}{4!0!} = 1.$$

Case A, which has the most uniform particle distribution, has the highest multiplicity and therefore is the most probable. Case C, which has the greatest particle segregation, is the least probable. If the degree of freedom is the extent of particle exchange, then the multiplicity will be greatest when the particles are distributed most uniformly. You can predict the tendencies of particles to mix and diffuse with the principle that the system tends toward the distribution that maximizes its multiplicity.

EXAMPLE 2.4 Why is rubber elastic? When you stretch a rubber band, it pulls back. This retractive force is due to the tendency of polymers to adopt conformations that maximize their multiplicities. Polymers have fewer *conformational* states when fully stretched (see Chapters 31–33). Figure 2.9 illustrates the idea. Fix the first monomer of a chain to a wall. The degree of freedom is the position of the other end, at distance $\ell = 1, 2,$ or 3 from the wall. $W(\ell)$ is the multiplicity of configurations that have the second chain end in layer ℓ.

This simple model shows that a polymer will tend to be found in a partly *retracted* state $\ell = 2$, rather than stretched to length $\ell = 3$, because of the greater number of conformations ($W = 4$) for $\ell = 2$ than for $\ell = 3$ ($W = 1$). For the same reason, the polymer also tends not to be *flattened* on the surface: $W = 2$ when $\ell = 1$. If you stretch a polymer molecule, it will retract on the basis of the principle that it will tend toward its state of maximum multiplicity. This is the nature of entropy as a force.

In Chapter 6 we will define the quantity called *entropy*, $S = \text{constant} \times \ln W$, which has a maximum for the values of the degrees of freedom that maximize W. The tendencies of systems toward states of maximum multiplicity are also tendencies toward states of maximum entropy.

Summary

Forces act on atoms and molecules. The degrees of freedom of a system will change until the system reaches a state in which the net forces are zero. An alternative description is in terms of extremum principles: systems tend toward states that have minimal energies or maximal multiplicities, or, as we will see in Chapter 8, a combination of the two tendencies. The tendency toward maximum multiplicity or maximum entropy accounts for the pressures of gases, the mixing of fluids, the retraction of rubber, and, as we'll see in the next chapter, the flow of heat from hot objects to cold ones.

Problems

1. A lattice gas.   How many arrangements are there of fifteen indistinguishable lattice gas particles distributed on:

   (a) $V = 20$ sites?

   (b) $V = 16$ sites?

   (c) $V = 15$ sites?

2. Maximum of binomial distribution.   Find the value $n = n^*$ that causes the function

$$W = \frac{N!}{n!(N-n)!} p^n (1-p)^{N-n}$$

to be at a maximum, for constants p and N. Use Stirling's approximation, $x! \simeq (x/e)^x$ (see page 56). Note that it is easier to find the value of n that maximizes $\ln W$ than the value that maximizes W. The value of n^* will be the same.

3. Finding extrema.   For the function

$$V(x) = \frac{x^3}{3} + \frac{5x^2}{2} - 24x :$$

   (a) Where is the maximum?

   (b) Where is the minimum?

4. The binomial distribution narrows as N increases. Flip a coin $4N$ times. The most probable number of heads is $2N$, and its probability is $p(2N)$. If the probability of observing N heads is $p(N)$, show that the ratio $p(N)/p(2N)$ diminishes as N increases.

5. De-mixing is improbable.   Using the diffusion model of Example 2.3, with $2V$ lattice sites on each side of a permeable wall and a total of $2V$ white particles and $2V$ black particles, show that perfect de-mixing (all white on one side, all black on the other) becomes increasingly improbable as V increases.

6. Stable states.   For the energy function $V(\theta) = \cos \theta$ for $0 \leq \theta \leq 2\pi$, find the values $\theta = \theta_s$ that identify stable equilibria, and the values $\theta = \theta_u$ that identify unstable equilibria.

Suggested Reading

HB Callen, *Thermodynamics and an Introduction to Thermostatistics*, Wiley, New York, 1985. In-depth discussion of stability and equilibrium.

3 Heat, Work & Energy

Heat Flows to Maximize Entropy

In Chapter 2, we noted that some forces on atoms and molecules are tendencies toward maximum multiplicity. However, molecules are also driven by energies. This chapter traces the history of thought that led to the modern view that atoms and molecules are particles that carry energies and move freely throughout space. Before 1800, a major mystery of physical science was why heat flows from hot objects to cold objects. Is heat like a fluid that equalizes its levels in two connected containers? This was the view until thermodynamics emerged. Heat was mistakenly thought to be a *conserved* property, like a mass of liquid that could flow between bodies. The correction to this misunderstanding was captured in the *First Law of thermodynamics*. We now know that heat flow is an energy transfer process in which some molecules exchange their energies with others. Why does heat flow? Why do molecules exchange energies? The answer is not found in an equalization principle or a conservation law like the First Law. Heat flow is a consequence of the tendency toward maximum multiplicity, which is the *Second Law of thermodynamics*.

Some Physical Quantities Obey Conservation Laws

The origin of ideas about molecules and their energies is deeply rooted in laws of conservation, which are central to physical science. The law of conservation of momentum was discovered in 1668 by J Wallis. Wallis observed that

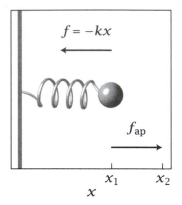

Figure 3.1 A spring force $f = -kx$ retracts to pull a ball to the left. To stretch the spring requires an applied force $f_{ap} = -f$ acting toward the right ($+x$ direction).

when objects collide, their velocities, positions, and accelerations can change, but the sum of the *mass × velocity* of all the objects does not. The quantity *mass × velocity*, called the momentum, is special because it is *conserved*.

Other functions of the position, time, velocity, and acceleration depend on the details of the collision process. For example, the quantity *velocity* or the quantity *mass × acceleration squared* will generally change when one particle collides with another. Such quantities are different for every situation and depend on the angles of the collisions and the shapes of the objects. However, conservation laws describe properties that are exceptionally simple owing to their invariance with respect to the particular details. The total momentum is the same before and after a collision, no matter how the collision occurs. Similar laws describe the conservation of mass, of angular momentum, and of energy. A property that is *conserved* is neither created nor destroyed as collisions take place. Because they are conserved, mass, momentum, and energy can only 'flow' from one place to another.

Conservation laws help to predict the behaviors of physical and chemical systems. For example, because the total momentum does not change in a collision, you can predict the final velocities of colliding objects on the basis of their masses and initial velocities. An important conserved quantity is the *energy*. Here's some terminology that we need for describing energies.

Force, Work, and Energy

FORCE. Newton's second law defines the force in terms of the acceleration of an object:

$$f = ma = m\frac{d^2x}{dt^2},$$ (3.1)

where the mass m of an object represents its resistance to acceleration, x is its position, the acceleration is $a = d^2x/dt^2$, and the velocity is $v = dx/dt$.

WORK. Consider a ball located at $x = 0$ that can move along the x-axis (see Figure 3.1). One force acting on the ball is from a spring attached to a wall pulling the ball to the left. Call this the intrinsic force f. For a spring we have $f = -k_s x$, where the minus sign indicates that it acts in the $-x$ direction. Suppose you apply an approximately equal opposing force to stretch the spring $f_{ap} = -f$. (It must be slightly greater, and not exactly equal, to stretch the spring. However the near equality ensures that there is approximately no net force or acceleration.) The work δw you perform *on* the system by moving the ball a distance dx to the right is

$$\delta w = f_{ap}\, dx = -f\, dx.$$ (3.2)

The total work w performed on the system in stretching the spring from x_1 to x_2 is the integral

$$w = \int_{x_1}^{x_2} f_{ap}\, dx = -\int_{x_1}^{x_2} f\, dx.$$ (3.3)

Examples 3.1 and 3.2 calculate the work of stretching a spring and the work of lifting a weight.

EXAMPLE 3.1 The work of stretching a spring. To stretch a spring from 0 to x_0, Equation (3.3) gives

$$w = \int_0^{x_0} k_s x \, dx = \frac{k_s x_0^2}{2}.$$

EXAMPLE 3.2 The work of lifting a weight. Gravity acts downward with a force $f = mg$, where m is the mass of the weight and g is a gravitational constant (see Appendix A). Suppose you lift a weight from the floor, $z = 0$, to a height $z = h$ (see Figure 3.2). The downward force defines the direction $x > 0$, but positive z is in the upward direction, so $dz = -dx$. Substituting $f = mg$ and $dx = -dz$ into Equation (3.3) gives the work of lifting the weight

$$w = -\int_0^h (mg)(-dz) = mgh.$$

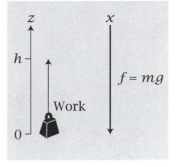

Figure 3.2 The intrinsic force f acts downward on a weight. The work to raise the weight to a height $z = h$ is positive.

CONSERVATIVE FORCES. The forces operating on the ball on the hill in Example 2.1 are *conservative forces*, because there are no frictional losses, no turbulence, and no other dissipations of energy. For a conservative force f no net work is performed in moving an object through any cycle that returns it to its starting point. For a force acting in the x direction the work performed in the cycle from A to B to A is

$$w(A, B) + w(B, A) = -\int_A^B f \, dx - \int_B^A f \, dx$$

$$= -\int_A^B f \, dx + \int_A^B f \, dx = 0. \qquad (3.4)$$

KINETIC ENERGY. The kinetic energy K of an object of mass m moving at velocity v is defined as

$$K = \frac{1}{2} m v^2. \qquad (3.5)$$

The kinetic energy is the work that an object can perform by virtue of its *motion*. As a moving object does work, it slows down and loses kinetic energy.

POTENTIAL ENERGY. In contrast, the potential energy V is the work that an object can perform by virtue of its *position*, or the positions of its subcomponents. If an object is raised above the earth, it gains the gravitational potential to do work when it comes back down. The potential to do work can be stored in a stretched or compressed spring, or in a battery through the separation of electrical charges. Like the kinetic energy, the potential energy of a system can change. The potential energy of a ball changes as it rolls downhill, as in Example 2.1. The potential energy of a mass attached to a spring changes as the mass bounces.

TOTAL ENERGY. The total energy E of a mechanical system is defined by the **law of conservation of energy**:

$$K + V = E = \text{constant}. \qquad (3.6)$$

Water In

Water Out

Figure 3.3 Heat doesn't flow in and out of a steam engine like water flows in and out of a water wheel.

The conservation of energy is a fundamental observation about nature. Although the kinetic and potential energies each change throughout a process, their sum, the total energy E, is constant. The total energy is an *invariant of the motion*. That is, the sum of all the kinetic energies plus all the potential energies of a system is the same at any stage before, after, and during a process. Any change in the kinetic energy of an object as it does work, or as work is done upon it, is balanced by an equal and opposite change in the potential energy of the system. Like momentum, the total energy cannot be created or destroyed: it can only flow from one place to another.

Heat Was Thought to Be a Fluid

The ability to *flow* is an aspect of conserved properties: mass, momentum, and energy can all flow. What about heat? When a hot object contacts a cold one, heat flows. Does it follow that heat is a conserved quantity? It does not. Heat is not a conserved quantity. All conserved properties can flow, but not all properties that flow are conserved.

Until the mid-1800s, heat was (incorrectly) thought to be a conserved form of matter, a fluid called *calorique*. The calorique model explained that materials expand upon heating because the added heat (calorique fluid) occupied space. The flow of heat from hot to cold objects was attributed to repulsions between the more crowded calorique particles in the hot material. Calorique particles escaped to cold materials, where they were less crowded. Because all known materials could absorb heat, it was thought that calorique particles were attracted to all materials, filling them in proportion to the volume of the material. The temperature of a substance was known to increase on heating, so it was thought that temperature directly measured the amount of heat in a material.

The misconception that calorique obeyed a law of conservation had important implications for industry and commerce. The industrial revolution began with J Watt's steam engine around 1780. During the late 1700s and early 1800s, power generated from steam gained major economic importance. The industrial quest for efficient steam engines drove the development of thermodynamics. A fundamental question was how heat is converted to work in steam engines. It was thought that the heat flow in a steam engine was like the water flow in a water wheel. In a water wheel, water goes in at the top, falls down to turn the wheel, and comes out the bottom (see Figure 3.3). The water is conserved: the flow of water from the bottom equals the water flow into the top. The amount of heat flowing out of the exhaust of a steam piston was believed to equal the amount of heat in the steam that enters the piston chamber.

The view of heat as a fluid began to change in the mid-1700s. The first step was the concept of *heat capacity*, developed by J Black around 1760: he heated mercury and water over the same flame and discovered that the temperature of the mercury was higher than the temperature of the water. He concluded that calorique could not be a simple fluid, because the amount taken up depended on the material that contained it. Different materials have different capacities to take up heat. The heat capacity of a material was defined as the amount of heat required to raise its temperature by 1 °C.

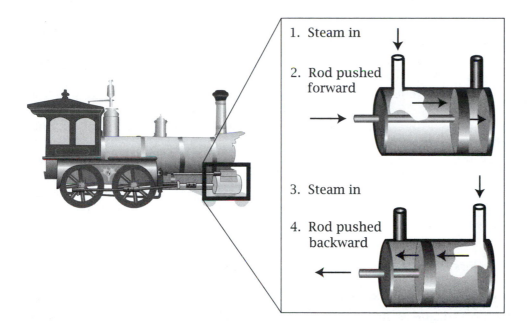

Figure 3.4 In a heat engine, hot gas enters a chamber, exerts pressure on a piston, and performs work. The gas cools on expansion and exits the chamber.

The next step was Black's discovery of *latent heat*, which showed that temperature and heat are different properties. Liquid water slightly below 100 °C can absorb heat and boil to become steam slightly above 100 °C. In this case the uptake of heat causes a change of phase, not a change of temperature. This heat is latent because it is 'stored' in the vapor and can be recovered in a condenser that converts the steam back to liquid water. The melting of ice also involves a latent heat. Although the discovery of latent heat showed that heat and temperature are different properties, heat was still regarded as a conserved fluid until the 1800s. And the question remained: What is temperature? We develop a conceptual model of temperature in Chapter 12, after laying the groundwork in thermodynamics and statistical mechanics.

The development of quantitative thermodynamics ultimately caused the downfall of the calorique theory of heat. The early 1800s were years of intensive development of engines driven by steam (see Figure 3.4), by hot air, and later by internal combustion. In 1853, J Ericsson launched a magnificent engine, designed using the calorique theory, with pistons 14 feet in diameter driven by hot air. Ericsson hoped the calorique engine would achieve 600 horsepower and compete with the steam engine. A set of 'regenerator' coils were supposed to collect exhaust heat and re-use it, on the basis of the assumption that heat was a conserved and recyclable fluid. But the calorique engine and the calorique theory ultimately failed because heat is not a conserved fluid. One of the main uses of the Second Law of thermodynamics in its early days was to debunk such 'perpetual motion' machines. The Second Law, now the centerpiece of statistical thermodynamics, is little needed for that purpose any longer.

Figure 3.5 The kinetic view of heat: matter is composed of molecules that exchange kinetic energies in collisions.

Other problems with the theory of calorique emerged at that time. First, it was shown that *radiant heat* could be transmitted through a vacuum by electromagnetic radiation. If heat is a form of matter, how could it pass through a vacuum that is devoid of matter?

Second, careful measurements showed that work could be converted to heat quantitatively. In 1798, Count Rumford showed that the mechanical work involved in boring cannons was converted to heat. In 1799, H Davy showed that the mechanical work involved in rubbing ice cubes together was sufficient to melt them. In the 1850s, JP Joule produced heat from work in many ways: a falling weight rotating a paddle wheel and imparting heat to a liquid, electrical work heating a resistor, and others. These experiments were difficult to reconcile with the view that heat is a conserved fluid. How could heat be created by work?

The First Law of Thermodynamics: the Conservation of Heat Plus Work

These experiments led to two paradigm-changing conclusions: (1) heat is *not* conserved, and (2) heat and work can be interconverted. This new understanding, first formulated by JR von Mayer in 1842, came to be known as the First Law of thermodynamics. It is not the heat (q) that is conserved ($q_{in} \neq q_{out}$). It is the *sum $q + w$* of the heat *plus* the work (w) that is conserved: this sum was called the *internal energy $U = q + w$*. Various forms of work were known before the 1800s, including mechanical, electrical, and magnetic. The advance embodied in the First Law was the recognition that heat can be added to this list, and that heat is a form of energy transfer, not a form of matter.

Atoms and Molecules Have Energies

The Kinetic Theory of Gases

The modern conception of particulate atoms and molecules developed in parallel with the modern view of heat. What is the microscopic nature of heat? This question was addressed in the kinetic theory of gases, a radically different view of heat that superseded the model of calorique, owing to the efforts of Clausius, Maxwell, Boltzmann, and others during the late 1800s. The concept of calorique had been intimately tied to the concept of the *ether*, a hypothetical pervasive medium that could transmit vibrations. Atoms were thought to be at fixed positions in the ether. The kinetic theory introduced three novel concepts. (1) Matter is composed of molecules that are not located at fixed positions in space, but are free to move through a space that is otherwise empty. (2) Heat is the exchange of energy that takes place due to the *motions and collisions of molecules.* In the kinetic theory, molecules collide and exchange energy like Newtonian billiard balls. (3) Electromagnetic radiation can influence the motions of molecules. This is the basis for radiant heat.

How does the kinetic theory of gases explain the conversion of heat to work? How does a gas lose heat and push the piston in a steam engine? The kinetic theory of gases is a mechanical model on a microscopic scale (see Figure 3.5). According to the kinetic theory, when molecules collide they exchange energy and momentum. At high temperature, gas particles move with high velocities. When a molecule collides with a piston, it imparts momentum, loses kinetic energy, moves the piston, and produces work. Alternatively, a high-energy particle colliding with its container can impart energy to the container wall, and the container wall can impart this kinetic energy to the surroundings as heat. Because the collisions between the gas and the piston are random, some collisions will perform work, owing to motion in the direction of the piston's trajectory. Other collisions will produce only heat, molecular motions in all other directions that result in a flow of energy out of the container.

The kinetic theory is remarkably successful. It predicts the ideal gas law and other properties of gases including diffusion rates, viscosities, thermal conductivities, and the velocities of sound in gases. It provides a model in which every molecule has its own energy, different molecules have different energies, and molecules can exchange their energies. The kinetic theory predicts that the temperature T of a gas is proportional to the average kinetic energy of the gas particles:

$$\frac{3}{2}kT = \frac{m\langle v^2 \rangle}{2}, \tag{3.7}$$

where m is the mass of one particle, $\langle v^2 \rangle$ is the square of the particle velocity averaged over all the particles, $m\langle v^2 \rangle/2$ is the average kinetic energy of the gas molecules, and k is Boltzmann's constant (see Appendix A, Table of Constants).

A Better Model: Energy is Quantized

Despite the tremendous advances that resulted from the kinetic theory of gases, it wasn't perfect either. The kinetic theory described gas particles as mechanical objects like billiard balls, with a continuum of possible energies. But twentieth century quantum theory, describing the motions of particles at the atomic level, showed that the energies of atoms and molecules are quantized. Each particle has discrete amounts of energy associated with each of its allowed degrees of freedom, some of which are translations, rotations, vibrations, and electronic excitations. For example, quantum mechanics might dictate that the molecules in a given system have energies of 0, 5, 16, 21, and 26 units, and no other energies are allowed. The allowed energies for a given system are indicated in *energy-level diagrams*, as shown in Figure 3.6. Although these diagrams seem to contain very little information, this information is sufficient to predict the thermodynamic properties.

For simple systems of independent particles such as ideal gases, we can express the total internal energy of a thermodynamic system as the sum of the particle energies:

$$U = \sum_i N_i \varepsilon_i, \tag{3.8}$$

where ε_i is the energy of any particle at level i, and N_i is the number of particles at energy level i. When the total internal energy of a system is increased by heating it, the energy levels do not change: the populations N_i change. Then more particles occupy higher energy levels.

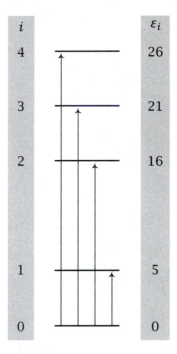

Figure 3.6 The quantum mechanical view: the quantized energies of particles are given by energy-level diagrams.

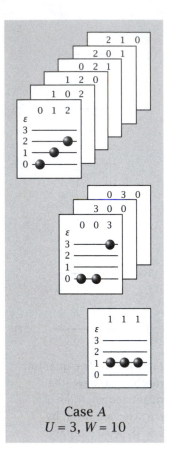

Case A
U = 3, W = 10

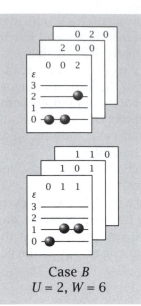

Case B
U = 2, W = 6

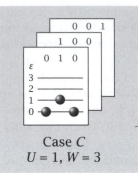

Case C
U = 1, W = 3

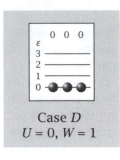

Case D
U = 0, W = 1

Figure 3.7 Each case represents a system with a given energy U. Each card represents a different distribution of the three particles over the four energy levels. The numbers shown at the top of each card are the individual energies of the particles. W is the number of configurations (cards).

Why Does Heat Flow?

In Example 2.2 we found that gases expand because the multiplicity $W(V)$ increases with volume V. The dependence of W on V defines the force called pressure. In Example 2.3 we found that particles mix because the multiplicity $W(N)$ increases as particle segregation decreases. This tendency defines the chemical potential. These are manifestations of the principle that systems tend toward their states of maximum multiplicity, also known as the Second Law of thermodynamics.

The following two examples illustrate that the flow of heat from hot to cold objects is also driven by a tendency to maximize multiplicity. First, Example 3.3 shows a model for how the multiplicity $W(U)$ of an object depends on its energy. Then Example 3.4 shows that heat flow maximizes multiplicity.

EXAMPLE 3.3 How does multiplicity depend on energy? Here is a miniaturized model of a material: it has just three distinguishable particles labeled x, y, and z. Each particle can be in any one of $i = 0, 1, 2, \ldots, t$ different energy levels ε_i. Just as you would for a sequence of dice rolls, list the energies for the particles in the order x, y, z. Different total energies U have different multiplicities $W(U)$. For illustration, suppose that $\varepsilon_i = i$, the energy equals the level number. (In Chapter 11 we'll derive this and other functional forms from quantum mechanics. For our present purposes, the functional form doesn't matter.)

Case A in Figure 3.7 shows ten different ways of achieving a total energy $U = \sum_{i=0}^{t} N_i \varepsilon_i = 3$. Case B shows there are six different ways to get $U = 2$. Case C shows the three ways of achieving $U = 1$: two particles must have zero energy and one particle must have $\varepsilon = 1$. And Case D shows that there is only one way to achieve $U = 0$: the energy level of each particle must be zero. Figure 3.8 shows that the number of configurations W is an increasing function of the total energy U, for the system shown in Figure 3.7. There are more arrangements of a system that have a high energy than a low energy.

Increasing the energy of a system increases its multiplicity of states. It follows that a tendency toward high multiplicity is a tendency for a system to take up heat from the surroundings. Now why does heat flow from hot objects to cold ones? Example 3.4 addresses this question.

EXAMPLE 3.4 Why does energy exchange? Consider the two systems, A and B, shown in Figure 3.9. Each system has ten particles and only two energy levels, $\varepsilon = 0$ or $\varepsilon = 1$ for each particle. The binomial statistics of coin flips applies to this simple model.

Systems A and B are identical except that A has less energy than B. System A has two particles in the 'excited state' ($\varepsilon = 1$), and eight in the 'ground state' ($\varepsilon = 0$) so the total energy is $U_A = 2$. System B has four in the excited state and six in the ground state so $U_B = 4$. In this case, heat flows from B (higher energy) to A (lower energy) because doing so increases the multiplicity. Let's see how.

The multiplicities of isolated systems A and B are

$$W_A = \frac{10!}{2!8!} = 45, \quad \text{and} \quad W_B = \frac{10!}{4!6!} = 210.$$

If A and B do not exchange energies, the total multiplicity is $W_{\text{total}} = W_A W_B = 9450$. Now suppose that you bring A and B into 'thermal contact' so that they can exchange energy. Now the system can change values of U_A and U_B subject to conservation of energy ($U_A + U_B$ will be unchanged). One possibility is $U_A = 3$ and $U_B = 3$. Then the total multiplicity W_{total} will be

$$W_{\text{total}} = \frac{10!}{3!7!} \frac{10!}{3!7!} = 14400.$$

This shows that a principle of maximum multiplicity predicts that heat will flow to equalize energies in this case. Consider the alternative. Suppose A were to lower its energy to $U_A = 1$ while B wound up with $U_B = 5$. Then the multiplicity of states would be

$$W_{\text{total}} = \frac{10!}{1!9!} \frac{10!}{5!5!} = 2520.$$

A maximum-multiplicity principle predicts that this inequitable distribution is unlikely. That is, heat will not flow from the cold to the hot object.

EXAMPLE 3.5 However, energy doesn't always flow downhill. Example 3.4 predicts that energies tend to equalize. But here's a more interesting case that shows that the tendency to maximize multiplicity does not always result in a draining of energy from higher to lower. Again system A has ten particles and an energy $U_A = 2$. However, now system B is smaller, with only four particles, and has energy $U_B = 2$. The energies of A and B are equal. The multiplicity is

$$W = W_A W_B = \frac{10!}{2!8!} \frac{4!}{2!2!} = 45 \times 6 = 270.$$

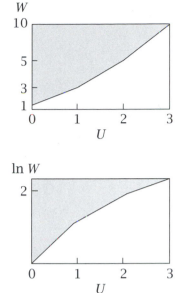

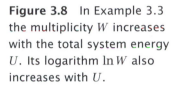

Figure 3.8 In Example 3.3 the multiplicity W increases with the total system energy U. Its logarithm $\ln W$ also increases with U.

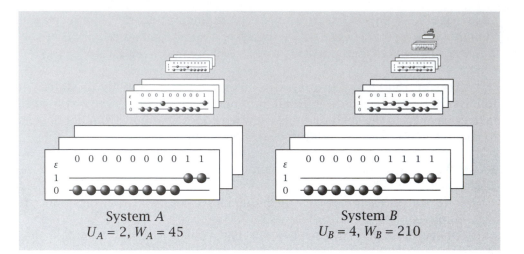

Figure 3.9 Energy-level diagrams for the two different systems in Example 3.4 with ten particles each. System A has total energy $U_A = 2$, and B has $U_B = 4$. System B has the greater multiplicity of states.

Now suppose that A and B come into thermal contact and the larger system absorbs energy from the smaller one, so $U_A = 3$ and $U_B = 1$. This causes the multiplicity to increase:

$$W = W_A W_B = \frac{10!}{3!7!} \frac{4!}{1!3!} = 120 \times 4 = 480.$$

The tendency for heat to flow is not always a tendency to equalize energies. It is a tendency to maximize multiplicity. We will see later that the concept of *temperature* describes the driving force for energy exchange. The tendency toward maximum multiplicity is a tendency toward equal temperatures, not equal energies. Temperature is the quantity that describes the tendency toward maximum multiplicity when energies can exchange. This is the topic of Chapter 12.

In this and the preceding chapter we have used simple models to predict that gases expand to fill the volume available to them, molecules will mix and diffuse to uniform concentrations, rubber retracts when pulled, and heat flows from hot bodies to colder ones. All these tendencies are predicted by a principle of maximum multiplicity: a system will change its degrees of freedom so as to reach the microscopic arrangement with the maximum possible multiplicity. This principle is the Second Law of thermodynamics, and is much broader than the simple models that we have used to illustrate it.

Summary

We have traced a little of the history of the laws of thermodynamics. The First Law says that heat is a form of energy exchange, and that the sum of heat plus work, called the internal energy, is a quantity that obeys a conservation law. The First Law is a bookkeeping tool. It catalogs the balance of heat and work. It doesn't tell us why heat flows. The Second Law says that systems tend toward their states of maximum multiplicity. Heat flows to maximize multiplicity. A simple example shows that transferring internal energy from one system to another can change the multiplicity.

To make these principles more useful, we now need some mathematical tools and then, beginning in Chapter 7, the definitions of thermodynamics.

Problems

1. The time dependence of a mass on a spring.

(a) For the harmonic motion of a mass on a spring, the kinetic energy is $K = (1/2)mv^2$, and the potential energy is $V = (1/2)k_s x^2$, where k_s is the spring constant. Using the conservation of energy find the time-dependent spring displacement, $x(t)$.

(b) Compute the force $f(t)$ on the mass.

2. Equalizing energies. For the two 10-particle two-state systems of Example 3.4, suppose the total energy to be shared between the two objects is $U = U_A + U_B = 4$. What is the distribution of energies that gives the highest multiplicity?

3. Energy conversion. When you drink a beer, you get about 100 Cal (1 food Cal = 1 kcal). You can work this off on an exercise bicycle in about 10 minutes. If you hook your exercise bicycle to a generator, what wattage of light bulb could you light up, assuming 100% efficiency?

4. Kinetic energy of a car. How much kinetic energy does a 1700 kg car have, if it travels 100 km h^{-1}?

Suggested Reading

RS Berry, JA Rice and J Ross, *Physical Chemistry*, Wiley, New York, 1980. Physical chemistry covered in great breadth and depth.

SG Brush, *The Kind of Motion We Call Heat*, volumes 1 and 2, North-Holland, Amsterdam, 1976. Scholarly history of thought and concepts in thermodynamics.

AP French, *Newtonian Mechanics*, WW Norton & Co, New York, 1971. Elementary treatment of forces, mechanics, and conservation laws.

D Halliday and R Resnick, *Fundamentals of Physics*, 3rd edition, John Wiley, New York, 1988. Broad survey of elementary physics, including forces and mechanics.

C Kittel, *Thermal Physics*, Wiley, New York, 1969. Excellent introduction to statistical thermodynamics for physicists. Introduces the lattice model for the ideal gas law.

F Reif, *Statistical Physics*, McGraw-Hill, New York, 1967. Extensive discussion of the kinetic theory of gases.

4 Math Tools: Series & Approximations

Physical Modelling
Often Involves Series Expansions

In this and the next chapter we describe some of the basic mathematical tools we need for the development of statistical thermodynamics. Here, we review the mathematics of series approximations. The importance of series expansions in physical models is that they provide a recipe for making systematic improvements. Among the most important is the Taylor series. As an illustration of the Taylor series, we introduce the model of random walks and show that the distance between the beginning and end of a random walk follows a Gaussian distribution.

Geometric Series

In mathematics, a *sequence* is a set of quantities ordered by integers. A *series* is the sum of a sequence. A series can have an infinite number of terms ($n \to \infty$), or it can be a *partial sum* of n terms. In a **geometric series**, each term is the product of the preceding term times x:

$$s_n = \sum_{k=1}^{n} ax^{k-1} = a + ax + ax^2 + \cdots + ax^{n-1}, \qquad (4.1)$$

where s_n is called the nth partial sum of the geometric series and a is a constant. This can be expressed in a simpler form. To see this, multiply Equation (4.1) by x:

$$s_n x = ax + ax^2 + ax^3 + \cdots + ax^n. \tag{4.2}$$

Now subtract Equation (4.2) from Equation (4.1) and rearrange:

$$s_n - s_n x = a - ax^n \quad \Rightarrow \quad s_n(1 - x) = a(1 - x^n)$$

$$\Rightarrow s_n = a\left(\frac{1 - x^n}{1 - x}\right), \tag{4.3}$$

if $x \neq 1$. If $|x| < 1$, then the limit of $n \to \infty$ gives $x^n \to 0$. It follows from Equation (4.3) that

$$s_\infty = \sum_{k=1}^{\infty} ax^{k-1} = \frac{a}{1 - x} \qquad \text{for } |x| < 1. \tag{4.4}$$

The quantity $a/(1 - x)$ is called the *closed form* of the series.

Here's an example of a geometric series.

EXAMPLE 4.1 A geometric series. What is the value of the series $1 + 1/2 + 1/4 + 1/8 + 1/16 + \cdots$? This is just Equation (4.4) with $a = 1$ and $x = 1/2$, so the sum is $1/(1 - 1/2) = 2$.

The geometric series is a special case of the **power series**:

$$f(x) = \sum_{k=0}^{\infty} c_k x^k = c_0 + c_1 x + c_2 x^2 + \cdots + c_n x^n + \cdots, \tag{4.5}$$

where the c_i's are constants.

Arithmetic–Geometric Series

Another common series is

$$t_n = \sum_{k=1}^{n} akx^k = ax + 2ax^2 + 3ax^3 + 4ax^4 + \cdots + nax^n. \tag{4.6}$$

Equation (4.6) is geometric because each term has an extra factor of x relative to its predecessor and it is arithmetic because each coefficient increases by one. This too is readily put into closed form. To simplify, notice that the derivative of Equation (4.1) (for $n = \infty$) is

$$\frac{ds_\infty}{dx} = a + 2ax + 3ax^2 + \cdots + (n - 1)ax^{n-2} + \cdots. \tag{4.7}$$

Multiply Equation (4.7) by x and you get t_∞:

$$t_\infty = x\frac{ds_\infty}{dx}. \tag{4.8}$$

Substituting $s_\infty = a/(1 - x)$ from Equation (4.4) into Equation (4.8) gives a closed form expression for t_∞:

$$t_\infty = x\frac{d}{dx}\left(\frac{a}{1 - x}\right) = \frac{ax}{(1 - x)^2}.$$

(4.9)

In many problems of statistical mechanics, such as in computing polymer length distributions (Example 4.2), energies (Chapter 11), fractional helicity in polymers (Chapter 26), or numbers of ligands bound to proteins or surfaces (Chapters 27 and 29), you will have a polynomial function such as

$$Q(x) = 1 + x + x^2 + x^3 + \cdots + x^n = \sum_{i=0}^{n} x^i,$$

(4.10)

and you will want the average exponent, $\langle i \rangle$:

$$\langle i \rangle = \frac{\sum_{i=0}^{n} ix^i}{\sum_{i=0}^{n} x^i} = \frac{x + 2x^2 + 3x^3 + \cdots + nx^n}{1 + x + x^2 + \cdots + x^n}.$$

(4.11)

As in Equation (4.8), you can express Equation (4.11) most simply by taking the derivative of Equation (4.10):

$$\frac{dQ}{dx} = 1 + 2x + 3x^2 + 4x^3 + \cdots + nx^{n-1} = \sum_{i=0}^{n} ix^{i-1}.$$

(4.12)

Now multiply Equation (4.12) by x and divide by Q to get $\langle i \rangle$:

$$\frac{x}{Q}\frac{dQ}{dx} = \frac{x + 2x^2 + 3x^3 + \cdots + nx^n}{1 + x + x^2 + \cdots + x^n} = \langle i \rangle.$$

(4.13)

For $n \to \infty$, Equation (4.13) can be expressed as the ratio $\langle i \rangle = t_\infty/s_\infty$.

$$\langle i \rangle = \frac{t_\infty}{s_\infty} = \frac{ax/(1 - x)^2}{a/(1 - x)} = \frac{x}{1 - x}.$$

(4.14)

For example, if $x = 0.1$, then Equation (4.14) gives $\langle i \rangle \approx 0.11$ (the *average* exponent need not be an integer). By using the notation $d\ln x = dx/x$ and $d\ln Q = dQ/Q$, Equation (4.13) can be expressed as

$$\langle i \rangle = \frac{x}{Q}\frac{dQ}{dx} = \frac{d\ln Q}{d\ln x}.$$

(4.15)

You can get the second moment $\langle i^2 \rangle$ by taking another derivative,

$$\langle i^2 \rangle = \frac{\sum_{i=0}^{n} i^2 x^i}{\sum_{i=0}^{n} x^i} = \frac{x + 4x^2 + 9x^3 + \cdots + n^2 x^n}{1 + x + x^2 + \cdots + x^n}$$

$$= \left(\frac{x}{Q}\right)\frac{d}{dx}\left(x\frac{dQ}{dx}\right).$$

(4.16)

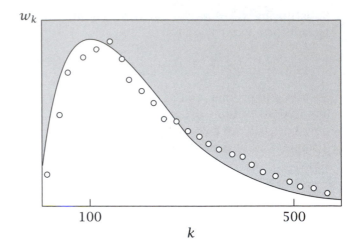

Figure 4.1 Shown here is the experimental molecular weight distribution w_k for nylon, as a function of chain length k. This quantity is given by $w_k = (\text{constant})\, k(1-p)n_k$ (see Problem 6) where n_k is given in Example 4.2. Points give experimental data for nylon. Source: PJ Flory, *Principles of Polymer Chemistry*, Cornell University Press, Ithaca, 1953. Data are from GB Taylor, *J. Am Chem Soc* **69**, 638 (1947).

Again, if $x = 0.1$, $\langle i^2 \rangle \approx 0.14$. Equations (4.13) and (4.16) also apply for the more general polynomial, $Q = a_0 + a_1 x + a_2 x^2 + \cdots + a_n x^n$. Example 4.2 shows why you might be interested in the average value of polynomial exponents.

EXAMPLE 4.2 The average molecular weights of polymers. To create polymer molecules—linear strings of monomer units linked together like beads on a necklace—chemists start with monomers in solution and initiate polymerization conditions. Monomers add to the ends of growing chains one unit at a time. Because the process is random, some chains are short and some are long. There is a distribution of molecular weights. Suppose that p is the probability that a monomer unit is reacted and connected in the chain. The statistical weight (unnormalized probability) for having a chain k monomers long is $n_k = p^{k-1}(1-p)$. The factor of $(1-p)$ arises because the terminal end unit must be unreacted for the molecule to be exactly k units long. For *condensation polymerization*, as for nylon shown in Figure 4.1, n_k is called the *most probable distribution*. The fraction of chains with length k is $P(k) = n_k / \sum_{k=1}^{\infty} n_k$. If you want the average chain length, compute $\langle k \rangle$,

$$\langle k \rangle = \sum_{k=1}^{\infty} kP(k) = \frac{\sum_{k=1}^{\infty} kp^{k-1}(1-p)}{\sum_{k=1}^{\infty} p^{k-1}(1-p)} = \frac{\sum_{k=1}^{\infty} kp^{k-1}}{\sum_{k=1}^{\infty} p^{k-1}}. \tag{4.17}$$

According to Equation (4.4), the denominator of Equation (4.17) is $1/(1-p)$. By using Equation (4.9), the numerator can be written as

$$\frac{1}{p} \sum_{k=1}^{\infty} kp^k = \frac{t_\infty}{p} = \frac{1}{(1-p)^2}, \quad \text{so}$$

$$\langle k \rangle = \frac{1/(1-p)^2}{1/(1-p)} = \frac{1}{1-p}. \tag{4.18}$$

It is not unusual for polymers to have $p = 0.999$ and average length $\langle k \rangle = 1000$. Figure 4.1 shows that this simple model predicts well an experimental polymerization distribution.

Making Approximations Involves Truncating Series'

A physical theory can involve many steps of mathematics. If one of the steps involves evaluating a complicated function $f(x)$, having exponentials, logarithms, sines or cosines, then it might not be possible to perform the subsequent mathematical steps. In such cases it is necessary to *approximate* the function by a simpler mathematical expression.

Taylor Series

Among the most important approximation tools in the physical sciences is the Taylor series, which expresses a function in terms of its derivatives. A function $f(x)$ near the point $x = a$ can be expressed as a power series:

$$f(x) = \sum_{k=0}^{\infty} c_k(x - a)^k = c_0 + c_1(x - a)$$

$$+c_2(x - a)^2 + \cdots + c_n(x - a)^n + \cdots. \tag{4.19}$$

The coefficients of this polynomial can be obtained from the derivatives of the function. Let $f^{(n)}(x)$ represent the nth derivative. $f^{(0)}(x)$ represents the function itself. The function and its first two derivatives are

$$f^{(0)}(x) = \sum_{k=0}^{\infty} c_k(x - a)^k, \qquad f^{(1)}(x) = \sum_{k=1}^{\infty} kc_k(x - a)^{k-1}, \qquad \text{and}$$

$$f^{(2)}(x) = \sum_{k=2}^{\infty} k(k - 1)c_k(x - a)^{k-2}.$$

For the nth derivative,

$$f^{(n)}(x) = \sum_{k=n}^{\infty} k(k - 1)(k - 2) \cdots (k - n + 1)c_k(x - a)^{k-n}. \tag{4.20}$$

We now want to turn these equations around to express the coefficients in terms of the derivatives. Assume that the derivatives of the function are known at one point, namely $x = a$. Putting $x = a$ into Equation (4.20) eliminates all except the first term ($k = n$) in the sum, so

$$f^{(n)}(a) = k(k - 1)(k - 2) \cdots (k - n + 1)c_n = (n!)c_n, \quad \text{where } k = n,$$

and you have

$$c_n = \frac{f^{(n)}(a)}{n!}. \tag{4.21}$$

(Note that $(x-a)^0 = 1$.) Substituting Equation (4.21) into (4.19) gives the Taylor series approximation of a function in terms of its derivatives:

$$f(x) \approx f(a) + \left(\frac{df}{dx}\right)_{x=a}(x - a) + \frac{1}{2}\left(\frac{d^2f}{dx^2}\right)_{x=a}(x - a)^2$$

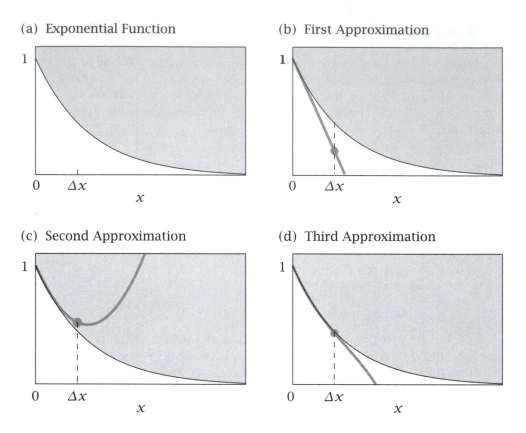

(a) Exponential Function

(b) First Approximation

(c) Second Approximation

(d) Third Approximation

Figure 4.2 Use the Taylor series to find values of $f(x) = e^{-x}$ for $x = \Delta x$ near $x = 0$. (a) The first term in the Taylor series is $f(0) = 1$. (b) The gray line is the first approximation, $f(x) \approx 1 - x$. (c) The second approximation is $f(x) \approx 1 - x + x^2/2$. (d) The third approximation is $f(x) \approx 1 - x + (x^2/2) - (x^3/6)$.

$$+ \frac{1}{6}\left(\frac{d^3 f}{dx^3}\right)_{x=a} (x - a)^3 + \cdots + \frac{1}{n!}\left(\frac{d^n f}{dx^n}\right)_{x=a} (x - a)^n + \cdots \tag{4.22}$$

Suppose that you know the function $f(a)$ at a point $x = a$, and you need a simple expression for $f(x)$ at a point x nearby, $x - a$ being small. The Taylor series gives a systematically improvable way to get better estimates for $f(x)$ as a polynomial expansion in $(x - a)$, $(x - a)^2$, $(x - a)^3$, etc.

Making approximations in physical theories is often a matter of truncating a series and keeping only the dominant terms. If x is near enough to the point a, where the derivatives are known, the quantity $(x - a)$ will be small. Then the Taylor series is convergent, meaning that the first- and second-derivative terms are larger than the subsequent terms. Often only one or two terms are needed to give useful approximations. If you keep only the constant and first-derivative terms, the approximation is called *first-order*. If you also keep the second-derivative term, it is called *second-order*.

Let's find the Taylor series approximation for the exponential function e^{-bx}.

EXAMPLE 4.3 The Taylor series expansion for $f(x) = e^{-bx}$ near $x = 0$. To find the Taylor series for this function, you evaluate $f(a)$ where $a = 0$, and the derivatives of the function at $x = 0$ according to Equation (4.22). For the first term, you have $f(0) = e^{-0} = 1$.

To compute the second term, find the derivative $f^{(1)}(x)$ and evaluate it at $x = 0$:

$$\frac{d}{dx}\left(e^{-bx}\right) = -be^{-bx}\Big|_{x=0} = -b.$$

So the second term of the Taylor series is $f^{(1)}(a)(x - a) = -bx$ (see Figure 4.2(b)). To compute the third term you need the second derivative $f^{(2)}(x)$ evaluated at $x = 0$:

$$\frac{d^2}{dx^2}(e^{-bx}) = b^2 e^{-bx}\Big|_{x=0} = b^2.$$

The third term of the Taylor series is $(1/2)f^{(2)}(a)(x - a)^2 = (bx)^2/2$.

To second-order approximation, the Taylor series expansion is $f(x) = e^{-bx} \approx 1 - bx + (bx)^2/2$, shown in Figure 4.2(c). Taylor series expansions for various functions are given in Appendix C, Useful Taylor Series Expansions.

For some approximations it is sufficient to keep first-order terms. In other cases, for example when all the first-order terms cancel, the second-order terms are the first non-zero contributions, so you need to keep both first-order and second-order terms. Here is an example.

EXAMPLE 4.4 A second-order approximation. Suppose you have a function $y(x) = bx + e^{-bx}$ and you want to know its dependence on x for small x, $bx \ll 1$. Since $e^{-bx} \approx 1 - bx + (bx)^2/2$, the first-order contributions cancel, so you need to keep the second-order term and $y(x) \approx 1 - (bx)^2/2$.

EXAMPLE 4.5 Two first-order approximations. Find a first-order approximation for $1/(1 + x)$. From Appendix C, Useful Taylor Series Expansions, use Equation (C.6) with $p = -1$ to get

$$\frac{1}{1 + x} \approx 1 - x.$$

To approximate $1/(1 + x)^{1/5}$, use Equation (C.6) with $p = -1/5$:

$$\frac{1}{(1 + x)^{1/5}} \approx 1 - \frac{1}{5}x.$$

The Newton Method for Finding Roots of Functions

Here is another application of the Taylor series method. Suppose that you want to find the values $x = x^*$ that cause a function to equal zero, $f(x) = 0$. Such values of x are called the *roots* of the equation. The Newton method is an iterative scheme that works best when you can make a reasonable first guess x^*, and when the function does not have an extremum between the guess and the correct answer. The Newton method assumes that the function $f(x)$ is given approximately by the first two terms of its Taylor series.

Suppose that your first guess is that the root is $x = x_1$; that is $f(x_1) = 0$. You want a systematic way to improve this guess. A better second estimate will be x_2; the third estimate will be x_3, etc. The Taylor series expansion for $f(x)$ in terms of the first guess x_1 is

$$f(x) = f(x_1) + f^{(1)}(x_1)(x - x_1) + \cdots$$

Given your first guess x_1, now choose x_2 to make $f(x) = 0$:

$$0 = f(x_1) + f^{(1)}(x_1)(x_2 - x_1).$$

That is, you should choose

$$x_2 = x_1 - \frac{f(x_1)}{f^{(1)}(x_1)},$$

on the basis of the function evaluated at x_1. This estimate for x_2 should be better than x_1, but still might not be exactly right because we have left out the higher terms in the Taylor series. So try again. In general, the $(n + 1)$th estimate is obtained from the nth estimate by using the expression

$$x_{n+1} = x_n - \frac{f(x_n)}{f^{(1)}(x_n)}. \tag{4.23}$$

If the function is well behaved, this procedure will converge to the correct value of x.

EXAMPLE 4.6 Finding the roots of a polynomial. Let's use the Newton method to compute the roots of an equation, $y = x^2 - 8x + 15$. We also know the answer by factorization: $y = (x - 3)(x - 5)$, and the roots are $x^* = (3, 5)$. Suppose that your first guess is $x_1 = 2$. Substitute $x_1 = 2$ into $y(x)$ to get

$$y_1 = f(x_1) = 4 - 16 + 15 = 3, \quad \text{and}$$

$$f^{(1)}(x_1) = (2x - 8)_{x_1=2} = -4.$$

Substituting these values into Equation (4.23) gives

$$x_2 = 2 - \left(\frac{3}{-4}\right) = 2.75.$$

Iterate again to get x_3:

$$f(x_2) = 0.5625, \quad f^{(1)}(x_2) = (2x - 8)_{x_2=2.75} = -2.5, \quad \text{so}$$

$$x_3 = 2.75 - \left(\frac{0.5625}{-2.5}\right) = 2.975.$$

This process is converging toward $x = 3$, one of the roots we seek.

Stirling's approximation

Combinatoric expressions are often given in terms of factorials, which can be difficult to manipulate mathematically. They are made more tractable by using

Stirling's approximation:

$$n! \approx (2\pi n)^{1/2} \left(\frac{n}{e}\right)^n. \tag{4.24}$$

You can derive Stirling's approximation using series methods. Equation (4.24) is very accurate for large n. We will usually be interested in the logarithm,

$$\ln n! \approx \frac{1}{2}\ln 2\pi + \left(n + \frac{1}{2}\right)\ln n - n. \qquad n! = (2\pi)^{1/2} n^{n+1/2} \exp(-n) \tag{4.25}$$

When n is larger than about 10, the first term $(1/2)\ln 2\pi = 0.92$ is negligible compared with the second term and $n + 1/2 \approx n$, so an even simpler approximation is

$$\ln n! \approx n\ln n - n \quad \Longrightarrow \quad n! \approx \left(\frac{n}{e}\right)^n. \tag{4.26}$$

(Try it. For $n = 100$, the better approximation is $\ln 100! \approx 0.92 + (100.5)\ln 100 - 100 = 363.74$, and the simpler approximation is $\ln 100! \approx 100\ln 100 - 100 = 360.5$.)

Now let's derive Equation (4.26). Since $n! = (1)(2)(3)\dots(n-1)(n)$, the logarithm of $n!$ is the sum

$$\ln n! = \ln 1 + \ln 2 + \dots + \ln n = \sum_{m=1}^{n} \ln m. \tag{4.27}$$

For large values of n, approximate the sum in Equation (4.27) with an integral:

$$\ln n! = \sum_{m=1}^{n} \ln m \approx \int_1^n \ln m \, dm = (m\ln m - m)\Big|_1^n. \tag{4.28}$$

For $n \gg 1$ this gives Equation (4.26).

The following section illustrates the Taylor series method, and also introduces an important model in statistical thermodynamics: the random *walk* (in two dimensions) or random *flight* (in three dimensions). In this example, we find that the Gaussian distribution function is a good approximation to the binomial distribution function (see page 15) when the number of events is large.

The Gaussian Distribution Describes a Random Walk

Random walks describe the diffusion of particles, the conformations of polymer chains, and the conduction of heat. The random-walker problem was first posed by the English mathematician K Pearson (1857–1936) in 1905. He was interested in how mosquitoes enter a cleared jungle. The solution (more or less the same one that we give here) was first provided for a different problem by the English physicist Lord J Rayleigh (1849–1919), who won the Nobel prize in physics in 1904 for the discovery of argon.

Let's start with a walk in one dimension. Each step has unit length in either the $+x$ or $-x$ direction, chosen randomly with equal probability. Steps are independent of each other. In a given walk of N total steps, m steps are in the $+x$ direction and $(N - m)$ steps are in the $-x$ direction. We aim to find the distribution of m values $P(m, N)$ for an N-step walk. Because the $(+)$ and $(-)$ steps are chosen just like coin flips, the result is the binomial probability distribution, Equation (1.28), with $p = 1 - p = 1/2$:

$$P(m, N) = \left(\frac{1}{2}\right)^N \frac{N!}{m!(N - m)!}. \tag{4.29}$$

We aim to approximate the function P, but owing to mathematical intractability of the factorials, it is easier to work with the logarithm, which can be obtained by a Taylor series expansion around the most probable end point of the walk m^* for the function $\ln P$ for fixed N,

$$\ln P(m) = \ln P(m^*) + \left(\frac{d \ln P}{dm}\right)_{m^*} (m - m^*)$$

$$+ \frac{1}{2} \left(\frac{d^2 \ln P}{dm^2}\right)_{m^*} (m - m^*)^2 + \cdots. \tag{4.30}$$

Taking the logarithm of Equation (4.29) and using Stirling's approximation gives the function for evaluating the derivatives:

$$\ln P = N \ln N - m \ln m - (N - m) \ln(N - m) + N \ln \left(\frac{1}{2}\right). \tag{4.31}$$

The value $m = m^*$ is defined as the point at which $P(m)$ or $\ln P(m)$ is at a maximum, so this is the point at which $(d \ln P/dm)_{m^*} = 0$. Taking the derivative of Equation (4.31), with N constant, gives

$$\left. \frac{d \ln P}{dm} \right|_{m^*} = -1 - \ln m^* + \ln(N - m^*) + 1 = 0$$

$$\implies -\ln \left(\frac{m^*}{N - m^*}\right) = 0 \implies m^* = \frac{N}{2}. \tag{4.32}$$

At the maximum of P, which occurs at $m = m^* = N/2$, the first derivative term in Equation (4.30) is zero. The second derivative, also evaluated at $m = m^*$, is

$$\left. \frac{d^2 \ln P}{dm^2} \right|_{m^*} = \left(-\frac{1}{m} - \frac{1}{N - m}\right)_{m^* = N/2} = -\frac{4}{N}. \tag{4.33}$$

Substituting Equation (4.33) into Equation (4.30) and exponentiating gives

$$P = P^* e^{-2(m - m^*)^2/N}. \tag{4.34}$$

This shows that the end points of the walk follow a Gaussian distribution.

Let's convert from the number of forward steps m to x. The net forward progress x equals $m - (N - m) = 2m - N$ (the number of forward minus reverse steps). Since $m^* = N/2$ from Equation (4.32), $x^* = 0$. That is, the most probable final destination is the origin, because it is most likely that the number of forward steps equals the number of backward steps. Since $x = 2m - N$, you have $m = (x + N)/2$. Since $m^* = N/2$, you have $m - m^* = x/2$. Substituting this into Equation (4.34) gives

$$P(x) = P^* e^{-x^2/(2N)}.$$

Now let's compute P^*. (If we had not made Stirling's approximation, our derivation above would have led to P^*, but mired in mathematical detail. The following is an easier way to get it.) To have a properly normalized probability distribution, you need the integral over all the probabilities to equal one. To find P^*, integrate:

$$\int_{-\infty}^{\infty} P(x)\,dx = \int_{-\infty}^{\infty} P^* e^{-x^2/(2N)}\,dx = 1.$$

Since $\int_{-\infty}^{\infty} e^{-ax^2}\,dx = \sqrt{\pi/a}$, you have

$$P^* = \left[\int_{-\infty}^{\infty} e^{-x^2/(2N)}\,dx\right]^{-1} = (2\pi N)^{-1/2}, \quad \text{so}$$

$$P(x) = (2\pi N)^{-1/2} e^{-x^2/(2N)}. \tag{4.35}$$

Equation (4.35) describes the distribution of ending points x, given that a one-dimensional walker started at the origin $x = 0$ and took N steps randomly in the $+x$ or $-x$ directions.

Finally, consider the *fluctuations*, a measure of the width of the distribution. For random processes such as random walks, the average $\langle x \rangle = 0$ is often less useful than the *mean square* $\langle x^2 \rangle$ or its square root $\langle x^2 \rangle^{1/2}$, called the *root-mean-square* (rms) distance. The second moment is

$$\langle x^2 \rangle = \int_{-\infty}^{\infty} x^2 P(x)\,dx = \left(\frac{1}{2\pi N}\right)^{1/2} \int_{-\infty}^{\infty} x^2 e^{-x^2/(2N)}\,dx.$$

Appendix D gives $\int_{-\infty}^{\infty} x^2 e^{-x^2/(2N)}\,dx = N(2\pi N)^{1/2}$, so you have

$$\langle x^2 \rangle = N, \tag{4.36}$$

and the rms distance traveled by the one-dimensional walker is

$$\langle x^2 \rangle^{1/2} = (N)^{1/2}.$$

This shows a characteristic feature of random walks. The *rms* displacement of the walker from the origin scales not linearly with the number of steps, as for a directed walker, but rather as the square root of the number of steps because of the random meandering.

Summary

Physical theories often require mathematical approximations. When functions are expressed as polynomial series, approximations can be systematically improved by keeping terms of increasingly higher order. One of the most important expansions is the Taylor series, an expression of a function in terms of its derivatives. These methods show that a Gaussian distribution function is a second-order approximation to a binomial distribution near its peak. We will find this useful for random walks, which are used to interpret diffusion, thermal conduction, and polymer conformations. In the next chapter we develop additional mathematical tools.

Problems

1. Taylor series. Derive Appendix Equation (C.6): $(1+x)^p \approx 1 + px + [p(p-1)/(2!)]x^2$, to second order using the Taylor series method, around $x \approx 0$.

2. Generalizing $(d \ln Q)/(d \ln x)$. Generalize the derivation in Equations (4.10)–(4.13). Show that $\langle i \rangle = (d \ln Q)/(d \ln x)$ for the function

$$Q = a_0 + a_1 x + a_2 x^2 + a_3 x^3 + \cdots + a_n x^n.$$

3. Roots of a polynomial. Example 4.6 shows that a first guess of $x = 2$ leads to the root $x = 3$ for the equation $y = x^2 - 8x + 15$. Toward what root does the Newton method converge if your first guess is $x = 6$?

4. Compute e. Computers and calculators use series expansions to get quantities such as e, the base of the natural logarithm, and π, etc. Compute e from a Taylor series expansion.

5. Derive a closed form for a series. Show that

$$\sum_{k=1}^{\infty} k^2 p^k = \frac{p(1+p)}{(1-p)^3}.$$

6. Molecular weight distribution in polymers. While the fraction of polymer chains having *length* k in Example 4.2 is $P(k) = n_k / \sum_{k=1}^{\infty} n_k$, the fraction of chains having molecular weight proportional to k is $w_k = k n_k / \sum_{k=1}^{\infty} k n_k$. This is the function plotted in Figure 4.1.

(a) Show that $w_k = k(1-p)n_k$.

(b) Compute the average molecular weight, $\langle k \rangle_{mw} = \sum_{k=1}^{\infty} k w_k$.

7. Taylor series for $\cos x$. Give the Taylor series expansion for $\cos x$ around $x = 0$, to the third non-zero term.

8. Average cluster sizes. Consider a one-dimensional lattice having V sites. n particles are randomly distributed on it. Let $q = n/V$ represent the probability that a given site is occupied.

(a) Compute the probability $p(m)$ that m adjacent sites are occupied by particles. Neglect end effects

(b) Compute the average cluster size $\langle m \rangle$.

9. Malthusian population growth. The rate of growth dp/dt of a biological population is often proportional to the population $p(t)$ itself, because the population size determines the number of parents that can beget offspring: $dp/dt = \alpha p$. Show that growth is exponential if $\alpha > 0$.

10. Predicting disaster. The probability that an airplane will be in an accident on any given flight is $p = 10^{-5}$, and the probability that the airplane will not be in an accident is $q = 1 - p$. As a function of p, calculate the average number of airplane trips you can take before you should expect to be in your first airplane accident.

Suggested Reading

HC Berg, *Random Walks in Biology*, Princeton University Press, Princeton, 1983. Elementary treatment of random walks, particularly as they pertain to diffusion.

BD Hughes, *Random Walks and Random Environments*, Clarendon Press, Oxford, 1995. An extensive discussion of random walks.

M Spiegel, *Mathematical Handbook*, Schaum's Outline Series, McGraw-Hill, New York, 1968. Tables of series expansions.

GB Thomas Jr, *Calculus and Analytic Geometry*, 8th edition, Addison-Wesley, Reading, 1992. Good discussion of series.

5 Math Tools: Multivariate Calculus

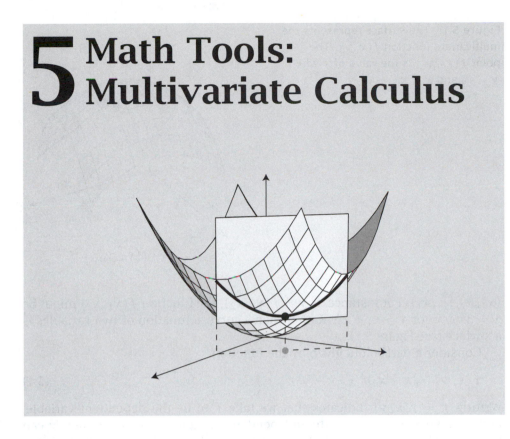

Multivariate Calculus Applies to Functions of Multiple Variables

So far, we've described model systems with only one degree of freedom: the number of heads n_H, the volume V, the number of particles N, or the amount of internal energy U. In each case we have considered the multiplicity as a function of only one variable: $W = W(V)$, $W = W(N)$, or $W = W(U)$. However, many problems in statistical thermodynamics involve simultaneous changes in multiple degrees of freedom: internal combustion engines burn gasoline, change particle composition, create heat, and expand against a piston all at the same time; metabolites enter cells, changing cell volume and particle number at the same time; a change in the number of bound ligand particles can change the permeability of cell membranes, and the shapes, activities, and stabilities of biomolecules.

Here we review the tools from multivariate calculus that we need to describe processes in which multiple degrees of freedom change together. We need these methods to solve two main problems: to find the extrema of multivariate functions, and to integrate them. Here we introduce the mathematics.

A function $y = f(x)$ assigns one number to y for each value of x. The function $f(x)$ may be $y = 3x^2$, or $y = \log(x - 14)$, etc. Now consider *multivariate* functions. Multivariate functions assign one number to each set of values of two or more variables. If y is a function of t variables, $y = f(x_1, x_2, \ldots, x_t)$, the function $f(x_1, x_2, \ldots, x_t)$ assigns one value to y for each set of values

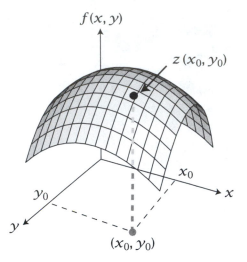

Figure 5.1 The surface represents the multivariate function $f(x, y)$. The point $f(x_0, y_0)$ is the value of f when $y = y_0$ and $x = x_0$.

$(x_1, x_2, \ldots, x_t)$. For instance, if $y = f(x_1, x_2)$, the function $f(x_1, x_2)$ might be $y = 3x_1/x_2$ or $y = x_1 + x_2^3$. Rather than a curve, a function of two variables is a surface (see Figure 5.1).

Consider a paraboloid given by the function

$$f(x, y) = (x - x_0)^2 + (y - y_0)^2. \tag{5.1}$$

Writing $f = f(x, y)$ indicates that we take f to be the dependent variable, and we take x and y to be the independent variables. Any x and any y can be specified independently, but then the result $f(x, y)$ is determined. If it is more convenient to specify x and f independently and observe the outcome y, Equation (5.1) is rearranged to $y(x, f)$. Figure 5.2 shows how f depends on x and y, according to Equation (5.1). The plane sliced out of Figure 5.2 parallel to the f and y axes shows how f depends on y alone, with x held constant.

Partial Derivatives Are the Slopes of the Tangents to Multivariate Functions

The *partial derivative* is the slope of one particular tangent to the curve at a given point.

Derivatives of functions of several variables are *partial* derivatives, taken with respect to only one variable, with all the other variables held constant. For example, the partial derivatives of $f(x, y)$ are defined by

$$\left(\frac{\partial f}{\partial x}\right)_y = \lim_{\Delta x \to 0} \frac{f(x + \Delta x, y) - f(x, y)}{\Delta x}, \quad \text{and}$$

$$\left(\frac{\partial f}{\partial y}\right)_x = \lim_{\Delta y \to 0} \frac{f(x, y + \Delta y) - f(x, y)}{\Delta y}, \tag{5.2}$$

where the symbol ∂ represents a partial derivative. $(\partial f/\partial y)_x$ is the partial derivative of f with respect to y while x is held constant (see Figures 5.2 and 5.3). To evaluate partial derivatives, use the rules for ordinary differentiation with respect to one variable, treating the other independent variables

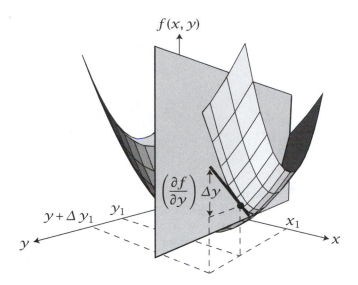

$f(x, y)$

$\left(\dfrac{\partial f}{\partial y}\right) \Delta y$

$y + \Delta y_1$ y_1

y

x_1

x

Figure 5.2 The partial derivative of f with respect to y at point (x_1, y_1) is defined by the slope of the tangent to f at point (x_1, y_1), according to Equation (5.2). This is the slope of $f(x, y)$ in the plane $x = x_1$.

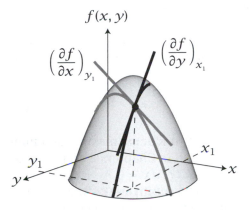

$f(x, y)$

$\left(\dfrac{\partial f}{\partial x}\right)_{y_1}$ $\left(\dfrac{\partial f}{\partial y}\right)_{x_1}$

y_1

y

x_1

x

Figure 5.3 The function $f(x, y)$ has two partial derivatives at every point. The partial derivative with respect to y, $(\partial f/\partial y)_{x_1, y_1}$, is the slope of the tangent to the curve parallel to the yf-plane, evaluated at the point (x_1, y_1). $(\partial f/\partial x)_{x_1, y_1}$ is the slope of the tangent parallel to the xf-plane, evaluated at (x_1, y_1).

as constants. For example, if the function is $f(x, y) = 3xy^2 + x + 2y^2$, then $(\partial f/\partial x)_{y=1} = (3y^2 + 1)_{y=1} = 4$, and $(\partial f/\partial y)_{x=2} = (6xy + 4y)_{x=2} = 16y$.

Just like ordinary derivatives, partial derivatives are functions of the independent variables. They can be further differentiated to yield second- and higher-order derivatives. The following example shows how these rules are applied to the *ideal gas law*. We explore the physics of this law later. For now our purpose is just to illustrate partial derivatives.

EXAMPLE 5.1 Partial derivatives and the ideal gas law.   The ideal gas law describes the pressure p of a gas as a function of the independent variables volume V and temperature T:

$$p = p(V, T) = \frac{RT}{V}, \tag{5.3}$$

where R is a constant. The first partial derivatives are

$$\left(\frac{\partial p}{\partial V}\right)_T = -\frac{RT}{V^2} \quad \text{and} \quad \left(\frac{\partial p}{\partial T}\right)_V = \frac{R}{V}.$$

The second partial derivatives are

$$\left(\frac{\partial^2 p}{\partial V^2}\right)_T = \frac{2RT}{V^3} \quad \text{and} \quad \left(\frac{\partial^2 p}{\partial T^2}\right)_V = 0,$$

$$\left(\frac{\partial^2 p}{\partial V \partial T}\right) = \left(\frac{\partial (\partial p/\partial T)_V}{\partial V}\right)_T = -\frac{R}{V^2}, \quad \text{and}$$

$$\left(\frac{\partial^2 p}{\partial T \partial V}\right) = \left(\frac{\partial (\partial p/\partial V)_T}{\partial T}\right)_V = -\frac{R}{V^2}.$$

Small Changes and the Total Differential

If you know the value of $f(x)$ at some point $x = a$, you can use the Taylor series expansion Equation (4.22) to compute $f(x)$ near that point:

$$\Delta f = f(x) - f(a)$$

$$= \left(\frac{df}{dx}\right)_{x=a} \Delta x + \frac{1}{2}\left(\frac{d^2 f}{dx^2}\right)_{x=a} \Delta x^2 + \frac{1}{6}\left(\frac{d^3 f}{dx^3}\right)_{x=a} \Delta x^3 + \cdots \quad (5.4)$$

For very small changes $\Delta x = (x - a) \rightarrow dx$, the terms involving Δx^2, Δx^3, etc., will become vanishingly small. The change in f will become small, and $\Delta f \rightarrow df$ will be given by the first term:

$$df = \left(\frac{df}{dx}\right)_{x=a} dx. \quad (5.5)$$

You can use derivatives to find the extrema of functions (maxima and minima), which predict equilibrium states (as described in Chapters 2 and 3). Maxima or minima in $f(x)$ occur at the points where $df/dx = 0$.

Now let's find the corresponding change for a multivariate function, say $f(x, y)$. In this case, too, the change Δf can be expressed as a Taylor series:

$$\Delta f = f(x, y) - f(a, b)$$

$$= \left(\frac{\partial f}{\partial x}\right)_{y=b} \Delta x + \left(\frac{\partial f}{\partial y}\right)_{x=a} \Delta y$$

$$+ \frac{1}{2}\left[\left(\frac{\partial^2 f}{\partial x^2}\right)(\Delta x)^2 + \left(\frac{\partial^2 f}{\partial y^2}\right)(\Delta y)^2 + 2\left(\frac{\partial^2 f}{\partial x \partial y}\right)\Delta x \Delta y\right] + \cdots$$

As $\Delta x = (x - a) \rightarrow dx$ and $\Delta y = (y - b) \rightarrow dy$, Δf becomes df and the higher-order terms in $\Delta x^2, \Delta y^2$, etc., become vanishingly small. Again, the Taylor series expansion reduces to the first-order terms:

$$df = \left(\frac{\partial f}{\partial x}\right)_y dx + \left(\frac{\partial f}{\partial y}\right)_x dy. \quad (5.6)$$

For a function of t variables such as $f(x_1, x_2, \ldots, x_t)$, the total differential is defined as

$$df = \sum_{i=1}^{t} \left(\frac{\partial f}{\partial x_i}\right)_{x_{j \neq i}} dx_i. \quad (5.7)$$

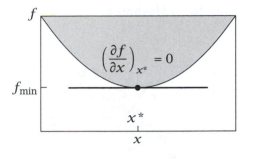

Figure 5.4 To find the extremum x^* of a function $f(x)$, find the point at which $df/dx = 0$.

The subscript $x_{j \neq i}$ indicates that all the other $t-1$ variables x_j are kept constant when you take the derivative with respect to x_i. The total differential df is the sum of contributions resulting from small increments in the independent variables x_i.

The Extrema of Multivariate Functions Occur Where the Partial Derivatives Are Zero

To use extremum principles, which can identify states of equilibrium, we find points at which derivatives are zero. In mathematics, a *critical point* is where a first derivative equals zero. It could be a maximum, a minimum, or a saddle point. In statistical thermodynamics, 'critical point' has a different meaning, but in this chapter critical point is used only in its mathematical sense.

How do we find an extremum of a multivariate function? First, recall how to find an extremum when there is only one variable. Figure 5.4 shows a single-variable function, $f(x) = x^2 + b$, where b is a constant. To locate the extremum, find the point at which the first derivative equals zero:

$$\left(\frac{df}{dx} \right)_{x^*} = 0. \tag{5.8}$$

The critical point x^* occurs at the value $f = f(x^*)$ that satisfies Equation (5.8). If the second derivative df^2/dx^2 is positive everywhere, the critical point is a minimum. If the second derivative is negative everywhere, the critical point is a maximum.

Now let's find an extremum of a multivariate function $f(x, y)$. A critical point is where small changes in both x and y together cause the total differential df to equal zero:

$$df = \left(\frac{\partial f}{\partial x} \right)_y dx + \left(\frac{\partial f}{\partial y} \right)_x dy = 0. \tag{5.9}$$

Since x and y are *independent* variables, the small non-zero quantities dx and dy are unrelated to each other. This means that we cannot suppose that the dx and dy terms in Equation (5.9) balance perfectly to cause df to equal zero. The independence of the variables x and y requires that *both* terms $(\partial f/\partial x)dx$ and $(\partial z/\partial y)dy$ equal zero simultaneously to satisfy Equation (5.9):

$$\left(\frac{\partial f}{\partial x} \right)_y dx = 0 \quad \textbf{and} \quad \left(\frac{\partial f}{\partial y} \right)_x dy = 0. \tag{5.10}$$

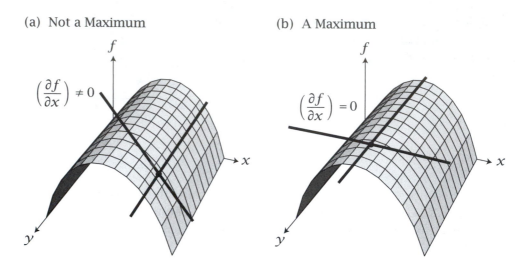

(a) Not a Maximum   (b) A Maximum

Figure 5.5 To identify the maximum of this function (Example 5.2), both $(\partial f/\partial x)$ and $(\partial f/\partial y)$ must equal zero as they do in (b), but not in (a).

Figure 5.6 For this paraboloid from Example 5.3 the critical point (x^*, y^*) equals (x_0, y_0). At (x^*, y^*) both partial derivatives equal zero.

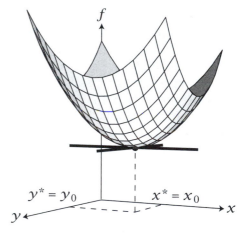

Furthermore, even though dx and dy are infinitesimal quantities, they are *not* equal to zero, so Equations (5.10) are satisfied only when the partial derivatives themselves equal zero:

$$\left(\frac{\partial f}{\partial x}\right)_y = 0 \quad \textbf{and} \quad \left(\frac{\partial f}{\partial y}\right)_x = 0. \tag{5.11}$$

To have $df = 0$, both Equations (5.11) must be satisfied. All the partial first derivatives must be zero for a point to be an extremum. Here is an example showing that it isn't sufficient for only one partial derivative to be zero.

EXAMPLE 5.2 Check all the partial derivatives. For the function f shown in Figure 5.5, f is independent of y so $(\partial f/\partial y)_x = 0$ everywhere. But $f(x, y)$ is only at a maximum where *both* $(\partial f/\partial x)_y$ *and* $(\partial f/\partial y)_x$ equal zero, only along the line at the top of the function (see Figure 5.5(b)).

The function $f(x_1, x_2, \ldots, x_t)$ of t independent variables will have extrema when $df(x_1, x_2, \ldots, x_t) = 0$, which means satisfying t simultaneous equations:

$$\left(\frac{\partial f}{\partial x_i}\right)_{x_{j\neq i}} = 0 \qquad \text{for all } i = 1, 2, \ldots, t. \tag{5.12}$$

Suppose that you want to reach the highest point on a hill. Traveling a path from east to west and finding the highest point on that path corresponds to setting one partial derivative to zero. However, the high point on the east-west path might not be the top of the hill. From your east-west path, perhaps traveling north takes you still farther uphill. Follow that route to the peak. At the summit, both derivatives are zero. The caveat is that this process might lead to a saddle point. The final tests that we need to ensure that we are at a maximum or a minimum are described in Examples 5.3 and 5.4.

EXAMPLE 5.3 Finding the minimum of a paraboloid. To find the minimum of the paraboloid $f(x, y) = (x - x_0)^2 + (y - y_0)^2$ in Figure 5.6, evaluate each partial derivative:

$$\left(\frac{\partial f}{\partial x}\right)_y = 2(x - x_0) \qquad \text{and} \qquad \left(\frac{\partial f}{\partial y}\right)_x = 2(y - y_0). \tag{5.13}$$

The critical point (x^*, y^*) is the point at which

$$\left(\frac{\partial f}{\partial x}\right)_y = 0 \quad \Rightarrow \quad 2(x^* - x_0) = 0 \quad \Rightarrow \quad x^* = x_0, \qquad \text{and}$$

$$\left(\frac{\partial f}{\partial y}\right)_x = 0 \quad \Rightarrow \quad 2(y^* - y_0) = 0 \quad \Rightarrow \quad y^* = y_0.$$

The critical point is at $(x^*, y^*) = (x_0, y_0)$. Check the signs of the partial derivatives at the critical point: $(\partial^2 f/\partial x^2)|_{x^*=x_0}$ and $(\partial^2 f/\partial y^2)_{y^*=y_0}$, which are both positive, so we might have a minimum. If those second partial derivatives were negative, we might have had a maximum. When some second partial derivatives are positive and some are negative, the function is at a saddle point and not at an extremum. For a function of two variables, the sign of the *Hessian* derivative expression,

$$\left(\frac{\partial^2 f(x, y)}{\partial x^2}\right)_{x^*,y^*} \left(\frac{\partial^2 f(x, y)}{\partial y^2}\right)_{x^*,y^*} - \left(\frac{\partial^2 f(x, y)}{\partial x \partial y}\right)_{x^*,y^*}^2,$$

evaluated at (x^*, y^*) must also be positive to ensure that we have an extremum. Here is an example.

EXAMPLE 5.4 A simple function that fails the Hessian test. A function that satisfies the test of the second partial derivatives $((\partial^2 f/\partial x^2)$ and $(\partial^2 f/\partial y^2)$ are both positive) but fails the Hessian test is $f(x, y) = x^2 + y^2 - 4xy$. The origin $(0, 0)$ is a critical point, and has positive second partials—both equal 2—but the origin is not an extremum, as Figure 5.7 shows.

We have considered the extrema of functions of *independent* variables. Now consider the extrema of functions of variables that are not independent, but are subject to *constraints*.

(a) $f(x, y) = x^2 + y^2 - 4xy$

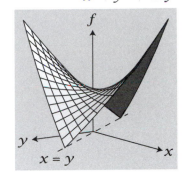

(b) Slice Through $x = 0$

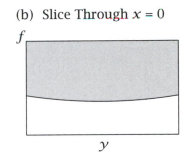

(c) Slice Through $y = 0$

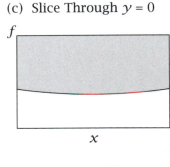

(d) Slice Through $x = y$

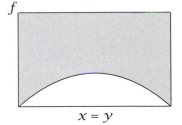

Figure 5.7 The origin is a saddle point, not a maximum or a minimum.

Finding the Extrema of Multivariate Functions Subject to Constraints

Constraints hold material systems at constant temperature or pressure, or at constant concentrations of chemical species, for example. Our aim in later chapters will be to find states of equilibrium by locating extrema of functions subject to such constraints. To find the extrema of a function $f(x, y)$ subject to constraints, you must find the set of values that satisfy *both* the extremum equation,

$$df = 0 = \left(\frac{\partial f}{\partial x}\right)_y dx + \left(\frac{\partial f}{\partial y}\right)_x dy, \tag{5.14}$$

and some constraint equation. A constraint is just a relationship between x and y. A constraint equation has the form $g(x, y) =$ constant. For example:

$$g(x, y) = x - y = 0$$
$$\Rightarrow \quad x = y. \tag{5.15}$$

Because x and y are related through the equation $g(x, y) =$ constant, they are not independent variables. To satisfy both Equations (5.14) and (5.15) simultaneously, put the constraint Equation (5.15) into differential form and combine it with the total differential Equation (5.14). Since $g =$ constant, the total differential dg is

$$dg = \left(\frac{\partial g}{\partial x}\right)_y dx + \left(\frac{\partial g}{\partial y}\right)_x dy = 0. \tag{5.16}$$

For the constraint Equation (5.15), the partial derivatives are $(\partial g/\partial x)_y = 1$ and $(\partial g/\partial y)_x = -1$. Equation (5.16) then gives $dx - dy = 0$, and $dx = dy$, which you can substitute into the extremum Equation (5.14):

$$df = 0 = \left(\frac{\partial f}{\partial x}\right)_y dx + \left(\frac{\partial f}{\partial y}\right)_x dx$$

$$\Rightarrow \quad \left[\left(\frac{\partial f}{\partial x}\right)_y + \left(\frac{\partial f}{\partial y}\right)_x\right] dx = 0.$$

Since dx is not zero we must have

$$\left(\frac{\partial f}{\partial x}\right)_y = -\left(\frac{\partial f}{\partial y}\right)_x. \tag{5.17}$$

Solving Equation (5.17) identifies the point that is both an extremum of f and also lies along the line $x = y$. This situation is different from that when x and y are independent. Then *each* of the partial derivatives equals zero (compare Equation (5.17) with Equation (5.12)).

EXAMPLE 5.5 Finding the minimum of a paraboloid subject to a constraint. Let's apply this reasoning to finding the minimum of the paraboloid $f(x, y) = x^2 + y^2$, subject to the constraint $g(x, y) = x + y = 6$ (see Figure 5.8). Look at this function geometrically: take a slice at $y = 6 - x$ through the paraboloid. Look for the lowest point on that slice (which is not necessarily the lowest point on the entire paraboloid).

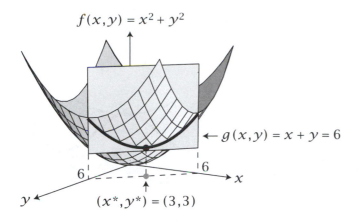

$f(x, y) = x^2 + y^2$

$\leftarrow g(x, y) = x + y = 6$

$(x^*, y^*) = (3, 3)$

Figure 5.8 The global minimum of $f(x, y) = x^2 + y^2$ is $(x^*, y^*) = (0, 0)$. But the minimum of $f(x, y)$ on the plane $g(x, y) = x + y = 6$ is $(x^*, y^*) = (3, 3)$.

To find the lowest point on $f(x, y)$ that satisfies $g(x, y)$, first evaluate the partial derivatives of g. In this case, $(\partial g / \partial x) = 1$ and $(\partial g / \partial y) = 1$. Equation (5.16) then gives $dx = -dy$. Substitute this into Equation (5.14) to get $(\partial f / \partial x)_{x^*, y^*} = (\partial f / \partial y)_{x^*, y^*}$. Taking these partial derivatives gives

$$2x^* = 2y^*. \tag{5.18}$$

The constraint equation requires $y^* = 6 - x^*$, so substitute y^* for x^* in Equation (5.18), with the result

$$2x^* = 2(6 - x^*) \quad \Rightarrow \quad x^* = y^* = 3.$$

This is the lowest point on the paraboloid that also lies in the plane $x + y = 6$. As shown in Figure 5.8, the constrained minimum is higher than the global minimum of the unconstrained function.

Example 5.5 shows one way to find the extrema of functions subject to constraints. A more general and powerful way is the method of Lagrange multipliers.

The Method of Lagrange Multipliers

Consider a function $f(x, y)$ for which we seek an extremum subject to the constraint $g(x, y) = $ constant.

Let's look at the problem graphically (Figure 5.9). In (a), $f(x, y)$ and $g(x, y)$ are intersecting surfaces. Because $g(x, y)$ relates to x and y, but does not relate to f, it is a ribbon-like surface parallel to f. (In Figure 5.8, g is a plane.) Curve h is the intersection of f and g, the set of points on the surface $f(x, y)$ that are also on the surface $g(x, y)$. h_{max} is the maximum value of f on the curve h. Figure 5.9(b) shows two tangent curves in the plane parallel to the xy plane through h_{max}: ℓ_g, the *level curve* of $g(x, y)$, and ℓ_f, the level curve of $f(x, y)$. The two level curves are tangent at h_{max}, the extremum of f that satisfies $g(x, y)$. The point $h_{max} = (x^*, y^*)$ must satisfy two conditions: the total differentials of df and dg are both zero,

$$\left(\frac{\partial f}{\partial x}\right)_y dx + \left(\frac{\partial f}{\partial y}\right)_x dy = 0, \quad \text{and} \tag{5.19}$$

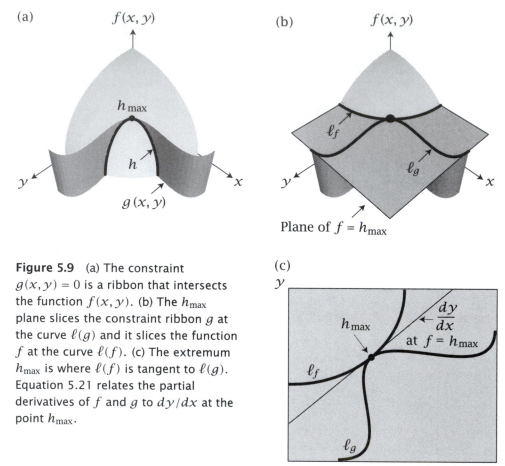

Figure 5.9 (a) The constraint $g(x, y) = 0$ is a ribbon that intersects the function $f(x, y)$. (b) The h_{max} plane slices the constraint ribbon g at the curve $\ell(g)$ and it slices the function f at the curve $\ell(f)$. (c) The extremum h_{max} is where $\ell(f)$ is tangent to $\ell(g)$. Equation 5.21 relates the partial derivatives of f and g to dy/dx at the point h_{max}.

$$\left(\frac{\partial g}{\partial x}\right)_y dx + \left(\frac{\partial g}{\partial y}\right)_x dy = 0. \tag{5.20}$$

We use the level curves to find the value of h_{max}. Figure 5.9(c) shows that at h_{max}, dy/dx is the slope of the tangent common to both ℓ_f and ℓ_g. The value of dy/dx at h_{max} is found by combining Equations (5.19) and (5.20):

$$\frac{dy}{dx} = \frac{-(\partial f/\partial x)_y}{(\partial f/\partial y)_x} = \frac{-(\partial g/\partial x)_y}{(\partial g/\partial y)_x}. \tag{5.21}$$

Because Equation (5.21) requires the equality of two ratios, the derivatives of f and g need only be the same to within an arbitrary constant λ, called the Lagrange multiplier:

$$\left(\frac{\partial f}{\partial x}\right)_y = \lambda \left(\frac{\partial g}{\partial x}\right)_y \quad \text{and} \quad \left(\frac{\partial f}{\partial y}\right)_x = \lambda \left(\frac{\partial g}{\partial y}\right)_x. \tag{5.22}$$

The values $x = x^*$ and $y = y^*$ that satisfy Equations (5.22) are at the extremum, and satisfy the constraint.

EXAMPLE 5.6 Finding the minimum of a paraboloid with Lagrange multipliers. Again let's find the minimum of the paraboloid $f(x, y) = x^2 + y^2$ subject to $g(x, y) = x + y = 6$. But now let's use the Lagrange multiplier method. We have $(\partial g / \partial x) = (\partial g / \partial y) = 1$. According to the Lagrange method, the solution will be given by

$$\left(\frac{\partial f}{\partial x}\right)_y = \lambda \left(\frac{\partial g}{\partial x}\right)_y \quad \Longrightarrow \quad 2x^* = \lambda, \tag{5.23}$$

$$\left(\frac{\partial f}{\partial y}\right)_x = \lambda \left(\frac{\partial g}{\partial y}\right)_x \quad \Longrightarrow \quad 2y^* = \lambda. \tag{5.24}$$

Combining Equations (5.23) and (5.24), and using $x + y = 6$, gives

$$2x^* = 2y^* \quad \Longrightarrow \quad y^* = x^* = 3. \tag{5.25}$$

The Lagrange multiplier result in Equation (5.25) is identical to our earlier result in Equation (5.18).

The Lagrange multiplier method doesn't simplify this particular example. However, it does simplify problems involving many variables and more than one constraint.

EXAMPLE 5.7 Lagrange multipliers again: finding a maximum. Find the maximum of the function $f(x, y) = 50 - (x - 3)^2 - (y - 6)^2$ subject to the constraint $g(x, y) = y - x = 0$, as shown in Figure 5.10. The global maximum is at $(x^*, y^*) = (3, 6)$, where both partial derivatives equal zero:

$$\left(\frac{\partial f}{\partial x}\right)_y = -2(x^* - 3) = 0 \quad \Longrightarrow \quad x^* = 3,$$

$$\left(\frac{\partial f}{\partial y}\right)_x = -2(y^* - 6) = 0 \quad \Longrightarrow \quad y^* = 6.$$

But since $(\partial g / \partial x) = -1$ and $(\partial g / \partial y) = 1$, the maximum along the constraint plane is at $(x^*, y^*) = (4.5, 4.5)$, according to Equation (5.22):

$$\left(\frac{\partial f}{\partial x}\right)_y = -2(x^* - 3) = -\lambda,$$

$$\left(\frac{\partial f}{\partial y}\right)_x = -2(y^* - 6) = \lambda.$$

Eliminating λ gives $(x^* - 3) = -(y^* - 6)$. Because the constraint is $y^* = x^*$, the maximum is at $y^* = x^* = 4.5$.

EXAMPLE 5.8 Lagrange multipliers find the maximum area/perimeter ratio. Suppose you have a fence 40 feet long, and want to use it to enclose a rectangle of the largest possible area. The area is

$$f(x, y) = xy.$$

Figure 5.10 The global maximum of the paraboloid, described in Example 5.7, $f(x, y) = 50 - (x-3)^2 - (y-6)^2$ is $f(x_0, y_0) = f(3, 6) = 50$. However, the maximum subject to the constraint $g(x, y) = x - y = 0$ is $f(x^*, y^*) = f(4.5, 4.5) = 45.5$.

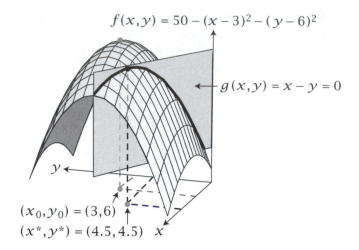

$$f(x, y) = 50 - (x-3)^2 - (y-6)^2$$

$$g(x, y) = x - y = 0$$

$$y$$

$$(x_0, y_0) = (3, 6)$$
$$(x^*, y^*) = (4.5, 4.5) \quad x$$

The perimeter is

$$g(x, y) = 2x + 2y = 40. \tag{5.26}$$

Use Lagrange multipliers to maximize f subject to g:

$$\left(\frac{\partial f}{\partial x}\right) = y, \qquad \left(\frac{\partial g}{\partial x}\right) = 2 \quad \Longrightarrow \quad y^* = 2\lambda, \tag{5.27}$$

$$\left(\frac{\partial f}{\partial y}\right) = x, \qquad \left(\frac{\partial g}{\partial y}\right) = 2 \quad \Longrightarrow \quad x^* = 2\lambda. \tag{5.28}$$

Substituting Equations (5.27) and (5.28) into Equation (5.26) and solving for λ gives $8\lambda = 40 \quad \Longrightarrow \quad \lambda = 5$ and $x^* = y^* = 10$. Therefore the rectangle of maximal area is a square, in this case with area $x^* y^* = 100$ square feet.

In general, to find the extremum of $f(x_1, x_2, \ldots, x_t)$ subject to a constraint $g(x_1, x_2, \ldots, x_t) = c$, where c is a constant, you have

$$\left(\frac{\partial f}{\partial x_1}\right) - \lambda \left(\frac{\partial g}{\partial x_1}\right) = 0,$$

$$\left(\frac{\partial f}{\partial x_2}\right) - \lambda \left(\frac{\partial g}{\partial x_2}\right) = 0,$$

$$\vdots$$

$$\left(\frac{\partial f}{\partial x_t}\right) - \lambda \left(\frac{\partial g}{\partial x_t}\right) = 0. \tag{5.29}$$

λ is eliminated by using the constraint equation

$$g(x_1, x_2, \ldots, x_t) = \text{constant}. \tag{5.30}$$

For the extremum of $f(x_1, x_2, \ldots, x_t)$ subject to more than one constraint, $g(x_1, x_2, \ldots, x_t) = c_1$ and $h(x_1, x_2, \ldots, x_t) = c_2$, etc., where the c_i's are constants, the Lagrange multiplier method gives the solutions

$$\left(\frac{\partial f}{\partial x_1}\right) - \lambda \left(\frac{\partial g}{\partial x_1}\right) - \beta \left(\frac{\partial h}{\partial x_1}\right) - \ldots = 0,$$

$$\left(\frac{\partial f}{\partial x_2}\right) - \lambda\left(\frac{\partial g}{\partial x_2}\right) - \beta\left(\frac{\partial h}{\partial x_2}\right) - \ldots = 0,$$

$$\vdots$$

$$\left(\frac{\partial f}{\partial x_t}\right) - \lambda\left(\frac{\partial g}{\partial x_t}\right) - \beta\left(\frac{\partial h}{\partial x_t}\right) - \ldots = 0. \tag{5.31}$$

where $\lambda, \beta, \ldots$ are the Lagrange multipliers for each constraint. Each Lagrange multiplier is found from its appropriate constraint equation.

Here is another perspective on the Lagrange multiplier strategy embodied in Equations (5.31). We intend to find the extremum

$$df = \sum_{i=1}^{t}\left(\frac{\partial f}{\partial x_i}\right)dx_i = 0, \tag{5.32}$$

subject to constraints

$$dg = \sum_{i=1}^{t}\left(\frac{\partial g}{\partial x_i}\right)dx_i = 0, \tag{5.33}$$

and

$$dh = \sum_{i=1}^{t}\left(\frac{\partial h}{\partial x_i}\right)dx_i = 0. \tag{5.34}$$

Incorporating Equations (5.33) and (5.34) into Equation (5.32) with Lagrange multipliers gives

$$d(f - \lambda g - \beta h) = \sum_{i=1}^{t}\left[\left(\frac{\partial f}{\partial x_i}\right) - \lambda\left(\frac{\partial g}{\partial x_i}\right) - \beta\left(\frac{\partial h}{\partial x_i}\right)\right]dx_i = 0. \tag{5.35}$$

Equations (5.31) show that each bracketed term in Equation (5.35) must equal zero, so the Lagrange multipliers have served to convert a problem involving constraints to a problem involving t independent quantities $[(\partial f/\partial x_i) - \lambda(\partial g/\partial x_i) - \beta(\partial h/\partial x_i)]$, each of which must equal zero.

Integrating Multivariate Functions Sometimes Depends on the Path of Integration, and Sometimes It Doesn't

Maxima or minima are not the only things that we need from multivariate functions. Sometimes we need to integrate them. We have considered only differential changes df so far. We now focus on larger changes $\Delta f = \int_A^B df$. Suppose you want to integrate

$$\int_{y_A}^{y_B}\int_{x_A}^{x_B}[s(x,y)\,dx + t(x,y)\,dy]. \tag{5.36}$$

You could first integrate over x holding $y = y_A$, then integrate over y. Or you could first integrate over y holding $x = x_A$, then integrate over x. The sequence of x and y values you choose is called a *pathway*. Some possible

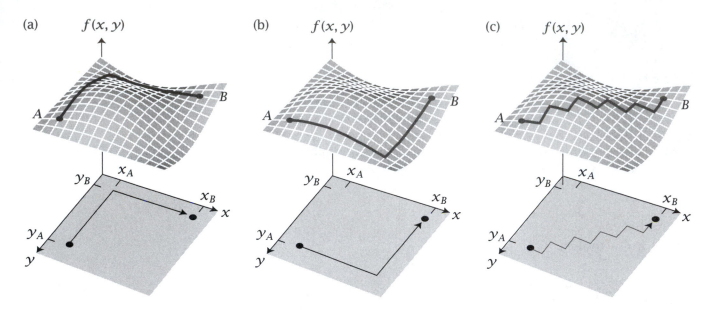

Figure 5.11 Different paths for integrating $f(x, y)$ from (x_A, y_A) to (x_B, y_B).

pathways for integration are shown as projections onto the xy-plane in Figure 5.11. Integrating a function $f(x)$ of a single variable does not involve pathways.

Which pathway should you use to integrate a multivariate function? Does it matter? It depends. Some functions are *state* functions, which are independent of pathways, and some are *path-dependent* functions.

If you can express $s\,dx + t\,dy$ as the differential of a function $f(x, y)$,

$$df = s\,dx + t\,dy, \tag{5.37}$$

then you can substitute Equation (5.37) into Equation (5.36) to get

$$\iint [s\,dx + t\,dy] = \int df = \Delta f = f(x_B, y_B) - f(x_A, y_A). \tag{5.38}$$

f is a state function because the integral depends only on the initial and final states (x_A, y_A) and (x_B, y_B). The integral will be the same no matter which path you use to integrate it.

> **EXAMPLE 5.9 State functions.** $x\,dy + y\,dx$ is a state function because it can be expressed as the differential of $f = xy$,
>
> $$df = \left(\frac{\partial f}{\partial x}\right)_y dx + \left(\frac{\partial f}{\partial y}\right)_x dy = y\,dx + x\,dy.$$
>
> Similarly, $6xy\,dx + 3x^2\,dy$ is a state function because it can be expressed as the differential of $f = 3x^2 y$.

df is called an *exact* differential. On the other hand, for differentials like $y^2\,dx + x^2\,dy$ or $x^2 y^3\,dx + 3x^2 y^2\,dy$, integration will give different answers depending on the integration pathway. $y^2\,dx + x^2\,dy$ cannot be written as d (some function of x and y) so Equation (5.38) does not apply, and the integral will depend on the integration pathway. These are *inexact* differentials. How do you determine whether any particular differential $s(x,y)dx + t(x,y)dy$ is exact or inexact? Use the Euler test.

The **Euler reciprocal relationship** is

$$\left(\frac{\partial^2 f}{\partial x \partial y}\right) = \left(\frac{\partial^2 f}{\partial y \partial x}\right),\tag{5.39}$$

for any state function f. To find out whether an expression $s(x,y)dx + t(x,y)dy$ is an exact differential, determine whether

$$\left(\frac{\partial s(x,y)}{\partial y}\right) = \left(\frac{\partial t(x,y)}{\partial x}\right).\tag{5.40}$$

If Equation (5.40) holds, then $s = (\partial f/\partial x)$, $t = (\partial f/\partial y)$, and $s(x,y)dx + t(x,y)dy = df$ is an exact differential.

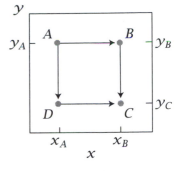

Figure 5.12 For proving the Euler reciprocal relationship, two paths from A to C: through B, or through D.

Proof of the Euler Reciprocal Relationship

In Chapter 9 we will use the Euler relationship to establish the Maxwell relations between the thermodynamic quantities. Here we derive the Euler relationship. Figure 5.12 shows four points at the vertices of a rectangle in the xy plane. Using a Taylor series expansion, Equation (4.22), compute the change in a function Δf through two different routes. First integrate from point A to point B to point C. Then integrate from point A to point D to point C. Compare the results to find the Euler reciprocal relationship. For $\Delta f = f(x+\Delta x, y+\Delta y) - f(x,y)$, the first terms of the Taylor series are

$$\Delta f = \left(\frac{\partial f}{\partial x}\right)\Delta x + \left(\frac{\partial f}{\partial y}\right)\Delta y$$

$$+ \frac{1}{2}\left[\left(\frac{\partial^2 f}{\partial x^2}\right)(\Delta x)^2 + \left(\frac{\partial^2 f}{\partial y^2}\right)(\Delta y)^2 + 2\left(\frac{\partial^2 f}{\partial x \partial y}\right)\Delta x \Delta y\right]$$

$$+ \cdots\tag{5.41}$$

Along the path from point A to point B in Figure 5.12, x changes, $\Delta x = x_B - x_A$, and y is constant. From point A to B, Equation (5.41) gives

$$f(B) = f(A) + \left(\frac{\partial f(A)}{\partial x}\right)\Delta x + \frac{1}{2}\left(\frac{\partial^2 f(A)}{\partial x^2}\right)(\Delta x)^2 + \cdots.\tag{5.42}$$

From B to C, only y changes, $\Delta y = y_C - y_B$, while x remains fixed:

$$f(C) = f(B) + \left(\frac{\partial f(B)}{\partial y}\right)\Delta y + \frac{1}{2}\left(\frac{\partial^2 f(B)}{\partial y^2}\right)(\Delta y)^2 + \cdots.\tag{5.43}$$

Substitute $f(B)$ given by Equation (5.42) into Equation (5.43), and keep terms up to second order because those are the lowest-order terms that do not cancel.

continued

Higher-order terms will be negligible. We find $f(C)$ in terms of $f(A)$:

$$f(C) = f(A) + \left(\frac{\partial f(A)}{\partial x}\right)\Delta x + \frac{1}{2}\left(\frac{\partial^2 f(A)}{\partial x^2}\right)(\Delta x)^2$$

$$+ \left(\frac{\partial f(A)}{\partial y}\right)\Delta y + \frac{\partial}{\partial y}\left(\frac{\partial f(A)}{\partial x}\right)\Delta y \Delta x$$

$$+ \frac{1}{2}\left(\frac{\partial^2 f(A)}{\partial y^2}\right)(\Delta y)^2 + \cdots. \tag{5.44}$$

Now evaluate the other path, from point A through point D to point C. Although the path differs, the values $\Delta x = x_C - x_D = x_B - x_A$ and $\Delta y = y_D - y_A = y_C - y_B$ are the same as for path ABC. Now you have

$$f(C) = f(A) + \left(\frac{\partial f(A)}{\partial y}\right)\Delta y + \frac{1}{2}\left(\frac{\partial^2 f(A)}{\partial y^2}\right)(\Delta y)^2$$

$$+ \left(\frac{\partial f(A)}{\partial x}\right)\Delta x + \frac{\partial}{\partial x}\left(\frac{\partial f(A)}{\partial y}\right)\Delta y \Delta x$$

$$+ \frac{1}{2}\left(\frac{\partial^2 f(A)}{\partial x^2}\right)(\Delta x)^2 + \cdots \tag{5.45}$$

Equations (5.44) and (5.45) both give $f(C)$, so you can set the right-hand sides equal and cancel identical terms. The remaining terms give the Euler reciprocal relationship (Equation (5.39)):

$$\left(\frac{\partial^2 f(A)}{\partial y \partial x}\right) = \left(\frac{\partial^2 f(A)}{\partial x \partial y}\right).$$

EXAMPLE 5.10 Applying the Euler test and integrating a state function. Is $6xy^3 dx + 9x^2 y^2 dy$ an exact differential? Apply the test of Equation (5.40):

$$\left(\frac{\partial(6xy^3)}{\partial y}\right) = 18xy^2, \quad \text{and} \quad \left(\frac{\partial(9x^2y^2)}{\partial x}\right) = 18xy^2.$$

Because the two second partial derivatives are equal, $6xy^3 dx + 9x^2 y^2 dy$ is an exact differential. The function $f(x, y)$ is given by

$$f(x, y) = 3x^2 y^3 + c, \tag{5.46}$$

where c is a constant of integration, as you can see by integrating $18xy^2$ once with respect to y then once with respect to x (or the other way around). That is, $d(3x^2 y^3) = 6xy^3 dx + 9x^2 y^2 dy$. To integrate from $(x_1, y_1) = (1, 1)$ to $(x_2, y_2) = (4, 5)$,

$$\int_{x=1}^{4} \int_{y=1}^{5} [6xy^3 \, dx + 9x^2 y^2 \, dy], \tag{5.47}$$

we need not be concerned with the pathway of integration. The integral depends only on the difference between the initial and final states, (x_1, y_1) and (x_2, y_2).

$$\Delta f = f(x_2, y_2) - f(x_1, y_1) = 3x^2 y^3 \Big|_{x_1, y_1}^{x_2, y_2}$$

$$= \left(3 \cdot 4^2 \cdot 5^3\right) - \left(3 \cdot 1^2 \cdot 1^3\right) = 5997. \tag{5.48}$$

EXAMPLE 5.11 Applying the Euler test again. Is $x^2 dx + xy\, dy$ an exact differential? Apply the test of Equation (5.40). You will find that it is not, because $(\partial x^2 / \partial y) = 0$, which does not equal $(\partial xy / \partial x) = y$. The value of the integral of this differential depends on the choice of pathway.

In physical systems, pathway-dependence means that it matters in what order, by what process, or how fast you do things. Here's an example [1]. Consider two quantities measured on a car trip. One quantity is the distance that you travel as measured by the odometer. This is a quantity that depends on the pathway. The distance traveled from San Francisco to Chicago depends on which route you take. It is *not* a state property. Another property of a trip is the change in altitude from sea level as measured by an altimeter. This property *is* a state function. The change in altitude during a trip from San Francisco to Chicago depends only on a characteristic (the *state*) of the two cities (their heights above sea level) and not on the route you take from one to the other.

In the rest of this chapter we give some useful rules for manipulating partial derivatives.

The Chain Rule
Applies to Functions of Several Variables

The chain rule is useful for *composite functions*, functions of variables that are themselves functions of other variables. Suppose that the variables x, y, and z in a function $f(x, y, z)$ are each a function of another variable u, that is $x = x(u)$, $y = y(u)$, and $z = z(u)$. Then $f(x, y, z)$ is a *composite* function $f = f(x(u), y(u), z(u))$. The total derivative df in terms of the variables (x, y, z) is

$$df = \left(\frac{\partial f}{\partial x}\right)_{y,z} dx + \left(\frac{\partial f}{\partial y}\right)_{x,z} dy + \left(\frac{\partial f}{\partial z}\right)_{x,y} dz. \tag{5.49}$$

The differentials of x, y, and z with respect to u are

$$dx = \frac{dx}{du} du, \qquad dy = \frac{dy}{du} du, \qquad dz = \frac{dz}{du} du. \tag{5.50}$$

Substitution of Equations (5.50) into Equation (5.49) produces the **chain rule for partial differentiation**:

$$df = \left[\left(\frac{\partial f}{\partial x}\right)_{y,z} \frac{dx}{du} + \left(\frac{\partial f}{\partial y}\right)_{x,z} \frac{dy}{du} + \left(\frac{\partial f}{\partial z}\right)_{x,y} \frac{dz}{du} \right] du, \text{ or} \tag{5.51}$$

$$\frac{df}{du} = \left(\frac{\partial f}{\partial x}\right)_{y,z} \frac{dx}{du} + \left(\frac{\partial f}{\partial y}\right)_{x,z} \frac{dy}{du} + \left(\frac{\partial f}{\partial z}\right)_{x,y} \frac{dz}{du}. \tag{5.52}$$

The symbol d/du indicates differentiation when a function depends only on a single variable, while $(\partial/\partial x)$ indicates partial differentiation of a multivariate function. In these expressions $x(u)$, $y(u)$, and $z(u)$ are functions of only one variable, while $f(x, y, z)$ is multivariate.

EXAMPLE 5.12 The chain rule. Let's return to the ball in the valley from Chapter 2, page 28. The height of the ball $z = x^2$ is a function of lateral position x. However, the energy is a function of height $U = mgz$, so the energy is a composite function $U(z(x))$. The chain rule gives the change in energy with lateral position, dU/dx:

$$\frac{dU}{dx} = \left(\frac{\partial U}{\partial z}\right)\frac{dz}{dx} = (mg)(2x) = 2mgx.$$

Dependent and Independent Variables Can Be Rearranged

We will show in Chapter 8 that the state of equilibrium in many laboratory experiments is defined by the total differential $dG = 0$, where the Gibbs free energy $G(T, p)$ is a function of temperature T and pressure p. However, suppose that you want to know the relationship between the measurable quantities T and p at equilibrium when $dG = 0$. This is the basis for *phase equilibria*. How is the relationship of T and p defined at constant G? We'll return to the physics later. For now let's look at the mathematics.

Let's look at a general case in which a function $f(x, y, z)$ is constrained to be constant and therefore the total differential equals zero:

$$df = 0 = \left(\frac{\partial f}{\partial x}\right)_{y,z} dx + \left(\frac{\partial f}{\partial y}\right)_{x,z} dy + \left(\frac{\partial f}{\partial z}\right)_{x,y} dz. \tag{5.53}$$

The variables x, y, and z are not independent of each other, owing to the constraint that $df = 0$. How are they related? Let's look at the relations between pairs of variables. If both f and z are held constant, df and dz equal zero, and you have

$$0 = \left(\frac{\partial f}{\partial x}\right)_{y,z} dx + \left(\frac{\partial f}{\partial y}\right)_{x,z} dy. \tag{5.54}$$

Divide Equation (5.54) by dx to get a useful relation:

$$0 = \left(\frac{\partial f}{\partial x}\right)_{y,z} + \left(\frac{\partial f}{\partial y}\right)_{x,z}\frac{dy}{dx} \quad \Longrightarrow \quad \frac{dy}{dx} = -\frac{(\partial f/\partial x)_{y,z}}{(\partial f/\partial y)_{x,z}}. \tag{5.55}$$

You can manipulate the differential quantities dx, dy, and dz in the same way as ordinary algebraic terms, but terms such as ∂f and ∂x **are not defined** as individual quantities.

Here is an example that demonstrates the rearrangement of dependent and independent variables.

EXAMPLE 5.13 Reshaping a cylinder while keeping its volume constant.
You have a cylinder of radius r and height h. Its volume is $V = \pi r^2 h$. If you double the length, how should you change the radius to keep the volume constant? The constraint is expressed by

$$dV = 0 = \left(\frac{\partial V}{\partial r}\right)_h dr + \left(\frac{\partial V}{\partial h}\right)_r dh. \tag{5.56}$$

Now divide Equation (5.56) by dr:

$$0 = \left(\frac{\partial V}{\partial r}\right)_h + \left(\frac{\partial V}{\partial h}\right)_r \frac{dh}{dr} \quad \Longrightarrow \quad \frac{dh}{dr} = -\frac{(\partial V/\partial r)_h}{(\partial V/\partial h)_r}. \tag{5.57}$$

Since $(\partial V/\partial r)_h = 2\pi rh$ and $(\partial V/\partial h)_r = \pi r^2$, Equation (5.57) gives

$$\frac{dh}{dr} = -\frac{2\pi rh}{\pi r^2} = -\frac{2h}{r}.$$

Rearrange to get the h's on the left and the r's on the right and integrate:

$$-\frac{1}{2}\int_{h_1}^{h_2}\frac{dh}{h} = \int_{r_1}^{r_2}\frac{dr}{r} \quad \Longrightarrow \quad -\frac{1}{2}\ln\frac{h_2}{h_1} = \ln\frac{r_2}{r_1} \quad \Longrightarrow \quad \frac{h_2}{h_1} = \left(\frac{r_1}{r_2}\right)^2.$$

Doubling the radius implies that the height must be reduced to one-quarter of its initial value to hold the volume constant.

Summary

Some functions depend on more than a single variable. To find extrema of such functions, it is necessary to find where all the partial derivatives are zero. To find extrema of multivariate functions that are subject to constraints, the Lagrange multiplier method is useful. Integrating multivariate functions is different from integrating single-variable functions: multivariate functions require the concept of a pathway. *State functions* do not depend on the pathway of integration. The Euler reciprocal relation relation can be used to distinguish state functions from path-dependent functions. In the next three chapters, we will combine the First and Second Laws with multivariate calculus to derive the principles of thermodynamics.

Problems

1. Applying the Euler test. How much kinetic energy does a 1700 kg car have, if it travels 100 km h^{-1}? Which of the following are exact differentials?

(a) $6x^5 dx + dy$

(b) $x^2 y^2 dx + 3x^2 y^3 dy$

(c) $(1/y)dx - (x/y^2)dy$

(d) $y dx + 2x dy$

(e) $\cos x dx - \sin y dy$

(f) $(x^2 + y)dx + (x + y^2)dy$

(g) $x dx + \sin y dy$

2. Differentiating multivariate functions. Compute the partial derivatives, $(\partial f/\partial x)_y$ and $(\partial f/\partial y)_x$, for the following functions.

(a) $f(x,y) = 3x^2 + y^5$

(b) $f(x,y) = x^{10} y^{1/2}$

(c) $f(x,y) = x + y^2 + 3$

(d) $f(x,y) = 5x$

3. Minimizing a single-variable function subject to a constraint. Given the function $y = (x-2)^2$, find x^*, the value of x that minimizes y subject to the constraint $y = x$.

4. Maximizing a multivariate function without constraints. Find the maximum (x^*, y^*, z^*) of the function $f(x,y,z) = d - (x-a)^2 - (y-b)^2 - (z-c)^2$.

5. Extrema of multivariate functions with constraints.

(a) Find the maximum of the function $f(x,y) = -(x-a)^2 - (y-b)^2$ subject to the constraint $y = kx$.

(b) Find the minimum of the paraboloid $f(x,y) = (x-x_0)^2 + (y-y_0)^2$ subject to the constraint $y = 2x$.

6. Composite functions.

(a) Given the functions $f(x,y(u)) = x^2 + 3y^2$ and $y(u) = 5u + 3$, express df in terms of changes dx and du.

(b) What is $\left(\dfrac{\partial f}{\partial u}\right)_{x, u=1}$?

7. Converting to an exact differential. Given the expression $dx + (x/y)dy$, show that dividing by x results in an exact differential. What is the function $f(x,y)$ such that df is $dx + (x/y)dy$ divided by x?

8. Propagation of error. Suppose that you can measure independent variables x and y and that you have a dependent variable $f(x,y)$. The average values are $\bar{x}$, $\bar{y}$, and $\bar{f}$, and the errors are $\varepsilon_x = x - \bar{x}$, $\varepsilon_y = y - \bar{y}$, and $\varepsilon_f = f - \bar{f}$.

(a) Use a Taylor series expansion to express the error in f, ε_f, as a function of the errors in x, ε_x, and y, ε_y.

(b) Compute the mean-squared error $\langle \varepsilon_f^2 \rangle$ as a function of $\langle \varepsilon_x^2 \rangle$ and $\langle \varepsilon_y^2 \rangle$.

(c) Consider the ideal gas law with $f = p$, $x = T$, and $y = V$. If T and V have 10% random errors, how big will be the error in p?

(d) How much increase in $\varepsilon(f)$ comes from adding n measurements that all have the same uncorrelated error $\varepsilon(x)$?

9. Small differences of large numbers can lead to nonsense. Using the results from problem 8, show that the propagated error is larger than the difference itself for $f(x,y) = x - y$, with $x = 20 \pm 2$ and $y = 19 \pm 2$.

10. Finding extrema. Find the point (x^*, y^*, z^*) that is at the minimum of the function

$$f(x,y,z) = 2x^2 + 8y^2 + z^2$$

subject to the constraint equation

$$g(x,y,z) = 6x + 4y + 4z - 72 = 0.$$

References

[1] R Dickerson. *Molecular Thermodynamics*. WA Benjamin, New York, 1969.

Suggested Reading

Good calculus references:

HB Callen, *Thermodynamics and an Introduction to Thermostatistics*, 2nd edition, Wiley and Sons, New York, 1985.

JE Marsden and AJ Tromba, *Vector Calculus*, 3rd edition, WH Freeman & Co, New York, 1988.

GB Thomas Jr, *Calculus and Analytic Geometry*, 8th edition, Addison-Wesley, Reading, 1992.

6 Entropy & the Boltzmann Law

$$S = k \log W$$

LVDWIG
BOLTZMANN
1844-1906

What Is Entropy?

Carved on the tombstone of Ludwig Boltzmann in the Zentralfriedhof (central cemetery) in Vienna is the inscription

$$S = k \log W. \tag{6.1}$$

This equation is the historical foundation of statistical mechanics. It connects the microscopic and macroscopic worlds. It defines the *entropy S*, a macroscopic quantity, in terms of the multiplicity W of the microscopic degrees of freedom of a system. For thermodynamics, $k = 1.380662 \times 10^{-23} \mathrm{J\,K^{-1}}$ is a quantity called Boltzmann's constant, and Boltzmann's inscription refers to the natural logarithm, $\log_e = \ln$.

In Chapters 2 and 3 we used simple models to illustrate that the composition of coin flips, the expansion of gases, the tendency of particles to mix, rubber elasticity, and heat flow can be predicted by the principle that systems tend toward their states of maximum multiplicity W. However, states that maximize W will also maximize W^2 or $15W^3 + 5$ or $k \ln W$, where k is any positive constant. Any monotonic function of W will have a maximum where W has a maximum. In particular, states that maximize W also maximize the entropy, $S = k \ln W$. Why does this quantity deserve special attention as a prediction principle, and why should it have this particular mathematical form?

In this chapter, we use a few general principles to show why the entropy must have this mathematical form. But first we switch our view of entropy from a multiplicity perspective to a probability perspective that is more general. In the probability perspective, the entropy is given as

$$\frac{S}{k} = -\sum_{i=1}^{t} p_i \ln p_i. \tag{6.2}$$

Let's see how Equation (6.2) is related to Equation (6.1). Roll a t-sided die N times. The multiplicity of outcomes is given by Equation (1.18) (see page 12),

$$W = \frac{N!}{n_1! n_2! \ldots n_t!},$$

where n_i is the number of times that side i appears face up. Use Stirling's approximation $x! \approx (x/e)^x$ (page 56), and define the probabilities $p_i = n_i/N$, to convert Equation (1.18) to

$$W = \frac{(N/e)^N}{(n_1/e)^{n_1}(n_2/e)^{n_2}\ldots(n_t/e)^{n_t}}$$

$$= \frac{N^N}{n_1^{n_1} n_2^{n_2} \ldots n_t^{n_t}} = \frac{1}{p_1^{n_1} p_2^{n_2} \ldots p_t^{n_t}}. \tag{6.3}$$

Take the logarithm and divide by N to get

$$\ln W = -\sum_{i=1}^{t} n_i \ln p_i \quad \Longrightarrow \quad \frac{1}{N} \ln W = -\sum_{i=1}^{t} p_i \ln p_i = \frac{S_N}{Nk}, \tag{6.4}$$

where S_N indicates that this is the entropy for N trials, and the entropy per trial is $S = S_N/N$. For this dice problem and the counting problems in Chapters 2 and 3, the two expressions for the entropy, Equations (6.2) and (6.1), are equivalent. The flattest distributions are those having maximum multiplicity W in the absence of constraints. For example, in N coin flips, the multiplicity $W = N!/[(n_H!)(N - n_H)!]$ is maximized when $n_H/N \approx n_T/N \approx 0.5$, that is, when the probabilities of heads and tails are as nearly equal as possible.

There are different types of entropy, depending on the degrees of freedom of the system. Examples 2.2 and 2.3 describe *translational* freedom due to the different positions of particles in space. In the next example we apply Equation (6.2) to the *rotational* or *orientational* entropy of *dipoles*. We show that flatter probability distributions have higher entropy than more peaked distributions.

EXAMPLE 6.1 Dipoles tend to orient randomly. Objects with distinguishable heads and tails such as magnets, chemically asymmetrical molecules, electrical dipoles with (+) charges at one end and (−) charges at the other, or even pencils with erasers at one end have rotational freedom as well as translational freedom. They can orient.

Spin a pencil on a table N times. Each time it stops, the pencil points in one of four possible directions: toward the quadrant facing north (n), east (e), south (s), or west (w). Count the number of times that the pencil points in each direction; label those numbers n_n, n_e, n_s, and n_w. Spinning a pencil and counting orientations is analogous to rolling a die with four sides labeled n, e, s or w. Each roll of that die determines the orientation of one pencil or

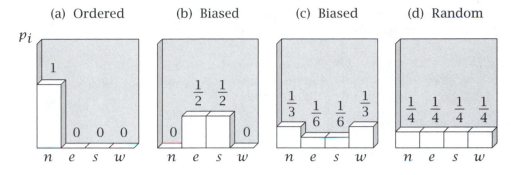

Figure 6.1 Spin a hundred pencils. Here are four (of a large number) of possible distributions of outcomes. (a) All pencils could point north (n). This is the most *ordered* distribution, $S/k = -1 \ln 1 = 0$. (b) Half the pencils could point east (e) and half could point south (s). This distribution has more entropy than (a), $S/k = -2(1/2 \ln 1/2) = 0.69$. (c) One-third of the pencils could point n, and one-third w, one-sixth e, and one-sixth s. This distribution has even more entropy, $S/k = -2(1/3 \ln 1/3 + 1/6 \ln 1/6) = 1.33$. (d) One-quarter of the pencils could point in each of the four possible directions. This is the distribution with highest entropy, $S/k = -4(1/4 \ln 1/4) = 1.39$.

dipole. N die rolls correspond to the orientations of N dipoles. The number of configurations for systems with N trials, distributed with any set of outcomes $\{n_1, n_2, \ldots, n_t\}$, where $N = \sum_{i=1}^{t} n_i$, is given by the multiplicity Equation (1.18): $W(n_1, n_2, \ldots, n_t) = N!/(n_1! n_2! \cdots n_t!)$. The number of different configurations of the system with a given composition n_n, n_e, n_s, and n_w is

$$W(N, n_n, n_e, n_s, n_w) = \frac{N!}{n_n! n_e! n_s! n_w!}.$$

The probabilities that a pencil points in each of the four directions are

$$p(n) = \frac{n_n}{N}, \qquad p(e) = \frac{n_e}{N}, \qquad p(s) = \frac{n_s}{N}, \qquad \text{and} \qquad p(w) = \frac{n_w}{N}.$$

Figure 6.1 shows some possible distributions of outcomes. Each distribution function satisfies the constraint that $p(n) + p(e) + p(s) + p(w) = 1$. You can compute the entropy per spin of the pencil of any of these distributions by using Equation (6.2), $S/k = -\sum_{i=1}^{t} p_i \ln p_i$. The absolute entropy is never negative, that is $S \geq 0$.

Flat distributions have high entropy. Peaked distributions have low entropy. When all pencils point in the same direction, the system is perfectly ordered and has the lowest possible entropy, $S = 0$. Entropy does not depend on being able to order the categories along an x-axis. For pencil orientations, there is no difference between the x-axis sequence *news* and *esnw*. To be in a state of low entropy, it does not matter which direction the pencils point in, just that they all point the same way. The flattest possible distribution has the highest possible entropy, which increases with the number of possible outcomes. In Figure 6.1 we have four states: the flattest distribution has $S/k = -4(1/4) \ln(1/4) = \ln 4 = 1.39$. In general, when there are t states, the flat distribution has entropy $S/k = \ln t$. Flatness in a distribution corresponds to *disorder* in a system.

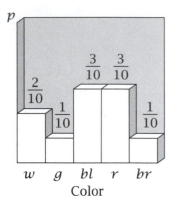

Figure 6.2 The entropy can be computed for *any* distribution function, even for colors of socks: white (*w*), green (*g*), black (*bl*), red (*r*), and brown (*br*).

The concept of entropy is broader than statistical thermodynamics. It is a property of any distribution function, as the next example shows.

EXAMPLE 6.2 Colors of socks. Suppose that on a given day, you sample 30 students and find the distribution of the colors of the socks they are wearing (Figure 6.2). The entropy of this distribution is

$$S/k = -0.2\ln 0.2 - 0.6\ln 0.3 - 0.2\ln 0.1 = 1.50.$$

For this example and Example 6.1, k should not be Boltzmann's constant. Boltzmann's constant is appropriate only when you need to put entropy into units that interconvert with energy, for thermodynamics and molecular science. For other types of probability distributions, k is chosen to suit the purposes at hand, so $k = 1$ would be simplest here. The entropy function just reports the relative flatness of a distribution function. The limiting cases are the most ordered, $S = 0$ (everybody wears the same color socks) and the most disordered, $S/k = \ln t = \ln 5 = 1.61$ (all five sock colors are equally likely).

Why should the entropy have the form of either Equation (6.1) or Equation (6.2)? Here is a simple justification. A deeper argument is given in 'Optional Material,' page 89.

The Simple Justification for $S = k \ln W$

Consider a thermodynamic system having two subsystems, A and B, with multiplicities W_A and W_B respectively. The multiplicity of the total system will be the product $W_{total} = W_A W_B$. Thermodynamics requires that entropies be *extensive*, meaning that the system entropy is the sum of subsystem entropies, $S_{total} = S_A + S_B$. The logarithm function satisfies this requirement. If $S_A = k \ln W_A$ and $S_B = k \ln W_B$, then $S_{total} = k \ln W_{total} = k \ln W_A W_B = k \ln W_A + k \ln W_B = S_A + S_B$. This argument illustrates why S should be a logarithmic function of W.

Let's use $S/k = -\sum_i p_i \ln p_i$ to derive the exponential distribution law, called the Boltzmann distribution law, that is at the center of statistical thermodynamics. The Boltzmann distribution law describes the energy distributions of atoms and molecules.

Underdetermined Distributions

In the rest of this chapter, we illustrate the principles that we need by concocting a class of problems involving die rolls and coin flips instead of molecules. How would you know if a die is biased? You could roll it N times and count the numbers of **1**'s, **2**'s, ..., **6**'s. If the probability distribution were perfectly flat, the die would not be biased. You could use the same test for the orientations of pencils or to determine whether atoms or molecules have biased spatial orientations or bond angle distributions. However the options available to molecules are usually so numerous that you could not possibly measure each one. In statistical mechanics you seldom have the luxury of knowing the full distribution, corresponding to all six numbers p_i for $i = 1, 2, 3, \ldots, 6$ on die rolls.

Therefore, as a prelude to statistical mechanics, let's concoct dice problems that are underdetermined in the same way as the problems of molecular science. Suppose that you do not know the distribution of all six possible outcomes. Instead, you know only the total score (or equivalently, the average score per roll) on the N rolls. In thousands of rolls, the average score per roll of an unbiased die will be $3.5 = (1 + 2 + 3 + 4 + 5 + 6)/6$. If you observe that the average score is 3.5, it is evidence (but not proof)[1] that the distribution is unbiased. In that case, your best guess consistent with the evidence is that the distribution is flat. All outcomes are equally likely.

However, if you observe the average score per roll is 2.0, then you must conclude that every outcome from 1 to 6 is *not* equally likely. You know only that low numbers are somehow favored. This one piece of data—the total score—is not sufficient to predict all six unknowns of the full distribution function. So we aim to do the next best thing. We aim to predict the least biased distribution function that is consistent with the known measured score. This distribution is predicted by the maximum-entropy principle.

Maximum Entropy Predicts Flat Distributions When There Are No Constraints

The entropy function is used to predict probability distributions. Here we show that the tendency toward the maximum entropy is a tendency toward maximum flatness of a probability distribution function when there are no constraints. Roll an unbiased t-sided die many times. Because the probabilities must sum to one,

$$\sum_{i=1}^{t} p_i = 1 \quad \Longrightarrow \quad \sum_{i=1}^{t} dp_i = 0. \tag{6.5}$$

We seek the distribution, $(p_1, p_2, \ldots, p_t) = (p_1^*, p_2^*, \ldots, p_t^*)$, that causes the entropy function, $S(p_1, p_2, \ldots, p_t) = -k \sum_i p_i \ln p_i$ to be at its maximum possible value, subject to the normalization Equation (6.5). For this problem, let $k = 1$. To solve it by the Lagrange multiplier method (see Equation (5.35)), with multiplier α for constraint Equation (6.5), we want

$$\sum_{i=1}^{t} \left[\left(\frac{\partial S}{\partial p_i} \right)_{p_{j \neq i}} - \alpha \right] dp_i = 0. \tag{6.6}$$

Set the term inside the brackets equal to zero for each i. The derivative of the entropy function gives $(\partial S / \partial p_i) = -1 - \ln p_i$ (since you are taking the derivative with respect to one particular p_i with all the other p's, $j \neq i$, held constant), so the solution is

$$-1 - \ln p_i - \alpha = 0 \quad \Longrightarrow \quad p_i^* = e^{(-1-\alpha)}. \tag{6.7}$$

To put this into a simpler form, divide Equation (6.7) by $\sum_i p_i^* = 1$ to get

$$\frac{p_i^*}{\sum_{i=1}^{t} p_i^*} = \frac{e^{(-1-\alpha)}}{t e^{(-1-\alpha)}} - \frac{1}{t}. \tag{6.8}$$

---

[1] For example, that score could also arise from 50% 2's and 50% 5's.

Maximizing the entropy predicts that when there is no bias, all outcomes are equally likely. However, what if there is bias? The next section shows how maximum entropy works in that case.

Maximum Entropy Predicts Exponential Distributions When There Are Constraints

Roll a die having t sides, with faces numbered $i = 1, 2, 3, \ldots, t$. You do not know the distribution of outcomes of each face, but you know the total score after N rolls. You want to predict the distribution function. When side i appears face up, the score is ε_i. The total score after N rolls will be $E = \sum_{i=1}^{t} \varepsilon_i n_i$, where n_i is the number of times that you observe face i. Let $p_i = n_i/N$ represent the fraction of the N rolls on which you observe face i. The average score per roll $\langle \varepsilon \rangle$ is

$$\langle \varepsilon \rangle = \frac{E}{N} = \sum_{i=1}^{t} p_i \varepsilon_i. \tag{6.9}$$

What is the expected distribution of outcomes $(p_1^*, p_2^*, \ldots, p_t^*)$ consistent with the observed average score $\langle \varepsilon \rangle$? We seek the distribution that maximizes the entropy, Equation (6.2), subject to two conditions: (1) that the probabilities sum to one and (2) that the average score agrees with the observed value $\langle \varepsilon \rangle$,

$$g(p_1, p_2, \ldots, p_t) = \sum_{i=1}^{t} p_i = 1 \quad \Longrightarrow \quad \sum_{i=1}^{t} dp_i = 0, \tag{6.10}$$

and

$$h(p_1, p_2, \ldots, p_t) = \langle \varepsilon \rangle = \sum_{i=1}^{t} p_i \varepsilon_i \quad \Longrightarrow \quad \sum_{i=1}^{t} \varepsilon_i dp_i = 0. \tag{6.11}$$

The solution is given by the method of Lagrange multipliers (pages 69–73):

$$\left(\frac{\partial S}{\partial p_i} \right) - \alpha \left(\frac{\partial g}{\partial p_i} \right) - \beta \left(\frac{\partial h}{\partial p_i} \right) = 0 \qquad \text{for } i = 1, 2, \ldots, t, \tag{6.12}$$

where α and β are the unknown multipliers. The partial derivatives are evaluated for each p_i:

$$\left(\frac{\partial S}{\partial p_i} \right) = -1 - \ln p_i, \quad \left(\frac{\partial g}{\partial p_i} \right) = 1, \quad \text{and} \left(\frac{\partial h}{\partial p_i} \right) = \varepsilon_i. \tag{6.13}$$

Substitute Equations (6.13) into Equation (6.12) to get t equations of the form

$$-1 - \ln p_i^* - \alpha - \beta \varepsilon_i = 0, \tag{6.14}$$

where the p_i^*'s are the values of p_i that maximize the entropy. Solve Equations (6.14) for each p_i^*:

$$p_i^* = e^{(-1 - \alpha - \beta \varepsilon_i)}. \tag{6.15}$$

To eliminate α in Equation (6.15), use Equation (6.10) to divide both sides by one. The result is an **exponential distribution law**:

$$p_i^* = \frac{p_i^*}{\sum_{i=1}^{t} p_i^*} = \frac{e^{(-1-\alpha)}e^{-\beta\varepsilon_i}}{\sum_{i=1}^{t} e^{(-1-\alpha)}e^{-\beta\varepsilon_i}} = \frac{e^{-\beta\varepsilon_i}}{\sum_{i=1}^{t} e^{-\beta\varepsilon_i}}. \qquad e^{-\varepsilon_i/KT} \qquad (6.16)$$

In statistical mechanics, this is called the **Boltzmann distribution law** and the quantity in the denominator is called the **partition function** q,

$$q = \sum_{i=1}^{t} e^{-\beta\varepsilon_i}. \qquad (6.17)$$

Using Equations (6.11) and (6.16) you can express the average score per roll $\langle\varepsilon\rangle$ (Equation (6.9)) in terms of the distribution,

$$\langle\varepsilon\rangle = \sum_{i=1}^{t} \varepsilon_i p_i^* = \frac{1}{q} \sum_{i}^{t} \varepsilon_i e^{-\beta\varepsilon_i}. \qquad (6.18)$$

The next two examples show how Equation (6.18) predicts all t of the p_i^*'s from the one known quantity, the average score.

EXAMPLE 6.3 Finding bias in dice by using the exponential distribution law.
Here we illustrate how to predict the maximum entropy distribution when an average score is known. Suppose a die has $t = 6$ faces and the scores equal the face indices, $\varepsilon(i) = i$. Let $x = e^{-\beta}$. Then Equation (6.17) gives $q = x + x^2 + x^3 + x^4 + x^5 + x^6$, and Equation (6.16) gives

$$p_i^* = \frac{x^i}{\sum_{i=1}^{6} x^i} = \frac{x^i}{x + x^2 + x^3 + x^4 + x^5 + x^6}. \qquad (6.19)$$

From the constraint Equation (6.18), you have

$$\langle\varepsilon\rangle = \sum_{i=1}^{6} i p_i^* = \frac{x + 2x^2 + 3x^3 + 4x^4 + 5x^5 + 6x^6}{x + x^2 + x^3 + x^4 + x^5 + x^6}. \qquad (6.20)$$

You have a polynomial, Equation (6.20), that you must solve for the one unknown x (a method for solving polynomials like Equation (6.20) is given on page 55). You begin with knowledge of $\langle\varepsilon\rangle$. Compute the value x^* that solves Equation (6.20). Then substitute x^* into Equations (6.19) to give the distribution function $(p_1^*, p_2^*, \ldots, p_t^*)$.

For example, if you observe the average score $\langle\varepsilon\rangle = 3.5$, then $x = 1$ satisfies Equation (6.20), predicting $p_i^* = 1/6$ for all i, indicating that the die is unbiased and has a flat distribution (see Figure 6.3(a)).

If, instead, you observe the average score is $\langle\varepsilon\rangle = 3.0$, then $x = 0.84$ satisfies Equation (6.20), and you have $q = 0.84 + 0.84^2 + 0.84^3 + 0.84^4 + 0.84^5 + 0.84^6 = 3.41$. The probabilities are $p_1 = 0.84/3.41 = 0.25$, $p_2 = 0.84^2/3.41 = 0.21$, $p_3 = 0.84^3/3.41 = 0.17$, and so on, as shown in Figure 6.3(b).

(a) $\langle \varepsilon \rangle = 3.5$

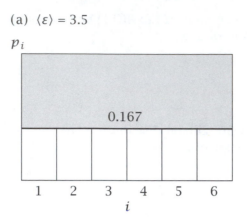

(b) $\langle \varepsilon \rangle = 3.0$

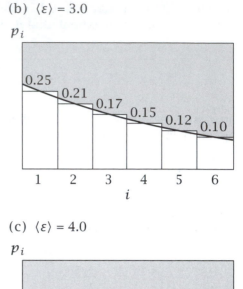

(c) $\langle \varepsilon \rangle = 4.0$

Figure 6.3 The probabilities of dice outcomes for known average scores. (a) If the average score per roll is $\langle \varepsilon \rangle = 3.5$, then $x = 1$ and all outcomes are equally probable, predicting that the die is unbiased. (b) If the average score is low ($\langle \varepsilon \rangle = 3.0$, $x = 0.84$), maximum entropy predicts an exponentially diminishing distribution. (c) If the average score is high ($\langle \varepsilon \rangle = 4.0$, $x = 1.19$), maximum entropy implies an exponentially increasing distribution.

If you observe $\langle \varepsilon \rangle < 3.5$, then the maximum-entropy principle predicts an exponentially decreasing distribution (see Figure 6.3(b)), with more **1**'s than **2**'s, more **2**'s than **3**'s, etc. If you observe $\langle \varepsilon \rangle > 3.5$, then maximum entropy predicts an exponentially increasing distribution (see Figure 6.3(c)): more **6**'s than **5**'s, more **5**'s than **4**'s, etc. For any given value of $\langle \varepsilon \rangle$, the exponential or flat distribution gives the most impartial distribution consistent with that score. The flat distribution is predicted either if the average score is 3.5 or if you have no information at all about the score.

EXAMPLE 6.4 Biased coins? The exponential distribution again. Let's determine a coin's bias. A coin is just a die with two sides, $t = 2$. Score tails $\varepsilon_T = 1$ and heads $\varepsilon_H = 2$. The average score per toss $\langle \varepsilon \rangle$ for an unbiased coin would be 1.5.

Again, to simplify, write the unknown Lagrange multiplier β in the form $x = e^{-\beta}$. In this notation, the partition function Equation (6.17) is $q = x + x^2$. According to Equation (6.16), the exponential distribution law for this two-state system is

$$p_T^* = \frac{x}{x + x^2} \quad \text{and} \quad p_H^* = \frac{x^2}{x + x^2}. \tag{6.21}$$

From the constraint Equation (6.18) you have

$$\langle \varepsilon \rangle = 1 p_T^* + 2 p_H^* = \frac{x + 2x^2}{x + x^2} = \frac{1 + 2x}{1 + x}.$$

Rearranging gives

$$x = \frac{\langle \varepsilon \rangle - 1}{2 - \langle \varepsilon \rangle}.$$

If you observe the average score to be $\langle \varepsilon \rangle = 1.5$, then $x = 1$, and Equation (6.21) gives $p_T^* = p_H^* = 1/2$. The coin is fair. If instead you observed $\langle \varepsilon \rangle = 1.2$, then $x = 1/4$, and you have $p_H^* = 1/5$ and $p_T^* = 4/5$.

There are two situations that will predict a flat distribution function. First, it will be flat if $\langle \varepsilon \rangle$ equals the value expected from a uniform distribution. For example, if you observe $\langle \varepsilon \rangle = 3.5$ in Example 6.3, maximum entropy predicts a flat distribution. Second, if there is no constraint at all, you expect a flat distribution. By the maximum-entropy principle, having no information is the same as expecting a flat distribution.

On page 84, we gave a simple rationalization for why S should be a logarithmic function of W. Now we give a deeper justification for the functional form, $S/k = -\sum_i p_i \ln p_i$. You might ask why entropy must be extensive, and why entropy, which we justified on thermodynamic grounds, also applies to a broad range of problems outside molecular science. Should $S = k \ln W$ also apply to interacting systems? The following section is intended to address questions such as these.

A Principle of Fair Apportionment Leads to the Function $-\sum p_i \ln p_i$

Here we derive the functional form of the entropy function, $S = -k \sum_i p_i \ln p_i$ from a Principle of Fair Apportionment. Coins and dice have intrinsic symmetries in their possible outcomes. In unbiased systems, heads is equivalent to tails, and every number on a die is equivalent to every other. The Principle of Fair Apportionment says that if there is such an intrinsic symmetry, and if there is no constraint or bias, then all outcomes will be observed with the same probability. That is, the system 'treats each outcome fairly' in comparison with every other outcome. The probabilities will tend to be apportioned between those outcomes in the most uniform possible way, if the number of trials is large enough. Throughout a long history, the idea that every outcome is equivalent has gone by various names. In the 1700s, Bernoulli called it the Principle of Insufficient Reason; in the 1920s, Keynes called it the Principle of Indifference [1].

However, the principle of the flat distribution, or maximum multiplicity, is incomplete. The Principle of Fair Apportionment needs a second clause that describes how probabilities are apportioned between the possible outcomes when there are constraints or biases. If die rolls give an average score that is convincingly different from 3.5 per roll, then the outcomes are not all equivalent and the probability distribution is not flat. The second clause that completes

Table 6.1 A possible distribution of outcomes for rolling a 30-sided die 1000 times. For example, a red **3** appears 18 times.

i	\multicolumn{6}{c}{j}
	1	**2**	**3**	**4**	**5**	**6**
red	16	18	18	17	15	16
blue	25	35	45	55	65	75
green	30	28	32	30	50	30
white	40	42	38	40	40	50
black	20	25	30	20	30	25

the Principle of Fair Apportionment says that when there are independent constraints, the probabilities must satisfy the multiplication rule of probability theory, as we illustrate here with a system of multicolored dice.

A Multicolored Die Problem

Consider a 30-sided five-colored die. The sides are numbered **1** through **6** in each of the five different colors. That is, six sides are numbered **1** through **6** and are colored red, six more are numbered **1** through **6** but they are colored blue, six more are green, six more are white, and the remaining six are black. If the die is fair and unbiased, a blue **3** will appear 1/30 of the time, for example, and a red color will appear 1/5 of the time.

Roll the die N times. Count the number of appearances of each outcome and enter those numbers in a table in which the six columns represent the numerical outcomes and the five rows represent the color outcomes. Table 6.1 is an example of a possible result for $N = 1000$. To put the table into the form of probabilities, divide each entry by N. This is Table 6.2. For row $i = 1, 2, 3, \ldots, a$ and column $j = 1, 2, 3, \ldots, b$, call the normalized entry p_{ij} ($a = 5$ and $b = 6$ in this case). The sum of probabilities over all entries in Table 6.2 equals one,

$$\sum_{i=1}^{a} \sum_{j=1}^{b} p_{ij} = 1. \tag{6.22}$$

If there were many trials, and no bias (that is, if blue **3** appeared with the same frequency as green **5**, etc.), then the distribution would be flat and every probability would be 1/30. The flat distribution is the one that apportions the outcomes most fairly between the 30 possible options, if there is no bias.

Now suppose that the system is biased, but that your knowledge of it is incomplete. Suppose you know only the sum along each row and the sum down each column (see Table 6.3). For tables larger than 2×2, the number of rows and columns will be less than the number of cells in the table, so that the probability distribution function will be *underdetermined*. For each row i the sum of the probabilities is

Table 6.2   The conversion of Table 6.1 to probabilities.

i	1	2	3	4	5	6
red	0.016	0.018	0.018	0.017	0.015	0.016
blue	0.025	0.035	0.045	0.055	0.065	0.075
green	0.030	0.028	0.032	0.030	0.050	0.030
white	0.040	0.042	0.038	0.040	0.040	0.050
black	0.020	0.025	0.030	0.020	0.030	0.025

Table 6.3   Suppose that you know only the row and column sums. In this case they are from Table 6.2. What is the best estimate of all the individual entries?

i	1	2	3	4	5	6	$u_i = \sum_{j=1}^{6} p_{ij}$
red	?	?	?	?	?	?	$u_1 = 0.10$
blue	?	?	?	?	?	?	$u_2 = 0.30$
green	?	?	?	?	?	?	$u_3 = 0.20$
white	?	?	?	?	?	?	$u_4 = 0.25$
black	?	?	?	?	?	?	$u_5 = 0.15$
$v_j = \sum_{i=1}^{5} p_{ij}$	v_1 0.131	v_2 0.148	v_3 0.163	v_4 0.162	v_5 0.200	v_6 0.196	

$$u_i = \sum_{j=1}^{b} p_{ij}, \tag{6.23}$$

where u_i represents the probability that a roll of the die will show a particular color, i = red, blue, green, white, or black. For example, u_2 is the probability of seeing a blue face with any number on it. Similarly, for each column j the sum of the probabilities is

$$v_j = \sum_{i=1}^{a} p_{ij}. \tag{6.24}$$

For example, v_4 is the probability of seeing a **4** of any color. If the die were unbiased, you would have $v_4 = 1/6$, but if the die were biased, v_4 might have a different value. Row and column sums constitute *constraints*, which are biases or knowledge that must be satisfied as you predict the individual table entries, the p_{ij}'s.

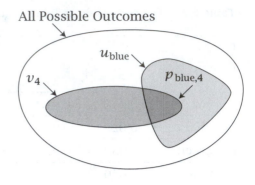

Figure 6.4 The darkest shaded region on this Venn Diagram represents a *blue* 4 outcome, the intersection of sets of all *blue* outcomes with all 4 outcomes.

All Possible Outcomes

How should you predict the p_{ij}'s if you know only the row and column sums? The rules of probability tell you exactly what to do. The joint probability p_{ij} represents the intersection of the set of number j and the set of color i (see Figure 6.4). Each p_{ij} is the product of u_i, the fraction of all possible outcomes that has the right color, and v_j, the fraction that has the right number (see Equation (1.6)),

$$p_{ij} = u_i v_j. \tag{6.25}$$

The multiplication rule was introduced in Chapter 1 (page 4). Table 6.4 shows this prediction: to get each entry p_{ij} in Table 6.4, multiply the row sum u_i by the column sum v_j from Table 6.3. With the multiplication rule, you can infer a lot of information (the full table) from a smaller amount (only the row and column sums). You have used only $a + b = 5 + 6 = 11$ known quantities to predict $a \times b = 30$ unknowns. The ability to make predictions for underdetermined systems is particularly valuable for molecular systems in which the number of entries in the table can be huge, even infinite, while the number of row and column sums might be only one or two.

However, if you compare Tables 6.4 and 6.2, you see that your incomplete information is not sufficient to give a perfect prediction of the true distribution function. Is there any alternative to the multiplication rule that would have given a better prediction? It can be proved that the multiplication rule is uniquely the only unbiased way to make consistent inferences about probabilities of the intersection of sets [2]. This is illustrated (but not proved) below, and is proved for a simple case in Problem 6 at the end of the chapter.

The Multiplication Rule Imparts the Least Bias

For simplicity, let's reduce our 5×6 problem to a 2×2 problem. Suppose you have a four-sided die, with sides red **1**, red **2**, blue **1**, and blue **2**. You do not know the individual outcomes themselves. You know only row and column sums: red and blue each appear half the time, **1** appears three-quarters of the time, and **2** appears one-quarter of the time (see Table 6.5). The data show no bias between red and blue, but show a clear bias between **1** and **2**. Now you want to fill in Table 6.5 with your best estimate for each outcome.

Because you know the row and column sums, you can express all the probabilities in terms of a single variable q, say the probability of a red **1**. Now Table 6.5 becomes Table 6.6.

Table 6.4 This table was created by using the multiplication rule Equation (6.26) and the row and column sums from Table 6.3. Compare with Table 6.2. The numbers are in good general agreement, including the large value of green 5 relative to the other green outcomes, and including the bias toward higher numbers in the blue series.

				j		
i	1	2	3	4	5	6
red	0.013	0.015	0.016	0.017	0.021	0.020
blue	0.040	0.044	0.049	0.050	0.062	0.059
green	0.026	0.030	0.033	0.033	0.041	0.039
white	0.033	0.037	0.041	0.042	0.052	0.049
black	0.020	0.022	0.024	0.025	0.031	0.029

Table 6.5 Assume the row and column constraints are known, but the probabilities of the individual outcomes are not, for this 2×2 case.

	Number		
Color	1	2	$u_i = \sum\limits_{j=1}^{2} p_{ij}$
red	?	?	$u_1 = 1/2$
blue	?	?	$u_2 = 1/2$
$v_j = \sum\limits_{i=1}^{2} p_{ij}$	$v_1 = 3/4$	$v_2 = 1/4$	

Table 6.6 The probabilities for Table 6.5 can be expressed in terms of a single variable q. All entries satisfy the row and column constraints.

	Number		
Color	1	2	
red	q	$1/2 - q$	$u_1 = 1/2$
blue	$3/4 - q$	$q - 1/4$	$u_2 = 1/2$
	$v_1 = 3/4$	$v_2 = 1/4$	

To ensure that each cell of Table 6.6 contains only a positive quantity (probabilities cannot be negative), the range of allowable values is from $q = 1/4$ to $q = 1/2$. At first glance, it would seem that you have four equations and four unknowns, so that you could solve directly for q. However, the four equations are not all independent, since $u_1 + u_2 = 1$ and $v_1 + v_2 = 1$. Tables 6.7(a), (b)

Table 6.7 These three tables are all consistent with the row and column constraints given in Table 6.5. Only (b) is unbiased. (b) is generated by the multiplication rule from the row and column constraints.

(a)

Color	Number	
	1	2
red	1/4	1/4
blue	1/2	0

(b)

Color	Number	
	1	2
red	3/8	1/8
blue	3/8	1/8

(c)

Color	Number	
	1	2
red	1/2	0
blue	1/4	1/4

and (c) show three possible probability distributions, all of which satisfy the given constraints.

Which of these tables is least biased? Notice that Tables 6.7(a) and (c) with $q = 1/4$ and $q = 1/2$ have unwarranted bias. The data tell us that there is a bias in favor of **1**'s relative to **2**'s but they give no evidence for a difference between the red and blue sides. Yet Tables 6.7(a) and (c) with $q = 1/4$ and $q = 1/2$ indicate that the red sides are different from the blue sides. These tables are perfectly consistent with the data, but they are not impartial toward each outcome. In short, Tables 6.7(a) and (c) are 'unfair' in their apportionment of probabilities to cells.

Only Table 6.7(b) with $q = 3/8$ offers a fair apportionment of the probabilities, and predicts that both colors are equivalent. This least-biased inference results from the multiplication rule. For example, the probability of a red **1** is the product of its row and column sums, $p(\text{red})p(\mathbf{1}) = (1/2)(3/4) = 3/8$. The multiplication rule applies because the constraints u_i and v_j are independent.

The Principle of Fair Apportionment of Outcomes

Equation (6.2) defines the entropy as $S/k = -\sum_i p_i \ln p_i$. Our aim in this section is to derive this expression by treating S as an unknown function of the p_i's using the requirement that S is maximal when the probabilities are fairly apportioned. Consider a table of probability cells indexed by labels $i = 1, 2, 3, \ldots, a$, and $j = 1, 2, 3, \ldots, b$. In typical dice problems, the score on face i is just equal to i. However, let us be more general here. Suppose you paste a score ε_i onto each face i. A score ε_{ij} is associated with each cell. Each cell represents an outcome having probability p_{ij}, the sum of which over all the cells equals one. There are many possible distributions of the p_{ij}'s, but only one distribution, $(p_{11}^*, p_{12}^*, p_{13}^*, \ldots, p_{ab}^*)$, is apportioned fairly. If there are no constraints, the distribution will be flat and the probabilities uniform. If there are row and column constraints, all the probabilities will obey the multiplication rule. We seek a function of all the probabilities, $S(p_{11}, p_{12}, p_{13}, \ldots, p_{ab})$, which we will call the *entropy*. The entropy function is maximal, $S = S_{\max}$, when the probabilities are distributed according to fair apportionment.

Why do we need a function S to do this? The multiplication rule already tells us how to fill in the table if the row and column constraints are known. We need S for two reasons. First, we need a way to generalize to situations

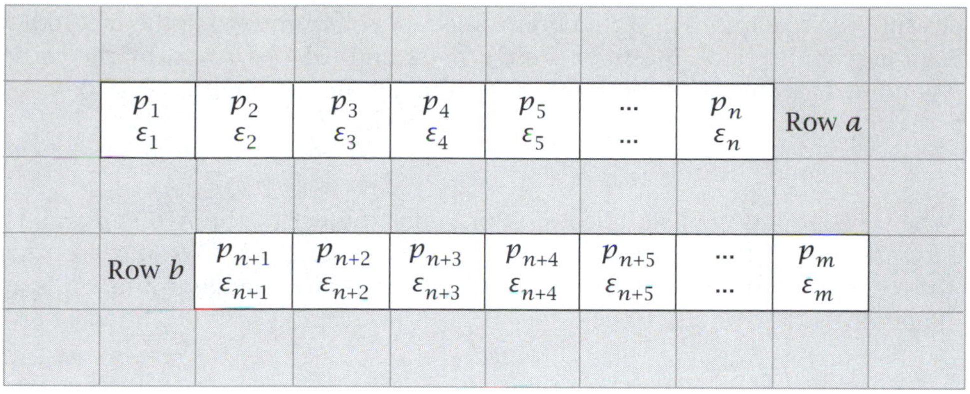

Figure 6.5 Two rows a and b in a grid of cells used to illustrate the notation for proving that entropy is extensive. Each row is subject to a constraint on the sum of probabilities and a constraint on the sum of the scores.

in which we have only a single constraint, or any set of constraints on only a part of the system. Second, we seek a measure of the driving force toward redistributing the probabilities when the constraints are changed.

The Principle of Fair Apportionment is all we need to specify the precise mathematical form that the entropy function must have. We will show this in two steps. First, we will show that the entropy must be *extensive*. That is, the function S that applies to the *whole system* must be a sum over *individual cell functions s*:

$$S(p_{11}, p_{12}, p_{13}, \ldots, p_{ab})$$
$$= s(p_{11}) + s(p_{12}) + s(p_{13}) + \cdots + s(p_{ab}). \tag{6.26}$$

Second, we will show that the *only* function that is extensive and at its maximum satisfies the multiplication rule is the function $-\sum_{i=1}^{t} p_i \ln p_i$.

The Entropy Is Extensive

Focus on two collections of cells a and b chosen from within the full grid. To keep it simple, we can eliminate the index j if a is a row of cells with probabilities that we label $(p_1, p_2, \ldots, p_n)$, and b is a different row with probabilities that we label $(p_{n+1}, p_{n+2}, \ldots, p_m)$ (see Figure 6.5). Our aim is to determine how the entropy function for the combined system a plus b is related to the entropy functions for the individual rows.

The entropy S is a function that we aim to maximize subject to two conditions. Two constraints, g and h, apply to row a: one is the normalization on the sum of probabilities, and the other is a constraint on the average score,

$$g(p_1, p_2, \ldots, p_n) = \sum_{i=1}^{n} p_i = u_a,$$

$$h(p_1, p_2, \ldots, p_n) = \sum_{i=1}^{n} \varepsilon_i p_i = \langle \varepsilon \rangle_a. \tag{6.27}$$

To find the maximum of a function subject to constraints, use the method of Lagrange multipliers. Multiply $(\partial g/\partial p_i) = 1$ and $(\partial h/\partial p_i) = \varepsilon_i$ by the corresponding Lagrange multipliers α_a and λ_a to get the extremum

$$\frac{\partial S(p_1, p_2, \ldots, p_n)}{\partial p_i} = \alpha_a + \lambda_a \varepsilon_i. \tag{6.28}$$

When this equation is satisfied, S is maximal subject to the two constraints. Express the entropy for row a as $S_a = S(p_1, p_2, \ldots, p_n)$. To find the total differential for the variation of S_a resulting from any change in the p's, sum Equation (6.28) over all dp_i's:

$$dS_a = \sum_{i=1}^{n} \left(\frac{\partial S_a}{\partial p_i} \right) dp_i = \sum_{i=1}^{n} (\alpha_a + \lambda_a \varepsilon_i) \, dp_i. \tag{6.29}$$

Equation (6.29) does not require any assumption about the form of $S_a = S(p_1, p_2, \ldots, p_n)$. It is just the definition of the total differential. Similarly, row b in a different part of the grid will be subject to constraints unrelated to the constraints on row a,

$$\alpha_b \sum_{i=n+1}^{m} dp_i = 0 \quad \text{and} \quad \lambda_b \sum_{i=n+1}^{m} \varepsilon_i dp_i = 0,$$

with Lagrange multipliers α_b and λ_b. For row b, index i now runs from $n + 1$ to m, rather than from 1 to n, and we will use the corresponding shorthand notation $S_b = S(p_{n+1}, p_{n+2}, \ldots, p_m)$. The total differential for S_b is

$$dS_b = \sum_{i=n+1}^{m} \left(\frac{\partial S_b}{\partial p_i} \right) dp_i = \sum_{i=n+1}^{m} (\alpha_b + \lambda_b \varepsilon_i) \, dp_i. \tag{6.30}$$

Finally, use the function $S_{\text{total}} = S(p_1, p_2, \ldots, p_n, p_{n+1}, \ldots, p_m)$ to apportion probabilities over both rows a and b at the same time, given exactly the same two constraints on row a and the same two constraints on row b as above. The total differential for S_{total} is

$$dS_{\text{total}} = \sum_{i=1}^{m} \left(\frac{\partial S_{\text{total}}}{\partial p_i} \right) dp_i$$

$$= \sum_{i=1}^{n} (\alpha_a + \lambda_a \varepsilon_i) \, dp_i + \sum_{i=n+1}^{m} (\alpha_b + \lambda_b \varepsilon_i) \, dp_i = dS_a + dS_b, \tag{6.31}$$

where the last equality comes from substitution of Equations (6.29) and (6.30). This says that if the entropy is to be a function having a maximum, subject to constraints on blocks of probability cells, then it must be *extensive*. The entropy of a system is a sum of subsystem entropies. The underlying premise is that the score quantities, $\langle \varepsilon \rangle_a$ and $\langle \varepsilon \rangle_b$, are extensive.

The way in which we divided the grid into rows a and b was completely arbitrary. We could have parsed the grid in many different ways, down to an individual cell at a time. Following this reasoning to its logical conclusion, and integrating the differentials, leads to Equation (6.26), which shows that the entropy of the whole grid is the sum of entropies of the individual cells.

The Form of the Entropy Function

To obtain the form of the entropy function, we use a table of probabilities with rows labeled by the index i and columns by the index j. Each row i is constrained to have an average score,

$$\sum_{j=1}^{b} \varepsilon_{ij} p_{ij} = \langle \varepsilon_i \rangle \quad \Longrightarrow \quad \lambda_i \sum_{j=1}^{b} \varepsilon_{ij} dp_{ij} = 0, \tag{6.32}$$

where λ_i is the Lagrange multiplier that enforces the score constraint for row i. Similarly, for column j,

$$\beta_j \sum_{i=1}^{a} \varepsilon_{ij} dp_{ij} = 0,$$

where β_j is the Lagrange multiplier that enforces the score constraint for column j. Finally,

$$\alpha \sum_{ij} dp_{ij} = 0$$

enforces the constraint that the probabilities must sum to one over the whole grid of cells. We use the extensivity property, Equation (6.26), to find the expression for maximizing the entropy subject to these three constraints,

$$\sum_{i=1}^{a} \sum_{j=1}^{b} \left[\left(\frac{\partial s(p_{ij})}{\partial p_{ij}} \right) - \lambda_i \varepsilon_{ij} - \beta_j \varepsilon_{ij} - \alpha \right] dp_{ij} = 0. \tag{6.33}$$

To solve Equation (6.33), each term in brackets must equal zero for each cell ij according to the Lagrange method (see Equation (5.35)).

Our aim is to deduce the functional form for $s(p_{ij})$ by determining how Equation (6.33) responds to changes in the row constraint u_i and column constraint v_j. To look at this dependence, let's simplify the notation. Focus on the term in the brackets, for one cell ij, and drop the subscripts ij for now. Changing the row or column sum u or v changes p, so express the dependence as $p(u, v)$.

Define the derivative $(\partial s_{ij}/\partial p_{ij}) = r_{ij}$, and express it as $r[p(u, v)]$. The quantities ε_{ij} differ for different cells in the grid, but they are fixed quantities and do not depend on u or v. Because the probabilities must sum to one over the whole grid no matter what changes are made in u and v, α is also independent of u and v. The value of the Lagrange multipliers $\lambda(u)$ for row i and $\beta(v)$ for column j will depend on the value of the constraints u and v (see Example 6.5). Collecting together these functional dependences, the bracketed term in Equation (6.33) can be expressed as

$$r[p(u, v)] = \lambda(u)\varepsilon + \beta(v)\varepsilon + \alpha. \tag{6.34}$$

Now impose the multiplication rule, $p = uv$. Let's see how r depends on u and v. This will lead us to specific requirements for the form of the entropy function. Take the derivative of Equation (6.34) with respect to v to get

$$\left(\frac{\partial r}{\partial v}\right) = \left(\frac{\partial r}{\partial p}\right)\left(\frac{\partial p}{\partial v}\right) = \left(\frac{\partial r}{\partial p}\right)\frac{p}{v} = \varepsilon\beta'(v). \tag{6.35}$$

where $\beta' = d\beta/dv$ and $(\partial p/\partial v) = u = p/v$. Now take the derivative of Equation (6.34) with respect to u instead of v to get

$$\left(\frac{\partial r}{\partial p}\right)\frac{p}{u} = \varepsilon\lambda'(u), \tag{6.36}$$

where $\lambda' = d\lambda/du$. Rearranging and combining Equations (6.35) and (6.36) gives

$$\left(\frac{\partial r}{\partial p}\right) = \frac{u\lambda'(u)\varepsilon}{p} = \frac{v\beta'(v)\varepsilon}{p}. \tag{6.37}$$

Notice that the quantity $u\lambda'(u)\varepsilon$ can only depend on (east–west) properties of the row i, particularly its sum u. Likewise, the quantity $v\beta'(v)\varepsilon$ can only depend on (north–south) properties of the column j, particularly its sum v. The only way that these two quantities can be equal in Equation (6.37) for any arbitrary values of u and v is if $u\lambda'(u)\varepsilon = v\beta'(v)\varepsilon = $ constant. Call this constant $-k$, and you have

$$\left(\frac{\partial r}{\partial p}\right) = \frac{-k}{p}. \tag{6.38}$$

Integrate Equation (6.38) to get

$$r(p) = -k\ln p + c_1,$$

where c_1 is the constant of integration. To get $s(p)$, integrate again:

$$s(p) = \int r(p)\,dp = \int (-k\ln p + c_1)\,dp$$
$$= -k(p\ln p - p) + c_1 p + c_2, \tag{6.39}$$

where c_2 is another constant of integration. Summing over all cells ij gives the total entropy:

$$S = \sum_{i=1}^{a}\sum_{j=1}^{b} s(p_{ij}) = -k\sum_{ij} p_{ij}\ln p_{ij} + c_3\sum_{ij} p_{ij} + c_2, \tag{6.40}$$

where $c_3 = k + c_1$. Since $\sum_{ij} p_{ij} = 1$, the second term on the right side of Equation (6.40) sums to a constant, and Equation (6.40) becomes

$$S = -k\sum_{ij} p_{ij}\ln p_{ij} + \text{constant}. \tag{6.41}$$

If you define the entropy of a perfectly ordered state (one p_{ij} equals one, all others equal zero) to be zero, the constant in Equation (6.41) equals zero. If you choose the convention that the extremum of S should be a maximum rather than a minimum, k is a positive constant. If there is only a single cell index i, you have $S/k = -\sum_i p_i\ln p_i$.

Equation (6.41) is a most remarkable result. It says that you can uniquely define a function, the entropy, that at its maximum will assign probabilities to the

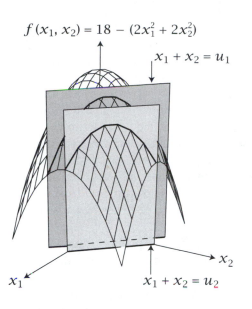

$$f(x_1, x_2) = 18 - (2x_1^2 + 2x_2^2)$$

$x_1 + x_2 = u_1$

x_2

x_1

$x_1 + x_2 = u_2$

Figure 6.6 The optimum of $f(x_1, x_2)$ changes if a row sum $g = x_1 + x_2 = u$ changes from a value u_1 to u_2.

cells in accordance with the Fair Apportionment Principle: (1) when there are no constraints, all probabilities will be equal, and (2) when there are independent constraints, the probabilities will satisfy the multiplication rule. Because S is the only function that succeeds at this task when it is maximal, S must also be the only function that gives the deviations from the Fair Apportionment Principle when it is not maximal.

EXAMPLE 6.5 Illustrating the dependence $\lambda(u)$. A Lagrange multiplier can depend on the value of a row (or column) constraint. Consider a function $f(x_1, x_2) = 18 - (2x_1^2 + 2x_2^2)$ subject to the constraint $g(x_1, x_2) = x_1 + x_2 = u$ (see Figure 6.6). The extremum is found where

$$\left(\frac{\partial f}{\partial x_1}\right) = -4x_1 = \lambda, \quad \text{and} \quad \left(\frac{\partial f}{\partial x_2}\right) = -4x_2 = \lambda \quad \Rightarrow \quad x_1 = x_2 = -\frac{\lambda}{4}. \quad (6.42)$$

Substitute Equation (6.42) into $g = x_1 + x_2 = u$ to find $\lambda(u) = -2u$.

Philosophical Foundations

For more than a century, entropy has been a controversial concept. One issue is whether entropy is a tool for making predictions based on the knowledge of an observer [1], or whether it is independent of an observer [3]. Both interpretations are useful under different circumstances. For dice and coin flip problems, constraints such as an average score define the limited knowledge that you have. You want to impart a minimum amount of bias in predicting the probability distribution that is consistent with that knowledge. Your prediction should be based on assuming you are maximally ignorant with respect to all else [1, 4]. Maximizing the function $-\sum_i p_i \ln p_i$ serves the role of making this prediction. Similar prediction problems arise in eliminating noise from spectroscopic signals or in reconstructing satellite photos. In these cases, the Principle we call Fair Apportionment describes how to reason and draw inferences and make the least-biased predictions of probabilities in face of incomplete knowledge.

However, for other problems, notably those of statistical mechanics, entropy describes a force of nature. In those cases, constraints such as averages over a row of probabilities are not only numbers that are known to an observer, they are also physical constraints that are imposed upon a physical system from outside (see Chapter 10). In this case, the Principle of Fair Apportionment is a description of nature's symmetries. When there is an underlying symmetry in a problem (such as t outcomes that are equivalent), then Fair Apportionment says that external constraints are shared equally between all the possible states that the system can occupy (grid cells, in our earlier examples). In this case, entropy is more than a description about an observer. It describes a tendency of nature that is independent of an observer.

A second controversy involves the interpretation of probabilities. The two perspectives are the frequency interpretation and the 'inference' interpretation [1]. According to the frequency interpretation, a probability $p_A = n_A/N$ describes some event that can be repeated N times. However, the inference interpretation is broader. It says that probabilities are fractional quantities that describe a state of knowledge. In the inference view, the rules of probability are simply rules for making consistent inferences from limited information. The probability that it might rain tomorrow, or the probability that Elvis Presley was born on January 8, can be perfectly well defined, even though these events cannot be repeated N times. In the inference view, probabilities can take on different values depending on your state of knowledge. If you do not know when Elvis was born, you would say the probability is 1/365 that January 8 was his birthday. That is a statement about your lack of knowledge. However, if you happen to know that was his birthday, the probability is one. According to the subjective interpretation, the rules of probability theory in Chapter 1 are laws for drawing consistent inferences from incomplete information, irrespective of whether or not the events are countable.

Historically, statistical mechanics has been framed in terms of the frequency interpretation. JW Gibbs (1839–1903), American mathematical physicist and a founder of chemical thermodynamics and statistical mechanics, framed statistical mechanics as a counting problem. He envisioned an imaginary collection of all possible outcomes, called an *ensemble*, which was countable and could be taken to the limit of a large number of imaginary repetitions. To be specific, for die rolls, if you want to know the probability of each of the six outcomes on a roll, the ensemble approach would be to imagine that you had rolled the die N times. You would then compute the number of sequences that you expect to observe. This number will depend on N. On the other hand, if the outcomes can instead be described in terms of probabilities, the quantity N never appears. For example, the grid of cells described in Table 6.2 describes probabilities that bear no trace of the information that $N = 1000$ die rolls were used to obtain Table 6.1.

These issues are largely philosophical fine points that have had little implication for the practice of statistical mechanics. We will sometimes prefer the counting strategy, and the use of $S/k = \ln W$, but $S/k = -\sum_i p_i \ln p_i$ will be more convenient in other cases.

Summary

The entropy $S(p_{11}, \ldots, p_{ij}, \ldots, p_{ab})$ is a function of a set of probabilities. The distribution of p_{ij}'s that cause S to be maximal is the distribution that most fairly apportions the constrained scores between the individual outcomes. That is, the probability distribution is flat if there are no constraints, and follows the multiplication rule of probability theory if there are independent constraints. If there is a constraint, such as the average score on die rolls, and if it is not equal to the value expected from a uniform distribution, then maximum entropy predicts an exponential distribution of the probabilities. In Chapter 10, this exponential function will define the Boltzmann distribution law. With this law you can predict thermodynamic and physical properties of atoms and molecules, and their averages and fluctuations. However, first we need the machinery of thermodynamics, the subject of the next three chapters.

Problems

1. Calculating the entropy of dipoles in a field. You have a solution of dipolar molecules with a positive charge at the head and a negative charge at the tail. When there is no electric field applied to the solution, the dipoles point north (n), east (e), west (w), or south (s) with equal probabilities. The probability distribution is shown in Figure 6.7(a). However when you apply a field to the solution, you now observe a different distribution, with more heads pointing north, as shown in Figure 6.7(b).

  (a) What is the polarity of the applied field? (In which direction does the field have its most positive pole?)

  (b) Calculate the entropy of the system in the absence of the field.

  (c) Calculate the entropy of the system in the applied field.

  (d) Does the system become more ordered or disordered when the field is applied?

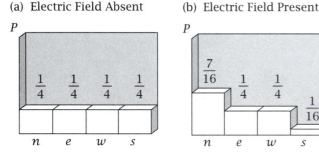

Figure 6.7

2. Comparing the entropy of peaked and flat distributions. Compute the entropies for the spatial concentration shown in Figures 6.8(a) and (b).

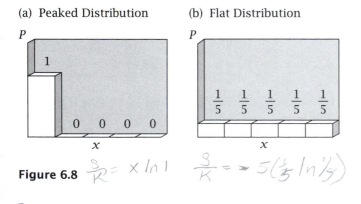

Figure 6.8 $\frac{S}{k} = x \ln 1$ $\frac{S}{k} = -5\left(\frac{1}{5}\ln\frac{1}{5}\right)$

3. Comparing the entropy of two peaked distributions. Which of the two distributions shown in Figure 6.9 has greater entropy?

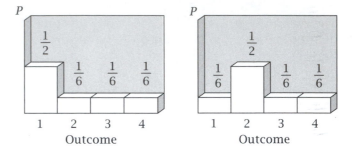

Figure 6.9

4. Calculating the entropy of mixing. Consider a lattice with N sites and n green particles. Consider another lattice, adjacent to the first, with M sites and m red particles. Assume that the green and red particles cannot switch lattices. This is state A.

  (a) What is the total number of configurations W_A of the system in state A?

  (b) Now assume that all $N + M$ sites are available to all the green and red particles. The particles remain distinguishable by their color. This is state B. Now what is the total number of configurations W_B of the system?

Now take $N = M$ and $n = m$ for the following two problems.

  (c) Using Stirling's approximation on page 56, what is the ratio W_A/W_B?

  (d) Which state, A or B, has the greatest entropy? Calculate the entropy difference given by

$$\Delta S = S_A - S_B = k \ln \left(\frac{W_A}{W_B} \right).$$

5. Proof of maximum entropy for two outcomes. Example 6.4 (page 88) is simple enough that you can prove the answer is correct even without using the maximum entropy principle. Show this.

6. Proving the multiplication rule maximizes the entropy. Show that if the entropy is defined by Equation (6.2), $S/k = -\sum_i p_i \ln p_i$, and if the 2×2 grids of Tables 6.4 through 6.7 have row sums u_1 and u_2, and column sums v_1 and v_2, then the function $p_{ij}(u_i, v_j)$ that maximizes the entropy is the multiplication rule $p_{ij} = u_i v_j$.

7. Other definitions of entropy do not satisfy the multiplication rule. In contrast to problem 6, show that if the entropy were defined by a least-squares criterion, as $S/k = \sum_{i=1}^{t} p_i^2$, the multiplication rule would not be satisfied when S is maximal.

8. Other definitions of entropy can predict the uniform distribution when there are no constraints. As in problem 7, assume that the definition of entropy was $S/k = \sum_{i=1}^{t} p_i^2$. Show that when there are no row or column constraints on the 2×2 grids of Tables 6.4 through 6.7, the definition will satisfactorily predict that the uniform distribution is an extremum of the entropy.

9. The maximum entropy distribution is Gaussian when the second moment is given. Prove that the probability distribution p_i that maximizes the entropy for die rolls subject to a constant value of the second moment $\langle i^2 \rangle$ is a Gaussian function. Use $\varepsilon_i = i$.

10. Maximum entropy for a three-sided die. You have a three-sided die, with numbers **1**, **2** and **3** on the sides. For a series of N dice rolls, you observe an average score α per roll using the maximum entropy principle.

(a) Write expressions that show how to compute the relative probabilities of occurrence of the three sides, n_1^*/N, n_2^*/N, and n_3^*/N, if α is given.

(b) Compute n_1^*/N, n_2^*/N, and n_3^*/N if $\alpha = 2$.

(c) Compute n_1^*/N, n_2^*/N, and n_3^*/N, if $\alpha = 1.5$.

(d) Compute n_1^*/N, n_2^*/N, and n_3^*/N if $\alpha = 2.5$.

11. Maximum entropy in Las Vegas. You play a slot machine in Las Vegas. For every \$1 coin you insert, there are three outcomes: (1) you lose \$1, (2) you win \$1, so your profit is \$0, (3) you win \$5, so your profit is \$4. Suppose you find that your average expected profit over many trials is \$0 (what an optimist!). Find the maximum entropy distribution for the probabilities p_1, p_2, and p_3 of observing each of these three outcomes.

12. Flat distribution, high entropy. For four coin flips, for each distribution of probabilities, $(p_H, p_T) = (0, 1), (1/4, 3/4), (1/2, 1/2), (3/4, 1/4), (1, 0)$, compute W, and show that the flattest distribution has the highest multiplicity.

References

[1] ET Jaynes. *The Maximum Entropy Formalism*, RD Levine and M Tribus eds., MIT Press, Cambridge, 1979, page 15.

[2] RT Cox. *Am J Phys* **14**, 1 (1946).

[3] KG Denbigh and JS Denbigh. *Entropy in Relation to Incomplete Knowledge*. Cambridge University Press, Cambridge, 1985.

[4] ET Jaynes. *Phys Rev* **106**, 620 (1957).

Suggested Reading

ET Jaynes, *Phys Rev* **106**, 620 (1957). This is the classic article that introduces the maximum entropy approach to statistical thermodynamics. It is well-written and includes a scholarly historical introduction.

JN Kapur and HK Kesavan, *Entropy Optimization Principles with Applications*, Academic Press, Boston, 1992. Excellent overview of maximum entropy applied to many fields of science.

RD Levine and M Tribus, *The Maximum Entropy Formalism*, MIT Press, 1979. This is an advanced edited volume that discusses applications and other aspects of maximum entropy, including an extensive review by Jaynes.

The following three articles show how the maximum entropy principle follows from the multiplication rule. A brief overview is given in J Skilling, *Nature*, **309**, 748 (1984).

AK Livesay and J Skilling, *Acta Crystallographica*, **A41**, 113–122 (1985).

JE Shore and RW Johnson, *IEEE Trans Inform Theory* **26**, 26 (1980).

Y Tikochinski, NZ Tishby, RD Levine, *Phys Rev Lett* **52**, 1357 (1984).

7 Thermodynamic Driving Forces

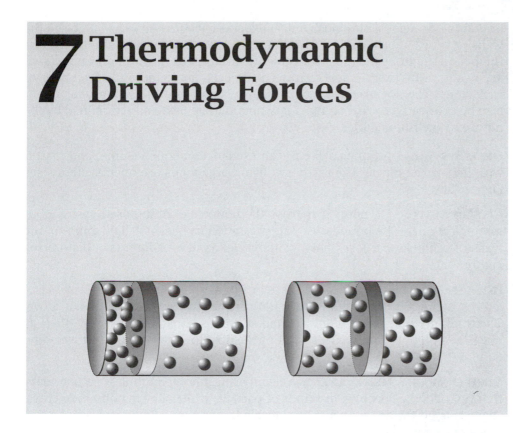

Thermodynamics Is Two Laws and a Little Calculus

Thermodynamics is a set of tools for reasoning about energies and entropies. It enables you to predict the tendencies of atoms and molecules to react; to bind, absorb, or partition; to dissolve or change phase; or to change their shapes or chemical bonds. The three basic tools of thermodynamics are: the First Law for the conservation of energy (see Chapter 3), the maximum entropy principle, also called the Second Law of thermodynamics (Chapters 2 and 6), and multivariate calculus (Chapter 5). Thermodynamics gives definitions of the forces that act in material systems—pressure, the tendency to exchange volume; temperature, the tendency to exchange energy; and chemical potential, the tendency to exchange matter. In this chapter, we combine these tools to create the machinery of thermodynamics and to explore the principles of equilibria.

What Is a Thermodynamic System?

A thermodynamic system is a collection of matter in any form, delineated from its surroundings by real or imaginary boundaries. The system may be a biological cell, the contents of a test tube, the gas in a piston, a thin film, or a can of soup. How you define a system depends on the problem that you want to solve. In thermodynamics, defining the boundaries is important, because boundaries determine what goes in or out. Much of thermodynamics is bookkeeping, accounting for energy or matter exchange across the boundaries,

Figure 7.1 An amoeba is an open system.

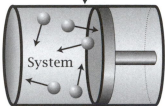

Figure 7.2 A piston can be a closed system.

or volume changes. For a biochemical experiment, the boundary might be the walls of a test tube, or a cell membrane. For a steam engine, the interior of the piston might be the most convenient definition of the system. The boundary need not be fixed in space. It need not even surround a fixed number of molecules. The system might contain subsystems, and the subsystems might also be delineated by real or imaginary boundaries. Systems are defined by the nature of their boundaries:

OPEN SYSTEM. An open system can exchange energy, volume, and matter with its surroundings. The earth and living organisms are open systems (see Figure 7.1).

CLOSED SYSTEM. Energy can cross the boundary of a closed system, but matter cannot. The boundaries can be stationary, as in a light bulb or the walls of a house that loses heat, or movable, as in a balloon or a piston (see Figure 7.2).

ISOLATED SYSTEM. Energy and matter cannot cross the boundaries of an isolated system. Also the volume does not change: the boundaries do not move. The total internal energy of an isolated system is constant. In reality, there is never complete isolation. This kind of idealization is common in thermodynamics.

SEMIPERMEABLE MEMBRANE. A semipermeable membrane is a boundary that restricts the flow of some kinds of particle, while allowing others to cross. Dialysis is performed with semipermeable membranes. Biological membranes are semipermeable.

ADIABATIC BOUNDARY. An adiabatic boundary allows no heat to flow between a system and its surroundings. Thermos bottles have adiabatic boundaries, more or less. Adiabatic walls permit you to measure work by prohibiting heat flow.

PHASE. A phase is a homogeneous part of a system that is mechanically separable from the rest of the system. Homogeneous means that the pressure, temperature, and concentration are uniform, or continuous functions of position within a phase. In a simple system, a phase can be solid, liquid, gaseous, liquid crystalline, or other.

SIMPLE SYSTEM. A simple system is defined as homogeneous, with only a single phase. It must be uncharged, large enough so that surface effects can be neglected, and not subject to changes due to electric, magnetic, or gravitational fields.

These definitions, like much of thermodynamics, involve idealizations and approximations to reality. In this chapter and Chapter 8, we will focus on simple systems.

Properties Are Extensive and Intensive

Some properties of matter are *extensive*. An extensive property P of a system is the sum of the properties $P_1 + P_2 + P_3 + \cdots$ of its component subsystems. For example, the volume of a glass of water is the sum of volumes of two half glasses of water. Alternatively, properties can be *intensive*, or independent of the size of a system. Temperature is intensive: the temperature of a glass of water is not the sum of the temperatures of two half glasses. Pressures and concentrations are also intensive quantities. Extensive properties include:

SPATIAL EXTENT. A system has volume, area, or length determined by the positions of its boundaries. We will usually consider the volume V, but with films or surfaces you might be interested in the area. When you stretch rubber bands, you might be interested in the length.

NUMBER OF PARTICLES. A system is composed of some number of particles such as atoms, molecules or other microscopic units. For a gas in a piston, a chemical reaction, or for ice in water, a natural microscopic unit is the atom or molecule. The particles of a system can belong to different species. For example, the gases in a piston can include O_2, N_2, CO, etc. You can have N_i molecules of species i, that is N_1 molecules of species 1, N_2 molecules of species 2, and so on. Rather than the number of molecules, we sometimes specify the numbers of moles, $n_i = N_i/\mathcal{N}$, where $\mathcal{N} = 6.022045 \times 10^{23}$ molecules per mole is Avogadro's number. We use bold type to denote the set of all species: $\mathbf{n} = n_1, n_2, \ldots, n_M$ or $\mathbf{N} = N_1, N_2, \ldots N_M$, where M is the number of species.

INTERNAL ENERGY. The internal energy U of a system is extensive. For a system of independent particles, the internal energy is the sum of the particle energies, according to Equation (3.8):

$$U = \sum_{i=1}^{t} N_i \varepsilon_i,$$

where t is the number of energy levels, N_i is the number of particles in energy level i, and ε_i is the energy at level i.

ENTROPY. The entropy of a system is the sum of the entropies of its independent subsystems (see Chapter 6).

The Fundamental Thermodynamic Equations, $S = S(U, V, \mathbf{N})$ or $U = U(S, V, \mathbf{N})$, Predict Equilibria

In Chapters 2 and 3 we used simple models to illustrate how the tendency toward maximum multiplicity W (and maximum entropy S) can account for some tendencies of physical systems. When the volume can change, the tendency toward maximum entropy $S(V)$ predicts the expansion of gases. When particle numbers can change, maximizing $S(N)$ predicts mixing. And when the energy can exchange, maximizing $S(U)$ predicts the tendency for heat to flow. However, many problems of interest involve multiple degrees of freedom. With the tools of multivariate calculus, you can predict more complex processes. If the energy, volume, and particle number are all free to change, then the **fundamental thermodynamic equation for entropy** is multivariate:

$$S = S(U, V, \mathbf{N}).$$

$S = S(U, V, \mathbf{N})$ is just a statement that we are taking $U, V, N_1, N_2, \ldots, N_M$ to be independent of each other, and that the entropy is dependent on them.

It is an inconvenient quirk of history (for us) that thermodynamics evolved instead with the arrangement of variables known as the **fundamental thermodynamic equation for energy**:

$$U = U(S, V, \mathbf{N}).$$

We will see that the fundamental definitions of pressure, chemical potential, and temperature are based on the form $U = U(S, V, \mathbf{N})$. Unfortunately, while the main definitions of thermodynamics originate from the energy equation, the microscopic driving forces are better understood in terms of the entropy equation $S = S(U, V, \mathbf{N})$. So we need to switch back and forth between $S = S(U, V, \mathbf{N})$ and $U = U(S, V, \mathbf{N})$. For example, in Chapter 2 we illustrated how $S(U)$ accounts for the tendency for heat to flow, but temperature is defined in terms of $U(S)$. This small inconvenience adds a few steps to our logic.

Either equation, $S = S(U, V, \mathbf{N})$ or $U = U(S, V, \mathbf{N})$, will completely specify the state of a simple system. Thermodynamics does not tell you the specific mathematical dependence of S on $(U, V, \mathbf{N})$ or U on $(S, V, \mathbf{N})$. *Equations of state*, such as the ideal gas law, specify interrelations among these variables. Equations of state must come either from experiments or from microscopic models. They are imported into thermodynamics, which gives additional useful relationships.

The Fundamental Equations Define the Thermodynamic Driving Forces

The Definitions

The fundamental equation for small changes in entropy $S(U, V, \mathbf{N})$ can be expressed as

$$dS = \left(\frac{\partial S}{\partial U}\right)_{V,\mathbf{N}} dU + \left(\frac{\partial S}{\partial V}\right)_{U,\mathbf{N}} dV + \sum_{j=1}^{M} \left(\frac{\partial S}{\partial N_j}\right)_{U,V,N_{i\neq j}} dN_j. \tag{7.1}$$

When you use the fundamental energy equation instead, you can describe small changes in energy $U(S, V, \mathbf{N})$ by

$$dU = \left(\frac{\partial U}{\partial S}\right)_{V,\mathbf{N}} dS + \left(\frac{\partial U}{\partial V}\right)_{S,\mathbf{N}} dV + \sum_{j=1}^{M} \left(\frac{\partial U}{\partial N_j}\right)_{S,V,N_{i\neq j}} dN_j. \tag{7.2}$$

Each of the partial derivatives in the fundamental energy Equation (7.2) corresponds to a measurable physical property. For now, take the following to be mathematical definitions: the *temperature* T, the *pressure* p, and the *chemical potential* μ_j, are *defined by* the partial derivatives given in Equation (7.2):

$$T = \left(\frac{\partial U}{\partial S}\right)_{V,\mathbf{N}}, \qquad p = -\left(\frac{\partial U}{\partial V}\right)_{S,\mathbf{N}}, \qquad \mu_j = \left(\frac{\partial U}{\partial N_j}\right)_{S,V,N_{i\neq j}}. \tag{7.3}$$

We will show, starting on page 110, that these mathematical definitions describe the physical properties that are familiar as temperature, pressure, and chemical potential. According to Equation (7.3), the quantities T, p, and μ are derivatives of the internal energy. In total derivative expressions, each partial derivative term $(\partial U / \partial x)$ is described as being *conjugate* to x. Temperature is conjugate to entropy, pressure is conjugate to volume, and chemical potential is conjugate to the number of particles.

Substituting the definitions given in Equation (7.3) into Equation (7.2) gives the **differential form of the fundamental energy equation:**

$$dU = T dS - p dV + \sum_{j=1}^{M} \mu_j dN_j. \tag{7.4}$$

Alternatively, you can rearrange Equation (7.4) to get a useful **differential form of the fundamental entropy equation:**

$$dS = \left(\frac{1}{T}\right) dU + \left(\frac{p}{T}\right) dV - \sum_{j=1}^{M} \left(\frac{\mu_j}{T}\right) dN_j. \tag{7.5}$$

Comparison of Equations (7.5) and (7.1) gives three more definitions:

$$\frac{1}{T} = \left(\frac{\partial S}{\partial U}\right)_{V,\mathbf{N}}, \qquad \frac{p}{T} = \left(\frac{\partial S}{\partial V}\right)_{U,\mathbf{N}}, \qquad \frac{\mu_j}{T} = -\left(\frac{\partial S}{\partial N_j}\right)_{U,V,N_{i \neq j}}. \tag{7.6}$$

Equations (7.4) and (7.5) are completely general statements that there are some functional dependences $S(U, V, \mathbf{N})$ and $U(S, V, \mathbf{N})$, and that these dependences define T, p, and μ. Equations (7.4) and (7.5) are *fundamental* because they completely specify all the changes that can occur in a simple thermodynamic system, and they are the bases for extremum principles that predict states of equilibria. An important difference between these fundamental equations and others that we will write later is that S and U are functions of only the extensive variables. Beginning on page 111 we will show how to use Equations (7.4) and (7.5) to identify states of equilibrium.

Sometimes thermodynamic reasoning enters a problem only by providing a definition. Here is an example in which the ideal gas law is derived from just two ingredients: the definition of pressure given in Equation (7.6), and the simple lattice model of a gas from Example 2.2 (page 32).

EXAMPLE 7.1 The ideal gas law derived from the lattice model. Take the definition of pressure, $p/T = (\partial S / \partial V)_{U,\mathbf{N}}$, from Equation (7.6). Into this expression, insert the function $S(V)$ from the lattice model in Example 2.2. For a lattice of M sites with N particles, use the Boltzmann expression for entropy and Equation (2.3) to get

$$\frac{S}{k} = \ln W(N, M) = \ln\left(\frac{M!}{N!(M-N)!}\right).$$

You can convert from $S(N, M)$ to $S(N, V)$ by using the chain rule (see page 77). There are M lattice sites per unit volume V, so $M = (\text{constant}) V$. This gives $(\partial M / \partial V) = M/V$, and

$$\left(\frac{\partial S}{\partial V}\right)_N = \left(\frac{\partial S}{\partial M}\right)_N \left(\frac{dM}{dV}\right) = \left(\frac{\partial S}{\partial M}\right)_N \left(\frac{M}{V}\right). \tag{7.7}$$

To get $S(M)$, apply Stirling's approximation (page 57) to Equation (2.3),

$$\frac{S}{k} \approx \ln\left(\frac{M^M}{N^N(M-N)^{(M-N)}}\right)$$

$$= M\ln M - N\ln N - (M-N)\ln(M-N). \tag{7.8}$$

Break the first term on the right side into two parts, $M\ln M = N\ln M + (M-N)\ln M$, and incorporate these into the second and third terms in Equation (7.8) to get

$$\frac{S}{k} = -N\ln\left(\frac{N}{M}\right) - (M-N)\ln\left(\frac{M-N}{M}\right). \tag{7.9}$$

Take the derivative of S in Equation (7.8) to get

$$\left(\frac{\partial S}{\partial M}\right)_N = k\left[1 + \ln M - \ln(M-N) - \left(\frac{M-N}{M-N}\right)\right]$$

$$= -k\ln\left(1 - \frac{N}{M}\right). \tag{7.10}$$

Equation (7.10) can be expressed by using the series expansion of the logarithm from Appendix C, $\ln(1-x) \approx -x - x^2/2 + \cdots$. If the density of molecules is small, $N/M \ll 1$, you need only the lowest order terms. Substituting Equation (7.10) and Equation (7.7) into the definition of pressure gives

$$p = -kT\left(\frac{M}{V}\right)\ln\left(1 - \frac{N}{M}\right)$$

$$\approx \left(-\frac{MkT}{V}\right)\left(-\frac{N}{M}\right)\left(1 + \frac{1}{2}\left(\frac{N}{M}\right) + \ldots\right) \approx \frac{NkT}{V}, \tag{7.11}$$

which is the ideal gas law.

We have derived the ideal gas law from a simple model for the dependence of S on V, using the thermodynamic definition of pressure. We will show in Chapter 24 that keeping the next higher order term in the expansion gives a refinement toward the van der Waals gas law.

$1/T$, p/T, and μ/T Behave as Forces

Now we'll show that T, p, and μ, mathematically defined by Equation (7.3), coincide with the physically measurable quantities temperature, pressure, and chemical potential. We will do this in two steps. First we will show that T, p, and μ are *intensive* quantities, not dependent on system size. Then we will show that $1/T$, p/T, and μ/T describe forces. $1/T$ describes a tendency for heat flow, p/T represents a tendency for volume change, and μ/T represents a tendency for particle exchange.

T, p, and μ are intensive because $U(S, V, \mathbf{N})$ is a *first-order homogeneous function* (see below). The fundamental energy equation has the form $U = U(x_1, x_2, x_3)$ where the x_i's are the extensive quantities, $S, V, \mathbf{N}$. If the system size is scaled up by a factor of λ, then each x_i is replaced by λx_i, and the energy, which is extensive, will increase from U to λU. But since T, p, and μ are partial derivatives $(\partial U / \partial x_i)$ (see Equation (7.3)), and since rescaling by λ does not change these partial derivatives, $(\partial(\lambda U)/\partial(\lambda x_i)) = (\partial U/\partial x_i)$, the quantities T, p, and μ are not dependent on the size of the system.

Homogeneous Functions

A first-order homogeneous function increases by a factor λ if the independent variables x_i are all increased by a factor λ:

$$f(\lambda x_1, \lambda x_2, \lambda x_3, \ldots, \lambda x_M) = \lambda f(x_1, x_2, x_3, \ldots, x_M), \tag{7.12}$$

where λ is a constant.

The volume of a liquid is a first-order homogeneous function of the number of molecules. If you double the number of molecules you double the volume of the liquid. In this case, x = number of molecules, f = volume, and $\lambda = 2$. In contrast, the volume of a sphere is *not* a first-order homogeneous function of the *radius*. If you double the radius, you do not double the volume.

An important property of first-order homogeneous functions is their relation to their partial derivatives. Differentiate both sides of Equation (7.12) with respect to λ and use the chain rule (see page 77) to get

$$\frac{df(\lambda x_1, \lambda x_2, \lambda x_3, \ldots, \lambda x_M)}{d\lambda} = f = \sum_{i=1}^{M} \left(\frac{\partial f}{\partial(\lambda x_i)}\right)\left(\frac{\partial(\lambda x_i)}{\partial \lambda}\right), \tag{7.13}$$

since $(\partial f(x_1, x_2, \ldots, x_n)/\partial \lambda) = 0$. The expression on the right side of Equation (7.13) can be simplified because $(\partial(\lambda x_i)/\partial \lambda)$ equals x_i:

$$\sum_{i=1}^{M} x_i \left(\frac{\partial f}{\partial(\lambda x_i)}\right) = f. \tag{7.14}$$

Since Equation (7.14) must hold for any value of λ, you can express the partial derivatives of the original function $f(x_1, x_2, x_3, \ldots, x_M)$ by setting λ equal to one, with the result

$$\sum_{i=1}^{M} x_i \left(\frac{\partial f}{\partial x_i}\right) = f. \tag{7.15}$$

Equation (7.15) does not apply to functions in general, only to first-order homogeneous functions. For example, for $U(S, V, \mathbf{N})$, Equation (7.15) gives $U = TS - pV + \sum_{i=1}^{M} \mu_i N_i$.

$dS = 0$ Defines Thermal, Mechanical, and Chemical Equilibria

In the next four examples we show that $1/T$ is a measure of a system's tendency for heat exchange, p/T is a measure of a system's tendency for volume change, and μ/T is a measure of a system's tendency for particle exchange. We follow

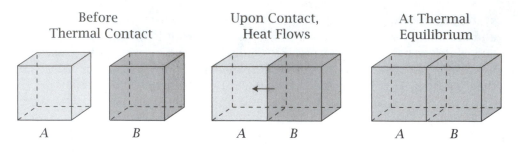

Before Thermal Contact   **Upon Contact, Heat Flows**   **At Thermal Equilibrium**

A   B   A   B   A   B

Figure 7.3 When objects A and B are brought into thermal contact, heat q (shown here as shading) can flow from one to the other. If there is no additional exchange with the surroundings, the process will conserve energy (see Equation (7.17)).

the same strategy in each case. We determine the relevant independent variables and apply the fundamental entropy equation, and the constraints. Then the maximum entropy principle ($dS_{\text{total}} = 0$) defines the state of equilibrium.

EXAMPLE 7.2 Temperature describes the tendency for energy exchange.
Temperature is the quantity that tells you when heat exchange will occur. Objects exchange heat to reach a state of maximum entropy (equilibrium), and in this state their temperatures are equal. Suppose two objects are brought into thermal contact. With *each other*, they can exchange energy but not volume or particles. Both objects are otherwise isolated from their surroundings. So with the *surroundings*, the two objects cannot exchange energy, particles, or volume. Object A has energy U_A, entropy S_A, and temperature T_A, where $1/T_A = (\partial S_A/\partial U_A)$. Object B has energy U_B, entropy S_B, and $1/T_B = (\partial S_B/\partial U_B)$. Although entropy is not experimentally measurable, temperature is, so temperature is a convenient surrogate quantity for determining the closeness to equilibrium. What temperatures will the two objects have at equilibrium? What state maximizes the entropy?

Entropy is an extensive property, so

$$S_{\text{total}} = S_A + S_B. \tag{7.16}$$

Equation (7.16) does not mean that the total entropy of the system remains fixed during the process. The total entropy can increase as a system approaches equilibrium. Equation (7.16) says only that the total entropy of the system is the sum of the entropies of the subsystems.

When the two subsystems are brought into contact, they can exchange energies U_A and U_B with each other (see Figure 7.3). U_A and U_B are the degrees of freedom and the entropy depends on them, $S_{\text{total}}(U_A, U_B)$. However, the system cannot exchange energy with the surroundings, so the total energy $U_A + U_B$ is constrained:

$$U_A + U_B = U_{\text{total}} = \text{constant.} \tag{7.17}$$

To find the state of thermal equilibrium, determine what variations in U_A and U_B will cause the total entropy of the system to be maximal, $dS = 0$. Write the fundamental entropy equation for the total differential dS_{total} in terms of the degrees of freedom, and set it equal to zero:

$$dS_{\text{total}} = dS_A + dS_B = \left(\frac{\partial S_A}{\partial U_A}\right)_{V,N} dU_A + \left(\frac{\partial S_B}{\partial U_B}\right)_{V,N} dU_B = 0. \tag{7.18}$$

The differential form of the constraint Equation (7.17) is

$$dU_A + dU_B = dU_{\text{total}} = 0 \quad \Longrightarrow \quad dU_A = -dU_B. \tag{7.19}$$

Substitute $-dU_A$ for dU_B in Equation (7.18) and rearrange:

$$dS_{\text{total}} = \left[\left(\frac{\partial S_A}{\partial U_A}\right)_{V,N} - \left(\frac{\partial S_B}{\partial U_B}\right)_{V,N}\right] dU_A = 0$$

$$\Longrightarrow \quad \left(\frac{\partial S_A}{\partial U_A}\right)_{V,N} = \left(\frac{\partial S_B}{\partial U_B}\right)_{V,N}. \tag{7.20}$$

Substituting the definition $1/T = (\partial S/\partial U)_{V,N}$ into Equation (7.20) gives the equilibrium condition

$$\frac{1}{T_A} = \frac{1}{T_B} \quad \Longrightarrow \quad T_A = T_B. \tag{7.21}$$

When two objects at different temperatures are brought into thermal contact, they exchange heat. They exchange heat because it leads to maximization of the entropy of the combined system ($dS_{\text{total}} = 0$). This results in the equalization of temperatures, $T_A = T_B$. The quantity $1/T$ describes a tendency to transfer energy. When the temperatures are equal, there is no heat exchange. The advantage of describing equilibrium in terms of the equality of temperatures, rather than in terms of the maximization of entropy, is that temperatures are measurable.

In what *direction* does energy exchange between subsystems A and B? Look at the signs on the differentials. Substituting $1/T = (\partial S/\partial U)_{V,N}$ into Equation (7.20) leads to

$$dS_{\text{total}} = \left(\frac{1}{T_A} - \frac{1}{T_B}\right) dU_A. \tag{7.22}$$

A change *toward* equilibrium *increases* the entropy: $dS_{\text{total}} \geq 0$. For dS_{total} to be positive, the signs on both factors dU_A and $((1/T_A) - (1/T_B))$ must be the same. If dU_A is negative, object A loses energy to object B in the approach to equilibrium. Then, in order that $dS_{\text{total}} > 0$, T_A must be greater than T_B. Thus if object A transfers energy to object B, A must have been the hotter object. The prediction that systems tend toward states of maximum entropy explains why heat flows from hot to cold objects.

What is Temperature? A First Look

Example 7.2 illustrates the relationships among energy, entropy, temperature, and heat. Energy is the capacity of a system to do work. That capacity can flow from one place to another as heat. The entropy describes the tendency of the work capacity (energy) to flow from one object to another. The entropy is a sort of potential to move energy from one place to another and $1/T$ represents the corresponding force. The higher the temperature of an object, the greater the tendency for energy to escape from it.

Figure 7.4 $1/T$ is the slope of S as a function of U for both systems A and B. Two systems at thermal equilibrium have the same temperature, T, not necessarily the same energy or entropy.

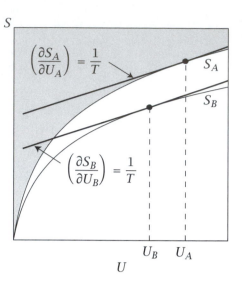

Is the flow of heat due to a tendency to equalize energies? No. Heat flows to maximize entropy. Temperatures are equalized, not energies. The First Law describes energy bookkeeping: the sum of heat plus work is constant. It does not tell us why, or when, or how much energy will be exchanged. For that we need a second principle, the principle that systems tend toward their states of maximum entropy, the Second Law of thermodynamics.

Consider money as an analog to energy. Money represents a capacity to produce things, just as energy represents the capacity for work. Money can flow between people, like energy between objects. When person A pays money to person B, the sum of their dollars is unchanged, just as when the two bodies exchange energy in Example 7.2. This conservation of dollars corresponds to the First Law, the conservation of energy.

However, the drive for energy to flow from hot objects to cold objects is not a tendency to equalize the distribution of energy, just as the rich do not pay the poor to equalize their wealth. This is where entropy and temperature come in. Energy flows only if it leads to an increase in the entropy of the whole system.

What is the analog of entropy in the money example? Suppose person A has money and wants widgets and person B has widgets to sell. Person A will benefit in some way by *paying* money to receive a widget. On the other hand, person B will benefit by *receiving* money for a widget. Money will be exchanged because both parties benefit. Economists call this 'maximizing a utility function' [1]. Money flows if it leads to an increase in the overall utility function for both parties, just as energy flows to maximize the entropy.

What is temperature in this analogy? Based on its definition, $1/T = (\partial S/\partial U)$, $1/T$ represents the incremental benefit to person A or B of getting a dollar or a widget. When these incremental benefits are equal, there is no further net redistribution of money and widgets. This analogy emphasizes that energy flows if it can increase another quantity, the entropy. When entropy can be increased no further by net changes in a degree of freedom, the system is in equilibrium. However, do not take this analogy too seriously. Money does not always flow to everyone's maximum benefit. For the limitations of the economics analogy, see reference [2].

Figure 7.4 shows that at thermal equilibrium, equality of temperatures T_A and T_B is not the same as equality of energies or entropies. We interpret temperature in more detail in Chapter 12.

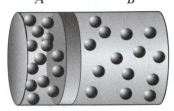

EXAMPLE 7.3 Pressure is a force for changing volume. How does volume change maximize entropy? Consider a cylinder partitioned into subsystems A and B by a movable piston as shown in Figure 7.5. The volumes of both subsystems change when the piston moves. In this problem, volume is the degree of freedom and there is no energy or matter exchange. You know from Example 7.2 that there will be no energy exchange if the temperatures on both sides are equal, $T_A = T_B$.

The total system is otherwise isolated, so it is not able to exchange energy, volume, or particles with the surroundings. Subsystem A has entropy $S_A(U_A, V_A)$, and subsystem B has entropy $S_B(U_B, V_B)$. Because the system is isolated, the volume of the total system is fixed, $V_A + V_B = V_{\text{total}} = \text{constant}$. Now allow subsystems A and B to exchange volume. For the moment, see what happens if they can also exchange energy. In differential form, the volume constraint is

After Equilibration

A B

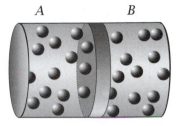

$$dV_A = -dV_B. \tag{7.23}$$

The state of maximum entropy, $dS = 0$, defines the equilibrium. Write the total differential dS_{total} in terms of the degrees of freedom:

$$dS = \left(\frac{\partial S_A}{\partial V_A}\right)dV_A + \left(\frac{\partial S_B}{\partial V_B}\right)dV_B + \left(\frac{\partial S_A}{\partial U_A}\right)dU_A + \left(\frac{\partial S_B}{\partial U_B}\right)dU_B = 0. \tag{7.24}$$

Because no energy is exchanged with the surroundings, $dU_A = -dU_B$. Substitute dU_A for $-dU_B$ and $1/T$ for (dS/dU):

$$dS = \left(\frac{\partial S_A}{\partial V_A}\right)dV_A + \left(\frac{\partial S_B}{\partial V_B}\right)dV_B + \left(\frac{1}{T_A} - \frac{1}{T_B}\right)dU_A = 0. \tag{7.25}$$

Since $T_A = T_B$, Equation (7.25) reduces to

$$dS = \left(\frac{\partial S_A}{\partial V_A}\right)dV_A + \left(\frac{\partial S_B}{\partial V_B}\right)dV_B = 0. \tag{7.26}$$

Substituting Equations (7.23) and (7.6) into Equation (7.26), the condition for equilibrium becomes

$$dS = \left(\frac{p_A}{T_A} - \frac{p_B}{T_B}\right)dV_A = 0. \tag{7.27}$$

Equation (7.27) is satisfied when

$$\frac{p_A}{T_A} = \frac{p_B}{T_B}. \tag{7.28}$$

Since $T_A = T_B$, *mechanical equilibrium* occurs when the pressures of the two subsystems are equal, $p_A = p_B$. Just as a system with $T_A > T_B$ increases its entropy and moves to equilibrium by transferring heat from subsystem A to subsystem B, a system with $p_A > p_B$ increases its entropy and moves toward equilibrium by increasing volume V_A and decreasing V_B. Volumes change to equalize the pressures.

Figure 7.5 Before equilibration, the piston in Example 7.3 is fixed in place and the pressures on the two sides are unequal. After the piston is freed, the volumes will change (subject to the constraint $V_A + V_B = V_{\text{total}} = \text{constant}$) until the pressures equalize.

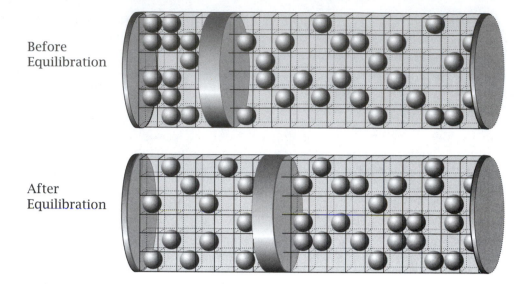

Before Equilibration

After Equilibration

Figure 7.6 Lattice model of volume changes ($T_A = T_B$) for Example 7.4. Particle numbers N_A and N_B are fixed on both sides of a piston. Volumes M_A and M_B change subject to the constraint $M_A + M_B$ = constant. Pressures become equalized, so according to lattice or ideal gas models, densities become equalized.

The pressures of gases are sometimes described as forces per unit area due to gas particles colliding with the walls of containers. Our definition of pressure in Equation (7.3) or Equation (7.6) is much more general. It applies to liquids and solids, and any other circumstance in which the energy changes with volume.

The thermodynamic reasoning in Example 7.3 is quite general, but it gives no insight into the microscopic nature of entropy. Let's return to the lattice model to see how equalizing the pressures maximizes the multiplicity of the arrangements of particles.

EXAMPLE 7.4 How does the equality of pressures maximize the multiplicity of states? Consider a gas contained on two sides of a piston as shown in Figure 7.6. The number N_A of particles on the left and the number N_B on the right are each fixed. The total volume is defined by M lattice sites, and the movable piston partitions the volume into M_A sites on the left and M_B sites on the right, with the constraint $M = M_A + M_B$ = constant.

The number of spatially distinguishable configurations on each side of the piston is given by Equation (2.3), $W(N, M) = M!/N!(M - N)!$. The total number of configurations of the system is the product

$$W = \left(\frac{M_A!}{N_A!(M_A - N_A)!} \right) \left(\frac{M_B!}{N_B!(M_B - N_B)!} \right). \tag{7.29}$$

Using Stirling's approximation (page 57) to replace the factorials, and using the constraint $M = M_A + M_B$ to replace M_B by $M - M_A$, you can express W in terms of one degree of freedom, M_A:

$$W \approx \left(\frac{M_A^{M_A}}{N_A^{N_A}(M_A - N_A)^{M_A - N_A}} \right) \left(\frac{(M - M_A)^{M - M_A}}{N_B^{N_B}(M - M_A - N_B)^{M - M_A - N_B}} \right). \qquad (7.30)$$

Take the logarithm of both sides of Equation (7.30) to find the entropy S/k:

$$S/k = \ln W = M_A \ln M_A - (M_A - N_A) \ln(M_A - N_A)$$

$$+ (M - M_A) \ln(M - M_A) - (M - M_A - N_B) \ln(M - M_A - N_B)$$

$$- N_A \ln N_A - N_B \ln N_B. \qquad (7.31)$$

Now find the value $M_A = M_A^*$ that maximizes $S/k = \ln W$. The constant terms $-N_A \ln N_A$ and $-N_B \ln N_B$ do not contribute to the derivative with respect to M_A. Taking M, N_A, and N_B as constants, solve

$$\frac{d \ln W}{d M_A} = 0$$

$$\Rightarrow 1 + \ln M_A^* - \ln(M_A^* - N_A) - 1 - \ln(M - M_A^*)$$

$$-1 + \ln(M - M_A^* - N_B) + 1 = 0$$

$$\Rightarrow \ln \left(\left(\frac{M_A^*}{M_A^* - N_A} \right) \left(\frac{M - M_A^* - N_B}{M - M_A^*} \right) \right) = 0$$

$$\Rightarrow \left(\frac{M_A^*}{M_A^* - N_A} \right) \left(\frac{M - M_A^* - N_B}{M - M_A^*} \right) = 1. \qquad (7.32)$$

$(M_A^* - N_A)/M_A^*$ is the fraction of A sites that are empty, and the fraction of B sites that are empty is $(M - M_A^* - N_B)/(M - M_A^*)$. Setting these expressions equal according to Equation (7.32) means that the fraction of filled sites must be the same on both sides,

$$\frac{N_B}{M_B^*} = \frac{N_A}{M_A^*}. \qquad (7.33)$$

The entropy is maximized and the system is at equilibrium when the density of particles N/M is the same on both sides. Because the pressure of an ideal gas is given by $p/T = Nk/M$, the equality of densities implies that $p_A/T_A = p_B/T_B$ at equilibrium, which is the same as the thermodynamic result in Example 7.3. However, the molecular model gives us information about densities that the thermodynamic model does not.

Now we turn our attention from systems that change volumes to systems that change particle numbers.

EXAMPLE 7.5 The chemical potential is a tendency for particle exchange.
The chemical potential describes a tendency for matter to move from one place to another. Suppose that two compartments (A and B) are separated by a fixed barrier. Consider N identical particles, N_A of which are on side A and N_B of which are on side B (see Figure 7.7). The degrees of freedom are N_A and N_B.

Figure 7.7 Before equilibration a permeation barrier prevents particle flow. After the barrier is removed, the particles exchange until the chemical potential is the same on both sides (see Example 7.5).

Before Equilibration

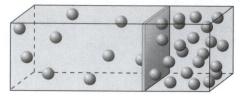

After Equilibration

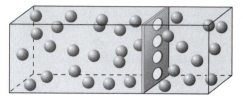

When particles are exchanged, they carry energy with them, so energy is also exchanged. However, if the temperatures and pressures of the subsystems are held equal, there will be no net exchange of energy or volume when particles exchange across the barrier.

The subsystems have entropies $S_A(N_A)$ and $S_B(N_B)$. Again write the differential form of the fundamental entropy equation in terms of the degrees of freedom, introduce the constraints, and then find the state that has the maximum entropy. When the permeation barrier is removed, the particle numbers on both sides change, subject to the constraint $N_A + N_B = N = $ constant. In differential form the number constraint is

$$dN_A = -dN_B. \tag{7.34}$$

The condition for equilibrium is

$$dS_{\text{total}} = \left(\frac{\partial S_A}{\partial N_A}\right) dN_A + \left(\frac{\partial S_B}{\partial N_B}\right) dN_B = 0. \tag{7.35}$$

Substitution of $-\mu/T$ for $(\partial S/\partial N)$ (Equation (7.6)) leads to the equilibrium condition

$$dS_{\text{total}} = \left(\frac{\mu_B}{T_B} - \frac{\mu_A}{T_A}\right) dN_A = 0. \tag{7.36}$$

The condition for material balance equilibrium at temperature $T = T_A = T_B$ is

$$\mu_A = \mu_B. \tag{7.37}$$

A change toward equilibrium must increase the entropy,

$$dS_{\text{total}} = \left(\frac{\mu_B}{T} - \frac{\mu_A}{T}\right) dN_A > 0. \tag{7.38}$$

For dS_{total} to be positive, the signs on the factors $((\mu_B/T) - (\mu_A/T))$ and dN_A must be the same. Therefore, if the particle number on side A increases as the system approaches equilibrium, μ_B must be greater than μ_A. The chemical potential μ is sometimes called the *escaping tendency* because particles tend to escape from regions of high chemical potential to regions of low chemical potential. The chemical potential pertains not just to particles in different spatial locations, say on different sides of a membrane. It also pertains to molecules in different phases or in different chemical states, as we will see in Chapters 13 and 16. Chemical potentials are found by measuring concentrations.

So far, we have used thermodynamics in two ways. First, in Example 7.1 we combined the thermodynamic definition of pressure with a molecular model to derive an equation of state, the ideal gas law. Second, in Examples 7.2, 7.3, and 7.5, we found that the condition for various equilibria ($dS = 0$) could be restated in terms of the experimentally accessible quantities, T, p, and μ.

$1/T$ is a tendency for exchanging energy, p/T is a tendency to change volume, and μ/T is a tendency for particles to exchange. For thermal, mechanical, and chemical equilibrium between subsystems A and B, $T_A = T_B$, $p_A = p_B$, and $\mu_A = \mu_B$.

Thermodynamic Logic Gives Unmeasurable Predictors of Equilibria from Measurable Properties

Thermodynamic logic often seems to be complex. This apparent complexity arises because the fundamental equilibrium functions cannot be measured directly. Equilibria are governed by energy and entropy through the First and Second Laws, but unfortunately there are no energy or entropy 'meters' to measure these quantities. Instead, inferences about equilibria are indirect and drawn from observations of quantities that can be measured, such as temperature, pressure, work, heat capacities, concentrations, or electrical potentials. Thermodynamics is a business of making clever inferences about unmeasurable quantities from observable ones, by various means.

In this and the next two chapters, we show six of the main methods of thermodynamics for drawing such inferences: (1) obtaining energies by measuring work, (2) obtaining entropies by measuring heat, (3) obtaining energies and entropies by measuring heat capacities, (4) using thermodynamic cycles to obtain some quantities from others, (5) obtaining partial derivative quantities from others by using Maxwell's relations, and (6) linking some partial derivative quantities to others by using the mathematics of homogeneous functions. Key concepts are *processes* and *cycles*, real or fictitious, as described below.

Cycles and Fictitious Processes

At the root of the classical thermodynamics of engines is the fact that work is an important and measurable quantity. However, work, while measurable, is rarely predictable. The work performed by a gas expanding in an engine piston can depend on pressure, volume, temperature, speed of the piston motion, frictional forces, and other factors in complex ways that are not fully understood.

Nevertheless, under some limited circumstances, predictions are possible. To describe those circumstances, we distinguish *state variables*, like p, V, T, N, U, and S that characterize the stable states of a system, from *process*

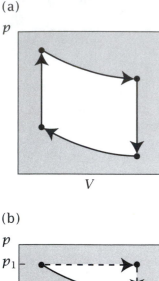

(a)

p

V

(b)

p

p_1

p_2

V_1 V_2

V

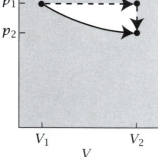

Figure 7.8 (a) Thermo-dynamic cycle for pressure–volume changes. (b) Modelling one step of the cycle (continuous line) by using a fictitious constant pressure process followed by a fictitious constant volume process (dashed lines).

variables, like piston velocity, thermal gradients, friction, diffusion, and other time-dependent quantities that characterize how the system gets from one state to another. Functions like energy $U(S, V, \mathbf{N})$ and entropy $S(U, V, \mathbf{N})$ have a special simplicity—they depend only on state variables such as $U, S, V, \mathbf{N}$, and not on process variables. In contrast, work w and heat q are more complex, because they depend on both state and process variables. If gas expansion is fast, it can generate much heat and little work. Slower expansion can generate more work and less heat. Just knowing state quantities like (p_1, V_1, T_1) and (p_2, V_2, T_2) without knowing the quantities that describe the intervening *process*, you cannot in general predict the work performed by a gas in a cylinder.

However, even if processes are performed infinitely slowly, so that time-dependent quantities are negligible, heat and work still depend on the process (or pathway) in ways that state functions do not. Recall from Chapter 5 (page 73) that for some functions $f(x, y)$, a change Δf, computed by integration, depends on the pathway of integration, while for others, called *state functions*, the integral Δf does not depend on the pathway of integration. Heat and work are path-dependent functions, while U and S are state functions. The work as a gas expands from (p_1, V_1) to (p_2, V_2) will be different if the system passes through an intervening state, say (p_x, V_x), than if it passes through a state (p_y, V_y). To compute the work, you need to know the sequence of intervening states in the physical process. However, to compute the difference in a state function, such as ΔU, you do not need to know the intervening states. That is, you get the same value of $\Delta U = U(p_2, V_2) - U(p_1, V_1)$ whether you assume that the system passed through state $x : U_2 - U_1 = (U_2 - U_x) + (U_x - U_1)$ or through state $y : U_2 - U_1 = (U_2 - U_y) + (U_y - U_1)$. The intervening term cancels in each of these expressions.

Two key principles follow from this logic. First, a thermodynamic change from state 1 to state 2 can be expressed as a sum of component steps, say from 1 to x followed by x to 2. Second, this logic is the basis for inventing *fictitious* processes and *imaginary* intervening states. Suppose that the pressure and volume of a gas change from initial state (p_1, V_1) to final state (p_2, V_2). To compute the difference in a state function, say ΔU, you are free to imagine that the real system passes through any state you wish to invent if it helps to make your calculation easier. You could model the overall change as a sum of two processes: a volume change at constant pressure $(p_1, V_1) \rightarrow (p_1, V_2)$ followed by a pressure change at constant volume $(p_1, V_2) \rightarrow (p_2, V_2)$, even if the real physical process never passes through such a state (p_1, V_2) (see Figure 7.8(b)). To compute differences in state functions such as ΔU, you can invent imaginary intervening states, even nonphysical ones, in whatever ways make it simpler to calculate. One important class of fictitious process is the *quasi-static process*.

Quasi-Static Processes

Quasi-static describes processes that are performed sufficiently slowly that properties are independent of time and independent of the speed of the process. To be quasi-static, a process must be significantly slower than a relaxation time of a system. Only quasi-static processes can be represented on state diagrams, because non-quasi-static processes involve more variables than just those that define state points. Non-quasi-static processes also depend on gradients or time-dependent quantities such as velocities and friction coefficients. The power of the idea of quasi-static processes is that it provides a relationship between the work w and the state variables, pressure and volume. For a gas in a piston expanding by a volume dV against an applied external pressure p_{ext}, the increment of work in a *quasi-static* process is defined to be

$$\delta w = -p_{ext}dV. \tag{7.39}$$

If a process is not quasi-static, Equation (7.39) is not valid.

Here are examples of different quasi-static processes or thermodynamic paths that could be taken when a gas expands in a piston at constant volume, constant pressure, or constant temperature.

EXAMPLE 7.6 Constant-volume work. Whenever there is no volume change in a quasi-static process in a simple system, no work is performed because $p\,dV = 0$.

EXAMPLE 7.7 Constant-pressure work. When the externally applied pressure p_{ext} is constant, the total work in a quasi-static process of expansion from volume V_A to V_B is

$$w = -\int_{V_A}^{V_B} p_{ext}\, dV = -p_{ext}(V_B - V_A). \tag{7.40}$$

According to our sign convention, δw is positive when the volume inside a piston decreases ($dV < 0$). That is, work is positive when it is done *on the system* to compress the gas, and negative when it is done *by the system* to expand.

EXAMPLE 7.8 Constant-temperature work. A gas does work when it expands quasi-statically in a piston at constant temperature. Now both the pressure and volume can change during the expansion, so you need to know the functional form of $p(V)$ for this process. A quasi-static process is slow enough for the gas pressure inside to equilibrate with the external pressure, $p_{int} = p_{ext}$. To compute the work of an ideal gas at constant T, integrate:

$$w = -\int_{V_A}^{V_B} p_{ext}\, dV = -\int_{V_A}^{V_B} p_{int}\, dV$$

$$= -\int_{V_A}^{V_B} \frac{NkT}{V}\, dV = -NkT \ln\left(\frac{V_B}{V_A}\right). \tag{7.41}$$

At the core of thermodynamics are *thermodynamic cycles*. A cycle is a series of steps that begin in one state, pass through other states, then return to the initial state to begin again. Engines take in fuel, which is combusted to form a vapor. The gas expands and performs work. The spent vapor exits. New fuel enters to begin another cycle. In refrigerators, a working fluid flows through cooling coils to cool a compartment, flows through other coils to dump heat outside the refrigerator, then flows around again to repeat the cycle. Muscle proteins convert energy to motion in repetitive cycles. Cycles are described by *state diagrams*, which show how certain state variables change throughout the cycle. Figure 7.8(a) illustrates a pressure–volume state diagram, a common way in which to describe the states of a gas in an engine. Figure 7.8(b) shows two fictitious processes that model a pV change in one step of the cycle.

We now aim to relate the two path-dependent quantities, q and w, to state functions like U and S, for two reasons. First, such relationships give ways of getting fundamental but unmeasurable quantities U and S from measurable quantities q and w. Second, there is an increase in predictive power whenever q and w depend only on state variables and not on process variables. The First Law of thermodynamics gives such a relationship.

The First Law Interrelates Heat, Work, and Energy

The **First Law of thermodynamics** relates a change in energy, dU, to increments of heat δq and work δw:

$$dU = \delta q + \delta w. \tag{7.42}$$

The use of δ as a differential element indicates that heat and work are path-dependent quantities, while the use of d indicates that their sum, the internal energy dU, is a state function (see pages 73–77). The signs of the quantities in the First Law are defined so that the internal energy increases when heat flows *into* a system, $\delta q > 0$, and when work is done *on* a system, $\delta w > 0$. Energy is a property of a *system* while heat and work are properties of a process of energy *transfer across a boundary*.

This difference is illustrated by an analogy: the level of water in a lake is like the amount of internal energy in a system, and the different modes of water exchange between the lake and its surroundings are like heat and work [3]. Say that rainfall corresponds to the process of heat going into the system ($q > 0$), evaporation corresponds to the process of heat going out of the system ($q < 0$), streams flowing into the lake correspond to work done on the system ($w > 0$), and streams flowing out of the lake correspond to work done *by* the system ($w < 0$). The change in internal energy $\Delta U = q_{in} - q_{out} + w_{in} - w_{out}$ is the sum of the heat and work *into* the system minus the heat and work *out of* the system, just as the change in the water level of the lake is the sum of rainfall plus river flows in, minus evaporation and river flows out. Once the water is in the lake, you cannot tell whether it came from rainfall or from streams. Similarly, you cannot tell whether the amount of internal energy in a system was acquired as heat or work. In this analogy, the internal energy corresponds to a property of the lake, while heat and work correspond to processes of the transfer across the boundaries of the lake.

Method (1)
Using Adiabatic Boundaries to Relate U and w

The First Law gives a way of relating w, a path-dependent quantity, to U, a state function. If you surround a working device with an insulating (adiabatic) boundary ($\delta q = 0$) it follows from the First Law that $dU = \delta w$. The point is to determine a change ΔU from an experimentally measurable quantity, the work done on the system. In the lake analogy, you could cover the lake with a tarp and measure the flow of streams to determine the change in water level. With an adiabatic boundary, work becomes a function of state properties only.

Heat, which is path-dependent, is related to the entropy, a state function, for quasi-static processes, as shown below.

Method (2)
Combine the First Law and the Fundamental Energy Equation to Find the Entropy

What experiment can determine entropy changes? Combining the First Law with the fundamental energy equation (Equation (7.4)) for a closed system (no loss or gain of particles), $dU = TdS - pdV$, gives

$$dU = TdS - pdV = \delta q + \delta w. \tag{7.43}$$

In a quasi-static process, the work is *Process that happens infinitely Slowly*

$$\delta w = -pdV. \tag{7.44}$$

Substituting Equation (7.44) into (7.43) gives $\delta q = TdS$, or

$$dS = \frac{\delta q}{T} \qquad \text{for quasi-static processes in closed systems.} \tag{7.45}$$

Equation (7.45) is sometimes called the **thermodynamic definition of entropy**. It shows how to obtain the entropy change (which you cannot otherwise measure directly) from experimentally observable quantities (heat transfer and temperature) by using a quasi-static process.

A central problem of the industrial revolution, understanding what makes steam engines efficient, was solved in the 1800s, establishing the importance of thermodynamics. Steam engines convert heat to work by the expansion of a hot gas in a piston. James Watt's original steam engine converted only 7% of the energy (heat) input to work. The remaining 93% was exhausted to the surroundings. The design of efficient engines had overwhelming economic importance for the industrial revolution. It was necessary to understand the nature of heat in order to convert heat to work efficiently. Is efficiency determined by the type of gas, its temperature, or the volume or shape of the piston? The concepts of thermodynamics emerged from the effort to solve this problem. To understand why an engine wastes heat, you need to know how the engine balances its energy books at the same time as it tends toward equilibrium. We need another idealization, called *reversibility*. We need three ideas: a reversible process, a reversible work source, and a reversible heat source.

Reversible Processes

Reversibility is an idealization that defines the maximum work that can be achieved. To achieve the maximum work requires more than just extreme slowness. Reversible processes are a subset of quasi-static processes. (Any process that is fast enough is irreversible.) Processes are reversible if they are both very slow and also if they can be undone by equilibrium physical changes that involve no entropy change, $\Delta S = 0$. (In Chapter 8 we will discuss processes

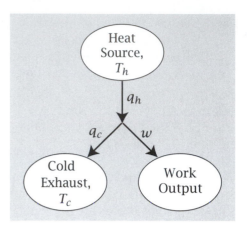

Figure 7.9 Schematic diagram of heat and work sources in which the hot gas in an engine partly converts to work, and partly converts to colder exhaust gas. The ellipses indicate reversible heat and work sources. The arrows indicate processes. A physical realization is shown in Figure 7.10.

at constant temperature; then reversibility will be defined as an unchanging *free energy*.) You can stretch clay very slowly, but it cannot be 'reversed' to its unstretched state by equilibrium steps. You can turn a paddle wheel clockwise in water to dissipate heat through frictional forces, but the process cannot be reversed: heat from the liquid cannot make the paddle wheel move counter-clockwise. You can put a droplet of dye in water and watch it spread, but the reverse does not happen. Dinosaurs decompose into hydrocarbons over the eons, but you cannot turn hydrocarbons back into dinosaurs. All these are examples of processes that can happen slowly, but are not reversible.

Reversible Heat and Work Sources

To compute the maximum work, we need two other idealizations. A *reversible work source* can change volume or perform work of any other kind quasi-statically, and is enclosed in an impermeable adiabatic wall, so $\delta q = TdS = 0$ and $dU = \delta w$. A *reversible heat source* can exchange heat quasi-statically, and is enclosed in a rigid wall that is impermeable to matter but not to heat flow, so $\delta w = -pdV = 0$ and $dU = \delta q = TdS$. A reversible *process* is different from a reversible heat or work *source*. A reversible heat source need not have $\Delta S = 0$. A reversible *process* refers to changes in a whole system, in which a collection of reversible heat plus work sources has $\Delta S = 0$. (Later we'll use $\Delta F = 0$ or $\Delta G = 0$.) The frictionless weights on pulleys and inclined planes of Newtonian mechanics are reversible work sources, for example. The maximum possible work is achieved when reversible processes are performed with reversible heat and work sources.

Now consider the problem of mechanical efficiency that motivated the development of thermodynamics.

EXAMPLE 7.9 Why do engines waste heat? Why can't you design an engine that converts 100% of the heat input to work output? This question was finally resolved by Sadi Carnot in the 1820s, when he determined the maximum efficiency of an idealized engine by the following logic. Treat the energy and entropy of the gas as though there were three compartments, two reversible heat sources and a reversible work source (see Figure 7.9). Heat and work can exchange among these imaginary compartments. The heat source loses energy,

$\Delta U_h = -q_h$ and the cold exhaust 'heat source' gains energy, $\Delta U_c = q_c$. The heat q_h from the hot gas parcels into two components: some of the heat converts to work w and the remaining heat q_c exits as cooler gas that leaves the engine: $q_h = q_c + w$. Rearranging gives

$$w = q_h - q_c. \tag{7.46}$$

Define the *efficiency* η of the system as the work output divided by the heat input:

$$\eta = \frac{w}{q_h} = 1 - \frac{q_c}{q_h}. \tag{7.47}$$

Figure 7.10 shows an alternative way of thinking about the compartments. Imagine a gas sealed in a cylinder with a piston. Bring a heat source at temperature T_h into contact, so that heat q_h flows into the cylinder. The heated gas in the cylinder expands and performs work w. Then bring the cylinder into thermal contact with another object at a lower temperature T_c, so that heat q_c flows out of the cylinder. No gas need ever enter or leave the cylinder, in this case.

The Second Law gives the entropy change toward equilibrium of the whole system, ΔS_{total}, as the sum of the three components, ΔS_h from the hot gas, ΔS_c from the cooled gas, and ΔS_w from the work source. $\Delta S_{\text{total}} = \Delta S_h + \Delta S_c + \Delta S_w \geq 0$. Because a reversible work source is adiabatic, $\Delta S_w = q_w/T = 0$, so

$$\Delta S_{\text{total}} = \Delta S_h + \Delta S_c \geq 0. \tag{7.48}$$

For reversible heat sources operating at high temperature T_h and low temperature T_c, the entropy changes are:

$$\Delta S_h = \frac{-q_h}{T_h}, \quad \text{and} \quad \Delta S_c = \frac{q_c}{T_c}. \tag{7.49}$$

Substitute Equations (7.49) into (7.48) and rearrange to get

$$\frac{q_c}{q_h} \geq \frac{T_c}{T_h}. \tag{7.50}$$

The maximum possible efficiency defines the reversible process, $\Delta S_{\text{total}} = 0$. All other processes cause larger changes in the entropy $\Delta S_{\text{total}} > 0$. The largest possible value of η is achieved by the smallest possible value of q_c/q_h in Equation (7.47), which is T_c/T_h (see Equation (7.50)), so

$$\eta \leq 1 - \frac{T_c}{T_h}. \tag{7.51}$$

Equation (7.51) implies that heat is most efficiently converted to work by engines that bring in heat at the highest possible temperature and exhaust it at the lowest possible temperature. The calculation of the efficiency of an idealized reversible engine answered two historically important questions. First, it said that the efficiency of an engine depends only on the temperatures of the heat intake and exhaust, and not on the type of gas or the volume of the piston. Second, the only way to reach 100% efficiency is to exhaust the waste heat at $T_c = 0\,\text{K}$.

(a) Heat In   (b) Work Performed   (c) Heat Out

Figure 7.10 (a) In this idealized heat engine, a piston containing cooled gas starts at rest. When heat q_h enters from the reversible heat source, the energies of the gas molecules increase. (b) Work is performed by the expansion of the heated gas (the reversible work source), extracting energies from the molecules. (c) Heat q_c flows out from the reversible heat source and the volume in the piston decreases.

The inability to achieve 100% efficiency in engines is *not* a consequence of friction or turbulence. We are assuming quasi-static processes here. Since T_c/T_h is usually greater than zero, a heat engine has an inherent efficiency. For a typical automobile engine, the Carnot (maximum) efficiency is about 50–60%, but the actual efficiency is closer to 25%. A considerable fraction of the heat input is lost to the environment as heat and is not converted to work. The problem is that perfect efficiency would require extracting all the motion and all the internal energy from the gas molecules, $T_c = 0$.

Engines can be reversed to make heat pumps. A heat pump creates a hot exhaust from a cold input when work is put in.

Why Is There an *Absolute* Temperature Scale?

How should we define a practical measure of temperature? Three scales are popular: Kelvin, Celsius, and Fahrenheit. The Kelvin scale is the most fundamental because it defines *absolute* temperature T. The idea of an absolute temperature scale derives from Equation (7.51), which relates the efficiency of converting heat to work to a ratio of temperatures, T_c/T_h. An absolute zero of the temperature scale, $T_c = 0$, is the point at which heat is converted to work with 100% efficiency, $\eta = 1$. This defines the zero of the Kelvin scale.

Other scales of temperature T' are 'shifted' so that their zero points coincide, for example, with the freezing point of water (Celsius scale) or with a point of coexistence of salt, water, and ice (Fahrenheit),

$$T' = aT - b,$$

where a and b are constants. For example, for the Celsius scale, $a = 1$ and $b = 273.15$. On these scales, the efficiency of a Carnot engine approaches some value other than one as the cold temperature approaches zero, $T'_c \rightarrow 0$:

$$\eta = 1 - \frac{T_c}{T_h} = 1 - \frac{(T_c' + b)/a}{(T_h' + b)/a} = 1 - \left(\frac{T_c' + b}{T_h' + b}\right)$$

so $\eta \to [1 - b/(T_h' + b)]$ as $T_c' \to 0$. For example, if the temperature of the hot reservoir is $T_h' = 100\,°$Celsius and the cold reservoir is $T_c' = 0\,°$Celsius, then $\eta = 1 - 273.15/373.15 = 0.268$. So the zero-point of the Celsius scale has no special meaning for thermodynamic efficiency.

Other Statements of the Second Law of Thermodynamics

Our statement of the Second Law is that isolated systems tend toward their states of maximum entropy. For example, heat does not spontaneously flow from cold to hot objects because that would *decrease* the entropy. The earliest statements of the Second Law of thermodynamics were based on such examples. One such statement was that heat will not flow from a colder body to a hotter one without the action of some external agent. We can express this quantitatively by modifying the system that we examined in Example 7.2. For a system in which two objects are brought into thermal contact, and are also allowed to exchange some other extensive quantity Y, in place of Equation (7.18), the direction toward equilibrium is given by

$$dS = \left(\frac{\partial S_A}{\partial U_A}\right) dU_A + \left(\frac{\partial S_B}{\partial U_B}\right) dU_B + \left(\frac{\partial S_A}{\partial Y_A}\right) dY_A + \left(\frac{\partial S_B}{\partial Y_B}\right) dY_B \geq 0$$

$$= \left(\frac{1}{T_A} - \frac{1}{T_B}\right) dU_A + \left(\frac{\partial S_A}{\partial Y_A}\right) dY_A + \left(\frac{\partial S_B}{\partial Y_B}\right) dY_B,$$

where $(\partial S/\partial Y)dY$ accounts for the entropy change due to changes in Y. The difference in $(1/T_A) - (1/T_B)$ can be less than or equal to zero if the additional work compensates. This implies that heat *can be* moved from cold objects to hot ones by the input of some kind of work or chemical change. This is what refrigerators do. Spontaneous processes lead to an entropy increase in isolated systems, but when heat or work can cross a boundary, the entropy of a system can be caused to decrease.

Summary

The First Law of thermodynamics is a bookkeeping tool that defines a conserved quantity, the internal energy, as a sum of heat plus work. The Second Law describes the tendencies of systems toward their states of maximum entropy. The fundamental equation $S(U,V,\mathbf{N})$ is $dS = (1/T)dU + (p/T)dV - \sum_{j=1}^{M}(\mu_j/T)dN_j$. The importance of this equation is twofold. It gives definitions of temperature, pressure, and chemical potential, and it is the basis for the maximum-entropy extremum principle. Set $dS = 0$ subject to the appropriate constraints for the problem at hand to find the condition for equilibrium in terms of experimental quantities. In the next chapter, we will introduce enthalpies and free energies, and move from heat engines and historical engineering problems to laboratory experiments in chemistry and biology.

Problems

1. The work of compression.
One mole of a van der Waals gas is compressed quasi-statically and isothermally from volume V_1 to V_2. For a van der Waals gas, the pressure is

$$p = \frac{RT}{V - b} - \frac{a}{V^2},$$

where a and b are material constants, V is the volume and RT is the gas constant × temperature.

(a) Write the expression for the work done.

(b) Is more or less work required than for an ideal gas in the low-density limit? What about the high-density limit? Why?

2. Deriving the ideal gas law in two dimensions.
Molecules at low density on a surface, such as surfactants at an interface of air and water, often obey a two-dimensional equivalent of the ideal gas law. The two-dimensional equivalent of p is π, where π is a lateral two-dimensional pressure. A is area. Using

$$\pi = T \left(\frac{\partial S}{\partial A} \right)_N \tag{7.52}$$

and assuming no energetic interactions, derive the two-dimensional equivalent of the ideal gas law by using a lattice model in the low-density limit.

3. The work of expansion in freezing an ice cube.
At 1 atm, freeze an amount of liquid water that is 2 cm × 2 cm × 2 cm in volume. The density (mass per unit volume) of liquid water at 0 °C is $1.000 \, \text{g cm}^{-3}$ and the density of ice at 0 °C is $0.915 \, \text{g cm}^{-3}$.

(a) What is the work of expansion upon freezing?

(b) Is work done *on* the system or *by* the system?

4. Compute $S(V)$ for an ideal gas.

(a) How does the entropy $S(V)$ depend on volume for an ideal gas, where $pV = NkT$?

(b) What is the entropy change if you double the volume from V to $2V$?

5. The work of expanding a gas.
Compute the total work performed when expanding an ideal gas, at constant temperature, from volume V to $2V$.

6. Pulling out the partition between two chambers of a gas.
A partition separates equal volumes containing equal numbers N of ideal gas molecules of the same species at the same temperature. Using a simple lattice model for ideal gases, evaluate the relevant multiplicity terms to show that the entropy of the composite system does not change when the partition is removed (hint: use Stirling's approximation from page 57).

7. The work in a thermodynamic cycle.
A thermodynamic cycle is a series of steps that ultimately returns to its beginning point. Compute the total work performed around the thermodynamic cycle of quasi-static processes in Figure 7.11.

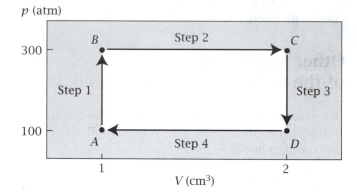

Figure 7.11

8. Engine efficiencies.
Consider a Carnot engine that runs at $T_h = 380 \, \text{K}$.

(a) Compute the efficiency if $T_c = 0 \, °\text{C} = 273 \, \text{K}$.

(b) Compute the efficiency if $T_c = 50 \, °\text{C} = 323 \, \text{K}$.

References

[1] PA Samuelson. *Economics*, 10th edition. McGraw-Hill, New York, 1976.

[2] P Mirowski. *More Heat than Light: Economics as Social Physics, Physics as Nature's Economics*. Cambridge University Press, Cambridge, 1989.

[3] HB Callen. *Thermodynamics and an Introduction to Thermostatistics*, 2nd edition. Wiley, New York, 1985.

Suggested Reading

CJ Adkins, *Equilibrium Thermodynamics*, 3rd edition, Cambridge University Press, Cambridge, 1983. Excellent introduction to thermodynamics and reversibility.

SG Brush, *The Kind of Motion We Call Heat*, North-Holland, New York, 1976. An excellent history of thermodynamics and statistical mechanics.

HB Callen, *Thermodynamics and an Introduction to Thermostatistics*, 2nd edition, Wiley, New York, 1985. The classic text on the axiomatic approach to thermodynamics. The logic of thermodynamics is explained with great clarity.

G Carrington, *Basic Thermodynamics*, Oxford University Press, Oxford, 1994. Good general text with many engineering examples.

K Denbigh, *The Principles of Chemical Equilibrium*, Cambridge University Press, Cambridge, 1981. General text on thermodynamics.

IM Klotz and RM Rosenberg, *Chemical Thermodynamics*, Wiley, New York, 1986. Excellent simple and general text.

HL Leff and AF Rex, *Maxwell's Demon: Entropy, Information, Computing*, Princeton Series in Physics, Princeton University Press, Princeton, 1990. Essays on the relationship between entropy and information theory.

EB Smith, *Basic Chemical Thermodynamics*, Clarendon Press, Oxford, 1990. Excellent simple and general text.

MW Zemansky and RH Dittman, *Heat and Thermodynamics*, 6th edition, McGraw-Hill, New York, 1981. Excellent general text; extensive comparisons with experiments and detailed discussion of reversibility.

8 Laboratory Conditions & Free Energies

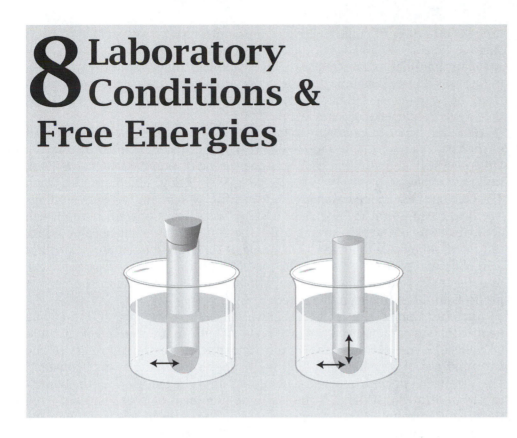

Switch from Maximum Entropy to Minimum Free Energy

In Chapter 7 we reasoned with the principle that systems tend toward states of maximum entropy. We considered systems with known or zero energy exchange across their boundaries. That logic helped explain gas expansion, particle mixing, and the interconversion of heat and work in engines. If we continued no further, we would miss much of the power of thermodynamics for physics, chemistry, and biology.

For processes in test tubes in laboratory heat baths, or processes open to the air, or processes in biological systems, it is not the *work* or *heat flow* that you control at the boundaries. It is the *temperature* and the *pressure*. This apparently slight change in conditions actually requires new thermodynamic quantities, the *free energy* and the *enthalpy*, and new extremum principles. Systems held at constant temperature do not tend toward their states of maximum entropy. They tend toward their states of *minimum free energy*.

Why Introduce New Independent Variables?

In Chapter 7, U, V, and $\mathbf{N}$ were the independent variables. Independent variables represent quantities that you can measure or control at the boundary of the system. You choose independent variables based on the type of boundaries enclosing the system. In Example 7.2 two otherwise isolated objects were

Constant Volume

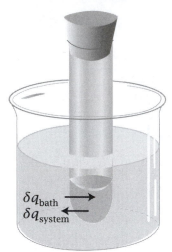

δq_{bath}
δq_{system}

Figure 8.1 A heat bath is a reservoir that holds the system (test tube, in this case) at constant temperature by allowing heat flow in or out, as required. The properties that do not change inside the system are temperature T, volume V, and particle number **N**, denoted $(T, V, \mathbf{N})$. The condition for equilibrium inside the test tube is that the Helmholtz free energy $F(T, V, \mathbf{N})$ is at a minimum.

brought into thermal contact. Because the whole system was isolated from its surroundings by an adiabatic boundary, we had $U_{\text{total}} = U_A + U_B = $ constant. Adiabatic boundaries constrain the whole system to undergo no energy change. In Carnot's engine (Example 7.9) the energy change of the system was known from the heat and work exchange across the boundaries. In these cases, U was the relevant independent variable.

However, when an *intensive* variable, such as T, p, or μ, is included as a part of the specification of a system, it means that the system is in contact with a large reservoir or 'bath' that can exchange the corresponding extensive quantity, $U, V,$ or N, respectively. Such exchanges are called 'fluctuations.' When T is constant, energy can exchange between the system and the surroundings (the heat bath), so the energy of the system fluctuates. Constant p implies an action like a piston stroke through which the system can exchange volume with the surroundings. In that case, the volume of the system fluctuates. Constant μ implies that a particle 'bath' is in contact with the system. Particles leave or enter the system to and from the particle bath. In this case, the number of particles in the system can fluctuate.

Consider a process in a *system* that we will call the *test tube*, immersed in a *heat bath*. The system need not be a real test tube in a water jacket. It could be molecules in a solvent or air in the atmosphere. 'Heat bath' refers to any surroundings of a system that hold the temperature of the system constant. This arrangement controls the temperature T, not the energy U, at the boundary around the subsystem. If the test tube plus heat bath are isolated from the greater surroundings, equilibrium will be the state of maximum entropy for the total system. However, we are not interested in the state of the total system. We are interested in what happens in the test tube itself. We need a new extremum principle that applies to the test tube, where the independent variables are $(T, V, \mathbf{N})$.

If the extremum of a function such as $S(U)$ predicts equilibrium, the variable U is called the *natural variable* of S. T is not a natural variable of S. Now we show that $(T, V, \mathbf{N})$ are natural variables of a function F, the *Helmholtz free energy*. An extremum in $F(T, V, \mathbf{N})$ predicts equilibria in systems that are constrained to constant temperature at their boundaries.

Free Energy Defines Another Extremum Principle

The Helmholtz Free Energy

Consider a process inside a test tube, sealed so that it has constant volume V and no interchange of its **N** particles with the surroundings (see Figure 8.1). A heat bath holds the test tube at constant temperature T. The process inside the test tube might be complex. It might vary in rate from a quasi-static process to an explosion. It might or might not involve chemical or phase changes. It might give off or absorb heat. Processes within the test tube will influence the heat bath only through heat exchange, because its volume does not change and no work is done.

Now we reason with the First and Second Laws of thermodynamics to find an expression that describes the condition for equilibrium in terms of the changes in the test tube alone.

Figure 8.2 A system moving *toward* its equilibrium value of a degree of freedom $x = x_{\text{equilibrium}}$ increases its entropy to $S(x) = S_{\text{max}}$.

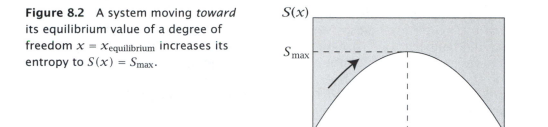

If the *combined system* (the subsystem plus the heat bath) is isolated from its surroundings, then equilibrium will be the state of maximum entropy $S(U, V, \mathbf{N})$ of the combined system. (Our strategy is to assume that the combined system is isolated. In the end, we will obtain properties of only the test tube system, so we will find that it does not matter how the bath interfaces with the greater surroundings.) Any change *toward* equilibrium must increase the entropy of the combined system, $dS_{\text{combined system}} \geq 0$ (see Figure 8.2).

Because the entropy is extensive, the entropy change of the total system is the sum of the entropy changes of its parts:

$$dS_{\text{combined system}} = dS_{\text{system}} + dS_{\text{bath}} \geq 0, \tag{8.1}$$

where the subscript 'system' indicates the test tube contents. Since the combined system is isolated,

$$dU_{\text{bath}} + dU_{\text{system}} = 0. \tag{8.2}$$

Our aim is to relate dS_{bath} to some property of the test tube system. Use the fundamental equation, $dS_{\text{bath}} = (1/T)dU + (p/T)dV - (\mu/T)dN = (1/T)dU_{\text{bath}}$ for a bath in which V and N are constant. Combine this with Equation (8.2) to get

$$dS_{\text{bath}} = -\frac{dU_{\text{system}}}{T}. \tag{8.3}$$

Substitute Equation (8.3) into Equation (8.1) to get

$$dS_{\text{system}} - \frac{dU_{\text{system}}}{T} \geq 0 \quad \Longrightarrow \quad dU_{\text{system}} - TdS_{\text{system}} \leq 0. \tag{8.4}$$

You now have an expression describing the approach to equilibrium in terms of the test tube subsystem alone. Define a quantity F, the **Helmholtz free energy**:

$$F = U - TS. \tag{8.5}$$

Its differential is

$$dF = dU - TdS - SdT$$

$$= dU - TdS \qquad \text{at constant temperature.} \tag{8.6}$$

Comparison of Equation (8.6) with (8.4) shows that when a system in which $(T, V, \mathbf{N})$ are constant is at equilibrium, the quantity F is at a minimum.

Figure 8.3 In the dimer state the two particles of a one-dimensional gas are adjacent.

Figure 8.4 In the dissociated state the two particles are not adjacent.

The definition $F = U - TS$ shows that F is determined by a balance between internal energy and entropy, and that the position of the balance is determined by the temperature. To minimize its Helmholtz free energy, the system in the test tube will tend toward *both* high entropy and low energy, depending on the temperature. At high temperatures, the entropy dominates. At low temperatures, the energy dominates. Let's return to simple lattice models to illustrate the Helmholtz free energy.

EXAMPLE 8.1 How to use free energies: a model of 'dimerization.' Suppose that $N = 2$ gas particles are contained in a test tube having a volume of V lattice sites arranged in a row, at temperature T. Under what conditions will the two particles associate into a dimer? Because systems at constant $(T, V, \mathbf{N})$ seek their states of minimum free energy, we compute the free energy of the two states—the dimer state and the dissociated monomer state—and we compare them. Whichever state has the lower free energy is the stable (equilibrium) state.

Dimer: Suppose that the dimer (shown in Figure 8.3) is held together by a 'bond energy' $U = -\varepsilon$ (where $\varepsilon > 0$). That is, when the two monomers are sitting on adjacent sites, there is an attractive energy between them. For now we will not be concerned about whether this bond is covalent or due to weaker interactions. At this level of simplicity, the model is equally applicable to many different types of bond.

On a linear lattice of V sites, there are $W_{\text{dimer}} = V - 1$ possible placements of a dimer (the first monomer could be in site 1, or 2, or 3,..., or $V - 1$, but not in site V because then the second monomer would be off the end of the lattice). Since $S = k \ln W$, the Helmholtz free energy for the dimer is

$$F_{\text{dimer}} = U_{\text{dimer}} - TS_{\text{dimer}} = -\varepsilon - kT \ln(V - 1). \tag{8.7}$$

Monomers: The two dissociated monomers (shown in Figure 8.4) have no energy of interaction. The number of ways in which two particles can be placed on V sites, *not adjacent* to each other (because adjacent placements are the dimer states) is

$$W_{\text{monomer}} = W_{\text{total}} - W_{\text{dimer}} = \frac{V!}{(2!)(V-2)!} - (V-1) = \left(\frac{V}{2} - 1\right)(V - 1),$$

and the Helmholtz free energy of the monomer-pair state is

$$F_{\text{monomer}} = U_{\text{monomer}} - TS_{\text{monomer}} = -TS_{\text{monomer}}$$

$$= -kT \ln\left[\left(\frac{V}{2} - 1\right)(V - 1)\right].$$

Figure 8.5 shows the free energies for monomers and dimer as a function of temperature. To determine which state is more stable, find the state with the lowest free energy. In this case, you choose between two options (associated or dissociated), rather than finding where a derivative is zero. Figure 8.5 shows that the dimer is the more stable state at low temperatures, while dissociation into monomers is favored at high temperatures. The monomers and dimer are equally stable (have the same free energy) at the temperature $T = T_0$:

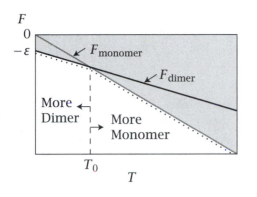

F
0
$-\varepsilon$

F_{monomer}

F_{dimer}

More Dimer   More Monomer

T_0

T

Figure 8.5 Free energy F versus temperature T for dimer association. The dotted line shows that more molecules are dimers at low temperature ($F_{\text{dimer}} < F_{\text{monomer}}$) and more are monomers at high temperature ($F_{\text{monomer}} < F_{\text{dimer}}$). For this simple model to correspond to the physical meanings of monomer and dimer, it must have more than four sites, $V > 4$.

$$-\varepsilon - kT_0 \ln(V-1) = -kT_0 \left[\ln\left(\frac{V}{2} - 1\right)(V-1) \right]$$

$$\Rightarrow \quad \varepsilon = kT_0 \ln\left(\frac{V}{2} - 1\right)$$

$$\Rightarrow \quad T_0 = \frac{\varepsilon}{k \ln((V/2) - 1)}. \tag{8.8}$$

The stronger the particle attraction (the more positive ε is), the higher is the dissociation temperature, and the more thermal energy is required to break the bond. We will develop methods in Chapter 10 to compute the relative amounts of monomer and dimer as functions of temperature.

On the one hand, we have said that equilibrium is the state of minimum free energy when T is constant. But in Figure 8.5 we have plotted the free energy versus temperature, $F(T)$. The latter is not a violation of the former. In Figure 8.5, the temperature of *each experiment* is fixed, and $F(T)$ shows the result of a series of different experiments.

Although this model is very simple, it represents the essentials of dimer dissociation and many other processes involving bond breakage, like boiling, melting, and the unfolding of polymers. The equilibrium is a balance between the energetic tendency of the particles to stick together, which dominates at low temperatures, and the entropic tendency of the particles to disperse, which dominates at high temperatures. If we had maximized the entropy instead of minimizing the free energy, we would have concluded that dimers should never form. We would have missed the importance of interaction energies in systems at lower temperatures. Here is another example of free energies.

EXAMPLE 8.2 Free energies again: a toy model of polymer collapse. When a polymer molecule is put into certain solvents (called 'poor' solvents) it collapses into a compact structure, owing to attractions similar to those that cause monomers to dimerize. For example, proteins fold to compact structures in water. Consider a two-dimensional model polymer having four monomers in a closed test tube solution in a water bath (Figure 8.6). The monomer units of the polymer chain are attracted to each other by a sticking energy $-\varepsilon$, where $\varepsilon > 0$. Suppose that

(a) Compact

ε

(b) Open

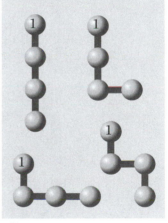

Figure 8.6 (a) The single compact (collapsed) conformation, and (b) the four open conformations of a toy model polymer. '1' indicates the first monomer. Other models distinguish mirror symmetric and rotated chains, leading to different numbers of conformations; the principles are the same.

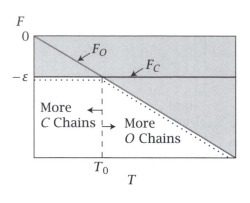

Figure 8.7 Free energy F versus temperature T for the collapse of the four-unit toy model polymer shown in Figure 8.6. Open (O) chains are more stable at high temperature, while the compact (C) chain is more stable at low temperature. T_0 is the collapse temperature.

experiments could distinguish between two conformational states: compact (Figure 8.6(a)) and open (Figure 8.6(b)). The single compact conformation ($W_C = 1$) is stabilized by the energy $U_C = -\varepsilon$ that attracts the ends to each other. The free energy of the compact state is $F_C = U_C - TS_C = -\varepsilon - kT \ln 1 = -\varepsilon$. The open state is the collection of the other 4 possible conformations, so $W_O = 4$. Open conformations have no pairs of monomers stuck together, $U_O = 0$, so $F_O = U_O - TS_O = -kT \ln 4$.

At low temperatures, the molecule is compact owing to the favorable sticking energy. At high temperatures the molecule is unfolded owing to the favorable entropy. Figure 8.7 shows the free energies of the lattice polymer versus temperature. There is a collapse temperature, $T_0 = \varepsilon/k \ln 4$, at which there are equal populations of open and compact chains. Strong sticking energy (large $\varepsilon > 0$) implies a high collapse temperature.

Now consider the collapse *process*, the transformation from open to compact states. The free energy difference for this change is $\Delta F_{\text{collapse}} = F_C(T) - F_O(T) = -\varepsilon + kT \ln 4$. The energy change for this process is $\Delta U_{\text{collapse}} = -\varepsilon$. The entropy change is $\Delta S_{\text{collapse}} = -k \ln 4$.

Because the volume of the test tube is constant, a quasi-static collapse process in this model involves no work, so the First Law gives $\Delta U = q = -\varepsilon < 0$.

According to this model, the collapse process causes heat to be given off from the test tube to the bath. Processes that give off heat are called *exothermic*. Processes that take up heat are called *endothermic*. This model collapse process is exothermic at every temperature.

The Fundamental Equation for the Helmholtz Free Energy

Just as the functional form $S(U, V, \mathbf{N})$ implies a fundamental entropy equation for dS, the form $F(T, V, \mathbf{N})$ implies a fundamental equation for dF:

$$dF = d(U - TS) = dU - TdS - SdT. \tag{8.9}$$

Substitute the fundamental energy Equation (7.4) for dU into Equation (8.9):

$$dF = \left(TdS - pdV + \sum_{j=1}^{M} \mu_j dN_j \right) - TdS - SdT$$

$$= -SdT - pdV + \sum_{j=1}^{M} \mu_j dN_j. \tag{8.10}$$

We will use Equation (8.10) later to develop Maxwell relations (see page 158). Because dF is also defined by its partial derivative expression,

$$dF = \left(\frac{\partial F}{\partial T}\right)_{V,N} dT + \left(\frac{\partial F}{\partial V}\right)_{T,N} dV + \sum_{j=1}^{M} \left(\frac{\partial F}{\partial N_j}\right)_{V,T,N_{i\neq j}} dN_j, \tag{8.11}$$

you get additional thermodynamic relations by comparing Equation (8.11) with (8.10):

$$S = -\left(\frac{\partial F}{\partial T}\right)_{V,N}, \qquad p = -\left(\frac{\partial F}{\partial V}\right)_{T,N}, \qquad \mu_j = \left(\frac{\partial F}{\partial N_j}\right)_{V,T,N_{i\neq j}}. \tag{8.12}$$

We derived $F(T,V,\mathbf{N})$ from $S(U,V,\mathbf{N})$ by physical arguments. You can also switch from one set of independent variables to another by purely mathematical arguments. These are called Legendre transforms.

Legendre Transforms

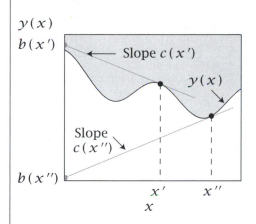

Figure 8.8 To create the Legendre transform, a function $y(x)$ is expressed as a tangent slope function $c(x)$, and a tangent intercept function $b(x)$. The tangent slopes and intercepts of points x' and x'' are shown here.

A function $y = f(x)$ can be described as a list of pairs (x_1, y_1), (x_2, y_2), ... Our aim here is to show that you can express the same function as a list of different pairs: the slopes $c(x)$ and the intercepts $b(x)$: (c_1, b_1), (c_2, b_2), ...

For a small change dx, the change dy in the function can be described by the slope $c(x)$ at that point,

$$dy = \left(\frac{dy}{dx}\right) dx = c(x)dx. \tag{8.13}$$

Figure 8.8 shows how the full function $y(x)$ (not just small changes) can be regarded as a set of slopes *and* intercepts—one slope $c(x)$ and one intercept $b(x)$ for each point x:

$$y(x) = c(x)x + b(x) \implies b(x) = y(x) - c(x)x. \tag{8.14}$$

continued

We are interested in the function that expresses the series of intercepts versus slopes, $b(c)$. To see how small changes in slope c lead to small changes in the intercept b, take the differential of Equation (8.14) and substitute Equation (8.13), $dy = c\,dx$, to get

$$db = dy - c\,dx - x\,dc = -x\,dc. \qquad (8.15)$$

EXAMPLE 8.3 Express $y(x)$ in terms of $b(c)$.   Suppose $y = 3x^2 + 5$. The slope is $c(x) = dy/dx = 6x$. From Equation (8.14), the intercept is $b(x) = 3x^2 + 5 - (6x)x = -3x^2 + 5$. Substitute $x = c/6$ into $b = -3x^2 + 5$ to get $b(c) = -c^2/12 + 5$. This is an alternative way to express $y = f(x)$. You can confirm Equation (8.15) by taking the derivative: $db = -(c/6)dc = -x\,dc$.

Now generalize to a multivariate function $y = y(x_1, x_2, x_3)$. The differential element is

$$dy = c_1 dx_1 + c_2 dx_2 + c_3 dx_3, \qquad (8.16)$$

where

$$c_1 = \left(\frac{\partial y}{\partial x_1}\right)_{x_2, x_3}, \qquad c_2 = \left(\frac{\partial y}{\partial x_2}\right)_{x_1, x_3}, \qquad \text{and} \qquad c_3 = \left(\frac{\partial y}{\partial x_3}\right)_{x_1, x_2}.$$

We want the intercept function b_1 along the x_1-axis,

$$b_1(c_1, x_2, x_3) = y - c_1 x_1. \qquad (8.17)$$

Take the differential of Equation (8.17) and substitute Equation (8.16) to get

$$db_1 = dy - c_1 dx_1 - x_1 dc_1 = -x_1 dc_1 + c_2 dx_2 + c_3 dx_3. \qquad (8.18)$$

From Equation (8.18) you can see that

$$x_1 = -\left(\frac{\partial b_1}{\partial c_1}\right)_{x_2, x_3}, \qquad c_2 = \left(\frac{\partial b_1}{\partial x_2}\right)_{c_1, x_3}, \qquad \text{and} \qquad c_3 = \left(\frac{\partial b_1}{\partial x_3}\right)_{c_1, x_2}. \qquad (8.19)$$

Any of the original independent variables x_i can be exchanged with their conjugate variables c_i in this way. The transformation can be performed on any combination of conjugate pairs, so there are a total of $2^n - 1$ possible transformations. Table 8.1 shows useful relations that can be derived from transformed fundamental equations.

The Enthalpy

We now have three useful fundamental functions for systems with fixed V and $\mathbf{N}$. Each is associated with an extremum principle: $S(U, V, \mathbf{N})$ is a maximum when U is controlled at the boundaries, $U(S, V, \mathbf{N})$ is a minimum when S is controlled at the boundaries (which we have not shown here), and $F(T, V, \mathbf{N})$ is a minimum when T is controlled at the boundaries. There are two other fundamental functions that are particularly important: the enthalpy $H(S, p, \mathbf{N})$ and the Gibbs free energy $G(T, p, \mathbf{N})$.

The enthalpy H is a function of the natural variables S, p, and $\mathbf{N}$. Enthalpy is seldom used as an extremum principle, because it is not usually convenient to control the entropy. However, the enthalpy is important because it can be obtained from calorimetry experiments, and it gives an experimental route to the Gibbs free energy, which is of central importance in chemistry and biology. To find the enthalpy, you could reason in the same way as we did for the Helmholtz free energy, but instead let's use the mathematics of Legendre transforms. Start with the internal energy $U(S, V, \mathbf{N})$. We seek to replace a dV term in the energy function with a dp term to get the enthalpy function dH. Add a pV term to the energy so that when you differentiate it, dV will be replaced by dp:

$$H = H(S, p, \mathbf{N}) = U + pV. \tag{8.20}$$

Now differentiate:

$$dH = dU + pdV + Vdp. \tag{8.21}$$

Substitute Equation (7.4), $dU = TdS - pdV + \sum_{i=1}^{M} \mu_i dN_i$, into Equation (8.21):

$$dH = TdS - pdV + \sum_{j=1}^{M} \mu_j dN_j + pdV + Vdp$$

$$\implies \quad dH = TdS + Vdp + \sum_{j=1}^{M} \mu_j dN_j. \tag{8.22}$$

If we had tried defining H with $H \overset{?}{=} U - pV$, the only other alternative for finding a function of dS, dp, and $d\mathbf{N}$, we would have failed because our alternative to Equation (8.22) would have contained a fourth variable dV. Therefore the only choice that will yield a function of only $(S, p, \mathbf{N})$ is $U + pV$. The enthalpy is a minimum at equilibrium when S, p, and $\mathbf{N}$ are the independent variables.

The Gibbs Free Energy

The *Gibbs free energy* is one of the most important fundamental functions. Constant temperature and pressure are the easiest constraints to impose in the laboratory, because the atmosphere provides them. T, p, and $\mathbf{N}$ are the natural variables for the Gibbs free energy $G = G(T, p, \mathbf{N})$, which has a minimum at equilibrium. To find the fundamental equation, start with the enthalpy, $H = H(S, p, \mathbf{N})$. Now we want to replace the dS term with a dT term in the equation $dH = TdS + Vdp + \sum_{j=1}^{M} \mu_j dN_i$. Define a function G:

$$G = H - TS. \tag{8.23}$$

The total differential dG is

$$dG = dH - TdS - SdT. \tag{8.24}$$

Substitute Equation (8.22) for dH into Equation (8.24) to get

$$dG = -SdT + Vdp + \sum_{j=1}^{M} \mu_j dN_j. \tag{8.25}$$

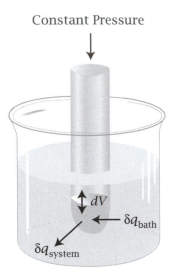

Constant Pressure

Figure 8.9 As in Figure 8.1, a heat bath holds the temperature constant. In addition, the pressure is now held constant and the volume fluctuates. In this case the quantities $(T, p, \mathbf{N})$ of the system are constant. Equilibrium results when the Gibbs free energy $G(T, p, \mathbf{N})$ is at a minimum.

The logic of the Gibbs free energy is similar to the logic of the Helmholtz free energy. If a process occurs in a test tube held at constant pressure and temperature (see Figure 8.9), it will be at equilibrium when the Gibbs free energy is at a minimum. Chemical reactions, phase changes, and biological or physical processes can take place in the test tube. The test tube exchanges energy with the surroundings by volume changes and heat transfer. Equilibrium is the state at which the entropy of the system *plus* surroundings is at a maximum. However, for the test tube system itself, which is at constant $(T, p, \mathbf{N})$, equilibrium occurs when the Gibbs free energy is at a minimum. dG can be expressed as

$$dG = \left(\frac{\partial G}{\partial T}\right)_{p,\mathbf{N}} dT + \left(\frac{\partial G}{\partial p}\right)_{T,\mathbf{N}} dp + \sum_{j=1}^{M} \left(\frac{\partial G}{\partial N_j}\right)_{p,T,N_{i\neq j}} dN_j. \tag{8.26}$$

Comparison of Equation (8.26) with (8.25) shows that:

$$S = -\left(\frac{\partial G}{\partial T}\right)_{p,\mathbf{N}}, \qquad V = \left(\frac{\partial G}{\partial p}\right)_{T,\mathbf{N}}, \qquad \mu_j = \left(\frac{\partial G}{\partial N_j}\right)_{p,T,N_{i\neq j}}. \tag{8.27}$$

For equilibrium phase changes at constant temperature, pressure, and particle number, the Gibbs free energy does not change, as shown in the next example.

EXAMPLE 8.4 Melting ice and freezing water. Consider a reversible phase change such as the melting of ice or the freezing of water in a test tube held at fixed constant pressure p_0. The phase change occurs at a temperature T_0. When the system is ice just below T_0, the free energy is at a minimum (T, p, $\mathbf{N}$ constant, $dG = 0$). Let us call this free energy G_{solid}. Now add heat. This can cause a change in phase with a negligible change in temperature. Ice melts to liquid water at a temperature just above 0 °C. The free energy of this new equilibrium is G_{liquid}. What is the free energy of melting, $\Delta G = G_{\text{liquid}} - G_{\text{solid}}$? Because the temperature, pressure, and mole numbers are unchanged ($dT = dp = dN_1 = dN_2, \ldots dN_M = 0$) inside the test tube, the free energy does not change: $\Delta G = 0$. The relative amounts of ice and water can change inside the test tube, but here $dN = 0$ means that no water molecules escape from the test tube. The enthalpy and entropy of melting balance:

$$\Delta H = T_0 \Delta S \qquad \text{when } \Delta G = 0. \tag{8.28}$$

Melting, boiling, and other phase changes involve an increase in entropy, $T_0 \Delta S$, that compensates for the increase in enthalpy ΔH that results from the breakage or weakening of interactions.

A Simple Way to Find Relationships

We now have several partial derivative relationships, such as Equations (8.27), (8.12), (7.3), and (7.6). How can you determine them when you need them? Suppose that you want to know $(\partial G/\partial T)_{p,N}$. The quantities in the denominator and in the subscripts are (T, p, N). Use Table 8.1 to find the fundamental function that applies to those constraints. In this case it is $G(T, p, N)$. Now identify the term you want in that equation. In this case, you see that $(\partial G/\partial T)_{p,N} = -S$.

Table 8.1 Fundamental equations and their natural variables.

Function	Extremum at Equilibrium	Fundamental Equation	Definition
$U(S, V, \mathbf{N})$	minimum	$dU = TdS - pdV + \sum_j \mu_i dN_i$	
$S(U, V, \mathbf{N})$	maximum	$dS = \left(\dfrac{1}{T}\right) dU + \left(\dfrac{p}{T}\right) dV - \sum_j \left(\dfrac{\mu_i}{T}\right) dN_j$	
$H(S, p, \mathbf{N})$	minimum	$dH = TdS + Vdp + \sum_j \mu_j dN_j$	$H = U + pV$
$F(T, V, \mathbf{N})$	minimum	$dF = -SdT - pdV + \sum_j \mu_j dN_j$	$F = U - TS$
$G(T, p, \mathbf{N})$	minimum	$dG = -SdT + Vdp + \sum_j \mu_j dN_j$	$G = H - TS$

The Limits on Constructing Thermodynamic Functions

What are the limits on constructing functions of T, S, p, V, $\mathbf{N}$, U, F, H, and G? You can divide thermodynamic functions into four categories:

FUNDAMENTAL AND USEFUL. Table 8.1 lists the main fundamental thermodynamic functions and their natural variables. The states of equilibrium are identified by extrema in these functions.

USEFUL BUT NOT FUNDAMENTAL. $U(T, V, \mathbf{N})$, $S(T, V, \mathbf{N})$, $H(T, p, \mathbf{N})$, and $S(T, p, \mathbf{N})$ are not functions of natural variables. These functions do not have corresponding extremum principles, but they are useful because they are components of $F(T, V, \mathbf{N})$ and $G(T, p, \mathbf{N})$.

COMPLETE BUT NOT USEFUL. Rearrangements of the dependent and independent variables from a fundamental thermodynamic function are possible, but not often useful. For example, $T(F, V, \mathbf{N})$ is a rearrangement of the fundamental Helmholtz free energy function, $F(T, V, \mathbf{N})$. This function is not very useful because you usually cannot constrain F at the system boundary.

INCOMPLETE. Additional functions could be constructed, such as $U(p, V, \mathbf{N})$ or $S(U, \mu, \mathbf{N})$ but because these involve conjugate pairs p and V, or μ and $\mathbf{N}$, and are missing other variables, they do not uniquely specify the state of a system. Such functions cannot be obtained by Legendre transforms of the fundamental equations.

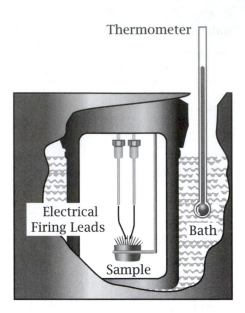

Figure 8.10 In some types of calorimeter, a measured electrical current heats a sample. A thermometer measures the corresponding change in temperature. The heat capacity is the ratio of heat input to temperature change.

Thermometer

Electrical Firing Leads

Bath

Sample

Method (3)
Functions of Non-natural Variables, Measurable Through Heat Capacities, Can Give Fundamental Functions

In this section, we describe a third way to obtain some nonmeasurable quantities from measurable ones. The first two were given in Chapter 7 (page 122–123). Although T, V, and $\mathbf{N}$ are natural variables for F, they are not natural variables for the components U and S in

$$F(T,V,\mathbf{N}) = U(T,V,\mathbf{N}) - TS(T,V,\mathbf{N}). \tag{8.29}$$

That is, minimizing $U(T,V,\mathbf{N})$ or maximizing $S(T,V,\mathbf{N})$ individually does not predict the state of equilibrium. Only their sum, in the form of F, has an extremum that identifies equilibrium if the constraints are $(T,V,\mathbf{N})$. Nevertheless, the functions $U(T,V,\mathbf{N})$ and $S(T,V,\mathbf{N})$ are useful because they are measurable by calorimetry. Similarly, the Gibbs free energy is composed of two functions of non-natural variables, $H(T,p,N)$ and $S(T,p,N)$:

$$G(T,p,\mathbf{N}) = H(T,p,\mathbf{N}) - TS(T,p,\mathbf{N}). \tag{8.30}$$

The enthalpy H has the same role in the Gibbs free energy G that the energy U has in the Helmholtz free energy F. The component quantities can be measured in calorimeters, as described in the next section.

The Heat Capacity

You can learn how the energy and entropy of a system depend on its temperature by measuring its heat capacity in a *calorimeter* (see Figure 8.10). Heat is introduced into a thermally insulated chamber, often via an electrical current, and the temperature change is measured. Some calorimeters, called *bomb calorime-*

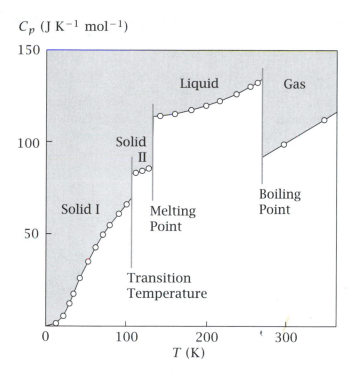

C_p (J K^{-1} mol^{-1})

Figure 8.11 The heat capacity of a material can have a complex dependence on temperature. Here is the heat capacity C_p (see Equation (8.34)) of *n*-butane in all its phases. Source: EB Smith, *Basic Chemical Thermodynamics*, 4th edition, Clarendon Press, Oxford, 1990.

ters, have fixed volume so no work occurs upon heating, $\delta w = -p\,dV = 0$, and $dU = \delta q = T\,ds$ for a quasi-static process.

The measured heat input divided by the temperature change defines the *heat capacity*, the amount of heat needed to raise the temperature of a given amount of a material by 1 K. At constant volume, the heat capacity C_V is

$$C_V = \left(\frac{\delta q}{dT}\right)_V = \left(\frac{\partial U}{\partial T}\right)_V = T\left(\frac{\partial S}{\partial T}\right)_V. \tag{8.31}$$

The definition of heat capacity given by Equation (8.31) is only an operational recipe. It does not say how the heat capacity of any particular material will depend on its volume or temperature. As shown in Figure 8.11, the ability of a material to absorb heat can be high over some ranges of temperature and low over others, or it can be flat over the temperature range of interest.

In general, to find the energy change ΔU from temperature T_A to T_B, you need to know the functional dependence of C_V on T for the material. Then you can integrate Equation (8.31):

$$\Delta U = \int_{T_A}^{T_B} C_V(T)\,dT. \tag{8.32}$$

Calorimetry can also be performed at constant pressure. Use the definition of enthalpy Equation (8.20) and the First Law to get

$$dH = d(U + pV) = dU + p\,dV + V\,dp = \delta q + \delta w + p\,dV + V\,dp.$$

For quasi-static processes ($\delta w = -p\,dV$) at constant pressure ($dp = 0$), this gives

$$dH = \delta q. \tag{8.33}$$

Table 8.2 Heat capacities at constant pressure of various substances near 25 °C. Source: I Tinoco, Jr, K Sauer, and JC Wang, *Physical Chemistry Principles and Applications in Biological Sciences*, Prentice-Hall, New York, 1978.

Gases	Molar Heat Capacities (cal deg^{-1} mol^{-1})	Liquids	Molar Heat Capacities (cal deg^{-1} mol^{-1})	Solids	Specific Heat Capacities (cal deg^{-1} g^{-1})
He	3.0	Hg	6.7	Au	0.0308
O_2	7.0	H_2O	18.0	Fe	0.106
N_2	7.0	Ethanol	27.0	C (diamond)	0.124
H_2O	7.9	Benzene	32.4	Glass (Pyrex)	0.2
CH_4	8.6	n-Heptane	53.7	Brick	~0.2
CO_2	9.0			Al	0.215
				Glucose	0.30
				Urea	0.50
				H_2O (0 °C)	0.50
				Wood	~0.5

The heat capacity C_p measured at constant pressure is:

$$C_p = \left(\frac{\delta q}{dT}\right)_p = \left(\frac{\partial H}{\partial T}\right)_p = T\left(\frac{\partial S}{\partial T}\right)_p. \tag{8.34}$$

Rearranging, you see that H and C_p have the same relation as U and C_V:

$$\Delta H = \int_{T_A}^{T_B} C_p(T)\, dT. \tag{8.35}$$

Table 8.2 shows the heat capacities of some materials. Within a given phase, heat capacities are often approximately independent of temperature.

Example 8.5 shows how you can compute the entropy change ΔS from the heat capacity.

EXAMPLE 8.5 The temperature dependence of the entropy. How can you determine $S(T)$? For a quasi-static process, Equation (8.31) gives

$$dS = \frac{C_V}{T} dT. \tag{8.36}$$

If the temperature dependence of the heat capacity is known from calorimetry, Equation (8.36) can be integrated to give the entropy at any temperature. If the heat capacity is independent of temperature, the entropy change is

$$\Delta S = \int_{T_A}^{T_B} \frac{C_V}{T} dT = C_V \ln\left(\frac{T_B}{T_A}\right). \tag{8.37}$$

If C_p is known rather than C_V, integrate $dS = (C_p/T)dT$ instead to get $\Delta S = C_p \ln(T_B/T_A)$. Notice that the entropy change in a system will be different depending on whether the pressure or the volume is held constant because in general $C_p \neq C_V$. Sometimes it is useful to know the entropy change integrated all the way from absolute zero, $T_A = 0 \, \text{K}$. The Third Law of thermodynamics defines the absolute entropy to be zero for a perfectly crystalline substance at a temperature of $0 \, \text{K}$.

Equations (8.32) and (8.37) give $\Delta U(T)$ and $\Delta S(T)$, so you can calculate $\Delta F(T)$:

$$\Delta F(T, V, \mathbf{N}) = \Delta U(T, V, \mathbf{N}) - T \Delta S(T, V, \mathbf{N}).$$

You can predict the free energy change ΔF of a system if you know ΔU and ΔS from heat capacity measurements. You can also use heat capacities to predict the equilibrium temperatures of objects in thermal contact. Let's revisit Example 7.2, in which two objects are brought into thermal contact.

EXAMPLE 8.6 The equilibrium temperature of objects in thermal contact. Suppose objects A and B have different constant-volume heat capacities, C_A and C_B, both independent of temperature. Initially, object A is colder with temperature T_A, and object B is hotter with temperature T_B. A and B are brought into thermal contact with each other, but they are isolated from the surroundings. At equilibrium, Example 7.2 shows that both objects will have the same temperature T. What is the final temperature? Because the objects are isolated from the surroundings, there is no net change in the energy of the total system:

$$\Delta U = \Delta U_A + \Delta U_B = 0. \tag{8.38}$$

The change in energy of A, ΔU_A, is

$$\Delta U_A = \int_{T_A}^{T} C_A \, dT = C_A(T - T_A), \tag{8.39}$$

because C_A is independent of temperature. Similarly, ΔU_B is

$$\Delta U_B = C_B(T - T_B). \tag{8.40}$$

A given energy change can be caused by either a smaller temperature change in a material with a larger heat capacity or a larger temperature change in a material with a smaller heat capacity. Substitute Equations (8.39) and (8.40) into (8.38) to get

$$C_A(T - T_A) + C_B(T - T_B) = 0. \tag{8.41}$$

Rearranging Equation (8.41) gives the final temperature T:

$$T = \frac{C_A T_A + C_B T_B}{C_A + C_B}.$$

If the two objects have the same heat capacity, $C_A = C_B$, the final temperature equals the average of the initial temperatures $T = (T_A + T_B)/2$. Note that heat capacities are usually given on a per mass or per volume basis, so such quantities need to be multiplied by the masses or volumes of the objects to give the total heat capacity of the object.

Method (4)
Thermodynamic Cycles
Give Some Quantities from Others

One way to get nonmeasurable quantities from measurable ones is to use thermodynamic cycles. Thermodynamic cycles are based on the principle that state functions (Chapter 5) must sum to zero around a complete cycle. (Path-dependent functions, such as work and heat, do not sum to zero around cycles. This is why engines can perform work.) Cyclic processes have great practical importance. Engines, muscles, catalysts and enzymes all undergo repeated cycles of events. However, thermodynamics also makes use of *imaginary* cycles. If you cannot determine a thermodynamic change directly, you can often compute it by constructing a fictitious cycle. Here is an example.

EXAMPLE 8.7 Measuring enthalpies under standard conditions and computing them for other conditions. Suppose you want to know the enthalpy of boiling water, $\Delta H_{\text{boil}(0\,°C)}$ at 0 °C and $p = 1$ atm. How would you obtain it? You cannot boil water at that temperature and pressure. Then why would you want to know it? We will see later that $\Delta H_{\text{boil}(0\,°C)}$ can tell us about the vapor pressure of water over a cold lake. You can get $\Delta H_{\text{boil}(0\,°C)}$ from the heat capacities of water and steam and the enthalpy of vaporization of water $\Delta H_{\text{boil}(100\,°C)}$ under more 'standard' boiling conditions ($T = 100$ °C, $p = 1$ atm) by using a simple thermodynamic cycle. The standard state enthalpy has been measured to be $\Delta H_{\text{boil}(100\,°C)} = 540$ cal g^{-1}. The heat capacity of steam is $C_p = 0.448$ cal K^{-1} g^{-1}, and the heat capacity of liquid water is $C_p = 1.00$ cal K^{-1} g^{-1}. To obtain $\Delta H_{\text{boil}(100\,°C)}$, construct the thermodynamic cycle shown in Figure 8.12.

With the directions of the arrows shown, summing to zero around a cycle means that $\Delta H_{\text{boil}(0\,°C)} = \Delta H_{\text{boil}(100\,°C)} - \Delta H_{\text{liquid}} + \Delta H_{\text{steam}}$. Because there is no phase change for the steam or liquid, and because the heat capacities are reasonably independent of temperature, you have

$$\Delta H_{\text{liquid}} = \int_{100}^{0} C_{p,\text{liquid}}\, dT = C_{p,\text{liquid}}\Delta T$$

$$= \left(1.00\,\frac{\text{cal}}{\text{K g}}\right)(-100\,\text{K}) = -100\,\text{cal g}^{-1}\ \text{and} \tag{8.42}$$

$$\Delta H_{\text{steam}} = \int_{100}^{0} C_{p,\text{steam}}\, dT = C_{p,\text{steam}}\Delta T$$

$$= \left(0.448\,\frac{\text{cal}}{\text{K g}}\right)(-100\,\text{K}) = -44.8\,\text{cal g}^{-1}.$$

Thus $\Delta H_{0\,°C} = (540 + 100 - 44.8)$ cal g^{-1} = 595.2 cal g^{-1}.

While state functions such as U, S, F, H, and G sum to zero around cycles, path-dependent functions such as q and w do not. The enormous utility of the concept of the state function is that even though a physical process can be so complex that you could not hope to know its actual physical pathway, you

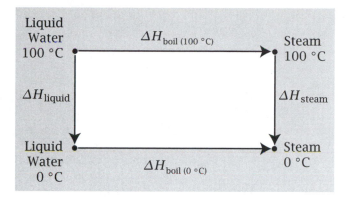

Figure 8.12 A thermodynamic cycle allows you to calculate the enthalpy for boiling water at the *freezing temperature* of water if you have measured the enthalpy at the *boiling temperature* (see Equation (8.42)).

can still often obtain exactly the quantities you need by using *any* pathway you choose in a thermodynamic cycle.

The workings of an internal combustion engine, described below, give another illustration of a thermodynamic cycle. Our first step is to focus on one part of the engine cycle: the adiabatic expansion of a gas in a piston to create work.

EXAMPLE 8.8 The quasi-static adiabatic expansion of an ideal gas. Let's start with an idealization, a gas expanding slowly in a cylinder with no heat flow, $\delta q = 0$. (Nearly adiabatic processes are common in real piston engines because the heat transfer processes are much slower than the volume changes within the cylinders.) What is the temperature change inside the cylinder as the gas expands?

In general, when the number of particles is constant, the internal energy of a gas can depend on the temperature and volume, $U = U(T, V)$, which can be expressed in differential form as

$$dU = \left(\frac{\partial U}{\partial V}\right)_T dV + \left(\frac{\partial U}{\partial T}\right)_V dT. \qquad (8.43)$$

What can be said about $U(T, V)$ for a gas? Experiments on ideal gases show that their internal energy depends on the temperature alone, not on the volume ($(\partial U/\partial V)_T = 0$). This is because at sufficiently low densities, where gases are ideal, changes in density do not much affect their weak intermolecular interactions. So we need to consider only the temperature change, and Equation (8.43) reduces to

$$dU = \left(\frac{\partial U}{\partial T}\right)_V dT = C_V dT \qquad \text{for an ideal gas at low density.} \qquad (8.44)$$

If the expansion is adiabatic, the First Law says that $dU = \delta w$. If the expansion is quasi-static, the work is given by $\delta w = -pdV$. Combining these relations with Equation (8.44) gives

$$C_V dT = dU = -pdV. \qquad (8.45)$$

Substitute the ideal gas law $p = NkT/V$ into Equation (8.45):

$$C_V dT = -\frac{NkT}{V} dV. \qquad (8.46)$$

Figure 8.13 The Otto cycle in Example 8.9.

(a) The fuel is ignited by a spark.

(b) Combustion of the fuel introduces heat into the cylinder. At this step, the combustion chamber has a small and constant volume. Burning the fuel converts it from a liquid to a gas and increases the pressure. $w_b = 0$, and $q_b = \Delta U_b = C_V(T_2 - T_1)$.

(c) The gas expands to perform work, increasing the volume and decreasing the pressure. The expansion is so rapid that it is approximately adiabatic. $q_c = 0$, and $w_c = \Delta U_c = C_V(T_3 - T_2)$.

(d) After the gas has expanded to the maximum volume V_2 allowed by the piston, the exhaust valve opens, and exhaust gases are forced out, reducing the volume and releasing heat.

(e) Fresh fuel is taken in. Little net work is performed on the cylinder to do this. For the combined steps (d) and (e), $w_{de} = 0$, and $q_{de} = \Delta U_{de} = C_V(T_4 - T_3)$.

(f) The fresh fuel is compressed by the action of the other cylinders. Like step (c), this step is so rapid that it is approximately adiabatic. $q_f = 0$, and $w_f = \Delta U_f = C_V(T_1 - T_4)$. Then the cycle repeats.

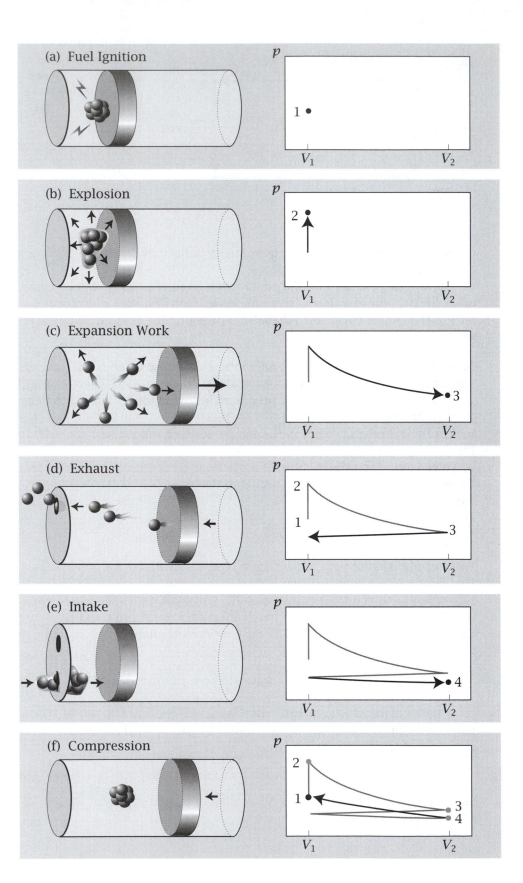

Rearrange and integrate to relate the changes in temperature and volume:

$$\int_{T_1}^{T_2} \frac{C_V dT}{T} = -\int_{V_1}^{V_2} \frac{Nk}{V} dV.$$ (8.47)

For ideal gases, heat capacities are independent of temperature, so

$$C_V \ln \left(\frac{T_2}{T_1} \right) = -Nk \ln \left(\frac{V_2}{V_1} \right).$$ (8.48)

Thus, for a quasi-static volume change in an ideal gas, the relation between temperature and volume is given by

$$\frac{T_2}{T_1} = \left(\frac{V_1}{V_2} \right)^{Nk/C_V}.$$ (8.49)

Equation (8.49) shows how the temperature decreases as the volume increases in an adiabatic ideal gas expansion.

Now let's introduce this adiabatic gas expansion into the Otto cycle, a model for a gasoline engine. We will compute the efficiency of the engine. Example 8.9 illustrates that heat and work are path-dependent functions, while energy is a state function.

EXAMPLE 8.9 How does an internal combustion engine work? Like a steam engine, an internal combustion engine is a heat engine, converting heat to work in a cyclic process. You can compute the maximum efficiency of an internal combustion engine from the intake and exhaust temperatures and the heat capacity of the fuel vapor, assuming certain idealizations: the fuel vapor is an ideal gas, and work is performed in adiabatic reversible quasi-static steps. The steps of the cycle are shown in Figure 8.13.

The efficiency η of the idealized internal combustion engine is the net work performed *by* the system $(-w)$ for the given heat influx:

$$\eta = \frac{-(w_f + w_c)}{q_b} = \frac{-C_V(T_1 - T_4) - C_V(T_3 - T_2)}{C_V(T_2 - T_1)} = 1 - \frac{T_3 - T_4}{T_2 - T_1}.$$ (8.50)

The subscripts b, c, and f in Equation (8.50) refer to the steps in the Otto cycle in Figure 8.13. The subscripts 1–4 refer to the points in Figure 8.13 at which the temperatures can be calculated. Assuming ideal adiabatic steps (c) and (f), we can compute the temperatures by using Equation (8.49),

$$\frac{T_3}{T_2} = \left(\frac{V_1}{V_2} \right)^{Nk/C_V} \quad \text{and} \quad \frac{T_4}{T_1} = \left(\frac{V_1}{V_2} \right)^{Nk/C_V},$$ (8.51)

so $T_3 = T_2(V_1/V_2)^{Nk/C_V}$ and $T_4 = T_1(V_1/V_2)^{Nk/C_V}$. Therefore the difference you need for Equation (8.50) is given by $T_3 - T_4 = (T_2 - T_1)(V_1/V_2)^{Nk/C_V}$. Substitution into (8.50) gives

$$\eta = 1 - \left(\frac{V_1}{V_2} \right)^{Nk/C_V}.$$ (8.52)

The efficiency of an Otto-cycle engine depends on V_2/V_1, known as the compression ratio, and on the heat capacity of the vaporized fuel. Typical compression ratios for gasoline engines range from four to ten. The Otto cycle is only an approximate model of internal combustion. In real engines, step (a) involves a change of mole numbers when the fuel burns, steps (b) and (d) are not quasi-static, and the expanding vapors are not ideal gases. Nevertheless, this crude model captures the general features of internal combustion engines.

Example 8.9 shows two main features of thermodynamic cycles. First, because the internal energy is a state function, it sums to zero around the cycle: You can check that $\Delta U_b + \Delta U_c + \Delta U_{de} + \Delta U_f = C_V[(T_2 - T_1) + (T_3 - T_2) + (T_4 - T_3) + (T_1 - T_4)] = 0$. Second, work and heat are not state functions and do not sum to zero around the cycle. The engine performs work on each cycle. This is evident either from computing the sum $w_b + w_c + w_{de} + w_f$, or from the graph of the pressure versus volume in Figure 8.13. Because work is the integral of $-p\,dV$, the total work is the area inside the cycle.

Summary

In this chapter, we have shown that different fundamental functions and extremum principles are required to define the state of equilibrium, depending on what quantities are known or controlled at the boundaries. Often you can control temperature T, rather than energy U, so the condition for equilibrium is that the free energy is at a minimum. Such systems have tendencies toward both low energies (or enthalpies) and high entropies, depending on T. Heat capacities, which are measurable quantities, can be used to obtain energies, enthalpies, and entropies, which in turn help to predict free energies. Thermodynamic cycles are valuable tools for computing unknown quantities from known quantities. The next chapter shows a few additional relationships for computing unknown quantities.

Problems

1. Finding a fundamental equation. While the Gibbs free energy G is the fundamental function of the natural variables $(T, p, \mathbf{N})$, growing biological cells often regulate not the numbers of molecules $\mathbf{N}$, but the chemical potentials μ_i. That is, they control *concentrations*. What is the fundamental function Z of natural variables (T, p, μ)?

2. Why does increasing temperature increase disorder? Systems become disordered as the temperature is increased. For example, liquids and solids become gases, solutions mix, adsorbates desorb. Why?

3. The difference between the energy and enthalpy changes in expanding an ideal gas. How much heat is required to cause the quasi-static isothermal expansion of one mole of an ideal gas at $T = 500\,K$ from $P_A = 0.42\,atm$, $V_A = 100$ liters to $P_B = 0.15\,atm$?

(a) What is V_B?

(b) What is ΔU for this process?

(c) What is ΔH for this process?

4. The work and the heat of boiling water. For the reversible boiling of five moles of liquid water to steam at $100\,°C$ and $1\,atm$ pressure, calculate q. Is w positive or negative?

5. The entropy and free energy of gas expansion. Two moles of an ideal gas undergo an irreversible isothermal expansion from $V_A = 100$ liters to $V_B = 300$ liters at $T = 300\,K$.

(a) What is the entropy change for this process?

(b) What is the Gibbs free energy change?

6. The free energy and entropy of membrane melting. Pure membranes of dipalmitoyl lecithin phospholipids are models of biological membranes. They melt at $T_m = 41$ °C. Reversible melting experiments indicate that $\Delta H_m = 9\,kcal\,mol^{-1}$. Calculate

(a) the entropy of melting ΔS_m, and

(b) the free energy of melting ΔG_m.

(c) Does the membrane become more or less ordered upon melting?

(d) There are 32 rotatable CH_2–CH_2 bonds in each molecule. What is the increase in multiplicity on melting one mole of bonds?

7. State- and path-dependent functions. Which quantities sum to zero around a thermodynamic cycle?

(a) q, heat

(b) w, work

(c) $-p\,dV$

(d) U

(e) G

8. Computing enthalpy and entropy with a temperature dependent heat capacity. The heat capacity for liquid *n*-butane depends on temperature:

$$C_p(T) = a + bT,$$

where $a = 100\,J\,K^{-1}\,mol^{-1}$ and $b = 0.1067\,J\,K^{-2}\,mol^{-1}$, from its freezing temperature $T_f \approx 140\,K$ to $T_b \approx 270\,K$, its boiling temperature.

(a) Compute ΔH for heating liquid butane from $T_A = 170\,K$ to $T_B = 270\,K$. $\Delta H = \int_{170}^{270}$

(b) Compute ΔS for the same process.

9. Cycle for determining the enthalpy of vaporization of water at 473 K. Suppose that you want to know how much heat it would take to boil water at $473\,K$, rather than $373\,K$. At $T = 373\,K$, $\Delta H_{boiling(100\,°C)} = 40.7\,kJ\,mol^{-1}$ is the enthalpy of vaporization. Assuming that the heat capacities of the liquid ($C_{p,liquid} = 75\,J\,K^{-1}mol^{-1}$) and the vapor ($C_{p,vapor} = 3.5\,J\,K^{-1}mol^{-1}$) are constant over this temperature range, calculate the enthalpy of vaporization at $473\,K$ using the thermodynamic cycle shown in Figure 8.14.

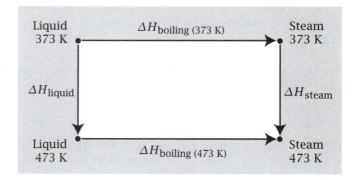

Figure 8.14 A thermodynamic cycle for calculating the enthalpy of boiling water at a temperature higher than the boiling point.

10. Heating a house. If your living room, having a volume of $6\,m \times 6\,m \times 3\,m \approx 100\,m^3$, were perfectly insulated, how much energy would be needed to raise the temperature inside the room from $T_{initial} = 0\,°C$ to $T_{final} = 25\,°C$? Note that $C_V = C_p - NR$ for an ideal gas.

11. Objects in thermal contact. Suppose two objects A and B, with heat capacities C_A and C_B and initial temperatures T_A and T_B, are brought into thermal contact. If $C_A \gg C_B$, is the equilibrium temperature T closer to T_A or to T_B?

12. ΔS for an adiabatic expansion of a gas. In an adiabatic quasi-static expansion of an ideal gas, how do you reconcile the following two facts: (1) the increase in volume should lead to an increase in entropy, but (2) in an adiabatic process, $\delta q = 0$ so there should be no change in entropy (since $dS = \delta q / T = 0$)?

13. A thermodynamic cycle for mutations in protein folding. Suppose that you can measure the stability of a wild-type protein, $\Delta G_1 = G_{\text{folded}} - G_{\text{unfolded}}$, the free energy difference between folded and unfolded states. A mutant of that protein has a single amino acid replacement. Design a thermodynamic cycle that will help you find the free energy difference $\Delta G_2 = G_{\text{unfolded, mutant}} - G_{\text{unfolded, wildtype}}$, the effect of the mutation on the unfolded state.

14. Ideal efficiency of a car engine. Suppose the compression ratio in your car engine is $V_2/V_1 = 8$. For a diatomic gas $C_V = (5/2)Nk$ and for ethane $C_V \approx 5Nk$.

(a) What is the efficiency of your engine if you burn a diatomic gas?

(b) Which is more efficient: a diatomic gas or ethane?

(c) Would your engine be more or less efficient with a higher compression ratio?

15. Free energy of an ideal gas.

(a) For an ideal gas, calculate $F(V)$, the free energy versus volume, at constant temperature.

(b) Compute $G(V)$.

16. Heat capacity of an ideal gas. The energy of an ideal gas does not depend on volume,

$$\left(\frac{\partial U}{\partial V} \right) = 0.$$

Use this fact to prove that the heat capacities $C_p = (\partial H / \partial T)_p$ and $C_v = (\partial U / \partial T)_V$ for an ideal gas are both independent of volume.

Suggested Reading

MM Abbott and HC VanNess, *Thermodynamics, Schaum's Outline Series*, McGraw-Hill, New York, 1972. Good discussion of Legendre transforms and many problems with solutions.

CJ Adkins, *Equilibrium Thermodynamics*, Cambridge University Press, Cambridge, 1983. Simple and concise.

HB Callen, *Thermodynamics and an Introduction to Thermostatics*, 2nd edition, Wiley, New York, 1985. This book is the classic development of the axiomatic approach to thermodynamics, in which energy and entropy, rather than temperature, are the primary concepts.

G Carrington, *Basic Thermodynamics*, Oxford University Press, Oxford, 1994. Describes the basics very clearly.

K Denbigh, *The Principles of Chemical Equilibrium*, Cambridge University Press, Cambridge, 1981. An excellent standard and complete text on thermodynamics.

EA Guggenheim, *Thermodynamics*, North-Holland, Amsterdam, 1978. An advanced treatment.

IM Klotz and RM Rosenberg, *Chemical Thermodynamics*, 4th edition, Benjamin-Cummings, New York, 1986. Well written, popular discussion of thermodynamics for chemists.

GN Lewis and M Randall, revised by KS Pitzer and L Brewer, *Thermodynamics*, McGraw-Hill, New York, 1961. A traditional classic text.

EB Smith, *Basic Chemical Thermodynamics*, 4th edition, Oxford University Press, Oxford, 1990. Simple and concise presentation of thermodynamics, with emphasis on experiments.

MW Zemansky and RH Dittman, *Heat and Thermodynamics*, 6th edition, McGraw-Hill, New York, 1981. A classic text for chemists and chemical engineers, with extensive discussion of experimental data.

9 Maxwell's Relations & Mixtures

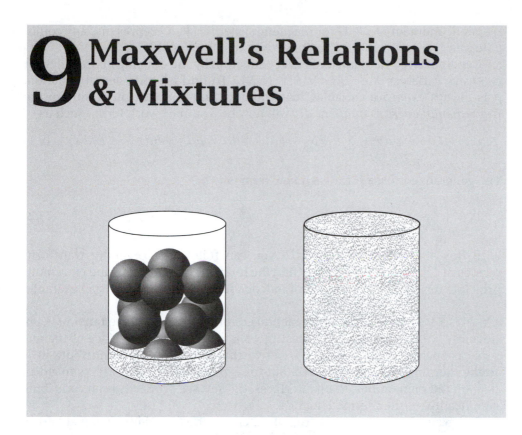

The Mathematics of Partial Derivatives Gives Two More Ways to Predict Unmeasurable Quantities

In this chapter, we introduce two more methods of thermodynamics. The first is Maxwell relations. With Maxwell relations you can find the entropic and energetic contributions to the stretching of rubber, the expansion of a surface or a film, or the compression of a bulk material. Second, we use the mathematics of *homogeneous functions* to develop the Gibbs–Duhem relationship. This is useful for finding the temperature and pressure dependences of chemical equilibria.

First we generalize our treatment of thermodynamics to allow for forces other than the pressures in pistons.

How to Design a Fundamental Equation

Here is the recipe for finding a fundamental equation and its corresponding extremum principle for any type of force. Each extensive degree of freedom in the fundamental energy equation is paired with its conjugate force. Pressure is paired with volume, temperature with entropy, and chemical potential with particle number. Other pairs include [force, length] = $[f, \ell]$ for solid or elastic materials, [surface tension, area] = $[\gamma, a]$ for liquids and interfaces, [electrical potential, charge] = $[\psi, q]$ for charged particles, and [magnetic field, magnetic moment] $= [B, I]$ for magnetic systems. $[P, X]$ represents any conjugate pair [generalized force, generalized extensive variable].

Generalize the fundamental energy equation by introducing all the relevant *extensive* variables X : $U = U(S, V, \mathbf{N}, X)$. To find a fundamental equation, it is always *extensive* variables that you add, and you add them only to the *fundamental energy equation*. dU will now be a sum of work terms such as

$$dU = TdS - pdV + \sum_j \mu_j dN_j + fdl + \gamma da + \psi dq + BdI + \sum_j P_j dX_j. \quad (9.1)$$

The generalized force P for each new term is

$$P_j = \left(\frac{\partial U}{\partial X_j} \right)_{S,V,\mathbf{N},X_{i \neq j}}. \quad (9.2)$$

With this augmented function dU, you can follow the Legendre transform recipes of Chapter 8 (page 137) to find the fundamental function and extremum principle for the appropriate set of independent variables. Here is an example.

EXAMPLE 9.1 What is the fundamental equation when surface tension is important? For experiments in test tubes and beakers, the surface of the material is such a small part of the system that surface forces contribute little to the thermodynamic properties. But for experiments on soap films in a dish pan, on cell membranes, or on interfaces, the surface properties can contribute significantly to the thermodynamics.

For example, if a spherical drop of water were distorted into a non-spherical shape with its volume held constant, its surface area would increase. This would increase its free energy. We know this because droplets spontaneously prefer to be spherical. The driving force is the surface tension.

To illustrate the design of a fundamental function, consider a process in which the surface tension is fixed and the area can vary. For example, area a can be varied at constant surface tension γ by pushing with constant force sideways on a floating bar in a dish pan (called a *Langmuir trough*) to keep a soap film corralled on one side. You want to find a function whose extremum identifies the state of equilibrium for constant T, p, $\mathbf{N}$, and γ. Begin by expressing $U = U(S, V, \mathbf{N}, a)$. The total differential dU is

$$dU = TdS - pdV + \sum_j \mu_j dN_j + \gamma da. \quad (9.3)$$

Now convert from $U(S, V, \mathbf{N}, a)$ to the independent variables of interest here, $(T, p, \mathbf{N}, \gamma)$, and find its natural function. To do this, add the following exact differentials to dU in Equation (9.3):

$$-d(TS) = -TdS - SdT,$$
$$+d(pV) = pdV + Vdp,$$
$$-d(\gamma a) = -\gamma da - ad\gamma,$$
$$\Rightarrow d(U - TS + pV - \gamma a) = -SdT + Vdp + \sum_j \mu_j dN_j - ad\gamma$$
$$= d(G - \gamma a). \quad (9.4)$$

To determine which terms are added and which are subtracted, you need to cancel terms in Equation (9.3) to give the appropriate independent variables. A system will be at equilibrium for constant T, p, $\mathbf{N}$, and y when the function $G - ya$ is at a minimum.

Here is another example. Suppose you keep the surface area constant and vary the surface tension instead.

EXAMPLE 9.2 What is the fundamental equation when *surface area* is the independent variable? Find the function whose extremum identifies the state of equilibrium for constant T, p, $\mathbf{N}$, and surface area a. In the Langmuir trough experiment, this corresponds to holding the bar at fixed position, and determining what force applied to the bar will hold it there. Again start with Equation (9.3), but do not subtract $d(ya)$ as you did in Equation (9.4).

$$dU = TdS - pdV + \sum_j \mu_j dn_j + yda, \tag{9.5}$$

$$-d(TS) = -TdS - SdT,$$
$$+d(pV) = pdV + Vdp,$$

$$\Rightarrow d(U - TS + pV) = -SdT + Vdp + \sum_j \mu_j dN_j + yda \tag{9.6}$$

$$= dG.$$

In this case the relevant function is just the Gibbs free energy itself, which now includes the term yda. For a system at constant $T, p, \mathbf{N}$, and a, the state of equilibrium is identified by a minimum in the Gibbs free energy, $G(T, p, \mathbf{N}, a)$. For example, for a water droplet held at constant $(T, p, \mathbf{N})$ to achieve its minimum free energy, $dG = yda = 0$, the droplet must adjust its shape to minimize its surface area. The shape that has a minimal surface, for a given volume of material, is a sphere.

In Chapters 7 and 8, we showed how the First Law, adiabatic boundaries, and thermodynamic cycles are used to predict nonmeasurable thermodynamic quantities from measurable ones. In the next section we show how the predictive power of thermodynamics is extended by the Maxwell relations.

Method (5)
Maxwell Relations Interrelate Partial Derivatives

We now derive the Maxwell relations. These are relationships between partial derivatives that follow from Euler's reciprocal relation, Equation (5.39) (page 75).

For example, consider the function $U(S, V)$ for fixed $\mathbf{N}$. According to the Euler Equation (5.39),

$$\left(\frac{\partial^2 U}{\partial V \, \partial S} \right) = \left(\frac{\partial^2 U}{\partial S \, \partial V} \right).$$

Equation (7.3) gives the definitions of temperature and pressure as

$$T = \left(\frac{\partial U}{\partial S}\right)_V \quad \text{and} \quad p = -\left(\frac{\partial U}{\partial V}\right)_S.$$

Substituting these definitions into the Euler relation gives

$$\left(\frac{\partial T}{\partial V}\right)_S = -\left(\frac{\partial p}{\partial S}\right)_V. \tag{9.7}$$

Equation (9.7) is a Maxwell relation.

A Recipe for Finding Maxwell Relations

Suppose that the quantity of interest to you is $(\partial S/\partial p)_{T,N}$. You might want to know this quantity because it describes how the multiplicity of microscopic states changes as you squeeze a material. This can be useful for constructing models of the forces between the atoms.

To find the Maxwell relation for any quantity, first identify what independent variables are implied. In this case the independent variables are (T, p, N) because these are the quantities that are either given as constraints in the subscript, or are in the denominator of the partial derivative. Second, find the *natural function* of these variables (see Table 8.1, page 141). For (T, p, N), the natural function is $G(T, p, N)$. Third, express the total differential of the natural function:

$$dG = -S\,dT + V\,dp + \sum_{j=1}^{M} \mu_j dN_j. \tag{9.8}$$

Fourth, based on the Euler relation, set the *cross-derivatives* equal. Set the derivative of S (from the first term on the righthand side of Equation (9.8)) with respect to p (from the second term) equal to the derivative of V (from the second term) with respect to T (from the first term). $-S$ in the first term on the right hand side of Equation (9.8) is $-S = (\partial G/\partial T)$. Take its derivative with respect to p, to get

$$-\left(\frac{\partial S}{\partial p}\right) = \left(\frac{\partial^2 G}{\partial p\,\partial T}\right). \tag{9.9}$$

The second term contains $V = (\partial G/\partial p)$. Take its derivative with respect to T, to get

$$\left(\frac{\partial V}{\partial T}\right) = \left(\frac{\partial^2 G}{\partial T\,\partial p}\right). \tag{9.10}$$

According to the Euler relation, the two second derivatives must be equal, so

$$\left(\frac{\partial S}{\partial p}\right)_{T,N} = -\left(\frac{\partial V}{\partial T}\right)_{p,N}. \tag{9.11}$$

This is the Maxwell relation you want. It gives you a quantity you cannot measure, $(\partial S/\partial p)$, from a quantity you can measure, $(\partial V/\partial T)$. Suppose that you have a block of metal or a liquid in a beaker. You measure its volume at constant pressure and plot how the volume changes with temperature. The slope $-(\partial V/\partial T)$ from that simple experiment will give $(\partial S/\partial p)_{T,N}$ through the Maxwell relationship. Or suppose that you have an ideal gas held at constant pressure. The gas volume increases with temperature, so $(\partial V/\partial T)_{p,N}$ is a positive quantity, and Equation (9.11) says that the entropy decreases with increasing pressure. Table 9.1 lists the Maxwell relations.

The next example illustrates how to get Maxwell relations when types of work other than pV changes are involved.

EXAMPLE 9.3 The thermodynamics of a rubber band. Is the retraction of a rubber band driven by a change in enthalpy or in entropy? The answer to this question will help us to construct a model for the microscopic behavior of polymeric materials in Chapter 29. Suppose you apply a quasi-static stretching force that increases the length ℓ of a rubber band. The force of retraction f exerted by the rubber band is equal and opposite to the applied stretching force. To deal with elastic forces when there is no particle exchange, we have $U = U(S, V, \ell)$ and

$$dU = TdS - pdV + fd\ell. \tag{9.12}$$

We are interested in experiments at constant T and p, so we want the Gibbs free energy $dG = d(H - TS) = d(U + pV - TS)$. Substitute Equation (9.12) into this expression to get

$$dG = -SdT + Vdp + fd\ell. \tag{9.13}$$

It follows from Equation (9.13) and the definition $G = H - TS$ that the force f can be defined in terms of enthalpic and entropic components,

$$f = \left(\frac{\partial G}{\partial \ell}\right)_{T,p} = \left(\frac{\partial H}{\partial \ell}\right)_{T,p} - T\left(\frac{\partial S}{\partial \ell}\right)_{T,p}. \tag{9.14}$$

To get a Maxwell relationship for $(\partial S/\partial \ell)_{T,p}$, take the cross-derivatives in Equation (9.13):

$$\left(\frac{\partial S}{\partial \ell}\right)_{T,p} = -\left(\frac{\partial f}{\partial T}\right)_{p,\ell}. \tag{9.15}$$

Equation (9.15) implies that you can get the entropic component of the force, $(\partial S/\partial \ell)_T$, from a very simple experiment. Hold the rubber band at a fixed stretched length ℓ (and constant pressure) and measure how the retractive force depends on the temperature (see Figure 9.1). The slope of that line, $(\partial f/\partial T)_\ell$, will give $-(\partial S/\partial \ell)_T$. The positive slope in Figure 9.1 indicates that the entropy decreases upon stretching.

How do you get the enthalpy component, $(\partial H/\partial \ell)_T$? Substituting Equation (9.15) into (9.14) and combining with Equation (9.15) gives

$$\left(\frac{\partial H}{\partial \ell}\right)_{T,p} = f - T\left(\frac{\partial f}{\partial T}\right)_{p,\ell}, \tag{9.16}$$

which you can determine from the same experiment. Figure 9.1 shows that the retraction of rubber is stronger at higher temperatures. This observation, first made in 1806 by J Gough, a British chemist, distinguishes rubber from

Table 9.1   Maxwell relations. Source: HB Callen, *Thermodynamics and an Introduction to Thermostatistics*, 2nd edition. Wiley, New York, 1985.

Fundamental Function	Variables to Relate	Maxwell Relation
$U(S,V,N)$	S,V	$(\partial T/\partial V)_{S,N} = -(\partial p/\partial S)_{V,N}$
$dU = TdS - pdV + \mu dN$	S,N	$(\partial T/\partial N)_{S,V} = (\partial \mu/\partial S)_{V,N}$
	V,N	$-(\partial p/\partial N)_{S,V} = (\partial \mu/\partial V)_{S,N}$
$F(T,V,N)$	T,V	$(\partial S/\partial V)_{T,N} = (\partial p/\partial T)_{V,N}$
$dF = -SdT - pdV + \mu dN$	T,N	$-(\partial S/\partial N)_{T,V} = (\partial \mu/\partial T)_{V,N}$
	V,N	$-(\partial p/\partial N)_{T,V} = (\partial \mu/\partial V)_{T,N}$
$H(S,p,N)$	S,p	$(\partial T/\partial p)_{S,N} = (\partial V/\partial S)_{p,N}$
$dH = TdS + Vdp + \mu dN$	S,N	$(\partial T/\partial N)_{S,p} = (\partial \mu/\partial S)_{p,N}$
	p,N	$(\partial V/\partial N)_{S,p} = (\partial \mu/\partial p)_{S,N}$
(S,V,μ)	S,V	$(\partial T/\partial V)_{S,\mu} = -(\partial p/\partial S)_{V,\mu}$
$TdS - pdV - Nd\mu$	S,μ	$(\partial T/\partial \mu)_{S,V} = -(\partial N/\partial S)_{V,\mu}$
	V,μ	$(\partial p/\partial \mu)_{S,V} = (\partial N/\partial V)_{S,\mu}$
$G(T,p,N)$	T,p	$-(\partial S/\partial p)_{T,N} = (\partial V/\partial T)_{p,N}$
$dG = -SdT + Vdp + \mu dN$	T,N	$-(\partial S/\partial N)_{T,p} = (\partial \mu/\partial T)_{p,N}$
	p,N	$(\partial V/\partial N)_{T,p} = (\partial \mu/\partial p)_{T,N}$
(T,V,μ)	T,V	$(\partial S/\partial V)_{T,\mu} = (\partial p/\partial T)_{V,\mu}$
$-SdT - pdV - Nd\mu$	T,μ	$(\partial S/\partial \mu)_{T,V} = (\partial N/\partial T)_{V,\mu}$
	V,μ	$(\partial p/\partial \mu)_{T,V} = (\partial N/\partial V)_{T,\mu}$
(S,p,μ)	S,p	$(\partial T/\partial p)_{S,\mu} = (\partial V/\partial S)_{p,\mu}$
$TdS + Vdp - Nd\mu$	S,μ	$(\partial T/\partial \mu)_{S,p} = -(\partial N/\partial S)_{p,\mu}$
	p,μ	$(\partial V/\partial \mu)_{S,p} = -(\partial N/\partial p)_{S,\mu}$
(T,p,μ)	S,p	$-(\partial S/\partial p)_{T,\mu} = (\partial V/\partial T)_{p,\mu}$
$-SdT + Vdp - Nd\mu$	T,μ	$(\partial S/\partial \mu)_{T,p} = (\partial N/\partial T)_{p,\mu}$
	p,μ	$(\partial V/\partial \mu)_{T,p} = -(\partial N/\partial p)_{T,\mu}$

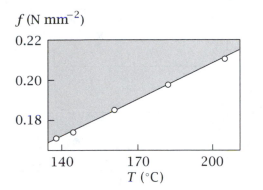

f (N mm^{-2})

Figure 9.1 The retractive force f of a rubbery polymer, amorphous polyethylene, held at constant length, as a function of temperature T. The slope is $(\partial f / \partial T)_\ell$. Source: JE Mark, A Eisenberg, WW Graessley, L Mandelkern, and JL Koenig, *Physical Properties of Polymers*, 2nd edition, American Chemical Society, Washington, DC (1993). The data are from JE Mark, *Macromo Rev*, **11**, 135 (1976).

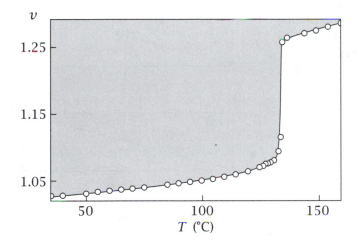

v

Figure 9.2 The specific volume v (cm^3 g^{-1}) of polyethylene versus temperature T. According to Equation (9.17), the thermal expansion coefficient α is proportional to the slope dv/dT. At low temperature, polyethylene is a hard crystalline plastic material. On melting at around 130 °C, the specific volume v increases sharply and α is large. Source: JE Mark, A Eisenberg, WW Graessley, L Mandelkern, and JL Koenig, *Physical Properties of Polymers*, 2nd edition, American Chemical Society, Washington, DC (1993). Data are from FA Quinn Jr and L Mandelkern, *J Am Chem Soc* **83**, 2857 (1961).

metal. The entropy of rubber *decreases* on stretching, while the entropy of metal *increases* on stretching. Stretching metal loosens the bonds, increasing the volume per atom, leading to translational disorder. Stretching rubber decreases the multiplicity of polymer conformations (see Example 2.4, page 34).

Changes in volume with temperature and pressure are readily measurable properties. These properties, in conjunction with the Maxwell relations, give useful insights into molecular behavior, as illustrated below.

Measuring Expansion and Compression

The thermal expansion coefficient α,

$$\alpha = \frac{1}{V} \left(\frac{\partial V}{\partial T} \right)_p, \tag{9.17}$$

is the fractional change in the volume of a system with temperature at constant pressure. For an ideal gas, Equation (9.17) gives $\alpha = (p/NkT)(Nk/p) = 1/T$.

Figure 9.2 shows the volume versus temperature for polyethylene. Thermal expansion coefficients are derived from the slopes of such plots by using

Figure 9.3 The top figure (a) shows the entropy for an ideal gas, $S(V) = Nk \ln V$. (Combining Equation (7.6) with the ideal gas law gives $(\partial S/\partial V)_{U,N} = p/T = Nk/V$. Integrate to get $S(V) = Nk \ln V$.) To get (b), take the slope of curve (a), $p/T = (\partial S/\partial V)_{U,N} = Nk/V$, and fix the pressure at a constant value, $p = p_0$. (b) is a plot of this function, $p_0/T = Nk/V$. To get from (b) to (c), invert the y-axis so you have $T(V) = p_0 V/Nk$. To get from (c) to (d), interchange the dependent and independent variables by interchanging x and y axes in (c): $V(T) = NkT/p_0$. Finally, the isothermal compressibility shown in (e) is the slope of (d), divided by V, $\alpha = V^{-1}(\partial V/\partial T)_p = (p_0/NkT)(Nk/p_0) = 1/T$.

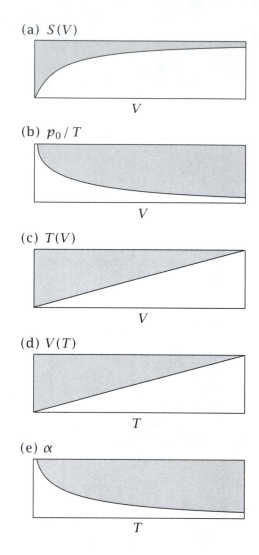

(a) $S(V)$

(b) p_0/T

(c) $T(V)$

(d) $V(T)$

(e) α

Equation (9.17). The slope of the curve below about 130 °C gives the thermal expansion coefficient for crystalline polyethylene, which is a hard plastic material. The volume expands sharply at the melting temperature. Above about 140 °C, the slope gives the thermal expansion coefficient of the plastic liquid. Thermal expansion coefficients are usually positive because increasing temperature causes a loosening up of the intermolecular bonds in the material.

The Tortuous Logic of Thermodynamics: a Graphical Perspective

We have noted in Chapters 7 and 8 that fundamental thermodynamic quantities are seldom measurable and measurable quantities are seldom fundamental. Here we use a graphical approach to illustrate the chain of thermodynamic logic starting from the fundamental but unmeasurable function $S(V)$, and working our way down to the thermodynamic expansion coefficient α, a measurable quantity. Figure 9.3 shows the steps.

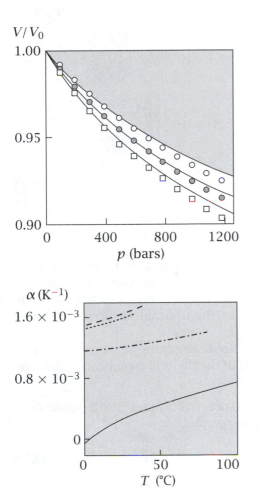

V/V_0

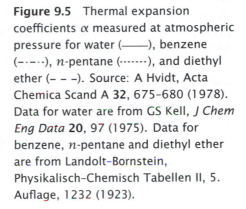

Figure 9.4 The relative volume V/V_0 of hexadecane decreases with applied pressure p (V_0 is the volume of the material extrapolated to zero pressure). The slopes of the curves give the isothermal compressibilities in the form of Equation (9.18). Compressibility increases with temperature (○ 60 °C, ◉ 90 °C, □ 120 °C). Source: RA Orwoll and PJ Flory, *J Am Chem Soc* **89**, 6814–6822 (1967).

Figure 9.5 Thermal expansion coefficients α measured at atmospheric pressure for water (——), benzene (–·–··), n-pentane (·······), and diethyl ether (– – –). Source: A Hvidt, Acta Chemica Scand A **32**, 675–680 (1978). Data for water are from GS Kell, *J Chem Eng Data* **20**, 97 (1975). Data for benzene, n-pentane and diethyl ether are from Landolt–Bornstein, Physikalisch–Chemisch Tabellen II, 5. Auflage, 1232 (1923).

The isothermal compressibility κ,

$$\kappa = -\frac{1}{V}\left(\frac{\partial V}{\partial p}\right)_T, \qquad (9.18)$$

is the fractional change in volume of a system as the pressure changes at constant temperature. For an ideal gas, $\kappa = (p/NkT)(NkT/p^2) = 1/p$.

Figure 9.4 shows the volume of hexadecane as a function of the applied pressure. Isothermal compressibilities are derived from the slopes of such plots. Because the slope of this curve is negative, Equation (9.18) shows that the isothermal compressibility is positive: increasing the applied pressure decreases the volume of hexadecane. Liquids and solids are sometimes modeled as *incompressible fluids*, $\kappa \approx 0$, where the volume is assumed to be approximately independent of the applied pressure. They are not perfectly incompressible, as Figure 9.4 shows. Nevertheless, gases are much more compressible.

Volume measurements give clues to the forces within materials. Figures 9.5 and 9.6 suggest that the bonding in liquid water is more 'rigid' than that in organic liquids. Figure 9.5 shows that organic liquids undergo larger volume expansions on heating than water does, while Figure 9.6 shows that water is

Figure 9.6 Compressibilities κ measured at atmospheric pressure for water (——), benzene (–·–··), n-pentane (·······), and diethyl ether (– – –). Source: A Hvidt, *Acta Chemica Scand A* **32**, 675–680 (1978). Data for water are from GS Kell, *J Chem Eng Data* **20**, 97 (1975). Data for benzene, n-pentane and diethyl ether are from Landolt-Bornstein, Physikalisch-Chemisch Tabellen II, 5. Auflage, 97 (1923).

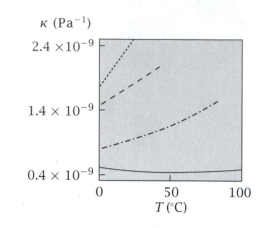

less compressible than organic liquids. Also, the compressibility κ of water changes quite unusually as a function of temperature: the compressibilities of the organics all increase between 0 and 100 °C, but water becomes less compressible between 0 and 46 °C. We examine the unusual properties of water more thoroughly in Chapters 29 and 30.

Here is an example showing how to compute an entropy change by combining a Maxwell relation and a measurement of the thermal expansion coefficient.

EXAMPLE 9.4 Entropy change with pressure. Consider the dependence of entropy on pressure, $S(p)$, at constant temperature:

$$dS = \left(\frac{\partial S}{\partial p}\right)_{T,N} dp. \tag{9.19}$$

Combine the Maxwell relation $(\partial S/\partial p)_{T,N} = -(\partial V/\partial T)_{p,N}$ with the definition of the thermal expansion coefficient given by Equation (9.17), $\alpha = V^{-1}(\partial V/\partial T)_{p,N}$, to get

$$dS = -\left(\frac{\partial V}{\partial T}\right)_{p,N} dp = -\alpha V dp. \tag{9.20}$$

Because the thermal expansion coefficients α of materials can be measured as a function of pressure, you can compute the entropy change from a series of thermal expansion experiments at different pressures:

$$\Delta S = -\int_{p_1}^{p_2} \alpha(p)V(p)\,dp. \tag{9.21}$$

Another Maxwell relation relates internal energy to volume to describe the attractive and repulsive forces within materials.

EXAMPLE 9.5 Energy change with volume. The function $(\partial U/\partial V)_T$ tells you about the cohesive forces in materials. How can you obtain it from experiments? Start with the functions $U(T,V)$ and $S(T,V)$. The differentials are

$$dU = \left(\frac{\partial U}{\partial V}\right)_T dV + \left(\frac{\partial U}{\partial T}\right)_V dT, \quad \text{and}$$

$$dS = \left(\frac{\partial S}{\partial V}\right)_T dV + \left(\frac{\partial S}{\partial T}\right)_V dT.$$

Substitute these into the fundamental energy equation $dU = TdS - pdV$ to get

$$\left(\frac{\partial U}{\partial V}\right)_T dV + \left(\frac{\partial U}{\partial T}\right)_V dT = T\left[\left(\frac{\partial S}{\partial V}\right)_T dV + \left(\frac{\partial S}{\partial T}\right)_V dT\right] - pdV.$$

When T is constant, this reduces to

$$\left(\frac{\partial U}{\partial V}\right)_T = T\left(\frac{\partial S}{\partial V}\right)_T - p.$$

Using the Maxwell relation $(\partial S/\partial V)_T = (\partial p/\partial T)_V$ (see Table 9.1) gives

$$\left(\frac{\partial U}{\partial V}\right)_T = T\left(\frac{\partial p}{\partial T}\right)_V - p.$$

The measurable quantity $(\partial p/\partial T)_V$ is called the thermal pressure coefficient. For an ideal gas, $(\partial p/\partial T)_V = Nk/V$, so $(\partial U/\partial V)_T = 0$. For typical liquids, $T(\partial p/\partial T)_V - p$ is negative at high densities and positive at low densities. This says that when you squeeze a material that is very compact, you are pushing against repulsive forces (the energy goes up as the volume decreases). And when you pull on a material having a lower density, you are pulling against attractive forces (the energy goes up as the volume increases).

Multicomponent Systems Have Partial Molar Properties

We now develop some useful thermodynamic relationships for systems having more than one chemical component, such as liquid mixtures or metal alloys. First, let's look at a simple property, the volume V of a system. Consider a system having n moles of just one type, a one-component system. The *molar volume* (the volume per mole) is $v = V/n$. Other molar properties, such as the molar free energy, $g = G/n$, are also found by dividing by n.

Partial Molar Volumes

But what if your system has more than one chemical species, with mole numbers $\mathbf{n} = n_1, n_2, \ldots, n_M$? Now each species of molecule can have a different molar volume. Equally important, the numbers of moles of each species can be specified independently, because different components can be present in different amounts. So a single quantity, the molar volume, is not sufficient to describe how the total volume depends on the components in multicomponent systems. Rather, you need the *partial molar volumes*,

$$v_j = \left(\frac{\partial V}{\partial n_j}\right)_{T,p,n_{i \neq j}}. \tag{9.22}$$

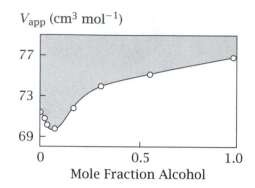

Figure 9.7 Adding alcohol to water at first decreases the total volume. But adding more alcohol increases the apparent volume V_{app}. Source: SG Bruun and A Hvidt, *Ber Bunsenges Phys Chem* **81**, 930–933 (1977).

The partial molar volume v_j describes how much the volume V of the system increases when you add dn_j moles of molecules of type j, with all other components held constant,

$$dV = \sum_{j=1}^{M} \left(\frac{\partial V}{\partial n_j} \right)_{T,p,n_{i \neq j}} dn_j = \sum_{j=1}^{M} v_j dn_j. \tag{9.23}$$

A related quantity is the *partial specific volume*, which is the derivative with respect to the *weights* w_j of species j, $(\partial V / \partial w_j)_{T,p,w_{i \neq j}}$, rather than with respect to mole numbers.

In simple cases, partial molar volumes are independent of the composition of the mixture, and are equal to the molar volumes of the individual pure components. But this is not true in general. For example, if you mix a volume V_a of alcohol with a volume V_w of water, the total volume of the mixture may not equal the sum, $V_a + V_w$. For small concentrations of alcohol, each alcohol molecule pulls water molecules to itself, so the solution volume is smaller than the sum of the volumes of the pure water and pure alcohol components.

This is illustrated in Figure 9.7 where there are two unknown quantities, the partial molar volume of the water and the partial molar volume of the alcohol. However, if you fix a quantity, say v_w, to equal the molar volume of pure water, then you can represent the volume of the alcohol in terms of a single variable, v_{app}, the *apparent* partial molar volume of the alcohol:

$$V_{app} = \frac{V - n_w v_w}{n_a},$$

where n_a is the number of moles of alcohol and n_w is the number of moles of water. At low concentrations in water, the apparent partial molar volumes of alcohols are smaller than for pure alcohols, indicating the shrinkage of the total volume when these two liquids are combined.

Ions such as magnesium sulfate have negative partial molar volumes in water. Water molecules are strongly attracted to them, through electrostatic interactions, so adding such salts to water can shrink the volume of the liquid. This is called *electrostriction*.

Figure 9.8 shows how the partial molar volume can depend on composition, using a barrel of bowling balls and adding sand. To begin, the barrel contains bowling balls and no sand. Now sand goes into the barrel. At first, the sand adds no volume to the system because it fills only the cavities between the

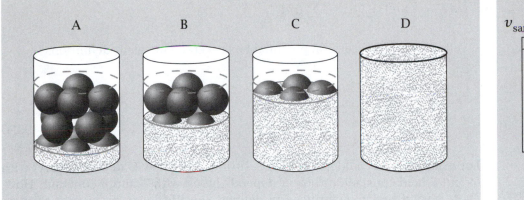
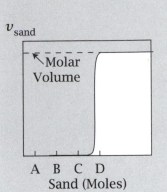

Figure 9.8 Adding sand to a barrel of bowling balls illustrates the idea of partial molar volume. At first, when the sand is at low 'concentration,' adding sand just fills in the holes between the bowling balls without increasing the barrel volume that is needed to contain the bowling balls and sand. However, when all the space between the bowling balls is filled, adding sand does add volume. At that point, D, the partial molar volume v_{sand} equals the molar volume v.

bowling balls. As long as there is not enough sand to go beyond the tops of the bowling balls, the partial molar volume of the sand equals zero because it does not change the volume of the system, which is defined by the floor and walls of the barrel and the height to which the barrel is filled. Once the sand fills all the cavities and extends above the bowling balls it begins to increase the volume of the system. Now the *partial molar* volume equals the *molar* volume of the sand. Figure 9.8(b) shows how the partial molar volume of sand depends on the amount of sand in the barrel.

Chemical Potentials Are Partial Molar Free Energies

Perhaps the most useful partial molar quantity is the chemical potential (see Chapters 13 and 16), which is the partial free energy, either per mole (if you prefer to work with numbers of moles n_j) or per molecule (if you prefer numbers of molecules N_j). Here we will use numbers of molecules N_j. The chemical potential is given in Table 8.1 in terms of each of the quantities U, F, H, and G:

$$dU = TdS - pdV + \sum_{j=1}^{M} \mu_j dN_j,$$

$$dF = -pdV - SdT + \sum_{j=1}^{M} \mu_j dN_j,$$

$$dH = TdS + Vdp + \sum_{j=1}^{M} \mu_j dN_j, \quad \text{and}$$

$$dG = -SdT + Vdp + \sum_{j=1}^{M} \mu_j dN_j.$$

It follows that the chemical potential has several equivalent definitions:

$$\mu_j = \left(\frac{\partial U}{\partial N_j}\right)_{V,S,N_{i \neq i}} = \left(\frac{\partial G}{\partial N_j}\right)_{T,p,N_{i \neq j}}$$

$$= \left(\frac{\partial F}{\partial N_j}\right)_{T,V,N_{i \neq j}} = \left(\frac{\partial H}{\partial N_j}\right)_{S,p,N_{i \neq j}}. \tag{9.24}$$

For example, you could get μ_j by measuring either how U depends on N_j with V and S held constant, or how G depends on N_j with T and p constant. However, *partial molar* quantities are defined specifically to be quantities measured *at constant T and p*. So only the derivative of the Gibbs free energy $\left(\partial G/\partial n_j\right)_{T,p,n_{i \neq j}}$ is called a *partial molar* quantity.

You can divide the partial molar free energy into its partial molar enthalpy and entropy components by using the definition $G = H - TS$:

$$\mu_j = \left(\frac{\partial G}{\partial N_j}\right)_{T,p,N_{i \neq j}}$$

$$= \left(\frac{\partial H}{\partial N_j}\right)_{T,p,N_{i \neq j}} - T\left(\frac{\partial S}{\partial N_j}\right)_{T,p,N_{i \neq j}} = h_j - Ts_j, \tag{9.25}$$

where h_j and s_j are the partial molar enthalpy and entropy, respectively.

When Equation (9.25) is used in expressions such as

$$dG = -SdT + Vdp + \sum_{j=1}^{M} (h_j - Ts_j)dN_j, \tag{9.26}$$

you see that entropy quantities appear in both the first and third terms. The first term, $-SdT$, describes how the free energy of a system changes if you change the temperature, without changing the pressure or the chemical constitution of the system. In contrast, the quantity $(h_j - Ts_j)dN_j$ describes how the free energy changes if dN_j moles of species j are transferred into the system, with both temperature and pressure held constant. Partial molar enthalpies and entropies pertain to changes in chemical composition, in the absence of thermal changes. Changes in one partial molar quantity can affect other partial molar quantities, when the quantities are 'linked.' To describe these interactions, we need to focus again on homogeneous functions.

Method (6)
Partial Molar Quantities Are Linked Through Homogeneous Functions

To describe the *linkage relationships* between partial molar quantities, we first focus on homogeneous functions. Extensive properties such as volume, energy, entropy, enthalpy, and free energy are first-order homogeneous functions of the mole numbers (see the box on page 111). Let's consider the volume. The mole numbers n_j are the independent variables. Assuming that T and p are constant, if all the mole numbers are uniformly increased by a factor λ, then the total volume of the system will increase by that factor,

$$V(\lambda n_1, \lambda n_2, \lambda n_3, \ldots, \lambda n_M) = \lambda V(n_1, n_2, n_3, \ldots, n_M). \tag{9.27}$$

Because volume is a first order homogeneous function, you have the expression (see page 111):

$$V = \sum_{i=j}^{M} n_i \left(\frac{\partial V}{\partial n_j} \right)_{T,p,n_{i \neq j}} = \sum_{j=1}^{M} n_j v_j. \tag{9.28}$$

In general, for any extensive property $Y(T, p, n_1, n_2, \ldots)$, multiplying the numbers of moles n_j or numbers of molecules N_j by a factor λ returns the function itself, multiplied by λ:

$$Y(T, p, \lambda n_1, \lambda n_2, \ldots) = \lambda Y(T, p, n_1, n_2, \ldots). \tag{9.29}$$

(T and p are intensive. They are not subject to the scaling by λ.) The *partial molar value* of Y for the component j is defined as

$$y_j(T, p, n_1, \ldots, n_M) = \left(\frac{\partial Y}{\partial n_j} \right)_{T,p,n_{i \neq j}}. \tag{9.30}$$

The partial molar quantity y_j is the change in Y that occurs when a small amount of the species j is added to a system, at constant pressure, temperature and number of moles of all other components. To derive the linkage between the partial molar volumes of two components of one system, we combine two results. First, dV is given in terms of its series expansion:

$$dV = \left(\frac{\partial V}{\partial n_1} \right)_{n_{i \neq 1}} dn_1 + \left(\frac{\partial V}{\partial n_2} \right)_{n_{i \neq 2}} dn_2 + \ldots + \left(\frac{\partial V}{\partial n_M} \right)_{n_{i \neq M}} dn_M$$

$$= \sum_{j=1}^{M} v_j dn_j. \tag{9.31}$$

Second take the derivative of Equation (9.28),

$$dV = \sum_{j=1}^{M} \left(n_j dv_j + v_j dn_j \right). \tag{9.32}$$

Setting Equations (9.31) and (9.32) equal gives a new relation:

$$\sum_{j=1}^{M} n_j dv_j = 0. \tag{9.33}$$

Equation (9.33) shows that partial molar volumes of different components, v_i and v_j, are linked; they are not independent of each other. The logic of Equations (9.31) through (9.33) also applies to chemical potentials, described in the next section.

The Gibbs–Duhem Equation

Because U is a first-order homogeneous function, you can use Equation (7.15), $\sum_{j=1}^{M} x_j (\partial f / \partial x_j) = f$. The independent variables are S, V, and $\mathbf{N}$, so the relation corresponding to Equation (7.15) is

$$U = \left(\frac{\partial U}{\partial S}\right)_{V,\mathbf{N}} S + \left(\frac{\partial U}{\partial V}\right)_{S,\mathbf{N}} V + \sum_{j=1}^{M} \left(\frac{\partial U}{\partial N_j}\right)_{S,V,N_{i \neq j}} N_j. \tag{9.34}$$

Substitute T, $-p$, and μ_j for the partial derivatives to get

$$U = TS - pV + \sum_{j=1}^{M} \mu_j N_j. \tag{9.35}$$

Taking the derivative of Equation (9.35) gives

$$dU = TdS + SdT - pdV - Vdp + \sum_{j=1}^{M} \mu_j dN_j + \sum_{j=1}^{M} N_j d\mu_j. \tag{9.36}$$

Subtract the fundamental energy Equation (7.4), $dU = TdS - pdV + \sum_{j=1}^{M} \mu_j dN_j$, from Equation (9.36) to get the **Gibbs–Duhem** equation:

$$\sum_{j=1}^{M} N_j d\mu_j = Vdp - SdT$$

$$\implies \sum_{j=1}^{M} N_j d\mu_j = 0 \qquad \text{at constant } T, p. \tag{9.37}$$

Equations (9.37) give the relations between the chemical potentials of the different species in a system. The Gibbs–Duhem equation helps in understanding how molecular interactions are coupled in osmosis, ligand binding, and phase transitions, and in describing how chemical potentials depend on temperature and pressure.

Summary

We have explored two methods of thermodynamics. The Maxwell relations derive from the Euler expression for state functions. They provide another way to predict unmeasurable quantities from measurable ones. For multicomponent systems, you get another relationship from the fact that many thermodynamic functions are homogeneous. In the following chapters we will develop microscopic statistical mechanical models of atoms and molecules.

Problems

1. How do thermodynamic properties depend on surface area? The surface tension of water is observed to decrease linearly with temperature (in experiments at constant p and a): $\gamma(T) = b - cT$, where T = temperature °C, $b = 75.6$ erg cm^{-2} (the surface tension at 0 °C) and $c = 0.1670$ erg cm^{-2}deg^{-1}.

  (a) If γ is defined by $dU = TdS - pdV + \gamma da$, where da is the area change of a pure material, give γ in terms of a derivative of the Gibbs free energy at constant T and p.

  (b) Using a Maxwell relation, determine the quantitative value of $(\partial S / \partial a)_{p,T}$ from the relationships above.

  (c) Estimate the entropy change ΔS from the results above if the area of the water/air interface increases by 4 Å^2 (about the size of a water molecule).

2. Water differs from simple liquids. Figures 9.5 and 9.6 show that the thermal expansion coefficient $\alpha = (1/V)(\partial V / \partial T)_p$ and isothermal compressibility $\kappa = -(1/V)(\partial V / \partial p)_T$ are both much smaller for water, which is hydrogen-bonded, than for simpler liquids like benzene, which are not. Give a physical explanation for what this implies about molecular packing and entropies in water versus simple liquids.

3. The heat capacity of an ideal gas. For an ideal gas $(\partial U / \partial V)_T = 0$. Show that this implies the heat capacity C_V of an ideal gas is independent of volume.

4. Using Maxwell relations. Show that $(\partial H / \partial p)_T = V - T(\partial V / \partial T)_p$.

5. Constant volume versus constant pressure: pV effects are usually small in liquids and solids.

  (a) Show that the free energy per molecule for inserting a water molecule into a liquid is about the same for a constant-volume process as for a constant-pressure process, $(\partial F / \partial N)_{T,p} \approx (\partial F / \partial N)_{T,V}$.

  (b) Show that for an ideal gas, these two quantities differ by kT.

6. Pressure dependence of the heat capacity.

  (a) Show that, in general, for quasi-static processes
  $$\left(\frac{\partial C_p}{\partial p} \right)_T = -T \left(\frac{\partial^2 V}{\partial T^2} \right)_p .$$

  (b) Show that $(\partial C_p / \partial p)_T = 0$ for an ideal gas.

7. Relating G to μ. Prove that $G = \sum_{i=1}^{N} \mu_i N_i$.

8. Piezoelectricity. Apply a mechanical force f along the x-axis of a piezoelectric crystal, such as quartz, which has dimension ℓ in that direction. The crystal will develop a polarization p_0, a separation of charge along the x-axis, positive on one face and negative on the other. Applying an electric field E along the x-axis causes a mechanical deformation. Such devices are used in microphones, speakers, and pressure transducers. For such systems the energy equation is $dU = TdS + fd\ell + Edp_0$. Find a Maxwell relation to determine $(\partial p_0 / \partial f)_{T,E}$.

9. Relating C_V and C_p. Show that $C_p = C_V + Nk$ for an ideal gas.

10. Rubber bands are entropic springs. Experiments show that the retractive force f of polymeric elastomers as a function of temperature T and expansion ℓ is approximately given by $f(T, \ell) = aT(\ell - \ell_0)$ where a and ℓ_0 are constants.

  (a) Use Maxwell relations to determine the entropy and enthalpy, $S(\ell)$ and $H(\ell)$, at constant T and p.

  (b) If you adiabatically stretch a rubber band by a small amount its temperature increases, but its volume does not change. Derive an expression for its temperature T as a function of ℓ, ℓ_0, a, and its heat capacity $C = (\partial U / \partial T)$.

11. Metal elasticity is due to energy, not entropy. Experiments show that the retractive force f of metal rods as a function of temperature T and extension ℓ relative to undeformed length ℓ_0 is given by $f(T, \ell) = Ea\Delta\ell/\ell_0$, where $\Delta\ell = \ell[1 - \alpha(T - T_0)] - \ell_0 = \ell - \ell\alpha(T - T_0) - \ell_0$. a is the cross-sectional area of the rod, E (which has the role of a spring constant) is called *Young's modulus*, and $\alpha \approx 10^{-5}$ is the linear expansion coefficient. Compute $H(\ell)$ and $S(\ell)$. Is the main dependence on ℓ due to enthalpy H or entropy S?

12. Graphical interpretation of compressibility. Following the same type of graphical procedure as in Figure 9.3, derive the isothermal compressibility κ from $S(V) = Nk \ln V$ for an ideal gas.

Suggested Reading

RS Berry, SA Rice, and J Ross, *Physical Chemistry*, Wiley, New York, 1980. A complete and advanced text on physical chemistry including thermodynamics.

HB Callen, *Thermodynamics and an Introduction to Thermostatistics*, 2nd edition, Wiley, New York, 1985. Very clearly written discussion of the Maxwell relationships.

G Carrington, *Basic Thermodynamics*, Oxford University Press, Oxford, 1994. Good discussion of the principles of thermodynamics, including many applications to models and equations of state.

RT DeHoff, *Thermodynamics in Materials Science*, McGraw-Hill, New York, 1993. A useful practical guide for finding thermodynamic relations by simple recipes.

10 The Boltzmann Distribution Law

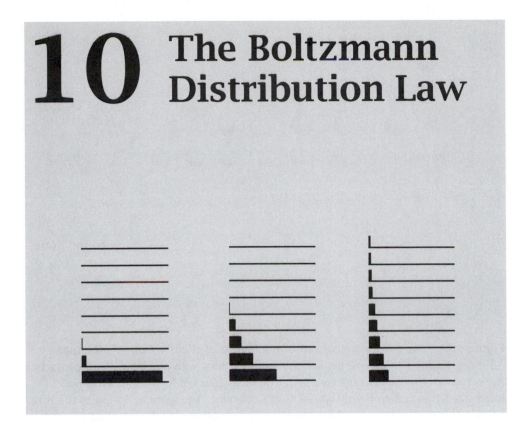

Statistical Mechanics Gives Probability Distributions for Atoms and Molecules

Now we begin statistical mechanics, the modeling and prediction of the properties of materials from the structures of the atoms and molecules that comprise them. The core of statistical mechanics is modeling the probability distributions of the energies of atoms and molecules, because the various averages over those distributions are what experiments measure. For example, to compute the properties of gases, you need the distributions of their energies and velocities (Chapter 11). You can predict chemical reaction equilibria if you know the distributions of the energies of the reactants and products (Chapter 13). And you can predict the average number of ligands bound to a DNA molecule if you know the distribution of energies of all the ligation states (Chapter 28).

The central result of this chapter is the Boltzmann distribution law, which gives probability distributions from the underlying energy levels. We derive this result by bringing together the two main threads from earlier chapters. First, the principles of thermodynamics in Chapters 7–9 describe how to predict the state of equilibrium. Second, Chapter 6 relates a macroscopic property of the equilibrium (the entropy) to a microscopic property, the probability distribution.

Here's the kind of problem we want to solve. Example 8.2 (page 135) describes a two-dimensional model of a four-bead polymer chain that has four open conformations and one compact conformation. In that example, we com-

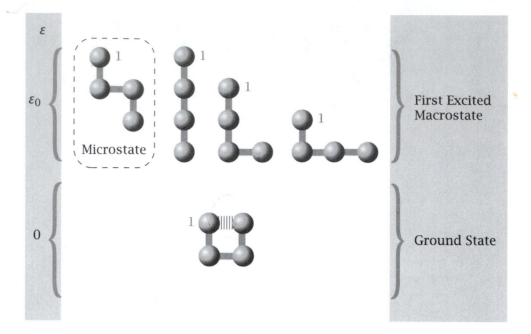

Figure 10.1 The five conformations of the four-bead chain in Example 8.2 (page 135) are grouped on an energy ladder. Conformations with one bead–bead contact, $n_C = 1$, are taken to have energy $\varepsilon = 0$. The other four conformations have no bead–bead contacts so $\varepsilon = \varepsilon_0$, where ε_0 is a constant. The number 1 next to the first bead indicates that the chain has a head and a tail; otherwise the counting of conformations is a little different.

puted the free energies of the open and compact states. Now we want to compute the probability distribution, the fraction of molecules that are in each conformation.

To begin, we need to define the system and its *energy levels*. Consider one four-bead chain. The four-bead chain has two energy levels (see Figure 10.1). Each energy level represents the number of bead–bead contacts that the chain can make. Let's use the convention that zero is the lowest energy and is the lowest rung on the ladder. Conformations with the maximum number of bead–bead contacts have zero energy. Breaking a contact increases the energy by an amount $\varepsilon = \varepsilon_0 > 0$. We use ε to indicate the energy in general and ε_0 to represent some particular constant value. We seek the distribution of probabilities, $p_1, p_2, \ldots, p_5$, that the four-bead chain is in each one of its conformations.

The state of lowest energy is called the *ground state*. States of higher energy are called *excited states*. Each of the five configurations is called a *microstate*, to distinguish it from a *state* or *macrostate*, which is a collection of microstates. Each microstate is a snapshot of the system. When you measure system properties, you measure averages over multiple microstates. In Figure 10.1, you can see two macrostates: the open macrostate, composed of four microstates, and the compact macrostate, composed of one microstate. For a lattice gas, a particular arrangement of N particles on M lattice sites is one microstate. A corresponding macrostate could include, for example, all such arrangements of particles that have the same density N/M.

We will compute the probability distribution for this four-bead chain after we derive the Boltzmann law. But first, let's look at a more general version of the problem that we are trying to solve. Figure 10.2 shows a complex system having N particles. Each particle has some *internal* energy due to its intrinsic properties. For example, different internal energies result from different rotational or vibrational states of molecules, or from the different conformations like the four-bead polymer molecule. In addition, there may be *interaction* energies between pairs or triples or larger collections of particles. The sum total of all the internal and interaction energies, taken for one particular arrangement j of the whole system, is E_j. The lowest energy over all the possible arrangements of the system is E_1. The next lowest energy arrangement will have energy E_2, etc. We will use two different notations for the energy: E_j for any system in general, no matter how complex, and ε_j for the specific simple case of independent particles (such as ideal gas atoms).

Now we derive the probability distribution for any system with known energy level spacings. We do this by combining the definition of entropy, $S/k = -\sum p_i \ln p_i$, with the definition of equilibrium. Our approach follows the dice problems of Chapter 6, but now instead of knowing an average score, we take the average energy of the system.

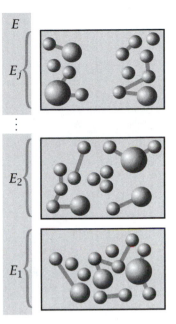

Figure 10.2 A complex system. In one configuration, it has the lowest possible energy E_1. Another configuration has energy E_2, etc.

The Boltzmann Distribution Law Describes the Equilibria Among Atoms and Molecules

Consider a system having N particles. (To make the math simple in this chapter, consider N particles of a single type, rather than a set of $\mathbf{N} = (N_1, N_2, \ldots, N_M)$ particles of multiple types.) Suppose the system has t different energy levels, labelled $E_j, j = 1, 2, \ldots, t$. The energy levels are defined by the physics of the system and the problem that you want to solve. Energies may be equally spaced, as in the polymer problem above, or they may be spaced differently. They may come from quantum mechanics, as we'll see in Chapter 11. Given the energies E_j, we aim to compute the probabilities p_j that the system is in each microstate j. Suppose (T, V, N) are held constant. Then the condition for equilibrium is $dF = dU - TdS = 0$. Apply the Lagrange multiplier method as in Chapter 6. We need dS and dU.

To get dS, use Equation (6.2) for the entropy as a function of the probabilities p_j:

$$\frac{S}{k} = -\sum_{j=1}^{t} p_j \ln p_j.$$

Differentiating with respect to p_j (holding the $p_{i \neq j}$'s constant) gives

$$dS = -k \sum_{j=1}^{t} (1 + \ln p_j) dp_j. \tag{10.1}$$

To get dU, we postulate that the internal energy U, which is the macroscopic quantity from thermodynamics, is the average over the microscopic energy levels (see Equation (1.35)),

$$U = \langle E \rangle = \sum_{j=1}^{t} p_j E_j. \tag{10.2}$$

Now take the derivative of Equation (10.2):

$$dU = d\langle E \rangle = \sum_{j=1}^{t} (E_j dp_j + p_j dE_j). \tag{10.3}$$

Like the macroscopic energy U, the microscopic energy levels $E_j = E_j(V, N)$ depend on V and N. But unlike U, the microscopic energy levels E_j do not depend on S or T. We take as a fundamental principle of quantum mechanics that only the populations $p_j(T)$, and not the energy levels E_j, depend on temperature. In this way, however, the *average energy* $\langle E \rangle = \sum p_j(T) E_j$ does depend on temperature. $dE_j = (\partial E_j / \partial V) dV + (\partial E_j / \partial N) dN = 0$ because both V and N are held constant here, and Equation (10.3) becomes

$$d\langle E \rangle = \sum_{j=1}^{t} E_j dp_j. \tag{10.4}$$

The First Law gives $dU = \delta q + \delta w$, which reduces to $d\langle E \rangle = dU = \delta q$ when V and N are constant. Because Equation (10.4) applies when V is constant, it follows that the term $\sum_j E_j dp_j$ is the heat, and $\sum_j p_j dE_j$ is the work.

We want the probability distribution that satisfies the equilibrium condition $dF = d\langle E \rangle - T dS = 0$ subject to the constraint that the probabilities sum to one, $\sum_{j=1}^{t} p_j = 1$. The constraint can be expressed in terms of a Lagrange multiplier α,

$$\alpha \sum_{j=1}^{t} dp_j = 0. \tag{10.5}$$

Substitute Equations (10.1), (10.3), (10.4), and (10.5) into $dF = dU - T dS = 0$ to get

$$dF = \sum_{j=1}^{t} \left[E_j + kT(1 + \ln p_j^*) + \alpha \right] dp_j^* = 0. \tag{10.6}$$

According to the Lagrange multiplier Equation (5.31), the term in the brackets in Equation (10.6) must equal zero for each value of j, so you have t equations of the form

$$\ln p_j^* = -\frac{E_j}{kT} - \frac{\alpha}{kT} - 1. \tag{10.7}$$

Exponentiate Equation (10.7) to find

$$p_j^* = e^{-E_j/kT} e^{(-\alpha/kT) - 1}. \tag{10.8}$$

To eliminate α from Equation (10.8), write the constraint equation

$$\sum_{j=1}^{t} p_j^* = 1 \quad \text{as} \quad 1 = \sum_{j=1}^{t} e^{-E_j/kT} e^{(-\alpha/kT)-1}.$$

Divide Equation (10.8) by this form of the constraint equation to get the **Boltzmann distribution law**:

$$p_j^* = \frac{e^{-E_j/kT}}{\sum_{j=1}^{t} e^{-E_j/kT}} = \frac{e^{-E_j/kT}}{Q}, \tag{10.9}$$

where Q is the **partition function**:

$$Q = \sum_{j=1}^{t} e^{-E_j/kT}. \tag{10.10}$$

The relative populations of particles in microstates i and j at equilibrium are given by

$$\frac{p_i^*}{p_j^*} = e^{-(E_i - E_j)/kT}. \tag{10.11}$$

Equation (10.9) gives an exponential distribution law, just as Equation (6.16) on page 87 does, but here the energy levels E_j replace the scores on die rolls.

The Boltzmann distribution says that more particles will have low energies and fewer particles will have high energies. Why? Particles don't have an intrinsic preference for lower energy levels. Fundamentally, all energy levels are equivalent. Rather, *there are more arrangements of the system that way.* It is extremely unlikely that one particle would have such a high energy that it would leave all the others no energy. There are far more arrangements in which most particles have energies that are relatively low, but non-zero. If each particle takes only a small fraction of the total energy, it leaves a great many more ways for the other particles to distribute the remaining energy.

Applications of the Boltzmann Law

Example 10.1 illustrates an application of the Boltzmann distribution law. We compute how the atmospheric pressure depends on the altitude above the earth's surface.

EXAMPLE 10.1 Barometric pressure of the atmosphere. The energy ε of a gas molecule in the earth's gravitational field is a function of altitude z:

$$\varepsilon(z) = mgz, \tag{10.12}$$

where g is the gravitational constant and m is the molecular mass. In this case, the energy is a continuous function (of z), not a discrete ladder, but Boltzmann's law still applies. We assume that the atmosphere is in equilibrium (valid only approximately) and is at constant temperature. The number

Figure 10.3 The temperature T (——) of the atmosphere is approximately constant up to about 100 km in altitude z. The pressure p (– – –) decreases exponentially with altitude, following the Boltzmann law. Source: ML Salby, *Fundamentals of Atmospheric Physics*, Academic Press, San Diego, 1996. Data are from: *US Standard Atmosphere*, NOAA, US Air Force, US Government Printing Office, NOAA-S/T 76-1562, Washington, DC, 1976.

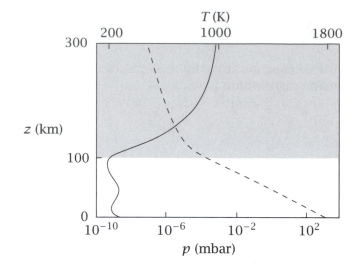

of molecules $N(z)$ at altitude z relative to the number $N(0)$ at sea level is given by the Boltzmann law,

$$\frac{N(z)}{N(0)} = e^{-\varepsilon(z)/kT} = e^{-mgz/kT}. \tag{10.13}$$

If the gas is ideal and the temperature is constant, the pressure $p(z)$ is proportional to the number of molecules per unit volume, so the pressure should decrease exponentially with altitude:

$$\frac{p(z)}{p(0)} = \frac{N(z)kT/V}{N(0)kT/V} = \frac{N(z)}{N(0)} = e^{-mgz/kT}. \tag{10.14}$$

Figure 10.3 shows experimental evidence that the temperature is reasonably constant for the earth's atmosphere up to about 100 km above the earth's surface, and it shows that the pressure decreases exponentially with altitude, as predicted. Above about 100 km, the equilibrium assumption no longer holds because the atmosphere becomes too thin for normal wind turbulence to mix the gases, and the temperature is no longer independent of altitude.

Example 10.2 gives another application of the Boltzmann distribution law, the distribution of the velocities of gas molecules. This is the basis for the kinetic theory of gases, a classical model of great historical importance.

EXAMPLE 10.2 The Maxwell–Boltzmann distribution of particle velocities. According to the kinetic theory, gases and liquids can be regarded as Newtonian particles having mass m, velocity v, and kinetic energy ε, where

$$\varepsilon(v) = \frac{1}{2}mv^2.$$

According to the Boltzmann law, the probability $p(v_x)$ that a particle in a container at constant volume and temperature will have velocity v_x in the x-direction is

$$p(v_x) = \frac{e^{-\varepsilon(v_x)/kT}}{\displaystyle\int_{-\infty}^{\infty} e^{-\varepsilon(v_x)/kT}\, dv_x} = \frac{e^{-mv_x^2/2kT}}{\displaystyle\int_{-\infty}^{\infty} e^{-mv_x^2/2kT}\, dv_x}$$

$$= \left(\frac{m}{2\pi kT}\right)^{1/2} e^{-mv_x^2/2kT}, \tag{10.15}$$

because $\int_{-\infty}^{\infty} e^{-ax^2}\, dx = \sqrt{\pi/a}$ (see Appendix D, Equation (D.1)) and $a = m/2kT$. This is called *the Maxwell–Boltzmann distribution*. Figure 10.4 shows the excellent agreement between this predicted distribution and experimental observations for the velocities of potassium atoms. The mean velocity is zero, because for any particular velocity v_x in the $+x$ direction, there is a corresponding velocity in the $-x$ direction.

The *mean-square* velocity $\langle v^2 \rangle$ defines the width of the distribution:

$$\langle v_x^2 \rangle = \int_{-\infty}^{\infty} v_x^2 p(v_x)\, dv_x = \left(\frac{m}{2\pi kT}\right)^{1/2} \int_{-\infty}^{\infty} v_x^2 e^{-mv_x^2/2kT}\, dv_x.$$

Because $\int_0^{\infty} x^2 e^{-ax^2}\, dx = (1/(4a))\sqrt{\pi/a}$ (see Appendix D, Equation (D.3)) and because $\int_{-\infty}^{\infty} = 2\int_0^{\infty}$, you have

$$\langle v_x^2 \rangle = \frac{kT}{m} \quad \Longrightarrow \quad \frac{1}{2} m \langle v_x^2 \rangle = \frac{1}{2} kT.$$

Now suppose that instead of measuring the x-component of velocity, you measure the three-dimensional velocity v, where $v^2 = v_x^2 + v_y^2 + v_z^2$. In an ideal gas, the components are independent of each other, so $\langle v^2 \rangle = \langle v_x^2 \rangle + \langle v_y^2 \rangle + \langle v_z^2 \rangle$, leading to:

$$\frac{1}{2} m \langle v^2 \rangle = \frac{3}{2} kT. \tag{10.16}$$

Because the velocity components are independent, the probabilities multiply, giving

$$p(v) = p(v_x)p(v_y)p(v_z) = \left(\frac{m}{2\pi kT}\right)^{3/2} e^{-m(v_x^2 + v_y^2 + v_z^2)/2kT}$$

$$= \left(\frac{m}{2\pi kT}\right)^{3/2} e^{-mv^2/2kT}. \tag{10.17}$$

Equations (10.16) and (10.17) are the central results of the kinetic theory of gases, describing the motions of atoms in terms of the kinetic energies of Newtonian particles. They provide a fundamental relationship between temperature and the velocities of the gas molecules.

What Does a Partition Function Tell You?

The partition function is the connection between macroscopic thermodynamic properties and microscopic models. It is a sum of *Boltzmann factors* $e^{-E_j/kT}$ that specify how *particles are partitioned* throughout the accessible states. For many typical problems, the ground state has zero energy, $E_1 = 0$. In those

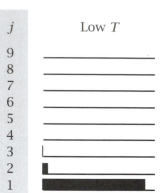

j	Low T
9	
8	
7	
6	
5	
4	
3	
2	
1	

Figure 10.4 Experiments showing $v^2 p(v) = v^2 \exp(-mv^2/2kT)$ versus average speed v, confirming the Maxwell–Boltzmann distribution of the speeds of potassium atoms in the gas phase. Source: DA McQuarrie, JD Simon, *Physical Chemistry a Molecular Approach*, University Science Books, Sausalito, 1997. Data are from RC Miller and P Kusch, *Phys Rev* **99**, 1314–1320 (1953).

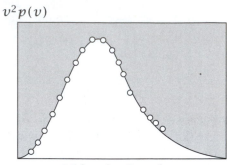

$v^2 p(v)$

Molecular Speed, v

cases, Equation (10.10) becomes

$$Q = \sum_{j=1}^{t} e^{-E_j/kT} = 1 + e^{-E_2/kT} + e^{-E_3/kT} + \cdots + e^{-E_t/kT}. \qquad (10.18)$$

The partition function Q accounts for the number of states that are *effectively* accessible to the system. To see this, look at the limiting values (see Figures 10.5 and 10.6). When the energies are small, or the temperature is high, all the states become equally populated:

$$\left. \begin{array}{c} E_j \to 0 \\ \text{or} \\ T \to \infty \end{array} \right\} \quad \Longrightarrow \quad \frac{E_j}{kT} \to 0 \quad \Longrightarrow \quad p_j^* \to \frac{1}{t} \quad \Longrightarrow \quad Q \to t. \qquad (10.19)$$

In this case, all t states become accessible.

At the other extreme, as the energy intervals become large or as the temperature approaches zero, the particles occupy only the ground state:

$$\left. \begin{array}{c} E_j \to \infty \\ \text{or} \\ T \to 0 \end{array} \right\} \quad \Longrightarrow \quad \frac{E_j}{kT} \to \infty \quad \Longrightarrow \quad \left\{ \begin{array}{c} (p_1^*) \to 1 \\ \text{and} \\ (p_j^*)_{j \neq 1} \to 0 \end{array} \right. \quad \Longrightarrow \quad Q \to 1. \qquad (10.20)$$

In this case, only the ground state becomes accessible.

The magnitude of E_j/kT determines whether or not the state j is effectively accessible. So kT is an important reference unit of energy. States that have energies higher than kT are relatively inaccessible and unpopulated at temperature T, while states having energies lower than kT are well-populated. Increasing the temperature makes the higher energy levels effectively more accessible.

The number of *effectively accessible states* is not the same as the number of accessible states, which is always t. The number t is fixed by the underlying physics of the system. In contrast, the effective accessibility Q also depends on the temperature.

j	Medium T
9	
8	
7	
6	
5	
4	
3	
2	
1	

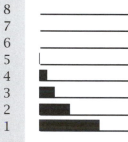

j	High T
9	
8	
7	
6	
5	
4	
3	
2	
1	

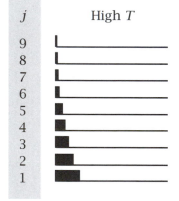

Figure 10.5 The Boltzmann distribution law. At low temperatures T, only the lowest energy states j are populated. Increasing T increases populations of higher energy states. The distribution is exponential.

The Density of States

In macroscopic systems, the number of microstates might be 10^{30}. Usually experiments cannot distinguish whether a system is in one microstate or another.

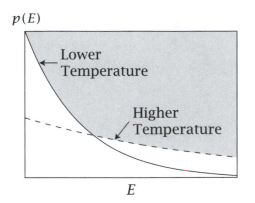

$p(E)$

Lower Temperature

Higher Temperature

E

Figure 10.6 According to the Boltzmann distribution, states of lower energy are more populated than states of higher energy. As temperature increases, higher energy states become more populated.

So it is often useful to switch your attention from microstates to macrostates. For example, for the four-bead polymer, each configuration shown in Figure 10.1 is one microstate. But a typical experiment might only tell you whether the chain is *open* (four microstates) or *compact* (one microstate). Then the terms 'open' and 'compact' define macrostates. The *density of states* $W(E)$ is the number of microstates in a given macrostate. When $W > 1$, an energy level is called *degenerate*.

For the four-bead chain, there is one compact conformation and four open conformations (see Figure 10.1), so the density of states is $W(0) = 1$ and $W(\varepsilon_0) = 4$. Since W is the number of microstates per level, the partition function can be expressed as a sum over the two levels (open and compact, in this case), rather than over the five microstates,

$$Q = 1e^{-0/kT} + 4e^{-\varepsilon_0/kT},$$

where the first term describes the one microstate in level 1 ($\varepsilon = 0$) and the second describes the four microstates in level 2 ($\varepsilon = \varepsilon_0$).

In general, when you prefer to focus on energy level E_ℓ, rather than on microstate j, you can express the partition function as a sum over energy levels $\ell = 1, 2, \ldots, \ell_{\max}$,

$$Q = \sum_{\ell=1}^{\ell_{\max}} W(E_\ell)e^{-E_\ell/kT}. \tag{10.21}$$

The probability that the system is in energy level ℓ is

$$p_\ell = Q^{-1}W(E_\ell)e^{-E_\ell/kT}. \tag{10.22}$$

You are free to choose whether to focus on microstates or energy levels, or some other alternative, depending on what is most convenient for the problem that you are trying to solve. Examples 10.3 and 10.4 illustrate the density of states with the lattice polymer collapse model.

EXAMPLE 10.3 The collapse distribution for the four-bead polymer chain. How does the population of collapsed conformations change as the temperature is changed? We focus here on the two macrostates, open and compact, rather than on the individual microstates, so the partition function for this two-state system is

Figure 10.7 In the polymer collapse model of Example 10.3, the collapsed population (probability p_C) diminishes with temperature T while the population of open conformations (probability p_O) grows.

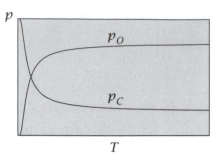

$$Q = 1 + 4e^{-\varepsilon_0/kT}.$$

At low temperatures, $T \to 0$, $Q \to 1$, so only the compact state is populated. But as the temperature increases, $T = \infty$, $Q \to 5$, indicating that all conformations become equally populated. Cooling collapses the chain. The fractions of the molecules p_O that are open, and p_C that are compact, are

$$p_C = \frac{1}{Q}, \quad \text{and} \quad p_O = \frac{4e^{-\varepsilon_0/kT}}{Q} = \frac{4e^{-\varepsilon_0/kT}}{1 + 4e^{-\varepsilon_0/kT}}.$$

Figure 10.7 shows how the populations change with temperature. This figure describes a series of different experiments, each one at a fixed temperature. This kind of curve is sometimes called an *unfolding* or *denaturation* profile for a polymer or protein [1].

To further illustrate the density of states, consider the collapse of a six-bead chain.

EXAMPLE 10.4 Collapse of the six-bead polymer chain. For the six-bead chain shown in Figure 10.8, there are three equally spaced energy levels because a chain can have 0, 1, or 2 bead–bead contacts, corresponding to energies $\varepsilon = 2\varepsilon_0$, $1\varepsilon_0$, and 0. The density of states is $W(0) = 4$; $W(\varepsilon_0) = 11$; and $W(2\varepsilon_0) = 21$. The partition function is

$$Q = 4 + 11e^{-\varepsilon_0/kT} + 21e^{-2\varepsilon_0/kT},$$

and the populations $p(\ell)$ of energy levels $\ell = 0, 1, 2$ are

$$p(0) = \frac{4}{Q}, \quad p(1) = \frac{11e^{-\varepsilon_0/kT}}{Q}, \quad \text{and} \quad p(2) = \frac{21e^{-2\varepsilon_0/kT}}{Q}.$$

Partition Functions for Independent and Distinguishable Particles

The Boltzmann distribution law applies to systems of any degree of complexity. The probabilities p_j can represent states $j = 1, 2, 3, \ldots, t$ of atoms in ideal gases. Or the probabilities can represent states with energies that depend on the complex intermolecular system configurations. Calculations are simplest whenever a system is composed of independent subsystems. For example, each

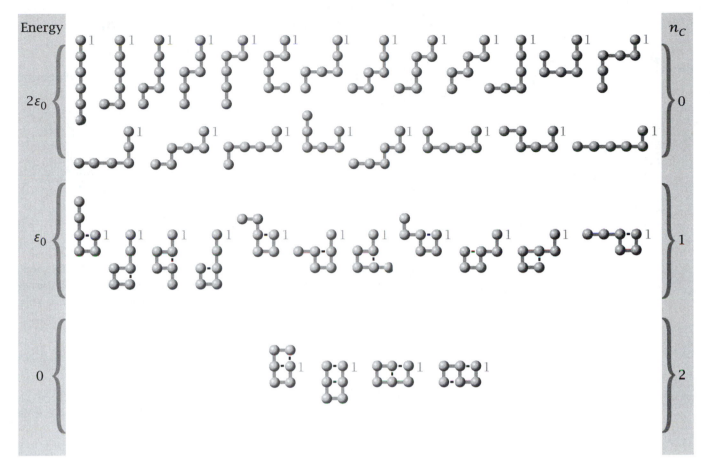

Figure 10.8 For the six-bead polymer chain, there are $W(0) = 4$ microstates having $n_C = 2$ contacts (contacts are shown as short dashes), $W(\varepsilon_0) = 11$ microstates having 1 contact, and $W(2\varepsilon_0) = 21$ microstates having 0 contacts.

molecule in an ideal gas is independent of every other molecule. Or suppose you have N four-bead polymer molecules that do not interact with each other. In such cases, the system energy is the sum of the energies of each of the particles, and the system partition function is the product of particle partition functions.

In this section, we show this, first for the case of *distinguishable*, then for *indistinguishable*, particles. What is the difference? The atoms in a crystal are spatially distinguishable because each one has its own private location in the crystal over the time scale of a typical experiment. Its location serves as a marker that distinguishes one particle from another. In contrast, according to quantum mechanics, the particles in a gas are indistinguishable from each other over the typical experimental time scale. Particles can interchange locations, so you cannot tell which particle is which.

First consider distinguishable particles. Consider a system with energy levels E_j. Suppose the system has two independent subsystems (two particles, for example), distinguishable by labels A and B, with energy levels ε_i^A and ε_m^B, respectively, $i = 1, 2, \ldots, a$ and $m = 1, 2, \ldots, b$. The system energy is

$$E_j = \varepsilon_i^A + \varepsilon_m^B.$$

Because the subsystems are independent, you can write the partition function q_A for subsystem A, and q_B for subsystem B according to Equation (10.9):

$$q_A = \sum_{i=1}^{a} e^{-\varepsilon_i^A/kT} \quad \text{and} \quad q_B = \sum_{m=1}^{b} e^{-\varepsilon_m^B/kT}. \tag{10.23}$$

The partition function Q for the entire system is the sum of Boltzmann factors $e^{-E_j/kT}$ over all $j = ab$ energy levels:

$$Q = \sum_{j=1}^{t} e^{-E_j/kT} = \sum_{i=1}^{a} \sum_{m=1}^{b} e^{-(\varepsilon_i^A + \varepsilon_m^B)/kT} = \sum_{i=1}^{a} \sum_{m=1}^{b} e^{-\varepsilon_i^A/kT} e^{-\varepsilon_m^B/kT}. \tag{10.24}$$

Because the subsystems are independent and distinguishable by their labels, the sum over the i levels of A has nothing to do with the sum over the m levels of B. The partition function Q in Equation (10.24) can be factored into subsystem partition functions q_A and q_B:

$$Q = \sum_{i=1}^{a} e^{-\varepsilon_i^A/kT} \sum_{m=1}^{b} e^{-\varepsilon_m^B/kT} = q_A q_B. \tag{10.25}$$

More generally, for a system having N independent and distinguishable particles, each with partition function q, the partition function Q for the whole system will be

$$Q = q^N. \tag{10.26}$$

Partition Functions for Independent and Indistinguishable Particles

Gas molecules are *indistinguishable*. They have *no labels A or B* that distinguish them from each other. For a system of two indistinguishable particles, the total energy is

$$E_j = \varepsilon_i + \varepsilon_m,$$

where $i = 1, 2, \ldots, t_1$, and $m = 1, 2, \ldots, t_2$. The system partition function is

$$Q = \sum_{j=1}^{t} e^{-E_j/kT} = \sum_{i=1}^{t_1} \sum_{m=1}^{t_2} e^{-(\varepsilon_i + \varepsilon_m)/kT}. \tag{10.27}$$

You cannot now factor the system partition function into particle partition functions as we did before. Here's the problem. If one particle occupied energy level 27 and other particle occupied energy level 56, you could not distinguish that from the reverse. Because of this indistinguishability, you would have overcounted by a factor of $2!$.[1]

---

[1] That's close, but not exactly right. Suppose *both* particles were in energy level 27: then you would need no indistinguishability factor correction, because there's only one way to have that happen. To compute Equation (10.27) correctly in general for indistinguishable particles is challenging. But fortunately a most important case is simple. Suppose you have a huge number

For this system, you have $Q = q^2/2!$ to a good approximation. For N indistinguishable particles, this argument generalizes to give a system partition function Q,

$$Q = \frac{q^N}{N!}. \tag{10.28}$$

We will use Equation (10.28) for gases.

If you know the partition function for a system or model, you can compute all the macroscopic thermodynamic properties.

Thermodynamic Properties Can Be Predicted from Partition Functions

Computing the Internal Energy from the Partition Function

Consider a system having fixed (T, V, N). To get the internal energy for a system with energies E_j, substitute Equation (10.9) for p_j^* into Equation (10.2):

$$U = \sum_{j=1}^{t} p_j^* E_j$$

$$= Q^{-1} \sum_{j=1}^{t} E_j e^{-\beta E_j}, \tag{10.29}$$

where $\beta = 1/kT$ is a useful quantity for simplifying the next few steps. Notice that the sum on the right-hand side of Equation (10.29) can be expressed as a derivative of the partition function in Equation (10.10):

$$\left(\frac{\partial Q}{\partial \beta} \right) = \frac{\partial}{\partial \beta} \sum_{j=1}^{t} e^{-\beta E_j} = - \sum_{j=1}^{t} E_j e^{-\beta E_j}. \tag{10.30}$$

Substituting Equation (10.30) into (10.29) simplifies it:

$$U = -\frac{1}{Q} \left(\frac{\partial Q}{\partial \beta} \right) = - \left(\frac{\partial \ln Q}{\partial \beta} \right). \tag{10.31}$$

Since $\beta = 1/kT$, you have

$$\left(\frac{\partial \beta}{\partial T} \right) = -\frac{1}{kT^2}. \tag{10.32}$$

So you can multiply the left side of Equation (10.31) by $-1/kT^2$ and the right side by $\partial \beta / \partial T$ to get

$$\frac{U}{kT^2} = \left(\frac{\partial \ln Q}{\partial T} \right). \tag{10.33}$$

of energy levels, say 100,000, and only two particles. The chance that those particles would have coincident energies is exceedingly small, so the 2! correction would be a very good approximation. In reality, this is often valid: the number of accessible states is often much larger than the number of particles. You will see in Chapter 11 that translational partition functions are of the order of 10^{30}, while usually the number of particles is much smaller, 10^{20} or less. Therefore you are often justified in neglecting this small correction.

Computing the Average Particle Energy

If particles are independent and distinguishable ($Q = q^N$), then Equation (10.33) gives the average energy $\langle \varepsilon \rangle$ per particle,

$$\langle \varepsilon \rangle = \frac{U}{N} = \frac{kT^2}{N} \left(\frac{\partial \ln q^N}{\partial T} \right)$$

$$= kT^2 \left(\frac{\partial \ln q}{\partial T} \right) = - \left(\frac{\partial \ln q}{\partial \beta} \right). \tag{10.34}$$

Computing the Entropy

The entropy of a system is defined by Equation (6.2):

$$\frac{S}{k} = - \sum_{j=1}^{t} p_j \ln p_j.$$

Substituting the Boltzmann distribution $p_j^* = Q^{-1} e^{-E_j/kT}$ from Equation (10.9) into Equation (6.2) gives

$$\frac{S}{k} = - \sum_{j=1}^{t} \left(\frac{1}{Q} e^{-E_j/kT} \right) \left(\ln \left(\frac{1}{Q} \right) - \frac{E_j}{kT} \right). \tag{10.35}$$

Substituting Equation (10.10) for Q and (10.29) for ΔU into (10.35) gives

$$S = k \ln Q + \frac{U}{T} = k \ln Q + kT \left(\frac{\partial \ln Q}{\partial T} \right). \tag{10.36}$$

For systems of N independent distinguishable particles, for which $Q = q^N$,

$$S = kN \ln q + \frac{U}{T}. \tag{10.37}$$

Because S increases linearly with N, the system entropy is the sum of the entropies of the independent particles.

Computing the Free Energy and Chemical Potential

From U and S in Equations (10.33) and (10.36), thermodynamic relationships can produce the rest—the Helmholtz free energy, chemical potential, and pressure, for example. Table 10.1 lists the main relationships.

Now we illustrate these relationships by computing the thermodynamic properties of a simple statistical mechanical model, the two-state model.

EXAMPLE 10.5 The Schottky two-state model. Consider a system that has N distinguishable particles with two energy levels for each particle: a ground state with energy zero, and an excited state with energy $\varepsilon = \varepsilon_0 > 0$. This model is useful for describing many different problems: our polymer or dimer lattice models in Chapter 8, the behaviors

Table 10.1 Thermodynamic quantities derived from the partition function for constant (T, V, N)

Internal Energy, U	$U = kT^2 \left(\dfrac{\partial \ln Q}{\partial T} \right)_{V,N}.$	(10.38)
Entropy, S	$S = k \ln Q + \dfrac{U}{T}.$	(10.39)
Helmholtz Free Energy, F	$F = U - TS = -kT \ln Q.$	(10.40)
Chemical Potential, μ	$\mu = \left(\dfrac{\partial F}{\partial N} \right)_{T,V} = -kT \left(\dfrac{\partial \ln Q}{\partial N} \right)_{T,V}.$	(10.41)
Pressure, p	$p = - \left(\dfrac{\partial F}{\partial V} \right)_{T,N} = kT \left(\dfrac{\partial \ln Q}{\partial V} \right)_{T,N}.$	(10.42)

of atoms or molecules that are excited by electromagnetic radiation, or the behavior of spins in magnetic fields (see Example 10.6).

Here we'll keep the model general and won't specify ε_0 in terms of any particular microscopic structure or property. We aim to find the average particle energy $\langle \varepsilon \rangle$, the heat capacity C_V, the entropy, and the free energy per particle from the partition function. The partition function for a two-level system is the sum of two Boltzmann factors, one for each level,

$$q = 1 + e^{-\beta \varepsilon_0}. \tag{10.43}$$

The partition function approaches one at low temperatures and two at high temperatures. The relative populations of the two states are given by the Boltzmann distribution, Equation (10.9):

$$p_1^* = \frac{1}{q}, \quad \text{and} \quad p_2^* = \frac{e^{-\beta \varepsilon_0}}{q}. \tag{10.44}$$

The average energy is found by taking the derivative in Equation (10.34):

$$\langle \varepsilon \rangle = -\frac{1}{q} \left(\frac{\partial q}{\partial \beta} \right) = \frac{\varepsilon_0 e^{-\beta \varepsilon_0}}{1 + e^{-\beta \varepsilon_0}} = \frac{\varepsilon_0 e^{-\varepsilon_0/kT}}{1 + e^{-\varepsilon_0/kT}}. \tag{10.45}$$

Figure 10.9(a) shows the energy of the two-state system as a function of temperature. At low temperatures, most molecules are in the ground state, so the system has low energy. As the temperature increases, the energy of the system increases and ultimately saturates at the value $\varepsilon_0/2$ per particle because energy levels 0 and ε_0 become equally populated.

To compute the heat capacity, use the definition $C_V = (\partial U/\partial T)$ from thermodynamics. Using Equation (10.34) to convert the total energy to the average energy per particle, $U = N \langle \varepsilon \rangle$, you have

(a) Average Energy

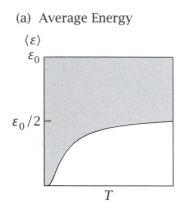

(b) Heat Capacity

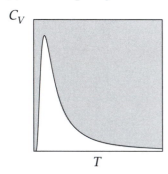

Figure 10.9 (a) The average energy per particle $\langle\varepsilon\rangle$ for the Schottky model (Equation (10.45)) saturates at $\varepsilon_0/2$ as temperature T approaches infinity. (b) The two-state model has a peak in the heat capacity C_V as a function of temperature (Equation (10.48)).

$$C_V = N\left(\frac{\partial\langle\varepsilon\rangle}{\partial T}\right) = N\left(\frac{\partial\langle\varepsilon\rangle}{\partial\beta}\right)\left(\frac{\partial\beta}{\partial T}\right) = -\frac{N}{kT^2}\left(\frac{\partial\langle\varepsilon\rangle}{\partial\beta}\right), \tag{10.46}$$

where the right-hand expressions convert from T to β to make the next step of the differentiation simpler. Take a derivative of the form $d(u/v) = (vu' - uv')/v^2$, where $u = \varepsilon_0 e^{-\beta\varepsilon_0}$ and $v = 1 + e^{-\beta\varepsilon_0}$, to get

$$\left(\frac{\partial\langle\varepsilon\rangle}{\partial\beta}\right) = \frac{(1 + e^{-\beta\varepsilon_0})(-\varepsilon_0^2 e^{-\beta\varepsilon_0}) - \varepsilon_0 e^{-\beta\varepsilon_0}(-\varepsilon_0 e^{-\beta\varepsilon_0})}{(1 + e^{-\beta\varepsilon_0})^2}$$

$$= \frac{-\varepsilon_0^2 e^{-\beta\varepsilon_0}}{(1 + e^{-\beta\varepsilon_0})^2}. \tag{10.47}$$

Substitute Equation (10.47) into the right-hand side of Equation (10.46) to find the heat capacity C_V in terms of the energy level spacing ε_0:

$$C_V = \frac{N\varepsilon_0^2}{kT^2}\frac{e^{-\beta\varepsilon_0}}{(1 + e^{-\beta\varepsilon_0})^2}. \tag{10.48}$$

The heat capacity is plotted in Figure 10.9(b), and is discussed in more detail in Chapter 12. Heat capacity peaks are characteristic of bond-breaking and melting processes. At low temperatures, the thermal energy kT from the bath is too small to excite the system to the higher energy level. At intermediate temperatures, the system can absorb heat from the bath and particles are excited into the higher energy state. At the highest temperatures, the system takes up no further energy from the bath because most particles already have their maximum energy.

To get the entropy, substitute Equation (10.43) for q and $Q = q^N$ into Equations (10.39) and (10.38):

$$\frac{S}{N} = \frac{\varepsilon_0 e^{-\beta\varepsilon_0}}{T(1 + e^{-\beta\varepsilon_0})} + k\ln(1 + e^{-\beta\varepsilon_0}). \tag{10.49}$$

To get the free energy, $F = U - TS$, use $U = N\langle\varepsilon\rangle$ with Equation (10.45) and and S from Equation (10.49) to get

$$\frac{F}{NkT} = -\ln q = -\ln(1 + e^{-\beta\varepsilon_0}). \tag{10.50}$$

As $\varepsilon_0 \to \infty$, the excited state becomes inaccessible so $S \to 0$, and $F \to 0$. On the other hand, as $\varepsilon_0 \to 0$, both states become accessible so $S \to Nk\ln 2$, and $F \to NkT\ln 2$.

Now let's consider a specific example of a two-level system, a magnetic material in a magnetic field.

EXAMPLE 10.6 Curie's law of paramagnetism, a two-level system. If you heat a magnet, it loses magnetization. The Schottky model illustrates this. Consider a simple paramagnetic material, in which every one of the N independent atoms has a magnetic dipole moment of magnitude $\mu_0 > 0$. The dipole moment is a measure of how much magnetization (alignment of each magnet) results from a given applied magnetic field B. (Don't confuse μ here with the chemical potential: the same symbol is used for both quantities, but they are unrelated.)

In a paramagnet, the dipole–dipole interaction between neighboring atoms is small compared with the interaction with an applied field. In such a field, the dipole moments align themselves either parallel or antiparallel to the applied field. If a magnetic field $B \geq 0$ is applied to an atom, the energy of the atom goes down by an amount $\varepsilon_1 = -\mu_0 B$, if the magnetic moment is parallel to the field. On the other hand, if the magnetic moment is antiparallel to the field, it leads to an unfavorable energy: $\varepsilon_2 = +\mu_0 B$. Using the convention that the ground state defines the zero of energy, the energy difference between these two states is $+2\mu_0 B$, which defines the excited state energy, so the partition function is

$$q = 1 + e^{-2\mu_0 B/kT}. \tag{10.51}$$

We want to calculate the average magnetic moment as a function of temperature. At equilibrium, the probability that an atom's magnetic moment is parallel to B is p_1^*. The probability that it is antiparallel is p_2^*:

$$p_1^* = \frac{1}{q}, \quad \text{and} \quad p_2^* = \left(\frac{1}{q}\right) e^{-2\mu_0 B/kT}. \tag{10.52}$$

Because the magnetic moment of the aligned state is $+\mu_0$ and the magnetic moment of the antialigned state is $-\mu_0$, the average magnetic moment is

$$\langle \mu \rangle = \sum_{i=1}^{2} \mu_0 p_j^*$$

$$= \mu_0 p_1^* + (-\mu_0) p_2^* = \frac{\mu_0}{q}(1 - e^{-2\mu_0 B/kT}) = \mu_0 \frac{1 - e^{-2\mu_0 B/kT}}{1 + e^{-2\mu_0 B/kT}}. \tag{10.53}$$

The last equality follows from the definition of q in Equation (10.51). A concise way to express this relationship is through use of the hyperbolic tangent function,

$$\tanh x = \frac{1 - e^{-2x}}{1 + e^{-2x}}. \tag{10.54}$$

Using the hyperbolic tangent, the average magnetic moment is given by

$$\langle \mu \rangle = \mu_0 \tanh \left(\frac{\mu_0 B}{kT}\right). \tag{10.55}$$

In weak magnetic fields or high temperatures, $\mu_0 B/kT \ll 1$. The Taylor series expansion for exponentials (see Appendix C, Equation (C.1)) gives $1 - e^{-2\mu_0 B/kT} \approx 2\mu_0 B/kT$ and $1 + e^{-2\mu_0 B/kT} \approx 2 - 2\mu_0 B/kT \approx 2$. At high temperatures or in weak fields, the total magnetic moment is inversely proportional to T, and Equation (10.55) becomes **Curie's law:**

$$\langle \mu \rangle = \frac{\mu_0^2 B}{kT}. \tag{10.56}$$

In contrast, at high fields ($B/kT \rightarrow \infty$) Equation (10.53) gives $\langle \mu \rangle = \mu_0$.

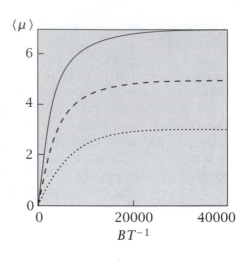

Figure 10.10 Average magnetic moment $\langle \mu \rangle$ saturates with increasing magnetic field B/kT. At low fields, $\langle \mu_0 \rangle$ is linear in B/kT. Data for gadolinium sulphate octahydrate (———), ferric ammonium alum (– – –), chromium potassium alum (········). Source: JD Fast, *Entropy; The Significance of the Concept of Entropy and Its Applications in Science and Technology*, McGraw-Hill, New York, 1962. Data are from CJ Gorter, WJ deHaas and J van den Handel, *Proc Kon Ned Akad Wetensch* **36**, 158 (1933) and WE Henry, *Phys Rev* **88**, 559 (1952).

Figure 10.10 shows experimental evidence that the magnetization of a material increases with increasing magnetic field. High fields cause all the spins to align with the field. Cooling leads to magnetization (alignment of the spins) and heating leads to demagnetization (random orientations of the spins).

What Is an Ensemble?

So far in this chapter, we have considered only systems in which (T, V, N) are constant. In statistical mechanics, such a system is called the *Canonical ensemble*. 'Ensemble' has two meanings. First, it refers to the collection of all the possible microstates, or snapshots, of the arrangements of the system. We have counted the number of arrangements W of particles on a lattice or configurations of a model polymer chain. The word 'ensemble' describes the complete set of all such configurations. 'Ensemble' is also sometimes used to refer to the constraints, as in the '(T, V, N) ensemble.' The (T, V, N) ensemble is so prominent in statistical mechanics that it is called 'canonical.' A system constrained by (T, p, N), is called the 'isobaric–isothermal' ensemble.

When the constraints are (T, V, μ), energy and particles can exchange across the boundary. This is called the *Grand Canonical ensemble*. It is important for processes of ligand binding, and is illustrated in Chapter 28. When (U, V, N) are held constant at the boundary, no extensive quantity exchanges across the boundary, so there is no bath, and maximum entropy $S(U, V, N)$ identifies the state of equilibrium (see Chapter 7). This is called the *Microcanonical ensemble*.

The Microcanonical Ensemble

The Microcanonical ensemble is qualitatively different from the Canonical and Grand Canonical ensembles because it involves no bath and no fluctuations of any extensive property across the boundary.

In the Canonical and Grand Canonical ensembles, a system has some probability, say p_3^*, of being in a state having energy E_3, and some probability p_{12}^* of being in a different state having energy E_{12}. The energy can *fluctuate*. But in the Microcanonical ensemble, the energy U is fixed, so every accessible state of the system must have precisely this same energy. In the Microcanonical ensemble, there are no energy fluctuations. To derive the probability distribution for the Microcanonical ensemble, maximize the entropy, subject only to the constraint on the sum of probabilities $\alpha \sum dp_j = 0$, to get

$$dS = -k \sum_{j=1}^{t} \left[(1 + \ln p_j) + \alpha \right] dp_j = 0. \tag{10.57}$$

Equation (10.57) is satisfied by $\ln p_j^* = -1 - \alpha$, or

$$p_j^* = \frac{1}{t}. \tag{10.58}$$

The system has the same probability of being in any one microstate as any other. This prediction is the same as the prediction in Chapter 6 for die rolls in the absence of knowledge of an average score, or for dipole orientations if all directions are equivalent.

The general definition of entropy is $S/k = -\sum_i p_i \ln p_i$. For the Microcanonical ensemble (U, V, N), the entropy is also given by the Boltzmann expression, Equation (6.1), $S = k \ln W(U, V, N)$. You can see this by substituting $p_j = 1/t = 1/W$ into Equation (6.2), to get

$$\frac{S}{k} = -\sum_{j=1}^{W} p_j \ln p_j = \ln W. \tag{10.59}$$

Summary

Boltzmann's law gives the equilibrium distribution of atoms and molecules over their energy levels. Starting with a model for the ladder of energies accessible to the system, you can compute the partition function. From the partition function, you can compute the thermodynamic and averaged physical properties. In the next chapter we illustrate how quantum mechanical models give the energy levels that can be used to predict the properties of gases and simple solids from their atomic structures.

Problems

1. Statistical thermodynamics of a cooperative system.
Perhaps the simplest statistical mechanical system having 'cooperativity' is the three-level system in Table 10.2.

Table 10.2

Energies	$2\varepsilon_0$	ε_0	0
Degeneracies	y	1	1

(a) Write an expression for the partition function q as a function of energy ε, degeneracy y, and temperature T.

(b) Write an expression for the average energy $\langle\varepsilon\rangle$ versus T.

(c) For $\varepsilon_0/kT = 1$ and $y = 1$, compute the populations, or probabilities, p_1^*, p_2^*, p_3^* of the three levels.

(d) Now if $\varepsilon_0 = 2\,\text{kcal mol}^{-1}$ and $y = 1000$, find the temperature T_0 at which $p_1 = p_3$.

(e) Under condition (d), compute p_1^*, p_2^*, and p_3^* at temperature T_0.

2. The speed of sound. The speed of sound in air is approximately the average velocity $\langle v_x^2 \rangle^{1/2}$ of the gas molecules. Compute this speed for $T = 0\,°\text{C}$, assuming that air is mostly nitrogen gas.

3. The properties of a two-state system. Given a two-state system in which the low energy level is $600\,\text{cal mol}^{-1}$, the high energy level is $1800\,\text{cal mol}^{-1}$, and the temperature of the system is 300 K,

(a) What is the partition function q?

(b) What is the average energy $\langle\varepsilon\rangle$?

4. Binding to a surface. Consider a particle that has two states: bonded to a surface, or non-bonded (released). The non-bonded state is higher in energy by an amount ε_0.

(a) Explain how the ability of the particle to bond to the surface contributes to the heat capacity, and why the heat capacity depends on temperature.

(b) Compute the heat capacity C_V in units of Nk if $T = 300\,\text{K}$ and $\varepsilon_0 = 1.2\,\text{kcal mol}^{-1}$ (which is about the strength of a weak hydrogen bond in water).

5. Entropy depends on distinguishability. Given a system of molecules at $T = 300\,\text{K}$, $q = 1 \times 10^{30}$, and $\Delta U = 3740\,\text{J mol}^{-1}$,

(a) What is the molar entropy if the molecules are distinguishable?

(b) What is the molar entropy if the molecules are indistinguishable?

6. The Boltzmann distribution of uniformly spaced energy levels. A system has energy levels uniformly spaced at $3.2 \times 10^{-20}\,\text{J}$ apart. The populations of the energy levels are given by the Boltzmann distribution. What fraction of particles is in the ground state at $T = 300\,\text{K}$?

7. The populations of spins in a magnetic field. The nucleus of a hydrogen atom, a proton, has a magnetic moment. In a magnetic field, the proton has two states of different energy; spin up and spin down. This is the basis of proton NMR. The relative populations can be assumed to be given by the Boltzmann distribution, where the difference in energy between the two states is $\Delta\varepsilon = g\mu B$, $g = 2.79$ for protons, and $\mu = 5.05 \times 10^{-24}\,\text{J Tesla}^{-1}$. For a 300 MHz NMR instrument, $B = 7\,\text{Tesla}$.

(a) Compute the relative population difference, $|N_+ - N_-|/(N_+ + N_-)$, at room temperature for a 300 MHz machine.

(b) Describe how the population difference changes with temperature.

(c) What is the partition function?

8. Energy and entropy for indistinguishable particles. Equations (10.34) for $\langle\varepsilon\rangle$ and (10.37) for S apply to distinguishable particles. Compute the corresponding quantities for systems of indistinguishable particles.

9. Computing the Boltzmann distribution. You have a thermodynamic system with three states. You observe the probabilities $p_1 = 0.9$, $p_2 = 0.09$, $p_3 = 0.01$ at $T = 300\,\text{K}$. What are the energies ε_2 and ε_3 of states 2 and 3 relative to the ground state?

10. The pressure reflects how energy levels change with volume. If energy levels $\varepsilon_i(V)$ depend on the volume of a system, show that the pressure is the average

$$p = -N\left\langle\frac{\partial\varepsilon}{\partial V}\right\rangle.$$

11. The end-to-end distance in polymer collapse. Use the two-dimensional four-bead polymer of Example 10.3. The distance between the chain ends is 1 lattice unit in the compact conformation, 3 lattice units in the extended conformation, and $\sqrt{5}$ lattice units in each of the other three chain conformations. Plot the average end-to-end distance as a function of temperature, if the energy is

(a) $\varepsilon = 1\,\text{kcal mol}^{-1}$.

(b) $\varepsilon = 3\,\text{kcal mol}^{-1}$.

12. The lattice model of dimerization. Use the lattice model for monomers bonding to form dimers, and assume large volumes $V \gg 1$.

(a) Derive the partition function.

(b) Compute $p_1(T)$ and $p_2(T)$, the probabilities of monomers and dimers as a function of temperature, and sketch the dependence on temperature.

(c) Compute the bond breakage temperature T_0 at which $p_1 = p_2$.

13. Equivalent premises for the Boltzmann distribution. Use Equation (10.6) to show that the distribution of probabilities p_j^* that minimizes the free energy F at constant (T, V, N) is the same distribution that maximizes the entropy S at constant $(U, V, N) = (\langle E \rangle, V, N)$.

References

[1] KA Dill, HS Chan, et al. *Prot Sci* **4**, 561–602 (1995).

Suggested Reading

D Chandler, *Introduction to Modern Statistical Mechanics*, Oxford University Press, Oxford, 1987. Concise and insightful.

HL Friedman, *A Course in Statistical Mechanics*, Prentice-Hall, New Jersey, 1985. Good description of the relationship between ensembles in thermodynamics and statistical mechanics.

TL Hill, *An Introduction to Statistical Thermodynamics*, Addison-Wesley, New York, 1960. A classic with an excellent discussion of ensembles.

K Huang, *Statistical Mechanics*, John Wiley, New York, 1963. Advanced text.

R Kubo, *Statistical Mechanics*, 2nd edition, North-Holland, New York, 1988. Advanced text with many excellent worked problems.

DA McQuarrie, *Statistical Mechanics*, Harper & Row, New York, 1973. A standard and complete text with a good discussion of ensembles.

11 The Statistical Mechanics of Simple Gases & Solids

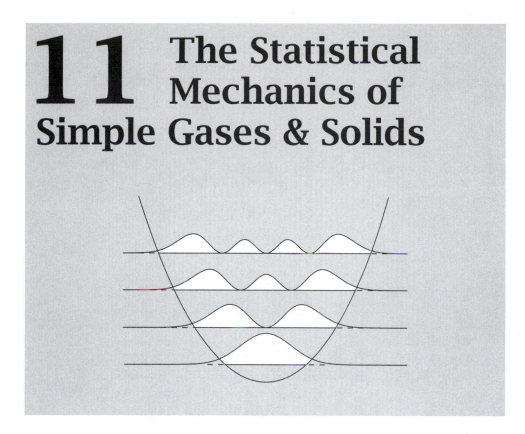

Statistical Mechanics Predicts Macroscopic Properties from Atomic Structures

To predict the properties of materials from the forces on the atoms that comprise them, you need to know the energy ladders. Energy ladders can be derived from spectroscopy or quantum mechanics. Here we describe some of the quantum mechanics that can predict the properties of ideal gases and simple solids. This will be the foundation for chemical reaction equilibria and kinetics in Chapters 13 and 19. Our discussion of quantum mechanics is limited. We just sketch the basic ideas with the *particle-in-a-box* model of translational freedom, the harmonic oscillator model for vibrations, and the rigid rotor model for rotations.

Evidence for the quantization of energies comes from atomic spectroscopy. Spectroscopy measures the frequency v of electromagnetic radiation that is absorbed by an atom, a molecule, or a material. Absorption of radiation by matter leads to an increase in its energy by an amount $\Delta\varepsilon = hv$, where $h = 6.626176 \times 10^{-34}$ Js is Planck's constant. This change, $\Delta\varepsilon$, is the difference from one energy level to another on an energy ladder.

Atoms and simple molecules absorb electromagnetic radiation only at certain discrete frequencies, rather than as continuous functions of frequency. For example, Figure 11.1 shows the infrared absorption spectrum of hydrogen bromide. Discrete absorption frequencies imply discrete energy spacings. This is the basis for quantum mechanics.

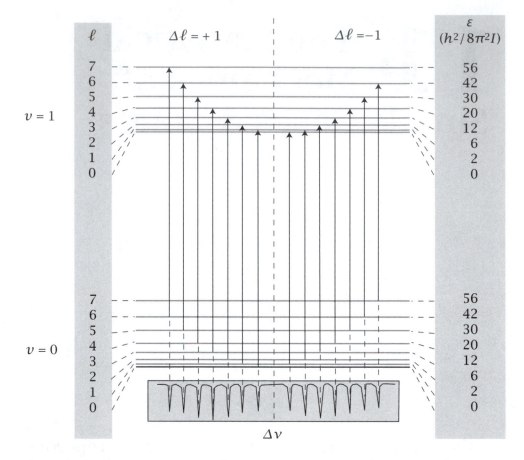

Figure 11.1 Electromagnetic radiation can excite transitions in molecules from one rotational energy level ℓ of a particular vibrational state v to a different level. The frequency of excitation radiation corresponds to the energy change of the transition, through the relation $\Delta\varepsilon = h\nu$. Shown here is the absorption band of gas-phase HBr in the infrared region. Source: GM Barrow, *The Structure of Molecules; an Introduction to Molecular Spectroscopy*, 2nd printing, WA Benjamin, New York, 1964. Series title: The general chemistry monograph series.

Energies form a hierarchy. Some energy changes are small, implying closely spaced energy levels, and others are large (see Figure 11.2). Energies fall into classes: *translations, rotations, vibrations, electronic excitations*, and others. Each step from one rung to the next on the ladder of electronic excitation energies typically contains many rungs on the vibrational ladder, each step on the vibrational ladder has many rotational rungs, and each step on the rotational ladder has many translational levels.

The basis for predicting quantum mechanical energy levels is the Schrödinger equation, which can be expressed in compact form:

$$\mathcal{H}\psi = E_j\psi, \tag{11.1}$$

where the *wavefunction* $\psi(x, y, z)$ is a function of spatial position (x, y, z) (ψ can also be a function of time but we don't consider that here). Solving Equation (11.1) gives ψ. The square of this function, ψ^2, is the spatial probability distribution of the particles for the problem of interest.

$\mathcal{H}$ is called the *Hamiltonian operator*. It describes the forces relevant for the problem of interest. In classical mechanics, the sum of kinetic plus potential energies is an invariant conserved quantity, the total energy (see Equation (3.6)). In quantum mechanics, the role of this invariant is played by the Hamiltonian operator. For example, for the one-dimensional translational motion of a particle having mass m and momentum p, the Hamiltonian operator is

$$\mathcal{H} = \frac{p^2}{2m} + \mathcal{V}(x), \qquad (11.2)$$

where $p^2/(2m)$ represents the kinetic energy and $\mathcal{V}(x)$ is the potential energy, as a function of the spatial coordinate x.

While classical mechanics regards p and $\mathcal{V}$ as functions of time and spatial position, quantum mechanics regards p and $\mathcal{V}$ as *mathematical operators* that create the right differential equation for the problem at hand. For example, the translational momentum operator is $p^2 = (-h^2/4\pi^2)(d^2/dx^2)$ in one dimension. The expression $\mathcal{H}\psi(x)$ in Equation (11.2) is a shorthand notation for taking the second derivative of ψ, multiplying by a collection of terms and adding $\mathcal{V}\psi$,

$$\mathcal{H}\psi = -\left(\frac{h^2}{8\pi^2 m}\right)\frac{d^2\psi(x)}{dx^2} + \mathcal{V}(x)\psi(x), \qquad (11.3)$$

so Equation (11.1) becomes

$$-\left(\frac{h^2}{8\pi^2 m}\right)\frac{d^2\psi(x)}{dx^2} + \mathcal{V}(x)\psi(x) = E_j\psi(x). \qquad (11.4)$$

According to Schrödinger's equation, the operator $\mathcal{H}$ acting on ψ (the left side) equals a constant E_j multiplied by ψ (the right side). Only certain functions $\psi(x)$ can satisfy Equation (11.4), if E_j is a constant. The quantities E_j are called the *eigenvalues* of the equation. They are the discrete energy levels that we seek. The index j for the eigenvalues is called the *quantum number*. Now, we solve the Schrödinger equation for a simple problem to show the quantum mechanical perspective on translational motion.

The Particle-in-a-Box Is the Quantum Mechanical Model for Translational Motion

The particle-in-a-box is a model for the freedom of a particle to move within a confined space. It applies to electrons contained within atoms and molecules, and molecules contained within test tubes. Let's first solve a one-dimensional problem.

A particle is free to move along the x-axis over the range $0 < x < L$. The idea of the 'box' is that at the walls ($x = 0$ and $x = L$) the potential is infinite ($\mathcal{V}(0) = \mathcal{V}(L) = \infty$) so the particle cannot escape (see Figure 11.3). But everywhere inside the box the molecule has free motion ($\mathcal{V}(x) = 0$ for $0 < x < L$). So we set $\mathcal{V}(x) = 0$ in Equation (11.4).

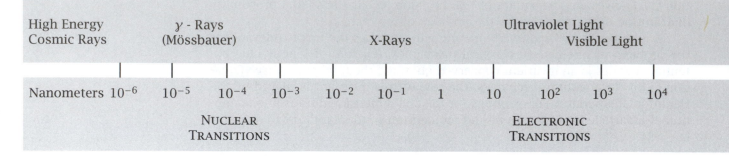

Figure 11.2 Electromagnetic energy spectrum. Rotational and vibrational motions occur in the infrared range. Ionization and covalent bond breakage occur at higher energies in the UV and X-ray range. Nuclear spins are affected at much lower energies (labeled NMR). Source: O Howarth, *Theory of Spectroscopy; an Elementary Introduction*, Wiley, New York, 1973.

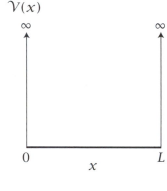

Figure 11.3 The potential energy $\mathcal{V}$ for the particle in a one-dimensional box is infinite at the walls, $x = 0$, and $x = L$. $\mathcal{V}(x) = 0$ inside.

Equation (11.4) then becomes a linear second-order differential equation,

$$\frac{d^2\psi}{dx^2} + K^2\psi = 0, \tag{11.5}$$

where

$$K^2 = \frac{8\pi^2 m\varepsilon}{h^2}. \tag{11.6}$$

Because we are dealing with a single particle, we represent the energies by ε rather than E. Equation (11.5) is satisfied by the expression

$$\psi(x) = A\sin Kx + B\cos Kx, \tag{11.7}$$

where A and B are constants.

To determine the constants A and B, use the boundary conditions (the values of $\mathcal{V}$ at the box walls). According to quantum mechanics, the probability of finding the particle in any region of space is the square of the wavefunction $\psi^2(x)$. Because the potential is infinite at the walls, there is no probability of finding the particle exactly at the wall, so

$$\psi^2(0) = 0 \implies \psi(0) = 0. \tag{11.8}$$

When $x = 0$ is substituted into Equation (11.7), you get $A\sin Kx = 0$ and $B\cos Kx = 1$. So you must have $B = 0$ to satisfy boundary condition Equation (11.8). Equation (11.7) reduces to

$$\psi(x) = A\sin Kx. \tag{11.9}$$

The second boundary condition implies that the wavefunction must equal zero also at $x = L$, the other wall of the box:

$$\psi(L) = A\sin KL = 0. \tag{11.10}$$

Near Infrared	Far Infrared	Microwave Electron Resonance & Radar				TV & FM Radio		AM Radio			

| meters | 10^{-6} | 10^{-5} | 10^{-4} | 10^{-3} | 10^{-2} | 10^{-1} | 1 | 10 | 10^2 | 10^3 | 10^4 | 10^5 |

MOLECULAR VIBRATION   MOLECULAR ROTATION   NUCLEAR SPIN FLIPS

1 kJ mol^{-1} = 1.195 × 10^{-4} m

kT at 300 K = 4.796 × 10^{-5} m

1 kcal mol^{-1} = 2.859 × 10^{-5} m

1 eV = 1.240 × 10^{-6} m

Equation (11.10) is satisfied when KL is any integer multiple of π,

$$KL = n\pi \qquad \text{for } n = 1, 2, 3, \dots . \tag{11.11}$$

$A = 0$ and $n = 0$ aren't useful solutions to Equation (11.10) because they imply that there's no particle in the box.

Substitute Equation (11.6) into Equation (11.11) to find the energy levels ε_n as a function of n:

$$\left(\frac{8\pi^2 m \varepsilon}{h^2} \right)^{1/2} L = n\pi,$$

$$\implies \varepsilon_n = \frac{(nh)^2}{8mL^2}. \tag{11.12}$$

Equation (11.12) is what you need for statistical mechanics—an expression for the energy level spacings ε_n as a function of the index n, the *quantum number*. Figure 11.4 shows the energy ladder for the one-dimensional particle in a box. Because ε_n increases as n^2, the energy-level spacings widen with increasing quantum number n.

Before we compute the partition function, let's complete the expression for the wavefunction $\psi(x)$. Probabilities must integrate to 1, $\int_0^L \psi_n^2(x)\, dx = 1$. So the normalized expression for the wavefunction in Equation (11.9) is

$$\psi_n(x) = \left(\frac{2}{L} \right)^{1/2} \sin\left(\frac{n\pi x}{L} \right).$$

Figure 11.5 shows $\psi(x)$, which can be negative, and $\psi^2(x)$, which is always positive. $\psi^2(x)$ represents the spatial distribution of the particle throughout the box. The particle is distributed differently in the ground state ($n = 1$) than in the higher excited states (larger n).

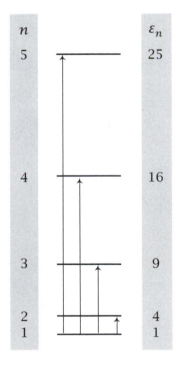

Figure 11.4 Energy ladder for the one-dimensional particle-in-a-box. The energy level ε_n increases with the square of the quantum number n, $\varepsilon_n = n^2 h^2 / (8mL^2)$.

(a) Classical Probability

(b) Wavefunction ψ, and Probability ψ^2

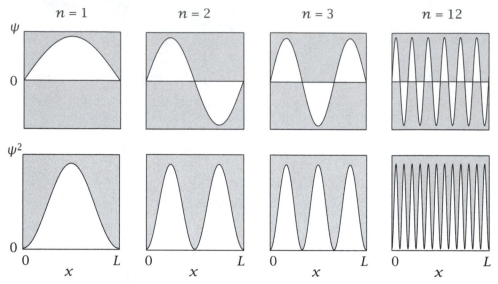

Figure 11.5 In contrast to the classical probability (a) of finding a particle distributed uniformly between $x = 0$ and L, the quantum mechanical particle in a box (b) has preferred locations, depending on n. The first row shows the wavefunction $\psi(x)$ and the second row shows $\psi^2(x)$, the probability distribution for the location of the particle.

EXAMPLE 11.1 Energy levels for argon in a macroscopic box. What are the energy levels for an argon atom contained in a one-dimensional box of length $L = 1\,\mathrm{cm}$? The mass of Ar is $m = 40\,\mathrm{g\,mol^{-1}} = (0.040\,\mathrm{kg\,mol^{-1}})/(6.02 \times 10^{23}\,\mathrm{atoms\,mol^{-1}}) = 6.64 \times 10^{-26}\,\mathrm{kg\,atom^{-1}}$. Equation (11.12) gives the energy spacings as a function of n:

$$\varepsilon = n^2 \frac{(6.626 \times 10^{-34}\,\mathrm{J\,s})^2}{8\,(6.64 \times 10^{-26}\,\mathrm{kg})\,(10^{-2}\,\mathrm{m})^2}$$

$$= \left(8.27 \times 10^{-39}\,\mathrm{J}\right) n^2 \approx \left(5 \times 10^{-15}\,\mathrm{J\,mol^{-1}}\right) n^2.$$

The energy spacings are exceedingly small, so very many energy levels are populated at room temperature. This is why gases in macroscopic containers can be treated classically, as if they were distributed continuously throughout space.

Now that you have the energy ladder, you can compute the translational partition function $q_{\text{translation}}$. Substitute $\varepsilon_n = n^2 h^2/8mL^2$ from Equation (11.12) into the definition of the partition function Equation (10.23). In this case for each particle,

$$q_{\text{translation}} = \sum_{n=1}^{\infty} e^{-\varepsilon_n/kT} = \sum_{n=1}^{\infty} e^{-n^2 h^2/(8mL^2 kT)}. \tag{11.13}$$

It is often useful to divide an energy by Boltzmann's constant to express the energy in units of temperature. The translational temperature is $\theta_{trans} = h^2/(8mL^2k)$, so the Boltzmann factors in Equation (11.13) can be expressed as $\exp(-n^2\theta_{trans}/T)$. For argon in Example 11.1, $\theta_{trans} = (8.27 \times 10^{-39}\,\text{J})/(1.38 \times 10^{-23}\,\text{J K}^{-1}) \approx 6 \times 10^{-16}\,\text{K}$, an extremely small temperature.

When $\theta_{trans}/T \ll 1$, as it is for argon in a 1 cm one-dimensional box at room temperature, many states are populated and the partition function is large. If the mass of the particle is large, if the size of the box is large, or if the temperature is high (even a few K for most systems) then $\theta/T \ll 1$, and the energy level spacings from Equation (11.12) are so small that the sum in Equation (11.13) can be approximated as an integral:

$$q_{\text{translation}} = \int_0^\infty e^{-(h^2/8mL^2kT)n^2}\,dn. \tag{11.14}$$

This integral, $\int_0^\infty e^{-ax^2}\,dx = (1/2)(\pi/a)^{1/2}$ where $a = (h^2/8mL^2kT)$, is given in Appendix D, Equation (D.1). Therefore the translational partition function for the particle in a one-dimensional box is

$$q_{\text{translation}} = \left(\frac{2\pi mkT}{h^2}\right)^{1/2} L. \tag{11.15}$$

Now let's generalize from one to three dimensions.

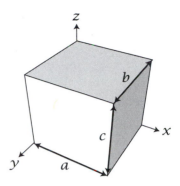

Figure 11.6 a, b, and c are the x, y, and z dimensions for the three-dimensional particle-in-a-box problem.

A Particle in a Three-Dimensional Box

Suppose the particle is confined within a three-dimensional box with dimensions $0 < x < a$, $0 < y < b$, and $0 < z < c$. The Schrödinger equation is

$$-\frac{h^2}{8\pi^2 m}\left(\frac{\partial^2}{\partial x^2} + \frac{\partial^2}{\partial y^2} + \frac{\partial^2}{\partial z^2}\right)\psi(x,y,z) + \mathcal{V}(x,y,z)\psi(x,y,z)$$
$$= \varepsilon\psi(x,y,z), \tag{11.16}$$

where $\mathcal{V}(x,y,z) = \mathcal{V}(x) + \mathcal{V}(y) + \mathcal{V}(z)$ and $\mathcal{V}(x) = \mathcal{V}(y) = \mathcal{V}(z) = 0$. Equation (11.16) can be factored into three independent equations:

$$\mathcal{H}_x\psi_x = \varepsilon_x\psi_x,$$

$$\mathcal{H}_y\psi_y = \varepsilon_y\psi_y, \quad \text{and}$$

$$\mathcal{H}_z\psi_z = \varepsilon_z\psi_z,$$

where $\mathcal{H}_x = -(h^2/8\pi^2 m)(\partial^2/\partial x^2)$, for example. The equations are *separable*—they can be solved separately. Then $\psi(x,y,z) = \psi(x)\psi(y)\psi(z)$ and the energies can be added together. The particle energy $\varepsilon_{n_x,n_y,n_z}$ is

$$\varepsilon_{n_x,n_y,n_z} = \frac{h^2}{8m}\left(\frac{n_x^2}{a^2} + \frac{n_y^2}{b^2} + \frac{n_z^2}{c^2}\right), \tag{11.17}$$

where a, b, and c are the x-, y-, and z-dimensions of the box (see Figure 11.6). Because the three solutions are independent, the partition function for particle translation in the three-dimensional case is just the product of three independent one-dimensional partition functions, each given by Equation (11.15):

$$q_{translation} = q_x q_y q_z = \left(\frac{2\pi mkT}{h^2}\right)^{3/2} abc = \left(\frac{2\pi mkT}{h^2}\right)^{3/2} V$$

$$= \frac{V}{\Lambda^3}, \tag{11.18}$$

where V is the volume of the box, and Λ^3, which has units of volume, is

$$\Lambda^3 = \left(\frac{h^2}{2\pi mkT}\right)^{3/2}. \tag{11.19}$$

This reference volume is of roughly molecular dimensions. For example $\Lambda = 0.714\,\text{Å}$ for H_2 at $T = 300\,\text{K}$.

Let's compute the partition function for a monatomic gas.

EXAMPLE 11.2 Calculate the translational partition function for argon. For temperature $T = 273\,\text{K}$, and standard state volume $V = 2.24 \times 10^{-2}\,\text{m}^3$ (the volume of one mole of gas at $p = 1\,\text{atm}$), the translational partition function is

$$q_{translation} = \left(\frac{2\pi mkT}{h^2}\right)^{3/2} V, \qquad \text{where}$$

$$m \approx 40\,\text{g mol}^{-1} = \frac{0.04}{6.02 \times 10^{23}}\,\frac{\text{kg mol}^{-1}}{\text{atoms mol}^{-1}}$$

$$= 6.64 \times 10^{-26}\,\text{kg per atom}.$$

$$kT = (1.38 \times 10^{-23}\,\text{J K}^{-1})(273\,\text{K}) = 3.77 \times 10^{-21}\,\text{J, so}$$

$$q_{translation} = \left(\frac{2\pi(6.64 \times 10^{-26}\,\text{kg per atom})(3.77 \times 10^{-21}\,\text{J})}{(6.63 \times 10^{-34}\,\text{J s})^2}\right)^{3/2}$$

$$\times (2.24 \times 10^{-2}\,\text{m}^3)$$

$$= (2.14 \times 10^{32})V = 4.79 \times 10^{30}\,\text{states per atom}.$$

The partition function $q_{translation}$ is the number of translational energy levels effectively accessible to an argon atom at this temperature. This number is certainly large enough to justify the replacement of the sum in Equation (11.13) by the integral in Equation (11.14).

Translational motion is only one of the types of degrees of freedom that atoms and molecules use to store and exchange energies. Quantum mechanics goes beyond the kinetic theory of gases in that it treats the internal degrees of freedom as well as translation. If you put all the forces between the nuclei and electrons into a Schrödinger equation to describe a particular atom or molecule, and if you are able to solve it, you can compute the orbitals (wavefunctions) and energy levels for that atom or molecule.

In the same way that we separated translational motion into three independent components, other Schrödinger equations can also be factored into simpler differential equations, which can be solved individually. Such component energies are additive. This is the basis for classifications such as *translations*, rotations and *vibrations*. For example, suppose you are interested in a diatomic gas molecule. The Hamiltonian operator can be expressed in terms of the distance between the two bonded atoms (vibration), their angular distribution (rotation), and the center-of-mass position of the molecule in space (translation). Because the differential operators are additive, you get a Schrödinger equation of the form $(\mathcal{H}_{translation} + \mathcal{H}_{vibration} + \mathcal{H}_{rotation})\psi = \varepsilon\psi$, which can be further simplified into three independent equations: $\mathcal{H}_{translation}\psi = \varepsilon_{translation}\psi$, $\mathcal{H}_{vibration}\psi = \varepsilon_{vibration}\psi$, and $\mathcal{H}_{rotation}\psi = \varepsilon_{rotation}\psi$.

Although quantum mechanics treats many kinds of problems, our main interest here is in the structures and properties of molecules, as a foundation for chemical reaction equilibria and kinetics. So we will focus on simple models of these types of degrees of freedom. The translational component was treated by the particle-in-a-box model. Now we consider vibrations, rotations, and electronic excitations.

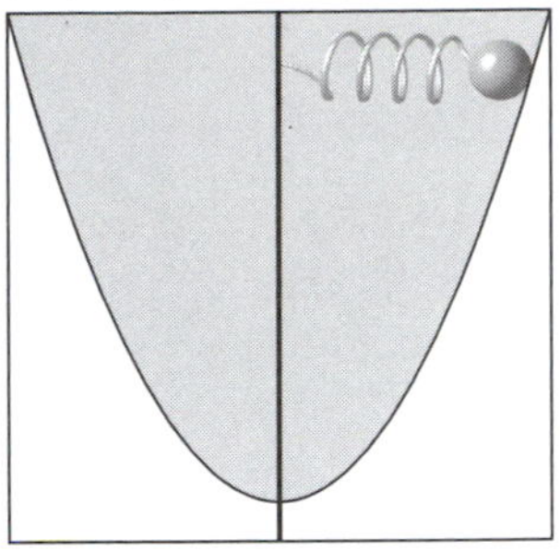

$\mathcal{V}(x) = k_s x^2 / 2$

0

x

Figure 11.7 The harmonic oscillator model for vibrations is based on a mass on a spring. The spring is attached at $x = 0$. The potential energy $\mathcal{V}$ increases as the square of the displacement x, with spring constant k_s.

Vibrations Can Be Treated by the Harmonic Oscillator Model

In many situations, atoms vibrate: inside molecules, in solids, and on surfaces, for example. To a first approximation, such atoms act as if they were connected to other atoms by springs. Here's the simplest model of such bonds.

Consider a particle of mass m that is free to move along the x-axis. It is held by a spring-like force around the position $x = 0$ (see Figure 11.7). If the spring constant is k_s, the potential energy is $\mathcal{V}(x) = k_s x^2 / 2$. This is called a *square-law*, or *parabolic* potential. Bonding potentials are often well approximated as a polynomial function, $\mathcal{V}(x) = V_0 + V_1 x + V_2 x^2 + \cdots$. For small motions near the point of minimum energy, where the first derivative is zero ($V_1 = (\partial \mathcal{V}/\partial)x = 0$), the lowest-order non-zero term is the square-law term, $V_2 x^2$. Therefore parabolic potentials are often good first approximations to bonding energy functions. The corresponding Schrödinger equation for the quantum harmonic oscillator is

$$-\frac{h^2}{8\pi^2 m}\frac{d^2\psi(x)}{dx^2} + \frac{1}{2}k_s x^2 \psi(x) = \varepsilon_v \psi(x), \tag{11.20}$$

where the subscript v indicates that these are vibrational energies. Rearranging gives the differential equation

$$\frac{d^2\psi}{dx^2} - \left(\frac{1}{a^4}\right)x^2\psi = -\left(\frac{2\varepsilon_v}{a^4 k_s}\right)\psi, \tag{11.21}$$

where $a^4 = h^2/(4\pi^2 m k_s)$. Wavefunctions $\psi(x)$ that satisfy Equation (11.20) are Gaussian functions multiplied by *Hermite polynomials*. The wavefunctions are given in Table 11.1 and shown in Figure 11.8.

The vibrational energy levels ε_v are

$$\varepsilon_v = \left(v + \frac{1}{2}\right)h\nu, \tag{11.22}$$

where the quantum number is $v = 0, 1, 2, 3, \ldots$. $\nu = (1/2\pi)(k_s/m)^{1/2}$ is the vibrational frequency of the harmonic oscillator. Notice that the lowest energy

Table 11.1 Energy eigenfunctions of the simple harmonic oscillator. Source: AP French and EF Taylor, *An Introduction to Quantum Physics*, WW Norton and Co, New York, 1978.

Quantum Number υ	Energy Eigenvalue ε_υ	Energy Eigenfunction $\psi_\upsilon(x) = \left(\dfrac{1}{\upsilon!2^\upsilon a\sqrt{\pi}}\right)^{1/2} H_\upsilon\left(\dfrac{x}{a}\right) e^{-x^2/2a^2}$
0	$\frac{1}{2}h\nu$	$\left(\dfrac{1}{a\sqrt{\pi}}\right)^{1/2} e^{-x^2/2a^2}$
1	$\frac{3}{2}h\nu$	$\left(\dfrac{1}{2a\sqrt{\pi}}\right)^{1/2} 2\left(\dfrac{x}{a}\right) e^{-x^2/2a^2}$
2	$\frac{5}{2}h\nu$	$\left(\dfrac{1}{8a\sqrt{\pi}}\right)^{1/2} \left[2 - 4\left(\dfrac{x}{a}\right)^2\right] e^{-x^2/2a^2}$
3	$\frac{7}{2}h\nu$	$\left(\dfrac{1}{48a\sqrt{\pi}}\right)^{1/2} \left[12\left(\dfrac{x}{a}\right) - 8\left(\dfrac{x}{a}\right)^3\right] e^{-x^2/2a^2}$
4	$\frac{9}{2}h\nu$	$\left(\dfrac{1}{348a\sqrt{\pi}}\right)^{1/2} \left[12 - 48\left(\dfrac{x}{a}\right)^2 + 16\left(\dfrac{x}{a}\right)^4\right] e^{-x^2/2a^2}$

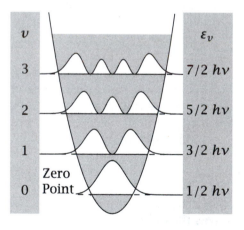

Figure 11.8 The parabola shows the harmonic oscillator potential. The horizontal lines show the energy levels, which are equally spaced for the quantum harmonic oscillator. The lowest level has energy $h\nu/2$. This is called the *zero-point energy*. The curve on each horizontal line shows $\psi^2(x)$, the particle distribution probability for each energy level.

level is not zero. The lowest level has energy $\varepsilon_0 = (1/2)h\nu$. This is called the *zero-point energy*. The system vibrates even at $T = 0$ K.

To apply the harmonic oscillator model to diatomic molecules, we need to generalize from a single particle on a spring to two masses, m_1 and m_2, connected by a spring (see Figure 11.9). This leads to only a small alteration of the theory. The vibrational energy levels are exactly as given in Equation (11.22), but now the vibrational frequency is

$$\nu = \left(\frac{1}{2\pi}\right)\left(\frac{k_s}{\mu}\right)^{1/2}, \tag{11.23}$$

where

$$\mu = \frac{m_1 m_2}{m_1 + m_2} \qquad (11.24)$$

is called the *reduced mass*.

You can get the single particle partition function by substituting the energy Equation (11.22) into the partition function Equation (10.23),

$$q_{\text{vibration}} = \sum_{v=0}^{\infty} e^{-(v+1/2)h\nu/kT} = e^{-h\nu/2kT}\left(1 + e^{-h\nu/kT} + e^{-2h\nu/kT} + \cdots\right)$$

$$= e^{-h\nu/2kT}(1 + x + x^2 + x^3 + \cdots), \qquad (11.25)$$

where $x = e^{-h\nu/kT}$. A vibrational temperature can be defined as $\theta_{\text{vibration}} = h\nu/k$. Using the series expansion in Appendix C, Equation (C.6), $(1-x)^{-1} = 1 + x + x^2 + \cdots$, for $0 < |x| < 1$, the vibrational partition function can be written more compactly as

$$q_{\text{vibration}} = \frac{e^{-h\nu/2kT}}{1 - e^{-h\nu/kT}}. \qquad (11.26)$$

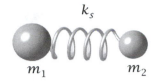

Figure 11.9 Vibrations in a diatomic molecule are modelled as two masses, m_1 and m_2, connected by a spring with constant k_s.

EXAMPLE 11.3 The vibrational partition function of O$_2$. Oxygen molecules have a vibrational frequency of $1580\,\text{cm}^{-1}$. To convert from cm^{-1} to ν, which has units of s^{-1}, multiply by the speed of electromagnetic radiation (light) to get $\nu = (1580\,\text{cm}^{-1})(2.997 \times 10^{10}\,\text{cm}\,\text{s}^{-1}) = 4.737 \times 10^{13}\,\text{s}^{-1}$. You have

$$\theta_{\text{vibration}} = \frac{h\nu}{k} = \frac{(6.626 \times 10^{-34}\,\text{J}\,\text{s})(4.737 \times 10^{13}\,\text{s}^{-1})}{1.38 \times 10^{-23}\,\text{J}\,\text{K}^{-1}} = 2274\,\text{K}.$$

At room temperature, $\theta_{\text{vibration}}/T = 2274\,\text{K}/300\,\text{K} = 7.58$. For the partition function at room temperature, Equation (11.26) gives $q_{\text{vibration}} = 1.0005$. Most oxygen molecules are in their vibrational ground states at this temperature. Even at $1000\,\text{K}$, most oxygen molecules are still in their vibrational ground states: $\theta_{\text{vibration}}/T = 2.27$ and $q_{\text{vibration}} = 1.11$.

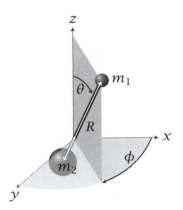

Figure 11.10 In the rigid rotor model of a diatomic molecule, masses m_1 and m_2 are separated by a rigid connector of length R. The origin is at the center of mass, and the angular degrees of freedom are θ and ϕ.

Rotations Can Be Treated by the Rigid Rotor Model

In classical mechanics, if a mass m orbits at a distance R around a center point and has tangential velocity v_t, its angular momentum is $L = mRv_t$ and its kinetic energy is $K = mv_t^2/2 = L^2/(2mR^2)$. For the corresponding quantum mechanical problem, L is an operator. Our interest here is in a related problem. The rotations of a diatomic molecule can be modelled as masses m_1 and m_2 separated by a rigid connector of length R (see Figure 11.10). The Hamiltonian operator is a function of the two angular degrees of freedom θ and ϕ,

$$\mathcal{H}\psi(\theta,\phi) = \frac{L^2\psi}{2\mu R^2} = \frac{-h^2}{8\pi^2\mu R^2}\left(\frac{1}{\sin\theta}\frac{\partial}{\partial\theta}\left(\sin\theta\frac{\partial}{\partial\theta}\right) + \frac{1}{\sin^2\theta}\frac{\partial^2}{\partial\phi^2}\right)\psi$$

$$= \varepsilon\psi, \qquad (11.27)$$

where μ is the reduced mass given by Equation (11.24), and $I = \mu R^2$ is the *moment of inertia*. The wavefunctions that satisfy Equation (11.27) are called the *spherical harmonics* $Y_{\ell,m}(\theta, \phi)$,

$$\psi(\theta, \phi) = A_{\ell,m} Y_{\ell,m}(\theta, \phi) = A_{\ell,m} P_{\ell,m}(\theta) e^{im\phi},$$

where $\ell = 0, 1, 2, \ldots$ and $m = -\ell, -\ell+1, \ldots 0, 1, \ldots \ell-1, \ell$ are the two quantum numbers, the $P(\theta)$'s are called *Legendre polynomials* [1], and the $A_{\ell,m}$'s are constants. The energy levels for the rigid rotor are

$$\varepsilon_\ell = \frac{\ell(\ell + 1)h^2}{8\pi^2 I}, \tag{11.28}$$

and the rotational temperature is $\theta_{\text{rotation}} = \varepsilon_\ell/k$. Rigid rotors have energy levels at $\ell(\ell + 1) = 0, 2, 6, 12, 20, 30, \ldots$.

For each value of ℓ, there is a degeneracy due to the $2\ell + 1$ values of m. So the rotational partition function is

$$q_{\text{rotation}} = \sum_{\ell=0}^{\infty} (2\ell + 1)e^{-\varepsilon_\ell/kT}. \tag{11.29}$$

For high temperatures, ($T \gg \theta_{\text{rotation}}$), Equation (11.29) can be approximated by an integral to give

$$q_{\text{rotation}} = \frac{T}{\sigma \theta_{\text{rotation}}} = \frac{8\pi^2 IkT}{\sigma h^2}, \tag{11.30}$$

where σ is a nuclear and rotation symmetry factor that accounts for the number of equivalent orientations of the molecule ($\sigma = 1$ for heteronuclear diatomic molecules; $\sigma = 2$ for homonuclear diatomic molecules; $\sigma = 2$ for H_2O; $\sigma = 12$ for CH_4, and $\sigma = 12$ for benzene, for example). Think of σ as a correction for overcounting.

For nonlinear molecules with three principal moments of inertia I_a, I_b, and I_c, the rotational partition function is

$$q_{\text{nonlinear rotation}} = \frac{(\pi I_a I_b I_c)^{1/2}}{\sigma} \left(\frac{8\pi^2 kT}{h^2}\right)^{3/2}. \tag{11.31}$$

In general, molecules with large moments of inertia have large rotational partition functions.

EXAMPLE 11.4 The rotational partition function of O_2. The length of the bond in an oxygen molecule is $R = 1.2074\,\text{Å}$, and the mass of each atom is $m = 16\,\text{g mol}^{-1}$, so the moment of inertia per molecule is

$$I = \mu R^2 = \frac{mR^2}{2} = \frac{\left(0.016\,\frac{\text{kg}}{\text{mol}}\right)(1.2074 \times 10^{-10}\,\text{m})^2}{2\left(6.02 \times 10^{23}\,\frac{\text{molecules}}{\text{mol}}\right)}$$

$$= 1.937 \times 10^{-46}\,\text{kg m}^2.$$

The rotational temperature is

Table 11.2 Properties of linear molecules, useful for computing partition functions.

Molecule	R^b (Å)	σ^a	I^b (amu Å^2)	$\theta_{rotation}^b$ (K)	$\theta_{vibration}^b$ (K)	$\theta_{electronic}^c$ (K)	g_0^a	D_0^a (kJ mol^{-1})
H_2	0.7415	2	0.2771	87.53	6338	129,000	1	432.1
HCl	1.2744	1	1.593	15.24	4302	61,800	1	427.8
N_2	1.0976	2	8.437	2.874	3395	97,400	1	941.4
CO	1.1282	1	8.733	2.777	3122	91,500	1	1070.1
NO	1.1506	1	9.888	2.453	2739	61,700	2	627.7
O_2	1.2074	2	11.663	2.080	2274	11,100	3	491.9
Cl_2	1.988	2	70.06	0.346	813	25,700	1	239.2
I_2	2.666	2	451.0	0.0538	308	16,700	1	148.8

Source: ^{a}RA Alberty and RJ Silbey, *Physical Chemistry*, 1st edition, Wiley, New York, 1992; Data are from M Chase et al. JANAF Thermochemical Tables, *J Phys Chem Ref Data* **14**, Supplement 1 (1985); ^{b}JH Knox, *Molecular Thermodynamics: an Introduction to Statistical Mechanics for Chemists*, Rev edition, New York, Wiley, 1978; ^{c}RS Berry, SA Rice and J Ross, with the assistance of GP Flynn and JN Kushick, *Physical Chemistry*, Wiley, New York, 1980.

$$\theta_{rotation} = \frac{h^2}{8\pi^2 I k} = \frac{(6.626 \times 10^{-34}\,\mathrm{J\,s})^2}{8\pi^2\,(1.937 \times 10^{-46}\,\mathrm{kg\,m^2})\,(1.38 \times 10^{-23}\,\mathrm{J\,K^{-1}})}$$

$$= 2.08\,\mathrm{K}.$$

Since $\sigma = 2$ for a homonuclear diatomic molecule, the rotational partition function at room temperature is (see Equation (11.30))

$$q_{rotation} = \frac{300\,\mathrm{K}}{2(2.08\,\mathrm{K})} = 72. \tag{11.32}$$

Therefore many rotational states are populated at room temperature. Table 11.2 gives rotational and vibrational constants for several small molecules.

The Electronic Partition Function

The partition function that accounts for the various states of electronic excitation is [1, 2]

$$q_{electronic} = g_0 + g_1 e^{-\Delta\varepsilon_1/kT} + g_2 e^{-\Delta\varepsilon_2/kT} + \dots, \tag{11.33}$$

where the g_i's are the electronic degeneracies and the $\Delta\varepsilon_i$'s are the electronic excited state energies. Table 11.2 shows that the first excited state energies for diatomic molecules are quite large relative to $T = 300\,\mathrm{K}$. Typically $\theta_e = \Delta\varepsilon_1/k \approx 10^4 - 10^5\,\mathrm{K}$. In those cases, the exponential terms approach zero and you can express Equation (11.33) as $q_{\text{electronic}} \approx g_0$. The ground-state electronic degeneracies g_0 are shown in Table 11.2.

For computing chemical equilibria (Chapter 13), you will need $q_{\text{electronic}}$ not just for molecules, but also for individual atoms. For atoms, $q_{\text{electronic}}$ depends on the electronic configuration of the atom (see Table 11.3). For example the $1s$ configuration of the H atom has degeneracy $g_0 = 2$, and the $2p$ configuration has $g_1 = 2$. But the $2p$ orbital has a high energy, so at temperatures T below about $kT = 10.20$ eV, the only contribution to the electronic partition function is $g_0 = 2$. Table 11.3 shows that for oxygen atoms, $g_0 = 5$, $g_1 = 3$, and $g_2 = 1$, and the energies of the first two excited states are small enough for you to retain all three terms except at the very lowest temperatures. Similarly the first two levels can contribute for fluorine. Atomic degeneracies can be found in [3]. Electronic configurations can be described as *term symbols*, where the lower right subscript ($1/2$ for H $1s$ for example) is J and the degeneracy is given by $g = 2J + 1$.

The Total Partition Function

Atoms and molecules can store energies using all their various degrees of freedom. To compute thermodynamic properties, you need the total partition function q, which is the product of terms,

$$q = q_{\text{translation}} q_{\text{rotation}} q_{\text{vibration}} q_{\text{electronic}}. \tag{11.34}$$

For example, for a linear diatomic molecule with $q_{\text{electronic}} = 1$, you will have a total partition function

$$q = \left[\frac{2\pi m k T}{h^2} \right]^{3/2} V \left[\frac{8\pi^2 I k T}{\sigma h^2} \right] \left[\frac{e^{-h\nu/2kT}}{1 - e^{-h\nu/kT}} \right]. \tag{11.35}$$

Ideal Gas Properties Are Predicted by Quantum Mechanics

Partition functions based on quantum mechanics can provide a more complete description of the ideal gas than our simple lattice model could. In this section, we compute the properties of ideal gases.

Ideal Gas Pressure

We derived the ideal gas law from a lattice model in Example 7.1. Now we derive the gas law from quantum mechanics instead. Equation (11.35) shows that the partition function for a gas is a product of volume V and a volume-independent factor we'll call q_0. Substituting $q = q_0 V$ into Equation (11.46) gives $F = -NkT \ln V - NkT \ln(eq_1/N)$. Taking the derivative that defines pressure in Equation (8.12) gives the ideal gas law,

$$p = -\left(\frac{\partial F}{\partial V} \right)_{T,N} = \frac{NkT}{V}.$$

Table 11.3 Atomic energy states used to obtain the electronic partition functions of atoms.

Atom	Electron Configuration	Term Symbol	Degeneracy $g = 2J + 1$	Energy (eV)
H	$1s$	$^2S_{1/2}$	2	0
	$2p$	$^2P_{1/2}$	2	10.20
	$2s$	$^2S_{1/2}$	2	10.20
	$2p$	$^2P_{3/2}$	4	10.20
He	$1s^2$	$^1S_0$	1	0
	$1s2s$	$^3S_1$	3	19.82
		$^1S_0$	1	20.62
Li	$1s^22s$	$^2S_{1/2}$	2	0
	$1s^22p$	$^2P_{1/2}$	2	1.85
		$^2P_{3/2}$	4	1.85
	$1s^23s$	$^2S_{1/2}$	2	3.37
O	$1s^22s^22p^4$	$^3P_2$	5	0
		$^3P_1$	3	0.02
		$^3P_0$	1	0.03
		$^1D_2$	5	1.97
		$^1S_0$	1	4.19
F	$1s^22s^22p^5$	$^2P_{3/2}$	4	0
		$^2P_{1/2}$	2	0.05
	$1s^22s^22p^43s$	$^4P_{5/2}$	6	12.70
		$^4P_{3/2}$	4	12.73
		$^4P_{1/2}$	2	12.75
		$^2P_{3/2}$	4	13.0
		$^2P_{1/2}$	2	13.0

Source: DA McQuarrie, *Statistical Mechanics*, Harper & Row, New York, 1976, and CE Moore, Atomic Energy States, *Natl Bur Standards, Circ* **1**, 467 (1949).

Ideal Gas Energy

We noted in Example 8.8 (page 147) that the internal energy of an ideal gas depends only on its temperature and not on its volume: $U = U(T)$. The lattice model doesn't give a basis for understanding the internal energy or its temperature dependence, but quantum mechanics does. Equation (10.34) expresses the internal energy U in terms of the partition function q and the number of particles N for either distinguishable or indistinguishable particles (see Problem 8, Chapter 10):

$$U = NkT^2 \left(\frac{\partial \ln q}{\partial T} \right).$$

For the purpose of taking the temperature derivative, express q as a product of its temperature-dependent and temperature-independent parts. For example, for translational freedom, Equation (11.18) would factor into $q = c_0 T^{3/2}$, where $c_0 = (2\pi m k/h^2)^{3/2} V$ is independent of temperature. Substitute this expression for q into Equation (10.34) to get

$$U = \frac{NkT^2}{q}\left(\frac{\partial q}{\partial T}\right) = \frac{NkT^2}{c_0 T^{3/2}}\left(\frac{3}{2}c_0 T^{1/2}\right) = \frac{3}{2}NkT. \tag{11.36}$$

The translational contribution to the average particle energy is $\langle \varepsilon \rangle = U/N = (3/2)kT$. This is the internal energy of an ideal gas.

Because the energy of an ideal gas does not depend on volume, the heat capacity does not either:

$$C_V = \left(\frac{\partial U}{\partial T}\right) = \frac{3}{2}Nk. \tag{11.37}$$

Rotations also contribute to the internal energy. For nonlinear gas molecules,

$$q_{\text{rotation}} = c_1 T^{3/2}, \tag{11.38}$$

where $c_1 = ((\pi I_a I_b I_c)^{1/2}/\sigma)(8\pi^2 k/h^2)^{3/2}$ is a temperature-independent constant. Substitute Equation (11.38) into (11.36) to find the component of the internal energy of nonlinear molecules that is due to rotation:

$$U_{\text{rotation (nonlinear)}} = \frac{3}{2}NkT. \tag{11.39}$$

Likewise for linear gas molecules the rotational internal energy is

$$U_{\text{rotation(linear)}} = NkT. \tag{11.40}$$

The total internal energy of an ideal gas is given as a sum of terms: $(1/2)kT$ for each of the three translational degrees of freedom, and $(1/2)kT$ per rotational degree of freedom. For weak vibrational modes of motion, $q \gg 1$, $NkT(\partial \ln q_{\text{vib}}/\partial T)$ contributes an energy of kT per vibrational mode.

Ideal Gas Entropy

A fundamental validation of the indistinguishability of gas molecules, and of the expression $Q = q^N/N!$ for the system partition function of a gas, is provided by the absolute entropies of gases.

The only contribution to the entropy of an ideal monatomic gas is from its translational freedom. Combining $Q = q^N/N!$, where $q = q_{\text{translation}}$ from Equation (11.18), with $U = (3/2)NkT$ and Stirling's approximation of Equation (11.39) gives the **Sackur–Tetrode** equation for the absolute entropy of a monatomic ideal gas:

$$S = k\ln\left(\frac{q^N}{N!}\right) + \frac{U}{T} = Nk\ln q - k(N\ln N - N) + \frac{3}{2}Nk$$

$$= Nk\left(\ln q - \ln N + \frac{5}{2}\right) = Nk\ln\left(\frac{qe^{5/2}}{N}\right)$$

$$= Nk \ln \left(\left(\frac{2\pi mkT}{h^2} \right)^{3/2} \left(\frac{e^{5/2}}{N} \right) V \right), \tag{11.41}$$

where V is the volume. Example 11.5 shows how the Sackur-Tetrode equation is applied.

EXAMPLE 11.5 Calculating the absolute entropy of argon from the Sackur–Tetrode equation.   Let's calculate the entropy of argon at $T = 300\,\text{K}$ and $p = 1\,\text{atm} = 1.01 \times 10^5\,\text{N m}^{-2}$. For argon, the Sackur–Tetrode Equation (11.41) is

$$S = Nk \ln \left(\frac{q_{\text{translation}} e^{5/2}}{N} \right). \tag{11.42}$$

We want $q_{\text{translation}}$ at $T = 300\,\text{K}$. Use $q_{\text{translation}}$ computed at $T = 273\,\text{K}$ from Example 11.2, and correct for the difference in temperature,

$$q_{\text{translation}}(300) = q_{\text{translation}}(273) \left(\frac{300\,\text{K}}{273\,\text{K}} \right)^{3/2}$$

$$= (2.14 \times 10^{32} V)(1.15)$$

$$= 2.47 \times 10^{32} V.$$

Substitute this into Equation (11.42) to get

$$\frac{S}{N} = k \ln \left(2.47 \times 10^{32} \left(\frac{V}{N} \right) e^{5/2} \right). \tag{11.43}$$

Substitute kT/p from the ideal gas law for V/N:

$$\frac{V}{N} = \frac{kT}{p} = \frac{(1.38 \times 10^{-23}\,\text{JK}^{-1}\ \text{per molecule})(300\,\text{K})}{1.01 \times 10^5\,\text{N m}^2}$$

$$= 4.10 \times 10^{-26},$$

$$\implies \frac{S}{N} = (8.31\,\text{J K}^{-1}\,\text{mol}^{-1}) \ln \left(2.47 \times 10^{32} (4.10 \times 10^{-26}) e^{5/2} \right)$$

$$= 154.8\,\text{J K}^{-1}\,\text{mol}^{-1}. \tag{11.44}$$

The value of S/N calculated here is likely to be more accurate than the thermodynamically measured value of the entropy for argon ($155\,\text{J K}^{-1}\text{mol}^{-1}$; see Table 11.4), because of the typical sizes of the errors in thermodynamic measurements.

The Sackur–Tetrode Equation (11.41) also predicts how the entropy changes with volume for an ideal gas at constant temperature:

$$\Delta S = S(V_B) - S(V_A) = Nk \ln \left(\frac{V_B}{V_A} \right), \tag{11.45}$$

which is the same result that you would get from the lattice model gas in Example 7.1, page 109.

Source: [a]RS Berry, SA Rice and J Ross, *Physical Chemistry*, Wiley, New York, 1980.
[b]W Kauzmann, *Thermodynamics and Statistics*, WA Benjamin Inc, New York, 1967. Data are from EF Westrum and JP McCullough, *Physics and Chemistry of the Organic Solid State*, D Fox, MM Labes, and A Weissberger, eds., Chapter 1, Interscience, 1963; DD Wagman et al., *Selected Values of Chemical Thermodynamic Properties*, National Bureau of Standards Technical Note 270-1, Washington, DC, 1965.

Table 11.4 Gas entropies determined calorimetrically and from statistical mechanics.

Substance	Statistical Entropy ($J\,K^{-1}\,mol^{-1}$)	Calorimetric Entropy ($J\,K^{-1}\,mol^{-1}$)
Ne^a	146.22	146.0
Ar^a	154.74	155.0
$N_2{}^b$	191.50	192.05
$O_2{}^b$	209.19	205.45
$Cl_2{}^a$	222.95	223.1
HCl^a	186.68	186.0
HBr^a	198.48	199.0
$NH_3{}^b$	192.34	192.09
$CO_2{}^a$	213.64	214.00
CH_3Cl^b	234.22	234.05
CH_3Br^a	243.0	242.0
$CH_3NO_2{}^b$	275.01	275.01
$C_2H_4{}^b$	219.33	219.58
cyclopropane[b]	227.02	226.65
benzene[b]	269.28	269.70
toluene[b]	320.83	321.21

Ideal Gas Free Energy

The Helmholtz free energy of an ideal gas is found by substituting $Q = q^N/N!$ into Equation (10.40):

$$F = -kT \ln Q = -kT \ln \frac{q^N}{N!} = -kT \ln \left(\frac{eq}{N} \right)^N$$
$$= -NkT \ln \left(\frac{eq}{N} \right), \tag{11.46}$$

where the factor of e comes from Stirling's approximation for $N!$

Ideal Gas Chemical Potential

Chapters 13 through 16 focus on the chemical potential, the quantity that describes the tendency for molecules to move from one place to another, or from one phase to another. It is the basis for treating boiling and freezing, for describing the partitioning of solute molecules between phases, and for treating chemical reaction equilibria and kinetics. To treat the gas phase in these processes, the key relationship that we will need is $\mu(p)$, the chemical potential as a function of pressure.

Combining the definition of chemical potential $\mu = (\partial F/\partial N)_{T,V}$ with Equation (11.46) gives the chemical potential of an ideal gas in terms of its partition function:

$$\mu = kT \ln\left(\frac{N}{eq}\right) + kT = -kT \ln\left(\frac{q}{N}\right). \tag{11.47}$$

Again we aim to factor the partition function, based on the derivative we will need to take, into a pressure-dependent term and a term for all the pressure-independent quantities. Express the partition function, $q = q_0 V$ as a product of the volume V and q_0, the partition function with the volume factored out. Use the ideal gas law $V = NkT/p$ to get

$$\frac{q}{N} = \frac{q_0 V}{N} = \frac{q_0 kT}{p}. \tag{11.48}$$

The quantity $q_0 kT$ has units of pressure,

$$p_{int}^\circ = q_0 kT = kT \left(\frac{2\pi mkT}{h^2}\right)^{3/2} q_{rotation} q_{vibration} q_{electronic}. \tag{11.49}$$

Our notation indicates that the quantity p_{int}° represents properties *internal* to the molecule. Substituting Equations (11.48) and (11.49) into Equation (11.47) gives

$$\mu = \mu^\circ + kT \ln p = kT \ln \frac{p}{p_{int}^\circ}, \tag{11.50}$$

where $\mu^\circ = -kT \ln p_{int}^\circ$. Equation (11.50) makes explicit how the chemical potential depends on the applied pressure p. Because p_{int}° has units of pressure, the rightmost side of Equation (11.50) shows that you are taking the logarithm of a quantity that is dimensionless, even when the chemical potential is written as $\mu^\circ + kT \ln p$. The quantities p_{int}° and μ° with superscripts $^\circ$ are called the *standard state* pressure and chemical potential. We will use Equation (11.50) throughout Chapters 13–16.

EXAMPLE 11.6 The standard state pressure of helium. To calculate p_{int}° for helium at $T = 300\,\text{K}$, first use Equation (11.19) to get

$$\Lambda^3 = \left[\frac{h^2}{2\pi mkT}\right]^{3/2}$$

$$= \left[\frac{(6.626 \times 10^{-34}\,\text{J s})^2}{2\pi(4 \times 1.66 \times 10^{-27}\,\text{kg})(1.38 \times 10^{-23}\,\text{J K}^{-1})(300\,\text{K})}\right]^{3/2}$$

$$= 1.28 \times 10^{-31}\,\text{m}^3 = 0.13\,\text{Å}^3.$$

Then

$$p_{int}^\circ = \frac{kT}{\Lambda^3} = \frac{(1.38 \times 10^{-23}\,\text{J K}^{-1})(300\,\text{K})}{1.28 \times 10^{-31}\,\text{m}^3}$$

$$= 3.2 \times 10^{10}\,\text{Pa} = 3.2 \times 10^5\,\text{atm}.$$

The Equipartition Theorem Says That Energies Are Uniformly Distributed to Each Degree of Freedom

On page 208 we noted that the average energies for gases are integral multiples of $(1/2)kT$ for rotational and translational degrees of freedom. This is a manifestation of the principle of equipartition of energy. Where does this principle come from?

When a form of energy depends on a degree of freedom x, the average energy $\langle \varepsilon \rangle$ of that component is

$$\langle \varepsilon \rangle = \frac{\displaystyle\sum_{\text{all } x} \varepsilon(x) e^{-\varepsilon(x)/kT}}{\displaystyle\sum_{\text{all } x} e^{-\varepsilon(x)/kT}}. \tag{11.51}$$

When quantum effects are unimportant so that a large number of states are populated, the sums in Equation (11.51) can be approximated by integrals and the average energy $\langle \varepsilon \rangle$ is

$$\langle \varepsilon \rangle = \frac{\displaystyle\int_{-\infty}^{\infty} \varepsilon(x) e^{-\varepsilon(x)/kT}\, dx}{\displaystyle\int_{-\infty}^{\infty} e^{-\varepsilon(x)/kT}\, dx}, \tag{11.52}$$

for $-\infty \leq x \leq \infty$. For many types of degrees of freedom, the energy is a square-law function, $\varepsilon(x) = cx^2$, where $c > 0$. Perform the integrations using Appendix D Equations (D.1) and (D.3). Equation (11.52) becomes

$$\langle \varepsilon \rangle = \frac{\displaystyle\int_{-\infty}^{\infty} cx^2 e^{-cx^2/kT}\, dx}{\displaystyle\int_{-\infty}^{\infty} e^{-cx^2/kT}\, dx} = \frac{1}{2}kT. \tag{11.53}$$

The constant c gives the scale for the energy level spacings, but note that $\langle \varepsilon \rangle$ does not depend on c.

Square-law relations hold for many types of degrees of freedom. For translations and rotations, the energy depends on the square of the appropriate quantum number:

$$\varepsilon_n^{\text{translation}} = \left(\frac{h^2}{8mL^2}\right) n^2, \quad \text{and} \quad \varepsilon_\ell^{\text{rotation}} = \left(\frac{h^2}{8\pi^2 I}\right) \ell(\ell + 1).$$

In classical mechanics, masses on springs, and balls in quadratic valleys, have square-law potentials. Even for more complex potentials, the lowest-order contribution to the energy is often a second-order, or square-law, dependence.

When equipartition applies and $kT/2$ is the average energy per degree of freedom, you need only count the number of degrees of freedom to compute the total average energy. Therefore you have energies of $3 \times (1/2)kT$ for translation in three dimensions, and $(1/2)kT$ for every rotational degree of freedom. Heat capacities, too, are additive according to equipartition, because for any two types of equipartitional energy ε_A and ε_B,

$$C_V = N\frac{\partial}{\partial T}\left(\langle \varepsilon_A \rangle + \langle \varepsilon_B \rangle\right) = N\left[\left(\frac{\partial \langle \varepsilon_A \rangle}{\partial T}\right) + \left(\frac{\partial \langle \varepsilon_B \rangle}{\partial T}\right)\right] = C_{V_A} + C_{V_B}.$$

The equipartition theorem says that energy in the amount of $kT/2$ partitions into each independent square-law degree of freedom, provided that the temperature is high enough for the sum in Equation (11.51) to be suitably approximated by the integral in Equation (11.52).

Equipartition of Vibrations

Quantum mechanical vibrations do not obey a square-law potential. Quantum vibrational energies depend only linearly on the vibrational quantum number, $\varepsilon_v = (v + 1/2)h\nu$. Therefore the equipartition theorem for vibrations is different from Equation (11.53). Using Appendix D, Equation (D.9), the vibrational equipartition theorem is

$$\langle \varepsilon \rangle = \frac{\int_0^\infty cx e^{-cx/kT}\,dx}{\int_0^\infty e^{-cx/kT}\,dx} = kT. \tag{11.54}$$

Energy in the amount of kT will partition equally into every vibrational degree of freedom at high temperatures. Equipartition does not apply when the temperature is low, because then the sum (Equation (11.51)) cannot be replaced by an integral (Equation (11.52)). At low temperatures, various degrees of freedom are 'frozen out', as shown in Figures 11.11, 11.12, and 11.13. Those degrees of freedom are unable to store energy because the energy difference from the ground state to the first excited state is too large for thermal energies to induce the system to populate the excited state.

The Einstein Model of Solids

Just like gases, solids can lose their ability to absorb energy at low temperatures. According to equipartition Equation (11.54), each vibration contributes kT to the energy. If there are N atoms in a solid, and each atom has three vibrational modes (in the x-, y-, and z-directions), the heat capacity will be $C_V = 3Nk$, independently of temperature. This is called the Law of Dulong and Petit, named after the experimentalists who first observed this behavior around 1819. But more recent experimental data, such as that shown in Figures 11.14 and 11.15, indicate that this law does not hold at low temperatures. As the temperature approaches zero, $C_V \to 0$. The Einstein model, developed in 1907, shows why. This work was among the first evidence for the quantum theory of matter.

Einstein assumed that there are $3N$ independent distinguishable oscillators. The vibrational partition function (Equation (11.26)) corrected for the 'zero-point energy' (see Equation (13.21)) is $q = (1 - e^{-\beta h\nu})^{-1}$. The average energy per vibrational mode is given by the derivative in Equation (10.34):

$$\langle \varepsilon \rangle = -\frac{1}{q}\left(\frac{\partial q}{\partial \beta}\right) = h\nu \left(\frac{e^{-\beta h\nu}}{1 - e^{-\beta h\nu}}\right). \tag{11.55}$$

Now take $3N$ times the temperature derivative of Equation (11.55) to get the

(a) Heat Capacity

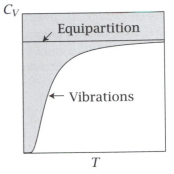

(b) Average Energy

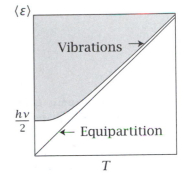

Figure 11.11 (a) Equipartition predicts heat capacities C_V that are constant and independent of temperature T. But at very low temperatures vibrational heat capacities become small because vibrational degrees of freedom 'freeze out.' (b) The corresponding average particle energy $\langle \varepsilon \rangle$ versus temperature.

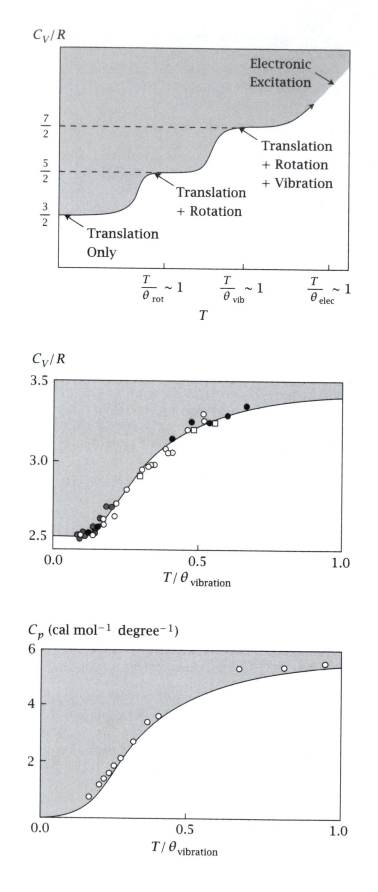

Figure 11.12 Dependence of the heat capacity on temperature for an ideal gas. Cooling a hot gas leads to freezing out of degrees of freedom: first electronic freedom, then vibrations, then rotations. Source: RS Berry, SA Rice and J Ross, with the assistance of GP Flynn and JN Kushick, *Physical Chemistry*, Wiley, New York, 1980.

Figure 11.13 Experimental data showing the freezing out of vibrations in diatomic gases, (●) CO, (◉) N_2, (□) Cl_2, and (○) O_2. Source: R Fowler and EA Guggenheim, *Statistical Thermodynamics*, Cambridge University Press, London, 1956. Data are from Henry, *Proc Roy Soc A* **133**, 492 (1931); Euken and von Lude, *Zeit Physikal Chem B* **5**, 413 (1929), Sharratt and Griffiths, *Proc Roy Soc A* **147**, 292 (1934).

Figure 11.14 Experimental values of the heat capacity of diamond (○) compared with values calculated by the Einstein model (——), using the characteristic temperature $\theta_{\text{vibration}} = h\nu/k = 1320\,\text{K}$. Vibrations are frozen out at low temperatures. Source: RJ Borg and GJ Dienes, *The Physical Chemistry of Solids*, Academic Press Inc, San Diego, 1992. The data are from A Einstein, *Ann Phys* **22**, 180 (1907).

Specific Heat $(\text{J g}^{-1}\text{ K}^{-1})$

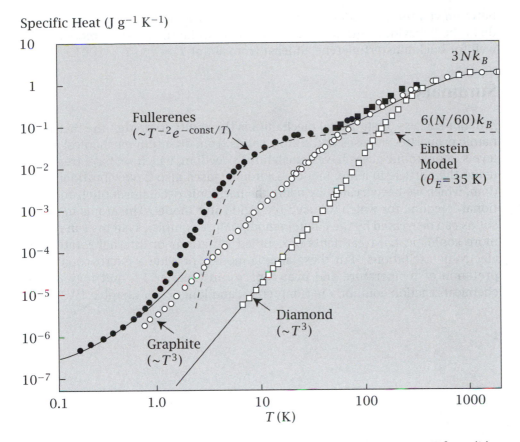

Figure 11.15 Specific heat (heat capacity per gram) versus temperature T for solids: diamond (□), graphite (○), and fullerenes (●). This log–log plot emphasizes the behavior at low temperatures. The Einstein model of independent oscillators $(\sim T^{-2}e^{-a/T})$ characterizes fullerenes from about $T = 5\,\text{K} - 100\,\text{K}$, but the more sophisticated Debye model of coupled oscillators $\sim T^3$ is a better model for diamond and graphite at low temperatures. Source: JR Olson, KA Topp and RO Pohl, *Science* **259**, 1145–1148 (1993).

heat capacity of the solid,

$$C_V = 3N\left(\frac{\partial\langle\varepsilon\rangle}{\partial T}\right) = 3Nk\left(\frac{h\nu}{kT}\right)^2\frac{e^{-h\nu/kT}}{\left(1 - e^{-h\nu/kT}\right)^2}. \tag{11.56}$$

(To see the steps of this derivative worked out, look at Equations (10.46) through (10.48) but replace the plus sign in the denominator of Equation (10.47) with a minus sign.)

Figure 11.14 shows that Equation (11.56) captures the freezing out of vibrational modes in diamonds at low temperatures. However, very precise measurements on solids show that the Einstein model is not quantitatively correct at extremely low temperatures. As $T \to 0$, Equation (11.56) predicts the temperature dependence $C_V \sim T^{-2}e^{-h\nu/kT}$. But very accurate experiments show that $C_V \sim T^3$ at extremely low temperatures. The T^3 dependence is predicted by the more sophisticated Debye theory of solids, which accounts for the coupling between vibrational modes [4]. Figure 11.15 shows that the Debye model predicts the behavior of graphite and diamonds, while the Einstein model works well for buckminsterfullerene, at least from about $T = 1$ to $T = 100 \, \text{K}$.

Summary

A major success of statistical mechanics is the ability to predict the thermodynamic properties of gases and simple solids from quantum mechanical energy levels. Monatomic gases have translational freedom, which we have treated by using the particle-in-a-box model. Diatomic gases also have vibrational freedom, which we have treated by using the harmonic oscillator model, and rotational freedom, for which we used the rigid-rotor model. The atoms in simple solids can be treated by the Einstein model. More complex systems can require more sophisticated treatments of coupled vibrations or internal rotations or electronic excitations. But these simple models provide a microscopic interpretation of temperature and heat capacity in Chapter 12, and they predict chemical reaction equilibria in Chapter 13, and kinetics in Chapter 19.

Problems

1. The heat capacity of an ideal gas. What is the heat capacity C_V of an ideal gas of argon atoms?

2. The statistical mechanics of oxygen gas. Consider a system of one mole of O_2 molecules in the gas phase at $T = 273.15\,K$ in a volume $V = 22.4 \times 10^{-3}\,m^3$. The molecular weight of oxygen is 32.

(a) Calculate the translational partition function $q_{translation}$.

(b) What is the translational component of the internal energy per mole?

(c) Calculate the constant-volume heat capacity.

3. The statistical mechanics of a basketball. Consider a basketball of mass $m = 1\,kg$ in a basketball hoop. To simplify, suppose the hoop is a cubic box of volume $V = 1\,m^3$.

(a) Calculate the lowest two energy states using the particle-in-a-box approach.

(b) Calculate the partition function at $T = 300\,K$. Show whether quantum effects are important or not. (Assume that they are important only if q is smaller than about 10.)

4. The statistical mechanics of an electron. Calculate the two lowest energy levels for an electron in a box of volume $V = 1\,\text{Å}^3$ (This is an approximate model for the hydrogen atom). Calculate the partition function at $T = 300\,K$. Are quantum effects important?

5. The translational partition function in two dimensions. When molecules adsorb on a two-dimensional surface, they have one less degree of freedom than in three dimensions. Write the two-dimensional translational partition function for an otherwise structureless particle.

6. The accessibility of rotational degrees of freedom. Diatomic ideal gases at $T = 300\,K$ have rotational partition functions of approximately $q = 200$. At what temperature would q become small (say $q < 10$) so that quantum effects become important?

7. The statistical thermodynamics of harmonic oscillations. Write the internal energy, entropy, enthalpy, free energy, and pressure for a system of N independent distinguishable harmonic oscillators.

8. Orbital steering in proteins. To prove that proteins do not require 'orbital steering,' a process once proposed to orient a substrate with high precision before binding, T. Bruice has calculated the dependence of the total energy on the rotational conformation of the hydroxymethylene group of 4-hydroxybutyric acid at $T = 300\,K$. Assume that the curve in Figure 11.16 is approximately parabolic, $\varepsilon = (1/2)k(\alpha - \alpha_0)^2$, where α is the dihedral angle of rotation. Use the equipartition theorem.

(a) Determine the spring constant k.

(b) What is the average energy $\langle \varepsilon \rangle$?

(c) What is the rms dihedral angle $\langle \alpha^2 \rangle^{1/2}$?

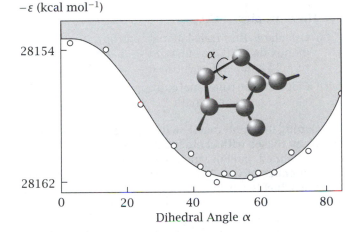

Figure 11.16 Source: TC Bruice, *Cold Spring Harbor Symposia on Quantitative Biology* **36**, 21–27 (1972).

9. The entropy of crystalline carbon monoxide at $T = 0\,K$. Carbon monoxide doesn't obey the 'third law' of thermodynamics: that is, its entropy is not zero when the temperature is zero. This is because molecules can pack in either the C≡O or O≡C direction in the crystalline state. For example, one packing arrangement of twelve CO molecules could be:

C≡O	C≡O	C≡O	C≡O
C≡O	C≡O	O≡C	O≡C
O≡C	O≡C	C≡O	C≡O

Calculate the partition function and the entropy of a carbon monoxide crystal at $T = 0\,K$.

10. Temperature-dependent quantities in statistical thermodynamics. Which quantities depend on temperature?

(a) Planck's constant h

(b) partition function q

(c) energy levels ε_j

(d) average energy $\langle \varepsilon \rangle$

(e) heat capacity C_V for an ideal gas.

11. Heat capacities of liquids.

(a) C_V for liquid argon (at $T = 100\,\text{K}$) is $18.7\,\text{J(K mol)}^{-1}$. How much of this heat capacity can you rationalize on the basis of your knowledge of gases?

(b) C_V for liquid water at $T = 10\,°\text{C}$ is about $75\,\text{J(K mol)}^{-1}$. Assuming water has three vibrations, how much of this heat capacity can you rationalize on the basis of gases? What is responsible for the rest?

12. The entropies of CO.

(a) Calculate the translational entropy for carbon monoxide CO (C has mass $m = 12$ amu, O has mass $m = 16$ amu) at $T = 300\,\text{K}$, $p = 1\,\text{atm}$.

(b) Calculate the rotational entropy for CO at $T = 300\,\text{K}$. The CO bond has length $R = 1.128 \times 10^{-10}\,\text{m}$.

13. Conjugated polymers: why the absorption wavelength increases with chain length.
Polyenes are linear double-bonded polymer molecules $(\text{C}=\text{C}-\text{C})_N$, where N is the number of $\text{C}=\text{C}-\text{C}$ monomers. Model a polyene chain as a box in which π-electrons are particles that can move freely. If there are $2N$ carbon atoms each separated by bond length $d = 1.4\,\text{Å}$, and if the ends of the box are a distance d past the end C atoms, then the length of the box is $\ell = (2N + 1)d$. An energy level, representing the two electrons in each bond, is occupied by two paired electrons. Suppose the N lowest levels are occupied by electrons, so the wavelength absorption of interest involves the excitation from level N to level $N + 1$. Compute the absorption energy $\Delta\varepsilon = \varepsilon_{N+1} - \varepsilon_N = hc/\lambda$ where c is the speed of light and λ is the wavelength of absorbed radiation, using the particle-in-a-box model.

14. Why are conjugated bonds so stiff?
As in problem 13, model polyene chain boxes of length $\ell \approx 2Nd$, where d is the average length of each carbon–carbon separation, and $2N$ is the number of carbons. There are $2N$ electrons in N energy levels, particles distributed throughout 'boxes,' according to the Pauli principle, with at most two electrons per level.

(a) Compute the total energy.

(b) Compute the total energy if the chain is 'bent,' that is if there are two boxes, each of length $\ell/2$ containing N electrons each.

15. Electrons flowing in wires carry electrical current.
Consider a wire 1 m long and $10^{-4}\,\text{m}^2$ in cross-sectional area. Consider the wire to be a box, and use the particle-in-a-box model to compute the translational partition function of an electron at $T = 300\,\text{K}$.

16. Fluctuations.
A stable state of a thermodynamic system can be described by the free energy $G(x)$ as a function of the degree of freedom x. Suppose G obeys a square law, with spring constant k_s, $G(x)/kT = k_s x^2$ as shown in Figure 11.17.

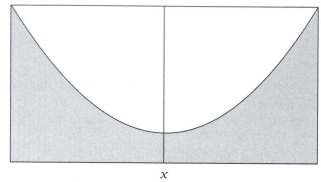

$G(x)$

x

Figure 11.17

(a) Compute the mean square thermal fluctuations $\langle x^2 \rangle$ in terms of k_s.

(b) Some systems have two single minima, with large spring constants k_1, and others have a single broad minimum with small spring constant k_2 as shown in Figures 11.18(a) and (b). For example, two-state equilibria may have two single minima, and the free energies near critical points have a single broad minimum. If $k_2 = 1/4k_1$, what is the ratio of fluctuations $\langle x_2^2 \rangle / \langle x_1^2 \rangle$ for individual energy wells?

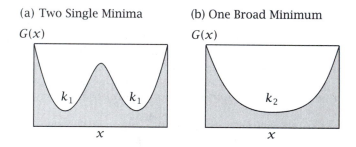

(a) Two Single Minima

$G(x)$

k_1 k_1

x

(b) One Broad Minimum

$G(x)$

k_2

x

Figure 11.18

17. Heat capacity for Cl_2.
What is C_V at $800\,\text{K}$ for Cl_2 treated as an ideal diatomic gas in the high-temperature limit?

18. Protein-in-a-box.
Consider a protein of diameter $40\,\text{Å}$ trapped in the pore of a chromatography column. The pore is a cubic box, $100\,\text{Å}$ on a side. The protein mass is $10^4\,\text{g mol}^{-1}$. Assume the box is otherwise empty and $T = 300\,\text{K}$.

(a) Compute the translational partition function. Are quantum effects important?

(b) If you deuterate all the hydrogens in the protein and increase the protein mass by 10%, does the free energy increase or decrease?

(c) By how much?

References

[1] RS Berry, SA Rice and J Ross. *Physical Chemistry*. Wiley, New York, 1980.

[2] RA Alberty and RJ Silbey. *Physical Chemistry*, 2nd Edition, John Wiley & Sons, New York 1992.

[3] CE Moore. Atomic Energy States, *Natl Bureau Standards Circ* **1**, 467 (1949).

[4] MW Zemansky and RH Dittman. *Heat and Thermodynamics: an Intermediate Textbook*, 7th ed., McGraw Hill, New York, 1997.

Suggested Reading

PW Atkins, *Molecular Quantum Mechanics*, 2nd edition, Oxford University Press, Oxford, 1983. Graphic and clear discussion of quantum mechanics.

PW Atkins, *Physical Chemistry*, 2nd edition, WH Freeman, San Francisco, 1982. Extensive elementary discussion of the principles of quantum mechanics.

HT Davis, *Statistical Mechanics of Phases, Interfaces, and Thin Films*, VCH Publishers, New York, 1996. Advanced statistical mechanics with a broad range of applications.

AP French and EF Taylor, *An Introduction to Quantum Physics*, 1st edition, Norton, New York, 1978. Excellent elementary textbook for the principles of quantum mechanics.

DA McQuarrie, *Statistical Mechanics*, Harper and Row, New York, 1976. Complete, extensive treatment of the subject.

Good treatment of the electronic partition function.

RA Alberty and RJ Silbey, *Physical Chemistry*, 2nd edition, J Wiley, New York, 1992.

Two classic texts on statistical mechanics:

RH Fowler and EA Guggenheim, *Statistical Thermodynamics; A Version of Statistical Mechanics for Students of Physics and Chemistry*, Cambridge University Press, Cambridge, 1939.

GS Rushbrooke, *Introduction to Statistical Mechanics*, Clarendon Press, Oxford, 1951.

Good elementary treatments of the statistical mechanics of gases:

JE House, *Fundamentals of Quantum Mechanics*, Academic Press, San Diego, 1998. Excellent elementary yet complete discussion of the principles of quantum mechanics including particle-in-a-box, rotors, vibrations, and simple atoms.

JH Knox, *Molecular Thermodynamics: An Introduction to Statistical Mechanics for Chemists*, Wiley, Chichester, 1978.

WGV Rosser, *An Introduction to Statistical Physics*, Halsted Press, New York, 1982.

NO Smith, *Elementary Statistical Thermodynamics: A Problems Approach*, Plenum Press, New York, 1982.

12 What Is Temperature? What Is Heat Capacity?

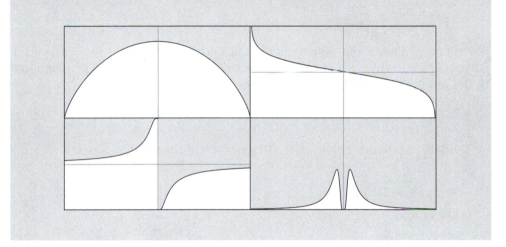

A Microscopic Perspective on Temperature and Heat Capacity

Temperature and heat capacity are two properties that are easy to measure, but not so easy to conceptualize. In this chapter, we develop a conceptual picture of temperature, heat capacity, and related quantities. We use two simple models—the ideal gas, and the two-state Schottky model of Chapter 10. Two-state systems can have *negative* temperatures (lower than $T = 0$ K). Negative temperatures help to illuminate the meanings of temperature and heat capacity in general.

Temperature is a property of a single object. Equation (7.6) defines the temperature in terms of the entropy change that results when a system takes up or gives off energy,

$$\frac{1}{T} = \left(\frac{\partial S}{\partial U}\right)_{V,N}.$$

The Temperature of a Two-State System

Let's use the Schottky two-state model of Chapter 10 to get $S(U)$, to see the meaning of T. The two-state model gives $S(U)$ as $S(W(n(U)))$, described below. In this model, there are two energy levels: the ground state (with energy 0) and the excited state (with energy $\varepsilon_0 > 0$). There are N particles, n of which are in the excited state, and $N - n$ of which are in the ground state (see Figure 12.1).

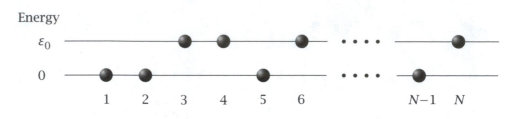

Energy

Figure 12.1 A two-state system with N particles, n of which are in the excited state. Each particle has energy ε_0 or zero.

When energy enters the system as heat, it excites particles to move from the ground state to the excited state. The energy U of the system is proportional to the number n of molecules in the excited state, so

$$U = n\varepsilon_0 \quad \Rightarrow \quad n = \frac{U}{\varepsilon_0}. \tag{12.1}$$

Coin–flip statistics Equation (1.19) gives the multiplicity of states:

$$W = \frac{N!}{n!(N-n)!},$$

so (see Example 7.1)

$$\frac{S}{k} = \ln W = -n \ln \frac{n}{N} - (N-n) \ln \left(\frac{N-n}{N} \right). \tag{12.2}$$

To get $S(U)$, replace n by U/ε_0 in Equation (12.2). A simple way to get $1/T$ is to express Equation (7.6) as

$$\frac{1}{T} = k \left(\frac{\partial \ln W}{\partial U} \right)_{V,N} = k \left(\frac{\partial \ln W}{\partial n} \right)_{V,N} \left(\frac{dn}{dU} \right). \tag{12.3}$$

Since $dn/dU = 1/\varepsilon_0$, Equation (12.3) becomes

$$\begin{aligned}
\frac{1}{T} &= \frac{k}{\varepsilon_0} \left[-1 - \ln \left(\frac{n}{N} \right) + 1 + \ln \left(\frac{N-n}{N} \right) \right] \\
&= -\frac{k}{\varepsilon_0} \ln \left(\frac{n/N}{1-(n/N)} \right) = -\frac{k}{\varepsilon_0} \ln \left(\frac{U/N\varepsilon_0}{1-U/N\varepsilon_0} \right) \\
&= \frac{k}{\varepsilon_0} \ln \left(\frac{f_{\text{ground}}}{f_{\text{excited}}} \right),
\end{aligned} \tag{12.4}$$

where $f_{\text{excited}} = (n/N)$ is the fraction of particles in the excited state and $f_{\text{ground}} = 1 - (n/N)$ is the fraction in the ground state.

Equation (12.4) shows that the temperature of a two-state system depends on the energy spacing ε_0 (the property that distinguishes one type of material from another), the number of particles N, and the total energy U through the quantity n.

Figure 12.2 shows $S(U)$ for a two-state system with $N = 3$ particles. For any value of U, $1/T$ is the slope of the curve $S(U)$. Figure 12.2 and Equation (12.4) show how the temperature of a two-state object depends on its total energy.

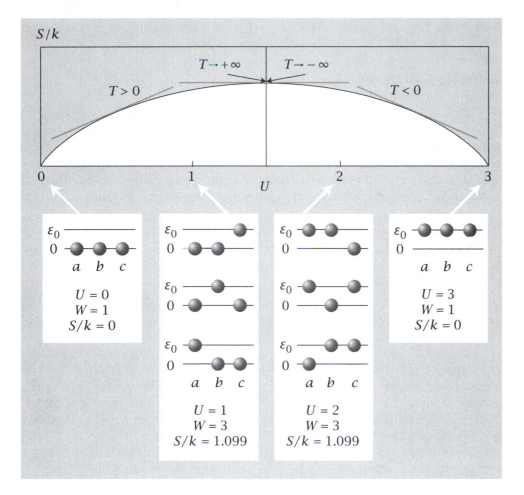

Figure 12.2 The entropy S/k of the two-state system as a function of its energy U. At low U, the multiplicity of states W is small because mostly the ground state is populated. Also at high U, W is small because mostly the excited state is populated. More states are populated for intermediate U. The slope of $S(U)$ everywhere is $1/T$, where T is the temperature. $T > 0$ on the left, $T < 0$ on the right.

Let's work through the figure from left to right.

Positive temperature, $T > 0$. For a two-state system that has low energy U (see the left side of Figure 12.2), most of the particles are in the ground state. As a consequence, $f_{ground}/f_{excited} > 1$, so $\ln(f_{ground}/f_{excited}) > 0$. According to Equation (12.4), the temperature of such a system is positive. If an external source of energy (a bath) is available, the system will tend to absorb energy from it. Particles will be excited from the ground state to the excited state because of the fundamental principle (the Second Law) that systems tend toward their states of maximum entropy. The system can increase its multiplicity of states by taking up energy. Systems having positive temperature absorb energy to gain entropy.

If our model consisted of six states instead of two, then the distribution function for dice (Figure 6.3(b)) would apply and populations would follow the

Boltzmann law, decreasing exponentially with increasing energy level. Lower energy levels would be more populated than higher energy levels.

Infinite temperature, $T = \infty$. For a two-state system having intermediate energy U (middle of Figure 12.2), with equal populations in the ground and excited states, $f_{\text{ground}}/f_{\text{excited}} = 1$, and $\ln(f_{\text{ground}}/f_{\text{excited}}) = 0$. According to Equation (12.4), this implies $1/T = 0$, which means that the temperature is infinite. This is the point at which $S(U)$ is maximal. Just as with coin flips, the maximum multiplicity occurs when half the particles are in each of the two states. Two-state systems at infinite temperature cannot gain additional entropy by absorbing more energy, so they have no tendency to take up energy from a bath.

In a system of six states, infinite temperature would correspond to equal populations of all the states, according to the Boltzmann law. This distribution corresponds to that of unbiased dice (Figure 6.3(a)) in which all outcomes have the same probability.

Negative temperature, $T < 0$. For a system that has high energy U (right side of Figure 12.2), most of the particles are in the excited state, so the ratio $f_{\text{ground}}/f_{\text{excited}} < 1$, and $\ln(f_{\text{ground}}/f_{\text{excited}}) < 0$. Equation (12.4) shows that the temperature at high U is a negative quantity. In this condition, if the system were to absorb additional energy, shifting the last few particles from the ground state to the excited state, it would lead to a *lower* multiplicity of states. The Second Law says that systems tend toward higher entropy, so systems at negative temperatures will tend to *give off* energy, not absorb it. In that regard, a system at negative temperature is *hotter* than a system at positive temperature, since 'hotness' corresponds to a tendency to give up energy.

In a six-state system, negative temperatures correspond to the dice distribution shown in Figure 6.3(c). Populations would *increase* exponentially with increasing energy level, according to the Boltzmann law. Higher energy levels would be more populated than lower levels.

The two-state model shows that the quantity $1/T$ represents the inclination of a system to absorb energy. When $1/T > 0$, a system tends to absorb energy. When $1/T < 0$, it tends to give off energy. When $1/T = 0$, the system has neither tendency. These inclinations result from the drive to maximize entropy.

Why is positive temperature so prevalent in nature and negative temperature so rare? Negative temperatures occur only in saturable systems that have finite numbers of energy levels. Ordinary materials have practically infinite ladders of energy levels and are not saturable, because their particles have translational, rotational, vibrational, and electronic freedom as described in Chapter 11. For such materials, energy absorption always leads to greater entropy. So most materials are energy absorbers and have only positive temperatures.

A system at negative temperature, with more excited than ground-state particles, is said to have a 'population inversion.' A population inversion cannot be achieved by equilibration with a normal heat bath because heat baths invariably have positive temperatures. Using a heat bath that has a positive temperature,

C.H

the most that increasing the bath temperature can achieve is equal populations of excited and ground state particles of the system. One way to achieve a population inversion is to use incident electromagnetic radiation to excite particles into a nonequilibrium state. The excited particles will typically spontaneously emit radiation to give off energy in order to reach a state of higher entropy. Lasers are based on the principle of population inversion.

The Relationship Between Entropy and Heat

What is the microscopic basis for the thermodynamic Equation (7.45), $dS = \delta q/T$? Consider a system with the degrees of freedom $(U, V, \mathbf{N})$. The Second Law of thermodynamics says that entropy $S(U, V, \mathbf{N})$ reaches a maximum at equilibrium. Hold V and $\mathbf{N}$ constant, so no work is done on the system. According to the First Law, the only way to increase the internal energy is through heat uptake, $dU = \delta q$. Consider the positive temperature region of a two-state system, where most molecules are in the ground state. Heat uptake leads to a conversion of ground-state molecules to excited-state molecules. If δq is the heat uptake, it causes an increase dU in the internal energy. Multiply $\delta q = dU$ by the slope $1/T$ to determine the entropy change $dS = \delta q/T$. Increasing the excited-state population, and correspondingly reducing the ground-state population, increases the multiplicity of states of a system at positive temperature.

Now let's look at the heat capacity and its relationship to temperature.

A Graphical Procedure Shows the Steps from Fundamental Functions to Experimental Measurables

To help untangle the relationships between S, U, T, and C_V, let's follow a series of graphical transformations beginning with $S(U)$, the relationship that is closest to the principle of equilibrium, and move toward $C_V(T)$, the most directly measurable quantity (see Figure 12.3). Let's compare the various thermal properties of an ideal gas and a two-state system, for constant V and $\mathbf{N}$.

First row, S versus U. For an ideal gas (Figure 12.3(a)), S is a monotonically increasing function of U because such systems have infinite ladders of energy levels. For an ideal gas at constant volume, $S = (3Nk/2) \ln U$. To see this, you can start with the equipartion expression $U = (3/2)NkT$ (Equation (11.36)) in the fourth row of the figure, and work up. Here, instead, we work from the top to the bottom. In contrast, for a two-state system (Figure 12.3(b)), $S(U)$ increases, reaches a maximum, then decreases, as described in Figure 12.2. The first row of Figure 12.3 gives the most fundamental description of either system in terms of the driving force, the entropy.

Second row, $1/T$ versus U. To get the second row of Figure 12.3 from the first row, take the slope $1/T = (\partial S/\partial U)$. This curve represents the driving force to absorb heat. For the ideal gas (Figure 12.3(a)), the driving force to take up heat is always positive, $1/T = 3Nk/(2U)$. For the two-state system (Figure 12.3(b)), the driving force changes sign if the energy U is large.

Figure 12.3 A graphical series showing the transformation from $S(U)$ (top row) to the heat capacity $C_V(T)$ (bottom row). The left column (a) describes the ideal gas, which is not saturable by energy because the ladder of energy levels is infinite. The right column (b) describes the two-state model, which is saturable.

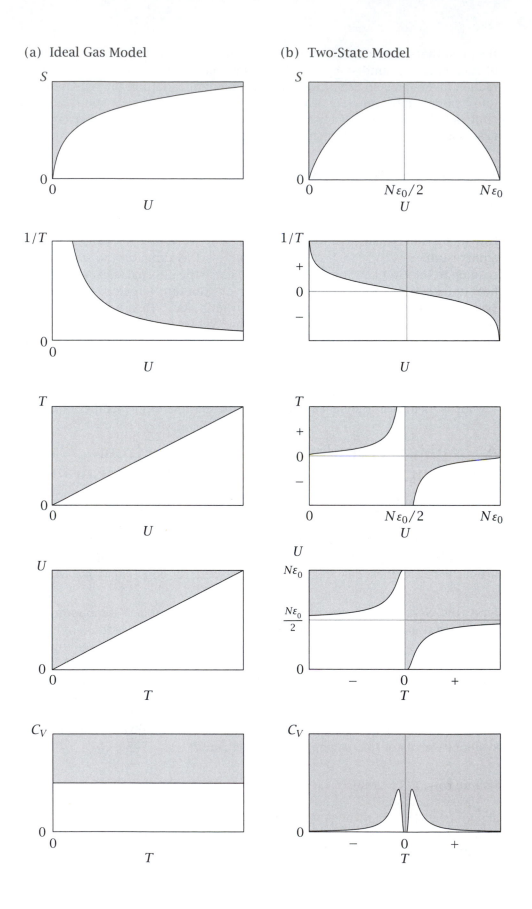

(a) Ideal Gas Model

(b) Two-State Model

Third row, T versus U. To get the third row of Figure 12.3 from the second row, take each y-axis value y_0 and invert it to get $1/y_0$. This gives the function $T(U)$. For ideal gases and many other systems, the temperature of the system is proportional to its energy (Figure 12.3(a)), $T = 2U/(3Nk)$. But you can see from Figure 12.3(b) that in general the temperature need not be proportional to energy. In general, temperature is simply a measure of the relative populations of energy levels: T is inversely proportional to $\ln(f_{ground}/f_{excited})$ in the two-state model.

Fourth row, U versus T. To get the fourth row of Figure 12.3, interchange the x- and y-axes, converting $T(U)$ into $U(T)$, a familiar function. For the ideal gas Figure 12.3(a), the energy is linearly proportional to the temperature, according to Equation (11.36), $U = (3/2)NkT$. The two-state model has a region of negative temperature, but the ideal gas does not. The energy of a two-state system can never exceed $N\varepsilon_0/2$, as long as the temperature is positive. At most, the two-state system can reach equal populations of the two states. To reach higher energies requires negative temperatures.

The fourth row of Figure 12.3(a) shows why you can normally speak of 'putting thermal energy into' a system. The higher the temperature of a bath, the greater is the internal energy of the equipartitioned system, such as an ideal gas that is in equilibrium with the bath. The ideal system absorbs energy in linear proportion to the temperature of the bath. On the other hand, the energy for the two-state model saturates (see Example 10.5, page 184). In this case you can't refer to 'putting in thermal energy,' because the system won't take up much energy from a bath in equilibrium at high temperatures.

Fifth row, C_V versus T. The fifth row of Figure 12.3 is found by taking the slope of the fourth-row figure to get the heat capacity, following the thermodynamic definition, $C_V = (\partial U/\partial T)$. C_V is the slope of the function $U(T)$. For an ideal gas (Figure 12.3(a)), the heat capacity is constant. For the two-state system (Figure 12.3(b)), there is a peak in $C_V(T)$. No thermal energy is absorbed either at high or low positive temperatures in the two-state system.

The series of steps from the top row to the bottom row of Figure 12.3 show that the heat capacity $C_V = (\partial U/\partial T)$ is very different from the tendency to absorb energy, $1/T = (\partial S/\partial U)$. $1/T$ describes the *driving force* for a system to soak up energy from its surroundings, while C_V describes how much energy will be soaked up as you change this driving force (by changing the bath temperature).

What Drives Heat Exchange?

Having considered the temperature of a single object, now consider *heat exchange*, a process that involves two objects. In Chapter 7, we used money as a metaphor to illustrate that the tendency of objects in thermal contact to approach the same temperature is not necessarily a tendency to equalize energies. Now let's use the two-state model to revisit thermal equilibrium. System A has energy U_A, energy spacing ε_A, and particle number N_A. System B has U_B, ε_B, and N_B.

At low positive temperatures, $U/N\varepsilon_0 \ll 1$, and Equation (12.4) becomes

$$\frac{1}{T} = \frac{k}{\varepsilon_0} \ln\left(\frac{N\varepsilon_0}{U}\right).$$

Thermal equilibrium between two two-state systems A and B leads to

$$\frac{1}{T_A} = \frac{1}{T_B} \quad \Longrightarrow \quad \frac{1}{\varepsilon_A} \ln\left(\frac{N_A\varepsilon_A}{U_A}\right) = \frac{1}{\varepsilon_B} \ln\left(\frac{N_B\varepsilon_B}{U_B}\right). \tag{12.5}$$

Here are some of the implications of Equation (12.5). Sometimes heat exchange equalizes energies. If the two objects are made of the same material ($\varepsilon_A = \varepsilon_B$), and have the same total number of particles ($N_A = N_B$), the tendency to maximize entropy and to equalize temperatures is a tendency to equalize their energies U_A and U_B, according to Equation (12.5).

But thermal equilibrium isn't a tendency of high-energy objects to unload their energy into low-energy objects. A large heat bath can have more energy than a system in a small test tube, but this doesn't force the small system to take up energy from the heat bath. If two two-state objects are made of identical material, $\varepsilon_A = \varepsilon_B$, but their sizes are different, $N_A \neq N_B$, then Equation (12.5) tells us that thermal equilibrium will occur when $U_A/N_A = U_B/N_B$, not when $U_A = U_B$. At equilibrium, the objects will share energies in proportion to their sizes.

The Heat Capacity Is Proportional to the Energy Fluctuations in a System

Hold a system at constant temperature with a heat bath. Although the temperature is fixed, the energy of the system will undergo fluctuations, usually very small, as heat flows to and from the bath. The heat capacity is a measure of the magnitude of these fluctuations. To derive this relationship, we switch our attention from microstates to energy levels (see Chapter 10), so we need the density of states.

The Density of States

There are often many different ways in which a system can have a given energy. For example, consider a particle in a cubic box of dimensions $a \times a \times a$. Equation (11.17) gives the translational energy as

$$\varepsilon = \varepsilon_{n_x,n_y,n_z} = \frac{h^2}{8ma^2}(n_x^2 + n_y^2 + n_z^2), \tag{12.6}$$

for $n_x, n_y, n_z = 1, 2, 3, \ldots$. We want to count the number of states $W(\varepsilon)$ that have an energy between ε and $\varepsilon + \Delta\varepsilon$.

Different combinations of integers n_x, n_y, and n_z can all lead to the same value of the sum $R^2 = n_x^2 + n_y^2 + n_z^2$. Substituting R^2 into Equation (12.6) gives $\varepsilon = (hR)^2/(8ma^2)$. Figure 12.4 shows that R can be regarded as a radius in the space of integers. The volume of integers in a sphere is $4\pi R^3/3$. In only

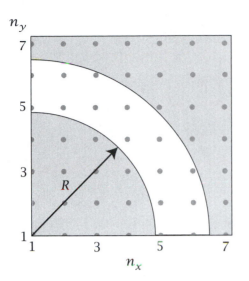

n_y

Figure 12.4 In this two-dimensional representation of translational quantum numbers n_x and n_y, the ring or shell of radius R includes all states having approximately equal values of $n_x^2 + n_y^2$ (•). The ring includes the states (points) with energies between ε and $\varepsilon + \Delta\varepsilon$. $W(\varepsilon)$ is the number of points in this shell.

n_x

1/8 of the sphere are n_x, n_y, and n_z all positive quantities, so the number of positive integer points M is

$$M = \frac{1}{8}\left(\frac{4\pi R^3}{3}\right) = \frac{\pi}{6}\left(\frac{8m\varepsilon}{h^2}\right)^{3/2} V, \tag{12.7}$$

where $V = a^3$. To get $W(\varepsilon)\Delta\varepsilon$, the number of positive integer points in the shell between energies ε and $\varepsilon + \Delta\varepsilon$, take the derivative,

$$W(\varepsilon)\Delta\varepsilon = M(\varepsilon + \Delta\varepsilon) - M(\varepsilon) \approx \frac{dM}{d\varepsilon}\Delta\varepsilon$$

$$= \left(\frac{\pi}{4}\right)\left(\frac{8m}{h^2}\right)^{3/2} V \varepsilon^{1/2} \Delta\varepsilon. \tag{12.8}$$

$W(\varepsilon)$ is the number of states of a single particle-in-a-box that have energy between ε and $\varepsilon + \Delta\varepsilon$. The following example shows that these degeneracies are very large numbers.

EXAMPLE 12.1 The degeneracy of states for the particle-in-a-box. Compute $W(\varepsilon)$ for an atom of argon, $m = 40\,\text{g mol}^{-1} = 6.64 \times 10^{-26}\,\text{kg per atom}$. At $T = 300\,\text{K}$, use $\varepsilon = 3kT/2 = 3(1.38 \times 10^{-23}\,\text{J K}^{-1})(300\,\text{K})/2 = 6.21 \times 10^{-21}\,\text{J}$. Take $a = 1\,\text{cm}$ and take a 1% band of energy, $\Delta\varepsilon = 0.01\varepsilon$. Equation (12.8) gives

$$W(\varepsilon)\Delta\varepsilon = \left(\frac{\pi}{4}\right)\left(\frac{8 \times 6.64 \times 10^{-26}\,\text{kg per atom}}{(6.63 \times 10^{-34}\,\text{J s})^2}\right)^{3/2} \times \left(10^{-2}\,\text{m}\right)^3$$

$$\times \left(6.21 \times 10^{-21}\,\text{J}\right)^{3/2} (0.01) = 5.11 \times 10^{24}\,\text{states}.$$

So far we have focused mostly on the probabilities $p(n_x, n_y, n_z) = q^{-1}e^{-\beta\varepsilon}$ that a particle is in a particular *microstate*, say (n_x, n_y, n_z) for the three-dimensional particle-in-a-box. But now we focus instead on the probability $p(\varepsilon) = q^{-1}W(\varepsilon)e^{-\beta\varepsilon}$ that a system is in a particular *energy level*.

Figure 12.5 The degeneracy of energy states $W(E)$ (– – –) increases with energy, while the Boltzmann factor $e^{-\beta\varepsilon}$ (——) decreases. The product is a peaked Gaussian probability distribution function $p(E)$ (——). This figure is computed for translations of independent particles, $W(\varepsilon) \sim \varepsilon^{1/2}$ for one particle; $W(\varepsilon) \sim \varepsilon^{(3N/2-1)}$ for N independent particles.

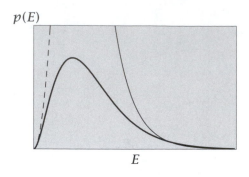

As a check on Equation (12.8), note that the partition function must be the same whether you sum over states or energy levels:

$$q = \sum_{\text{states}} e^{-\beta\varepsilon} = \sum_{\substack{\text{energy} \\ \text{levels}}} W(\varepsilon)e^{-\beta\varepsilon}. \tag{12.9}$$

Substituting Equation (12.8) for $W(\varepsilon)$ into the right-hand side of Equation (12.9) and converting to an integral gives

$$q = \int_0^\infty \left(\frac{\pi}{4}\right)\left(\frac{8m}{h^2}\right)^{3/2} V\varepsilon^{1/2}e^{-\beta\varepsilon}\,d\varepsilon. \tag{12.10}$$

This integral is of the form $\int_0^\infty x^{1/2}e^{-\beta x}\,dx$, where $x = \varepsilon$. Using $n = 1/2$ in Equation (D.9) of Appendix D gives this integral as $\Gamma(3/2)/\beta^{3/2}$, where $\Gamma(3/2) = \sqrt{\pi}/2$. Replace the integral with $\sqrt{\pi}(kT)^{3/2}/2$ in Equation (12.10) to get

$$q = \left(\frac{\pi}{4}\right)\left(\frac{8m}{h^2}\right)^{3/2} V\left(\frac{1}{2}\right)\left(\sqrt{\pi}\right)(kT)^{3/2} = \left(\frac{2\pi mkT}{h^2}\right)^{3/2} V, \tag{12.11}$$

which is the same result we found when summing over states in Equation (11.18).

Energy Fluctuations

How large are the energy fluctuations of a system that is held at constant temperature? Just as you can describe the probability that a system is in a particular microstate, $p(n_x, n_y, n_z) = Q^{-1}e^{-\beta E}$, you can also predict the probability that it is in a particular energy level, $p(E) = Q^{-1}W(E)e^{-\beta E}$. Here we use the distribution over energy levels because we are interested in the fluctuations of the energy. We replace ε and q with E and Q to indicate that the systems of interest are not restricted to independent particles. We show that

$$p(E) = Q^{-1}W(E)e^{-\beta E} \tag{12.12}$$

is a highly peaked function because $W(E)$ grows sharply with E while $e^{-\beta E}$ diminishes exponentially with E (see Figure 12.5).

We use E to represent any particular energy the system might have at any instant, and U to represent the equilibrium value of the energy, the quantity that appears in thermodynamic expressions. Near the peak, the function $p(E)$

is approximately Gaussian, and its width is proportional to C_V. To see this, express $\ln p(E)$ as a Taylor series (see page 53) around the mean value $\langle E \rangle = U$,

$$\ln p(E) = \ln p(U) + \left(\frac{\partial \ln p(E)}{\partial E}\right)_{E=U}(E - U)$$

$$+ \frac{1}{2}\left(\frac{\partial^2 \ln p(E)}{\partial E^2}\right)_{E=U}(E - U)^2 + \cdots. \qquad (12.13)$$

The peak of this function defines the equilibrium state: the fluctuations are deviations from it. We are interested in the variations away from equilibrium. At the peak, $E = U$, and the entropy $S(E)$ equals its equilibrium value $S(U)$. At the peak, the temperature $T(E)$ equals its equilibrium value T_0, which is the temperature of the bath.

In evaluating Equation (12.13), you can use Equation (12.12) to get $\ln p(E) = \ln[Q^{-1}W(E)] - \beta E$. Then take its first and second derivatives. Q is not a function of E (it is a sum over E), so Q does not contribute to such derivatives. Use the expressions $S(E) = k \ln W(E)$ and $(\partial S(E)/\partial E) = 1/T(E)$ to evaluate the terms in Equation (12.13):

$$\left(\frac{\partial \ln p(E)}{\partial E}\right) = \left(\frac{\partial \ln W(E)}{\partial E}\right) - \beta = \left(\frac{\partial(S/k)}{\partial E}\right) - \beta = \frac{1}{kT(E)} - \frac{1}{kT_0}. \qquad (12.14)$$

At the peak, this first derivative of the Taylor series equals zero, so $T(E) = T_0$. To get the second derivative for the Taylor series, take another derivative of the last term in Equation (12.14) using the relation $(\partial E/\partial T) = C_V$ to get

$$\left(\frac{\partial^2 \ln p(E)}{\partial E^2}\right) = -\left(\frac{1}{kT^2}\right)\left(\frac{\partial T}{\partial E}\right)_{E=U} = -\left(\frac{1}{kT_0^2 C_V}\right). \qquad (12.15)$$

Substituting Equation (12.15) into (12.13) and exponentiating shows that $p(E)$ is a Gaussian function (see Equation (1.48)),

$$p(E) = p(U)e^{-(E-U)^2/(2kT_0^2 C_V)}$$

$$= e^{U - T_0 S(U)}e^{-(E-U)^2/(2kT_0^2 C_V)}. \qquad (12.16)$$

Equation (12.16) is based on using $S = k \ln W$ and Equation (12.12). Comparing Equation (12.16) with Gaussian Equation (1.48) and the definition of the variance in Equation (1.40) shows that the width of the energy distribution is characterized by its variance σ^2,

$$\sigma^2 = \langle(E - U)^2\rangle = \langle E^2\rangle - (U)^2 = kT_0^2 C_V. \qquad (12.17)$$

Equation (12.17) shows that you can determine the magnitude of the energy fluctuations if you know C_V. Example 12.2 shows that the width of this Gaussian function is usually exceedingly narrow.

EXAMPLE 12.2 The width of the energy distribution is usually very narrow. A useful dimensionless measure of the width of the distribution in Equation (12.16) is the ratio of the standard deviation to the mean value, σ/U. For an ideal gas $U = (3/2)NkT$, and $C_V = (3/2)Nk$, so using Equation (12.17) you have

$$\frac{\sigma}{U} = \frac{(kT^2 C_V)^{1/2}}{\frac{3}{2}NkT} = \frac{\left[\frac{3}{2}N(kT)^2\right]^{1/2}}{\frac{3}{2}NkT} = \left(\frac{3}{2}N\right)^{-1/2}.$$

If the number of particles in a system is of order 10^{23}, then σ/U is of order 10^{-12}, implying that $p(E)$ has an extremely sharp peak, and the fluctuations are usually exceedingly small. There are exceptions, however. In Chapter 25, we will discuss phase transitions and conditions called critical points where the heat capacity and the fluctuations both become large.

The fact that $p(E)$ is so sharply peaked answers an important question. How do you know whether a system should be considered to be at constant energy (U, V, N) and should be treated by the microcanonical ensemble or whether it should be considered to be at constant temperature (T, V, N) and treated by the Canonical ensemble? If $p(E)$ is very sharply peaked, and if you are interested in thermodynamic average properties such as U, then fixing the temperature is tantamount to fixing the energy U to within a very narrow range. For averaged properties such as U, you would make very little error by using the Microcanonical ensemble and assuming that the preponderance of accessible states have the same energy. Although the choice of ensemble doesn't matter for averaged properties, it does matter when you are interested in fluctuations. Fluctuations in U where T is fixed (the Canonical ensemble) are given by Equation (12.17). There are no fluctuations in U in the Microcanonical ensemble.

Summary

$1/T$ is the driving force for the uptake of energy. For two bodies A and B in thermal contact, energy will flow from one to the other until the driving forces are equal, $1/T_A = 1/T_B$. Setting a bath temperature T_B fixes the system temperature $T_A = T_B$ at equilibrium. C_V describes how much energy uptake results from changing the bath temperature, and is a measure of the magnitude of energy fluctuations. But the amount of energy taken up for a given change in temperature depends on the type of system. A two-state system is very different from an equipartitioned system like an ideal gas. The two-state system has a finite number of energy levels, which can be saturated. In such systems, increasing the bath temperature may not substantially increase the energy of the system. In contrast, the ideal gas does not saturate, and tends to absorb more energy at any temperature.

Problems

1. The heat capacity peaks of phase transitions. The peak heat capacity in Figure 12.6 shows that helium gas adsorbed on graphite undergoes a transition from an ordered state at low temperature to a disordered state at high temperature. Use Figure 12.6 to estimate the energy associated with the transition.

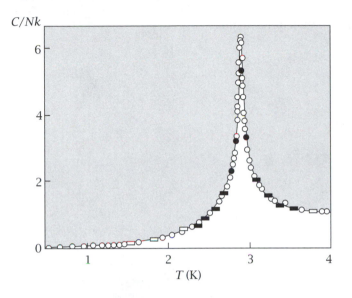

Figure 12.6 Specific heat for helium on graphite in the vicinity of the lattice-gas ordering transition. Source: HJ Kreuzer and ZW Gortel, *Physisorption Kinetics*, Springer Verlag, Heidelberg, 1986. Data are from RL Elgin and DL Goodstein, *Phys Rev A* **9**, 2657–2675 (1974); *Proceedings of the 13th International Conference on Low Temperature Physics*, edited by KD Timmerhaus, WJ O'Sullivan and EF Hammel, Plenum, New York.

2. A two-state model of a hydrogen bond. Suppose a hydrogen bond in water has an energy of about 2 kcal mol^{-1}. Suppose a 'made' bond is the ground state in a two-state model and a 'broken' bond is the excited state. At $T = 300\,\mathrm{K}$, what fraction of hydrogen bonds are broken?

3. A random energy model of glasses. Glasses are materials that are disordered—and not crystalline—at low temperatures. Here's a simple model [1]. Consider a system that has a Gaussian distribution of energies E, according to Equation (12.16):

$$p(E) = p_0 \exp\left[-(E - \langle E \rangle)^2/2\Delta E^2\right],$$

where $\langle E \rangle$ is the average energy and ΔE characterizes the magnitude of the fluctuations, that is, the width of the distribution.

(a) Show that the entropy $S(E)$ is an inverted parabola,

$$S(E) = S_0 - \frac{k(E - \langle E \rangle)^2}{2\Delta E^2}.$$

(b) An *entropy catastrophe* happens (left side of the curve in Figure 12.7) where $S = 0$. For any physical system, the minimum number of accessible states is $W = 1$, so $S = k \ln W$ implies that the minimum entropy is $S = 0$. At the point of the entropy catastrophe, the system has no states that are accessible below an energy $E = E_0$. Compute E_0.

(c) The *glass transition temperature* T_g is the temperature of the entropy catastrophe. Compute T_g from this model.

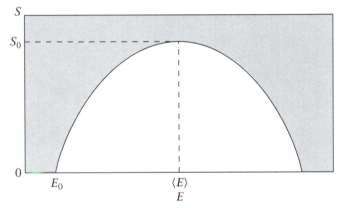

Figure 12.7

4. Fluctuations in enthalpy. The mean square fluctuations in enthalpy for processes at constant pressure are given by an expression similar to Equation (12.17) for processes at constant pressure:

$$\langle \delta H^2 \rangle = \left\langle (H - \langle H \rangle)^2 \right\rangle = kT^2 C_p.$$

What is the root-mean-square enthalpy fluctuation $\langle \delta H^2 \rangle^{1/2}$ for water at $T = 300\,\mathrm{K}$?

5. Oxygen vibrations. Oxygen molecules have a vibrational frequency of 1580 cm^{-1} (see Example 11.3). If the relative populations are $f_{\mathrm{gnd}}/f_{\mathrm{exc}} = 100$ in the Schottky model, what is the temperature of the system?

References

[1] JN Onuchic, Z Luthey-Schulten, and PG Wolynes. *Ann Rev Phys Chem* **48**, 545–600 (1997).

Suggested Reading

P Hakonen and OV Lounasmaa, *Science* **265**, 1821 (1994). Describes experiments on spins in metals showing negative temperatures.

TL Hill, *An Introduction to Statistical Thermodynamics*, Addison-Wesley, 1960. The classic text for discussion of ensembles and fluctuations, and many applications of statistical mechanics.

C Kittel and H Kroemer, *Thermal Physics*, 2nd edition, WH Freeman, New York, 1980. Outstanding general text on statistical thermodynamics. Good discussion of negative temperatures.

RK Pathria, *Statistical Mechanics*, Pergamon, Oxford, 1972. Excellent advanced treatment of statistical mechanics, including a good description of fluctuations.

JR Waldram, *The Theory of Thermodynamics*, Cambridge University Press, Cambridge, 1985. Excellent discussion of fluctuations.

13 Chemical Equilibria

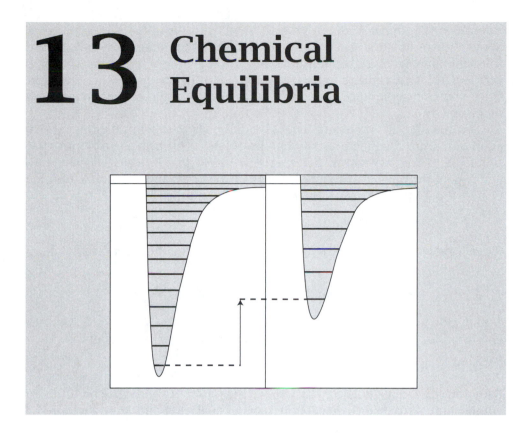

Simple Chemical Equilibria Can Be Predicted from Atomic Structures

A major goal in chemistry is to predict the equilibria of chemical reactions, the relative amounts of the reactants and products, from their atomic masses, bond lengths, moments of inertia, and their other structural properties. In this chapter, we consider reactions in the gas phase, which are simpler than reactions in liquid phases. In Chapter 16 we will consider solvation effects.

The Condition for Chemical Equilibrium

First consider the simplest equilibrium between two states, A and B,

$$A \xrightarrow{K} B. \tag{13.1}$$

Two-state equilibria include: chemical isomerization, the folding of a biopolymers from an open to a compact state, the binding of a ligand to a surface or a molecule, the condensation of a vapor to a liquid or the freezing of a liquid to a solid.

The *equilibrium constant K* is the ratio of the numbers or concentrations of particles in each of the two states at equilibrium. To make it unambiguous whether K is the ratio of the amount of A divided by the amount of B or B divided by A, the arrow in Equation (13.1) has a direction. It points to the *final state* from the *initial state*. The terms 'initial' and 'final' have no implication about time or kinetics. They imply only which is the numerator and which is the denominator in the equilibrium constant K. With the direction shown in Equation (13.1), K is the ratio B/A. You can point the arrow in either direction. Once you choose the direction, the sign of every thermodynamic quantity is determined.

The quantity that predicts chemical equilibria is the chemical potential. For equilibria at fixed temperature and pressure, the appropriate extremum function is the Gibbs free energy: $dG = -S dT + V dp + \mu_A dN_A + \mu_B dN_B$, where N_A and N_B are the numbers of particles in the two states, and μ_A and μ_B are their chemical potentials. At constant T and p, the condition for equilibrium is

$$dG = \mu_A dN_A + \mu_B dN_B = 0. \tag{13.2}$$

If every molecule is in either state A or B, the total number of molecules N_{total} is constant:

$$N_A + N_B = N_{\text{total}} = \text{constant} \quad \Longrightarrow \quad dN_A + dN_B = 0. \tag{13.3}$$

A molecule lost as A is converted to a B molecule, so $dN_A = -dN_B$, and the condition for equilibrium Equation (13.2) can be written in terms of dN_A:

$$(\mu_A - \mu_B) dN_A = 0. \tag{13.4}$$

Equation (13.4) must hold for any non-zero variation dN_A, so the condition for equilibrium is that the chemical potentials of species A and B are the same:

$$\mu_A = \mu_B. \tag{13.5}$$

Now we want to get μ_A and μ_B from microscopic models, to predict how chemical equilibria depend on atomic structures. We need to relate each chemical potential to its partition function.

Partition Functions for Chemical Reactions

First, to simplify our notation for the rest of the chapter, we express the partition function (Equation (10.18) as q', with an added $'$,

$$q' = \sum_{j=0}^{t} e^{-\varepsilon_j/kT} = e^{-\varepsilon_0/kT} + e^{-\varepsilon_1/kT} + \ldots + e^{-\varepsilon_t/kT}. \tag{13.6}$$

This allows us to redefine q, without the $'$, as a reduced partition function with the ground-state term factored out,

$$q = e^{\varepsilon_0/kT} q'$$
$$= 1 + e^{-(\varepsilon_1 - \varepsilon_0)/kT} + e^{-(\varepsilon_2 - \varepsilon_0)/kT} + \ldots + e^{-(\varepsilon_t - \varepsilon_0)/kT}. \tag{13.7}$$

In terms of q', the chemical potential μ_A for species A is given by Equation (11.47):

$$\mu_A = -kT \ln\left(\frac{q'_A}{N_A}\right). \tag{13.8}$$

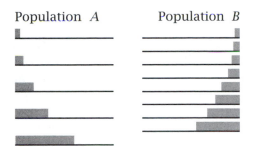

Population A Population B

Figure 13.1 Hypothetical energy ladders for the two-state equilibrium $A \to B$. Two factors contribute to the equilibrium populations of A and B: low energies and high densities of states. A has the lowest energy ground state, but B has the higher density of states.

Similarly for B,

$$\mu_B = -kT \ln \left(\frac{q'_B}{N_B} \right). \tag{13.9}$$

At equilibrium, all the energy levels of both species A and B are accessible to any molecule. A given molecule may be species A for a while at one energy level, then it may change to species B, at some other energy level. At equilibrium, the individual molecules of species A and B will be distributed according to the Boltzmann distribution, with the largest number of molecules in the lowest energy state. The number of molecules populating a given state is determined only by the energy, and not by whether the molecule is in the form of A or B.

Figure 13.1 shows a pair of possible energy ladders for A and B, illustrating a case in which the state of lowest energy (and therefore highest population) is the ground state of A. This does not mean that more particles will be A's than B's. The relative numbers of A's and B's is a sum of Boltzmann factors over all the energy levels. If the energy level spacings are closer for B than for A, B may be the species with the greater total population.

Substituting Equations (13.9) and (13.8) into the equilibrium condition $\mu_A = \mu_B$ (Equation (13.5)) gives $q'_B / N_B = q'_A / N_A$. Defining the **equilibrium constant** K as the ratio N_B / N_A gives

$$K = \frac{N_B}{N_A} = \left(\frac{q'_B}{q'_A} \right) = \left(\frac{q_B}{q_A} \right) e^{-(\varepsilon_{0B} - \varepsilon_{0A})/kT}. \tag{13.10}$$

Equation (13.10) gives a way to compute K from the atomic properties of A and B through their partition functions. Equation (13.10) takes no account of the interactions of one molecule with another, and therefore it applies only to isolated particles such as those in the gas phase.

More Complex Equilibria

Let's generalize to the reaction

$$aA + bB \xrightarrow{K} cC, \tag{13.11}$$

where a, b, and c indicate the stoichiometries of species A, B, and C. What is the appropriate way to define an equilibrium constant for this reaction? At constant T and p the condition for equilibrium is

$$dG = \mu_A dN_A + \mu_B dN_B + \mu_C dN_C = 0. \tag{13.12}$$

This equilibrium is subject to two stoichiometric constraints:

$$\frac{N_A}{a} + \frac{N_C}{c} = \text{constant}, \quad \text{and} \quad \frac{N_B}{b} + \frac{N_C}{c} = \text{constant}. \tag{13.13}$$

Here is the logic behind these constraints. Because N_A is the number of particles of type A, and because a is the number of A-type particles required for each stoichiometric conversion to C, N_A/a is the number of stoichiometric conversions that would consume the N_A particles. Similarly, N_C/c is the number of A-type particles that *have already been* stoichiometrically converted to C. The sum $N_A/a + N_C/c$ is the total number of possible stoichiometric conversions of all the A-type particles that were originally put into the reaction vessel. Example 13.1 may make this clearer.

EXAMPLE 13.1 Stoichiometry. Suppose 4 tires T and 6 windows W are needed to construct one car C. Then the 'equilibrium' for the 'reaction' that transforms tires and windows into cars can be written as

$$4T + 6W \rightarrow C.$$

There are two constraints. The first is

$$\frac{T}{4} + C = \text{constant},$$

so $T + 4C$ is a constant. The number of tires not yet on cars is T. The number of tires already on cars is $4C$. $T + 4C$ is the total number of tires, including those that are on cars and those that are not. ($T/4$ is the number of 'car-units' of tires not yet on cars and C is the number of car-units of tires already on cars.) The other constraint,

$$\frac{W}{6} + C = \text{constant},$$

describes the conservation of windows, on cars and not yet installed.

Now let's return to chemical equilibria. Taking the differential elements of constraint Equations (13.13), you have

$$dN_A = -\frac{a}{c} dN_C \quad \text{and} \quad dN_B = -\frac{b}{c} dN_C. \tag{13.14}$$

Substituting these constraints into Equation (13.12) leads to

$$\left(\mu_C - \frac{a}{c}\mu_A - \frac{b}{c}\mu_B \right) dN_C = 0, \tag{13.15}$$

so at equilibrium,

$$c\mu_C = a\mu_A + b\mu_B. \tag{13.16}$$

Substituting the definitions of chemical potential from Equations (13.8) and (13.9) into equilibrium Equation (13.16) gives

$$c\left[-kT \ln\left(\frac{q'_C}{N_C} \right) \right] = a\left[-kT \ln\left(\frac{q'_A}{N_A} \right) \right] + b\left[-kT \ln\left(\frac{q'_B}{N_B} \right) \right],$$

$$\Rightarrow \quad \left(\frac{q'_C}{N_C}\right)^c = \left(\frac{q'_A}{N_A}\right)^a \left(\frac{q'_B}{N_B}\right)^b. \tag{13.17}$$

To express the relative numbers of particles of each species present at equilibrium, a natural definition of the equilibrium constant K arises from rearranging Equation (13.17):

$$K = \frac{N_C^c}{N_A^a N_B^b} = \left(\frac{(q'_C)^c}{(q'_A)^a (q'_B)^b}\right)$$

$$= \left(\frac{q_C^c}{q_A^a q_B^b}\right) e^{-(c\varepsilon_{0C} - a\varepsilon_{0A} - b\varepsilon_{0B})/kT}. \tag{13.18}$$

To determine K from experiments, you need some way of detecting the numbers of particles of types A, B, and C at equilibrium, and you need to know the stoichiometric coefficients, a, b, and c. Equation (13.18) allows you to predict chemical equilibria from atomic structures.

Finding Ground-State Energies

To compute equilibrium constants using Equation (13.18), you need to know $\Delta\varepsilon_0 = c\varepsilon_{0C} - a\varepsilon_{0A} - b\varepsilon_{0B}$, the difference in ground state energies. For $A \rightarrow B$, both states A and B must contain the same atoms, so the fully dissociated state of A must be identical to the fully dissociated state of B. Therefore there is a common zero of energy, the fully dissociated state (see Figure 13.2). To define the ground state energies further, we must resolve a matter of the vibrational ground state. Vibrational energies are given by Equation (11.22) as $\varepsilon_v = (v + 1/2)h\nu$. The zero of energy $\varepsilon_v = 0$ is at the bottom of the energy well. However, because there is no energy level there, systems cannot access this state and it cannot be measured experimentally. The lowest energy level accessible to the system is $\varepsilon_0 = (1/2)h\nu$, the zero-point energy. However, we previously used the well-bottom as the zero of energy to derive the vibrational partition function, Equation (11.26)

$$q_{\text{vibration}} = \frac{e^{-h\nu/2kT}}{1 - e^{-h\nu/kT}}. \tag{13.19}$$

In order to put chemical equilibria in terms of measurable quantities, we now switch to the zero-point, instead of the well-bottom, as our reference state. This leads to a corresponding change in how we express the vibrational partition function. Figure 13.2 shows the *dissociation energy* $D = -\varepsilon_0 - (1/2)h\nu$ and its relationship to the well-bottom energy. D can be determined from spectroscopic or calorimetric experiments on the dissociation of diatomic molecules, extrapolated to $T = 0\,\text{K}$. We combine the vibrational part of the partition function Equation (11.26) with the expression for D to get

$$q_{\text{vibration}} e^{-\varepsilon_0/kT} = \left(\frac{e^{-h\nu/2kT}}{1 - e^{-h\nu/kT}}\right) e^{(D+h\nu/2)/kT}$$

$$= \left(\frac{1}{1 - e^{-h\nu/kT}}\right) e^{D/kT} = q_{\text{vz}} e^{D/kT}, \tag{13.20}$$

Figure 13.2 Definition of dissociation energies. Energy versus the separation r of the atoms in systems A and B. The dissociation energy $D = -\varepsilon_0 - h\nu/2$ has the opposite sign of ε_0 and is smaller in magnitude by $h\nu/2$. The diagram shows the difference in dissociation energies, $\Delta D = D_A - D_B$.

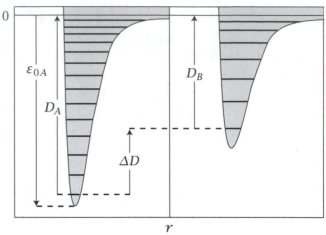

where q_{vz} is the vibrational partition function defined with its zero of energy at the zero-point instead of at the well-bottom,

$$q_{vz} = \frac{1}{1 - e^{-h\nu/kT}}. \tag{13.21}$$

To use Equation (13.18), replace products of the vibrational partition function that appear in the full partition function q, and the Boltzmann factor, $q_{vibration} \exp(-\varepsilon_0/kT)$ with products $q_{vz} \exp(D/kT)$, for each species in the chemical equilibrium. It is these quantities q_{vz} that are implied in the partition functions that follow in this chapter. The next example illustrates how to compute K from partition functions.

EXAMPLE 13.2 Predicting K in a change-of-symmetry reaction. Consider the gas-phase exchange reaction,

$$H_2 + D_2 \xrightarrow{K} 2HD. \tag{13.22}$$

The equilibrium constant is (see Equation (13.18))

$$K = \left(\frac{q_{HD}^2}{q_{H_2} q_{D_2}}\right) e^{\Delta D/RT}, \tag{13.23}$$

where ΔD is the difference between the molar dissociation energies of all the products and all the reactants. Because these energies are in molar units, divide them by RT rather than kT, to get Boltzmann factors with dimensionless exponents. The dissociation energies are $431.8\,\mathrm{kJ\,mol^{-1}}$ for H_2, $439.2\,\mathrm{kJ\,mol^{-1}}$ for D_2, and $435.2\,\mathrm{kJ\,mol^{-1}}$ for HD. Therefore,

$$\Delta D = 2D_{HD} - D_{H_2} - D_{D_2} = -0.6\,\mathrm{kJ\,mol^{-1}}, \tag{13.24}$$

so $e^{\Delta D/RT} = 0.79$ at $T = 300\,\mathrm{K}$.

You can write the equilibrium constant as a product of its component translational (t), rotational (r), and vibrational (v) factors, $K = K_t K_r K_v e^{\Delta D/RT}$, where

$$K_t = \frac{\left[(2\pi m_{HD}kTh^{-2})^{3/2}\right]^2}{(2\pi m_{H_2}kTh^{-2})^{3/2}(2\pi m_{D_2}kTh^{-2})^{3/2}}$$

$$= \left(\frac{m_{HD}^2}{m_{H_2}m_{D_2}}\right)^{3/2} = \left(\frac{3^2}{2\times 4}\right)^{3/2} = 1.19, \tag{13.25}$$

and

$$K_r = \frac{\left((8\pi^2 I_{HD}kT)/(\sigma_{HD}h^2)\right)^2}{\left((8\pi^2 I_{H_2}kT)/(\sigma_{H_2}h^2)\right)\left((8\pi I_{D_2}kT)/(\sigma_{D_2}h^2)\right)}$$

$$= \left(\frac{\sigma_{H_2}\sigma_{D_2}}{\sigma_{HD}^2}\right)\left(\frac{I_{HD}^2}{I_{H_2}I_{D_2}}\right) = 4\left(\frac{(6.13)^2}{(4.60)(9.19)}\right) = 3.56. \tag{13.26}$$

K_t is the ratio of the appropriate powers of translational partition functions from Equation (11.18), K_r is the ratio of rotational partition functions from Equation (11.30), where $\sigma_{H_2} = \sigma_{D_2} = 2$ and $\sigma_{HD} = 1$. Because the vibrational energies are large for all three species at room temperature, $(h\nu)/kT \gg 1$ and $q_{\text{vibration}} = 1$,

$$K_v = \frac{(1 - e^{-h\nu_{HD}/kT})^{-2}}{(1 - e^{-h\nu_{H_2}/kT})^{-1}(1 - e^{-h\nu_{D_2}/kT})^{-1}} = 1 \tag{13.27}$$

is the ratio of vibrational partition functions q_{vz}, from Equations (13.20) and (13.21). Combining the factors $K = K_t K_r K_v e^{\Delta D/RT}$ gives $K = (1.19)(3.56)(1)(0.79) = 3.35$. As $T \to \infty$, $K \to 4$. The change in rotational symmetry is the main contributor to this equilibrium. The reaction is driven by the gain in entropy due to the rotational asymmetry of the products.

Pressure-Based Equilibrium Constants

Because pressures are easier to measure than particle numbers for gases, it is often more convenient to use equilibrium constants K_p based on pressures rather than equilibrium constants K based on the numbers of molecules. Combining Equation (13.18) with the ideal gas law $N = pV/kT$ gives

$$K = \frac{N_C^c}{N_A^a N_B^b} = \frac{(p_C V/kT)^c}{(p_A V/kT)^a (p_B V/kT)^b} = \frac{q_C^c}{q_A^a q_B^b}e^{\Delta D/kT}. \tag{13.28}$$

Multiply the two terms on the right-hand side of Equation (13.28) by $(V/kT)^{a+b-c}$ to get the pressure-based equilibrium constant K_p:

$$K_p = \frac{p_C^c}{p_A^a p_B^b} = (kT)^{c-a-b}\frac{(q_{0C})^c}{(q_{0A})^a (q_{0B})^b}e^{\Delta D/kT}, \tag{13.29}$$

where $q_0 = q/V = \left[(2\pi mkT)/h^2\right]^{3/2}q_r q_v \ldots$ is the partition function with the volume V factored out.

You can also express chemical potentials in terms of the partial pressures of each gas. Using $\mu = -kT\ln(q'/N)$ from Equation (13.8), factoring out the volume $q' = q_0'V$, and using the ideal gas law $V = NkT/p$, you have

$$\mu = -kT \ln \left(\frac{q_0' kT}{p} \right) = kT \ln \left(\frac{p}{p_{int}^\circ} \right) = \mu^\circ + kT \ln p, \qquad (13.30)$$

where $p_{int}^\circ = q_0' kT$ and $\mu^\circ = -kT \ln q_0' kT$.

Equation (13.30) will be useful in the following chapters. It divides the chemical potential into a part that depends on pressure ($kT \ln p$) and a part that does not (μ°). μ° is called the *standard-state* chemical potential. Although the right-most expression ($\mu^\circ + kT \ln p$) would appear to imply that you should take the logarithm of a quantity that is not dimensionless, the preceding expression ($-kT \ln(p/p_{int}^\circ)$) shows that the argument of the logarithm is indeed dimensionless. μ° depends on temperature.

The pressure-based equilibrium constant K_p is computed for a dissociation reaction in Example 13.3.

EXAMPLE 13.3 Dissociation reaction. For the dissociation of iodine, $I_2 \rightarrow 2I$, compute the equilibrium constant at $T = 1000\,K$. The mass of an iodine atom is $m_I = 0.127\,kg\,mol^{-1}/(6.02 \times 10^{23}\,molecules\,mol^{-1}) = 2.109 \times 10^{-25}\,kg$. Table 11.2 and the associated discussion give the ground-state electronic degeneracies as $q_{elec,I} = 4$, $q_{elec,I_2} = 1$, $\theta_{rotation} = 0.0537\,K$, $\theta_{vibration} = 308\,K$, and $\Delta D = -35.6\,kcal\,mol^{-1}$. From Equation (13.29), the pressure-based equilibrium constant is

$$K_p = kT \frac{(q_{0I})^2}{q_{0I_2}} e^{\Delta D/RT}. \qquad (13.31)$$

You have

$$kT = RT/\mathcal{N} = \frac{(8.206 \times 10^{-5}\,m^3\,atm\,K^{-1}\,mol^{-1})(1000\,K)}{6.02 \times 10^{23}\,molecules\,mol^{-1}}$$

$$= 1.363 \times 10^{-25}\,m^3\,atm, \quad \text{and}$$

$$e^{\Delta D/RT} = e^{-35,600/(1.987)(1000)} = 1.66 \times 10^{-8}.$$

Factor the partition functions q_0 into translational (q_t), rotational (q_r), vibrational (q_{vz}), and electronic (q_e) components:

$$\frac{q_{0I}^2}{q_{0I_2}} = \left(\frac{q_{tI}^2}{q_{tI_2}} \right) \left(\frac{1}{q_{rI_2}} \right) \left(\frac{1}{q_{vzI_2}} \right) \left(\frac{q_{eI}^2}{q_{eI_2}} \right). \qquad (13.32)$$

For the rotations, $\sigma = 2$, and

$$q_{rI_2} = \frac{T}{2\theta_{rotation}} = \frac{1000}{2(0.0537)} = 9310.$$

For the vibrations (Equation (13.21)),

$$q_{vzI_2} = \frac{1}{1 - e^{-\theta_{vibration}/T}} = \frac{1}{1 - e^{-308/1000}}$$

$$= 3.77.$$

For the electronic terms, $q_{eI}^2/q_{eI_2} = 16$. For the translations (which have units of m^{-3} because volume has been factored out), $m_{I_2} = 2m_I$, and

$$\frac{q_{tI}^2}{q_{tI_2}} = \frac{[(2\pi m_I kT/h^2)^{3/2}]^2}{(2\pi m_{I_2} kT/h^2)^{3/2}} = \left(\frac{\pi m_I kT}{h^2}\right)^{3/2}$$

$$= \left[\frac{\pi (2.109 \times 10^{-25}\,\text{kg})(1.38 \times 10^{-23}\,\text{J K}^{-1})(1000\,\text{K})}{(6.626 \times 10^{-34}\,\text{J s})^2}\right]^{3/2}$$

$$= 3.01 \times 10^{33}\,\text{m}^{-3}.$$

Combine all these terms using Equations (13.31) and (13.32) to get

$$K_p = \left(1.363 \times 10^{-25}\,\text{m}^3\,\text{atm}\right)\left(1.66 \times 10^{-8}\right)\left(\frac{1}{9310}\right)\left(\frac{1}{3.77}\right) \quad (16)$$

$$\times \left(3.01 \times 10^{33}\,\text{m}^{-3}\right) = 3.1 \times 10^{-3}\,\text{atm}.$$

Figure 13.3 shows that this prediction is quite good. The measured dissociation constant of $I_2 \rightarrow 2I$ shown in Figure 13.3 is highly dependent on temperature. This is mainly because of the exponential dependence of the Boltzmann factor $e^{\Delta D/RT}$ on temperature. Increasing the temperature dissociates I_2 into iodine atoms.

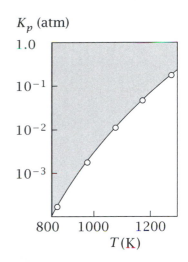

Figure 13.3 Theoretical curve and experimental points for the dissociation of iodine as a function of temperature. Source: ML Perlman and GK Rollefson, *J Chem Phys* **9**, 246 (1941).

Le Chatelier's Principle Describes the Response to Perturbation from Equilibrium

Any perturbation away from a stable equilibrium state must increase the free energy of the system. The system will respond by moving back toward the state of equilibrium. To illustrate, let's return to the general two-state equilibrium,

$$A \xrightarrow{K} B.$$

Suppose a fluctuation changes the number of B molecules by an amount dN_B. The resulting change dG in free energy is given by Equations (13.2) and (13.4),

$$dG = (\mu_B - \mu_A)dN_B. \quad (13.33)$$

Define a *reaction coordinate* $\xi = N_B/(N_A + N_B)$, the fractional degree to which the system has proceeded to B (see Figure 13.4). The total number of particles is fixed: $N_A + N_B = N$ so $N_B = N\xi$ and $dN_B = Nd\xi$. Substituting $Nd\xi$ into Equation (13.33) gives

$$dG = (\mu_B - \mu_A)Nd\xi. \quad (13.34)$$

To move toward equilibrium, $dG \leq 0$, implying that the quantities $(\mu_B - \mu_A)$ and $d\xi$ must have opposite signs. If the system fluctuates into a state in which B happens to have the higher chemical potential, $\mu_B > \mu_A$, then the direction toward equilibrium is $d\xi < 0$, meaning that N_B will decrease. The chemical potential μ is sometimes called the **escaping tendency** because the higher the value of μ_B, the greater is the tendency of the system to escape from the state B to enter the state A. *Le Chatelier's principle* is the term that refers to the tendency of a system to return to equilibrium by moving in a direction opposite to that caused by an external perturbation.

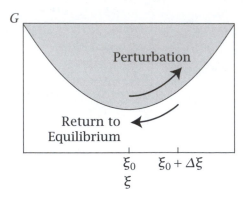

Figure 13.4 For the reaction $A \to B$ at equilibrium, the extent of reaction is $\xi = N_B/(N_A + N_B) = \xi_0$. A fluctuation that increases B leads to $\xi = \xi_0 + \Delta\xi$. A stable system returns to equilibrium by reducing the amount of B.

The Temperature Dependence of Equilibrium Is Described by the van 't Hoff Equation

If you measure an equilibrium constant $K(T)$ at different temperatures, you can learn the enthalpy and entropy of the reaction, which are often useful for constructing or testing microscopic models. To illustrate, we return to the two-state equilibrium,

$$A \xrightarrow{K} B.$$

At constant T and p, the condition for equilibrium is $\mu_A = \mu_B$. The pressure-based equilibrium constant is $K_p = p_B/p_A$. Using Equation (13.30) you have, at equilibrium:

$$\mu_A^\circ + kT \ln p_A = \mu_B^\circ + kT \ln p_B$$

$$\Rightarrow \ln K_p = \ln \left(\frac{p_B}{p_A} \right) = \frac{-(\mu_B^\circ - \mu_A^\circ)}{kT} = -\frac{\Delta\mu^\circ}{kT}. \tag{13.35}$$

$\Delta\mu^\circ$ depends only on temperature (see Equation (13.30), not on pressure, and can be expressed in terms of components (see Equation (9.25)).

$$\Delta\mu^\circ = \Delta h^\circ - T\Delta s^\circ.$$

The temperature dependence in Equation (13.35) is given by

$$\left(\frac{\partial \ln K_p}{\partial T} \right) = -\frac{\partial}{\partial T} \left(\frac{\Delta\mu^\circ}{kT} \right) = -\frac{\partial}{\partial T} \left(\frac{\Delta h^\circ - T\Delta s^\circ}{kT} \right). \tag{13.36}$$

If $\Delta h^\circ = h_B^\circ - h_A^\circ$ and $\Delta s^\circ = s_B^\circ - s_A^\circ$ are independent of temperature, then Equation (13.36) gives

$$\left(\frac{\partial \ln K_p}{\partial T} \right) = \frac{\Delta h^\circ}{kT^2}. \tag{13.37}$$

An alternative form of Equation (13.37), called the **van 't Hoff** relation, provides a useful way to plot data to obtain Δh°. Use $d(1/T) = -(1/T^2)dT$:

$$\left(\frac{\partial \ln K_p}{\partial (1/T)} \right) = -\frac{\Delta h^\circ}{k}. \tag{13.38}$$

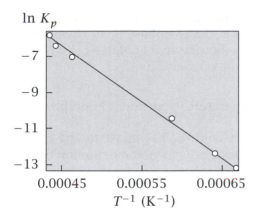

Figure 13.5 van 't Hoff plot for the dissociation of water in the gas phase, $H_2O \rightarrow H_2 + (1/2)O_2$. Higher temperatures cause more dissociation than lower temperatures. Source: MW Zemansky and RH Dittman, *Heat and Thermodynamics: an Intermediate Textbook*, 6th edition, McGraw-Hill, New York, 1981.

van 't Hoff plots show $\ln K$ versus $1/T$: the slope is $-\Delta h^\circ/k$. Figure 13.5 shows a van 't Hoff plot for the dissociation of water vapor, $H_2O \rightarrow H_2 + (1/2)O_2$. Water is more dissociated at high temperatures than at lower temperatures. For dissociation, $\Delta h^\circ > 0$, which is characteristic of bond-breaking processes. Because this enthalpy change is positive, dissociation must be driven by entropy. Figure 13.5 illustrates a common, but not universal, feature of van 't Hoff plots: they are often linear, implying that Δh° is independent of temperature.

When Δh° is independent of temperature, integration of Equation (13.38) gives

$$\ln\left(\frac{K_p(T_2)}{K_p(T_1)}\right) = \frac{-\Delta h^\circ}{k}\left(\frac{1}{T_2} - \frac{1}{T_1}\right). \tag{13.39}$$

Equation (13.39) can be used to find how K_p depends on temperature if you know Δh°, or to determine Δh° if you measure $K_p(T)$.

EXAMPLE 13.4 Getting Δh° from a van 't Hoff plot. To get the enthalpy of dissociation of water, take two points $(1/T, \ln K)$ from Figure 13.5. At temperatures $T = 1500\,K$ and $2257\,K$, you have $1/T = 6.66 \times 10^{-4}$ and $1/T = 4.43 \times 10^{-4}$, respectively. The corresponding points are $(6.66 \times 10^{-4}, -13.147)$ and $(4.43 \times 10^{-4}, -6.4)$. Use Equation (13.39):

$$\Delta h^\circ = \frac{-R[\ln K_p(T_2) - \ln K_p(T_1)]}{(1/T_2) - (1/T_1)}$$

$$= -(8.314\,J\,K^{-1}\,mol^{-1})\frac{[-13.1 - (-6.4)]}{(6.64 \times 10^{-4} - 4.43 \times 10^{-4})}$$

$$= 249\,kJ\,mol^{-1}.$$

Other thermodynamic quantities are also readily obtained from Figure 13.5. For example, at $T = 1500\,K$,

$$\Delta\mu^\circ = -RT\ln K_p = -(8.314\,J\,K^{-1}\,mol^{-1})(1500\,K)(-13.147) = 164\,kJ\,mol^{-1}$$

and

$$\Delta s^\circ = \frac{\Delta h^\circ - \Delta\mu^\circ}{T} = \frac{(249 - 164)\,kJ\,mol^{-1}}{1500\,K} = 56.7\,J\,K^{-1}\,mol^{-1}.$$

Equations (13.37) through (13.39) are quite general and apply to any two-state equilibria, not just to those that are in the gas phase or to chemical reactions. Here is a more general derivation showing that Equations (13.37) and (13.38) hold even when $\Delta h°$ is dependent on temperature.

The Gibbs–Helmholtz Equation for Temperature–Dependent Equilibria

Let's generalize beyond chemical equilibria and the gas phase to any dependence of a free energy $G(T)$ on temperature. Rearrange the definition of free energy $G = H - TS$:

$$H = G + TS. \tag{13.40}$$

Substitute Equation (8.27), $S = -(\partial G/\partial T)_p$, into Equation (13.40) to get

$$H = G - T\left(\frac{\partial G}{\partial T}\right)_p. \tag{13.41}$$

Use $d(u/v) = (vu' - uv')/v^2$ with $v = T$ and $u = G$ to express the temperature derivative of G/T as

$$\left(\frac{\partial(G/T)}{\partial T}\right)_p = \frac{1}{T}\left(\frac{\partial G}{\partial T}\right)_p - \frac{G}{T^2} = -\frac{1}{T^2}\left(G - T\left(\frac{\partial G}{\partial T}\right)_p\right). \tag{13.42}$$

Substituting Equation (13.41) into the last parentheses of Equation (13.42) leads to the **Gibbs–Helmholtz equation**:

$$\left(\frac{\partial(G/T)}{\partial T}\right)_p = -\frac{H(T)}{T^2}. \tag{13.43}$$

Similarly, for the Helmholtz free energy, you have:

$$\left(\frac{\partial(F/T)}{\partial T}\right)_V = -\frac{U(T)}{T^2}. \tag{13.44}$$

Here's an illustration of the Gibbs–Helmholtz equation.

EXAMPLE 13.5 Obtaining $S(T)$ and $H(T)$ from $G(T)$. Suppose the free energy has the temperature dependence $G(T) = aT^3$, where a is a constant. Equation (8.27) then gives $S(T) = -3aT^2$. From Equation (13.41), you get $H(T) = aT^3 - 3aT^3 = -2aT^3$. This is consistent with the result from Equation (13.43): $(d(aT^2)/dT) = 2aT = -H(T)/T^2$. This shows that the Gibbs–Helmholtz equation does *not* mean that the temperature dependence of the Gibbs free energy is solely due to the enthalpy.

Pressure Dependence of the Equilibrium Constant

The dependence on pressure can be derived in much the same way as the dependence on temperature. The pressure dependence $K(p)$ reflects a volume change. The slope of the equilibrium constant with pressure is

$$\frac{\partial \ln K(p)}{\partial p} = \frac{\partial}{\partial p}\left[-\frac{(\mu_B° - \mu_A°)}{kT}\right] = -\frac{1}{kT}\left(\frac{\partial \Delta\mu°}{\partial p}\right). \tag{13.45}$$

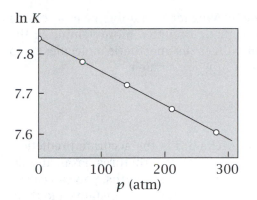

$\ln K$

Figure 13.6 Applying pressure drives anesthetic molecules into water from lipid bilayer membranes. K is the partition coefficient from water into the bilayer. Source: S Kaneshina, H Kamaya, I Ueda, *J Coll Interf Sci* **93**, 215–224 (1983).

To determine the pressure dependence $\Delta\mu(p)$, use either the Gibbs–Duhem equation (9.37) or Maxwell's relations. Here we use the Maxwell relation

$$\left(\frac{\partial\mu}{\partial p}\right)_{T,N} = \left(\frac{\partial V}{\partial N}\right)_{T,P} = v, \qquad (13.46)$$

where v is the volume per molecule, or per mole, depending on the units of μ. Substituting Equation (13.46) into (13.45) gives

$$\left(\frac{\partial(\mu_B^\circ - \mu_A^\circ)}{\partial p}\right)_T = v_B - v_A = \Delta v. \qquad (13.47)$$

Substituting Equation (13.47) into (13.45) gives

$$\left(\frac{\partial \ln K}{\partial p}\right)_T = -\frac{\Delta v}{kT}. \qquad (13.48)$$

Therefore, if B is the state with the smaller volume, $v_B - v_A < 0$, then increasing the pressure will shift the equilibrium from A towards B. Equations (13.37), (13.38), and (13.48) are applicable to complex processes, as illustrated by the following example.

EXAMPLE 13.6 Pressure affects a two-state equilibrium. A common anesthetic drug molecule is halothane (2-bromo-2-chloro-1,1,1-trifluoroethane). Its mode of action is presumed to involve partitioning from water (state A) into lipid bilayer membranes (state B). Figure 13.6 shows how the partitioning equilibrium depends on pressure. Putting values of $(p, \ln K) = (0, 7.84)$ and $(280, 7.6)$ into Equation (13.48) gives $\Delta v = v_{\text{bilayer}} - v_{\text{water}}$:

$$\Delta v = -RT\frac{(\ln K_2 - \ln K_1)}{p_2 - p_1}$$

$$= \frac{(8.205 \times 10^{-5}\,\text{m}^3\,\text{atm}\,\text{K}^{-1}\,\text{mol}^{-1})(300\,\text{K})(7.84 - 7.6)}{280\,\text{atm}}\left(\frac{10^2\,\text{cm}}{\text{m}}\right)^3$$

$$= 21\,\text{cm}^3\,\text{mol}^{-1}.$$

Multiplying by $(10^8 \text{ Å/cm})^3$ and dividing by Avogadro's number gives a volume change of $35\,\text{Å}^3$ per molecule. The anesthetic occupies more volume in the membrane phase than in water. Pressure forces the anesthetic molecules to go into the water, where the volumes they occupy are smaller.

Summary

A remarkable achievement of statistical mechanics is the accurate prediction of gas-phase chemical reaction equilibria from atomic structures. From atomic masses, moments of inertia, bond lengths, and bond strengths, you can calculate partition functions. You can then calculate equilibrium constants and their dependence on temperature and pressure. In Chapter 19, we will apply these ideas to chemical kinetics, which pertains to the *rates* of reactions. Reactions can be affected by the medium they are in. Next we will develop models of liquids and other condensed phases.

Problems

1. Iodine dissociation. Compute the dissociation constant K_p for iodine at $T = 300$ K.

2. Temperature dependence of a simple equilibrium. In a reaction

$$A \xrightarrow{K} B,$$

the equilibrium constant is $K = 10$ at $T = 300$ K.

(a) What is $\Delta\mu°$?

(b) If $\Delta h° = 10$ kcal mol^{-1}, what is K at $T = 310$ K?

(c) What is $\Delta s°$ at $T = 300$ K?

3. Dissociation of oxygen, $O_2 \rightarrow 2O$. Compute K_p, the pressure-based equilibrium constant for this dissociation reaction at $T = 3000$ K. The electronic ground-state degeneracy for O is $g_0(0) = 9$.

4. Temperature dependence of K_p. For the dissociation of oxygen, derive an expression for $d \ln K_p / dT$ near $T = 300$ K from the expression for K_p that you used in problem 3.

5. Polymerization. Consider a polymerization reaction in the gas phase in which n moles of identical monomers are in equilibrium with one mole of chains of n-mers:

(a) Do you expect typical polymerization processes to be driven, or opposed, by enthalpy? By entropy? What are the physical origins of these enthalpies and entropies?

(b) Do you expect polymerizations to be exothermic (giving off heat) or endothermic (taking up heat)? Explain the explosions in some polymerization processes.

(c) Are polymerization equilibrium constants for long chains more or less sensitive to temperature than for shorter chains?

6. Hydrogen dissociation. A hydrogen atom can dissociate in the gas phase:

$$H \xrightarrow{K} H^+ + e^-.$$

Calculate the equilibrium constant K for temperature $T = 5000$ K. There is no rotational or vibrational partition function for H, H$^+$, or e$^-$, but there are spin partition functions: $q_s = 1$ for H$^+$, and $q_s = 2$ for e$^-$. $\Delta D = -311$ kcal mol^{-1}.

7. Free energy along the reaction coordinate. Consider the ideal gas-phase equilibrium $2A \rightarrow B$, at constant temperature and a constant pressure of 1 atm. Assume that there is initially 1 mole of A and no B present, and

that $\mu_A° = 5$ kcal mol^{-1} and $\mu_B° = 10$ kcal mol^{-1} at this temperature.

(a) Show that G, at any time during the reaction, can be written as

$$G = \left(5 + RT \ln \left[\left(\frac{1-2\xi}{1-\xi} \right)^{1-2\xi} \left(\frac{\xi}{1-\xi} \right)^{\xi} \right] \right) \frac{\text{kcal}}{\text{mol}},$$

where ξ is the extent of reaction.

(b) What is the value of the extent of reaction ξ at equilibrium?

8. Pressure denaturation of proteins. For a typical protein, folding can be regarded as involving two states, native (N) and denatured (D),

$$N \xrightarrow{K} D.$$

At $T = 300$ K, $\Delta\mu° = 10$ kcal mol^{-1}. Applying about 10,000 atm of pressure can denature a protein at $T = 300$ K. What is the volume change Δv for the unfolding process?

9. Clusters. Sometimes isolated molecules of type A can be in a two-state equilibrium with a cluster of m monomers,

$$mA \xrightarrow{K} A_m,$$

where A_m represents an m-mer cluster.

(a) At equilibrium, what is the relationship between μ_1, the chemical potential of the monomer, and μ_m, the chemical potential of the molecule in the cluster?

(b) Express the equilibrium constant K in terms of the partition functions.

Suggested Reading

Elementary and detailed discussions of chemical equilibria:

JH Knox, *Molecular Thermodynamics*, Wiley, New York, 1978.

NO Smith, *Elementary Statistical Thermodynamics: A Problems Approach*, Plenum Press, New York, 1982.

Excellent advanced texts with worked examples:

TL Hill, *Introduction to Statistical Thermodynamics*, Addison-Wesley, New York, 1960.

R Kubo, *Statistical Mechanics, an Advanced Course with Problems and Solutions*, North Holland, Amsterdam, 1965.

DA McQuarrie, *Statistical Mechanics*, Harper and Row, New York, 1976.

CL Tien and JH Lienhard, *Statistical Thermodynamics*, Hemisphere Publishing, New York, 1979.

14 Equilibria Between Liquids, Solids, & Gases

Phase Equilibria Are Described by the Chemical Potential

In this chapter, we explore some properties of pure liquids and solids. We model the vapor pressures over liquids and the processes of boiling and sublimation. This is the foundation for treating solvation effects in chemical and physical equilibria in Chapters 15 and 16.

In Chapter 13, we found that the chemical potential describes the tendency for particles to interconvert from one chemical species to another. Here we find that the chemical potential also describes the tendency for particles to move between two phases, between a liquid and a vapor phase for example. The surface tension of a liquid is described in terms of the tendency for particles to move between the bulk (interior) and the surface of a material.

Why Do Liquids Boil?

At room temperature and pressure, water is mostly liquid, but some water molecules escape to the vapor phase. Above the boiling point, water vaporizes. (*Vapor* usually refers to the gas phase of a material that is mostly liquid or solid at room temperature, such as water, while *gas* refers to a material that is mostly gaseous at room temperature, such as helium.) How is vaporization controlled by temperature and pressure? The essence is a balance of two forces. Attractions hold the molecules together in the liquid phase. On the other hand, molecules gain translational entropy when they escape to the vapor phase. Increasing the temperature boils water because the entropy gain at high temperatures is greater than the loss of bond energies, and therefore dominates the free energy, $\Delta F = \Delta U - T\Delta S$. At lower temperatures, energies dominate, and systems prefer the liquid state.

Consider a beaker of liquid (the condensed phase is indicated with the subscript c) in equilibrium with its vapor (indicated with the subscript v). Hold T and p constant. The degree of freedom is the number of particles N_c in the condensed phase, or the number N_v in the vapor phase. The free energy change of the combined system (vapor plus condensed phase) is expressed in terms of the chemical potentials μ_c and μ_v. Because $dT = dp = 0$, the free energy depends only on the chemical potentials and the numbers of particles in the two phases:

$$dG = -S dT + V dp + \sum_{j=1}^{2} \mu_j dN_j = \mu_v dN_v + \mu_c dN_c. \qquad (14.1)$$

If the total number of molecules is conserved in the exchange between vapor and condensed phases, the constraint is

$$N_v + N_c = N = \text{constant}$$
$$\implies \quad dN_v + dN_c = dN = 0 \quad \implies \quad dN_c = -dN_v, \qquad (14.2)$$

so you can express the change in free energy as

$$dG = (\mu_v - \mu_c)dN_v. \qquad (14.3)$$

The condition for equilibrium at constant T and p is $dG = 0$, so

$$\mu_v = \mu_c. \qquad (14.4)$$

To compute the vaporization equilibrium from Equation (14.4), you need μ_c and μ_v. If the vapor is dilute enough to be an ideal gas, then according to Equations (11.47) and (11.50), you have

$$\mu_v = kT \ln \left(\frac{p}{p_{\text{int}}^\circ} \right), \qquad (14.5)$$

where $p_{\text{int}}^\circ = kT/\Lambda^3$ has units of pressure, describes the internal degrees of freedom in the molecule, and is a constant that does not depend on the vapor pressure p.

Now you need a model for μ_c, the chemical potential for the molecules in the condensed phase. The particles in liquids and solids are close together so they interact significantly with each other. Predicting the properties of condensed phases can be complex. Instead of proceeding toward sophisticated theory, we return to the lattice model, which gives simple insights.

Let's model a liquid or solid as a lattice of particles of a single type (see Figure 14.1). Because gases are roughly a thousandfold less dense than liquids, and liquids are only about 10% less dense than solids, we will neglect the distinction between liquids and solids here. We will model a liquid as if its particles occupied a crystalline lattice, with every site occupied by one particle.

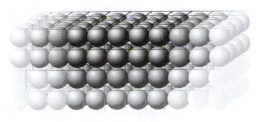

Figure 14.1 Lattice model of liquids and solids. For practical purposes, the lattice is infinite.

This is an approximation in two respects: liquids are slightly less dense than solids, and solids have longer ranged repeating periodicities, called *long-range order*. Solids are more regular. Every atom or molecule in a crystalline solid has the same number of nearest-neighbor atoms or molecules as every other. But in a liquid, there is greater variation—one molecule may have five nearest neighbors for awhile, then six for awhile, etc. The main insight represented by the lattice liquid model is that the most important energetic interactions for holding liquids together are the *short-ranged* interactions of each particle with its nearest neighbors, and that the number of such neighbors has a relatively well-defined average (see Chapter 24). A particle interacts only very weakly with other particles that are more distant.

Because liquids and solids are much less compressible than gases, it won't matter whether we treat the liquid as being at constant pressure (T, p, N) or constant volume (T, V, N). When the pressure is held constant, the volumes of liquids and solids don't change much. In microscopic models, it is often simpler to treat constant V than constant p. Here we use constant V, and we compute the free energy $F = U - TS$ to determine which phase—vapor or liquid—is most stable under various conditions.

The Lattice Model for the Entropy and Energy of a Condensed Phase

The translational entropy of the lattice of particles shown in Figure 14.1 is zero. $W = 1$ and $S = k \ln W = 0$, because if pairs of particles trade positions, the rearrangement can't be distinguished from the original arrangement.

Now we need the energy of the lattice model liquid. At equilibrium, the exchange of particles between the liquid and the vapor phases involves little or no change in the internal quantum mechanical state of the particles—their rotations, vibrations, and electronic states do not change, to first approximation. So the energies that are relevant for vaporization are those between pairs of particles, not within each particle. In the gas phase, if it is ideal, the particles do not interact with each other.

Uncharged particles are attracted to each other (see Chapter 24). Let's represent the attraction between two particles of type A in terms of a 'bond' energy: $w_{AA} < 0$. This energy applies to every pair of particles that occupy neighboring lattice sites. Each such pair has one bond, or contact. w_{AA} is negative, indicating that the particles attract. The energies that hold uncharged atoms and molecules together in liquids and solids are less than a few kilocalories per mole of bonds. They are much weaker than the covalent and ionic bonds that hold atoms in molecules (tens to hundreds of kilocalories per mole). A reasonable first approximation is that the bond energies are independent of temperature.

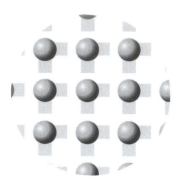

Figure 14.2 Focus on an interior particle in an infinite lattice of particles. Each particle has z neighbors ($z = 4$ in this two-dimensional case). To avoid double-counting, assign $z/2 = 2$ bonds shown as shaded bars to each particle. For N particles, the total number of bonds is $m = Nz/2$.

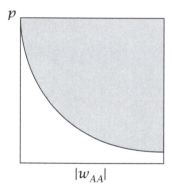

Figure 14.3 The vapor pressure p of A molecules versus the strength of the AA bond, $|w_{AA}|$.

To compute the interaction energy of the lattice model liquid, we need to count the number of bonds among the N particles and multiply by the energy per bond w_{AA}. Each particle on a lattice has z nearest neighbors. z is called the *coordination number* of the lattice. Even though each particle has z neighbors, and therefore z interactions, the total energy is *not* $U = Nzw_{AA}$ because that would count every bond twice. Figure 14.2 shows that you can avoid double-counting by assigning $z/2$ of the interactions to each particle. So, because the outer edges contribute negligibly to an infinite lattice, the number of bonds is $m = Nz/2$, and the total bond energy U is

$$U = mw_{AA} = \left(\frac{Nzw_{AA}}{2} \right). \tag{14.6}$$

Free Energy and Chemical Potential

The free energy of the lattice model liquid is

$$F = U - TS = U = N\left(\frac{zw_{AA}}{2} \right), \tag{14.7}$$

and the chemical potential μ_c is

$$\mu_c = \left(\frac{\partial F}{\partial N} \right)_{T,V} = \frac{zw_{AA}}{2}. \tag{14.8}$$

The Vapor Pressure

For the condensed phase chemical potential μ_c, you can use lattice model Equation (14.8). Use Equation (14.5) for μ_v and set the chemical potentials equal, $\mu_c = \mu_v$, to get the condition for equilibrium in terms of the vapor pressure p,

$$kT \ln\left(\frac{p}{p_{\text{int}}^{\circ}} \right) = \frac{zw_{AA}}{2} \quad \Longrightarrow \quad p = p_{\text{int}}^{\circ} e^{zw_{AA}/2kT}. \tag{14.9}$$

Equation (14.9) describes the vapor pressure p of the A molecules that escape the liquid. The vapor pressure is a measure of the density of the vapor-phase molecules (because $p = (N/V)kT$ from the ideal gas law). As the AA bonds are made stronger (w_{AA} becomes more negative), the vapor pressure decreases because the molecules prefer to bond together in the liquid rather than to escape to the vapor phase (see Figure 14.3). If w_{AA} is fixed, increasing the temperature increases the vapor pressure. Heating a lake and the air above it increases the water vapor pressure in the air. When the air is then cooled, the process reverses, and the result is rain or fog.

Cavities in Liquids and Solids

Molecular processes in liquids and solids often involve creating or filling cavities. Figure 14.4(a) shows that the energy cost ΔU_{remove} of removing one particle, leaving behind an open cavity, is

$$\Delta U_{\text{remove}} = -zw_{AA}, \tag{14.10}$$

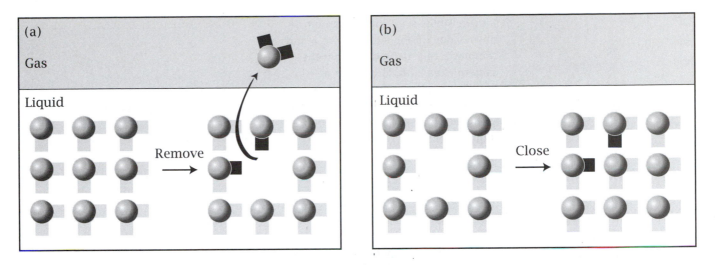

Figure 14.4 (a) Removing a particle and leaving a cavity behind breaks the $z/2$ bonds of the escaping particle, and the $z/2$ bonds of the cavity molecules, so $\Delta U_{\text{remove}} = -(z/2 + z/2)w_{AA} = -zw_{AA}$. (b) Closing a cavity makes $z/2$ bonds, so $\Delta U_{\text{close}} = zw_{AA}/2$.

if the cavity doesn't distort or adjust after the particle is removed. The energy goes up when the particle is removed ($\Delta U_{\text{remove}} > 0$) because this process involves bond breaking. Of the z bonds that are broken, $z/2$ of them are associated with the molecule that leaves (all its bonds are broken), and $z/2$ of the broken bonds are associated with the molecules that line the cavity that is left behind.

Another process of interest is the removal of one particle and the subsequent closure of the cavity. $\Delta U_{\text{remove+close}}$ must equal $U(N-1) - U(N)$, the energy difference between a system of $N-1$ particles and N particles. Using Equation (14.6) for $U(N)$ gives

$$\Delta U_{\text{remove+close}} = U(N-1) - U(N) = -\frac{zw_{AA}}{2}. \qquad (14.11)$$

By taking the difference between Equation (14.11) and (14.10), you can also see that the energy costs of opening and closing a cavity are

$$\Delta U_{\text{close}} = \frac{zw_{AA}}{2}, \quad \text{and}$$

$$\Delta U_{\text{open}} = -\frac{zw_{AA}}{2} \qquad (14.12)$$

(see Figure 14.4(b)).

In Chapters 15 and 16 we will find Equations (14.10) to (14.12) useful for interpreting solvation and partitioning processes.

Is it better to think of particles escaping from the surface or from the interior of the liquid? It doesn't matter because the chemical potential of a molecule must be the same at the surface or in the bulk. Otherwise the system and its surface would not be in equilibrium.

Figure 14.5 A pT phase diagram. A given value of applied pressure p and temperature T defines whether the system is gas, liquid, or solid at equilibrium. The phase boundary lines represent boiling (liquid–gas), sublimation (solid–gas), or melting (liquid–solid).

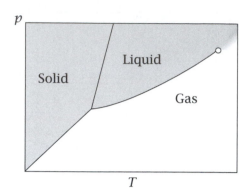

Figure 14.6 Gas–liquid phase boundary for water. Water's vapor pressure is 1 atm at $T = 100\,°C$ and is 0.03 atm at $T = 25\,°C$. At the phase boundary the applied pressure equals the vapor pressure.

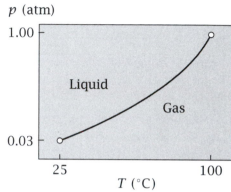

The Clapeyron Equation Describes $p(T)$ at Phase Equilibrium

In this section we describe phase equilibria such as boiling, and how they change with pressure and temperature. Figure 14.5 shows a phase diagram. Each line on the diagram (called a phase boundary) represents a set of (p, T) points at which two phases are equally stable: solid–gas (sublimation), solid–liquid (melting or freezing), or liquid–gas (boiling or condensation). For example, the line separating the liquid from the gas indicates the combinations of pressure and temperature (p, T) at which the system boils. Figure 14.6 illustrates that water boils at $T = 100\,°C$ and $p = 1\,atm$, or that you can boil water at room temperature ($T = 25\,°C$) if you reduce the applied pressure to $p = 0.03\,atm$. In general, increasing the pressure increases the boiling temperature. When (p, T) is above the line, the system is liquid, and when (p, T) is below the line, the system is gas. For example, under conditions $p = 1\,atm$ and $T = 25\,°C$, water is liquid.

In Figures 14.5 and 14.6, p is the applied pressure, which need not equal the vapor pressure. When the applied pressure equals the vapor pressure, the system changes phase. When the applied pressure is greater (say 1 atm) than the vapor pressure (say 0.03 atm at $T = 25\,°C$), water is a liquid, but some molecules are in the vapor phase. The positive slope of $p(T)$ is a general feature of many types of phase equilibria.

Here's a general derivation of the slope of a phase boundary, illustrated for the equilibrium between gas (G) and liquid (L) (see Figure 14.7). We want

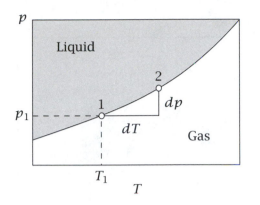

Figure 14.7 The slope of the boiling equilibrium line is given by the Clausius–Clapeyron equation.

to compute how the boiling pressure changes with boiling temperature. You have two points (p_1, T_1) and (p_2, T_2) at which the liquid and the vapor are in equilibrium, so

$$\mu_L(T_1, p_1) = \mu_G(T_1, p_1), \quad \text{and} \quad \mu_L(T_2, p_2) = \mu_G(T_2, p_2). \tag{14.13}$$

The chemical potential at point 2 involves a small perturbation from its value at point 1,

$$\mu_L(T_2, p_2) = \mu_L(T_1, p_1) + d\mu_L(T, p), \quad \text{and}$$

$$\mu_G(T_2, p_2) = \mu_G(T_1, p_1) + d\mu_G(T, p). \tag{14.14}$$

Substituting Equations (14.13) into Equations (14.14) gives

$$d\mu_G(T, p) = d\mu_L(T, p). \tag{14.15}$$

We are regarding μ as a function of (T, p), so

$$d\mu(T, p) = \left(\frac{\partial \mu}{\partial T}\right)_{p,N} dT + \left(\frac{\partial \mu}{\partial p}\right)_{T,N} dp. \tag{14.16}$$

If you use free energies *per mole* (rather than *per molecule*) as units for μ, the derivatives in Equation (14.16) are the partial molar entropy s and volume v (see the Maxwell relations in Table 9.1, page 158),

$$\left(\frac{\partial \mu}{\partial T}\right)_{p,N} = -\left(\frac{\partial S}{\partial N}\right)_{T,p} = -s, \quad \text{and}$$

$$\left(\frac{\partial \mu}{\partial p}\right)_{T,N} = \left(\frac{\partial V}{\partial N}\right)_{T,p} = v. \tag{14.17}$$

Substitute Equations (14.17) into (14.16) and (14.15) to get

$$d\mu_G = -s_G dT + v_G dp = d\mu_L = -s_L dT + v_L dp. \tag{14.18}$$

Rearranging Equation (14.18) gives

$$\frac{dp}{dT} = \frac{s_G - s_L}{v_G - v_L} = \frac{\Delta s}{\Delta v}, \tag{14.19}$$

where $\Delta s = s_G - s_L$ is the partial molar change of entropy, and $\Delta v = v_G - v_L$ is

Figure 14.8 Vapor pressure of benzene versus $1/T$. When $\ln p$ is a linear function of $1/T$, as it is here, Equation (14.23) shows that Δh is independent of temperature. Source: JS Rowlinson and FL Swinton, *Liquids and Liquid Mixtures* in Butterworth's Monographs in Chemistry, 3rd edition, Butterworth Science, London, 1982.

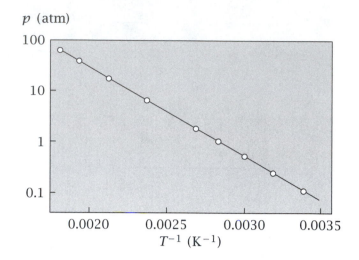

the partial molar change of volume from liquid to gas. At the phase transition, $\Delta\mu = \Delta h - T\Delta s = 0$, so

$$\Delta h = T\Delta s, \tag{14.20}$$

where $\Delta h = h_G - h_L$ is the enthalpy difference between the two phases. Substituting Equation (14.20) into (14.19) yields the **Clapeyron equation**:

$$\frac{dp}{dT} = \frac{\Delta h}{T\Delta v}. \tag{14.21}$$

The molar volume of the gas phase is much larger than the molar volume of the liquid or solid phase, so $\Delta v = v_{gas} - v_{condensed} \approx v_{gas} = RT/p,$[1] if the ideal gas law applies. Substituting this expression into Equation (14.21) gives the **Clausius–Clapeyron equation**,

$$\frac{d\ln p}{dT} = \frac{\Delta h}{RT^2}. \tag{14.22}$$

To put Equation (14.22) into a more useful form, integrate it. When Δh is independent of p and T,

$$\int_{p_1}^{p_2} d\ln p = \int_{T_1}^{T_2} \frac{\Delta h}{RT^2}\, dT$$

$$\Rightarrow \quad \ln\frac{p_2}{p_1} = -\frac{\Delta h}{R}\left(\frac{1}{T_2} - \frac{1}{T_1}\right). \tag{14.23}$$

Figure 14.8 shows the vapor pressure of benzene versus $1/T$, illustrating that $\ln p$ is linearly proportional to $1/T$, and therefore that Δh is independent of T. Values of Δh can be obtained from tables (see Table 14.1) and used to predict a boiling point (p_2, T_2) if another boiling point (p_1, T_1) is known, as shown in Example 14.1.

---

[1] Volumes per mole are given by RT/p, while volumes per molecule are given by kT/p. Use kT to express energies per molecule, or multiply by Avogadro's number ($\mathcal{N} = 6.022 \times 10^{23}$ molecules mol^{-1}) to get $RT = \mathcal{N}kT$, if you want energies per mole.

Table 14.1 Enthalpies of fusion (Δh_f) and boiling (Δh_{vap}), and the corresponding transition temperatures (T_f and T_b) at various pressures. Source: PW Atkins, *Physical Chemistry*, 6th edition, WH Freeman, San Francisco, 1997.

| Substance | Fusion[a] | | Evaporation[b] | |
	T_f (K)	Δh_f (kJ mol^{-1})	T_b (K)	Δh_{vap} (kJ mol^{-1})
He	3.5	0.021	4.22	0.084
Ar	83.81	1.188	87.29	6.506
H_2	13.96	0.117	20.38	0.916
N_2	63.15	0.719	77.35	5.586
O_2	54.36	0.444	90.18	6.820
Cl_2	172.1	6.41	239.1	20.41
Br_2	265.9	10.57	332.4	29.45
I_2	386.8	15.52	458.4	41.80
Hg	234.3	2.292	629.7	59.30
Ag	1234	11.30	2436	250.6
Na	371.0	2.601	1156	98.01
CO_2	217.0	8.33	194.6	25.23[c]
H_2O	273.15	6.008	373.15	40.656[d]
NH_3	195.4	5.652	239.7	23.35
H_2S	187.6	2.377	212.8	18.67
CH_4	90.68	0.941	111.7	8.18
C_2H_6	89.85	2.86	184.6	14.7
C_6H_6	278.61	10.59	353.2	30.8
CH_3OH	175.2	3.16	337.2	35.27

[a] Various pressures; [b] at 1 atm; [c] sublimation; [d] 44.016 at 298.15 K.

EXAMPLE 14.1 Boiling water at high altitudes   Water boils at $T_1 = 373$ K and $p_1 = 1$ atm. At a high altitude where $p_2 = 1/2$ atm, what is the boiling temperature T_2? Substituting the enthalpy of vaporization for water, $\Delta h_{vap} = 40.66$ kJ mol^{-1} from Table 14.1 into Equation (14.23) gives

$$\frac{1}{T_2} = \frac{1}{T_1} - \frac{R}{\Delta h_{vap}} \ln\left(\frac{p_2}{p_1}\right)$$

$$\Rightarrow \quad \frac{1}{T_2} = \frac{1}{373} - \left(\frac{8.314\,\mathrm{J\,K^{-1}\,mol^{-1}}}{40{,}660\,\mathrm{J\,mol^{-1}}}\right)\ln\frac{1}{2} \quad \Rightarrow \quad T_2 = 354\,\mathrm{K} = 81\,°\mathrm{C}.$$

Water boils at lower temperatures at higher altitudes.

Equation (14.19) says that the phase with the greater entropy has the greater volume per molecule, if the measured slope dp/dT is positive. Because the vapor has a greater volume than the liquid or solid, it must have greater entropy. Similarly a typical liquid has more volume than its solid. The correlation of increasing volume with increasing entropy is a general property of materials, although an interesting exception is water (see Chapter 29). For water, the solid–liquid phase boundary has a negative slope, $dp/dT < 0$. Because

Figure 14.9 The thermodynamic cycle of a refrigerator. The spheres indicate the relative densities of the refrigerant molecules as they move through the system. There are four steps in the cycle. (1) Heat is absorbed from inside the refrigerator compartment to boil the working fluid. (2) The fluid is compressed and pumped. (3) Heat is dumped outside. The gas refrigerant re-condenses into a liquid. (4) The fluid is expanded until it is ready to boil.

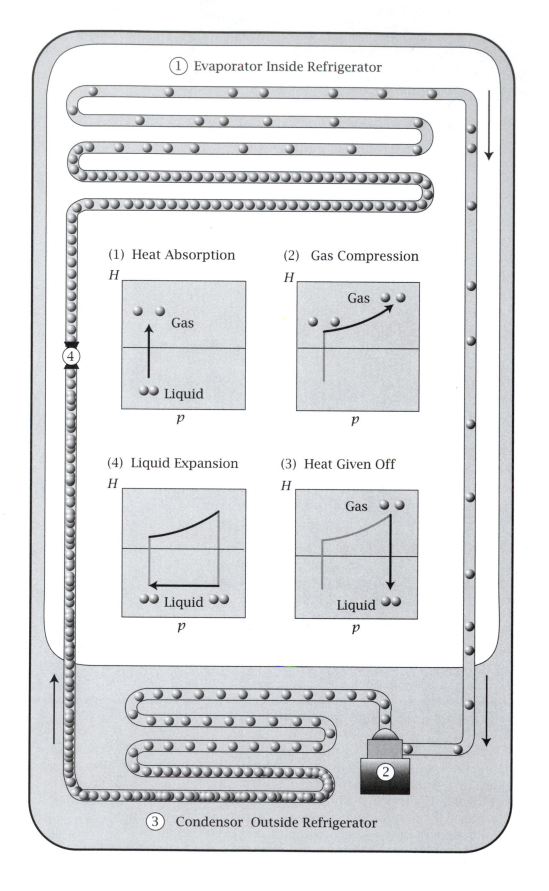

liquid water has greater entropy than solid water (ice), liquid water must have a *smaller* volume per molecule than ice. Indeed, ice floats on water.

Comparing Equation (14.9) with Equation (14.23) shows that the vaporization enthalpy is related to the lattice model pair interaction energy w_{AA} through

$$\Delta h_{\text{vap}} = -\frac{z w_{AA}}{2}. \tag{14.24}$$

Figure 14.4 illustrates that the enthalpy of vaporization comes from breaking bonds, removing a particle, and closing a cavity in a liquid.

To illustrate the principles of vapor–liquid equilibria, let's look at refrigerators and heat pumps.

How Do Refrigerators and Heat Pumps Work?

A refrigerator uses external energy to absorb heat from a cold place (inside the refrigerator box) and dump the heat in a warmer place (outside the refrigerator). A heat pump does the same. It pumps heat from a cold place (outside a house in winter, for example) to a warmer place (inside the house). In both types of device, a 'working fluid' is pumped around a system of tubes and undergoes repeated thermodynamic cycles of vaporization and condensation.

Refrigerators and heat pumps are based on two principles: (1) that boiling stores energy (breaks bonds) and condensation gets it back, and (2) that a fluid can be boiled at low temperature and recondensed at high temperature by controlling the pressure. Figure 14.9 shows the four steps in an idealized refrigeration cycle. The graphs of enthalpy versus pressure show when the noncovalent bonds of the working fluid break (high enthalpy) and form (low enthalpy).

(1) Absorbing heat from inside the cold refrigerator. Before step 1, the working fluid is in the liquid state at low pressure, poised to boil. In step 1, the working fluid flows through the coils inside the refrigeration box, absorbing heat from the food compartment and cooling the inside of the refrigerator. This heat uptake boils the fluid, breaking the liquid–liquid bonds, increasing the enthalpy. The choice of fluid is based on its ability to boil at an appropriately low temperature and pressure. A typical refrigerant (working fluid) is 1,1,1,2-tetrafluoroethane, which boils at $T = -26.1\,^\circ\text{C}$ and has the molecular structure shown in Figure 14.10. Older refrigerants such as freon contained chlorines, which react and deplete the ozone layer of the earth, so they are now less common.

(2) Compressing the working fluid. The low-pressure gas is then compressed to a gas at higher pressure. The gas is now under sufficiently high pressure that it is poised to condense to the liquid state.

(3) Dumping the heat outside the refrigerator. The working gas now flows through coils mounted on the outside of the refrigeration box. Heat is given off to the surroundings, condensing the working gas into a liquid under high pressure. This step results in a decrease in enthalpy, because the liquefaction

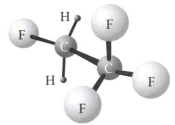

Figure 14.10
1,1,1,2-tetrafluoroethane, a modern refrigerant, circulates through refrigerators, undergoing repeated cycles of boiling and condensation.

Figure 14.11 Lattice model of surface tension. Of N total molecules, n are at the surface. The surface tension is the free energy of moving molecules from the bulk to the surface, per unit area.

Surface
n molecules

Bulk
$N-n$ molecules

reforms liquid–liquid bonds. The heat captured from inside the refrigerator in step 1 is now released to the surrounding room in step 3.

(4) Reducing the pressure of the working fluid. The fluid flows through an expansion valve where its pressure is reduced without much enthalpy change. The liquid is again at low pressure, poised for vaporization, and the cycle repeats. This is how vaporization cycles can be used to pump heat from cold places to hot places.

In this chapter we have focused on liquid–vapor equilibria. The same ideas can be applied to other phase equilibria. For example, the sublimation pressure of a gas over a solid can be computed from Equation (14.23) by replacing the enthalpy of vaporization with the enthalpy of sublimation. Now we consider another process of particle exchange, not between condensed phase and vapor, but between the interior and the surface of a liquid.

Surface Tension Describes the Equilibrium Between Molecules at the Surface and in the Bulk

A *surface* is defined as the boundary between a condensed phase and a gas or vapor. More generally, an *interface* is defined as the boundary between any two media. *Surface tension* is the free energy cost of increasing the surface area of the system. For example, when a water droplet is spherical, it has the smallest possible ratio of surface to volume. When the droplet changes shape, its surface gets larger relative to its volume. Water tends to form spherical droplets because deviations away from spherical shapes are opposed by the surface tension. Here is a model.

Consider a lattice condensed phase of N molecules. If n of the molecules are at the surface, then $N - n$ are in the 'bulk' (the interior of the phase, away from the surface; see Figure 14.11). Bulk particles have z nearest neighbors. Surface particles have only $z - 1$ nearest neighbors because they have one side exposed at the surface. Taking this difference into account, the total energy of the surface plus the bulk can be computed according to Equation (14.6):

$$U = \left(\frac{zw_{AA}}{2}\right)(N - n) + \left(\frac{(z - 1)w_{AA}}{2}\right)n = \frac{w_{AA}}{2}(Nz - n). \tag{14.25}$$

The surface tension is defined as the derivative of the free energy with respect to the total area $\mathcal{A}$ of the surface, $\gamma = (\partial F/\partial\mathcal{A})_{T,V,N}$. Because the lattice liquid

Table 14.2 The surface tension γ of some solids and liquids.

Source: G Somorjai, *Introduction to Surface Chemistry and Catalysis*, J Wiley, New York, 1994.

[a] JM Blakeley and PS Maiya, *Surfaces and Interfaces*, JJ Burke et al., editors, Syracuse University Press, Syracuse, 1967.
[b] AW Adamson, *Physical Chemistry of Surfaces*, J Wiley, New York, 1967.
[c] GC Benson and RS Yuen, *The Solid-Gas Interface*, EA Flood, editor, Marcel Dekker, New York, 1967.

Material	γ (ergs cm^{-2})	T (°C)
W(solid)[a]	2900	1727
Au (solid)[a]	1410	1027
Ag (solid)[a]	1140	907
Ag (liquid)[b]	879	1100
Fe (solid)[a]	2150	1400
Fe (liquid)[b]	1880	1535
Pt (solid)[a]	2340	1311
Hg (liquid)[b]	487	16.5
NaCl (solid)[c]	227	25
KCl (solid)[c]	110	25
CaF$_2$ (solid)[c]	450	−195
He (liquid)[b]	0.308	−270.5
Ethanol (liquid)[b]	22.75	20
Water (liquid)[b]	72.75	20
Benzene (liquid)[b]	28.88	20
n-Octane (liquid)[b]	21.80	20
Carbon tetrachloride[b] (liquid)	26.95	20
Nitrobenzene[b] (liquid)	25.2	20

has $S = 0$ and $F = U$,

$$\gamma = \left(\frac{\partial F}{\partial \mathcal{A}}\right)_{T,V,N} = \left(\frac{\partial F}{\partial n}\right)_{T,V,N} \left(\frac{dn}{d\mathcal{A}}\right) = \left(\frac{\partial U}{\partial n}\right)_{T,V,N} \left(\frac{dn}{d\mathcal{A}}\right). \tag{14.26}$$

Use the definition of U in Equation (14.25) to take the derivative:

$$\left(\frac{\partial U}{\partial n}\right) = \frac{-w_{AA}}{2}. \tag{14.27}$$

The total area of the surface is $\mathcal{A} = na$, where a is the area per particle. So the second part of the derivative that you need for Equation (14.26) is

$$\left(\frac{dn}{d\mathcal{A}}\right) = \frac{1}{a}.$$

Therefore, the lattice model gives

$$\gamma = \frac{-w_{AA}}{2a}, \tag{14.28}$$

for the surface tension.

Figure 14.12 Vaporization and surface tension γ both involve breaking bonds, modelled through the quantity w_{AA}. The slope of $\gamma \times$ area (a) versus ΔH_{vap} indicates that 6.6-fold more bonds are broken per atom in vaporization than in creating surface, over a range of liquid metals. Source: redrawn from SH Overbury, PA Bertrand and GA Somorjai, *Chem Rev* **75**, 547–560 (1975).

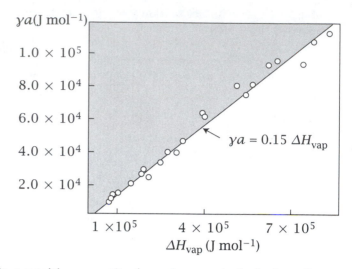

The surface tension is a positive quantity (w_{AA} is negative). It describes the free energy cost of transferring particles from the bulk to the surface to increase the surface area. Surface tensions are greatest for the materials with the strongest intermolecular attractive forces. Table 14.2 shows that the surface tensions are very high for liquid metals, such as silver, iron, or mercury, in which free–flowing electrons bind atoms tightly together. This is why a mercury droplet is so spherical: it strongly seeks the shape of minimum possible surface. For water, which is hydrogen-bonded, the surface tension is $72.8\,\mathrm{dyn\,cm^{-1}}$ at 20 °C. For *n*-hexane, which is bound together only by dispersion forces, the surface tension is $18.4\,\mathrm{dyn\,cm^{-1}}$ at 20 °C. Adding surfactants to water decreases the surface tension, resulting in the increased surface area that you see in the foams on beer and dishwater.

Measuring various equilibria can give information about the intermolecular interaction energies w_{AA}. Different measures are related to each other. For example, Δh_{vap} and γ are both linear functions of w_{AA}, so Δh_{vap} and γ should be linearly related to each other. From the lattice model, the surface tension of a liquid is predicted to be $1/z$ times the energy of vaporization per unit area of the particle. Figure 14.12 confirms this approximate relationship between surface tensions and vaporization enthalpies for liquid metals.

Summary

Vaporization equilibria involve a balance of forces. Particles stick together at low temperatures but they vaporize at high temperatures to gain translational entropy. The chemical potential describes the escaping tendency, μ_v from the vapor phase, and μ_c from the condensed phase. When these escaping tendencies are equal, the system is at equilibrium and no net particle exchange takes place. We used the statistical mechanical model of Chapter 11 for the chemical potential of the gas phase. For the liquid or solid phase, we used a lattice model. Vapor pressure and surface tension equilibria, combined with models, give information about intermolecular interactions. In the next chapter we will go beyond pure phases and consider the equilibria of mixtures.

Problems

1. Applying the Clausius–Clapeyron equation.

(a) The vapor pressure of water is 23 mm Hg at $T = 300$ K, and 760 mm Hg at $T = 373$ K. Calculate the enthalpy of vaporization, Δh_{vap}.

(b) Assuming that each water has $z = 4$ nearest neighbors, calculate the interaction energy, w_{AA}.

2. How does surface tension depend on temperature?
If the surface tension of a pure liquid is due entirely to energy (and no entropy), then will the surface tension *increase*, *decrease*, or *not change* with increasing temperature?

3. Why do spray cans get cold?
Explain why an aerosol spray can gets cold when you spray the contents.

4. The surface tension of water.
The surface tension of water $\gamma(T)$ decreases with temperature as shown in Figure 14.13.

(a) As a first-order approximation, estimate the water–water pair attraction energy w_{AA}, assuming that the interaction is temperature independent.

(b) Explain the basis for this attraction.

(c) Explain why γ decreases with T.

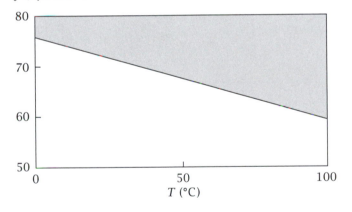

γ (dyn cm^{-1})

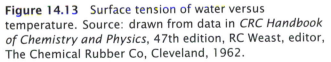

T (°C)

Figure 14.13 Surface tension of water versus temperature. Source: drawn from data in *CRC Handbook of Chemistry and Physics*, 47th edition, RC Weast, editor, The Chemical Rubber Co, Cleveland, 1962.

5. Applying the Clausius–Clapeyron equation again.
Suppose you want to add a perfuming agent to a liquid aerosol spray. You want its vapor pressure to double if the temperature increases from 25 °C to 35 °C. Calculate the pair interaction energy that this agent should have in the pure liquid state. Assume that the number of nearest-neighbor molecules in the liquid is $z = 6$.

6. How does the boiling temperature depend on the enthalpy of vaporization?
A fluid has a boiling temperature $T = 300$ K at $p = 1$ atm; $\Delta h_{vap} = 10$ kcal mol^{-1}. Suppose you make a molecular modification of the fluid that adds a hydrogen bond so that the new enthalpy is $\Delta h_{vap} = 15$ kcal mol^{-1} at $p = 1$ atm. What is the new boiling temperature?

7. Trouton's rule.

(a) Using Table 14.1, show that the entropy of vaporization, $\Delta s_{vap} = \Delta h_{vap}/T_b$, is relatively constant for a broad range of different substances. Δh_{vap} is the enthalpy of vaporization and T_b is the boiling temperature. This is called Trouton's rule.

(b) Why are the values of Δs_{vap} for water and CO_2 in Table 14.1 larger than for other substances?

8. Sublimation of graphite.
The heat of sublimation of graphite is $\Delta h_{sub} = 716.7$ kJ mol^{-1}. Use this number to estimate the strength of a carbon–carbon bond.

Suggested Reading

RS Berry, SA Rice and J Ross, *Physical Chemistry*, Wiley, New York, 1980. A complete and advanced text on physical chemistry including thermodynamics.

TD Eastop and A McConkey, *Applied Thermodynamics for Engineering Technologists*, 5th edition, Longman Scientific and Technical, Essex, 1993. An engineering perspective on refrigerators and heat pumps.

Good treatments of phase boundaries and the Clausius–Clapeyron equation:

P Atkins, *Physical Chemistry*, 6th edition, WH Freeman, New York, 1997.

DA McQuarrie and JD Simon, *Physical Chemistry, A Molecular Approach*, University Science Books, Sausalito, 1997.

Surface tension is discussed in:

PC Hiemenz, *Principles of Colloid and Surface Chemistry*, 3rd edition, Marcel Dekker, New York, 1997.

J Lyklema, *Fundamentals of Interface and Colloid Science*, Vol I, Academic Press, San Diego, 1991.

15 Solutions & Mixtures

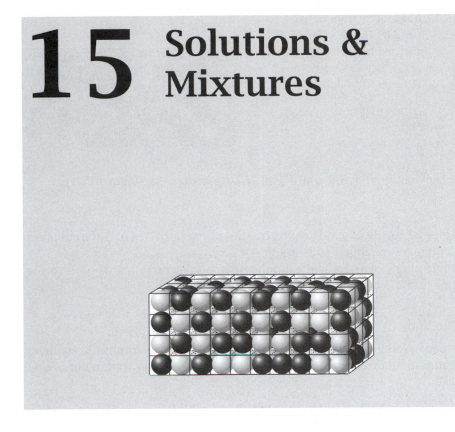

A Lattice Model Describes Mixtures

In Chapter 14 we considered pure liquids or solids composed of a single chemical species. Here we consider solutions, mixtures of more than one component. (A solution is a mixture that is homogeneous.) The fundamental result that we derive in this chapter, $\mu = \mu^\circ + kT \ln \gamma x$, is a relationship between the chemical potential μ and the concentration x of one of the components in the solution. This relationship will help address questions in Chapters 16, and 25 to 30: When does one chemical species dissolve in another? When is it insoluble? How do solutes lower the freezing point of a liquid, elevate the boiling temperature, and cause osmotic pressure? What forces drive molecules to partition differently into different phases? We continue with the lattice model because it gives simple insights and because it gives the foundation for treatments of polymers, colloids, and biomolecules.

We use the $(T, V, \mathbf{N})$ ensemble, rather than $(T, p, \mathbf{N})$, because it allows us to work with the simplest possible lattice model that captures the principles of solution theory. The appropriate extremum principle is based on the Helmholtz free energy, $F = U - TS$, where S is the entropy of mixing and U accounts for the interaction energies between the lattice particles.

The Entropy of Mixing

Suppose there are N_A molecules of species A, and N_B molecules of species B. Particles of A and B are the same size—each occupies one lattice site—and

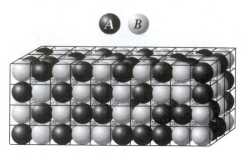

Figure 15.1 A lattice mixture of two components A and B. The number of A's is N_A and the number of B's is N_B. The total number of lattice sites is $N = N_A + N_B$. All sites are filled.

together they completely fill a lattice of N lattice sites (see Figure 15.1),

$$N = N_A + N_B.$$

(See Chapter 31 for the Flory–Huggins treatment for particles of different sizes, such as polymers in simple solvents.)

The multiplicity of states is the number of spatial arrangements of the molecules:

$$W = \frac{N!}{N_A! N_B!}.$$

The translational entropy of the mixed system can be computed by using the Boltzmann Equation (6.1), $S = k \ln W$, and Stirling's approximation, Equation (4.26):

$$\Delta S_{\mathrm{mix}} = k \left(N \ln N - N_A \ln N_A - N_B \ln N_B \right)$$

$$= k \left(N_A \ln N + N_B \ln N - N_A \ln N_A - N_B \ln N_B \right). \tag{15.1}$$

It is most useful to express the mixing entropy in terms of the *mole fraction* x. For a mixture of A and B, $x_A = N_A/N$ and $x_B = N_B/N$. Equation (15.1) becomes

$$\Delta S_{\mathrm{mix}} = -k \left(N_A \ln x_A + N_B \ln x_B \right). \tag{15.2}$$

To simplify further, let $x = x_A$, so $1 - x = x_B$. Divide by Nk. The entropy of mixing a binary (two-component) solution is

$$\frac{\Delta S_{\mathrm{mix}}}{Nk} = -x \ln x - (1 - x) \ln(1 - x). \tag{15.3}$$

Figure 15.2 shows how the entropy of mixing $\Delta S_{\mathrm{mix}}/Nk$ depends on x, the mole fraction of A, according to Equation (15.3). The process is illustrated in Figure 15.3. Put N_A A molecules into a container that has N_B B molecules. The composition x is now fixed. If the two materials did not mix with each other, there would be no change in the entropy, relative to the two pure fluids, $\Delta S_{\mathrm{mix}} = 0$. If the two materials do mix in the random way that we have assumed, then the entropy will increase to the value shown in Figure 15.2 because of the multiplicity of ways of intermingling the A's with the B's. This is the driving force for mixing.

Don't confuse x with a degree of freedom. The system doesn't change its composition toward $x = 1/2$ to reach the highest entropy. x is the composition that is fixed by the number of A and B molecules put into the solution.

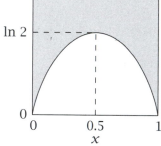

Figure 15.2 The entropy of mixing as a function of the mole fraction x.

EXAMPLE 15.1 Mixing entropy. For a solution containing a mole fraction of 20% methanol in water, compute the mixing entropy. Equation (15.3) gives

$$\frac{\Delta S_{mix}}{N} = k(-0.2 \ln 0.2 - 0.8 \ln 0.8)$$

$$= \left(1.987 \frac{cal}{mol\,K}\right)(0.5) \approx 1.0 \frac{cal}{mol\,K}.$$

If there were no interaction energy, then the free energy of mixing at $T = 300$ K would be

$$\frac{\Delta F_{mix}}{N} = \frac{-T\Delta S_{mix}}{N}$$

$$= -300 \frac{cal}{mol}.$$

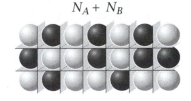

Figure 15.3 Mixing is a process that begins with N_A molecules of pure A, N_B molecules of pure B, and combines them into a solution of $N_A + N_B$ molecules.

Ideal Solutions

A solution is called *ideal* if its free energy of mixing is given by $\Delta F_{mix} = -T\Delta S_{mix}$ with ΔS_{mix} taken from Equation (15.3). Mixing an ideal solution involves no change in energy. And it involves no other entropies due to changes in volume, or structuring, or ordering in the solution. Chapter 25 describes solubility and insolubility in more detail. Now we move to a model of mixtures that is more realistic than the ideal solution model.

The Energy of Mixing

In practice, few solutions are truly ideal. They involve energies of mixing. In the lattice model, the total energy of mixing is the sum of the contact interactions of noncovalent bonds of all the pairs of nearest neighbors in the mixture. For a lattice solution of A and B particles, Figure 15.4 shows the three possible types of contact: an AA bond, a BB bond, or an AB bond. There are no other options, because the lattice is completely filled by A's and B's.

The total energy of the system is the sum of the individual contact energies over the three types of contact:

$$U = m_{AA}w_{AA} + m_{BB}w_{BB} + m_{AB}w_{AB}, \tag{15.4}$$

where m_{AA} is the number of AA bonds, m_{BB} is the number of BB bonds, m_{AB} is the number of AB bonds, and w_{AA}, w_{BB}, w_{AB} are the corresponding contact energies. As noted in Chapter 14, the quantities w are negative.

In general, the numbers of contacts, m_{AA}, m_{BB}, and m_{AB}, are not known. To put Equation (15.4) into a more useful form, you can express the m quantities in terms of N_A and N_B, the known numbers of A and B particles. Each lattice site has z 'sides.' Figure 15.5 shows that every contact involves two sides. The total number of sides of type A particles is zN_A, which can be expressed in terms of the numbers of contacts as

$$zN_A = 2m_{AA} + m_{AB}, \tag{15.5}$$

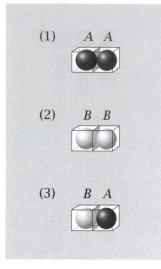

Figure 15.4 Three types of contact (or bond) occur in a lattice mixture of components A and B.

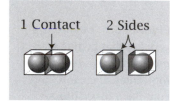

Figure 15.5 One contact between lattice particles involves two lattice site sides.

because the total number of A sides equals

$$(\text{number of } AA \text{ bonds}) \times \left(\frac{2A \text{ sides}}{AA \text{ bond}}\right) + (\text{number of } AB \text{ bonds})\left(\frac{1A \text{ side}}{AB \text{ bond}}\right).$$

Similarly, for B particles,

$$zN_B = 2m_{BB} + m_{AB}. \tag{15.6}$$

Solve Equations (15.5) and (15.6) for the number of AA bonds m_{AA}, and for the number of BB bonds m_{BB}:

$$m_{AA} = \frac{zN_A - m_{AB}}{2}, \quad \text{and} \quad m_{BB} = \frac{zN_B - m_{AB}}{2}. \tag{15.7}$$

Substitute Equations (15.7) into (15.4) to arrive at an expression for the total interaction energy, in which the only unknown term is now m_{AB}, the number of AB contacts:

$$U = \left(\frac{zN_A - m_{AB}}{2}\right)w_{AA} + \left(\frac{zN_B - m_{AB}}{2}\right)w_{BB} + m_{AB}w_{AB}$$

$$= \left(\frac{zw_{AA}}{2}\right)N_A + \left(\frac{zw_{BB}}{2}\right)N_B + \left(w_{AB} - \frac{w_{AA} + w_{BB}}{2}\right)m_{AB}. \tag{15.8}$$

Now we use the 'Bragg–Williams,' or *mean-field*, approximation to evaluate m_{AB} [1, 2, 3].

The Mean-Field Approximation

The reason that we don't know m_{AB}, the count of AB contacts, is that there are many possible configurations of the system, some of which have many AB contacts, and some of which have few. In principle, we should consider each configuration of the system, and account for its appropriate Boltzmann weight (for more discussion, see page 275). This would lead to sophisticated models. Here we explore a much simpler approach that gives many of the same insights. We make an assumption, called the mean-field approximation, that for any given numbers N_A and N_B, the particles are mixed as randomly and uniformly as possible. This gives us a way to estimate m_{AB}.

Consider a specific site next to an A molecule. What is the probability that a B occupies that neighboring site? In the Bragg–Williams approximation, you assume that the B's are randomly distributed throughout all the sites. The probability p_B that any site is occupied by B equals the fraction of all sites that are occupied by B's,

$$p_B = \frac{N_B}{N} = x_B = 1 - x.$$

Because there are z nearest-neighbor sites for each A molecule, the average number of AB contacts made by that particular A molecule is $zN_B/N = z(1-x)$. The total number of A molecules is N_A, so

$$m_{AB} \approx \frac{zN_A N_B}{N} = zNx(1 - x). \tag{15.9}$$

Now compute the total contact energy of the mixture from the known quantities N_A and N_B by substituting Equation (15.9) into Equation (15.8):

$$U = \left(\frac{zw_{AA}}{2}\right)N_A + \left(\frac{zw_{BB}}{2}\right)N_B + z\left(w_{AB} - \frac{w_{AA} + w_{BB}}{2}\right)\frac{N_A N_B}{N}$$

$$= \left(\frac{zw_{AA}}{2}\right)N_A + \left(\frac{zw_{BB}}{2}\right)N_B + kT\chi_{AB}\frac{N_A N_B}{N}, \tag{15.10}$$

where we define a dimensionless quantity called the **exchange parameter** χ_{AB},

$$\chi_{AB} = \frac{z}{kT}\left(w_{AB} - \frac{w_{AA} + w_{BB}}{2}\right) \tag{15.11}$$

(discussed on page 273).

How does the Bragg–Williams approximation err? If AB interactions are more favorable than AA and BB interactions, then B's will prefer to sit next to A's more often than the mean-field assumption predicts. Or, if the self-attractions are stronger than the attractions of A's for B's, then A's will tend to cluster together, and B's will cluster together, more than the random mixing assumption predicts. Nevertheless, the Bragg–Williams mean-field expression is often a reasonable first approximation.

The Free Energy of Mixing

Now combine terms to form the free energy, $F = U - TS$, using Equation (15.2) for the entropy, and Equation (15.10) for the energy:

$$\frac{F(N_A, N_B)}{kT} = N_A \ln\left(\frac{N_A}{N}\right) + N_B \ln\left(\frac{N_B}{N}\right)$$

$$+ \left(\frac{zw_{AA}}{2kT}\right)N_A + \left(\frac{zw_{BB}}{2kT}\right)N_B + \chi_{AB}\frac{N_A N_B}{N}. \tag{15.12}$$

$F(N_A, N_B)$ is the free energy of a mixed solution of N_A A's and N_B B's, totalling $N = N_A + N_B$ particles.

Sometimes we are interested in the *free energy difference* between the mixed final state and the initial pure states of A and B, ΔF_{mix} (see Figure 15.3):

$$\Delta F_{mix} = F(N_A, N_B) - F(N_A, 0) - F(0, N_B). \tag{15.13}$$

$F(N_A, 0) = zw_{AA}N_A/2$ is the free energy of a pure system of N_A A's, found by substituting $N = N_A$ and $N_B = 0$ into Equation (15.12). Similarly, $F(0, N_B) = zw_{BB}N_B/2$ is the free energy of a pure system of N_B B's. $F(N_A, N_B)$, the free energy of the mixed final state, is given by Equation (15.12). Substitute these three free-energy expressions into Equation (15.13) and divide by N to get the free energy of mixing in terms of the mole fraction x and the interaction parameter χ_{AB}:

$$\frac{\Delta F_{mix}}{NkT} = x\ln x + (1 - x)\ln(1 - x) + \chi_{AB}x(1 - x). \tag{15.14}$$

This model was first described by JH Hildebrand in 1929, and is called the *regular solution model* [4]. Solutions with free energies of mixing that are well described by Equation (15.14) are called regular solutions. Here's an application of Equation (15.14).

EXAMPLE 15.2 Oil and water don't mix. Show that oil and water don't mix unless one component is very dilute. A typical value is $\chi \approx 5$ for hydrocarbon/water interactions. If $x_{oil} = 0.3$ and $T = 300\,K$, Equation (15.14) gives the free energy of mixing as

$$\frac{\Delta F_{mix}}{N} = [0.3 \ln 0.3 + 0.7 \ln 0.7 + 5(0.3)(0.7)](8.314\,JK^{-1}\,mol^{-1})(300\,K)$$

$$= 1.1 \frac{kJ}{mol}.$$

Because this value is positive, it predicts that oil and water won't mix to form a random solution of this composition. If the oil is very dilute, $x_{oil} = 10^{-4}$, then

$$\frac{\Delta F_{mix}}{N} = \left[10^{-4} \ln 10^{-4} + 0.9999 \ln 0.9999 + 5(10^{-4})(0.9999)\right]$$

$$\times (8.314\,JK^{-1}\,mol^{-1})(300\,K) = -1.3\,J\,mol^{-1},$$

which is negative, so mixing is favorable.

In general, determining solubility is not as simple as these calculations imply, because another option is available to the system—it may separate into phases with different compositions. Phase separation, for which this model is a starting point, is described in Chapter 25.

The Chemical Potentials

The chemical potential for A is found by taking the derivative of F (Equation (15.12)) with respect to N_A, holding N_B (not N) constant:

$$\frac{\mu_A}{kT} = \left[\frac{\partial}{\partial N_A}\left(\frac{F}{kT}\right)\right]_{N_B,T}$$

$$= \ln \frac{N_A}{N} + 1 - \frac{N_A}{N} - \frac{N_B}{N} + \frac{zw_{AA}}{2kT} + \chi_{AB}\frac{(N_A + N_B)N_B - N_A N_B}{(N_A + N_B)^2}$$

$$= \ln x_A + \frac{zw_{AA}}{2kT} + \chi_{AB}(1 - x_A)^2. \tag{15.15}$$

Similarly, the chemical potential for B is

$$\frac{\mu_B}{kT} = \left[\frac{\partial}{\partial N_B}\left(\frac{F}{kT}\right)\right]_{N_A,T} = \ln x_B + \frac{zw_{BB}}{2kT} + \chi_{AB}(1 - x_B)^2. \tag{15.16}$$

The main result is that $\mu = kT \ln x +$ other terms. We will rely heavily on this expression in the next few chapters.

The lattice model leads to a small ambiguity. In principle, our degrees of freedom are (T, V, N_A, N_B). But the volume is not independent of the total particle number because the lattice is fully filled and constrained by $N_A + N_B = N$. Because the lattice contains no empty sites, this model does not treat pV effects. We regard the relevant constraints as (T, N_A, N_B) and neglect the pV term. This is why you hold N_B constant, rather than N, when taking the derivative with respect to N_A.

Equation (15.15) gives the chemical potential for particle types A and B in a mixture of components, according to the lattice model. More generally, the

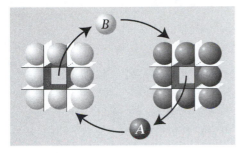

Figure 15.6 The quantity $2\chi_{AB}$ is the energy divided by kT for the process that begins with pure components A and B, and exchanges an A for a B.

chemical potential can be expressed as

$$\mu = \mu^\circ + kT \ln \gamma x, \tag{15.17}$$

where γ is called the *activity coefficient*, and μ° is the *standard state chemical potential.* (Don't confuse this γ with the surface tension.)

In Chapter 16 we will apply both Equations (15.15) and (15.17) to problems of mixing, solubility, partitioning, solvation, and colligative properties. We will find that the chemical potential describes the escaping tendency for particles to move from one phase or one state to another. Particles are driven by at least two tendencies. First, A particles tend to leave regions of high A concentration and move toward regions of low A concentration to gain mixing entropy (this is described by the term $kT \ln x$). Second, A particles are attracted to regions or phases for which they have high chemical affinity (described by μ°).

Now let's look at the quantity χ_{AB}. This quantity appears for many different processes; in solution, or at interfaces, or for conformational changes in polymers. The exchange parameter χ_{AB} describes the energetic cost of beginning with the pure states of A and B and transferring one B into a medium of pure A's and one A into a medium of pure B's (see Figure 15.6):

$$\frac{1}{2}z(AA) + \frac{1}{2}z(BB) \longrightarrow z(AB), \quad \text{and} \quad \chi_{AB} = -\ln K_{\text{exch}}, \tag{15.18}$$

where K_{exch} is the equilibrium constant for the exchange process.

If this process is favorable in the direction of the arrow in Equation (15.18) and Figure 15.6, $K_{\text{exch}} > 1$. In that case, χ_{AB}, which has units of energy divided by kT, is negative. According to *Hildebrand's principle*, for most systems, the AB affinity is weaker than the AA and BB affinities, so usually $\chi_{AB} > 0$. The quantity χ_{AB} also contributes to the interfacial free energy between two materials, which we take up in the next section.

Interfacial Tension Describes the Free Energy of Creating Surface Area

The boundary between two condensed phases is an *interface.* The interfacial tension γ_{AB} is the free energy cost of increasing the interfacial area between phases A and B. If γ_{AB} is large, the two media will tend to minimize their interfacial contact. Let's determine γ_{AB} by using the lattice model for molecules of types A and B that are identical in size.

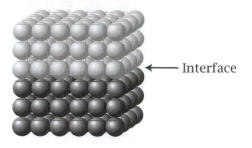

Figure 15.7 Lattice model of interfacial tension.

Interface

Suppose there are N_A molecules of A, n of which are at the interface, and there are N_B molecules of B, n of which are at the interface in contact with A (see Figure 15.7). (Because the particles have the same size in this model, there will be the same number n of each type for a given area of interfacial contact). The total energy of the system is treated as it was for surface tension (see Equation (14.25)), with the addition of n AB contacts at the interface:

$$U = (N_A - n)\left(\frac{zw_{AA}}{2}\right) + n\left(\frac{(z-1)w_{AA}}{2}\right) + nw_{AB}$$

$$+ (N_B - n)\left(\frac{zw_{BB}}{2}\right) + n\left(\frac{(z-1)w_{BB}}{2}\right). \tag{15.19}$$

Because the entropy of each bulk phase is zero according to the lattice model, the interfacial tension is defined by

$$\gamma_{AB} = \left(\frac{\partial F}{\partial \mathcal{A}}\right)_{N_A,N_B,T} = \left(\frac{\partial U}{\partial \mathcal{A}}\right) = \left(\frac{\partial U}{\partial n}\right)\left(\frac{dn}{d\mathcal{A}}\right), \tag{15.20}$$

where $\mathcal{A}$ is the total area of the surface in lattice units. We have $dn/d\mathcal{A} = 1/a$, where a is the area per molecule exposed at the surface. Take the derivative $(\partial U/\partial n)$ using Equation (15.19):

$$\left(\frac{\partial U}{\partial n}\right)_{N_A,N_B,T} = w_{AB} - \frac{w_{AA} + w_{BB}}{2}. \tag{15.21}$$

You can assemble an expression for γ_{AB} from Equations (15.11), (15.20), and (15.21):

$$\gamma_{AB} = \frac{1}{a}\left(w_{AB} - \frac{w_{AA} + w_{BB}}{2}\right) = \left(\frac{kT}{za}\right)\chi_{AB}. \tag{15.22}$$

When there are no B molecules, $w_{AB} = w_{BB} = 0$, and γ_{AB} reduces to the surface tension given by Equation (14.28), $-w_{AA}/2a$.

EXAMPLE 15.3 Estimating χ_{AB} from interfacial tension experiments. The interfacial tension between water and benzene at 20 °C is 35 dyn cm^{-1}. Suppose $z = 6$ and the interfacial contact area of a benzene with water is about 3 Å $\times$ 3 Å = 9 Å^2. Determine χ_{AB} at 20 °C. Equation (15.22) gives

$$\chi_{AB} = \frac{za\gamma_{AB}}{kT} = \frac{(6)(9\,\text{Å}^2)(35 \times 10^{-7}\,\text{J cm}^{-2})\left(\dfrac{1\,\text{cm}}{10^8\,\text{Å}}\right)^2}{(1.38 \times 10^{-23}\,\text{J K}^{-1})(293\,\text{K})} = 4.67.$$

χ_{AB} is a dimensionless quantity, and $RT\chi_{AB} = 11.3\,\text{kJ mol}^{-1} = 2.7\,\text{kcal mol}^{-1}$ is the exchange energy.

What Have We Left Out?

There are two main ways in which our lattice model is a simplification. First, the partition function Q should be a sum over all the possible states of the system. It should be a sum of the number of states with small numbers of AB contacts, m_{AB}, and states with large m_{AB}, rather than a mean-field estimate of the number of uniformly mixed conformations (see Figure 15.8). In this way, we have made the approximation that

$$Q = \sum_{m_{AB}} W(N_A, N_B, m_{AB}) e^{-E(N_A, N_B, m_{AB})/kT}$$

$$\approx \frac{N!}{N_A! N_B!} e^{-U/kT}, \tag{15.23}$$

where E is the energy of a configuration having m_{AB} contacts and W is the density of states (see page 178), the number of configurations having the given value of m_{AB}. U is the mean-field average energy from Equation (15.10).

The second approximation that we made was leaving out the quantum mechanical degrees of freedom: rotations, vibrations, electronic configurations. Those degrees of freedom normally are the same in the pure phase as in the mixed phase. In *differences* of thermodynamic quantities, like ΔF_{mix}, such terms cancel, so you are at liberty to leave them out in the first place. It is only the quantities that *change* in a process that need to be taken into account. Only the intermolecular interactions and the translational entropy change in simple mixing processes. But for chemical reactions in solution, quantum mechanics contributes too (see Chapter 16). Taking internal degrees of freedom into account gives

$$Q = q_A^{N_A} q_B^{N_B} \sum_{m_{AB}} W(N_A, N_B, m_{AB}) e^{-E(N_A, N_B, m_{AB})/kT}$$

$$\approx q_A^{N_A} q_B^{N_B} \frac{N!}{N_A! N_B!} e^{-U/kT}, \tag{15.24}$$

where q_A and q_B are the partition functions for the rotations, vibrations and electronic states of molecules A and B, respectively. Using $F = -kT \ln Q$ (Equation (10.40)) with Equation (15.24), you get a general expression for the free energy of mixing that includes the quantum mechanical contributions in the mean-field lattice model:

$$\frac{F}{kT} = N_A \ln\left(\frac{N_A}{N}\right) + N_B \ln\left(\frac{N_B}{N}\right) + \left(\frac{z w_{AA}}{2kT}\right) N_A + \left(\frac{z w_{BB}}{2kT}\right) N_B$$

$$+ \chi_{AB} \frac{N_A N_B}{N} - N_A \ln q_A - N_B \ln q_B. \tag{15.25}$$

Similarly, the generalizations of Equation (15.15) for the chemical potential are

$$\frac{\mu_A}{kT} = \ln\left(\frac{x_A}{q_A}\right) + \frac{z w_{AA}}{2kT} + \chi_{AB}(1 - x_A)^2, \quad \text{and}$$

$$\frac{\mu_B}{kT} = \ln\left(\frac{x_B}{q_B}\right) + \frac{z w_{BB}}{2kT} + \chi_{AB}(1 - x_B)^2. \tag{15.26}$$

Unmixed
m_{AB} is Small

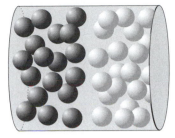

Slightly Mixed
m_{AB} is Intermediate

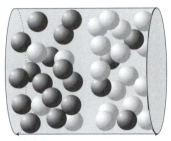

Uniformly Mixed
m_{AB} is Large

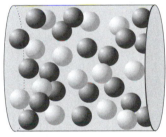

Figure 15.8 *A* and *B* particles can mix to different degrees to have different numbers m_{AB} of AB contacts. The partition function Q is a sum over all these states, but the mean-field model assumes uniform mixing.

The quantum mechanical components cancel in quantities such as ΔF_{mix} when the internal degrees of freedom are unaffected by the mixing process. To see this, use Equation (15.25) to get the pure state components for Equation (15.13): $F(N_A, 0) = N_A[(zw_{AA}/2) - \ln q_A]$ and $F(0, N_B) = N_B[(zw_{BB}/2) - \ln q_B]$, and subtract. You get the same result as in (Equation (15.14)):

$$\frac{\Delta F_{\text{mix}}}{NKT} = x \ln x + (1 - x) \ln(1 - x) + \chi_{AB} x (1 - x).$$

Summary

We have developed a model for the thermodynamic properties of ideal and regular solutions. Two components A and B will tend to mix because of the favorable entropy resulting from the many different ways of interspersing A and B particles. The degree of mixing also depends on whether the AB attractions are stronger or weaker than the AA and BB attractions. In the next chapters we will apply this model to the properties of solutions.

Problems

1. Ternary lattice mixtures. Consider a lattice model liquid mixture of three species of spherical particles, A, B, and C. As with binary mixtures, assume that all $N = n_A + n_B + n_C$ sites are filled.

(a) Write an expression for the entropy of mixing.

(b) Using the Bragg–Williams approximation, write an expression for the energy of mixing U in terms of the binary interaction parameters χ.

(c) Write an expression for the chemical potential μ_A of A.

2. Enthalpy of mixing. For a mixture of benzene and n-heptane having equal mole fractions, $x = 1/2$, and temperature $T = 300\,\mathrm{K}$, the enthalpy of mixing is $\Delta H_{\mathrm{mix}} = 220\,\mathrm{cal\,mol}^{-1}$. Compute χ_{AB}.

3. Plot $\mu(x)$. Plot the chemical potential versus x for:

(a) $\chi_{AB} = 0$,

(b) $\chi_{AB} = 2$,

(c) $\chi_{AB} = 4$.

4. Limit of $x \ln x$ terms that contribute to mixing entropies and free energies. What is the value of $x \ln x$ as $x \to 0$?

5. Hydrophobic entropy. The entropy of dissolving benzene in water at high dilution is approximately $14\,\mathrm{cal\,mol}^{-1}\,\mathrm{K}^{-1}$ at $T = 15\,°\mathrm{C}$.

(a) How does this compare with the mixing entropy?

(b) Speculate on the origin of this entropy.

6. Solubility parameters. The quantity χ_{AB} describes AB interactions relative to AA and BB interactions. If you have N different species $A, B, \ldots$, and if you need the pairwise quantities χ_{ij} for all of them, it will involve $\sim N^2$ experiments involving mixtures of components $i = 1, 2, \ldots, N$ with components $j = 1, 2, \ldots, N$. However, if all the species are nonpolar, then you can make the approximation $w_{AB} \approx \sqrt{w_{AA} w_{BB}}$ (see Chapter 24). Show how this reduces the number of experiments to $\sim N$. Each experiment finds a *solubility parameter*, $\delta_A(w_{AA})$ or $\delta_B(w_{BB})$, by a measurement on a pure system, so no experiments are needed on mixtures.

References

[1] PJ Flory. *Principles of Polymer Chemistry*. Cornell University Press, New York, 1953.

[2] TL Hill. *Introduction to Statistical Thermodynamics*. Wiley, New York, 1960.

[3] R Kubo. *Satistical Mechanics, An Advanced Course with Problems and Solutions*. North-Holland, New York, 1965.

[4] JH Hildebrand and RL Scott. *Regular Solutions*. Prentice-Hall, Englewood Cliffs, 1962.

Suggested Reading

Excellent discussions of the lattice model of mixing:

PJ Flory, *Principles of Polymer Chemistry*, Cornell University Press, New York, 1953.

TL Hill, *Introduction to Statistical Thermodynamics*, Wiley, New York, 1960.

R Kubo, *Statistical Mechanics, An Advanced Course with Problems and Solutions*, North-Holland, New York, 1965.

Good treatments of liquids, regular solutions, and surfaces:

JS Rowlinson and FL Swinton, *Liquids and Liquid Mixtures*, 3rd edition, Butterworth Scientific, London, 1982. A complete and extensive discussion of mixing, including much experimental data.

JH Hildebrand and RL Scott, *Regular Solutions*, Prentice-Hall, Englewood Cliffs, 1962. The regular solution model is described in detail.

JS Rowlinson and B Widom, *Molecular Theory of Capillarity*, Clarendon Press, Oxford, 1982. An excellent advanced discussion of surfaces and interfacial tensions.

WE Acree, Jr, *Thermodynamic Properties of Nonelectrolyte Solutions*, Academic Press, San Diego, 1984.

16 The Solvation & Transfer of Molecules Between Phases

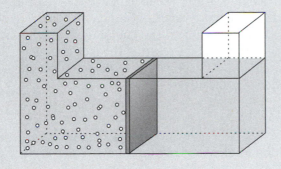

The Chemical Potential Describes the Tendency of Molecules to Exchange and Partition

Now we consider the solvation and partitioning of one type of molecule into a medium of another type. For example, toxins dissolve in one medium, such as water, and partition into another, such as the oil components of fish. Metabolites and drugs cross cell membranes. Many noncovalent equilibria in chemistry and biology are the result of such solvation forces. We also consider how molecules of one type can affect the phase equilibria of other types of molecules. For example, salt melts ice and condenses boiling water. These are examples of *colligative properties*: vapor pressure depression by solutes, boiling temperature elevation, freezing temperature depression, and osmotic pressure. In this chapter we use the liquid mixture lattice model to illustrate, in a simple approximate way, the molecular basis for colligative processes.

What drives these processes? On the one hand, atoms and molecules tend to transfer from regions where their concentrations are high to where they are low, to increase the entropy of the system. On the other hand, particles tend to move to regions for which they have high chemical affinity. In this chapter we explore these driving forces through the use of the chemical potential. While the free energy describes the tendency of a whole system toward equilibrium, the chemical potential describes the tendency toward equilibrium for each individual chemical component. In a mixture, the low-concentration component is called the *solute* and the high-concentration component is the *solvent*.

Solvation Is the Transfer of Molecules Between Vapor and Liquid Phases

Consider a liquid mixture of two components, A and B. Suppose B is volatile but A is not. That is, B exchanges freely between the liquid and the vapor phase, but A stays in the liquid phase. The degree of freedom is the number of B molecules in the gas phase, $N(\text{gas})$, or the number of B molecules in the liquid phase, $N(\text{liquid})$. The total number of B molecules is fixed, $N(\text{gas}) + N(\text{liquid}) = \text{constant}$. No degree of freedom is associated with A because it does not exchange.

For example, you might be interested in gas (B) dissolved in water (A). Since the gas is more volatile than the water, it is often reasonable to assume that water doesn't exchange. On the other hand, if you were interested in salt in water, water would be the more volatile component. Then you would take water to be the volatile component B and salt the nonexchangeable component A.

At constant temperature and pressure, the free energy will be at a minimum when the exchangeable B molecules have the same chemical potential in the gas phase as in the solution phase,

$$\mu_B(\text{gas}) = \mu_B(\text{liquid}). \tag{16.1}$$

The Lattice Solution Model for Solvation and Vapor Pressure Depression

To find the relationship between the vapor pressure p_B of the B molecules and the solution concentration x_B, use Equation (11.50), $\mu_B(\text{gas}) = kT \ln p_B/p_{B,\text{int}}^\circ$, for the chemical potential of B in the vapor phase. For the chemical potential of B in the solution, use lattice model Equation (15.15), $\mu_B(\text{liquid}) = kT \ln x_B + zw_{BB}/2 + kT\chi_{AB}(1 - x_B)^2$. Substitute these equations into the equilibrium Equation (16.1) and exponentiate to get

$$\frac{p_B}{p_{B,\text{int}}^\circ} = x_B \exp\left(\chi_{AB}(1 - x_B)^2 + \frac{zw_{BB}}{2kT}\right). \tag{16.2}$$

Equation (16.2) can be expressed more compactly as

$$p_B = p_B^\circ x_B \exp[\chi_{AB}(1 - x_B)^2], \tag{16.3}$$

where p_B° is the vapor pressure of B over a pure solvent B,

$$p_B^\circ = p_{B,\text{int}}^\circ \exp\left[\frac{zw_{BB}}{2kT}\right].$$

Figure 16.1 plots Equation (16.3) for different values of χ_{AB}, the quantity that characterizes the chemical affinity of the A's for the B's. The vapor pressure p_B of the B molecules increases linearly with their concentration x_B in an 'ideal solution' ($\chi_{AB} = 0$). The linear dependence is called *Raoult's law*. Start at the right end of the $\chi = 0$ curve in Figure 16.1, where $x_B \approx 1$ and the volatile component B is the predominant species. For example, think about a little salt (A) in a lot of water (B). Adding a little salt to pure water (moving to the left from the right end of the figure, reducing x_B to less than one) reduces the vapor pressure of the water. Why should a few salt molecules in the liquid reduce

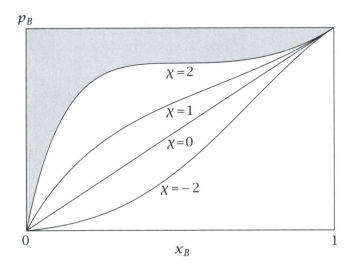

p_B

$\chi = 2$

$\chi = 1$

$\chi = 0$

$\chi = -2$

0   x_B 1

Figure 16.1 The vapor pressure p_B of B molecules over a liquid solution having composition x_B, according to the lattice model. Increasing the concentration x_B of B in the liquid increases the vapor pressure of B. $\chi = 0$ represents an ideal solution. B has more escaping tendency when $\chi > 0$ and less when $\chi < 0$, for a given x_B.

the tendency of the water molecules to escape to the vapor phase, particularly if their interaction energy plays no role?

Adding A's to a liquid of pure B reduces B's tendency to escape from the liquid phase. Like ideality in gases, ideality in solution ($\chi = 0$) describes processes that are driven by the translational entropy, and not by energies of interaction. For a liquid of pure B molecules, the number of lattice configurations is $W = 1$, so $S = k \ln W = 0$ (see Figure 16.2). Because $x_B = 1$, you have $\mu = \mu° + RT \ln x_B = \mu°$. But when A molecules are present in the liquid, $W > 1$, so $S > 0$. Then $x_B < 1$, so $\mu = \mu° + RT \ln x_B < \mu°$. A mixture of A's and B's has more translational entropy than a liquid of pure B. This mixing entropy draws B molecules from pure B phases into AB mixtures. The exchangeable component is drawn into the phase containing the non-exchangeable component.

The chemical potential is sometimes called the *escaping tendency*. The more positive the value of the chemical potential of a particular component in a particular phase, the greater is its tendency to escape from that phase.

What is the additional influence of energetic interactions between A and B? When A and B 'dislike' each other ($\chi_{AB} > 0$), B has greater tendency to escape than it would from an ideal solution. But when the affinity of A for B is greater than the combined AA and BB affinities ($\chi_{AB} < 0$), B will escape less than it would from an ideal solution. Most nonpolar molecules have more affinity for 'like' molecules than for different molecules, so usually $\chi_{AB} > 0$. This is captured in the general rule that 'like dissolves like' (see Chapter 24).

Now look at the left end of Figure 16.1, where the volatile species B is the minority component ($x_B \to 0$). This corresponds to a small amount of gas (B) dissolved in a large amount of water (A), for example. The higher the vapor pressure of the gas, the higher is the gas concentration in water. For example, high pressures of CO_2 gas over water create carbonated beverages. At low concentrations, the vapor pressure is a linear function of composition x_B,

$$\frac{p_B}{p°_{B,\text{int}}} = x_B \exp\left(\chi_{AB} + \frac{z w_{BB}}{2kT}\right) = x_B \exp\left[\frac{z}{kT}\left(w_{AB} - \frac{w_{AA}}{2}\right)\right], \qquad (16.4)$$

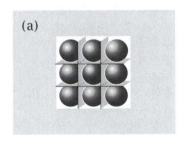

(a)

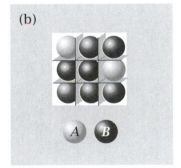

(b)

A B

Figure 16.2 (a) For pure B, the number of arrangements is $W = 1$. (b) Mixing in A's gives $W > 1$, increasing the entropy and reducing the escaping tendency μ of B.

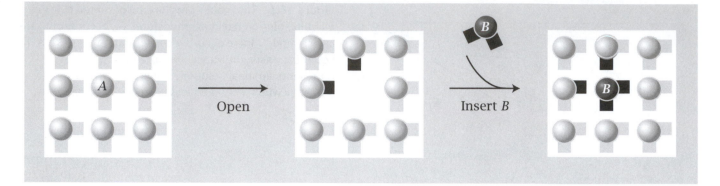

Figure 16.3 The microscopic process underlying the Henry's law coefficient k_H and μ_B° in the solute convention: opening a cavity in pure A (breaking $(z/2)$ AA contacts), then inserting a B (making z AB contacts).

Table 16.1 Henry's law constants for gases in water at 25 °C. Source: I Tinoco, K Sauer, and JC Wang, *Physical Chemistry: Principles and Applications in Biological Sciences*, Prentice-Hall, Englewood Cliffs, 1978.

Gas	k_H (atm)
He	131×10^3
N_2	86×10^3
CO	57×10^3
O_2	43×10^3
CH_4	41×10^3
Ar	40×10^3
CO_2	1.6×10^3
C_2H_2	1.3×10^3

which can be expressed in terms of p_B°, the vapor pressure over pure B,

$$p_B = \left(p_B^\circ e^{\chi_{AB}}\right) x_B. \tag{16.5}$$

This linear region is known as the Henry's law region. The slope, $k_H = p_B^\circ e^{\chi_{AB}}$, which has units of pressure, is the Henry's law constant. When p_B° is known from independent measurements, k_H gives χ_{AB}. Henry's law constants for gases in aqueous solutions at 25 °C are given in Table 16.1.

The exponential factor $\exp[(z/kT)(w_{AB} - w_{AA}/2)]$ in Henry's law Equation (16.4) represents the process shown in Figure 16.3: a cavity is opened in A and a B molecule is inserted. Equation (16.4) shows that B's will concentrate in liquid A if either the AB affinity is strong or the AA affinity is weak. A weak AA affinity means that the A medium readily accommodates a B molecule because it costs little energy to open a cavity in A.

The exchange parameter χ_{AB} is common to various solution properties, giving relationships between them. For example, χ_{AB} for the transfer of a solute B into a liquid A is often proportional to the interfacial tension between the two materials (see Figure 16.4).

A Thermodynamic Model Defines the Activity and Activity Coefficient

Now we switch to a thermodynamic model to look at the same process. The advantage is that the thermodynamic model is more general and not dependent on lattices or the Bragg–Williams approximation. The disadvantage is that it gives less physical insight into microscopic interactions. In this approach, experimentally measurable quantities are used in place of a lattice model.

Start with the condition for equilibrium Equation (16.1), which must apply to any model. Now instead of making a detailed model for the functions $\mu_B(p_B)$ and $\mu_B(x_B)$, we will only assume that these functions have the same logarithmic form as above, $\mu(\text{gas}) \sim \ln p$ and $\mu(\text{liquid}) \sim \ln x_B$:

$$\mu_B(\text{gas}) = \mu_B^\circ(\text{gas}) + kT \ln p_B, \tag{16.6}$$

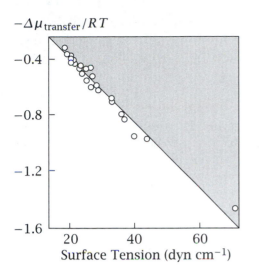

$-\Delta\mu_{\text{transfer}}/RT$

Surface Tension (dyn cm^{-1})

Figure 16.4 The free energy of transferring argon into various hydrocarbons at 25 °C correlates with surface tension, indicating how various solution properties are related to each other through χ_{AB}. Source: A Ben-Naim, *Hydrophobic Interactions*, Plenum Press, New York, 1980. Data were reproduced with changes from R Battino and HL Clever, *Chem Rev* **66**, 395 (1966).

(neglecting nonidealities) in the gas phase and

$$\mu_B(\text{liquid}) = \mu_B^\circ(\text{liquid}) + kT \ln \gamma_B x_B \qquad (16.7)$$

in the liquid phase. The superscript ° in $\mu_B^\circ(\text{gas})$ simply means that this is the chemical potential when the gas pressure is $p_B = 1$, in whatever units you are using. The superscript ° in $\mu_B^\circ(\text{liquid})$ has either of two different meanings described below, depending on whether the volatile component B is in higher or lower concentration than A. For equilibrium, set the chemical potential in Equation (16.6) equal to that in Equation (16.7), to get

$$p_B = \gamma_B x_B \exp\left[\frac{\Delta\mu_B^\circ}{kT}\right], \qquad (16.8)$$

where $\Delta\mu_B^\circ = \mu_B^\circ(\text{liquid}) - \mu_B^\circ(\text{gas})$. The quantity $x_B \gamma_B(x_B)$ in Equation (16.7) is the *activity*. The quantity $\gamma_B(x_B)$, called the *activity coefficient*, can be a function of x_B in general, and accounts for the nonideality of mixing. In a perfectly ideal mixture, $\gamma_B = 1$ for all values of x_B.

Here's how to use these quantities. Suppose you measure the vapor pressure of B, $p_B(x_B)$, over a series of solutions of different compositions x_B (see Figure 16.5). Dividing the observed p_B by x_B gives the 'nonideality' factor as a product of $\gamma_B(x_B)$ and $\exp(\Delta\mu_B^\circ/kT)$. We have not yet defined μ_B° (liquid) or γ_B, so there are many different choices of $\Delta\mu_B^\circ$ and $\gamma_B(x_B)$ that would give the observed value of that product.

So, to be definitive, for uncharged molecules we choose one of two standard conventions. The **solvent convention** is used when the exchangeable component B is the majority component. Then ideality ($\gamma_B(x_B) \to 1$) is defined in the limit as B becomes pure solvent, $x_B \to 1$. In this case, $\mu^\circ(\text{liquid})$ is the chemical potential for a B molecule completely surrounded by other B molecules.

Alternatively, the **solute convention** is used when the exchangeable component B is the minority component. In this case, ideality ($\gamma_B(x_B) \to 1$) is reached as B approaches infinite dilution, $x_B \to 0$. Then $\mu^\circ(\text{liquid})$ is the chemical potential for a B molecule completely surrounded by A molecules.

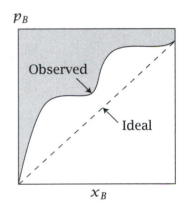

p_B

Observed

Ideal

x_B

Figure 16.5 In an ideal solution, p_B is a linear function of x_B. A real solution can be more complex. The degree of nonideality is determined by the ratio p_B/x_B.

Suppose you are interested in the case in which the exchangeable component is the solvent. Use the solvent convention, $y_B = 1$ for $x_B = 1$. The vapor pressure over pure B gives you the quantity $p_B^\circ = \exp[\Delta\mu_B^\circ/kT]$, so Equation (16.8) becomes

$$p_B(x_B) = p_B^\circ x_B y_B(x_B). \qquad (16.9)$$

The measured vapor pressure of B over a series of solutions that have different compositions gives the function $p_B(x_B)$. According to Equation (16.9), dividing $p_B(x_B)$ by $p_B^\circ x_B$ gives $y_B(x_B)$. If you find that $y_B(x_B) \approx 1$, then you have an ideal solution,

$$p_B = p_B^\circ x_B.$$

The measured function $y_B(x_B)$ or its logarithm $kT \ln y_B(x_B)$, called the *excess free energy*, can be used to test microscopic models of solutions.

Interpreting the Thermodynamic Quantities by Using the Lattice Model

Here's a microscopic interpretation of μ_B° and y_B, using the lattice model. Comparing Equation (15.17) with Equation (15.16) shows that

$$\frac{\mu_B^\circ}{kT} + \ln y_B = \frac{zw_{BB}}{2kT} + \chi_{AB}(1 - x_B)^2. \qquad (16.10)$$

SOLVENT CONVENTION. To find the correspondence between the lattice and thermodynamic models in the solvent convention, let $y_B(x_B) \to 1$ as $x_B \to 1$ to reduce Equation (16.10) to

$$\frac{\mu_B^\circ}{kT} = \frac{zw_{BB}}{2kT}. \qquad (16.11)$$

In this case, μ_B° represents the free energy cost of opening a cavity in pure B, and inserting a B.

To interpret the activity coefficient, substitute Equation (16.11) into Equation (16.10) and exponentiate:

$$y_B = \exp[\chi_{AB}(1 - x_B)^2]. \qquad (16.12)$$

In this lattice model, the nonideal concentration dependence is due to the energies of the AB interactions.

SOLUTE CONVENTION. If B is the solute, substitute $y_B \to 1$ as $x_B \to 0$ into Equation (16.10) to get

$$\frac{\mu_B^\circ}{kT} = \frac{zw_{BB}}{2kT} + \chi_{AB}$$

$$= \frac{z}{kT}\left(w_{AB} - \frac{w_{AA}}{2}\right), \qquad (16.13)$$

which describes the process shown in Figure 16.3. μ_B° represents the free energy cost of inserting a B molecule at infinite dilution into a cavity created in otherwise pure solvent A. Substitute Equation (16.13) into Equation (16.10) to find the definition of γ_B in the solute convention:

$$\gamma_B = \exp\left(\chi_{AB}x_B(x_B - 2)\right). \tag{16.14}$$

Now let's see how the solute A can affect the phase equilibrium of B.

Adding Solute Can Raise the Boiling Temperature of the Solvent

At $p = 1$ atm, pure water boils at 100 °C. But salt water boils at a higher temperature. This is a another manifestation of the colligative principle that salt reduces the vapor pressure of water. Adding salt reduces water's tendency to vaporize, so salt water remains a liquid at 100 °C. To model this, start with the condition for the boiling equilibrium of a solution containing a concentration x_B of B at temperature T and at vapor pressure p,

$$\mu_B(\text{gas}, T, p) = \mu_B(\text{liquid}, T, p, x_B), \tag{16.15}$$

where B is the water, the more volatile component.

We are interested in how liquid and gas chemical potentials depend on x_B and T, at fixed pressure p. For the liquid solution, we have $\mu_B(\text{liquid}, T, x_B) = \mu_B^\circ(\text{liquid}, T) + kT \ln \gamma_B x_B$, where the superscript $^\circ$ for the liquid state in the solvent convention means that the liquid B is pure and contains no solute.

For the gas at standard pressure, $p^\circ = 1$, say in units of atmospheres, $\mu_B(\text{gas}, T, p^\circ) = \mu_B^\circ(\text{gas}, T) + kT \ln p^\circ = \mu_B^\circ(\text{gas}, T)$. For the gas, the superscript $^\circ$ means it is at standard pressure. Substituting these expressions into Equation (16.15) and rearranging gives

$$\ln \gamma_B x_B = \frac{1}{kT}[\mu_B^\circ(\text{gas}, T) - \mu_B^\circ(\text{liquid}, T)]. \tag{16.16}$$

$\mu_B^\circ(\text{gas}, T)$ will depend on temperature because gases have heat capacity. To express the temperature dependence, use Equation (9.25), $\mu^\circ = h^\circ - Ts^\circ$.

Now consider two different situations: (1) a mixture having a concentration x_B of B that boils at temperature T, and (2) pure B ($x_B = 1$), which has a boiling temperature T_b. For the mixture Equation (16.16) gives

$$\ln \gamma_B x_B = \frac{1}{kT}[h^\circ(\text{gas}) - h^\circ(\text{liquid})] - \frac{1}{k}[s^\circ(\text{gas}) - s^\circ(\text{liquid})], \tag{16.17}$$

and for pure B,

$$\ln 1 = \frac{1}{kT_b}[h^\circ(\text{gas}) - h^\circ(\text{liquid})] - \frac{1}{k}[s^\circ(\text{gas}) - s^\circ(\text{liquid})]. \tag{16.18}$$

Subtracting Equation (16.18) from Equation (16.17) gives

$$\ln \gamma_B x_B = \frac{\Delta h^\circ}{k}\left(\frac{1}{T} - \frac{1}{T_b}\right), \tag{16.19}$$

where $\Delta h^\circ = h^\circ(\text{gas}) - h^\circ(\text{liquid})$ is the enthalpy change for boiling the pure solvent. Because the increase in boiling temperature will be small, $T \approx T_b$,

$TT_b \approx T_b^2$, and

$$\ln \gamma_B x_B \approx \frac{\Delta h^\circ}{k} \left(\frac{T_b - T}{T_b^2} \right) = \frac{\Delta h^\circ}{kT_b^2} (T_b - T). \tag{16.20}$$

You can further simplify Equation (16.20) for small solute concentrations $x_A \ll 1$. Use the lattice model Equation (16.12), $\ln \gamma_B = \chi_{AB}(1 - x_B)^2$, and expand the logarithm (see Appendix C, Equation (C.4)) $\ln x_B = \ln(1 - x_A) \approx -x_A - x_A^2/2 - x_A^3/3 - \dots$ to get

$$\ln \gamma_B x_B = \ln \gamma_B + \ln(1 - x_A)$$

$$= \chi_{AB} x_A^2 + \left(-x_A - \frac{x_A^2}{2} - \frac{x_A^3}{3} - \dots \right)$$

$$= - \left[x_A + \left(\frac{1}{2} - \chi_{AB} \right) x_A^2 + \frac{x_A^3}{3} \dots \right]. \tag{16.21}$$

To second-order approximation in x_A, the elevation ΔT of the boiling temperature due to component A is

$$\Delta T = T - T_b$$

$$= \frac{kT_b^2}{\Delta h^\circ} \left[x_A + \left(\frac{1}{2} - \chi_{AB} \right) x_A^2 \right]. \tag{16.22}$$

For ideal solutions ($\chi_{AB} = 0$), $\Delta T = (kT_b^2/\Delta h^\circ)x_A$, and the boiling point elevation, like the vapor pressure depression, depends linearly on the concentration of the nonvolatile solute A. Then the increase in the boiling temperature depends on a product of only two terms: (1) the solute concentration x_A, and (2) quantities that depend only on the pure state of B, its enthalpy of vaporization and boiling temperature. Because ideal solution properties depend only on the concentration, and not on the chemical character, of the solute, the colligative laws are the analogs of the ideal gas law for dilute solutions.

The interaction energy χ_{AB} appears only in the second-order and higher terms of Equation (16.21). The linear dependence on x_A can occur for either small x_A or, if $\chi_{AB} = 1/2$, for somewhat higher concentrations of A. By the definition of χ_{AB} from Equation (15.11), such ideality occurs at a temperature $T = \theta$, where

$$\theta = \frac{2z}{k} \left(w_{AB} - \frac{w_{AA} + w_{BB}}{2} \right).$$

At $T = \theta$, the coefficient of the second-order concentration term in Equation (16.21) is zero (higher-order terms are not necessarily zero). The concept of a 'theta temperature' is particularly useful for polymers (see Chapter 31).

Units of Concentration

You may sometimes prefer concentration units other than mole fractions. The *molality* m_A of A is the number of moles of solute A dissolved in 1000 grams of solvent. The *molarity* c_A is the number of moles of A per liter of solution. To compute ΔT from Equation (16.22), let's convert x_A to other units. If n_A

is the number of moles of the solute A and n_B is the number of moles of the solvent B, then in a dilute solution

$$x_A = \frac{n_A}{n_A + n_B} \approx \frac{n_A}{n_B}. \tag{16.23}$$

In a solution containing 1000 g of solvent B, having solute molality m_A, n_A is given by

$$n_A \text{ moles of } A = \left(m_A \frac{\text{moles of } A}{1000 \text{ g of } B} \right) (1000 \text{ g of } B). \tag{16.24}$$

Since the solution is nearly pure B, the number of moles, n_B, of solvent in this system can be computed simply from its molecular weight. If M_B is the molecular weight of B in grams per mole, then $1/M_B$ is the number of moles of B contained in one gram of solvent. So, in 1000 grams of solvent B, the number of moles is

$$n_B \text{ moles of } B = \left(\frac{1}{M_B} \right) (1000 \text{ g of } B). \tag{16.25}$$

Substituting Equations (16.24) and (16.25) into (16.23) gives

$$x_A \approx m_A M_B. \tag{16.26}$$

Expressed in terms of the molality of A in a dilute solution, the boiling point elevation is

$$T - T_b = K_b m_A, \tag{16.27}$$

where K_b is a constant for boiling that depends only on the pure solvent (for example, $K_b = 0.51$ for water).

Example 16.1 shows a calculation of K_b.

EXAMPLE 16.1 Compute the boiling point elevation constant K_b for water.
Combine Equations (16.22) and (16.27) with Equation (16.26), and take $\Delta h_b^\circ = 40,656 \text{ J mol}^{-1}$ and $T_b = 373.15 \text{ K}$ from Table 14.1, to get

$$K_b = \frac{RT_b^2 M_B}{\Delta h_b^\circ}$$

$$= \frac{(8.314 \text{ J K}^{-1} \text{ mol}^{-1})(373.15 \text{ K})^2 (18 \text{ g mol}^{-1})}{(40,656 \text{ J mol}^{-1})}$$

$$= 0.513 \text{ K kg mol}^{-1}.$$

The molality is often preferred as a concentration measure because it is expressed in terms of weight, which is independent of temperature. Molarities are expressed in terms of volumes, which depend on temperature. In an ideal solution, the solute contributes to colligative properties only through its concentration, and not through its interaction with the solvent. Solutes affect freezing points in much the same way that they affect boiling points.

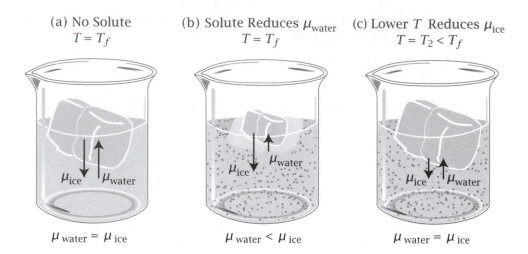

(a) No Solute
$T = T_f$

(b) Solute Reduces μ_{water}
$T = T_f$

(c) Lower T Reduces μ_{ice}
$T = T_2 < T_f$

μ_{ice} μ_{water}

μ_{ice} μ_{water}

μ_{ice} μ_{water}

$\mu_{water} = \mu_{ice}$

$\mu_{water} < \mu_{ice}$

$\mu_{water} = \mu_{ice}$

Figure 16.6 The freezing temperature is lowered by adding solute. (a) The freezing temperature of pure solvent is T_f. (b) Adding solute reduces the tendency of the water to escape from the liquid mixture to ice. At T_f, this melts the ice. (c) Lowering the temperature to $T < T_f$ lowers the tendency of water to escape from the ice, to reach a new freezing point.

Adding Solute Can Lower the Freezing Temperature of a Solvent

Adding salt lowers the freezing temperature of water below 0 °C, following the same principles as boiling point elevation. The melting process is an equilibrium of an exchangeable component, such as water, between its liquid and solid phases. At the melting temperature of pure water, the escaping tendency of water molecules from the liquid to the solid is the same as the escaping tendency from the solid to the liquid. Salt preferentially dissolves in the liquid, so to first approximation, salt is not an exchangeable component in this equilibrium. Salt reduces the chemical potential of the water to escape from the liquid to the solid, drawing water from the pure phase (ice) to the mixture (liquid), and melting the ice (see Figure 16.6).

The lowering of the freezing temperature ΔT_f by solute can be treated in the same way as the boiling point elevation in Equations (16.15) to (16.22), with the result

$$\Delta T_f = T_f - T = \left(\frac{kT_f^2}{\Delta h_{\text{fus}}^\circ} \right) \left[x_A + \left(\frac{1}{2} - \chi_{AB} \right) x_A^2 + \cdots \right], \tag{16.28}$$

where $T_f > T$. T_f is the freezing temperature of the pure liquid B, and $\Delta h_{\text{fus}}^\circ = h_{\text{liquid}}^\circ - h_{\text{solid}}^\circ$ is the enthalpy of *fusion* (this is the enthalpy of melting, with sign opposite to the enthalpy of freezing). Concentration units other than mole fraction x_A may be more convenient. In terms of molality,

$$T_f - T = K_f m_A,$$

where K_f is defined by the freezing properties of the solvent (for example, $K_f =$

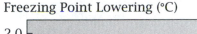
Freezing Point Lowering (°C)

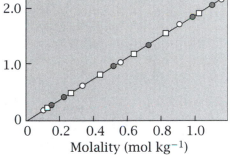

Molality (mol kg^{-1})

Figure 16.7 The depression of the freezing temperature of water is a linear function of the concentration of glycerol (○), dextrose (●), and sucrose (□). Source: CH Langford and RA Beebe, *The Development of Chemical Principles*, Dover Publications, New York, 1969.

1.86 K kg mol^{-1} for water). Figure 16.7 shows that the freezing temperature of water is depressed in linear proportion to concentration for glycerol and some sugars, in good agreement with this ideal solution model.

In a few instances, a solute preferentially dissolves in the solid, rather than in the liquid. For example, some long-chain alkanes are more concentrated in solid lipid bilayer membranes than they are in liquid-state bilayers. Then such solutes must increase, rather than decrease, the freezing temperature.

The next section treats another colligative effect. Adding solutes to solvents can lead to osmotic pressures.

Adding Solute on One Side of a Semipermeable Membrane Causes an Osmotic Pressure

Suppose you have a pure liquid B separated from a mixture of A with B by a *semipermeable membrane*, through which B molecules can pass freely, but A cannot (see Figure 16.8). The compartment on the right contains pure liquid B. The compartment on the left contains B as a solvent with A as a solute.

B molecules will be drawn from the right (pure liquid) into the left compartment (the mixture) because of the favorable entropy of mixing with the A component. But the net flow of solvent B can be stopped by applying more pressure to the left-hand compartment than to the right-hand compartment. An alternative way to counter the leftward solvent flow is to accumulate extra volume on top of the left side. This added volume will itself exert a hydrostatic pressure because of its weight. In either case, the extra pressure that is required on the left to hold the system in equilibrium is called the *osmotic pressure*.

What is the relationship between the composition x_B of the solution and the osmotic pressure π that is required to hold the solvent flow in equilibrium when the temperature T is the same on both sides? The relationship we want is between x_B and the extra pressure π added to the left compartment. At equilibrium, the chemical potential for the exchangeable component B must be the same on both sides of the membrane,

$$\mu_B(\text{pure}, p) = \mu_B(\text{mixture}, p + \pi, x_B). \tag{16.29}$$

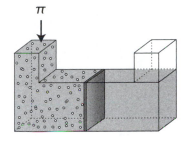

Figure 16.8 Osmotic pressure. The right compartment contains pure liquid B. The left compartment contains a mixture of A and B. B can exchange freely across the membrane, but A cannot. At equilibrium, the net flow of B to the left is halted by a pressure π applied to the left compartment to hold the system in equilibrium. The pressure π that is needed to halt the net flow increases linearly with the concentration of solute A in the left compartment.

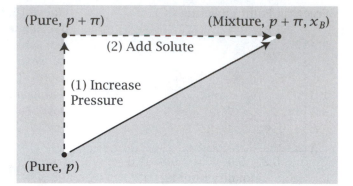

Figure 16.9 Thermodynamic cycle for osmotic pressure in the left compartment. Begin with pure solvent B at pressure p. Step 1: increase its pressure to $p + \pi$. Step 2: add solute A until the composition x_B of B is reached.

To find the relationship between π and x_B, use the thermodynamic cycle shown in Figure 16.9. For the left compartment, imagine starting with pure liquid B at pressure p, then increasing its pressure to $p + \pi$ in step 1, then add the solute at constant pressure $p + \pi$ in step 2. How does μ_B change in these two steps?

To see how μ_B depends on pressure in a pure liquid B at constant T, integrate $(\partial \mu_B / \partial p)$.

$$\mu_B(\text{pure}, p + \pi) = \mu_B(\text{pure}, p) + \int_p^{p+\pi} \frac{\partial \mu_B}{\partial p}\, dp. \tag{16.30}$$

Using the Maxwell relation $\partial \mu_B / \partial p = \partial V / \partial N_B = v_B$, where v_B is the molar volume of B, you get

$$\mu_B(\text{pure}, p + \pi) = \mu_B(\text{pure}, p) + \int_p^{p+\pi} v_B\, dp$$

$$= \mu_B(\text{pure}, p) + \pi v_B. \tag{16.31}$$

We are assuming that v_B is independent of pressure.

For the second step—adding the solute—the chemical potential is given by $\mu_B(\text{mixture}, p + \pi, x_B) = \mu_B(\text{pure}, p + \pi) + RT \ln y_B x_B$. Use Equation (16.31) to replace $\mu_B(\text{pure}, p + \pi)$ in this expression to get

$$\mu_B(\text{mixture}, p + \pi, x_B) = \mu_B(\text{pure}, p) + \pi v_B + RT \ln y_B x_B. \tag{16.32}$$

Substitute Equation (16.29) into Equation (16.32) to get

$$-\pi v_B = RT \ln y_B x_B. \tag{16.33}$$

To get the osmotic pressure π in terms of the composition of a dilute solution, substitute Equation (16.21) into Equation (16.33):

$$\pi = \frac{RT}{v_B}\left[x_A + \left(\frac{1}{2} - \chi_{AB}\right) x_A^2 + \cdots \right]. \tag{16.34}$$

When $x_A \ll 1$, Equation (16.34) describes ideal solution behavior, $\pi v_B = RT x_A$.

To convert to molarity c_A, use $x_A \approx n_A / n_B$ and note that $n_B v_B \approx V$ is approximately the volume of the solution. Then Equation (16.34) becomes

$$\pi \approx \frac{n_A RT}{n_B v_B} = \frac{n_A RT}{V} = c_A RT. \tag{16.35}$$

EXAMPLE 16.2 Determining molecular weights from osmotic pressure. Find the molecular weight of a compound that has a weight concentration $w = 1.2\,\text{g L}^{-1}$ in solution and an osmotic pressure of $\pi = 0.2\,\text{atm}$ at $T = 300\,\text{K}$. Combine Equation (16.35), $\pi = cRT$, with $c = w/M$ to get $\pi = wRT/M$. Now rearrange to get

$$M = \frac{wRT}{\pi} = \frac{(1.2\,\text{g L}^{-1})(8.2 \times 10^{-2}\,\text{L atm K}^{-1}\,\text{mol}^{-1})(300\,\text{K})}{0.2\,\text{atm}}$$

$$= 147\,\text{g mol}^{-1}.$$

Solutes Can Transfer and Partition from One Medium to Another

In many processes, including chromatographic separations, the adsorption of drugs to surfaces, the equilibration of solutes between oil and water phases, and diffusion across polymeric or biological membranes, molecules start in one environment and end in another. Nucleation, aggregate or droplet formation, membrane and micelle formation, and protein folding have been modeled in terms of such processes.

So far we've concentrated on colligative processes. Often they involve the transfer of *solvent* molecules from one place to another. Now we focus on the transfer of *solutes*, so we will use the solute convention (see page 284).

Solvents A and B are immiscible or otherwise isolated from each other in some way. A solute species s dissolves in both solvents A and B, and can exchange between them. If, at equilibrium, the concentration of s in A is x_{sA}, and the concentration of s in B is x_{sB}, then the **partition coefficient** is defined as $K_A^B = x_{sB}/x_{sA}$ at equilibrium. (Partition coefficients can also be expressed in other concentration units.)

The degree of freedom is the number of solute molecules in each medium. The total number of solute molecules is fixed. The chemical potential for s in A is $\mu_s(A)$ and the chemical potential for s in B is $\mu_s(B)$. To determine the partition coefficient, use the equilibrium condition that the chemical potential of the solute must be the same in both phases,

$$\mu_s(A) = \mu_s(B). \tag{16.36}$$

To compute the partition coefficient, you can use the lattice model for the chemical potential of s in each liquid. According to Equations (15.15) and (15.16),

$$\frac{\mu_s(A)}{kT} = \frac{zw_{ss}}{2kT} + \ln x_{sA} + \chi_{sA}(1 - x_{sA})^2, \quad \text{and} \tag{16.37}$$

$$\frac{\mu_s(B)}{kT} = \frac{zw_{ss}}{2kT} + \ln x_{sB} + \chi_{sB}(1 - x_{sB})^2. \tag{16.38}$$

Setting the chemical potential for s in A equal to the chemical potential of s in B gives the partition coefficient K_A^B,

$$\ln K_A^B = \ln\left(\frac{x_{sB}}{x_{sA}}\right) = \chi_{sA}(1 - x_{sA})^2 - \chi_{sB}(1 - x_{sB})^2. \tag{16.39}$$

Figure 16.10 $\Delta\mu°$ for the transfer of hydrocarbons from aqueous solution to pure liquid hydrocarbon at 25 °C, based on solubility measurements of C McAuliffe, *J Phys Chem* **70**, 1267 (1966). The transfer free energy is linearly proportional to the number of solute–solvent contacts, represented by z in the lattice model. Source: C Tanford, *The Hydrophobic Effect: Formation of Micelles and Biological Membranes*, 2nd edition, Wiley, New York, 1980.

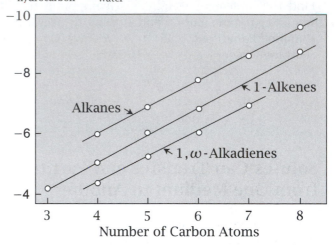

The corresponding result for the thermodynamic model comes from combining Equations (16.36) and (16.7):

$$\ln K_A^B = \ln\left(\frac{x_{sB}}{x_{sA}}\right) = -\frac{\mu_s°(B) - \mu_s°(A)}{kT} - \ln\left(\frac{\gamma_{sB}}{\gamma_{sA}}\right). \tag{16.40}$$

The quantity $\Delta\mu° = \mu_s°(B) - \mu_s°(A)$ is called the *free energy of transfer*. At infinite dilution of solute in both phases, $x_{sA}, x_{sB} \ll 1$. In the solute convention ($\gamma_{sA} \to 1$ as $x_{sA} \to 0$ and $\gamma_{sB} \to 1$ as $x_{sB} \to 0$), you have

$$\ln K_A^B = \frac{\mu_s°(A) - \mu_s°(B)}{kT} = \chi_{sA} - \chi_{sB}. \tag{16.41}$$

EXAMPLE 16.3 Partition coefficient for transferring butane into water. Figure 16.10 shows $\Delta\mu° = \mu_{water}° - \mu_{hydrocarbon}° = 6\,\text{kcal mol}^{-1}$ for butane. To compute the partition coefficient for this transfer, $K_{hc}^w = x_{s,w}/x_{s,hc}$ at $T = 300\,\text{K}$, use Equation (16.40),

$$K_{hc}^w = e^{-\Delta\mu°/RT} = \exp\left[\frac{-6000}{(1.987)(300)}\right]$$

$$= 4.3 \times 10^{-5}.$$

One particularly simple process of partitioning is the transfer of s from pure liquid s to a mixture with A. By definition, $\chi_{ss} = 0$ (see Equation (15.11)). In that case, Equation (16.39) gives

$$\ln K_{sA} = -\chi_{sA}.$$

Partitioning experiments provide a direct means for measuring the binary interaction quantities χ_{AB}. For butane and water in Example 16.3, $-kT\chi_{\text{butane,water}} = -6\,\text{kcal mol}^{-1}$, so at $T = 300\,\text{K}$, $\chi_{\text{butane,water}} \approx 10$.

The exchange contribution to the transfer process at infinite dilution is shown in Figure 16.11. Partitioning is driven by the energetic cost of removing a solute molecule from medium A, closing the cavity in A, opening a cavity in

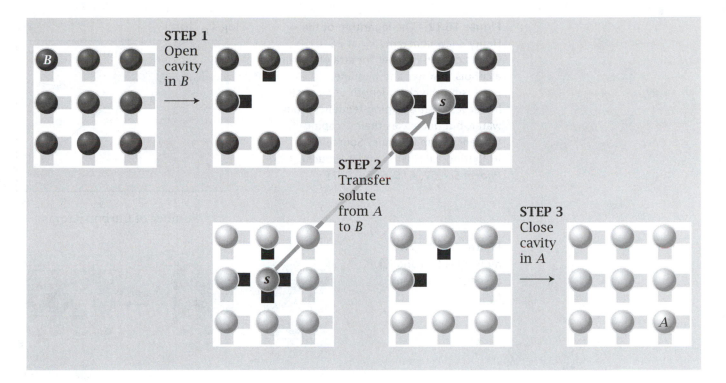

Figure 16.11 The solute partitioning process: (1) open a cavity in B (breaking $z/2$ BB contacts), (2) transfer solute s from medium A to B (breaking $z\,sA$ contacts and making $z\,sB$ contacts), then (3) close the cavity in A (making $z/2$ AA contacts).

B, and inserting the solute into that cavity. This is the difference

$$\chi_{sB} - \chi_{sA} = \left(\frac{z}{kT}\right)\left[\left(w_{sB} - \frac{w_{ss} + w_{BB}}{2}\right) - \left(w_{sA} - \frac{w_{ss} + w_{AA}}{2}\right)\right]$$

$$= \left(\frac{z}{kT}\right)\left[w_{sB} - w_{sA} + \frac{w_{AA}}{2} - \frac{w_{BB}}{2}\right]. \qquad (16.42)$$

If the solute were not sufficiently dilute, say in medium A, then some of the neighbors of s would be other s molecules, rather than exclusively A's. This would change the driving force for exchange. Exchange experiments are simplest to interpret when partition coefficients are measured at, or extrapolated to, infinite dilution.

Partitioning Experiments

In the lattice model, the driving force for partitioning depends on the lattice coordination number z, which serves as a measure of the surface area of contact between the solute molecule and surrounding medium. It follows that $\mu_s^\circ(A) - \mu_s^\circ(B)$ should be a linear function of the sizes of solutes that partition between different media. Figure 16.10 shows examples in which the partitioning free energy increases linearly with the solute size. For the same reason, the Henry's law constant should increase with solute size. Figure 16.12 shows this result.

Figure 16.12 The logarithm of the Henry's law constants k_H at 25 °C versus carbon number for straight-chain alcohols in water and heptane. Increasing the chain length of alcohols increases their escaping tendency from water, but decreases their escaping tendency from heptane. Source: JH Rytting, LP Huston and T Higuchi, *J Pharm Sci*, **67**, 615–618 (1978).

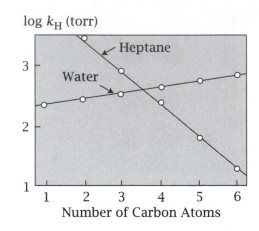

Figure 16.13 Two molecules (◐) and (●) associate in solvent (●).

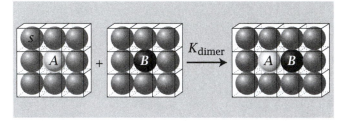

Solvation forces can drive not only partitioning processes, but also association and dimerization in solution.

Noncovalent Dimerization in Solution Involves Desolvation

In this section we compute the equilibrium constant for the noncovalent association of a molecule of type A with a molecule of type B to form species AB (see Figure 16.13). All three species are at infinite dilution in a solvent s.

The degrees of freedom are the numbers N_A of monomer particles of type A, N_B of type B, and the number N_{AB} of dimers of type AB. Pressure and temperature are constant. Changes in free energy are given by

$$dG = \mu_A dN_A + \mu_B dN_B + \mu_{AB} dN_{AB}. \tag{16.43}$$

Conservation of A and B particles leads to the constraints

$$N_A + N_{AB} = N_{A,\text{total}} = \text{constant, and}$$

$$N_B + N_{AB} = N_{B,\text{total}} = \text{constant}$$

$$\implies \quad dN_A = dN_B = -dN_{AB}. \tag{16.44}$$

The equilibrium condition $dG = 0$ can be expressed as

$$(-\mu_A - \mu_B + \mu_{AB})dN_{AB} = 0$$

$$\implies \quad \mu_{AB} = \mu_A + \mu_B. \tag{16.45}$$

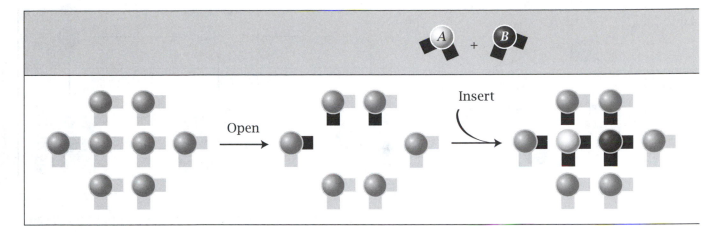

Figure 16.14 For dimerization, a two-site cavity opens in solvent s. There are $2(z-1)$ first-neighbor s molecules (six first shell neighbors in this two-dimensional example in which $z = 4$). The number of ss bonds broken in this process is half this number ($z - 1 = 3$ indicated by the three bonds pointed inward in this case). Inserting one A makes $z - 1$ contacts between s and A. Inserting one B makes $z - 1$ contacts between s and B. There is one AB contact.

Assume A, B, and AB are at infinite dilution in the solvent s. Substituting the definition of μ from Equation (16.7) into Equation (16.45), you have

$$\frac{\mu^{\circ}_{AB}}{kT} + \ln x_{AB} = \left(\frac{\mu^{\circ}_A}{kT} + \ln x_A \right) + \left(\frac{\mu^{\circ}_B}{kT} + \ln x_B \right). \qquad (16.46)$$

From Equation (16.46), the dimerization equilibrium constant K_{dimer} is

$$\ln K_{\text{dimer}} = \ln \left(\frac{x_{AB}}{x_A x_B} \right) = -\frac{1}{kT} (\mu^{\circ}_{AB} - \mu^{\circ}_A - \mu^{\circ}_B). \qquad (16.47)$$

The lattice model gives μ°_A and μ°_B (see Equation (15.26)):

$$\frac{\mu^{\circ}_A}{kT} = \frac{z w_{AA}}{2kT} + \chi_{sA} - \ln q_A = \frac{z}{kT} \left(w_{sA} - \frac{w_{ss}}{2} \right) - \ln q_A, \quad \text{and} \qquad (16.48)$$

$$\frac{\mu^{\circ}_B}{kT} = \frac{z w_{BB}}{2kT} + \chi_{sB} - \ln q_B = \frac{z}{kT} \left(w_{sB} - \frac{w_{ss}}{2} \right) - \ln q_B. \qquad (16.49)$$

Now you need to include the internal partition functions, q_A, q_B, and q_{AB} because the internal degrees of freedom, such as the rotational symmetry, can change upon dimerization (see Equation (15.26)).

You need μ°_{AB}, the free energy cost of opening a two-site cavity in s and inserting an A and a B adjacent to each other (see Chapter 15). The simplest way to get this quantity is to consider a two-site cavity surrounded by a first shell of $2(z-1)$ solvent molecules. The number of bonds that are broken upon opening the cavity is half the number of first-shell neighbors, $2(z-1)/2 = z-1$ (see Figure 16.14). So the energy cost of opening a two-site cavity is $-w_{ss} \times (z-1)$. The energy gained upon inserting one A into the cavity is $(z-1)w_{sA}$, and upon inserting one B the energy gain is $(z-1)w_{sB}$. The energy gained in forming

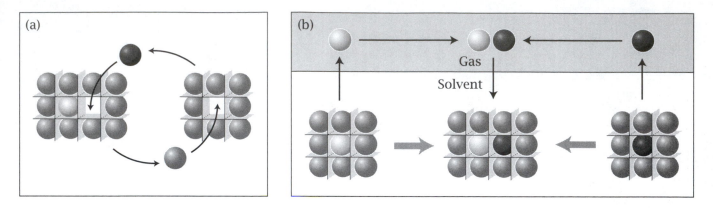

Figure 16.15 (a) The association of two molecules, (⚪) and (⚫), in solvent (⚫) can be represented as an exchange process, or (b) as a series of steps: first A and B are transferred from solvent to the vapor phase, then A associates with B, then the AB dimer is solvated.

the one AB contact is w_{AB}, so

$$\frac{\mu_{AB}^{\circ}}{kT} = \left(\frac{z-1}{kT}\right)(w_{sA} + w_{sB} - w_{ss}) + \frac{w_{AB}}{kT} - \ln q_{AB}, \qquad (16.50)$$

where q_{AB} is the partition function for the dimer. To find K_{dimer}, substitute Equations (16.48), (16.49) and (16.50) into Equation (16.47):

$$\ln K_{\text{dimer}} = -\frac{\Delta\mu^{\circ}}{kT} = \frac{w_{sA} + w_{sB} - w_{ss} - w_{AB}}{kT} + \ln\left(\frac{q_{AB}}{q_A q_B}\right) \qquad (16.51)$$

$$= \left(\frac{1}{z}\right)[\chi_{sA} + \chi_{sB} - \chi_{AB}] + \ln\left(\frac{q_{AB}}{q_A q_B}\right). \qquad (16.52)$$

Dimerization in solution can be described by the thermodynamic cycle shown in either Figure 16.15(a) or (b).

AB association in solution can be driven by the affinity of A for B, or by the affinity of s for s ($w_{ss} \ll 0$), or by the disaffinity of s for A or B. The hydrophobic interaction is an example: hydrocarbon molecules in water at 25 °C are driven to associate mainly because the water-water affinity is high owing to hydrogen bonding, not because there is a disaffinity between hydrocarbon molecules and water (see Chapter 30).

If you are interested in AA dimerization, then substitute A for B in Equation (16.51):

$$\ln\left(\frac{x_{AA}}{x_A^2}\right) = \frac{2w_{sA} - w_{ss} - w_{AA}}{kT} + \ln\left(\frac{q_{AA}}{q_A^2}\right) = \frac{2}{z}\chi_{sA} + \ln\left(\frac{q_{AA}}{q_A^2}\right). \quad (16.53)$$

As for dimerization in the gas phase, the rotational partition function favors the dimer, because there are more distinguishable orientations of AB than of A or B spheres.

Summary

We have considered mixtures of two components in which one can exchange into the gas phase, or into other liquids or solids. The degree to which a species exchanges depends on its concentration in the solution. Such processes can be described by thermodynamic models, which do not describe the microscopic details, or statistical mechanical models such as the lattice model. Chapters 27 and 28 treat dimerization and other binding processes in more detail.

Problems

1. The mechanism of anesthetic drugs.
Anesthetic drug action is thought to involve the solubility of the anesthetic in the hydrocarbon region of the lipid bilayer of biological membranes. According to the classical 'Meyer–Overton hypothesis,' anesthesia occurs whenever the concentration of drug is greater than $0.03 \, \text{mol kg}^{-1}$ membrane, no matter what the anesthetic.

(a) For gaseous anesthetics like nitrous oxide or ether, how would you determine what gas pressure of anesthetic to use in the inhalation mix for a patient in order to achieve this membrane concentration?

(b) Lipid bilayers 'melt' from a solid-like state to a liquid-like state. Do you expect introduction of the anesthetic to increase, decrease, or not change the melting temperature? If the melting temperature changes, how would you predict the change?

2. Divers get the bends.
Divers can get the 'bends' from nitrogen gas bubbles in their blood. Assume that blood is largely water. The Henry's law constant for N_2 in water at 25 °C is 86,000 atm. The hydrostatic pressure is 1 atm at the surface of a body of water and increases by approximately 1 atm for every 33 feet in depth. Calculate the N_2 solubility in the blood as a function of depth in the water, and explain why the bends occur.

3. Hydrophobic interactions.
Two terms describe the hydrophobic effect: (i) hydrophobic hydration, the process of transferring a hydrophobic solute from the vapor phase into a very dilute solution in which the solvent is water, and (ii) hydrophobic interaction, the process of dimer formation from two identical hydrophobic molecules in a water solvent.

(a) Using the lattice model chemical potentials, and the solute convention, write the standard state chemical potential differences for each of these processes, assuming that these binary mixtures obey the regular solution theory.

(b) Describe physically, or in terms of diagrams, the driving forces and how these two processes are similar or different.

4. Solutes partition into lipid bilayers.
Robinson Crusoe and his trusty friend Friday are stranded on a desert island and need to conserve their whiskey. Crusoe, an anesthesiologist, realizes that the effect of alcohol, as with other anesthetics, is felt when the alcohol reaches a particular critical concentration, c_0, in the membranes of neurons. Crusoe and Friday have only a supply of propanol, butanol, and pentanol, and a table of their free energies of transfer for $T = 300 \, \text{K}$.

(a) If a concentration of c_1 of ethanol in water is needed to produce a concentration c_0 in the membrane, a hydrocarbon-like environment. Use Table 16.2 to predict what concentrations in water of the other alcohols would produce the same degree of anesthesia.

Table 16.2 Partitioning quantities in cal mol^{-1}.

	$\mu_w^\circ - \mu_{hc}^\circ$	$h_w^\circ - h_{hc}^\circ$	$s_w^\circ - s_{hc}^\circ$
ethanol	760	-2430	-10.7
propanol	1580	-2420	-13.4
butanol	2400	-2250	-15.6
pentanol	3222	-1870	-17.1

(b) The natives find Crusoe and Friday and throw them into a pot of boiling water. Will the alcohol stay in their membranes and keep them drunk at 100 °C?

(c) Which alcohol has the greatest tendency to partition into the membrane per degree of temperature rise?

5. Global warming.
CO_2 in the earth's atmosphere prevents heat from escaping, and is responsible for roughly half of the 'greenhouse' effect, the putative origin of global warming. Would global warming cause a further increase in atmospheric CO_2 through vaporization from the oceans? Assume that the ocean is a two-component solution of water plus CO_2, and that CO_2 is much more volatile than water. Give an algebraic expression for the full temperature dependence of Henry's law constant k_H for the CO_2 in water; that is, derive an equation for $\partial k_H / \partial T$.

6. Modelling cavities in liquids.
Assume that you have a spherical cavity of radius r in a liquid. The surface tension of the liquid is γ, in units of energy area^{-1}.

(a) Write an expression for the energy $\varepsilon(r)$ required to create a cavity of radius r.

(b) Write an expression for the probability $p(r)$ of finding a cavity of radius r.

(c) What is the average energy $\langle \varepsilon \rangle$ for cavities in the liquid?

(d) Write an expression for the average cavity radius.

(e) If $kT = 600 \, \text{cal mol}^{-1}$, and $\gamma = 25 \, \text{cal mol}^{-1}\text{Å}^2$, then compute $\langle r \rangle$.

7. Sparkling drinks.
CO_2 in water has a Henry's law constant $k_H = 1.25 \times 10^6 \, \text{mm Hg}$. What mole fraction of CO_2 in water will lead to 'bubbling up' and a vapor pressure equal to 1 atm?

8. Modelling binding sites.
You have a two-dimensional molecular lock and key in solvent s, as shown in Figure 16.16. Different parts of each molecule have different chemical characters, A, B, or C.

(a) In terms of the different pair interactions, $(w_{AB}, w_{AC}, w_{AS}, \ldots etc.)$ write an expression for the binding constant K (i.e., for association).

(b) Which type of pair interaction (AB, AC, BC) will dominate the attraction?

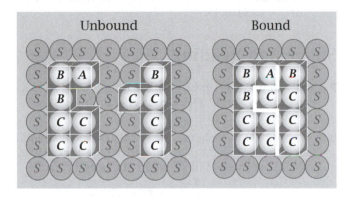

Figure 16.16

9. Oil/water partitioning of drugs. In the process of partitioning of a drug from oil into water, $\Delta s° = -50$ cal (mol deg)$^{-1}$ and $\Delta h° = 0$ at $T = 300$ K.

(a) What is $\Delta \mu°$ at $T = 300$ K?

(b) What is the partition coefficient from oil to water, $K_{\text{oil}}^{\text{water}}$ at $T = 300$ K?

(c) Assume that $\Delta s°$ and $\Delta h°$ are independent of temperature. Calculate K_0^w at $T = 320$ K.

10. Another oil/water partitioning problem. Assume that a drug molecule partitions from water to oil with a partition coefficients $K_1 = 1000$ at $T_1 = 300$ K, and $K_2 = 1400$ at $T_2 = 320$ K.

(a) Calculate the free energy of transfer, $\Delta \mu°$ at $T_1 = 300$ K.

(b) Calculate the enthalpy of transfer, $\Delta h°$ (water to oil).

(c) Calculate the entropy of transfer, $\Delta s°$ (water to oil).

11. Oil/water partitioning of benzene. You put the solute benzene into a mixture containing the two solvents oil and water. You observe the benzene concentration in water to be $x_w = 2.0 \times 10^{-6}$ M and in oil to be $x_o = 5.08 \times 10^{-3}$ M.

(a) What is the partition coefficient $K_{\text{water}}^{\text{oil}}$ (from water into oil)?

(b) What is $\Delta \mu°$ for the transfer of benzene from water into oil at $T = 300$ K?

12. Balancing osmotic pressures. Consider a membrane-enclosed vesicle that contains within it a protein species A that cannot exchange across the membrane.

This causes an osmotic flow of water into the cell. Could you reverse the osmotic flow by a sufficient concentration of a different nonexchangeable protein species B in the external medium?

13. Vapor pressures of large molecules. Why do large molecules, such as polymers, proteins, and DNA, have very small vapor pressures?

14. Osmosis in plants. Consider the problem of how plants might lift water from ground level to their leaves. Assume that there is a semipermeable membrane at the roots, with pure water on the outside, and an ideal solution inside a small cylindrical capillary inside the plant. The solute mole fraction inside the capillary is $x = 0.001$. The radius of the capillary is 10^{-2} cm. The gravitational potential energy that must be overcome is mgh, where m is the mass of the solution, and g is the gravitational constant, 980 cm s^{-2}. The density of the solution $= 1$ g cm^{-3}. What is the height of the solution at room temperature? Can osmotic pressure account for raising this water?

15. Polymerization in solution. Using the lattice dimerization model (see Figure 16.17), derive the equilibrium constant for creating a chain of m monomers of type A in a solvent s.

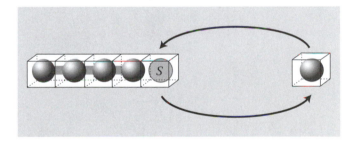

Figure 16.17

16. Ethane association in water. The association of ethane molecules in water is accompanied by an enthalpy of 2.5 kcal mol^{-1} at $T = 25$ °C. Calculate $(\partial \ln K_{\text{assoc}} / \partial T)$ at this temperature. Does the 'hydrophobic effect' get stronger or weaker as temperature is increased?

17. Freezing point depression by large molecules. Freezing temperature depression is a useful way of measuring the molecular weights of some molecules. Is it useful for proteins? One g of protein of molecular weight $100,000$ g mol^{-1} is in 1 g of water. Calculate the freezing temperature of the solution.

18. Partitioning into small droplets is opposed by interfacial tension. When a solute molecule (s) partitions from water (A) into a small oil droplet (B), the droplet will grow larger, creating a larger surface of contact between the droplet and the water. Thus partitioning the solute into the droplet will be opposed by the droplet's interfacial tension with water. Derive an expression to show how much the partition coefficient will be reduced as a function of the interfacial tension and radius of the droplet.

19. Alternative description of Henry's law. Show that an alternative way to express Henry's law of gas solubility is to say that the volume of gas that dissolves in a fixed volume of solution is independent of pressure at a given temperature.

20. Benzene transfer into water. At $T = 25\,°C$, benzene dissolves in water up to a mole fraction of 4.07×10^{-4} (its solubility limit).

(a) Compute $\Delta\mu°$ for transferring benzene to water.

(b) Compute $\chi_{benzene,water}$.

(c) Write an expression for the temperature dependence of $\Delta\mu°$ as a function of $\Delta h°$, $\Delta s°$, the molar enthalpy and entropy at 25 °C, and ΔC_p.

(d) Using the expression you wrote for (c), calculate $\Delta\mu°$ for transferring benzene to water at $T = 100$ °C, if $\Delta h° = 2\,kJ\,mol^{-1}$, $\Delta s° = -58\,J\,mol\,K^{-1}$, and $\Delta C_p = 225\,J\,mol\,K^{-1}$.

Suggested Reading

P Atkins, *Physical Chemistry*, 6th edition, WH Freeman and Co, New York, 1997. Good general physical chemistry text.

IM Klotz and RM Rosenberg, *Chemical Thermodynamics*, 5th edition, Benjamin Cummings, Menlo Park, 1986. Excellent discussion of standard states and activities.

17 Mathematical Toolkit: Some Vector Calculus

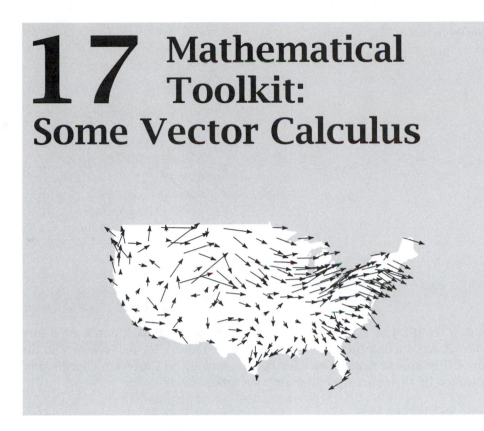

Vectors Describe Forces and Flows

Vector algebra and calculus are useful for describing the forces and flows of fluids and particles as a function of spatial position and time. Chapters 18–23 describe the forces and flows of neutral and electrically charged molecules. In this chapter we just develop the background vector mathematics.

Vectors Add and Subtract by Components

A *vector* is a quantity that has both a magnitude and a direction. An example is a force **f**. In contrast, a *scalar* quantity, such as work, is a single number. Vectors are represented as arrows having a length, which represents the magnitude, and a direction. Vectors are indicated by **bold** type, such as **x** and **v**.

Moving the origin does not change the magnitude or direction of a vector, so vectors can be *added*, $\mathbf{v}_1 + \mathbf{v}_2 = \mathbf{v}_2 + \mathbf{v}_1 = \mathbf{v}_3$, by placing them head to tail as in Figure 17.1. Vectors are subtracted by reversing their direction and adding, so that $\mathbf{v}_1 - \mathbf{v}_1 = 0$, as shown in Figure 17.2. Multiplication by a scalar only changes the magnitude of a vector, and not the direction. A *unit vector* is derived by dividing a vector by its magnitude as in $\mathbf{v}/v$.

The magnitude and direction of a vector do not depend on what the coordinate system is, but the components do. To define the Cartesian coordinate system, for example, let **i** be a vector that is directed along the x-axis, and

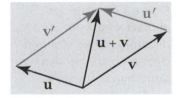

Figure 17.1 Summing vectors **u** and **v** can be done by putting the tail of **v** at the head of **u** (vector **v'**), or vice versa (vector **u'**).

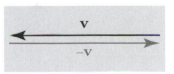

Figure 17.2 Multiplying a vector **v** by -1 reverses its direction.

Figure 17.3 Vector **v** is the sum of the three scalar magnitudes v_x, v_y, and v_z, multiplied by their corresponding unit vectors **i**, **j**, and **k**.

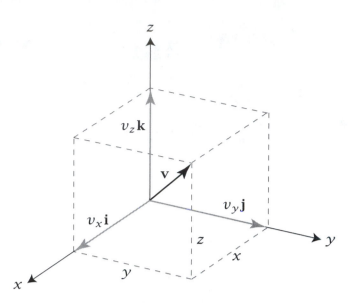

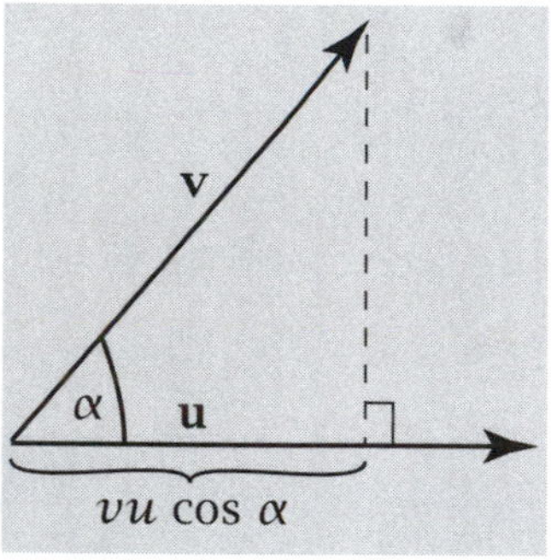

has unit length, **j** be a unit vector directed along the y-axis, and **k** be a unit vector directed along the z-axis. As shown in Figure 17.3, any vector **v** is the sum of the three scalar magnitudes v_x, v_y, and v_z (in Cartesian coordinates) multiplied by their corresponding unit vectors:

$$\mathbf{v} = v_x\mathbf{i} + v_y\mathbf{j} + v_z\mathbf{k}. \tag{17.1}$$

The *magnitude v* of **v** is

$$v = \sqrt{v_x^2 + v_y^2 + v_z^2}. \tag{17.2}$$

In this notation, adding vector $\mathbf{u} = u_x\mathbf{i} + u_y\mathbf{j} + u_z\mathbf{k}$ to **v** is expressed as

$$\mathbf{v} + \mathbf{u} = (v_x + u_x)\mathbf{i} + (v_y + u_y)\mathbf{j} + (v_z + u_z)\mathbf{k}. \tag{17.3}$$

Subtracting **u** from **v** is expressed as

$$\mathbf{v} - \mathbf{u} = (v_x - u_x)\mathbf{i} + (v_y - u_y)\mathbf{j} + (v_z - u_z)\mathbf{k}, \tag{17.4}$$

and multiplying **v** by the scalar c is expressed as

$$c\mathbf{v} = cv_x\mathbf{i} + cv_y\mathbf{j} + cv_z\mathbf{k}. \tag{17.5}$$

Sometimes it is useful to know the *projection* of one vector onto another, the magnitude of one of them in the direction of the other one. Projections are computed using the dot product.

The Dot Product Gives the Length of One Vector Projected Onto Another

For two vectors **v** and **u** with magnitudes v and u and an angle α between them (see Figure 17.4), the **dot product** is a scalar quantity given by

$$\mathbf{u} \cdot \mathbf{v} = \mathbf{v} \cdot \mathbf{u} = vu\cos\alpha. \tag{17.6}$$

Figure 17.4 The dot product $\mathbf{v} \cdot \mathbf{u} = vu\cos\alpha$ is the projection of **v** onto **u** (shown here). It is also the projection of **u** onto **v**.

If two vectors are perpendicular to each other, the angle between them is $\alpha = \pi/2$, so $\cos\alpha = 0$, and the dot product equals zero. For parallel vectors, $\alpha = 0$, so $\cos\alpha = 1$ and the dot product equals vu. For example, the work along the direction $\mathbf{r}$ is the dot product of force $\mathbf{f}$ with direction $\mathbf{r}$. The work is maximal if the force acts in the direction of $\mathbf{r}$ ($\cos\alpha = 1$), and zero if the force acts in a direction perpendicular to $\mathbf{r}$ ($\cos\alpha = 0$).

To find the dot product for any two vectors in terms of their components, apply the definition of the dot product (Equation (17.6)) to the coordinate unit vectors. The dot product of a vector with itself involves an angle $\alpha = 0$, so the products of the components are found with

$$\mathbf{i}\cdot\mathbf{i} = \mathbf{j}\cdot\mathbf{j} = \mathbf{k}\cdot\mathbf{k} = 1. \qquad (17.7)$$

Because the three unit vectors along the coordinate axes are perpendicular to one another, $\alpha = \pi/2$, you have

$$\mathbf{i}\cdot\mathbf{j} = \mathbf{i}\cdot\mathbf{k} = \mathbf{j}\cdot\mathbf{k} = 0. \qquad (17.8)$$

Using Equations (17.7) and (17.8), you can express the dot product of any two vectors $\mathbf{v}$ and $\mathbf{u}$ in terms of their components:

$$\mathbf{v}\cdot\mathbf{u} = (v_x\mathbf{i} + v_y\mathbf{j} + v_z\mathbf{k})\cdot(u_x\mathbf{i} + u_y\mathbf{j} + u_z\mathbf{k})$$

$$= v_x u_x + v_y u_y + v_z u_z. \qquad (17.9)$$

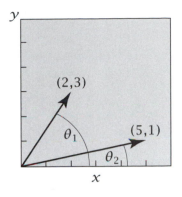

Figure 17.5 Vectors $(2, 3)$ and $(5, 1)$ for Example 17.1.

EXAMPLE 17.1 Two ways to compute a dot product. Find the dot product of the two-dimensional vector $\mathbf{v}_1 = (2, 3)$ with $\mathbf{v}_2 = (5, 1)$, shown in Figure 17.5. From Equation (17.9), you have

$$\mathbf{v}_1 \cdot \mathbf{v}_2 = (2 \times 5) + (3 \times 1) = 13.$$

Equation (17.6) shows that the dot product can also be expressed as $|\mathbf{v}_1||\mathbf{v}_2|\cos\alpha$. The magnitudes are $|\mathbf{v}_1| = \sqrt{2^2 + 3^2} = \sqrt{13}$ and $|\mathbf{v}_2| = \sqrt{5^2 + 1^2} = \sqrt{26}$. The angle between these vectors is $\alpha = 45°$. (To see this, note that $\theta_1 = \tan^{-1}(3/2) = 56.3°$ and $\theta_2 = \tan^{-1}(1/5) = 11.3°$, so $\alpha = \theta_1 - \theta_2 = 45°$.) This method gives the same result:

$$\mathbf{v}_1 \cdot \mathbf{v}_2 = \sqrt{13}\sqrt{26}\cos 45° = 13.$$

To describe the forces and flows when multiple particles are involved, you need to consider more than two vectors at a time. This purpose is served by the concept of a *field*.

Spatial Distributions Can Be Described by Scalar and Vector Fields

A field is a physical quantity that can take on different values at different positions in space. A weather map of the temperatures (a scalar quantity) is a *scalar field* (see Figure 17.6). An example of a *vector field* is the set of velocities of a flowing fluid at all points throughout the fluid. Another vector field is the

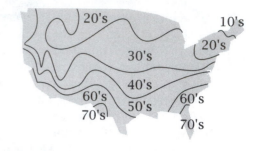

Figure 17.6 Temperature is a scalar field, as indicated on a weather map. Temperature is a function of coordinates (x, y).

Figure 17.7 A vector field: Average January wind velocities in the USA. Each vector has an origin, indicating where the wind velocity was measured, and a magnitude and direction, indicating how strongly and in which direction the wind blows there. Source: Supercomputer Institute Research Bulletin Newsletter of the University of Minnesota **13**, 6 (1997). Data are from Professor Katherine Klunk of the Department of Geography, University of Minnesota.

set of wind velocities in the USA, shown in Figure 17.7. To find the forces and gradients of scalar or vector fields, we need a calculus for vectors.

Defining the Gradient Vector and the ∇ Operator

Consider a weather map with a distribution of temperatures $T(x, y, z)$ where x and y are the east–west and north–south coordinates on a map. To be general, let's also consider the altitude z. We want to know the difference in temperatures from one point $\mathbf{r} = (x, y, z)$ to another $\mathbf{r} + d\mathbf{r}$, where $d\mathbf{r}$ is given by the vector

$$d\mathbf{r} = \mathbf{i}\,dx + \mathbf{j}\,dy + \mathbf{k}\,dz. \tag{17.10}$$

Recall from Chapter 5 that a change such as dT can be computed as

$$dT = \left(\frac{\partial T}{\partial x}\right) dx + \left(\frac{\partial T}{\partial y}\right) dy + \left(\frac{\partial T}{\partial z}\right) dz. \tag{17.11}$$

You can write Equation (17.11) more simply using the vector dot product defined in Equation (17.9). Let one vector be the temperature gradient, written as ∇T, where '∇' is the gradient operator called 'del' or 'grad':

$$\nabla T = \mathbf{i}\left(\frac{\partial T}{\partial x}\right) + \mathbf{j}\left(\frac{\partial T}{\partial y}\right) + \mathbf{k}\left(\frac{\partial T}{\partial z}\right). \tag{17.12}$$

For example, suppose $T(x, y, z) = 1 - ax$: the temperature decreases linearly along the x-axis, and is independent of y and z. Then $\nabla T = -\mathbf{i}a$ is the slope of the function along the x-axis. Now dT in Equation (17.11) is the dot product of the temperature gradient with the change of position vector $d\mathbf{r}$

given by Equation (17.10):

$$dT = \nabla T \cdot d\mathbf{r}. \tag{17.13}$$

The gradient of a scalar function is a vector that tells how the function varies in every direction in the neighborhood of a point. For example, a vector pointing from Dallas to Fort Worth Texas might have a magnitude of -10, indicating that the temperature is ten degrees lower in Fort Worth than in Dallas. The gradient is perpendicular to lines of constant temperature.

∇ is a *vector operator*, which itself can be manipulated as a vector. The **gradient vector** ∇ is defined by

$$\nabla = \mathbf{i}\left(\frac{\partial}{\partial x}\right) + \mathbf{j}\left(\frac{\partial}{\partial y}\right) + \mathbf{k}\left(\frac{\partial}{\partial z}\right). \tag{17.14}$$

∇ operates on the quantity that follows it. The notation ∇T means to perform the ∇ operation on T. How the ∇ operator is applied depends on whether it is operating on a scalar or on a vector function. For a scalar function, Equation (17.14) defines ∇ as the following operation: take the partial derivative of the function with respect to each Cartesian coordinate and multiply it by the unit vector in the direction of the coordinate, then sum the products. The result is the *gradient* vector. Now let's find the gradient of a vector field.

Defining the Divergence

Consider a field of vectors, rather than a field of a scalar quantity. A vector field is a set of vectors distributed throughout space, pointing in various directions. For example, Figure 17.7 shows the vector field of wind velocities throughout the USA. Each vector on this diagram conveys three pieces of information: (1) the location (indicated by the spatial position of the vector), (2) the magnitude of the wind velocity at that location (indicated by the vector length), and (3) the direction in which the wind blows (indicated by the vector direction).

Now we want to know the rates of change at any point for a field of vectors. This function is called the *divergence*. The divergence, which is a scalar quantity, is the dot product of the vector operator ∇ with the vector on which it operates. In Cartesian coordinates, the divergence of the vector $\mathbf{u}$ is

$$\nabla \cdot \mathbf{u} = \left[\mathbf{i}\left(\frac{\partial}{\partial x}\right) + \mathbf{j}\left(\frac{\partial}{\partial y}\right) + \mathbf{k}\left(\frac{\partial}{\partial z}\right)\right] \cdot \left(\mathbf{i}u_x + \mathbf{j}u_y + \mathbf{k}u_z\right)$$

$$= \left(\frac{\partial u_x}{\partial x}\right) + \left(\frac{\partial u_y}{\partial y}\right) + \left(\frac{\partial u_z}{\partial z}\right). \tag{17.15}$$

Figures 17.8 and 17.9 illustrate the concept of divergence. In a constant vector field (Figure 17.8), all the vectors point in the same direction and have the same magnitude. The divergence is zero, because there is no change in the vectors from one position in space to another. This might represent the water flow in the middle of a river. In contrast, as shown in Figure 17.9, an idealized point source of fluid (for example, a wellspring or fountainhead) or point sink (for example, a drainhole) causes the velocity vectors to diverge, to point in different directions from one point in space to the next. Near a point source

$\nabla \cdot \mathbf{u} = 0$

Figure 17.8 The divergence is zero for a constant vector field.

(a) $\nabla \cdot \mathbf{u} > 0$

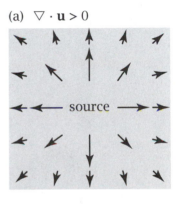

(b) $\nabla \cdot \mathbf{u} < 0$

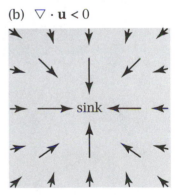

Figure 17.9 The divergence of a vector field is (a) positive for a point source, and (b) negative for a point sink.

or sink, the divergence is not zero. The divergence describes the change in a vector field from one point in space to another.

> **EXAMPLE 17.2 Computing the divergence.** Compute the divergence $\nabla \cdot \mathbf{u}$ for the vector $\mathbf{u} = x^2\mathbf{i} + y^2\mathbf{j} + z^2\mathbf{k}$. Equation (17.15) gives
>
> $$\nabla \cdot \mathbf{u} = 2x + 2y + 2z.$$

In Equation (17.12), we operated with the ∇ operator on a scalar field to find the first derivative, called the gradient. Often we also need the second derivative, for which we use the *Laplacian* operator.

The Laplacian Operator

The Laplacian operator is the dot product of ∇ with itself. It is written ∇^2 and pronounced 'del squared.' For Cartesian coordinates, the Laplacian operator is

$$\nabla \cdot \nabla = \left(\mathbf{i}\frac{\partial}{\partial x} + \mathbf{j}\frac{\partial}{\partial y} + \mathbf{k}\frac{\partial}{\partial z}\right) \cdot \left(\mathbf{i}\frac{\partial}{\partial x} + \mathbf{j}\frac{\partial}{\partial y} + \mathbf{k}\frac{\partial}{\partial z}\right)$$

$$\Rightarrow \quad \nabla^2 = \frac{\partial^2}{\partial x^2} + \frac{\partial^2}{\partial y^2} + \frac{\partial^2}{\partial z^2}. \tag{17.16}$$

If $T(x, y, z)$ is a scalar field, $\nabla^2 T$ is another scalar field given by:

$$\nabla^2 T = \left(\mathbf{i}\frac{\partial}{\partial x} + \mathbf{j}\frac{\partial}{\partial y} + \mathbf{k}\frac{\partial}{\partial z}\right)\left(\mathbf{i}\frac{\partial T}{\partial x} + \mathbf{j}\frac{\partial T}{\partial y} + \mathbf{k}\frac{\partial T}{\partial z}\right)$$

$$= \frac{\partial^2 T}{\partial x^2} + \frac{\partial^2 T}{\partial y^2} + \frac{\partial^2 T}{\partial z^2}.$$

> **EXAMPLE 17.3 Computing the Laplacian.** For the function $f(x, y, z) = x^2 + y^3 + z^4$, compute $\nabla^2 f$. Equation (17.16) gives
>
> $$\nabla^2 f = 2 + 6y + 12z^2.$$

Operators in Different Coordinate Systems

Choosing a coordinate system appropriate to the symmetry of the problem at hand can make problems easier to solve. Cartesian coordinates are usually easiest for problems involving planes and flat surfaces. For problems involving spheres or cylinders, spherical or cylindrical coordinates are most convenient. The advantage of the operator notation, ∇, $\nabla\cdot$, and ∇^2, is that it is independent of the coordinate system. But when it comes to performing specific calculations, you can use Table 17.1 (pages 312 and 313) to translate from the vector notation, such as ∇T, to the specific expressions that you need for a particular problem.

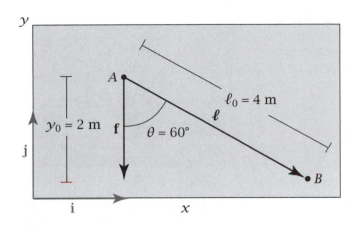

Figure 17.10 An object is on a 4-meter-long plane inclined at $60°$ from vertical. The gravitational force f_g acts vertically downwards. The work exerted by the force in moving the object from A to B is calculated in Example 17.4.

Integrating Vector Quantities: The Path Integral

When you have a function $f(x)$, you may need to integrate over x, $\int f(x)dx$. Similarly, you sometimes need to integrate vector quantities. For vectors, you integrate by components. A *path integral* involves a dot product of two vectors over a particular vector pathway. For example, if you have a vector $\mathbf{u} = (u_x, u_y, u_z)$ to be integrated along $d\mathbf{r} = (dx, dy, dz)$, then the *path integral* is

$$\int \mathbf{u} \cdot d\boldsymbol{r} = \int u_x dx + u_y dy + u_z dz. \tag{17.17}$$

Here's an example.

EXAMPLE 17.4 A simple path integral: the work performed over a path from A to B. The work w equals the path integral of the vector force $\mathbf{f}$ acting on an object, projected onto each differential element of the path $d\boldsymbol{\ell}$:

$$w = \int_A^B \mathbf{f} \cdot d\boldsymbol{\ell}. \tag{17.18}$$

The work of the force on the object is positive, $w > 0$, if the force acts in the direction of positive $d\boldsymbol{\ell}$, and negative if the force acts in the opposite direction (see Equation (3.3)). In Figure 17.10, the path is diagonally downward.

Suppose that gravity acts downward with a force $f_g = 2$ newtons. The vector force is

$$\mathbf{f} = f_x \mathbf{i} + f_y \mathbf{j} = -f_g \mathbf{j}.$$

Suppose the downward force is pushing an object along an inclined plane at an angle $\theta = 60°$ (see Figure 17.10) over a distance of $\ell_0 = 4$ m. The work is

$$w = \int \mathbf{f} \cdot d\boldsymbol{\ell} = \int \left(f_x \mathbf{i} + f_y \mathbf{j} \right) (dx \mathbf{i} + dy \mathbf{j})$$

$$= -\int_0^{-y_0} f_g \, dy = f_g y_0 = f_g \ell_0 \cos \theta = (2\,\text{N})(4\,\text{m})(\cos 60°)$$

$$= 4\,\text{N\,m}. \tag{17.19}$$

Figure 17.11 In a fountain, the water flux through an imaginary hemispherical surface is highest at the top and smallest at the sides.

In this way, work is computed as a path integral of a force acting over a distance. In general, the path integral also treats cases where the force varies along the trajectory.

The Flux of a Vector Field Is Its 'Flow' Through a Surface

Vectors are useful for describing the flows of fluids and particles. The central concept that you need is the *flux*, which is a familiar scalar quantity. The flux of water through a garden hose is the volume of water per unit time that flows through an imaginary plane of unit area that cuts perpendicularly through the hose. For fluid flow, flux has units of (volume of fluid)/[(time)(unit area)].

You can also define the flux in more complex situations, such as a water fountain in a pool (Figure 17.11). You first need to define a surface through which there is flux. For the fountain, invent an imaginary hemispherical surface, or 'balloon', that encloses it. The flux is the amount of water that flows through this surface per unit time, per unit area. Because water fountains are directed upwards, the flux is greatest through the 'north pole' of the hemispherical balloon, and is nearly zero through the 'equator.' To compute the total flux J you need to integrate the different amounts of flux passing through the many different infinitesimal elements of the imaginary surface.

Let's describe the total water flux through the imaginary bounding surface mathematically. Consider the flow of a fluid through a small element of a surface. Two vectors define the liquid movement. One is the fluid velocity vector **v**: its magnitude represents the speed of the fluid and its orientation indicates the direction of fluid flow. The second vector is $d\mathbf{s}$, which defines the size and orientation of the imaginary small plane element through which the fluid is flowing. $d\mathbf{s}$ is an infinitesimal quantity but it obeys the same mathematical rules as any other vector. The magnitude of $d\mathbf{s}$ represents the infinitesimal area of the imaginary surface through which the fluid flows. The orientation of $d\mathbf{s}$ is normal to the imaginary plane, indicating how the planar element is oriented in space. Because there are two possible normals to the surface, the convention for a closed surface is that $d\mathbf{s}$ is positive when it points *outward*. Flux is positive for flow *out* of a closed surface and negative for flow *into* a closed surface.

The dot product of these two vectors—the flow velocity vector **v** and the area element vector $d\mathbf{s}$—gives the infinitesimal flux of fluid $dJ = \mathbf{v} \cdot d\mathbf{s}$ through

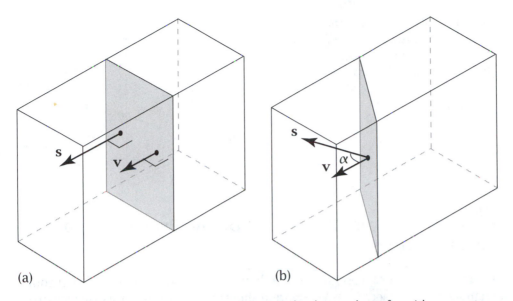

Figure 17.12 The flux of a vector quantity **v** is the dot product of **v** with **s**, a vector having direction normal to a planar element and magnitude equal to its area. The flux (a) is maximal if **v** is parallel to **s**, and (b) is reduced by $\cos \alpha$ if the vectors are oriented at an angle α with respect to each other.

the area element. If the surface area is rotated so that the vector $d\mathbf{s}$ forms an angle α with respect to the direction of fluid flow (Figure 17.12), then the flux is $dJ = \mathbf{v} \cdot d\mathbf{s} = v \, ds \cos \alpha$, where v is the scalar magnitude of the flow velocity and ds is the scalar magnitude of the area of the small element. If the area element is normal to the flow direction, then $dJ = v \, ds$. Here, the flux is a scalar quantity.

To find the total flow out through the balloon surface, integrate the flux over all the surface elements. Represent the entire surface as a large number of small flat areas each with its own normal vector given by $d\mathbf{s}_i$ as shown in Figure 17.13. The total flux J is the sum over each of the individual fluxes. Because the individual fluxes are infinitesimal, this sum becomes an integral, and the total flux is

$$J = \int_{\text{surface}} \mathbf{v} \cdot d\mathbf{s}. \tag{17.20}$$

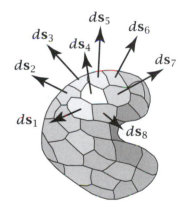

Figure 17.13 The flux out of a surface is the sum of the fluxes out of infinitesimal patches of the surface.

EXAMPLE 17.5 Computing the flux. Suppose you have a field of vectors that all point radially outward from an origin at $r = 0$. Each vector has magnitude $v(r) = a/r$. What is the flux through a spherical balloon with radius R centered at $r = 0$? All the volume elements have identical fluxes, so in this case Equation (17.20) becomes

$$J = v(R)4\pi R^2 = \left(\frac{a}{R}\right)(4\pi R^2) = 4\pi aR.$$

The next section describes *Gauss's theorem*, a mathematical result that sometimes simplifies the calculation of flux.

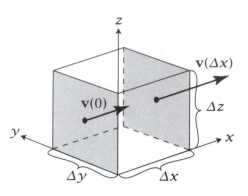

Figure 17.14 To prove Gauss's theorem, construct a cube with its left face at $x = 0$, its right face at $x = \Delta x$, its front face at $y = 0$, its back face at $y = \Delta y$, its bottom face at $z = 0$, and its top face at $z = \Delta z$.

Gauss's Theorem Relates a Volume Property to a Surface Property

Flux is a *surface* property. It tells you how much of something passes through an imaginary surface. Gauss's theorem shows how to replace this surface property with a *volume* property, the divergence of the field inside the surface.

Consider a field of velocities **v**, for example the velocities of fluid flow through an infinitesimal cube having volume $\Delta V = \Delta x \Delta y \Delta z$ (see Figure 17.14). Our balloon is now a surface with a cubical shape. Suppose the cube has its left face at $x = 0$, its front face at $y = 0$, and its bottom face at $z = 0$. To compute the flux you need the dot products of the vector field and the six unit normal surface vectors for the cube. The flux through each surface is the product of the surface area multiplied by the component of the vector field that is perpendicular to it.

First compute the flux through the two surfaces parallel to the yz plane (see Figure 17.14). Because the unit normal vectors are in the directions of the coordinate axes, you need only to keep track of their direction in the plus or minus sense. The area parallel to the yz plane is $\Delta y \Delta z$. The outward flux through the right-hand face at $x = \Delta x$ is $v_x(\Delta x)\Delta y \Delta z$. Through the left-hand face at $x = 0$, the outward flux is is $-v_x(0)\Delta y \Delta z$. In each case v_x is the x component of the field at the center of the face. Because Δx is taken to be very small, spatial variations of v_x can be expanded as a Taylor series. To first-order approximation v_x varies linearly with x:

$$v_x(\Delta x) = v_x(0) + \left(\frac{\partial v_x}{\partial x}\right)\Delta x. \tag{17.21}$$

Multiply both sides of Equation (17.21) by $\Delta y \Delta z$ to get the fluxes at $x = 0$ and Δx, and rearrange:

$$(v_x(\Delta x) - v_x(0))\,\Delta y \Delta z = \left(\frac{\partial v_x}{\partial x}\right)\Delta x \Delta y \Delta z. \tag{17.22}$$

You can compute the flux J in two different ways. First, from Equation (17.20), you have $J = \int_s \mathbf{v} \cdot d\mathbf{s}$. Second, Equation (17.22) shows that the x-direction flux is $(\partial v_x/\partial x)\Delta x \Delta y \Delta z$. Because the contributions along the y and z axes are calculated in the same way, the total flux through all six faces is

$$\int_s \mathbf{v} \cdot d\mathbf{s} = \left(\left(\frac{\partial v_x}{\partial x} \right) + \left(\frac{\partial v_y}{\partial y} \right) + \left(\frac{\partial v_z}{\partial z} \right) \right) \Delta x \Delta y \Delta z$$

$$= (\nabla \cdot \mathbf{v}) \Delta V. \tag{17.23}$$

Equation (17.23) gives the flux through a small volume element.

Now integrate over all the infinitesimal volume elements. The left-hand side of Equation (17.23) must be integrated over the whole surface S, and the right-hand side over the whole volume V. Because the flux into one volume element equals the flux out of an adjacent element, the only component of the surface integral that is non-zero is the one that represents the outer surface. This integration gives **Gauss's theorem**:

$$\int_S \mathbf{v} \cdot d\mathbf{s} = \int_V \nabla \cdot \mathbf{v} \, dV. \tag{17.24}$$

Gauss's theorem equates the *flux* of a vector field *through a closed surface* with the *divergence* of that same field *throughout its volume*. This result will be useful in the following chapters.

Summary

Forces and flows have magnitudes and directions, and can be described by vectors. Vectors are added and subtracted by components. Vector calculus is a collection of methods and notations for finding spatial gradients and integrals of scalar and vector fields. Gauss's theorem shows how to relate the flux through a surface to the divergence of a vector throughout a volume. These methods are useful for computing flows, diffusion, and electrostatic forces, the subjects of the next few chapters.

Table 17.1 For a scalar f or a vector $\mathbf{u}$, this table gives the gradient Δf, the divergence $\nabla \cdot \mathbf{u}$, and the Laplacian $\nabla^2 f$ in Cartesian, cylindrical, and spherical coordinates.

Cartesian Coordinates x, y, z

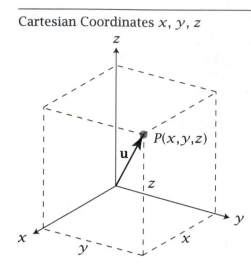

$$\nabla f = \frac{\partial f}{\partial x}\mathbf{i} + \frac{\partial f}{\partial y}\mathbf{j} + \frac{\partial f}{\partial z}\mathbf{z} \tag{17.25}$$

$$\nabla \cdot \mathbf{u} = \frac{\partial u_x}{\partial x} + \frac{\partial u_y}{\partial y} + \frac{\partial u_z}{\partial z} \tag{17.26}$$

$$\nabla^2 f = \frac{\partial^2 f}{\partial x^2} + \frac{\partial^2 f}{\partial y^2} + \frac{\partial^2 f}{\partial z^2} \tag{17.27}$$

Cylindrical Coordinates r, θ, z

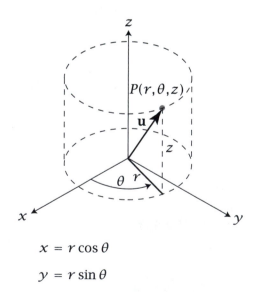

$$x = r\cos\theta$$

$$y = r\sin\theta$$

$$z = z$$

$$\nabla f = \left(\frac{\partial f}{\partial r}\right)\mathbf{r} + \frac{1}{r}\left(\frac{\partial f}{\partial \theta}\right)\boldsymbol{\theta} + \left(\frac{\partial f}{\partial z}\right)\mathbf{z} \tag{17.28}$$

$$\nabla \cdot \mathbf{u} = \frac{1}{r}\left(\frac{\partial (ru_r)}{\partial r}\right) + \frac{1}{r}\left(\frac{\partial u_\theta}{\partial \theta}\right) + \left(\frac{\partial u_z}{\partial z}\right) \tag{17.29}$$

$$\nabla^2 f = \frac{1}{r}\left(\frac{\partial}{\partial r}\left(r\frac{\partial f}{\partial r}\right)\right) + \frac{1}{r^2}\left(\frac{\partial^2 f}{\partial \theta^2}\right) + \left(\frac{\partial^2 f}{\partial z^2}\right) \tag{17.30}$$

Spherical Coordinates r, θ, ϕ

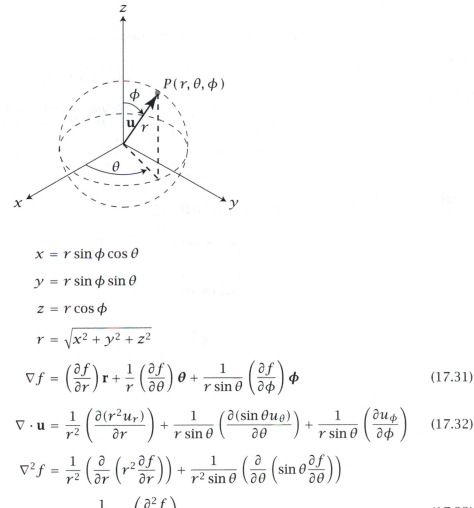

$$x = r\sin\phi\cos\theta$$

$$y = r\sin\phi\sin\theta$$

$$z = r\cos\phi$$

$$r = \sqrt{x^2 + y^2 + z^2}$$

$$\nabla f = \left(\frac{\partial f}{\partial r}\right)\mathbf{r} + \frac{1}{r}\left(\frac{\partial f}{\partial \theta}\right)\boldsymbol{\theta} + \frac{1}{r\sin\theta}\left(\frac{\partial f}{\partial \phi}\right)\boldsymbol{\phi} \tag{17.31}$$

$$\nabla \cdot \mathbf{u} = \frac{1}{r^2}\left(\frac{\partial (r^2 u_r)}{\partial r}\right) + \frac{1}{r\sin\theta}\left(\frac{\partial (\sin\theta u_\theta)}{\partial \theta}\right) + \frac{1}{r\sin\theta}\left(\frac{\partial u_\phi}{\partial \phi}\right) \tag{17.32}$$

$$\nabla^2 f = \frac{1}{r^2}\left(\frac{\partial}{\partial r}\left(r^2\frac{\partial f}{\partial r}\right)\right) + \frac{1}{r^2\sin\theta}\left(\frac{\partial}{\partial \theta}\left(\sin\theta\frac{\partial f}{\partial \theta}\right)\right)$$

$$+ \frac{1}{r^2\sin^2\theta}\left(\frac{\partial^2 f}{\partial \phi^2}\right) \tag{17.33}$$

Problems

1. Adding, subtracting, and multiplying vectors. You have three vectors:

$$\mathbf{v} = 2\mathbf{i} + 0.5\mathbf{j} + 3\mathbf{k},$$

$$\mathbf{w} = \mathbf{i} - 5\mathbf{j} + 6\mathbf{k}, \text{ and}$$

$$\mathbf{y} = -4\mathbf{i} - 2\mathbf{k}.$$

Calculate $\mathbf{v} + \mathbf{w}$, $\mathbf{v} - \mathbf{w}$, $\mathbf{v} \cdot \mathbf{w}$, $\mathbf{v} + \mathbf{w} + \mathbf{y}$, $\mathbf{v} \cdot \mathbf{y}$, and $\mathbf{w} \cdot \mathbf{y}$.

2. Equivalent expressions for the Laplacian. Show for spherical coordinates, where a function $\phi(r)$ depends only on r, that the following two forms are equivalent:

$$\nabla^2 \phi = \frac{1}{r}\frac{d^2(r\phi)}{dr^2} = \frac{1}{r^2}\frac{d}{dr}\left(r^2\frac{d\phi}{dr}\right).$$

3. Finding gradients in Cartesian coordinates. What is ∇f for the following functions of Cartesian coordinates?

(a) $f = x^4 + z^2$

(b) $f = xy + yz + zx$

(c) $f = xyz$

4. Finding gradients in cylindrical coordinates. What is the gradient of f for the following functions of cylindrical coordinates?

(a) $f = 3r^2$

(b) $f = 5rz + z^2$

(c) $f = r^2\theta + z\theta^2$

5. Finding the divergence in Cartesian coordinates. You have vector functions of Cartesian coordinates:

(a) $\mathbf{f} = 3x\mathbf{i} - 2y\mathbf{j} - z\mathbf{k}$

(b) $\mathbf{f} = y^2\mathbf{i} - x^2\mathbf{k}$

(c) $\mathbf{f} = 2x^2y\mathbf{i} + y\sqrt{(z)}\mathbf{j} + 3xy^3\mathbf{k}$

Give the divergence of these functions.

6. Finding the divergence in spherical coordinates. You have vector functions in spherical coordinates:

(a) $f = (1/r^2)\mathbf{r} + (\theta/r)\boldsymbol{\theta} + \phi r\boldsymbol{\phi}$,

(b) $f = (e^{-C_0 r}/r^2)\mathbf{r}$, and

(c) $f = (1/r)\mathbf{r}$,

where C_0 is a constant. Derive the divergence of f.

7. Computing Laplacians. Apply the Laplacian operator to the functions in problems 3 and 4. What is $\nabla^2 f$?

8. Finding a flow field. The three-dimensional velocity $\mathbf{v}$ of a fluid around a point source is radially symmetric. It is given by $\mathbf{v}(r) = r^p$, where p is a constant. Determine p from the condition that the divergence of the flow must vanish.

Suggested Reading

HM Shey, *Div, Grad, Curl, and All That*, WW Norton, New York, 1973.

MR Spiegel, *Vector Analysis*, Schaum's Outline Series, McGraw-Hill, New York, 1959.

18 Physical Kinetics: Diffusion, Permeation & Flow

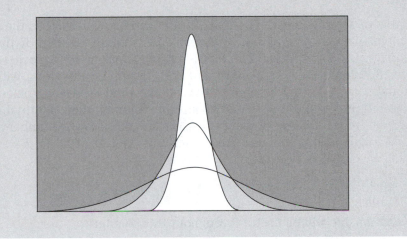

Forces Drive Molecules to Flow

Molecules diffuse, transporting matter and heat. For example, soft drinks lose their fizz when CO_2 diffuses out of their plastic containers. Drugs diffuse out of clever encapsulation devices into the body according to tailored time schedules. Metabolites flow in and out of cells through biological membranes. External forces also drive particles to move and flow. The electrical currents that power household appliances, modern electronics, and neurons result from the flows of ions and electrons that are driven by gradients of electrical potentials. What are the rates at which molecules flow from one place to another? What forces drive them?

Defining the Flux

The flow of particles or fluids is described by their flux (see Chapter 17, page 308). To avoid vector arithmetic, let's just consider a flow along a single direction, which we'll choose to be the x-axis. The flux J is defined as the amount of material passing through a unit area per unit time. Sometimes the flux is defined as a quantity that is not divided by a unit area (see Chapter 17, for example), but we do divide by the unit area in this chapter. You are free to define the amount of material in terms of either the number of particles, or their mass, or their volume, whichever is most convenient for the problem at hand.

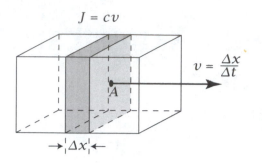

$J = cv$

Figure 18.1 The flux of particles is $J = cv$, where c is the concentration, A is the cross-sectional area, and the fluid moves a distance Δx in time Δt at velocity v.

$v = \dfrac{\Delta x}{\Delta t}$

A

Δx

(a)

$c(x)$

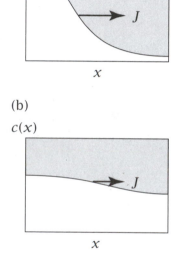

J

x

(b)

$c(x)$

J

x

Figure 18.2 Flow results from concentration gradients. (a) A steep gradient causes large flux, while (b) a shallow gradient causes a small flux.

Think of a garden hose. The flux of water can be expressed as the number of gallons of water coming out of the hose per minute, divided by the cross-sectional area of the hose. The flux is proportional to the velocity of flow. Figure 18.1 shows a rectangular element of the fluid, Δx units long, with cross-sectional area A, having a volume $A\Delta x$. If the fluid carries particles having concentration c = (number of particles)/(unit volume), then the total number of particles in the volume element is $cA\Delta x$. The flux J is the number of particles, $cA\Delta x$, divided by the unit area A per unit time Δt. That is, $J = cA\Delta x/A\Delta t$. If the flow velocity is $v = \Delta x/\Delta t$, the flux is

$$J = \frac{c\Delta x}{\Delta t} = cv. \tag{18.1}$$

This relationship will be useful throughout this chapter.

Fick's Law, Ohm's Law, and Fourier's Law Are Linear Laws that Relate Forces to Flows

A single-phase system that is in thermodynamic equilibrium and not subject to external forces will have particles uniformly distributed in space. But a system having concentration *gradients* may not be in equilibrium. The thermodynamic tendency toward equilibrium is a tendency toward uniform concentrations.

Just as thermodynamics is governed by fundamental empirical laws—the First and Second Laws—kinetics too is governed by fundamental empirical laws. Among them are *Fick's law* relating the forces and flows of particles, *Ohm's law* relating the forces and flows of electrical current, and *Fourier's law* relating the forces and flows of heat. **Fick's first law** captures the general observation that the flux J of particles is proportional to the gradient of concentration, dc/dx,

$$J = -D\frac{dc}{dx}, \tag{18.2}$$

if the flow is one-dimensional. In three dimensions the flux **J** is a vector (c is always a scalar), and flows can have components in all directions, so

$$\mathbf{J} = -D\nabla c. \tag{18.3}$$

The proportionality constant D is called the *diffusion coefficient* and the minus sign indicates that particles flow *down* their concentration gradients, from regions of high concentration to regions of low concentration (see Figure 18.2). The flow rate of particles is proportional to the particle concentration gradient.

Table 18.1 Diffusion constants D for various molecules in air or water, as a function of their molecular weights M. Source: RW Baker, *Controlled Release of Biologically Active Agents*, Wiley, New York, 1987.

Molecule	Medium	T (°C)	M (g mol^{-1})	D (cm^2 s^{-1})
Hydrogen	Air	0	2	6.11×10^{-1}
Helium	Air	3	4	6.24×10^{-1}
Oxygen	Air	0	32	1.78×10^{-1}
Benzene	Air	25	78	9.60×10^{-2}
Hydrogen	Water	25	2	4.50×10^{-5}
Helium	Water	25	4	6.28×10^{-5}
Oxygen	Water	25	32	2.10×10^{-5}
Urea	Water	25	60	1.38×10^{-5}
Benzene	Water	25	78	1.02×10^{-5}
Sucrose	Water	25	342	5.23×10^{-6}
Ribonuclease	Water	20	13,683	1.19×10^{-6}
Hemoglobin	Water	20	68,000	6.90×10^{-7}
Catalase	Water	20	250,000	4.10×10^{-7}
Myosin	Water	20	493,000	1.16×10^{-7}
DNA	Water	20	6,000,000	1.30×10^{-8}
Tobacco mosaic virus	Water	20	50,000,000	3.00×10^{-8}

Steep concentration gradients cause large fluxes of particles. Small gradients cause small fluxes. Also, the larger the diffusion constant, the higher the flux, for a given gradient. Table 18.1 lists some molecular diffusion constants.

Two other relationships have the same form as Fick's first law: Fourier's law says that heat flow is proportional to the gradient of the temperature, and Ohm's law says that the flow of electrical current is proportional to the gradient of the voltage (see Chapter 22).

These laws are based on experimental observations of kinetics and are outside the scope of equilibrium thermodynamics. According to these empirical laws, the flux J of particles, of electrical current, or of heat is linearly proportional to the driving force f that causes it,

$$J = Lf, \tag{18.4}$$

where L is a proportionality constant.

What is the 'force' on a particle in a concentration gradient? If particle concentration $c(x)$ varies with x, the drive toward equalizing the chemical potential everywhere in space tends to flatten the concentration gradients. This is not a force that acts on each particle directly. It results, at least in part, from mixing entropy. Just as Newton's laws give forces as derivatives of energies, the force from a concentration gradient can be expressed as a derivative of the chemical potential. If c depends on x, then the chemical potential can be ex-

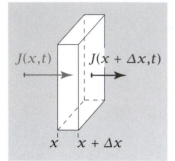

Figure 18.3 The flux $J(x + \Delta x, t)$ out of a volume element at $x + \Delta x$ can be different from the flux $J(x, t)$ into the element at (x, t) because the material can be depleted or accumulated in the volume element.

pressed as $\mu(x) = \mu° + kT \ln c(x)$ and the average force f from a concentration gradient is

$$f = -\frac{d\mu}{dx} = -kT\frac{d\ln c}{dx} = -\frac{kT}{c}\frac{dc}{dx}. \tag{18.5}$$

Substituting Equation (18.5) into Equation (18.4) gives

$$J = -\frac{LkT}{c}\frac{dc}{dx}. \tag{18.6}$$

Comparing Equation (18.6) with Equation (18.2) gives a useful relationship: $D = ukT$, where $u = L/c$ is called the *mobility*, and $1/u$ is called the *friction coefficient*. In the next section we combine Fick's law with a conservation relationship to predict how particle concentrations vary in time and space.

The Diffusion Equation Describes How Concentration Gradients Change over Time and Space

To combine Fick's law with the constraint due to the conservation of particles, focus on a one-dimensional process. Figure 18.3 shows the flux $J(x, t)$ into a small element of volume, and the flux $J(x + \Delta x, t)$ out of that element. The flow in and out of the volume element need not be the same at a given instant of time because particles cannot traverse the volume element instantaneously, and because particles can be accumulated or depleted within the volume. The increase in the number of particles in the volume element at time t is the flow in at time t minus the flow out at time t, $A\Delta t[J(x, t) - J(x + \Delta x, t)]$.

The total accumulation of particles can also be computed in a different way, as (volume) × (midpoint concentration) at two different times, t and $t + \Delta t$, or $A\Delta x[c(x + \Delta x/2, t + \Delta t) - c(x + \Delta x/2, t)]$. Combining these two calculations of the same property gives

$$A\Delta t\,[J(x, t) - J(x + \Delta x, t)]$$
$$= A\Delta x\left[c\left(x + \frac{\Delta x}{2}, t + \Delta t\right) - c\left(x + \frac{\Delta x}{2}, t\right)\right]. \tag{18.7}$$

Divide both sides of Equation (18.7) by $\Delta x\Delta t$ and simplify by using the definition of the derivative, for example $\{[J(x, t) - J(x + \Delta x, t)]/\Delta x\} \rightarrow (\partial J/\partial x)$ as $\Delta x \rightarrow dx \rightarrow 0$. Then take the limits $\Delta x \rightarrow dx$ and $\Delta t \rightarrow dt$ to get

$$\left(\frac{\partial c}{\partial t}\right) = -\left(\frac{\partial J}{\partial x}\right). \tag{18.8}$$

Substituting Equation (18.2) into Equation (18.8) gives **Fick's second law**, also called the **diffusion equation**,

$$\left(\frac{\partial c}{\partial t}\right) = \left(\frac{\partial}{\partial x}\left[D\left(\frac{\partial c}{\partial x}\right)\right]\right) = D\left(\frac{\partial^2 c}{\partial x^2}\right), \tag{18.9}$$

where the last equality applies if the diffusion coefficient D does not depend

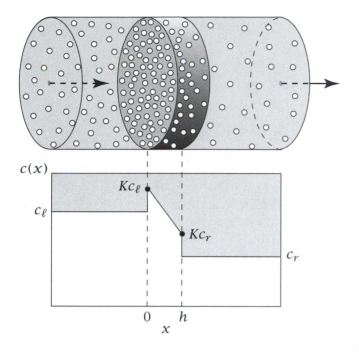

Figure 18.4 Particles flow from a reservoir at concentration c_ℓ on the left, through a slab of material of thickness h, to a reservoir having a lower concentration c_r on the right. The partition coefficient from the solutions into the slab is K. At steady state, there is a linear gradient of particle concentration along the x-axis through the slab (see Example 18.1).

on x. The generalization to three dimensions is

$$\left(\frac{\partial c}{\partial t}\right) = -\nabla \cdot \mathbf{J} = D\nabla^2 c. \tag{18.10}$$

The diffusion equation is a partial differential equation. You solve it to find $c(x, y, z, t)$, the particle concentration as a function of spatial position and time. To solve it, you need to know two *boundary conditions* and one *initial condition*, that is, two pieces of information about the concentration at particular points in space and one piece of information about the concentration distribution at some particular time. Example 18.1 shows how to solve the diffusion equation for the permeation of particles through a planar film or membrane.

EXAMPLE 18.1 Diffusion through a slab or membrane. Drugs and metabolites pass through cell membranes or skin, and gases are forced through thin polymer membranes in industrial processes of gas purification or recovery. For example, forcing air through polymeric films separates O_2 from N_2 gases, because of the faster permeation of O_2.

Here's a model for the permeation of atoms and molecules through films and membranes. Figure 18.4 shows a slab of material having thickness h, through which particles flow. To the left of the membrane is a higher concentration c_ℓ of particles, and to the right is a lower concentration c_r. The difference in particle concentrations, $\Delta c = c_\ell - c_r$, drives particles to flow from left to right through the slab. We want to know the concentration profile $c(x, t)$ of particles inside the slab.

For simplicity, consider *steady-state* flow, which results from reservoirs of particles so large that their particle concentration does not change with time, $(\partial c/\partial t) = 0$. Then the diffusion Equation (18.9) reduces to

$$\left(\frac{\partial^2 c}{\partial x^2}\right) = 0. \tag{18.11}$$

Integrating Equation (18.11) once gives

$$\left(\frac{\partial c}{\partial x}\right) = A_1, \tag{18.12}$$

where A_1 is a constant of integration. Integrating again gives

$$c(x) = A_1 x + A_2, \tag{18.13}$$

where A_2 is the second constant of integration.

Equation (18.13) says that the concentration profile in the slab is linear in x. You can get A_1 and A_2 if you know the partition coefficient K for the molecule from the external solvent into the slab. Just inside the left wall of the slab at $x = 0$, the concentration is $c(0) = Kc_\ell$. Substituting this condition and $x = 0$ into Equation (18.13) gives

$$A_2 = Kc_\ell. \tag{18.14}$$

Inside the right wall at $x = h$, the concentration is $c(h) = Kc_r$. Substituting these boundary conditions into Equation (18.13) gives

$$A_1 = \frac{K(c_r - c_\ell)}{h}. \tag{18.15}$$

With these two boundary conditions, Equation (18.13) becomes

$$c(x) = \frac{K(c_r - c_\ell)}{h} x + Kc_\ell. \tag{18.16}$$

The gradient of the concentration in the slab is shown in Figure 18.4.

What is the flux J of particles through the slab? Take the derivative of Equation (18.16), as prescribed by Fick's law equation (18.2), to get

$$J = \frac{KD}{h}(c_\ell - c_r) = \frac{KD}{h}\Delta c. \tag{18.17}$$

The flow of particles through the membrane is driven by the difference in concentration on the two sides. The flux increases with the diffusion constant for the particles in the slab, with the partition coefficient for particles going into the slab, and inversely with the thickness of the slab.

The *permeability* P of a membrane is defined as the flux divided by the driving force Δc,

$$P = J/\Delta c = KD/h. \tag{18.18}$$

The *resistance* of the membrane to the particle flux is defined as $1/P$, the inverse of the permeability. Figure 18.5 shows that the permeabilities of various solutes through lipid bilayer membranes increase approximately linearly with the partition coefficient, as predicted. Flow into biological cells is called *passive transport* if it is driven by such concentration gradients. *Active transport* is flow that goes against concentration gradients, driven by some energy source.

Now we switch from one-dimensional diffusion to a three-dimensional problem.

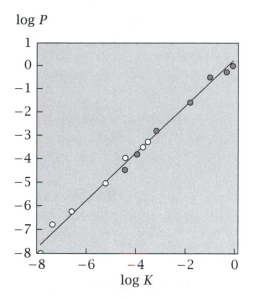

log P

log K

Figure 18.5 The permeability P is proportional to the oil-water partition coefficient K for p-methylhippuric acids (○) and p-toluic acids (●) through lipid bilayer membranes. Source: P Mayer, T-X Xiang, and BD Anderson, *AAPS Pharm Sci* **2**, article 14 (2000).

Diffusion of Particles Toward a Sphere

Consider a spherical particle of radius a such as a protein or micelle, toward which small molecules or ligands diffuse (see Figure 18.6). At what rate do the small molecules collide with the sphere? To compute the rate, solve the diffusion Equation (18.10) in spherical coordinates (see page 313 and problem 2, page 314). At steady state $\partial c/\partial t = 0$. Assume that the flow depends only on the radial distance away from the sphere, and not on the angular variables. Then you can compute $c(r)$ using

$$\nabla^2 c = \frac{1}{r}\frac{d^2(rc)}{dr^2} = 0. \tag{18.19}$$

To solve for $c(r)$, integrate Equation (18.19) once to get

$$\frac{d(rc)}{dr} = A_1, \tag{18.20}$$

where A_1 is the first constant of integration. Integrate again to get

$$rc = A_1 r + A_2, \tag{18.21}$$

where A_2 is the second constant of integration. Rearranging Equation (18.21) gives $c(r)$,

$$c(r) = A_1 + \frac{A_2}{r}. \tag{18.22}$$

Two boundary conditions are needed to determine A_1 and A_2. First, you know the concentration c_∞ of small molecules at a large distance from the sphere, $r \to \infty$, so Equation (18.22) gives $A_1 = c_\infty$. Second, assume that when each small molecule contacts the sphere, it is absorbed, dissipated, or transformed (for example, by a chemical reaction and dispersal of the products). This boundary condition gives an upper limit on the rate at which collisions lead to conversion to product. This condition, called an *absorbing boundary*

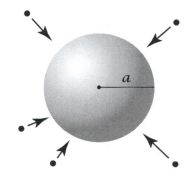

Figure 18.6 Small molecules (●) diffuse toward a sphere of radius a.

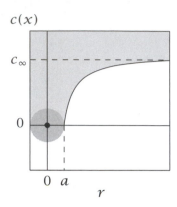

$c(x)$

c_∞

0

0 a

r

Figure 18.7 Concentration profile $c(r) = c_\infty(1 - a/r)$ of molecules as a function of distance r away from an absorbing sphere of radius a.

condition, gives the concentration at the surface of the sphere, $c(a) = 0$. Substituting $c(a) = 0$ and $A_1 = c_\infty$ into Equation (18.22) gives $A_2 = -c_\infty a$. With these boundary conditions, Equation (18.22) becomes

$$c(r) = c_\infty \left(1 - \frac{a}{r}\right). \tag{18.23}$$

Figure 18.7 shows this distribution of small molecules near the sphere. To get the flux, use

$$J(r) = -D\frac{dc}{dr} = \frac{-Dc_\infty a}{r^2}. \tag{18.24}$$

To get the number of collisions per second at $r = a$, called the *current* $I(a)$, multiply the flux (the number of particles colliding per unit area per second, from Equation (18.24)) by the area of the sphere's surface:

$$I(a) = J(a)4\pi a^2 = -4\pi Dc_\infty a. \tag{18.25}$$

The minus sign in Equation (18.25) for $I(a)$ indicates that the current is in the direction $-r$, toward the sphere. The current is large if D or c_∞ or a are large.

For spherical particles, **diffusion-controlled** reaction rates are given by Equation (18.25). Diffusion control implies that reactions are fast. Other reactions involve additional rate-limiting steps that occur after the reacting molecules have come into contact with the sphere. Diffusion control defines an upper limit on the speed of reactions. Any other kind of process must be slower than the diffusion-controlled process because the reaction must take additional time to complete after contact. Association rates are often expressed in terms of a rate coefficient k_a defined by $I(a) = -k_a c_\infty$, where

$$k_a = 4\pi Da. \tag{18.26}$$

EXAMPLE 18.2 Diffusion-controlled rate. Compute the diffusion-controlled collision rate with a protein sphere having radius $a = 10\,\text{Å}$ for benzene, which has a diffusion constant $D = 1 \times 10^{-5}\,\text{cm}^2\,\text{sec}^{-1}$. Equation (18.26) gives

$$k_a = 4\pi \left(1 \times 10^{-5}\frac{\text{cm}^2}{\text{s}}\right)\left(10^{-7}\,\text{cm}\right)\left(6.02 \times 10^{23}\frac{\text{molecule}}{\text{mol}}\right)$$

$$\times \left(\frac{1\,\text{L}}{1000\,\text{cm}^3}\right) \approx 7.6 \times 10^9\,\text{M}^{-1}\,\text{s}^{-1},$$

where M is moles per liter.

Let's look at another diffusion problem.

EXAMPLE 18.3 Diffusion from a point source. Put a drop of dye in water and observe how the dye spreads. To make the math simple, consider the spreading in one dimension. Begin by putting the dye at $x = 0$. To determine the dye profile $c(x, t)$, you can solve Equation (18.9) by various methods. The solutions of this equation for many different geometries and boundary conditions are given in two classic texts on diffusion [1], and on heat conduction [2].

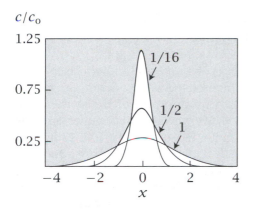

c/c_0

Figure 18.8 Concentration–distance curves for molecules that begin at $x = 0$ and diffuse in one dimension along x. The numbers on the curves are the values of Dt. Source: J Crank, *The Mathematics of Diffusion*, 2nd edition, Clarendon Press, Oxford, 1993.

Heat conduction obeys the same differential equation as diffusion: if you replace particle concentration $c(x,y,z,t)$ with temperature $T(x,y,z,t)$, and replace the diffusion constant D with the *thermal diffusivity κ*, which expresses how the flow of heat depends on the temperature gradient, then the distribution of temperature in space and time is given by

$$\left(\frac{\partial T}{\partial t}\right) = \kappa \nabla^2 T. \tag{18.27}$$

So, one way in which to solve a diffusion equation is to solve instead the corresponding heat flow equation having the same boundary conditions, and replace T with c, and κ with D. You can find solutions to diffusion problems in books of either diffusion or heat conduction and in problems having the given boundary conditions. For the present problem, see Equation (2.6) in [1]. The solution to this one-dimensional problem is

$$c(x,t) = \frac{n_0}{(4\pi Dt)^{1/2}} \exp(-x^2/4Dt), \tag{18.28}$$

where n_0 is the initial amount of dye at the point $x = 0$. You can verify this solution by substituting Equation (18.28) into Equation (18.9) and taking the appropriate derivatives. Figure 18.8 shows how the dye spreads out with time. The larger the diffusion constant D, the faster the dye diffuses away from the site of the initial droplet.

To generalize to two or three dimensions, you can regard diffusion along the x, y, and z axes as independent of each other. For any dimensionality, $d = 1, 2, 3, \ldots$, Equation (18.28) becomes

$$c(r,t) = \frac{n_0}{(4\pi Dt)^{d/2}} \exp(-r^2/4Dt), \tag{18.29}$$

where r represents the distance in d-dimensional space. In three dimensions, for example, $r^2 = x^2 + y^2 + z^2$.

So far, we have considered how particles diffuse when there are spatial concentration gradients. But diffusive flows can also be affected by additional sources or sinks of particles, such as chemical reactions. Sources and sinks are considered in the next section. At the same time we generalize to show how diffusion modeling applies to other applications.

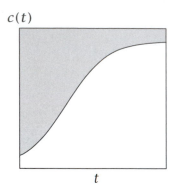

$c(t)$

t

Figure 18.9 The Verhulst model in which the population $c(t)$ grows until it saturates, according to Equations (18.32) and (18.33). Source: JD Murray, *Mathematical Biology*, 2nd edition, Springer-Verlag, Berlin, 1993.

Sources and Sinks Also Contribute to Flows: Examples from Population Biology

In population biology, plants and animals evolve in time and space in ways that can often be described by diffusion equations. Births, deaths, migration, and competition are sources, sinks, and fluxes into or out of the system that are subject to 'diffusion' in time and space. We develop such equations one step at a time. We start with the simplest time dependence for the growth of populations, then add the effects of competition, and finally we introduce the spatial dependence, following the discussion in reference [3].

Consider a population of some species $c(t)$, a function of time. In 1798, TR Malthus first modelled situations in which a change in population, dc/dt, is proportional to the population itself,

$$\frac{dc}{dt} = ac, \qquad (18.30)$$

because the number of offspring is proportional to the number of parents. The constant a equals the number of births minus the number of deaths per member of the population. Chemical reactions obey similar equations: amounts of product are proportional to amounts of reactant (see Chapter 19). Rearranging to $dc/c = a\,dt$ and integrating both sides gives

$$c(t) = c_0 e^{at}, \qquad (18.31)$$

where c_0, the population at $t = 0$, is a constant. Equation (18.31) predicts that populations grow exponentially with time if $a > 0$, and that populations die out exponentially if $a < 0$.

But real populations can't sustain exponential growth indefinitely. In 1836, PF Verhulst introduced the term $-bc^2(t)$ to account for how high populations are reduced by competition or crowding. For high populations, c^2 becomes large in relation to c, and the minus sign says that high populations reduce the growth rate. This model for population growth gives

$$\frac{dc}{dt} = ac(t) - bc^2(t). \qquad (18.32)$$

Integrating Equation (18.32) gives the solution

$$c(t) = \frac{c_0 e^{at}}{1 + (c_0 b/a)\,(e^{at} - 1)}. \qquad (18.33)$$

In this case $c(t)$ saturates with time, as Figure 18.9 shows.

To add another level of refinement and account for geographic migrations of species, RA Fisher in 1937 included a spatial dependence by introducing the diffusion equation,

$$\frac{\partial c(x, y, t)}{\partial t} = D\nabla^2 c + ac - bc^2. \qquad (18.34)$$

Figure 18.10 shows a solution of this model. The population spreads by diffusion, but it reaches a limit to growth.

These examples illustrate that fluxes and diffusion are not limited to atoms and molecules, and they show how sources and sinks can be added to the

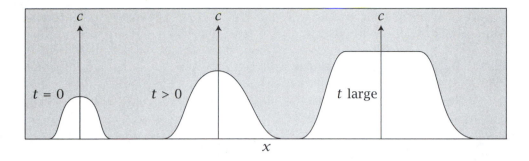

Figure 18.10 Diffusion and growth of a population in one dimension, subject to limits on growth due to crowding (see Equation (18.34)). Source: JD Murray, *Mathematical Biology*, 2nd edition, Springer-Verlag, Berlin, 1993.

diffusion equation. Example 18.4 applies the same principles to molecules that diffuse and react.

EXAMPLE 18.4 Diffusion coupled with a chemical reaction. Suppose that drug molecules diffuse out of a tablet (which we model as a planar wall) into a solution (see Figure 18.11). The drug undergoes a chemical reaction in the solution. The chemical reaction rate is k_{rx} and the diffusion constant of the drug in the solution is D. Let's find the concentration profile for the drug as a function of distance x away from the tablet wall in the solution. If the reaction causes the drug to be depleted in proportion to its concentration c, then

$$\frac{dc}{dt} = -k_{rx}c. \tag{18.35}$$

Including this reaction process (sink) in the general diffusion Equation (18.34) gives

$$\left(\frac{\partial c}{\partial t}\right) = D\left(\frac{\partial^2 c}{\partial x^2}\right) - k_{rx}c. \tag{18.36}$$

To simplify the problem, consider steady state conditions, $(\partial c/\partial t) = 0$. c now depends only on x. You can get $c(x)$ by solving

$$\frac{d^2 c}{dx^2} - \frac{k_{rx}}{D}c = 0. \tag{18.37}$$

The general solution is

$$c(x) = A_1 \exp(-ax) + A_2 \exp(ax), \tag{18.38}$$

where $a = \sqrt{k_{rx}/D}$ and A_1 and A_2 are constants that must satisfy the boundary conditions. You can check this solution by substituting it into Equation (18.37). Because the concentration at $x \to \infty$ in a large container must be $c(\infty) = 0$, you have $A_2 = 0$ and

$$c(x) = c(0) \exp\left(-x\sqrt{k_{rx}/D}\right), \tag{18.39}$$

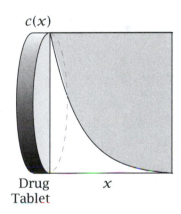

$c(x)$

Drug Tablet x

Figure 18.11
Concentration profile of a drug from a tablet, $c(x)$ at steady state. Drug is depleted in solution with rate constant k_{rx} (see Example 18.4).

where $c(0)$ is the concentration at the wall, $x = 0$. The drug concentration diminishes exponentially with distance from the tablet. When $x = \sqrt{D/k_{rx}}$, $c(x)/c(0) = 1/e$. For comparison, if you had no chemical reaction ($k_{rx} = 0$ in Equation (18.37)) with the same boundary conditions, the drug concentration would diminish as a linear function of distance (Equation (18.13)).

At the surface ($x = 0$), the steady-state flux of the drug out of the tablet can be computed from Equation (18.39):

$$J = -D \left(\frac{dc}{dx} \right)_{x=0} = c(0) a D e^{-ax} |_{x=0} = c(0)\sqrt{Dk_{rx}}. \tag{18.40}$$

The drug is drawn out of the tablet rapidly if it has a high diffusion constant or has a high reaction rate in solution.

Diffusing Particles Can Be Subject to Additional Forces

So far, we have considered how gradients drive the flows of particles, and how chemical reactions can affect those flows by serving as sources or sinks. Diffusing particles can also be subjected to directional external forces. Unlike sources and sinks, forces and fluxes are vectorial.

For example, gravity draws particles toward the bottom of a beaker of water. Forces accelerate particles according to Newton's laws. But for particles in liquids, acceleration is short-lived. The velocity v times a proportionality constant called the *friction coefficient* ξ equals the applied force f,

$$f = \xi v. \tag{18.41}$$

For proteins or colloids, this frictional regime is reached in nanoseconds or less (see Example 18.2), so you are justified in using Equation (18.41) except for the very fastest processes. Substituting Equation (18.41) into Equation (18.1) gives the flux of particles J_{ap} that results from an applied force f,

$$J_{ap} = cv = \frac{cf}{\xi}. \tag{18.42}$$

If particles are subject to both a concentration gradient and an applied force, then the two fluxes will sum and Fick's First Law Equation (18.2) generalizes to

$$J = -D \left(\frac{\partial c}{\partial x} \right) + \frac{cf}{\xi}. \tag{18.43}$$

The corresponding generalization of the diffusion equation, called the **Smoluchowski equation**, is given by combining Equations (18.8) and (18.43):

$$\left(\frac{\partial c}{\partial t} \right) = D \left(\frac{\partial^2 c}{\partial x^2} \right) - \frac{f}{\xi} \left(\frac{\partial c}{\partial x} \right). \tag{18.44}$$

Now we use Equation (18.44) to derive a relationship between the frictional drag experienced by a particle moving through a fluid, and the particle's diffusion constant.

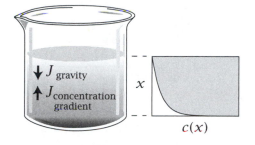

Figure 18.12 Particles in a beaker flow downward due to gravity, building up a high concentration at the bottom, leading to an opposing flux upward at equilibrium.

The Einstein–Smoluchowski Equation Relates Diffusion and Friction

The Einstein–Smoluchowski equation relates the diffusion constant D to the friction coefficient ξ. To derive this relationship, consider the equilibrium illustrated in Figure 18.12. Put some molecules in a beaker of liquid. Gravity forces the molecules toward the bottom. Particles begin to concentrate at the bottom, forming a concentration gradient. This concentration gradient acts to drive particles upward, opposing the force of gravity. At equilibrium, the two fluxes (from gravity, forcing the particles downward, and from the concentration gradient, forcing the particles upward) are balanced so $J = 0$ in Equation (18.43). c does not depend on time and you have

$$D\left(\frac{dc}{dx}\right) = \frac{cf}{\xi}. \tag{18.45}$$

Rearranging Equation (18.45) to put the c's on one side and x on the other gives

$$D\frac{dc}{c} = \frac{f\,dx}{\xi}. \tag{18.46}$$

Integrate both sides and use Equation (3.3) for the reversible work $w = -\int f\,dx$. The minus sign here is because the force acts downwards while the positive x direction is upwards. This gives

$$D\ln\frac{c(x)}{c(0)} = \frac{-w}{\xi}. \tag{18.47}$$

Exponentiate Equation (18.47) to get

$$\frac{c(x)}{c(0)} = e^{-w(x)/\xi D}. \tag{18.48}$$

Because the system is in equilibrium, the Boltzmann distribution law also applies, and the quantity on the right side of Equation (18.48) must also equal $\exp(-w/kT)$. Equating the exponents of these two expressions gives the **Einstein–Smoluchowski** equation,

$$D = \frac{kT}{\xi}. \tag{18.49}$$

This relationship shows how to determine the friction coefficient ξ if you know D. Because ξ depends on the sizes and shapes of particles, measuring the diffusion constant D gives some information about the structures of particles.

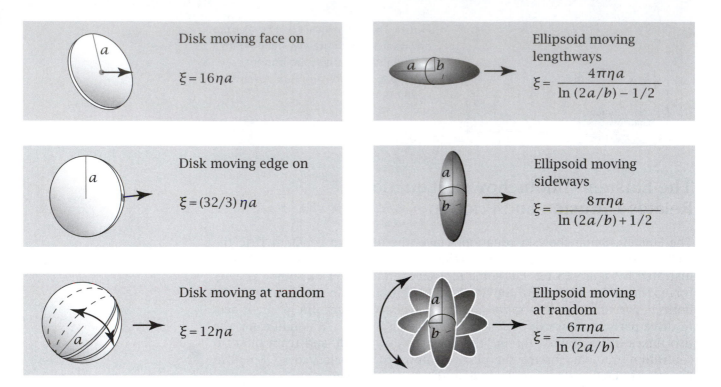

Figure 18.13 Friction coefficients ξ for objects in flow. η is the solvent viscosity, a and b are the dimensions shown, and the straight arrow indicates the flow direction. The objects include a disk moving face-on, edge-on, or averaged over all possible orientations, and an ellipsoid moving lengthways, sideways, or averaged over all orientations. Source: HC Berg, *Random Walks in Biology*, Princeton University Press, Princeton, 1993.

For example, solving fluid flow equations (which we don't do here; see [4]) gives *Stokes's law* for spherical particles,

$$\xi = 6\pi\eta a, \tag{18.50}$$

where η is the viscosity of the solvent and a is the radius of the sphere.

Stokes's law is known from the following type of experiment. Put a ball of radius a in a liquid. The gravitational force is $\Delta m g$, where Δm is the mass difference between the object and displaced liquid. Measure its velocity. The ratio of gravitational force to velocity is ξ. Figure 18.13 gives friction factors for particles that aren't spherical. Combining Equations (18.49) and (18.50) gives the **Stokes–Einstein** law of diffusion for spherical particles,

$$D = \frac{kT}{6\pi\eta a}, \tag{18.51}$$

which predicts that larger spheres have greater friction and lower diffusion rates than smaller spheres. Viscosity has units of *poise*, dyn s cm^{-2} = g (s cm)$^{-1}$. The viscosity of water at 20 °C is 1.000 centipoise.

Because a sphere has a mass that is proportional to its volume $\Delta m \sim a^3$, the Stokes–Einstein law implies $D \sim m^{-1/3}$. Figure 18.14 confirms that diffusion

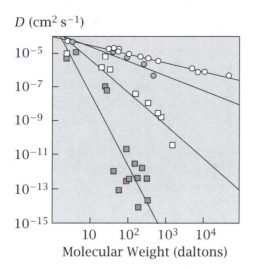

Figure 18.14 Diffusion coefficient as a function of solute molecular weight in water (○) and in three polymeric solvents: silicone rubber (⊙), natural rubber (□), and polystyrene (■). The regression lines through the measurements have slopes of -0.51 (water), -0.86 (silicone rubber), -1.90 (natural rubber), and -4.20 (polystyrene). Source: adapted from RW Baker, *Controlled Release of Biologically Active Agents*, Wiley, New York, 1987.

coefficients decrease with particle size, but it also shows that Equation (18.51) is not quantitatively accurate for particles in polymeric solvents, where the one-third-power law does not always hold.

So far, our perspective on diffusion has been *macroscopic*, focused on an experimentally observable quantity D, the diffusion constant. Now we develop a more microscopic model of diffusion.

A Microscopic Perspective on Diffusion

Diffusion results from Brownian motion, the random battering of a molecule by the solvent. Let's apply the one-dimensional random walk model of Chapter 4 (called *random flight*, in three dimensions) to see how far a particle is moved by Brownian motion in a time t. A molecule starts at position $x = 0$ at time $t = 0$. At each time step, assume that the particle randomly steps either one unit in the $+x$ direction or one unit in the $-x$ direction. Equation (4.34) gives the distribution of probabilities (which we interchangeably express as a concentration) $c(x, N)$ that the particle will be at position x after N steps,

$$c(x, N) = \frac{n_0}{(2\pi N)^{1/2}} e^{(-x^2/2N)}, \tag{18.52}$$

where n_0 is the initial amount at $x = 0$.

Comparing the microscopic random walk Equation (4.34) with the diffusion Equation (18.28) gives the number N of steps in terms of the mean-square distance traversed in a time t,

$$\langle x^2 \rangle = N = 2Dt. \tag{18.53}$$

Generalizing to d dimensions, you have

$$\langle r^2 \rangle = \langle x^2 \rangle + \langle y^2 \rangle + \langle z^2 \rangle + \ldots = N = 2dDt. \tag{18.54}$$

For example in $d = 3$ dimensions, $\langle r^2 \rangle = 6Dt$.

Equation (18.53) or (18.54) tells you how to convert D, a measurable quantity, into microscopic information about the root-mean-square distance $\langle x^2 \rangle^{1/2}$

that the particle moves in time t. For example, suppose you perform computer simulations using an atomic model of a particle undergoing Brownian motion. From that model, you simulate the particle distribution and compute the mean-square particle displacement $\langle x^2 \rangle$. Equation (18.53) or (18.54) provides a way to relate your microscopic model to experimental measurements of D.

Example 18.5 calculates the friction coefficient, the diffusion coefficient, and the mean-square displacement of a particle.

EXAMPLE 18.5 A diffusing protein. Let's compute the dynamic properties of a small spherical protein having a radius $a = 20\,\text{Å} = 2$ nm, in water, which has a viscosity $\eta = 10^{-3}\,\text{kg m}^{-1}\,\text{s}^{-1}$. According to Stokes's law Equation (18.50), this particle will have a friction coefficient

$$\xi = 6\pi \left(10^{-3}\frac{\text{kg}}{\text{m s}}\right)(2 \times 10^{-9}\,\text{m}) = 3.77 \times 10^{-11}\,\text{kg s}^{-1}. \tag{18.55}$$

The Stokes–Einstein law gives the diffusion constant,

$$D = \frac{kT}{\xi} = \frac{1.38 \times 10^{-23}\,\text{J K}^{-1} \times 300\,\text{K}}{3.77 \times 10^{-11}\,\text{kg s}^{-1}} = 1.1 \times 10^{-10}\,\text{m}^2\,\text{s}^{-1}$$

$$= 1.1 \times 10^{-6}\,\text{cm}^2\,\text{s}^{-1}. \tag{18.56}$$

In one hour, the protein diffuses an average x-axis distance

$$\langle x^2 \rangle^{1/2} = \sqrt{2Dt} = (2 \times 1.1 \times 10^{-10}\,\text{m}^2\,\text{s}^{-1} \times 3600\,\text{s})^{1/2} \tag{18.57}$$

$$= 8.9 \times 10^{-4}\,\text{m} \approx 0.9\,\text{mm}. \tag{18.58}$$

Now we explore a model of how molecular machines such as proteins can combine Brownian motion with binding and release processes to create directed motion.

Brownian Ratchets Convert Binding and Release Events into Directed Motion

Inside biological cells are motor proteins. Examples include kinesin, which walks along microtubules, myosin, which walks along actin in muscle, helicases, which unwind DNA, translocases, which pull proteins across membranes, and other protein machines that convert chemical energy into motions in some particular direction. How do molecular machines produce directed motion? So far in this chapter we have considered how gradients or external forces can cause flows. But some molecular machines use neither gradients nor external forces. Moreover, Brownian motion, by itself, cannot be the source of directed motion, because it is random. Figure 18.15 illustrates the **Brownian ratchet** model [5, 6] of how random diffusion, coupled with energy-driven but nondirectional binding and release events, can lead to directed motion.

Consider a ligand molecule L that moves itself along some 'molecular rope,' a partner molecule P having a chain of binding sites. Suppose there is an asymmetric binding free energy as a function of spatial coordinate x, which is aligned with the rope axis. Assume that this function is shaped like sawteeth

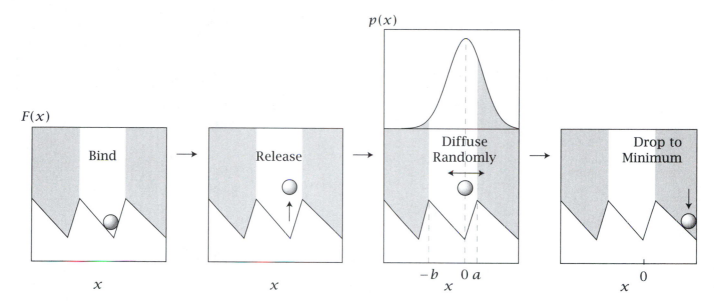

Figure 18.15 A simple Brownian ratchet. $F(x)$ is the free energy versus x, the one-dimensional coordinate of motion. There are four steps, as follows. (1) The ligand L binds to a low-energy well. (2) L is released. (3) L diffuses randomly in $+x$ and $-x$ directions for a time τ_{off}. Because $F(x)$ is asymmetric, the particle is more likely to move to the right of $x = a$ than to the left of $x = -b$, as indicated by the shading in the diagram of probability $p(x)$. (4) At the final diffusional value of x, L binds again, and rolls downhill to the nearest energy minimum. The cycle repeats.

(see Figure 18.15). Before time $t = 0$, the system is stable, and L is bound at a location where the binding free energy $F(x)$ is a minimum. At time $t = 0$, energy is put into the system in some way to release L from its binding site on P. The ligand, no longer in contact with P, then diffuses freely along the x-axis with no bias favoring either the $+x$ or $-x$ direction. The ligand remains unbound to P and diffuses for a time τ_{off}. Diffusion leads to a Gaussian distribution along x. During that time some of the ligand molecules will diffuse to $x \geq a$, where $x = a$ is the location of the next maximum to the right. At that time, those ligand molecules will rebind and slide energetically downhill to the next energy well to the right of the original binding site. A smaller number of molecules will diffuse to the left to $x \leq -b$, where they can fall into a well to the left of the original binding site.

Even though diffusion is symmetrical in x, the ligand binding potential $F(x)$ is not. At time τ_{off} more particles fall into the energy well on the right than fall into the well on the left. Repeated cycles of release, diffusion, and rebinding lead to a net hopping from one well to the next toward the right. If the time interval τ_{off} is too short, most of the ligands will return to the same well from which they started, so there will be little net motion. If τ_{off} is too long, the particles will have time to spread so broadly in both directions that again there will be no net directed motion. So a key parameter that determines the average velocity is the off-rate of the ligand from P, $k_{\text{off}} = 1/\tau_{\text{off}}$.

Let's make the model more quantitative. L diffuses along the x-axis for a time τ_{off}. p_{right} is the probability that the ligand diffuses to beyond $x = a$ in

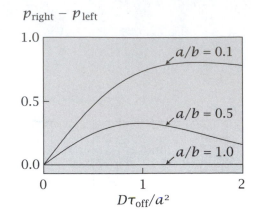

Figure 18.16 Brownian ratchet particle velocity to the right, v_{right}, is proportional to $p_{\text{right}} - p_{\text{left}}$. This velocity increases as D increases, as τ_{off} increases, or as a decreases.

that time. p_{right} is given by integrating Equation (18.28):

$$p_{\text{right}} = (4\pi D\tau_{\text{off}})^{-1/2} \int_a^\infty e^{-x^2/4D\tau_{\text{off}}} \, dx. \tag{18.59}$$

Equation (18.59) can be expressed more compactly in terms of the complementary error function,

$$\text{erfc}(z) = \frac{2}{\sqrt{\pi}} \int_z^\infty e^{-u^2} \, du. \tag{18.60}$$

Comparison of the exponents in Equations (18.59) and (18.60) shows that u is a dimensionless quantity given by

$$u^2 = \frac{x^2}{4D\tau_{\text{off}}} = \left(\frac{x}{\ell_0}\right)^2 \tag{18.61}$$

where $\ell_0 = 2\sqrt{D\tau_{\text{off}}}$ has units of length. ℓ_0 is the root-mean-square distance travelled in time τ_{off} by a particle having diffusion constant D (see Equation (18.54) for one dimension, $d = 1$). z is the lower limit of the integral in these reduced units. That is, when $x = a$, Equation (18.61) gives $u = a/\ell_0$, so the lower limit of the integral Equation (18.60) is $z = a/\ell_0$. Substituting $z = a/\ell_0$ and Equation (18.61) into Equation (18.60) allows you to express Equation (18.59) as

$$p_{\text{right}} = \frac{1}{2}\text{erfc}\left(\frac{a}{2\sqrt{D\tau_{\text{off}}}}\right). \tag{18.62}$$

Similarly, the probability p_{left} that the particle diffuses to the left and reaches the next well at a distance $|x| = b$ is $p_{\text{left}} = (1/2)\text{erfc}(b/2\sqrt{D\tau_{\text{off}}})$. Because the ligand travels a unit distance $a + b$ in a unit time $\tau_{\text{off}} + \tau_{\text{on}}$, the net velocity to the right, v_{right}, is

$$v_{\text{right}} = \left(\frac{a + b}{\tau_{\text{off}} + \tau_{\text{on}}}\right) (p_{\text{right}} - p_{\text{left}}). \tag{18.63}$$

Equation (18.63) is an approximation based on the assumption that the root-mean-square diffusion distance, $2\sqrt{D\tau_{\text{off}}}$, is not large compared with a, because otherwise some particles might jump two or more units in the time $\tau_{\text{off}} + \tau_{\text{on}}$.

Figure 18.16 is a plot of $p_{right} - p_{left}$ versus $(D\tau_{off})/a^2$ for various ratios a/b. It shows that the net velocity toward the right increases as D increases, as τ_{off} increases or as the asymmetry, a/b, of the potential increases. If the potential is symmetric, $a = b$, this model shows that there is no net motion. This model shows how Brownian motion, which is nondirectional, and energy-driven binding and release events, which are also nondirectional, can combine to create directed motion if there is an asymmetric potential.

The Fluctuation-dissipation Theorem Relates Equilibrium Fluctuations to the Rate of Approach to Equilibrium

An important theorem of statistical mechanics relates a property of kinetics to a property of equilibrium. At equilibrium, a system undergoes thermal fluctuations. Remarkably, the magnitudes of these equilibrium fluctuations are related to how fast the system approaches equilibrium. This theorem is quite general, and applies to many different physical and chemical processes. It allows you to determine the diffusion constant, viscosity, and other transport properties from knowledge of the equilibrium fluctuations.

To illustrate the idea, we develop the *Langevin model* of a particle moving in one dimension, subject to Brownian motion. The particle—whether a protein molecule, a colloidal particle, or a biological cell—is large relative to the solvent molecules. The solvent molecules bombard the particle rapidly from all directions. According to Newton's laws, a particle's mass m multiplied by its acceleration dv/dt, where v is the particle velocity, equals the sum of all the forces acting on the particle. In the Langevin model, two forces act on the particle: friction, and a random force $f(t)$ that fluctuates rapidly as a function of time t, representing Brownian motion. Every time a Brownian bombardment causes the particle to move in the $+x$ direction, frictional drag acts in the $-x$ direction to slow it down. So for this situation, Newton's law gives the **Langevin equation**:

$$m\frac{dv}{dt} = f(t) - \xi v, \tag{18.64}$$

where ξ is the friction coefficient. Because the fluctuating force acts just as often in the $+x$ direction as in the $-x$ direction, the *average force* is zero,

$$\langle f(t) \rangle = 0. \tag{18.65}$$

We use this model to illustrate the idea of a *time correlation function*.

The Time Correlation Function Describes How Quickly Brownian Motion Erases the Memory of the Initial Particle Velocity

Here's how to construct a time-correlation function. Take the velocity $v(0)$ of the particle at time $t = 0$. Multiply by the velocity of the particle $v(t)$ at a later time t. Take the equilibrium ensemble average of this product over many different collisions to get $\langle v(0)v(t) \rangle$. This is the *velocity autocorrelation*

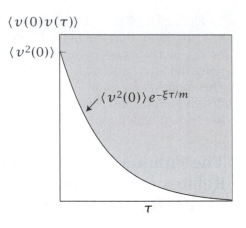

Figure 18.17 Autocorrelation function, showing the ensemble average of the product of the velocity at two different times, $t = 0$ and $t = \tau$. If τ is large enough, the velocity $v(\tau)$ becomes uncorrelated with $v(0)$.

function. It is an example of a more general quantity called a time correlation function. For example, you can do this with positions $\langle x(0)x(t) \rangle$, or forces $\langle f(0)f(t) \rangle$.

The velocity autocorrelation function tells you how fast a particle 'forgets' its initial velocity, owing to Brownian randomization. When the time t is short relative to the correlation time of the physical process, the particle velocity will be nearly unchanged from time 0 to t, and $v(t)$ will nearly equal $v(0)$ so $\langle v(0)v(t) \rangle \approx \langle v^2(0) \rangle$. But when t is much greater than the system's correlation time, Brownian motion will have had time to randomize the particle velocity relative to its initial velocity, so $v(t)$ will be *uncorrelated* with $v(0)$. This means that $v(0)v(t)$ will be negative just as often as it is positive, so the ensemble average will be zero, $\langle v(0)v(t) \rangle = 0$.

Let's compute the velocity autocorrelation function from the Langevin model. Multiply both sides of the Langevin Equation (18.64) by $v(0)$, which is a constant independent of t. Then take the ensemble average to get:

$$m \left\langle v(0) \frac{dv(t)}{dt} \right\rangle = \langle v(0)f(t) \rangle - \xi \langle v(0)v(t) \rangle. \qquad (18.66)$$

Because the fluctuating force is uncorrelated with the initial velocity of the particle, you have $\langle v(0)f(t) \rangle = \langle v(0) \rangle \langle f(t) \rangle$. This product is zero since $\langle f(t) \rangle = 0$, so

$$\frac{d}{dt} \langle v(0)v(t) \rangle + \frac{\xi}{m} \langle v(0)v(t) \rangle = 0$$

$$\implies \quad \langle v(0)v(t) \rangle = \langle v^2(0) \rangle e^{-\xi t/m}. \qquad (18.67)$$

Now since $\langle mv^2(0)/2 = kT/2 \rangle$ (see equipartition Equation (10.16)), Equation (18.67) can be expressed as

$$\langle v(0)v(t) \rangle = \frac{kT}{m} e^{-\xi t/m}. \qquad (18.68)$$

This function is shown in Figure 18.17. This figure shows that the longer the delay between two collisions, the less correlation there is between the velocities $v(0)$ and $v(t)$.

In the Langevin model, there are two time scales. First, each bombardment of the large particle by a solvent molecule is very fast. There is a second, much slower, time scale over which the particle's velocity at one time becomes uncorrelated with its velocity at another time. Because the exponent in Equation (18.68) is dimensionless, the *time constant* for this slower process is m/ξ.

EXAMPLE 18.6 Autocorrelation times. The *correlation time m/ξ* is the time required for the velocity autocorrelation function to reach $1/e$ of its initial value. The correlation time is short when the mass is small or when the friction coefficient is large. The correlation time for the diffusional motion of a small protein of mass $m = 10,000\,\text{g mol}^{-1}$ is in the picosecond time range (see Equation (18.56)):

$$\frac{m}{\xi} = \frac{(10,000\,\text{g/mol})(1\,\text{kg}/1000\,\text{g})}{(6.023 \times 10^{23}\,\text{molecules/mol})(3.77 \times 10^{-11}\,\text{kg s}^{-1})}$$

$$= 4.4 \times 10^{-13}\,\text{s}.$$

A particle's velocity is correlated with its earlier velocity over times shorter than m/ξ, but is uncorrelated over times much longer than this.

Now let's see how $\langle v(0)v(t)\rangle$, a property of the fluctuations in equilibrium, is related to a kinetic property, in this case the diffusion coefficient D. Integrate the time correlation function Equation (18.68) over all the possible time lags t,

$$\int_0^\infty \langle v(0)v(t)\rangle\,dt = \frac{kT}{m}\int_0^\infty e^{-\xi t/m}\,dt = \frac{kT}{\xi} = D, \tag{18.69}$$

because $\int_0^\infty e^{-\xi t/m}\,dt = m/\xi$. In three dimensions, the velocities are vectors $\mathbf{v}(0)$ and $\mathbf{v}(t)$ and Equation (18.69) becomes

$$\int_0^\infty \langle \mathbf{v}(0)\mathbf{v}(t)\rangle\,dt = 3D. \tag{18.70}$$

Equations (18.69) and (18.70) are examples of **Green–Kubo** relationships [7, 8] between an equilibrium correlation function and a *transport coefficient*, the diffusion constant D in this case. There are several such relationships. Without going into the details, we just note that another such relationship gives the friction coefficient in terms of the **force correlation function**:

$$\xi = \frac{1}{2kT}\int_{-\infty}^\infty \langle f(0)f(t)\rangle\,dt. \tag{18.71}$$

In addition, the viscosity is the autocorrelation function of momentum transfer in liquid flows. Green–Kubo relationships equate a kinetic property (such as the diffusion coefficient or the friction coefficient) with an ensemble average of an equilibrium property, such as $\langle v(0)v(t)\rangle$.

Onsager Reciprocal Relations Describe Coupled Flows

So far, we have considered a single type of force f and a single type of flow J, related by $J = Lf$, where $L = cu$ (see Equations (18.4) and (18.6)). Now consider a more complex situation, e.g., a material in which a temperature

Table 18.2 Phenomenological flow coefficients L_{ij} for the system NaCl/KCl/water. The subscript 1 is NaCl and 2 is KCl. Sources: [a]H Fujita and LJ Gostling, *J Phys Chem* **64**, 1256-1263 (1960); [b]PJ Dunlop and LJ Gostling, *J Phys Chem* **63**, 86-93 (1959).

	Salt Concentrations (M)			
C_{NaCl}	0.25	0.5	0.25	0.5
C_{KCl}	0.25	0.25	0.5	0.5
	$L_{ij} \times 10^9 RT$			
L_{11} [b]	2.61	4.76	2.79	5.15
L_{12} [b]	−0.750	−1.03	−0.99	−1.52
L_{21} [a]	−0.729	−1.02	−0.97	−1.45
L_{22} [b]	3.50	3.83	6.36	7.02
L_{12}/L_{21}	1.03	1.01	1.02	1.05

gradient drives a heat flow, and a voltage difference also drives an electrical current. If such processes were independent of each other, then you would have

$$J_1 = L_1 f_1 \quad \text{and} \quad J_2 = L_2 f_2, \tag{18.72}$$

where the subscript 1 denotes the heat flux and 2 denotes the electrical flux.

But in general, multiple flow processes in a system are not independent; they are coupled. The heat flow can affect the electrical current and vice versa. (Such observations were first made by Lord Kelvin in 1854.) So the simplest relationships between two forces and two flows are

$$J_1 = L_{11} f_1 + L_{12} f_2 \quad \text{and} \quad J_2 = L_{21} f_1 + L_{22} f_2, \tag{18.73}$$

where L_{12} and L_{21} are the 'coupling' coefficients that describe how the temperature gradient affects the electrical current and vice versa. A most remarkable experimental observation is that the coupling coefficients obey a very simple reciprocal relationship: $L_{21} = L_{12}$.

Here are two examples of coupling. Suppose you solder together two different metals to make a junction. If you apply a voltage to drive a current flow, it can cause heating or cooling of the junction. This is called the *Peltier effect*. But the existence of a reciprocal relation means that if you heat or cool a bimetallic junction instead, an electrical current will flow. This is the principle of operation for thermocouples, which convert temperature changes to electrical signals.

Here is another example. The diffusion of one salt, say NaCl, affects the diffusion of another, such as KCl. Table 18.2 shows careful measurements of such coupled diffusion processes, and shows that at least in this case the reciprocal relations hold to within experimental error.

Such couplings are called Onsager reciprocal relations, after L Onsager, (1903–1976), a Norwegian physical chemist who explained this symmetry as arising from *microscopic reversibility* [9, 10]. (Although the Norges Tekniske Hogskole judged Onsager's work unacceptable and did not award him a Ph.D. degree, they later awarded him an honorary doctorate. This work was awarded the Nobel Prize in Chemistry in 1968.)

The principle of microscopic reversibility can be described in terms of time correlation functions. In the Langevin example, microscopic reversibility says that the average probability of observing a velocity $v_2(t)$ at a time t after the particle has a velocity $v_1(0)$ is the same as the average probability of observing $v_1(t)$ after observing $v_2(0)$, that is, $\langle v_1(0)v_2(t)\rangle = \langle v_2(0)v_1(t)\rangle$. Onsager showed that this principle of microscopic reversibility implies the reciprocal relationship of forces and flows, $L_{12} = L_{21}$.

Summary

Equilibrium is a state of matter that results from spatial uniformity. In contrast, when there are concentration differences or gradients, particles will flow. In these cases, the rate of flow is proportional to the gradient. The proportionality constant between the flow rate and the gradient is a transport property: for particle flow, this property is the diffusion constant. Diffusion can be modelled at the microscopic level as a random flight of the particle. The diffusion constant describes the mean square displacement of a particle per unit time. The fluctuation-dissipation theorem describes how transport properties are related to the ensemble-averaged fluctuations of the system in equilibrium.

Problems

1. Diffusion-controlled dimerization reactions.

(a) Using the Stokes–Einstein relation, show that the rate of diffusion of a particle having radius a to a sphere of radius a can be expressed in terms that depend only on temperature and solvent viscosity.

(b) Using the expression from (a), compute the diffusion-controlled dimerization rate k_a of two identical spheres in water ($\eta = 0.01$ poise) at $T = 300\,\text{K}$.

(c) For the diffusion of two particles of radii a_A and a_B and diffusion constants D_A and D_B, the generalization of Equation (18.29) for the dimerization rate is

$$k_a = 4\pi(D_A + D_B)(a_A + a_B). \qquad (18.74)$$

Show that Equation (18.74) reduces to $k_a = 8kT/3\eta$ when $a_A = a_B$ and $D_A = D_B$.

2. Diffusion of drug from a delivery tablet.

Suppose a drug is encapsulated between two planes at $x = 0$ and $x = h$. The drug diffuses out of both planes at a constant rate R, so the diffusion equation is

$$D\frac{d^2c}{dx^2} = R.$$

(a) Solve for $c(x)$ inside the tablet, subject to boundary conditions $c(0) = c(h) = 0$, that is, the drug is used up the instant it is released.

(b) Compute the flux of the drug out of the tablet.

3. Diffusion of light from the center of the sun.

Neutrino particles fly from the sun's center to its surface in about 2.3 s because they travel at the speed of light ($c \approx 3 \times 10^{10}\,\text{cm s}^{-1}$), and they undergo little interaction with matter. But a photon of light takes much longer to travel from the sun's center to its surface because it collides with protons and free electrons and undergoes a random walk to reach the sun's surface. A photon's step length is approximately 1 cm per step [11]. The sun's radius is $7 \times 10^{10}\,\text{cm}$. How long does it take a photon to travel from the center to the surface of the sun?

4. How far does a protein diffuse?

Consider a protein that has a diffusion constant in water of $D = 10^{-6}\,\text{cm}^2\,\text{s}^{-1}$.

(a) What is the time that it takes the protein to diffuse a root-mean-square distance of 10 μm in a three-dimensional space? This distance is about equal to the radius of a typical cell.

(b) If a cell is spherical with radius $r = 10\,\mu m$, and if the protein concentration outside the cell is 1 μM (micromolar), what is the number of protein molecules per unit time that bombard the cell at the diffusion-limited rate?

5. Diffusion of long rods is faster than diffusion of small spheres.

If ligand molecules have diffusion coefficient D and concentration c_∞ in solution, the current $I(L, a)$ of their diffusional encounters with a rod of length L and radius a is

$$I(L, a) \approx -\frac{4\pi DLc_\infty}{\ln(2L/a)}.$$

For a long rod, such as DNA, for which $L/a = 10^6$, compute the ratio of $I(L, a)$ to $I(a)$, the current to a sphere of radius a.

6. HIV growth kinetics.

According to [12], the following model is sufficient to describe the time-dependent amount $V(t)$ of virus in the body of an HIV-infected patient. Two factors contribute to the rate dV/dt: (1) a constant rate P of viral production, and (2) a clearance rate cV at which virus is removed from the body.

(a) What differential equation gives $V(t)$?

(b) Solve that equation to show the function $V(t)$.

7. Einstein's estimate of Brownian motion and Avogadro's number.

Einstein assumed that a particle of radius 5×10^{-5} cm could be observed to undergo Brownian motion under a microscope [13]. He computed the root-mean-square distance $\langle x^2 \rangle^{1/2}$ that the particle would move in one minute in a one-dimensional walk in water, $\eta \approx 1$ centipoise at about $T = 300\,\text{K}$.

(a) Compute $\langle x^2 \rangle^{1/2}$.

(b) Show how the argument can be turned around to give Avogadro's number.

8. $\langle x^2 \rangle$ from the Langevin model.

(a) Derive an expression for $\langle x^2 \rangle(t)$ from the Langevin model.

(b) Find limiting expressions for $\langle x^2 \rangle(t)$ as $t \to 0$, and as $t \to \infty$.

9. Permeation through bilayers.

Acetamide has an oil/water partition coefficient of $K_w^o = 5 \times 10^5$. It has a permeability $P = 5 \times 10^4\,\text{cm s}^{-1}$ through oil-like bilayers, which are 30 Å thick. What is the diffusion constant D for acetamide across the bilayer?

10. Computing the radius of hemoglobin.

Assume that hemoglobin is a sphere. Compute its radius from its diffusion constant in Table 18.1.

11. Rotational diffusion.

Consider particles that are oriented at an angle θ with respect to the x-axis. The distribution of particles at different angles is $c(\theta)$. The flux

$J(\theta)$ of particles through different angles depends on the gradient as in Fick's law,

$$J(\theta) = -\Theta\left(\frac{\partial c}{\partial \theta}\right),$$

where Θ is the orientational diffusion coefficient.

(a) Write an expression for the rotational diffusion equation.

(b) If the orientational friction coefficient for a sphere of radius r is $f_{or} = 8\pi\eta r^3$, write an expression for $\Theta(r)$.

12. Friction of particles.

(a) A sphere and disk have the same radius. For a given applied force, which moves faster in flow?

(b) Does a long thin ellipsoid move faster lengthwise or sideways?

13. Diffusion/reaction length from a drug tablet. A drug that has diffusion constant $D = 10^{-6} \text{cm}^2 \text{ s}^{-1}$ reacts at a rate of $k_{rx} = 10^2 \text{ s}^{-1}$ at the aqueous surface of a tablet. What is the 'decay length' at which the concentration of drug is $1/e$ of its value inside the tablet?

References

[1] J Crank. *The Mathematics of Diffusion*. 2nd edition, Clarendon Press, Oxford, 1975.

[2] HS Carslaw and JC Jaeger. *Conduction of Heat in Solids*. 2nd edition, Clarendon Press, Oxford, 1959.

[3] JD Murray. *Mathematical Biology*. 2nd edition, Springer-Verlag, New York, 1993.

[4] GK Batchelor. *An Introduction to Fluid Dynamics*. Cambridge University Press, Cambridge, 1973.

[5] Y Okada and N Hirokawa. *Science* **283**, 1152 (1999).

[6] RD Astumian. *Science* **276**, 917 (1997).

[7] D Chandler. *Introduction to Modern Statistical Mechanics*. Oxford University Press, New York, 1987.

[8] DA McQuarrie. *Statistical Mechanics*. Harper & Row, New York, 1976.

[9] L Onsager. *Phys Rev* **37**, 405 (1931).

[10] L Onsager. *Phys Rev* **37**, 2265 (1931).

[11] M Harwit. *Astrophysical Concepts*. 3rd edition, Springer, New York, 1998.

[12] DD Ho et al., *Nature* **373**, 123 (1995).

[13] A Einstein. *Investigations on the Theory of Brownian Movement*. Dover Publications, New York, 1956.

Suggested Reading

HC Berg, *Random Walks in Biology*, Princeton University Press, Princeton, 1993. Elementary and concise overview of diffusion, flow, and random walks.

TF Weiss, *Cellular Biophysics*, Vol 1, MIT Press, Cambridge, 1996. Detailed and extensive treatment of forces and flows through biological membranes.

Excellent treatments of the Langevin equation and time correlation functions:

RS Berry, SA Rice and J Ross, *Physical Chemistry* (Chapter 10) Wiley, New York, 1980.

D Chandler, *Introduction to Modern Statistical Mechanics*, Oxford University Press, New York, 1987.

N Wax (editor), *Selected Papers on Noise and Stochastic Processes*, Dover Publications, New York, 1954.

Excellent treatments of Onsager relations and coupling include:

R Haase, *Thermodynamics of Irreversible Processes*, Dover Publications, New York, 1969.

A Katchalsky and PF Curran, *Nonequilibrium Thermodynamics in Biophysics*, Harvard University Press, Cambridge, 1965.

19 Chemical Kinetics & Transition States

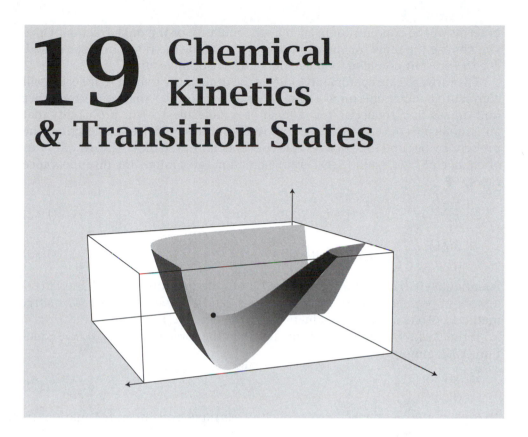

Diffusion and Chemical Reaction Rates Depend on Temperature

In this chapter we focus on the kinetics of processes, such as chemical reactions, that speed up strongly as temperature increases. Remarkably, to model the rates of these processes, you can use the same statistical thermodynamics approach that we used in Chapter 13 to model equilibria. You need only one additional concept: the *transition state* or *activation barrier*.

Reaction Rates Are Proportional to Concentrations

We start with a simple kinetic process, the interconversion between two states, as in Equation (13.1),

$$A \underset{k_r}{\overset{k_f}{\rightleftharpoons}} B, \tag{19.1}$$

where k_f and k_r indicate the forward and reverse rate coefficients, which are described below. How do the amounts of A and B change with time t, given the initial amounts at time $t = 0$?

The rate of increase of A is $d[A]/dt$, and the rate of increase of B is $d[B]/dt$, where '[ ]' indicates the amount of each species, for which you can use any units you wish: $[A(t)]$ can be given in terms of the numbers of molecules, or in terms of the concentration, such as the molarity or the mole fraction. Once you choose the units for the amounts, the units for the rates are determined. For the present problem, we'll use the numbers of molecules.

The forward **rate coefficient** k_f gives the probability that an A molecule will convert to a B, per unit time. The number of A molecules that convert to B per unit time is the product of (the number of A molecules) $\times$ (the probability that an A molecule converts to a B per unit time). A rate coefficient k_r describes the probability per unit time of a reverse conversion, from B to A. The overall rates of change of $[A(t)]$ and $[B(t)]$ are the creation rates minus the disappearance rates,

$$\frac{d[A(t)]}{dt} = -k_f[A(t)] + k_r[B(t)], \qquad \text{and} \tag{19.2}$$

$$\frac{d[B(t)]}{dt} = k_f[A(t)] - k_r[B(t)]. \tag{19.3}$$

Equations (19.2) and (19.3) are *coupled*: both the quantities $[A(t)]$ and $[B(t)]$ appear in both equations. Such equations can be solved by standard matrix methods. Texts on chemical kinetics give the details [1, 2].

If the backward rate is much smaller than the forward rate, $k_r \ll k_f$, Equation (19.2) simplifies to

$$\frac{d[A(t)]}{dt} = -k_f[A(t)]. \tag{19.4}$$

To express the time dependence explicitly, rearrange and integrate Equation (19.4) from time $t = 0$ to t to get

$$\int \frac{dA'}{A'} = -\int_0^t k_f \, dt' \quad \Longrightarrow \quad \ln \frac{[A(t)]}{[A(0)]} = -k_f t$$

$$\Longrightarrow \quad [A(t)] = [A(0)]e^{-k_f t}, \tag{19.5}$$

where $[A(0)]$ is the concentration of A at time 0, the start of the reaction. If $k_f > 0$, the amount of A diminishes exponentially with time. If $[A(t)] + [B(t)] = $ constant, then $[B(t)]$ increases with time, $[B(t)] = \text{constant} - [A(0)] \exp(-k_f t)$.

At Equilibrium, Rates Obey Detailed Balance

The *principle of detailed balance* says that the forward and reverse rates must be identical for an elementary reaction at equilibrium:

$$k_f[A]_{\text{eq}} = k_r[B]_{\text{eq}}, \tag{19.6}$$

where $[\ ]_{\text{eq}}$ is the equilibrium concentration. To see that this is a condition of equilibrium, substitute Equation (19.6) into Equations (19.2) and (19.3) to get $d[A]/dt = 0$ and $d[B]/dt = 0$. Note that it is the rates, not the rate coefficients, that are equal at equilibrium.

Equation (19.6) relates rate coefficients k_f and k_r to the equilibrium constant K:

$$K = \frac{[B]_{\text{eq}}}{[A]_{\text{eq}}} = \frac{k_f}{k_r}. \tag{19.7}$$

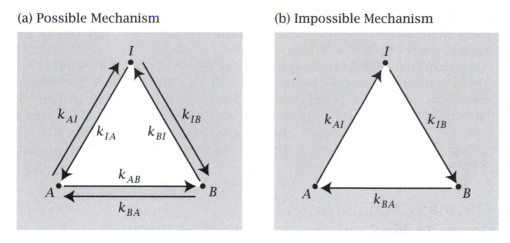

(a) Possible Mechanism

(b) Impossible Mechanism

Figure 19.1 The principle of detailed balance is satisfied by mechanism (a) between three states A, I, and B, but violated by mechanism (b). Forward rates must equal reverse rates for each pair of states.

For more complex systems, the principle of detailed balance gives more information beyond the statement of equilibrium. For a system having more than one elementary reaction, 'detailed' balance means that the forward and reverse rates must be equal for every elementary reaction. Figure 19.1(a) shows a three-state equilibrium between states A, B, and I. Detailed balance says that each individual forward rate equals its corresponding individual backward rate,

$$\frac{[I]_{eq}}{[A]_{eq}} = \frac{k_{AI}}{k_{IA}}, \qquad \frac{[B]_{eq}}{[I]_{eq}} = \frac{k_{IB}}{k_{BI}}, \qquad \text{and} \qquad \frac{[A]_{eq}}{[B]_{eq}} = \frac{k_{BA}}{k_{AB}}. \tag{19.8}$$

Here's an implication of detailed balancing. Suppose that for the three-state process in Figure 19.1(a) you hypothesized a mechanism in which $k_{IA} \approx 0$, $k_{BI} \approx 0$, and $k_{AB} \approx 0$ (see Figure 19.1(b)). In the forward direction A converts to B through an intermediate state I, while in the reverse direction, B converts to A, either through a different intermediate state, or through none at all, as indicated here. The condition for equilibrium tells you only that the sum of fluxes into each state must equal the sum of fluxes out. For mechanism (b) this implies that

$$[A]k_{AI} = [I]k_{IB} = [B]k_{BA}. \tag{19.9}$$

Equation (19.9) is two independent equations in three unknown concentrations. It has an infinite number of solutions.

Detailed balancing, Equations (19.8), gives additional information. It says that the mechanism shown in Figure 19.1(b) is impossible. Multiplying Equations (19.8) gives

$$1 = \frac{k_{AI}k_{IB}k_{BA}}{k_{IA}k_{BI}k_{AB}}. \tag{19.10}$$

Equation (19.10) cannot be satisfied in mechanism (b) because the denominator would be zero if the reverse rates were zero. The principle of detailed balancing says that forward and backward reactions at equilibrium cannot have different intermediate states. That is, if the forward reaction is $A \to I \to B$, the backward reaction cannot be $B \to A$.

The principle of detailed balancing can be derived from the *principle of microscopic reversibility* (see Chapter 18), which says that Newton's laws of motion for the collisions between atoms and molecules are symmetrical in time—they would look the same if time ran backward. The proof of detailed balance from microscopic reversibility is given in [3]. Detailed balance is helpful in understanding the *mechanisms* of chemical reactions, the molecular steps from reactants to products.

The Mass Action Laws Describe Mechanisms in Chemical Kinetics

Suppose you have a chemical reaction in which a product P is produced from reactants A, B, and C, with stoichiometric coefficients a, b, and c:

$$aA + bB + cC \to P. \tag{19.11}$$

In general, the initial reaction rate depends on the concentrations of the reactants, the temperature and pressure, and on the coefficients a, b, and c. Chemical kinetics experiments often measure how reaction rates depend on the concentrations of the reactants. Such experiments can provide valuable information about the mechanism of the reaction.

The kinetic *law of mass action*, first developed by CM Guldberg and P Waage in 1864, says that reaction rates should depend on stoichiometry in the same way that equilibrium constants do. According to this law, the initial rate of product formation, $d[P]/dt$ for the reaction in Equation (19.11), depends on reactant concentrations:

$$\frac{d[P]}{dt} = k_f[A]^a[B]^b[C]^c, \tag{19.12}$$

where k_f is the rate coefficient for the forward reaction. Suppose that you performed an experiment and found that the initial rate of formation of product P depends, say, on the seventh power of the concentration $[A]$, and not on B or C, $(d[P]/dt = k_f[A]^7)$. Then the conclusion from the law of mass action would be that $a = 7$ is the stoichiometric coefficient: seven molecules of A must come together at a time to form one P molecule. This is the type of mechanistic information that is available from kinetics experiments.

However, although mass action summarizes a large body of experimental data, there are exceptions. Kinetic mechanisms do not always follow the thermodynamic stoichiometries. In general, the prediction of the rate law (the dependence of the reaction speed on the reactant concentrations) from the atomic structures of the reactants and products requires quantum mechanics at a level that is only currently possible for the simplest reactions. Reaction mechanisms are described in chemical kinetics textbooks such as [1, 2].

Now we switch attention from the concentration dependence of reaction rates to the temperature dependence of rate coefficients. This, too, can give insights into the mechanism of the reaction.

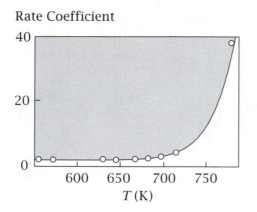

Rate Coefficient

Figure 19.2 Rate coefficients k_f for the reaction $2HI \rightarrow H_2 + I_2$ in the gas phase as a function of temperature T. These data show the exponential dependence described by Equation (19.16). Source: H Eyring and EM Eyring, *Modern Chemical Kinetics*, Reinhold Publishing Corp, New York, 1963.

Reaction Rates Depend on Temperature

Consider a binary reaction in the gas phase,

$$A + B \xrightarrow{k_2} P. \tag{19.13}$$

Suppose that the initial rate of appearance of P can be expressed as

$$\frac{d[P]}{dt} = k_2[A][B], \tag{19.14}$$

where $k_2[A][B]$ is the rate of the reaction, and the subscript 2 on the rate coefficient indicates that two reactants are involved. By definition, the rate coefficient k_2 is independent of the concentrations of A and B. However, k_2 can depend strongly on temperature.

The typical temperature dependence of a rate coefficient is shown in Figure 19.2. This observed dependence of the reaction rate on the temperature is much greater than you would find from just the enhanced thermal motions of the molecules. Increasing the temperature from 700 K to 800 K would increase the average thermal energy kT by only 16%, but the rate coefficient increases 40-fold (see Figure 19.2).

In 1889, S Arrhenius (1859–1927), a Swedish chemist and physicist, proposed a simple explanation for the strong temperature dependence of reaction rates. Based on the van't Hoff Equation (13.37) for the strong dependence of the equilibrium constant K on temperature,

$$\frac{d \ln K}{dT} = \frac{\Delta h^{\circ}}{kT^2},$$

Arrhenius proposed that the forward and reverse rate coefficients k_f and k_r also have the van 't Hoff form:

$$\frac{d \ln k_f}{dT} = \frac{E_a}{kT^2} \quad \text{and} \quad \frac{d \ln k_r}{dT} = \frac{E'_a}{kT^2}, \tag{19.15}$$

where E_a and E'_a have units of energy that are chosen to fit the experimental data. E_a and E'_a are called *activation energies*.

According to Arrhenius, it is not the average energy of the reactants that determines the reaction rates but only the high energies of the 'activated' molecules. Figure 19.3 shows this idea in a diagram. There are two plateaus

Figure 19.3 An activation energy diagram relates three quantities: the activation energy E_a for the forward reaction, the activation energy E_a' for the reverse reaction, and the enthalpy Δh° for the equilibrium. ξ is the reaction coordinate. A more microscopic interpretation of the reaction coordinate is given on page 348.

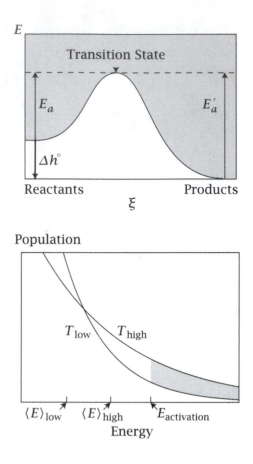

Figure 19.4 Two Boltzmann distributions, one for higher temperature T_{high} and the other for lower temperature T_{low}. Increasing the temperature increases the average energy E a small amount from $\langle E \rangle_{low}$ to $\langle E \rangle_{high}$, but it increases the population of activated (high-energy) molecules by a larger amount. The shaded area between the curves where $E \geq E_{activation}$ indicates the enhanced population of high-energy molecules with temperature.

in energy, representing reactants and products. It shows a maximum, called the *transition state* or *activation barrier*, which is the energy that activated molecules must have to proceed from reactants to products. All three of the energy differences between minima or maxima shown on this diagram are obtainable from experiments. Measuring the forward rate k_f as a function of temperature and using Equation (19.15) gives E_a. Measuring the reverse rate gives E_a'. Measuring the equilibrium constant versus temperature and using Equation (13.37) gives Δh°. Because the equilibrium constant is related to the rates by $K = k_f / k_r$, the three energy quantities are related by $\Delta h^\circ = E_a - E_a'$. Figure 19.4 shows how activation is interpreted according to the Boltzmann distribution law: a small increase in temperature can lead to a relatively large increase in the population of high-energy molecules.

Integrating Equation (19.15) and exponentiating gives the temperature dependence of the forward reaction,

$$k_f = Ae^{-E_a/kT}, \tag{19.16}$$

where A is a constant.

Just as equilibria can be represented by a van't Hoff plot of $\ln K$ versus $1/T$, kinetics can be represented by an *Arrhenius plot* of $\ln k$ versus $1/T$. Example 19.1 shows how Equations (19.15) can account for the temperature dependence of reaction rates, taking E_a and E_a' to be constants.

Rate Coefficient

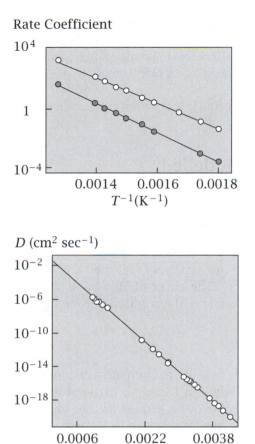

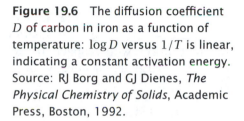

Figure 19.5 Arrhenius plots of the hydrogen iodide reactions: (○) $H_2 + I_2 \rightarrow 2HI$; (●) $2HI \rightarrow H_2 + I_2$. Source: H Eyring and EM Eyring, *Modern Chemical Kinetics*, Reinhold Publishing Corp, New York, 1963.

D (cm^2 sec^{-1})

Figure 19.6 The diffusion coefficient D of carbon in iron as a function of temperature: $\log D$ versus $1/T$ is linear, indicating a constant activation energy. Source: RJ Borg and GJ Dienes, *The Physical Chemistry of Solids*, Academic Press, Boston, 1992.

EXAMPLE 19.1 Calculate E_a for $H_2 + I_2 \rightarrow 2HI$. Use Figure 19.5 and the Arrhenius Equation (19.16) to get

$$\ln\left(\frac{k_{f2}}{k_{f1}}\right) = -\frac{E_a}{k}\left(\frac{1}{T_2} - \frac{1}{T_1}\right)$$

$$\Rightarrow E_a = \frac{-k[\ln k_{f2} - \ln k_{f1}]}{(1/T_2) - (1/T_1)} \approx \left(-2\frac{\text{cal}}{\text{mol K}}\right)\left(\frac{\ln 10^4 - \ln 10^{-1}}{0.0012 - 0.0018}\right)$$

$$\approx 38\,\frac{\text{kcal}}{\text{mol}}.$$

Arrhenius kinetics applies to many physical processes, not just to chemical reactions. One example is the diffusion of atoms in solids. Figure 19.6 shows that the diffusion rates of carbon atoms through solid iron metal follow Arrhenius behavior over a remarkable 14 orders of magnitude. This evidence supports the *interstitial model*, in which the carbon atoms occupy the interstices in the iron lattice, and 'jump' over energy barriers to travel from one interstitial site to another.

When should you treat a process as activated? If a small increase in temperature gives a large increase in rate, then a good first step is to try the Arrhenius model. The making or breaking of bonds is often, but not always, well modeled as an activated process. Highly reactive radicals are counterexamples. Reactions such as $H_3^+ + HCN \rightarrow H_2 + H_2CN^+$ can be much faster than typical activated processes, and they slow down with increasing temperature, although only slightly. Such 'negative activation energies' are usually small, 0 to -2 kcal mol^{-1} [4, 5].

We now describe a more microscopic approach to reaction rates, called transition state theory.

Activated Processes Can Be Modelled by Transition State Theory

The Energy Landscape of a Reaction

An *energy landscape* defines how the energy of a reacting system depends on its degrees of freedom—the positions and orientations of all the reactant and product atoms with respect to each other. The cover of this book shows a complex energy landscape. A simpler example is the reaction between atoms $A, B,$ and C,

$$A + BC \rightarrow AB + C. \tag{19.17}$$

This reaction can be described by two degrees of freedom: the distance R_{AB} between A and B, and the distance R_{BC} between B and C, if the attack and exit of the isolated atom are collinear with the bond. Otherwise, the angles of attack and exit are also relevant degrees of freedom. Figure 19.7 shows the energy landscape computed from quantum mechanics as a function of these two degrees of freedom for the reaction of deuterium D with a hydrogen molecule, $D + H_2 \rightarrow HD + H$.

When D is far away from H_2, the H_2 bond length is constant: D approaches along an energy valley of constant depth. When D gets close to H_2, it causes the H-H bond to stretch and the energy increases to a maximum, which is indicated as a saddle point resembling a mountain pass (see Figure 19.7). At the maximum, the central hydrogen is weakly bound to the incoming D and is also weakly bound to the outgoing H. After the H-H bond is broken and the lone H exits, the D-H bond length R_{HD} shortens and approaches constant length and energy, indicated by the outgoing valley. The width of each valley represents the vibrations of the H-H and D-H bonds.

The reaction trajectory for any one set of reacting atoms typically involves some excursions up the walls of the valleys, corresponding to the central H oscillating between the incoming and outgoing atoms. But when averaged over multiple trajectories, the reaction process can be described as following the lowest energy route, along the entrance valley over the saddle point and out of the exit valley, because the Boltzmann populations are highest along that average route.

The reaction coordinate ξ defines the position of the reaction along this average route. Even when energy landscapes involve many degrees of freedom, if there is a single lowest-energy route involving valleys and mountain passes, that route defines the reaction coordinate. The saddle point, which is the highest point along the reaction coordinate, is called the *transition state*

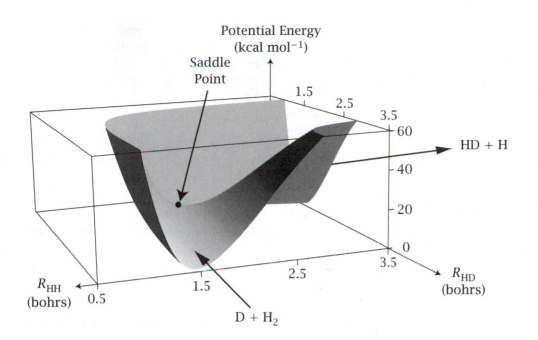

Figure 19.7 The energy surface for the reaction $D + H_2 \rightarrow HD + H$ as a function of the bond distances R_{HH} and R_{HD} (1 bohr = 0.53×10^{-10} m) when the three atoms are collinear. The arrows indicate the path of the reaction. The point of highest energy on the minimum-energy path is the saddle point. Source: S Borman, *Chem Eng News*, June 4 (1990). Data are from DG Truhlar, University of Minnesota.

and is denoted by the symbol ‡. The transition state is unstable: a ball placed at the saddle point will roll downhill along the reaction coordinate in one direction or the other. Because the transition state has high energy, it has low population.

Calculating Rate Coefficients from Transition State Theory

Let's calculate the rate coefficient k_2 for the reaction $A + B \xrightarrow{k_2} P$, by means of *transition state theory*, also called *absolute rate theory*. The approach is to divide the reaction process into two stages. The first stage is an equilibrium between the reactants and the transition state $(AB)^\ddagger$ with 'equilibrium constant' $K^\ddagger$. The second stage is a direct step downhill from the transition state to form the product with rate coefficient $k^\ddagger$:

$$A + B \xrightarrow{K^\ddagger} (AB)^\ddagger \xrightarrow{k^\ddagger} P. \tag{19.18}$$

A key assumption of transition state theory is that the first stage can be expressed as an equilibrium between the reactants A and B and the transition state $(AB)^\ddagger$, with equilibrium constant $K^\ddagger$,

$$K^\ddagger = \frac{[(AB)^\ddagger]}{[A][B]}, \tag{19.19}$$

(a) Potential Energy Surface

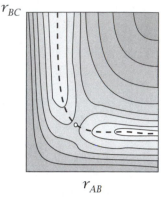

r_{BC}

r_{AB}

(b) Three-dimensional Transition State

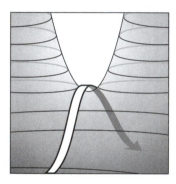

Figure 19.8 (a) Contour plot of a reaction pathway (– – –) on an energy landscape for the reaction $A + BC \rightarrow AB + C$. The shading shows the energy contours, and the broken line shows the lowest-energy path between reactants and products. (b) The transition state is an unstable point along the reaction pathway (indicated by the arrow), and a stable point perpendicular to the reaction pathway. Source: A Pross, *Theoretical and Physical Principles of Organic Reactivity*, Wiley, New York, 1995.

even though the transition state $(AB)^{\ddagger}$ is not a true equilibrium state. It is an unstable state because it is at the top of an energy hill. The overall rate is expressed as the number of molecules in the transition state, $[(AB)^{\ddagger}]$, multiplied by the rate coefficient $k^{\ddagger}$ for converting transition state molecules to product,

$$\frac{d[P]}{dt} = k^{\ddagger}[(AB)^{\ddagger}] = k^{\ddagger}K^{\ddagger}[A][B]. \tag{19.20}$$

The second equality in Equation (19.20) follows from Equation (19.19). Combining Equations (19.20) and (19.14) gives the measurable rate coefficient k_2,

$$k_2 = k^{\ddagger}K^{\ddagger}. \tag{19.21}$$

Because the quantity $K^{\ddagger}$ is regarded as an equilibrium constant, you can express it in terms of the molecular partition functions developed in Chapter 13 (see Equation (13.18)):

$$K^{\ddagger} = \left(\frac{q_{(AB)^{\ddagger}}}{q_A q_B}\right) e^{\Delta D^{\ddagger}/kT}, \tag{19.22}$$

where $\Delta D^{\ddagger}$ is the dissociation energy of the transition state minus the dissociation energy of the reactants.

$q_{(AB)^{\ddagger}}$ is the partition function for the transition state. At the transition state the AB bond is ready to form, but it is stretched, weak, and unstable. The system will move rapidly downhill in energy, either forward along the reaction coordinate forming the bond, or backward breaking it. Although the system is *unstable* (at a maximum of energy) *along* the reaction coordinate, it is *stable* (at a minimum) of energy in all other directions that are normal to the reaction coordinate (see Figure 19.8(b)). For the stable degrees of freedom, the system is presumed to act like any other equilibrium system. Those degrees of freedom are described by equilibrium partition functions.

Provided that the unstable degrees of freedom are independent of the stable degrees of freedom, $q_{(AB)^{\ddagger}}$ is factorable into two components:

$$q_{(AB)^{\ddagger}} = \overline{q^{\ddagger}} q_{\xi}, \tag{19.23}$$

where $\overline{q^{\ddagger}}$ represents the partition function for all of the ordinary thermodynamic degrees of freedom of the transition state structure, and q_{ξ} represents the one nonequilibrium vibrational degree of freedom of the bond along the reaction coordinate. Thus $\overline{q^{\ddagger}}$ is computed as an ordinary equilibrium partition function, but the 'final state' of this 'equilibrium' is the transition state. For example, for the translational partition function component $q_t^{\ddagger}$, use the sum of the masses of molecules A and B. For the rotational partition function $q_r^{\ddagger}$, use the moment of inertia given by the masses and bond separations of the transition state structure.

Because the bond along ξ is only partly formed in the transition state structure, we treat it as a weak vibration (see page 203). The discussion on page 201 shows that the frequency ν of a vibration depends on the spring constant k_{ξ} and the mass. For the frequency ν_{ξ} of the reacting bond,

$$\nu_{\xi} = \frac{1}{2\pi}\left(\frac{k_{\xi}}{\mu}\right)^{1/2}, \tag{19.24}$$

where μ is the reduced mass, $\mu = m_A m_B/(m_A + m_B)$. The vibration is weak, so k_ξ and ν_ξ are small, and

$$\frac{h\nu_\xi}{kT} \ll 1. \tag{19.25}$$

(In contrast, for stable covalent bonds, $h\nu/kT \gg 1$.) The partition function q_{vz} for vibrations is given by Equation (13.21). When a vibration is weak, you can approximate the exponential in the partition function by $e^{-x} \approx 1 - x$ for small x (see Appendix C, Equation (C.1)), so

$$q_\xi = \frac{1}{1 - e^{-h\nu_\xi/kT}} \approx \frac{kT}{h\nu_\xi}. \tag{19.26}$$

Once the system has reached the transition state, it is assumed to proceed to product as rapidly as the system permits, namely at the frequency of the reaction coordinate vibration. The rate coefficient for the downhill step from transition state to product is

$$k^\ddagger = \nu_\xi. \tag{19.27}$$

A factor κ, called the *transmission coefficient*, is often introduced ($k^\ddagger = \kappa\nu_\xi$) to account for observed deviations from the simple rate theory. When the reaction occurs in a condensed-phase medium, or in complex systems, $\kappa < 1$.

Now substitute Equations (19.22), (19.23), (19.26) and (19.27) into Equation (19.21) to get the rate coefficient k_2 from the partition functions,

$$k_2 = \left(\frac{kT}{h}\right)\left(\frac{\overline{q^\ddagger}}{q_A q_B}\right) e^{\Delta D^\ddagger/kT}$$

$$= \left(\frac{kT}{h}\right)\overline{K^\ddagger}, \tag{19.28}$$

where the overbars indicate that the unstable reaction coordinate degree of freedom has been factored out (into the term kT/h). Example 19.2 shows how you can predict chemical reaction rates if you know the transition state structure and its energy.

EXAMPLE 19.2 The rate of the reaction $F + H_2 \rightarrow HF + H$. Compute the rate of this reaction at $T = 300\,K$. The masses are $m_{H_2} = 2$ atomic mass units (amu), and $m_F = 19$ amu. For the transition state $F \cdots H - H$, the mass is $m = 21$ amu. Quantum mechanical modelling shows that the transition state has an F–H bond length of 1.602 Å, an H–H bond length of 0.756 Å, moment of inertia $I^\ddagger = 7.433$ amu Å², and an electronic degeneracy $g_e = 4$. There are two bending vibrations, each with frequency 397.9 cm^{-1}. The reactant H_2 has bond length 0.7417 Å, moment of inertia $I_{H_2} = 0.277$ amu Å², symmetry factor $\sigma = 2$, and electronic degeneracy $g_e = 1$. It is in its vibrational ground state. The F atom has no rotational or vibrational partition function and its electronic degeneracy is $g_e = 4$. The activation energy is $\Delta D^\ddagger = -657$ kJ mol^{-1}. Because the experimental rate data are in units of molar volume per second, you have to multiply Equation (19.28) by volume V and Avogadro's number $\mathcal{N}$ to put the reaction rate coefficient into those same units,

$$\text{rate coefficient} = k_2 V \mathcal{N} = \left(\frac{RT}{h}\right) \frac{q^{\ddagger}/V}{(q_F/V)(q_{H_2}/V)} e^{-\Delta D^{\ddagger}/kT}. \tag{19.29}$$

The contributions to the ratio of the partition functions are:

Electronic: $\dfrac{4}{4 \cdot 1} = 1,$

Rotational: $\dfrac{8\pi^2 I^{\ddagger} kT/h^2}{8\pi^2 I_{H_2} kT/\sigma_{H_2} h^2} = \dfrac{\sigma_{H_2} I^{\ddagger}}{I} = \dfrac{2(7.433)}{0.277} = 53.67,$ and

Each vibration:
$$\frac{h\nu}{kT} = \frac{(6.626 \times 10^{-34}\,\text{J s})(397.9\,\text{cm}^{-1})(2.99 \times 10^{10}\,\text{cm s}^{-1})}{(1.38 \times 10^{-23}\,\text{J K}^{-1})(300\,\text{K})}$$
$$= 1.904.$$

So the contribution of the two vibrational modes is

$$(1 - e^{-h\nu/kT})^{-2} = (1 - e^{-1.904})^{-2} = (1 - 0.149)^{-2} = 1.38.$$

Translations:

$$\left(\frac{m_F + m_{H_2}}{m_F m_{H_2}}\right)^{3/2} \left(\frac{2\pi kT}{h^2}\right)^{-3/2} = \left[\left(\frac{1 + m_{H_2}/m_F}{m_{H_2}}\right)\left(\frac{h^2}{2\pi kT}\right)\right]^{3/2}$$

$$= \left[\left(\frac{1 + 2/19}{(0.002\,\text{kg mol}^{-1})/(6.02 \times 10^{23}\,\text{mol}^{-1})}\right)\right.$$

$$\left. \times \left(\frac{(6.626 \times 10^{-34}\,\text{J s})^2}{2\pi(1.38 \times 10^{-23}\,\text{J K}^{-1})(300\,\text{K})}\right)\right]^{3/2}$$

$$= \left[(3.33 \times 10^{26})(1.688 \times 10^{-47})\right]^{3/2}$$

$$= 4.21 \times 10^{-31}\,\text{m}^3.$$

Substituting these terms into Equation (19.29) gives

$$\text{rate coefficient} = \left(\frac{(8.314\,\text{J K}^{-1})(300\,\text{K})}{6.626 \times 10^{-34}\,\text{J s}}\right)(53.67)(1.38)$$

$$\times (4.21 \times 10^{-31}\,\text{m}^3) e^{-6570/(8.314)T}$$

$$= 1.17 \times 10^8 \times e^{-790/T}\,\text{m}^3\,\text{s}^{-1}\,\text{mol}^{-1}$$

The experimental value is $2.0 \times 10^8 \times e^{-800/T}\,\text{m}^3\,\text{s}^{-1}\,\text{mol}^{-1}$ [2]. The 1.6-fold discrepancy at $T = 300\,\text{K}$ is probably due to the limitations of the transition state theory: the assumed separability of the reaction coordinate from other degrees of freedom, and the assumption that each trajectory crosses the transition state only once, $\kappa = 1$.

We have focused here on reactions involving multiple reactants, such as atoms A and B. Unimolecular reactions involving only a single reactant atom are somewhat different because there is no vibration or bond formation that defines the reaction coordinate. Such reactions are discussed in [6]. In the next section we show how the transition-state theory explains *isotope effects*, which are used experimentally for determining reaction mechanisms.

The Primary Kinetic Isotope Effect

Isotope substitution is useful for studying chemical reaction mechanisms. For our purposes, an isotope is an atom that is identical to another except for its mass. In general, chemical reaction rates do not change much if one isotope is substituted for another. However, if the isotope substitution occurs at a reacting position in the molecule, it can change the reaction kinetics. For example, at room temperature, a C–H bond cleaves about 8 times faster than a C–D bond, and about 60 times faster than a carbon-tritium bond. A ^{12}C–H bond cleaves about 1.5 times faster than a ^{14}C–H bond. Isotope effects are called primary when the isotope is substituted at a reacting position, and secondary when it is substituted elsewhere in the molecule.

Here is the basis for kinetic isotope effects. Compare the rate k_H for breaking a carbon–hydrogen bond,

$$CH \rightarrow (CH)^{\ddagger} \rightarrow C + H. \tag{19.30}$$

with the rate k_D for breaking a carbon–deuterium bond,

$$CD \rightarrow (CD)^{\ddagger} \rightarrow C + D. \tag{19.31}$$

From Equation (19.28), the ratio of the rates for breaking these two types of bond is

$$\frac{k_H}{k_D} = \frac{\left(\dfrac{q_{(CH)^{\ddagger}}}{q_{CH}}\right) e^{\Delta D_{CH}^{\ddagger}/kT}}{\left(\dfrac{q_{(CD)^{\ddagger}}}{q_{CD}}\right) e^{\Delta D_{CD}^{\ddagger}/kT}} \approx e^{(\Delta D_{CH}^{\ddagger} - \Delta D_{CD}^{\ddagger})/kT}, \tag{19.32}$$

where $q_{(CH)}$ and $q_{(CD)}$ are the partition functions for the respective molecules before bond breakage, and $q_{(CH)^{\ddagger}}$ and $q_{(CD)^{\ddagger}}$ are the partition functions for the respective transition states. Changing the mass mainly affects the ground-state vibrational energy of the bond, and thus the energy difference, $\Delta D_{CH}^{\ddagger} - \Delta D_{CD}^{\ddagger}$, giving the right-hand expression in Equation (19.32).

The transition state for this reaction is the point at which the bond breaks and the atoms dissociate, so its energy $D^{\ddagger}$ is assumed to be the same for both isotopes. Only the zero-point vibrational energy of the *reactant* bond is affected by the isotope substitution (see Figure 19.9), so

$$\Delta D_{CH}^{\ddagger} - \Delta D_{CD}^{\ddagger} = -\frac{h}{2}(\nu_{CD} - \nu_{CH}). \tag{19.33}$$

The vibrational frequencies ν_{CD} and ν_{CH} depend on the reduced masses μ_{CH} and μ_{CD},

$$\mu_{CH} = \frac{m_C m_H}{m_C + m_H} \approx m_H \quad \text{and}$$

$$\mu_{CD} = \frac{m_C m_D}{m_C + m_D} \approx m_D = 2m_H = 2\mu_{CH}, \tag{19.34}$$

Figure 19.9 Interpretation of primary kinetic isotope effects. The bond energy ε is shown as a function of R, the bond length. Most molecules are in their lowest vibrational energy state at ordinary temperatures. This state is lower by about 4.8 kJ mol^{-1} in a C–D bond than in a C–H bond. So the C–D bond requires about 4.8 kJ mol^{-1} more energy to reach the same transition state. Source: A Cornish-Bowden, *Fundamentals of Enzyme Kinetics*, Portland Press, London, 1995.

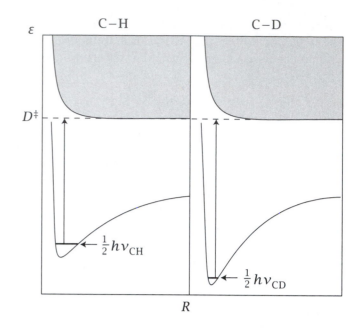

where m_C, m_D and m_H are the masses of carbon, deuterium, and hydrogen, respectively. If the spring constant is k_s, the vibrational frequencies are (see Equation (11.23)):

$$\nu_{CD} = \frac{1}{2\pi}\left(\frac{k_s}{\mu_{CD}}\right)^{1/2} = \frac{1}{2\pi}\left(\frac{k_s}{2\mu_{CH}}\right)^{1/2} = \frac{\nu_{CH}}{\sqrt{2}}$$

$$\Rightarrow \quad \nu_{CD} - \nu_{CH} = \left(\frac{1}{\sqrt{2}} - 1\right)\nu_{CH}. \tag{19.35}$$

Combine Equations (19.32), (19.33), and (19.35) to get

$$\frac{k_H}{k_D} = \exp\left[\frac{-h\nu_{CH}}{2kT}\left(\frac{1}{\sqrt{2}} - 1\right)\right]. \tag{19.36}$$

EXAMPLE 19.3 C–H bonds break about 8 times faster than C–D bonds. Compute the kinetic isotope rate effect at $T = 300$ K if the C–H stretch mode observed in the infrared is 2900 cm^{-1}. Use Equation (19.36):

$$\frac{k_H}{k_D} = \exp\left[\frac{(-6.626 \times 10^{-34}\,\text{J s})(2900\,\text{cm}^{-1})(3.0 \times 10^{10}\,\text{cm s}^{-1})}{2(1.38 \times 10^{-23}\,\text{J K}^{-1})(300\,\text{K})}\left(\frac{1}{\sqrt{2}} - 1\right)\right]$$

$$= 7.68$$

at $T = 300$ K. In general, bonds of lighter isotopes cleave more rapidly than those of heavier isotopes, because the lighter isotopes have higher zero-point energies.

Transition state theory gives a microscopic basis for the Arrhenius model. To see this, let's convert to a thermodynamic notation.

Migration Rate (s^{-1})

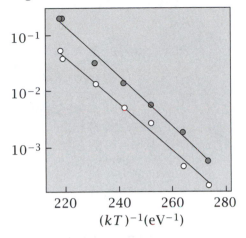

Figure 19.10 The migration rate of CO monomers (○) and dimers (●) on copper surfaces shows an Arrhenius temperature dependence. Source: BG Briner, MD Doering, HP Rust and AM Bradshaw, *Science* **278**, 257–260 (1997).

The Thermodynamics of the Activated State

We are treating $\overline{K^{\ddagger}}$ in Equation (19.28) as an equilibrium constant for the stable degrees of freedom (the overbar indicates that the unstable reaction coordinate degree of freedom ξ has been factored out). So you can use $\overline{K^{\ddagger}}$ as you would any other equilibrium constant, and express it in terms of thermodynamic quantities, which are called the **activation free energy** $\Delta G^{\ddagger}$, **activation enthalpy** $\Delta H^{\ddagger}$, and **activation entropy** $\Delta S^{\ddagger}$:

$$-kT \ln \overline{K^{\ddagger}} = \Delta G^{\ddagger} = \Delta H^{\ddagger} - T\Delta S^{\ddagger}. \tag{19.37}$$

Substituting Equation (19.37) into (19.28) gives

$$k_2 = \left(\frac{kT}{h}\right) e^{-\Delta G^{\ddagger}/kT} = \left(\frac{kT}{h}\right) e^{-\Delta H^{\ddagger}/kT} e^{\Delta S^{\ddagger}/k}. \tag{19.38}$$

Measurements of the temperature dependence of a chemical reaction rate can be fitted to Equation (19.38) to give the two parameters $\Delta H^{\ddagger}$ and $\Delta S^{\ddagger}$. The quantity $\Delta H^{\ddagger}$ is related to the activation energy E_a of Arrhenius. The factor $(kT/h) \exp(\Delta S^{\ddagger}/k)$ gives the front factor A in the Arrhenius expression. For any reaction that is activated (i.e., has a barrier, $\Delta G^{\ddagger} > 0$), Equation (19.38) says that the fastest possible rate, which is achieved as $\Delta G^{\ddagger} \to 0$, is $kT/h = (1.38 \times 10^{-23} \text{ J K}^{-1})(300 \text{ K})/(6.626 \times 10^{-34} \text{ J s}) = 6.24 \times 10^{12} \text{ s}^{-1}$ or about one conversion per 0.16 ps.

Figure 19.10 shows the rates for two processes: (1) the migration of individual carbon monoxide (CO) molecules on the surface of metallic copper, and (2) the migration of CO dimers on the same surface. Figure 19.11 shows a model of this activated kinetics as a landscape of periodic energy barriers that must be surmounted by the CO molecules as they move across the surface. $\Delta H^{\ddagger}$ can be determined from the slope and $\Delta S^{\ddagger}$ can also be found from Figure 19.10. Figure 19.10 shows that the activation enthalpy barrier is about the same for monomers and dimers. The dimers migrate faster for entropic reasons. The dimers gain orientational freedom at the point where they just begin to detach from the surface, at the transition state. The monomers don't, so the migration rate is greater for the dimers.

Figure 19.11 Atoms and molecules can hop over energy or enthalpy barriers to migrate laterally on solid surfaces. Figure 19.10 shows data that can be interpreted in this way.

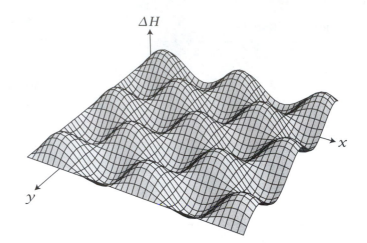

Figure 19.12 Free energy barrier $\Delta G^{\ddagger}$ is reduced by a catalyst. A and B are reactants, P is product, and C is the catalyst. $(AB)^{\ddagger}$ is the transition state for the uncatalyzed reaction, and $(ABC)^{\ddagger}$ is the transition state for the catalyzed reaction.

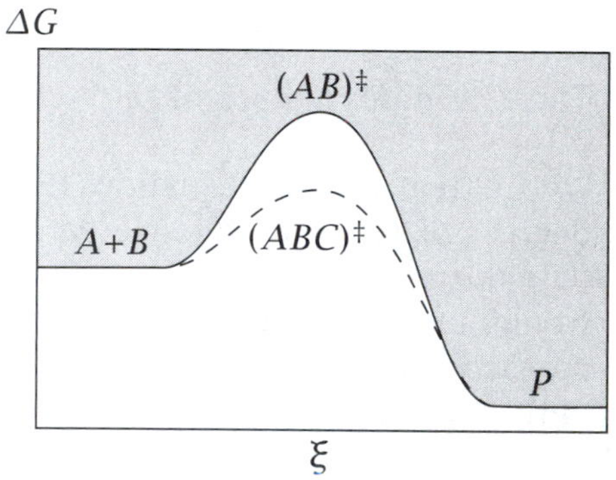

In the following section we look at the reduction of transition-state barriers by catalysts.

Catalysts Speed Up Chemical Reactions

Catalysts affect the rates of chemical reactions; for example, biochemical reactions are accelerated by enzyme catalysts. Enzymes can achieve remarkably large accelerations, by a factor of 2×10^{23} for orotidine 5′-phosphate decarboxylase, for example [7]. Petroleum is converted to gasoline and natural gas by using platinum–nickel catalysts. An important industrial method for manufacturing ammonia from hydrogen and nitrogen gases ($3H_2 + N_2 \rightarrow 2NH_3$) is the Haber process. If you put nitrogen and hydrogen gases together in a container, they won't react, even though ammonia is the more stable state. But add a little iron dust, and you'll get ammonia. Iron oxide or chromium oxide are catalysts that speed up the reaction. A catalyst affects only the rate of a reaction, and not the equilibrium ratio of products to reactants.

American chemist L Pauling proposed in 1946 that catalysts work by *stabilizing the transition state* [8, 9] (see Figure 19.12). Suppose that an uncatalyzed reaction has a forward rate coefficient k_0,

$$A + B \xrightarrow{k_0} P. \tag{19.39}$$

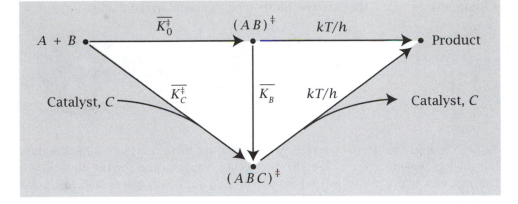

Figure 19.13 Across the top, the uncatalyzed reaction goes from reactant to transition state ( )‡, with equilibrium constant $\overline{K_0^{\ddagger}}$, then to product, with rate kT/h. The bottom arrows show the catalyzed reaction. Reactant and catalyst form a transition state with equilibrium constant $\overline{K_c^{\ddagger}}$, and on to product with rate kT/h. $\overline{K_B}$ is the 'binding constant' of the catalyst to the transition state.

k_0 is given by transition state theory Equation (19.28) as a product of a rate factor (kT/h) and an equilibrium factor $\overline{K_0^{\ddagger}}$,

$$k_0 = \left(\frac{kT}{h}\right)\overline{K_0^{\ddagger}} = \left(\frac{kT}{h}\right)\frac{[\overline{AB^{\ddagger}}]}{[A][B]}. \tag{19.40}$$

$\overline{K_0^{\ddagger}}$ is the equilibrium ratio of the concentration of the transition state $[\overline{AB^{\ddagger}}]$ to the product of the concentrations of the reactants (see Equation (19.19)). The overbar indicates that the unstable degree of freedom has been factored out into the term (kT/h). The corresponding process for the same reaction when a catalyst C is involved is shown by the bottom arrows on Figure 19.13. For the catalyzed reaction, the rate is $k_c = (kT/h)\overline{K_c^{\ddagger}}$, where

$$\overline{K_c^{\ddagger}} = \frac{[\overline{ABC^{\ddagger}}]}{[A][B][C]}. \tag{19.41}$$

The rate enhancement due to the catalyst is given by the ratio

$$\frac{k_c}{k_0} = \frac{(kT/h)\overline{K_c^{\ddagger}}}{(kT/h)\overline{K_0^{\ddagger}}} = \frac{\overline{K_c^{\ddagger}}}{\overline{K_0^{\ddagger}}} = \overline{K_B} = \frac{[(\overline{ABC})^{\ddagger}]}{[(\overline{AB})^{\ddagger}][C]}, \tag{19.42}$$

where $\overline{K_B}$ represents the 'binding constant' of the catalyst to the transition state (see Figure 19.13). Pauling's principle, expressed in Equation (19.42), is that the rate enhancement by the catalyst is proportional to the binding affinity of the catalyst for the transition state. This has two important implications. First, to accelerate a reaction, Pauling's principle says to design a catalyst that binds tightly to the transition state (and not to the reactants or products, for example). The second is that a catalyst that reduces the transition state free energy for the forward reaction is also a catalyst for the backward reaction (see Figure 19.12).

(a) Reactants Polarize, so Water Reorganizes   **(b)** Enzyme Binds Ions in Prepolarized Pocket

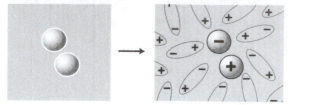
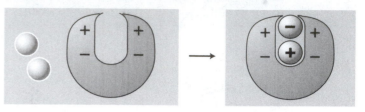

Figure 19.14 (a) Two neutral reactants become charged in the transition state. Creating this charge separation costs free energy because it orients the solvent dipoles. (b) Enzymes can reduce the activation barrier by having a site with pre-organized dipoles [10, 11].

Table 19.1 The relative reaction rates k_{rel} for two reactants; A is COO^- and B is $COOC_6H_4Br$, either as isolated molecules ($k_{rel} = 1$) or covalently linked together in various configurations in different molecules. Source: FC Lightstone and TC Bruice, *J Am Chem Soc* **118**, 2595–2605 (1996).

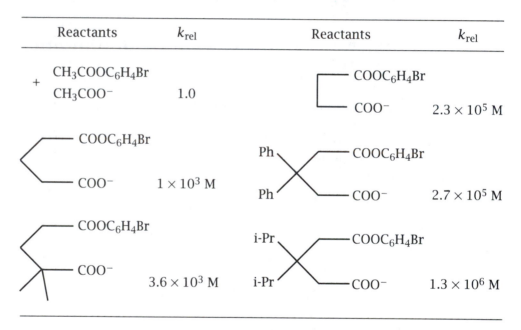

Reactants	k_{rel}	Reactants	k_{rel}
$CH_3COOC_6H_4Br$ + CH_3COO^-	1.0	$COOC_6H_4Br$ / COO^-	2.3×10^5 M
$COOC_6H_4Br$ / COO^-	1×10^3 M	Ph, Ph $COOC_6H_4Br$ / COO^-	2.7×10^5 M
$COOC_6H_4Br$ / COO^-	3.6×10^3 M	i-Pr, i-Pr $COOC_6H_4Br$ / COO^-	1.3×10^6 M

**Speeding Up Reactions
by Intramolecular Localization or Solvent Preorganization**

What are the physical mechanisms of catalysis? One possibility is that the slow step in the reaction of A with B is their diffusion together in the appropriate mutual orientation to react. In that case, an appropriate catalyst would be a molecule or surface that provides a common binding site for bringing reactants A and B together in an appropriate orientation. To test the importance of this kind of spatial and orientational localization, experiments have been performed in which reactants A and B are contained within a single molecule, sometimes in a particular orientation. The reaction rates of A with B are then

$Cl^- + CH_3Br \rightarrow ClCH_3 + Br^-$	
Solvent	Relative rate $\left(cm^3\, molecule^{-1}s^{-1} \right)$
None (gas phase)	1
MeCO	10^{-10}
Dimethylformamide	10^{-11}
CH_3OH	10^{-15}
H_2O	10^{-16}

Table 19.2 Reaction rates in various solvents (top, least polar; bottom, most polar). Source: WR Cannon and SJ Benkovic, *J Biol Chem* **273**, 26257–27260 (1998).

compared with reference reactions in which A and B are not covalently linked. Table 19.1 shows an example. Intramolecular localization can lead to large rate enhancements. Such localization can involve both entropic and enthalpic factors.

Another mechanism has been proposed for reducing activation barriers in enzyme-catalyzed reactions [10, 11, 12]. The enzyme may act as a *preorganized 'solvent'* cavity. Table 19.2 shows that some reactions can be extremely slow, particularly in polar solvents. In such cases, the reactants A and B may be neutral but the transition state may involve charge separation, $A^+ B^-$. A big energy cost in creating this charge separation involves orienting the surrounding polar solvent molecules. That is, water molecules near the transition state molecule re-orient so that the partial negative charges on their oxygen atoms will associate with A^+ and the partial positive charges on their hydrogen atoms will associate with B^-. Figure 19.14 shows the idea. A way to reduce the activation energy is to put the reactants into a cavity that is appropriately polarized in advance. An enzyme may be able to provide an environment of this type to stabilize the charge distribution on the transition state.

The Brønsted Law of Acid and Base Catalysis: The Stronger the Acid, the Faster the Reaction It Catalyzes

Some chemical reactions are catalyzed by acids or bases. For example, a reactant R converts to a product P, assisted by a coupling to an acid or base equilibrium, according to the reaction

$$
AH \xrightarrow{K_a} H^+ + A^-
$$

$$
R \xrightarrow{k_a} P.
$$

(19.43)

AH is an acid catalyst having an acid dissociation constant

$$
K_a = \frac{[H^+][A^-]}{[AH]}.
$$

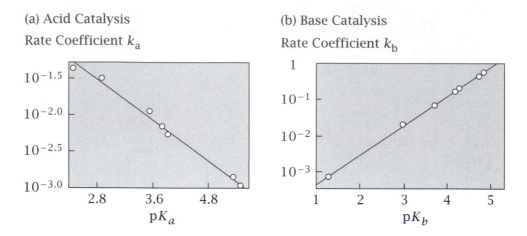

(a) Acid Catalysis
Rate Coefficient k_a

(b) Base Catalysis
Rate Coefficient k_b

Figure 19.15 Brønsted plots. (a) Acid catalyzed hydrolysis of ethyl vinyl ether in H_2O at 25 °C. (b) Base catalysis of the decomposition of nitramide in H_2O at 15 °C. Data are from AJ Kresge, HL Chen, Y Chiang, E Murrill, MA Payne, and DS Sagatys, *J Am Chem Soc* **93**, 413 (1971).

Under some conditions, the rate of disappearance of the reactant is

$$\frac{d[R]}{dt} = -k_a[AH][R],$$

where k_a is the rate coefficient for the reaction.

The **Brønsted law** is the observation that the stronger the acid (K_a large), the faster the reaction $R \rightarrow P$ that it catalyzes. That is,

$$\log k_a = \alpha \log K_a + c_a, \tag{19.44}$$

where c_a is a constant and $\alpha > 0$. Using the notation $pK_a = -\log K_a$, the Brønsted law is expressed as $\log k_a = -\alpha \, pK_a + c_a$. The acid dissociation equilibrium matters because the rate-limiting step in the reaction $R \rightarrow P$ is often the removal of a proton from the catalyst, AH. Base catalysis often follows a similar linear relationship, but with a slope of opposite sign: $\log k_b = \beta \, pK_b + c_b$, where K_b is the base dissociation constant, $\beta > 0$, and c_b is a constant. Examples of acid and base catalysis are given in Figure 19.15. Such observations imply a linear relationship between the activation energy E_a for the reaction and the free energy ΔG of acid dissociation:

$$E_a = a\Delta G + b, \tag{19.45}$$

where a and b can be positive or negative constants. Equation (19.45) is the **Brønsted relationship**. It is an example of a broader class of *linear free energy relationships* in which the logarithm of a rate coefficient is proportional to the logarithm of an equilibrium constant, as in Equation (19.44) (see Figure 19.16).

Equation (19.43) describes two coupled processes in which the equilibrium of one ($AH \rightarrow H^+ + A^-$) is coupled to the kinetics of the other ($R \rightarrow P$). In other cases, the coupling of equilibria and kinetics can occur within the same reaction, such as

$$R^- + AH \rightarrow RH + A^-, \tag{19.46}$$

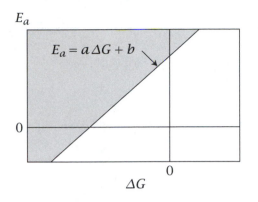

$$E_a = a\,\Delta G + b$$

Figure 19.16 According to the Brønsted relation, the activation energy E_a of a process is a linear function of ΔG, the stability of the product relative to the reactant. As ΔG becomes more negative, E_a is reduced (see Figure 19.17).

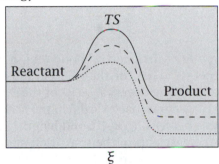

Figure 19.17 The Brønsted relationship shown schematically. The more stable is the product, the lower the barrier is, so the forward reaction is faster.

Increasing the strength of the acid in this case (increasing K_a) increases the rate of the reaction.

What is the basis for the Brønsted law? Figure 19.17 shows the reaction coordinate diagrams for a series of related reactions, indicating that the more stable the product, the faster the reaction. For acid catalysis, the figure implies that the more weakly bonded the proton is to the catalyst, the faster it comes off, and the faster is the catalyzed reaction. Here is a model.

A First Approximation: The Evans–Polanyi Model of the Brønsted Law and Other Linear Free-Energy Relationships

In 1936, MG Evans and M Polanyi developed a simple model to explain Equation (19.45) [13]. Consider the reaction

$$AB + C \rightarrow A + BC. \tag{19.47}$$

This reaction can be described in terms of two degrees of freedom: r_{AB}, the bond length between A and B, and r_{BC}, the bond length between B and C. To simplify, Evans and Polanyi considered the distance between A and C to be fixed, so r_{BC} diminishes as r_{AB} increases. Now the problem can be formulated in terms of a single reaction coordinate $r = r_{AB} = \text{constant} - r_{BC}$. This is usually a good approximation for the range of bond lengths in which one bond is being made while the other is being broken.

Figure 19.18 The Evans-Polanyi model for linear free-energy relationships. ΔG is the stability of the product relative to the reactant, E_a is the activation energy, r_1 and r_2 are the stable bond lengths of AB and BC, and $r^{\ddagger}$ is the A-B bond length in the activated state.

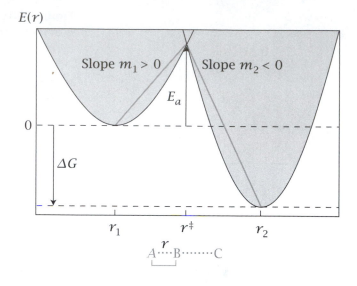

Figure 19.18 shows how the bond energy $E(r)$ depends on bond length r. The left parabola in the figure gives the energy $E_{AB}(r)$ of the A-B bond as a function of its length r. The right parabola gives the energy $E_{BC}(r)$ of the B-C bond. The right parabola as shown on this diagram is 'backwards' in the sense that increasing r (from $r^{\ddagger}$ to r_2) represents a decreasing B-C bond length.

As the A-B bond stretches, the energy increases up to the transition state, where the bond breaks, then the energy decreases as the B-C bond forms. Evans and Polanyi approximated the two energy functions between reactants, transition state, and products, by two straight lines that intersect at the transition state. For the AB molecule, the energy is

$$E_{AB}(r) = m_1(r - r_1), \tag{19.48}$$

where m_1 is the slope of the straight line from reactants to transition state and r_1 is the equilibrium length of the A-B bond. At the transition state, $E_a = m_1(r^{\ddagger} - r_1)$ so

$$r^{\ddagger} = \frac{E_a}{m_1} + r_1. \tag{19.49}$$

Similarly, for the BC molecule:

$$E_{BC}(r) = m_2(r - r^{\ddagger}) + E_a \quad \Longrightarrow \quad \Delta G = m_2(r_2 - r^{\ddagger}) + E_a, \tag{19.50}$$

where $\Delta G = G_{\text{products}} - G_{\text{reactants}}$ is the equilibrium difference in free energy (see Figure 19.18), and the slope of the straight line from transition state to product is negative $m_2 < 0$. To express the activation energy E_a in terms of the free energy difference ΔG, substitute Equation (19.49) into (19.50) to get

$$\Delta G = m_2 \left[r_2 - \left(\frac{E_a}{m_1} + r_1 \right) \right] + E_a. \tag{19.51}$$

Rearranging Equation (19.51) gives a linear relationship between E_a and ΔG:

$$E_a = \frac{m_1}{m_1 - m_2} \left[\Delta G - m_2(r_2 - r_1) \right]. \tag{19.52}$$

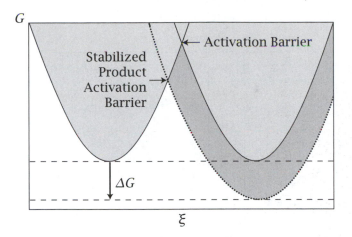

Figure 19.19 When the product is stabilized, the right-hand parabola shifts from (——) to (······). The activation barrier is also lowered and shifted to the left along the reaction coordinate ξ.

This simple model shows how linear free energy relationships can arise from reactant and product energy surfaces with a transition state between them that is defined by the *curve-crossings*. Figure 19.19 shows how shifting the equilibrium to stabilize the products can speed up the reaction. It also illustrates that such stabilization can shift the transition state to the left along the reaction coordinate, to 'earlier' in the reaction. If $|m_1| \gg |m_2|$, the transition state will be closer to the reactants than to the products, and if $|m_1| \ll |m_2|$, the reverse will hold. The Evans–Polanyi model rationalizes why stabilities should correlate linearly with rates.

Activation energies are not always linearly proportional to product stabilities. Some experiments show an 'inverted region,' where further increasing the product stability can *slow down* the reaction. This was explained by R Marcus, who was awarded the 1992 Nobel Prize in Chemistry. Marcus theory replaces Equation (19.48) with a parabola, $E_{AB}(r) = m_1(r - r_1)^2 + b$, and a similar parabola replaces Equation (19.50). The result is a square-law dependence $E_a \sim (\text{constant} + \Delta G)^2$ between activation energy and product stability instead of Equation (19.52) (see [13]). Figure 19.20 indicates how this leads to the inverted behavior.

Funnel Landscapes Describe Diffusion and Polymer Folding Processes

All the processes described in this chapter so far involve well-defined reactants and products, and a well-defined reaction coordinate. But diffusional processes and polymer conformational changes often cannot be described in this way. Consider a solution of mobile molecules diffusing toward a sphere, or an ensemble of polymer conformations put into folding conditions (conditions that strongly favor some compact polymer conformations). Now the starting state is not a single point on an energy landscape. It is a broad distribution—the many different spatial locations of the starting positions of the mobile molecules, or the many different open conformations of the polymer chain.

Figure 19.21 shows an example of such an energy landscape. The observed kinetics may often still be describable in terms of an activation barrier. But in

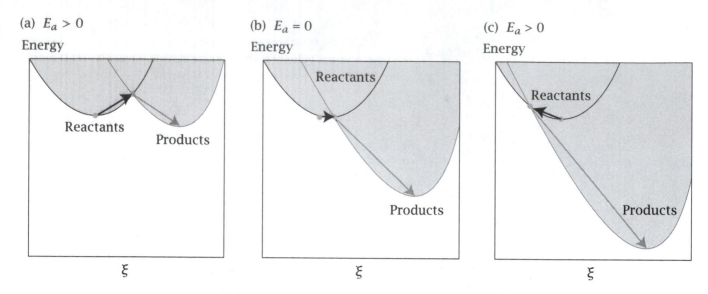

(a) $E_a > 0$

Energy

Reactants

Products

ξ

(b) $E_a = 0$

Energy

Reactants

Products

ξ

(c) $E_a > 0$

Energy

Reactants

Products

ξ

Figure 19.20 Increasing stability of products, from (a) to (c), leads first to a reduction of the activation barrier to $E_a = 0$ in (b), then to an increase. Reactant and product curves against the reaction coordinate ξ, as in Figure 19.19.

Figure 19.21 A bumpy energy landscape, such as occurs in some diffusion processes, polymer conformational changes, and biomolecule folding. A single minimum in the center may represent 'product,' but there can be many different 'reactants,' such as the many open configurations of a denatured protein. KA Dill and HS Chan, *Nat Struc Biol* **4**, 10–19 (1997).

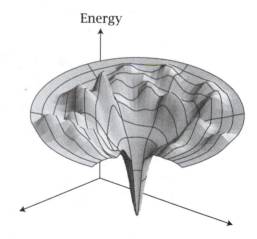

Energy

these cases, there may be no single structure or single mountain-pass enthalpy barrier that defines the kinetic bottleneck. Instead, the activation barrier may be entropic. The rate limit is just the diffusional meanderings from the different starting points as each molecule progresses toward the stable state.

Summary

Chemical reactions and diffusion processes usually speed up with temperature. This can be explained in terms of a transition state or activation barrier and an equilibrium between reactants and a transient and unstable species, called the transition state. For chemical reactions, the transition state involves an unstable weak vibration along the reaction coordinate, and an equilibrium between all other degrees of freedom. You can predict kinetics by using microscopic structures and the equilibrium partition functions of Chapter 13 if the transition state structure is known. Catalysts act by binding to the transition state structure. They can speed reactions by forcing the reactants into transition-state-like configurations.

Problems

1. Isotope substitution experiments can detect hydrogen tunnelling. Isotope substitution experiments can sometimes be used to determine whether hydrogens are cleared from molecules through mechanisms that involve tunnelling. To test this, two isotopes are substituted for hydrogen: (1) deuterium (D, mass = 2) is substituted and the ratio of rate coefficients k_H/k_D is measured, and (2) tritium (T, mass = 3) is substituted and k_H/k_T is measured.

(a) Using the isotope substitution model in this chapter, show that

$$\frac{k_H}{k_T} = \left(\frac{k_H}{k_D}\right)^\alpha.$$

(b) compute the numerical value of α.

2. Relating stability to activation barriers. Using the Evans–Polanyi model, with $r_1 = 5$, $r_2 = 15$, $m_1 = 1$, and $m_2 = -2$:

(a) Compute the activation barriers E_a for three systems having product stabilities $\Delta G = -2$ kcal mol^{-1}, -5 kcal mol^{-1}, and -8 kcal mol^{-1}.

(b) Plot the three points E_a versus ΔG, to show how the activation barrier is related to stability.

3. Reduced masses. Equations (19.40) and (19.41) give the reduced masses for C–H and C–D bonds as $\mu_{CH} \approx m_H$ and $\mu_{CD} \approx 2m_H$. The approximation is based on assuming the mass of carbon is much greater than of H or D. Give the more correct values of these reduced masses if you don't make this assumption.

4. Classical collision theory. According to the kinetic theory of gases, the reaction rate k_2 of a sphere of radius r_A with another sphere of radius r_B is

$$k_2 = \pi R^2 \left(\frac{8kT}{\pi \mu_{AB}}\right)^{1/2} e^{-\Delta\epsilon_0^\ddagger/kT},$$

where $R = r_A + r_B$ is the distance of closest approach, μ_{AB} is the reduced mass of the two spheres, and $\Delta\epsilon_0^\ddagger$ is the activation energy. Derive this from transition state theory.

5. The pressure dependence of rate constants.

(a) Show that the pressure dependence of the rate constant k for the reaction

$$A \xrightarrow{k_f} B$$

is proportional to an *activation volume* $v^\ddagger$,

$$\left(\frac{\partial \ln k_f}{\partial p}\right) = -\frac{(v^\ddagger - v_A)}{kT}.$$

(b) Show that the expression in (a) above is consistent with Equation (13.48), $K = k_f/k_r$, where k_r is the rate of the reverse reaction.

6. Relating the Arrhenius and activated-state parameters. Derive the relationship of the activation parameter $\Delta H^\ddagger$ in Equation (19.38) to the Arrhenius activation energy E_a in Equation (19.16) for a gas-phase reaction.

7. Enzymes accelerate chemical reactions. Figure 19.22 shows an Arrhenius plot for the uncatalyzed reaction of 1-methylorotic acid (OMP) [7].

(a) Estimate $\Delta H^\ddagger$ from the figure.

(b) Estimate $\Delta S^\ddagger$ at $T = 300$ K.

(c) At $T = 25\,°C$, the enzyme OMP decarboxylase accelerates this reaction 1.4×10^{17}-fold. How fast is the catalyzed reaction at $25\,°C$?

(d) What is the binding constant of the enzyme to the transition state of the reaction at $T = 300$ K?

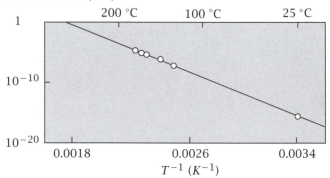

Figure 19.22 Source: A Radzicka and R Wolfenden, *Science* **267**, 90–93 (1995).

8. Rate increase with temperature. A rule-of-thumb used to be that chemical reaction rates would roughly double for a ten-degree increase in temperature, say from $T_1 = 300$ K to $T_2 = 310$ K. For what activation energy E_a would this be exactly correct?

9. Catalytic rate enhancement. The reaction rate of an uncatalyzed reaction is $k_0 = 10^3$ M^{-1} s^{-1} at $T = 300$ K. A catalyst C binds to the transition state with free energy $\Delta G = -5$ kcal mol^{-1}. What is the rate k_0 of the catalyzed reaction at $T = 300$ K?

10. Negative Activation Energies. Activation energy barriers often indicate a process that is rate-limited by the need to break a critical bond, or achieve some particular strained state. In contrast, some processes involve *negative* activation energies—their rates decrease with increasing temperature (see Figures 19.23, 19.24). Table 19.23 shows some Arrhenius parameters for the formation of duplex DNA from single strands of various nucleotide sequences. Figure 19.24 shows an example of the kind of data from which the parameters in the table were derived. Which molecules have negative activation energies?

Table 19.3 Relaxation kinetics of oligonucleotides (21° to 23 °C). Source: CR Cantor and PR Schimmel, Biophysical Chemistry, Volume 3, WH Freeman, San Francisco, 1980.

Ordered form	k_1 $(M^{-1}s^{-1})$	E_a (kcal mol^{-1})
$A_9 \cdot U_9$	5.3×10^5	-8
$A_{10} \cdot U_{10}$	6.2×10^5	-14
$A_{11} \cdot U_{11}$	5.0×10^5	-12
$A_{14} \cdot U_{14}$	7.2×10^5	-17.5
$A_4U_4 \cdot A_4U_4$	1.0×10^6	-6
$A_5U_5 \cdot A_5U_5$	1.8×10^6	-4
$A_6U_6 \cdot A_6U_6$	1.5×10^6	-3
$A_7U_7 \cdot A_7U_7$	8.0×10^5	$+5$
$A_2GCU_2 \cdot A_2GCU_2$	1.6×10^6	$+3$
$A_3GCU_3 \cdot A_3GCU_3$	7.5×10^5	$+7$
$A_4GCU_4 \cdot A_4GCU_4$	1.3×10^5	$+8$

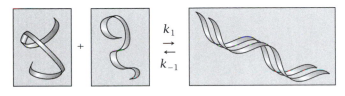

Figure 19.23

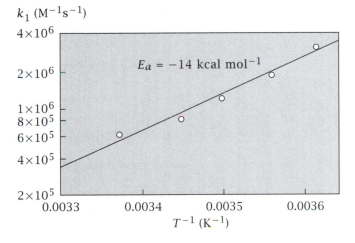

Figure 19.24 Source: CR Cantor and PR Schimmel, Biophysical Chemistry, Volume 3, WH Freeman, San Francisco, 1980.

References

[1] WC Gardiner Jr. *Rates and Mechanisms of Chemical Reactions*. WA Benjamin, New York, 1969.

[2] JI Steinfeld, JS Francisco and WL Hase. *Chemical Kinetics and Dynamics*. Prentice-Hall, Englewood Cliffs, 1989.

[3] RC Tolman. *The Principles of Statistical Mechanics*. Dover Publications, New York, 1979.

[4] SW Benson and O Dobis. *J Phys Chem A* **102**, 5175–5181 (1998).

[5] DC Clary. *Annu Rev Phys Chem* **41**, 61-90 (1990).

[6] T Baer and WL Hase. *Unimolecular Reaction Dynamics: Theory and Experiments*. Oxford University Press, New York, 1996.

[7] A Radzicka and R Wolfenden. *Science* **267**, 90-93 (1995).

[8] J Kraut. *Science* **242**, 533-540 (1988).

[9] WR Cannon, SF Singleton and SJ Benkovic. *Nat Struc Biol*, **3**, 821–833 (1996).

[10] A Warshel. *J Biol Chem* **273**, 27035-27038 (1998).

[11] A Warshel. *Computer Modeling of Chemical Reactions in Enzymes and Solutions*. Wiley, New York, 1991.

[12] WR Cannon and SJ Benkovic. *J Biol Chem* **273**, 26257-26260 (1998).

[13] RI Masel. *Principles of Adsorption and Reaction on Solid Surfaces*. Wiley, New York, 1996.

Suggested Reading

Excellent texts on isotope effects, the Brønsted law, and other aspects of physical organic chemistry:

ML Bender, RJ Bergeron, M Komiyama, *The Bio-organic Chemistry of Enzymatic Catalysis*, Wiley, New York, 1984.

H Eyring and EM Eyring, *Modern Chemical Kinetics*, Reinhold Publishing Corp, New York, 1963.

JE Leffler and E Grunwald, *Rates and Equilibria of Organic Reactions*, Dover Publications, New York, 1963.

H Maskill, *The Physical Basis of Organic Chemistry*, Oxford University Press, New York, 1990.

A Pross, *Theoretical and Physical Principles of Organic Reactivity*, Wiley, New York, 1995.

20 Coulomb's Law of Electrostatic Forces

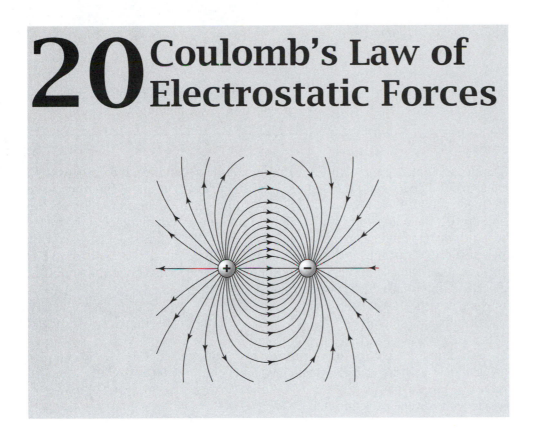

Charge Interactions Are Described by Coulomb's Law

Electrical interactions between charges govern much of physics, chemistry, and biology. They are the basis for chemical bonding, weak and strong. Salts dissolve in water to form the solutions of charged ions that cover three-quarters of the Earth's surface. Salt water forms the working fluid of living cells. pH and salts regulate the associations of proteins, DNA, cells, and colloids, and the conformations of biopolymers. Nervous systems would not function without ion fluxes. Electrostatic interactions are also important in batteries, corrosion, and electroplating.

Charge interactions obey Coulomb's law. When more than two charged particles interact, the energies are sums of coulombic interactions. To calculate such sums, we introduce the concepts of the *electric field* and Gauss's law, and, in the next chapter, the *electrostatic potential*. With these tools you can determine the electrostatic force exerted by one charged object on another, as when an ion interacts with a protein, DNA molecule, or membrane, or when a charged polymer changes conformation.

Coulomb's law was discovered in careful experiments by H Cavendish (1731–1810), J Priestley (1733-1804), and CA Coulomb (1736-1806) on macroscopic objects such as magnets, glass rods, charged spheres, and silk cloths. But Coulomb's law applies to a wide range of size scales, including atoms, molecules and biological cells. It states that the interaction energy $u(r)$ between two charges in a vacuum is

$$u(r) = C\frac{q_1 q_2}{r},$$ (20.1)

where q_1 and q_2 are the magnitudes of the two charges, r is the distance separating them, and $C = 1/4\pi\varepsilon_0$ is a proportionality constant (see box below).

Units

The proportionality constant C in Equation (20.1) depends on the units used to measure the charge and the distance. In the older cgs system, the units were defined so that $C = 1$. We use the SI system in which the unit of charge is the coulomb C, the unit of energy is the joule J, and the unit of distance is the meter m. The corresponding constant C equals $(4\pi\varepsilon_0)^{-1}$. The factor ε_0 is called the permittivity of vacuum. In SI units $\varepsilon_0 = 8.85 \times 10^{-12}$ farad m^{-1}. The farad F is the SI unit of capacitance, equal to 1 coulomb volt^{-1}. The volt is the SI unit of electrical potential, which is equal to one J C^{-1}. So $\varepsilon_0 = 8.85 \times 10^{-12}$ C^2 (J m)$^{-1}$. In cgs units, the unit of charge is 1 statcoulomb or 1 esu $= 3.00 \times 10^9$ C, the unit of potential is 1 statvolt $= (1/300)$ V, and the unit of capacitance is 1 statfarad $= 9 \times 10^{11}$ F.

In SI units, the charge on a proton is e $= 1.60 \times 10^{-19}$ C or $\mathcal{N}$e $= 9.647 \times 10^4$ C mol^{-1} and the charge on an electron is $-$e $= -1.60 \times 10^{-19}$ C. A useful quantity for calculations is $Ce^2\mathcal{N} = e^2\mathcal{N}/4\pi\varepsilon_0 = 1.386 \times 10^{-4}$ J m mol^{-1}, where $\mathcal{N}$ is Avogadro's number.

EXAMPLE 20.1 The coulombic attraction between Na$^+$ and Cl$^-$. Compute the pair interaction $u(r)$ between a sodium ion and a chloride ion at $r = 2.8$ Å, the bond length of NaCl, in a vacuum. Equation (20.1) gives

$$u(r) = \frac{-Ce^2\mathcal{N}}{r} = \frac{-1.39 \times 10^{-4}\,\text{J m mol}^{-1}}{(2.8 \times 10^{-10}\,\text{m})}$$

$$= -496\,\text{kJ mol}^{-1} \approx -119\,\text{kcal mol}^{-1}.$$

Charge Interactions Are Long-Ranged

In previous chapters we considered only the interactions between uncharged atoms. Uncharged atoms interact with each other through *short-ranged* interactions. That is, attractions between uncharged atoms are felt only when the two atoms are so close to each other that they nearly contact. Short-ranged interactions typically diminish with distance as r^{-6}. We will explore this in more detail in Chapter 24. For now, it suffices to note that charge interactions are very different: they are long-ranged. Coulombic interactions diminish with distance as r^{-1}. Counting lattice contacts is no longer sufficient.

The difference between short-ranged and long-ranged interactions is much more profound than it may seem. The mathematics and physics of long-ranged and short-ranged interactions are very different. A particle that interacts through short-ranged interactions feels only its nearest neighbors, so

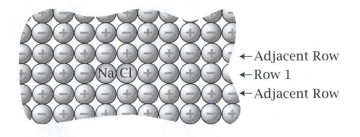

Figure 20.1 Crystalline sodium chloride packs in a cubic array of alternating sodium and chloride ions.

system energies can be computed by counting the nearest-neighbor contacts, as we did in previous chapters. But when interactions involve charges, more distant neighbors contribute to the energies, so our methods of summing energies must become more sophisticated.

Example 20.2 shows how long-ranged electrostatic interactions hold a salt crystal together.

EXAMPLE 20.2 Why are sodium chloride crystals stable? Let's assume that a crystal of sodium chloride is composed of hard spherical ions that interact solely through electrostatics. To compute the total energy, we sum all the pairwise coulombic interaction energies of one ion with every other ion in the crystal. Crystalline NaCl is arranged in a simple cubic lattice, with interionic spacings $a = 2.81$ Å at 25 °C.

First sum the interactions of a sodium ion in row 1 with all the other ions in the same row, shown in Figure 20.1. Because there are two negative ions at distance a with unit charge $-e$, two positive ions at distance $2a$ with unit charge e, and two more negative ions at distance $3a$, etc., the total coulombic interaction along the row $u_{\text{row 1}}$ is given by the series

$$u_{\text{row 1}} = \frac{\mathcal{C}e^2}{a}\left(-\frac{2}{1} + \frac{2}{2} - \frac{2}{3} + \frac{2}{4} - \cdots\right)$$

$$= \frac{2\mathcal{C}e^2}{a}\left(-1 + \frac{1}{2} - \frac{1}{3} + \frac{1}{4} - \cdots\right). \tag{20.2}$$

This series is slow to converge, implying that two charges can have substantial interaction even when they are far apart. The series converges to $-\ln 2$ per row (see Appendix C, Equation (C.4) with $a = x = 1$), so the energy of interaction of one sodium ion with all other ions in the same row is

$$u_{\text{row 1}} = -\frac{2\mathcal{C}e^2}{a}\ln 2 = -\frac{1.386\mathcal{C}e^2}{a}. \tag{20.3}$$

Next, sum the interactions of the same sodium ion with all the ions in the four adjacent rows above, below, in front of, and behind row 1. In these four rows, the closest ions are the four negative charges at distance a, the eight positive charges at distance $a\sqrt{2}$, and the eight negative charges at distance $a\sqrt{5}$. The pattern of pairwise distances gives a jth term involving eight charges at distance $a(1 + j^2)^{1/2}$. These four adjacent rows contribute energies $u_{\text{four adjacent rows}}$,

$$u_{\text{four adjacent rows}} = \frac{4\mathcal{C}e^2}{a}\left(-1 + \frac{2}{\sqrt{2}} - \frac{2}{\sqrt{5}} + \frac{2}{\sqrt{10}} - \cdots\right). \tag{20.4}$$

When the pairwise energies are summed over all the rows of charges, the total coulombic energy of the interaction between one ion and all the others in the crystal is found to be [1]:

$$U = -1.747 \frac{\mathcal{C}e^2 \mathcal{N}}{a}$$

$$= \frac{(-1.747)(1.386 \times 10^{-4} \, \mathrm{J\,m\,mol^{-1}})}{2.81 \times 10^{-10} \, \mathrm{m}}$$

$$= -862 \, \mathrm{kJ\,mol^{-1}} = -206 \, \mathrm{kcal\,mol^{-1}}. \qquad (20.5)$$

The energy holding each positive charge in the crystal is $U/2$, where the factor of 2 corrects for the double-counting of interactions. To remove a NaCl molecule, both a Na^+ ion and a Cl^- ion, the energy is $U = 206 \, \mathrm{kcal\,mol^{-1}}$, according to this calculation.

The experimentally determined energy of vaporization (complete ionization) of crystalline NaCl is $183 \, \mathrm{kcal\,mol^{-1}}$. The $23 \, \mathrm{kcal\,mol^{-1}}$ discrepancy between the value we calculated and the experimental value is due mainly to the assumption that the ions are hard incompressible particles constrained to a separation of exactly $a = 2.81$ Å. In reality, ions are somewhat compressible. Nevertheless, we can conclude that the coulombic interactions are the dominant attractions that hold ionic crystals together.

The quantity -1.747 in Equation (20.5) is called the *Madelung* constant. The magnitude of the Madelung constant depends on the symmetry of the crystal lattice.

Charges that interact with each other over long spatial distances contribute substantially to the energy of an electrostatic system. To see this, suppose that we had summed only the contributions of the six nearest neighbors, as we have done for systems dominated by short-range interactions. All six nearest neighbors are negative charges, so our estimate for the Madelung constant would have been -3 (correcting for double counting), rather than -1.747, leading to a huge error, about $148 \, \mathrm{kcal\,mol^{-1}}$!

Electrostatic interactions are strong as well as long-ranged. The electrostatic repulsion is about 10^{36} times stronger than the gravitational attraction between two protons. Strong, short-ranged nuclear forces are required to hold the protons together against their electrostatic repulsions in atomic nuclei. When charge builds up on a macroscopic scale, violent events may result, such as lightning and thunder storms, and explosions in oil tankers. Because charge interactions are so strong, matter does not sustain charge separations on macroscopic scales: macroscopic matter is neutral. The transfer of less than a thousand charges is enough to cause the static electricity that you observe when you rub your shoes on a carpet. In solution, too, imbalances are usually less than thousands of charges, but because this is one part in 10^{21}, it means that you can assume that solutions have charge neutrality—no net charge of one sign—to a very high degree of approximation. Although macroscopic charge separations aren't common in ordinary solutions, charge separations on microscopic scales are important and will be described in Chapters 22 and 23.

Charge Interactions Are Weaker in Media

Charges interact more weakly when they are in liquids than when they are in a vacuum. To varying degrees, liquids can be *polarized*. This means that if two fixed charges, say a negative charge A and a positive charge B, are separated by a distance r in a liquid, then the intervening molecules of the liquid tend to reorient their charges. The positive charges in the medium orient toward A, and the negative charges toward B, to shield and weaken the interaction between A and B (see Figure 20.2). This weakening of the coulombic interactions between A and B is described by the *dielectric constant* of the medium. Media that can be readily polarized shield charge interactions strongly: they have high dielectric constants.

In this and the next three chapters, we model the medium as a *polarizable isotropic continuum*. In treating a medium as a continuum, we neglect its atomic structure and focus on its larger-scale properties. In treating a medium as *isotropic*, we assume that its polarizability is the same in all directions. By treating a medium as *polarizable*, we assume that the charge redistributes in response to an electric field, even if the medium is neutral overall.

The more polarizable the medium, the greater is the reduction of an electrostatic field across it. The reduction factor is the *dielectric constant D*, a scalar dimensionless quantity. When charges q_1 and q_2 are separated by a distance r in a medium of dielectric constant D, their coulombic interaction energy is

$$u(r) = \frac{\mathcal{C}q_1 q_2}{Dr}. \tag{20.6}$$

For air at 0 °C, $D = 1.00059$. Table 20.1 gives the dielectric constants of some liquids. Some polar liquids have dielectric constants larger than $D = 100$, implying that an electrostatic interaction in such a liquid has less than 1% of its strength in a vacuum.

Coulombic interactions are traditionally described as an energy $u(r)$, but when dielectric constants are involved, $u(r)$ is a free energy, because the dielectric constant is temperature dependent. Figure 20.2 illustrates that polarization often involves orientations, and therefore entropies.

The polarizability of a medium arises from several factors. First, if the medium is composed of molecules that have a *permanent dipole moment*, a positive charge at one end and a negative charge at the other, then applying an electric field tends to orient the dipoles of the medium in a direction that opposes the field.

Second, the polarizability of a medium can arise from the polarizabilities of the atoms or molecules composing it, even if they have no permanent dipole moment. Atoms or molecules that lack permanent dipoles have electronic *polarizability*, a tendency of nuclear or electronic charge distributions to shift slightly within the atom, in response to an electric field (see Chapter 24). The electronic polarizabilities of hydrocarbons and other nonpolar substances are the main contributors to their dielectric constants ($D \approx 2$).

Third, polarizabilities can also arise from networks of hydrogen bonds in liquids such as water. If there are many alternative hydrogen-bond donors and acceptors, a shift in the pattern of hydrogen bonding is not energetically costly,

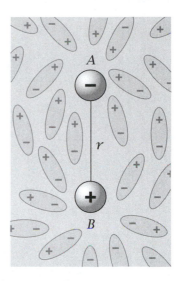

Figure 20.2 Fixed charges A and B cause dipoles in the surrounding liquid to orient partly, which causes some *shielding*, a reduction of the interaction between A and B. The dielectric constant D describes the degree of shielding.

Table 20.1 The dielectric constants D of several organic liquids. Source: RC Weast, editor, *Chemical Rubber Company Handbook of Chemistry and Physics*, 53rd edition, CRC, Cleveland, 1972.

Liquid	T (°C)	D
Heptane	0	1.958
Heptane	30	1.916
Methanol	25	33
Formamide	20	109
Formic acid	16	58
Nitrobenzene	25	35
HCN	0	158
HCN	20	114
Glycol	25	37
Water	0	88.00
Water	25	78.54

so the networks of hydrogen bonds in hydrogen-bonded liquids are labile when electric fields are applied.

The Bjerrum Length

To show the effect of the polarizability of the medium, let's first compute a useful quantity called the *Bjerrum length* ℓ_B, defined as the charge separation at which the coulomb energy between ions $u(r)$ just equals the thermal energy RT (see Figure 20.3). Substitute ℓ_B for r and RT for u in $u(r) = \mathcal{C}q_1q_2/Dr$ (Equation (20.6)), and solve for ℓ_B for single charges $q_1 = q_2 = $ e:

$$\ell_B = \frac{\mathcal{C}e^2\mathcal{N}}{DRT}. \tag{20.7}$$

For a temperature $T = 298.15$, in a vacuum or air ($D = 1$),

$$\ell_B = \frac{1.386 \times 10^{-4}\,\mathrm{J\,m\,mol^{-1}}}{(8.314\,\mathrm{J\,K^{-1}\,mol^{-1}} \times 298.15\,\mathrm{K})} = 560\,\text{Å}.$$

560 Å is a relatively long distance, many times the diameter of an atom. When two charges are much closer together than this distance, they feel a strong repulsive or attractive interaction, much larger than the thermal energy RT. But when two charges are much further apart than this distance, their interactions are weaker than RT, and they are more strongly directed by Brownian motion than by their coulombic interactions.

Dividing $u(r) = \mathcal{C}q_1q_2/Dr$ by ($RT = \mathcal{C}q_1q_2/D\ell_B$) gives

$$\frac{u}{RT} = \frac{\ell_B}{r}. \tag{20.8}$$

Example 20.3 shows an effect of the polarizability of a medium on charge interactions. Although Na^+ is strongly attracted to Cl^- in a vacuum (119 kcal mol^{-1}; Example 20.1), the attraction is much weaker in water, where $D = 78.54$ at 25 °C.

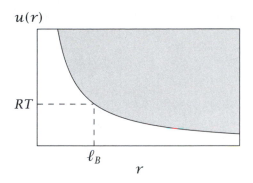

$u(r)$

RT

ℓ_B

r

Figure 20.3 The Bjerrum length ℓ_B is defined as the distance r at which two unit charges interact with an energy $u(r)$ equal to RT.

EXAMPLE 20.3 Why do salts ionize in water?   In a vacuum or in oil, the bonds between oppositely charged ions are strong.

But NaCl readily ionizes in water, that is, it falls apart into individual ions, indicating that its coulombic bonds are weak in water. In water at 25 °C, $D = 78.54$ so the Bjerrum length is $\ell_B = 7.13$ Å (substitute $D = 78.54$ into Equation (20.7)). At a distance of 2.81 Å, Equation (20.8) shows that the ionic interaction in water is $u(r, D = 78.54) = -(7.13/2.81) \times RT$ per mole, or -1.5 kcal mol^{-1} at $T = 300$ K, only a little more than twice RT.

Because ionic bonds are so weak in water, salts readily dissociate into ions, even though their interactions in air or nonpolar media are as strong as covalent bonds.

So far, our calculations for ionic interactions have involved simple arrangements of charges. When several charges are involved, or when charges have complex spatial arrangements, more sophisticated methods are needed to treat them. We now develop some useful tools that lead to Gauss's law and Poisson's equation.

Electrostatic Forces Add Like Vectors

Vectors and Coulomb's Law

Coulomb's law may be expressed either in terms of energy $u(r) = Cq_1q_2/Dr$, or force $f = -(\partial u/\partial r) = Cq_1q_2/Dr^2$. The difference is that energies are scalar quantities, which add simply, while forces are vectors that add in component form. Either way, an important fundamental law is the superposition principle: electrostatic energies and forces are both additive.

The total electrostatic force on a particle is the vector sum of the electrostatic forces from all other particles. You can express the force on particle B from particle A in Figure 20.4 as

$$\mathbf{f} = \frac{Cq_Aq_B}{Dr^2} \cdot \frac{\mathbf{r}}{r}, \tag{20.9}$$

where $\mathbf{r}/r$ indicates a vector pointing in the direction of vector $\mathbf{r}$, along the line connecting the particles, but with unit length. $\mathbf{f}$ is the force acting in the same direction. The component of the force $\mathbf{f}$ in the x-direction is described by using

Figure 20.4 The electrostatic force **f** of fixed charge A on test charge B is a vector with components f_x and f_y along the x- and y-axes, respectively.

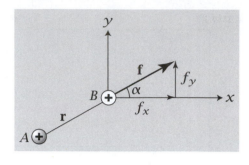

Figure 20.5 The force on the test charge B is the vector sum from the two fixed charges A and C.

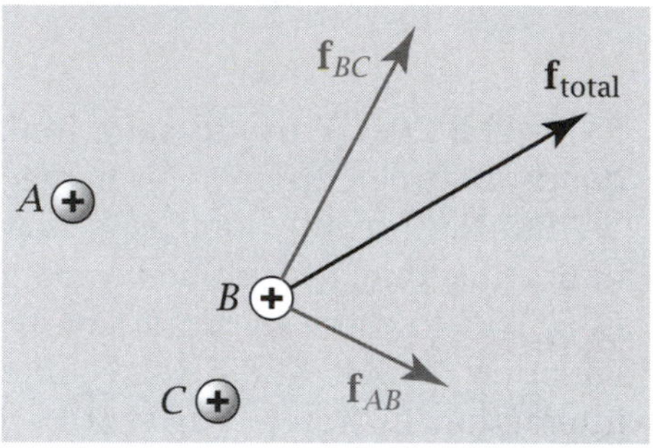

the component of the vector **r** between the charges, $r_x = r \cos \alpha$, where α is the angle between **f** and the x-axis. The x-direction component of the force is

$$f_x = \left(\frac{q_A q_B}{4\pi\varepsilon_0 D r^2} \right) \cos \alpha. \tag{20.10}$$

The y-direction component is

$$f_y = \left(\frac{q_A q_B}{4\pi\varepsilon_0 D r^2} \right) \sin \alpha. \tag{20.11}$$

Figure 20.5 shows how electrostatic forces add vectorially.

What Is an Electrostatic Field?

Now we want a general way to describe the force that would act on a charged particle if it were placed at *any position* in space containing other fixed charges. This task is simplified by the concept of the *electrostatic field*. An electrostatic field is a vector field (see page 303). Suppose that charges A and B in Figure 20.6 are in fixed positions.

Figure 20.6 diagrams the force vector that would act on a charge C at either of the two positions shown. When C is at position 1, the force from A and B on C is large and in the $+x$-direction. When C is at position 2 the force on C is in a different direction, and it is weaker because of the r^{-2} dependence of force on distance. The force $\mathbf{f}(x, y, z)$ that acts on a particle C can vary with the spatial position (x, y, z) of the particle C.

To describe how the force acting on C depends on spatial position, you compute the electrostatic field. Consider C to be a *test particle*, having a charge q_u. By test particle, we mean a charge that you are free to place at any position to probe the force there. The test particle is acted upon by the fixed charges.

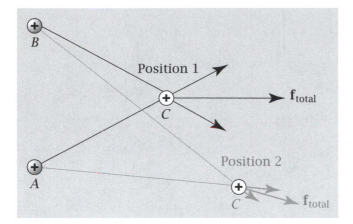

Figure 20.6 *A* and *B* are charges at fixed positions in space. *C* is *a test particle*. When *C* is placed at position 1, the net force on *C* is large and along the +*x*-direction. When *C* is placed at position 2, the force is weaker and in a different direction. The *electrostatic field* is the set of vectors that describes the force per unit charge that the test particle would feel at any position in space.

You need not be concerned with how the fixed charges act on each other, because that isn't a degree of freedom. The electric (or electrostatic) field **E** is the vector located at position $\mathbf{r} = (x, y, z)$ in space that indicates the direction and magnitude of the electrostatic force that acts on a charged particle *C* at that point, due to all the fixed charges. Wherever you place *C* in space, you will find a vector there that gives the direction and magnitude of the force that will act on the particle at that position. Suppose the vector force that would act on a test particle *C* in Figure 20.6 at position **r** is **f**(**r**). Dividing the force in Equation (20.9) by the test charge q_{test} defines the field:

$$\mathbf{E}(\mathbf{r}) = \frac{\mathbf{f}(\mathbf{r})}{q_{\text{test}}} = \frac{q_{\text{fixed}}}{4\pi\varepsilon_0 D r^2}\frac{\mathbf{r}}{r},\tag{20.12}$$

the force per unit charge on the test particle. The SI units of electrostatic field are N C^{-1} or V m^{-1}.

EXAMPLE 20.4 The field from a point charge. A point charge is radially symmetric so you can dispense with vector notation to compute its field, and just treat the dependence on radial distance r. The field from a fixed point charge acting on a test particle is

$$E(r) = \frac{Cq_{\text{fixed}}}{D r^2}.\tag{20.13}$$

Figure 20.7 shows the electrostatic field around a positive charge, and around a negative charge. It shows discrete vectors at specific positions in space, because that's the best you can do in a diagram. The actual fields are continuous functions of spatial position. The figure shows that a positive test charge is pulled toward a fixed negative charge, or repelled from a fixed positive charge. The relative lengths of the vector arrows indicate that the force is greatest near the charge, and diminishes with the distance from the fixed charge.

Figure 20.8 uses field lines instead of field vectors to show the force that a positive test charge would experience near a dipolar arrangement of one positive and one negative charge. In this type of diagram, each line represents only the direction of the force at any position in space, and not its magnitude.

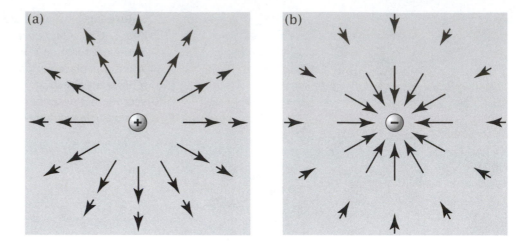

Figure 20.7 The electric field is indicated by vectors pointing (a) away from a positive charge and (b) toward a negative charge. The field represents the force per unit charge that acts on a positive test charge put at each particular position in space.

Figure 20.8 The electric field near a dipole is represented by field lines pointing away from the positive charge and toward the negative charge. This diagram shows the direction that a positive test charge would be moved if it were placed at a given position.

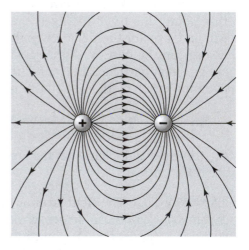

Now suppose that you are interested in the electrical force acting on a charged particle, from a line of charges like the phosphate groups in DNA or other charged groups on a polyelectrolyte molecule, or from a plane of charge like a membrane or surface. You will need to sum many Coulomb force vectors. If the charges on the line or plane are close together, you could approximate this sum as an integral over force vectors. But there is a much easier way to do all this, called Gauss's law. Gauss's law is nothing but Coulomb's law in a different form. To derive Gauss's law you need a quantity called the electric field flux.

Electric Fields Have Fluxes

Chapter 17 describes the flux of a flowing fluid. You can also define the flux for electric field vectors. Why is it useful to define a flux for electric fields? The electric field flux has the important general property that it is independent of the size and shape of any balloon that fully surrounds all the charges, as we will show. This generality leads to some important methods for simplifying the problem of computing complex electrostatic fields.

Positive and negative charges are the sources and sinks of electrical fields. Consider some constellation of positive and negative charges fixed at certain positions in space. Just as we did for fluids (page 308), we invent an imaginary balloon that we place around the charges. We want to determine the flux of the electric field through the imaginary balloon surface. The electric field flux Φ is defined as the integral of D times the electric field $\mathbf{E}$ over all the surface area elements $d\mathbf{s}$ of the balloon,

$$\Phi = \int_{\text{surface}} D\mathbf{E} \cdot d\mathbf{s}, \tag{20.14}$$

where D is the dielectric constant. Because $\mathbf{E}$ is proportional to $1/D$ in a uniform medium (see Equation (20.12)), the flux Φ is independent of D.

Equation (20.14) is general and applies to any constellation of fixed charges, and to any shape and position of an imaginary balloon around them. Many problems of interest have a special simplicity or symmetry that allows you to simplify Equation (20.14). For example, suppose that the charge constellation is just a single point charge q and that the imaginary balloon is just a sphere of radius r centered on the point charge (see Figure 20.9). We are interested in computing the electric field flux from the point charge through the imaginary spherical surface.

The electrostatic field of a point charge is spherically symmetrical—the magnitude of the field $\mathbf{E}$ is the same through the north pole as it is through the south pole or through the equator, or through any other patch of surface of a given area. So you can drop the vector notation and express the field as $E(r)$, which depends only on r. Because E doesn't differ from one surface patch to another, the integral in Equation (20.14) becomes a simple product. That is, $\mathbf{E}$ is parallel to $d\mathbf{s}$ everywhere, so $\mathbf{E} \cdot d\mathbf{s} = E\,ds$ and the flux integral becomes $\Phi = DE \int ds = DE(r) \times$ (total area of sphere). Use $E(r) = q/(4\pi\varepsilon_0 D r^2)$ from Equation (20.13), and $4\pi r^2$ for the surface area of a sphere, to get

$$\Phi = DE(r)(4\pi r^2)$$

$$= \left(\frac{q}{4\pi\varepsilon_0 r^2}\right)(4\pi r^2)$$

$$= \frac{q}{\varepsilon_0}, \tag{20.15}$$

where q is the charge at the center of the sphere.

Two profound implications of Equation (20.15) lead to a general method for calculating electrostatic forces and fields. First, the right-hand side of Equation (20.15) does not depend on r. For the electric flux through a spherical container, the r^{-2} dependence of E cancels with the r^2 dependence of the spherical surface area. Because of this cancelation, the electric field flux out of a sphere is independent of its radius. It doesn't matter how large you choose your imaginary balloon to be, the flux is the same through spherical balloons of any radius. Second, we show now that the *shape* of the balloon doesn't matter either.

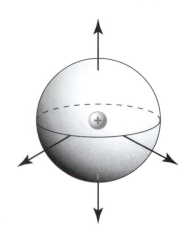

Figure 20.9 The vector electric field $\mathbf{E}$ is normal to a sphere surrounding a point charge located at the center of the sphere.

Figure 20.10 The flux through *any* closed surface around a charge q is the same as the flux through the small sphere. R is the distance from the charge to the outer surface element. $d\mathbf{s}$ indicates the outer surface area element. E indicates the field at distance R from the charge.

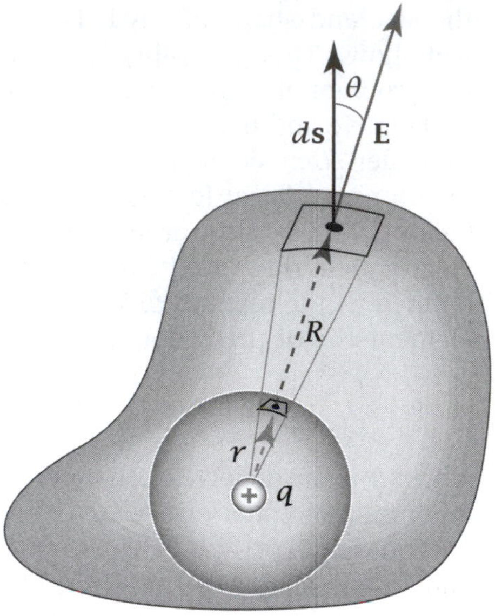

The Electric Field Flux is Independent of Balloon Shape

Now construct another imaginary balloon (the outer balloon) at radius $R > r$ fully containing the original spherical balloon (the inner balloon) that encloses the point charge q (see Figure 20.10). The outer balloon may have any arbitrary shape: it need not be spherical. Consider a radial cone emanating from the point charge. The flux through a small area element $s(r)$ of the inner balloon is $\Phi_{\text{inner}} = DE(r)s(r)$. The same cone cuts through a patch of area $s(R)$ on the outer balloon that is oriented at an angle θ with respect to the radial vector. Because of this angle, the flux through the outer balloon element is reduced by $\cos\theta$, so

$$\Phi_{\text{outer}} = DE(R)s(R)\cos\theta.$$

To relate Φ_{outer} to Φ_{inner}, compute the field through the outer element,

$$E(R) = \frac{\mathcal{C}q}{DR^2} = E(r)\left(\frac{r}{R}\right)^2,$$

and notice that the area of the outer balloon element is greater than the area of the inner balloon element owing to two factors: the area is larger by a factor $(R/r)^2$ owing to the greater distance from the origin; the area is larger by a factor of $1/\cos\theta$ owing to its tilt. Hence $s(R) = s(r)(R/r)^2/\cos\theta$. Combining these factors gives

$$\Phi_{\text{outer}} = DE(r)\left(\frac{r}{R}\right)^2\left(\frac{s(r)}{\cos\theta}\right)\left(\frac{R}{r}\right)^2\cos\theta = \Phi_{\text{inner}}.$$

The flux is the same through the outer balloon area element as through the inner balloon element.

Because this argument applies to all surface elements at any angle θ or distance R, it means that the electric field flux does not depend on the size or shape of the bounding surface. Therefore, because we know that the flux through the sphere of radius r is $\Phi = q/\varepsilon_0$, it must be true that for *any closed surface* of any radius or shape that encloses the point charge q,

$$\Phi = \int_{\text{surface}} D\mathbf{E} \cdot d\mathbf{s} = \frac{q}{\varepsilon_0}. \tag{20.16}$$

Before we show how Equation (20.16) is useful, consider one last generalization. Because this argument applies to *any* bounding surface, it surely must apply to a spherical balloon that is shifted so that its center does not coincide with the charge point.

These generalizations now lead to a remarkable result called Gauss's law. Because the electric field $\mathbf{E}(\mathbf{r})$ from any constellation of fixed charges is always the vector sum of the fields from the component charges, $\mathbf{E} = \mathbf{E}_1 + \mathbf{E}_2 + \mathbf{E}_3 + \cdots + \mathbf{E}_n$, the flux through any closed surface around any constellation of charges is

$$\Phi = \int_{\text{surface}} D\mathbf{E} \cdot d\mathbf{s} = \int_{\text{surface}} D(\mathbf{E}_1 + \mathbf{E}_2 + \mathbf{E}_3 \cdots) \cdot d\mathbf{s} = \frac{1}{\varepsilon_0} \sum_{i=1}^{n} q_i, \tag{20.17}$$

where $\sum_{i=1}^{n} q_i$ is just the net charge contained inside the boundary surface. The vector quantity $D\mathbf{E}$ is called the electric displacement.

Equation (20.17) is **Gauss's law**. It says that a very complex quantity can be computed by a very simple recipe. If you want to find the flux of the electrostatic field through any bounding balloon, no matter how complex its shape, you don't need to compute the electrostatic field vector at each point in space, find its dot product with all the surface elements, and integrate. Instead, you can compute the flux simply by counting up the total net charge contained within the bounding surface, and divide by ε_0, a constant.

We have derived Gauss's law for a uniform medium. But Equation (20.17) is also valid in a medium with spatial variations of the dielectric constant D. In that case the local value of D is used in the surface integrals of Equation (20.17). This will be useful for treating dielectric boundaries in Chapter 21.

If the charges are many or densely distributed, it is sometimes more convenient to represent them in terms of a continuous spatial distribution function $\rho(x, y, z)$, the charge per unit volume. Then the net charge is the integral over the charge distribution, and Gauss's law is

$$\int D\mathbf{E} \cdot d\mathbf{s} = \frac{1}{\varepsilon_0} \int_V \rho \, dV. \tag{20.18}$$

The real power of Gauss's law is in the third equality in Equation (20.17), not the first. It prescribes a very simple way to compute the electrostatic field of force from a constellation of fixed charges. Here are some examples.

EXAMPLE 20.5 The field from a line charge. Let's model a charged wire or a polyelectrolyte molecule such as DNA as a long charged line of length L, a cylinder of zero radius. Suppose that this line has λ charges per unit length and is so long that you can neglect end effects. Then the only electrostatic field is *radial*, perpendicular to the line of charge, because the cylindrical symmetry of the line causes the nonradial components to cancel each other. What is the radial force on a charged test particle near the line charge? One way to compute the field is to sum the vector forces

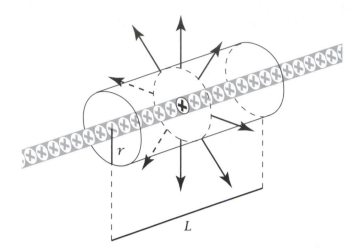

Figure 20.11 To compute the electric field at a distance r from a charged line, construct an imaginary cylinder of radius r and length L and use Gauss's law.

on the test charge from each small piece of the line charge. A much simpler way is to use Gauss's law. Use a cylindrical bounding surface with its axis along the line of charges to enclose a length L of the line of charges (see Figure 20.11). The flux that exits perpendicular to the cylinder surface at a distance r is $D \times E(r) \times$ (area) $= DE(2\pi rL)$. Because the total charge is (number of charges per unit length) $\times$ (length) $= \lambda L$, Gauss's law for this system is given by $D \times E(r) \times$ (area) $=$ (total charge)$/\varepsilon_0 = \lambda L/\varepsilon_0$, so

$$DE(r)(2\pi rL) = \frac{\lambda L}{\varepsilon_0}. \tag{20.19}$$

Rearranging Equation (20.19) shows that the field $E(r)$ at a radial distance r from a line of charge is

$$E(r) = \frac{\lambda}{2\pi\varepsilon_0 Dr}. \tag{20.20}$$

Equation (20.20) gives the force per unit charge that acts on a test charge at any distance r from the line charge.

EXAMPLE 20.6 The field from a charged planar surface. Let's compute the field due to a charged planar surface, such as the plate of a capacitor, an electrode, or a cell membrane in an isotropic dielectric medium. Suppose that the charge is distributed uniformly on the planar surface. Assume that the plane is so large that you can neglect edge effects, and that it has a uniform surface charge σ per unit surface area.

To use Gauss's law, let a bounding surface take the shape of a cylinder of radius R sandwiching the plane, as shown in Figure 20.12. By symmetry, the magnitude of the electric field must be equal on the two opposite sides of the sheet, so the vectors **E** point in opposite directions. The vectors on the flat ends of the bounding cylinder point in opposite directions on the two sides of the sheet. The electric field vectors pass only in the x-direction, perpendicular to the sheet, through the imaginary bounding cylinder at the flat ends because for every vector of force that is parallel to the plane, there is an equal and opposite one that cancels it. If $E(x)$ is the magnitude of the field on one side

(a) Electric Field from a Plane Charge   (b) Gauss's Law Cylinder

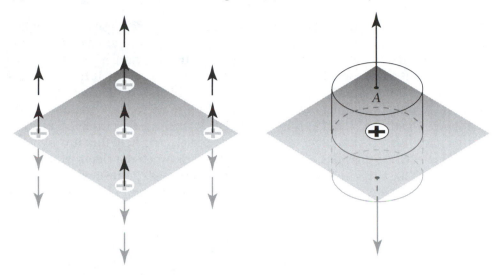

Figure 20.12 (a) A field emanates in both directions perpendicular to a charged plane. (b) To compute the field from a plane, construct a cylindrical box having end area A and containing the plane, and use Gauss's law.

of the plane, then the total flux out (both ends of) the bounding surface is $2DE(x)$ multiplied by the area A of each end of the cylinder. Because the total charge enclosed by the surface is $A\sigma$, Gauss's law gives

$$2DE(x)A = \frac{\text{total charge}}{\varepsilon_0} = \frac{A\sigma}{\varepsilon_0}. \tag{20.21}$$

Rearrange this to find E for a plane of charges:

$$E = \frac{\sigma}{2\varepsilon_0 D}. \tag{20.22}$$

The height or volume of the cylinder you choose doesn't matter because the field strength is independent of the distance x from the plane.

E changes sign from one side of the charged surface to the other. The total difference in electrostatic field from one side of the plane to the other, from E to $-E$, equals $\sigma/\varepsilon_0 D$. In general (the normal component of) the electric field changes discontinuously upon passing through a charged surface or upon passing through a dielectric boundary (see Chapter 21).

We have found that the electrostatic field from a point charge is shorter-ranged ($1/r^2$) than the field from a line charge ($1/r$), and that the field from a plane charge is independent of distance. These relations are a matter of scale. If you are interested in a field on a small scale very close to a large charged sphere, the charged surface acts like a plane, and the field is independent of r. Very far away from a charged plane, much farther than the size of the plane, the plane appears to be a point charge, and the field decays as $1/r^2$.

Summary

Electrostatic interactions are governed by Coulomb's law. They are long-ranged: $u(r)$ for two point charges diminishes with charge separation as $1/r$. While charge interactions in a vacuum can be very strong, charges interact more weakly in polarizable media, because the medium arranges itself to counteract the field set up by the charges. The degree to which a medium can polarize and shield charge interactions is described by the dielectric constant of the medium. Electrostatic forces sum like vectors, and are described by electrostatic fields. Gauss's law, defined in terms of the electric field flux through any bounding surface, provides a useful relationship between the electrostatic field and the net charge for any constellation of fixed charges. In the next chapter we define the electrostatic potential and the Poisson equation for more complex problems than we can treat with Gauss's law.

Problems

1. NaCl in a gas or crystal. In Example 20.1 we calculated the electrostatic interaction energy between Na^+ and Cl^- ions in a vacuum at 2.81 Å distance. How is this energy related to the crystal energy calculated in Example 20.2?

2. Divalent ion attraction. A Mg^{2+} ion and a Cl^- ion are 6 Å apart in water at 25 °C. What is their electrostatic interaction energy in units of kT?

3. Charges in alcohol. The dielectric constant of methanol is $D = 33$. Two charges are separated by a distance ℓ in methanol. For what value of ℓ is the electrostatic interaction equal to kT at $T = 25$ °C?

4. Gauss's law by summing coulombic interactions. Derive Equations (20.20) and (20.22) from Coulomb's law by integrating the appropriate field component.

5. The field around DNA. Consider a line charge with the linear charge density of the DNA double helix, $\lambda = 2e$ per 3.4 Å. What is the electric field in V m^{-1} at a radial distance of 30 Å in water (dielectric constant $D = 80$)?

6. The field inside a spherical shell. What is the electric field inside a uniformly charged spherical shell?

7. The field of a sphere. Consider a uniformly charged spherical volume with radius R and space charge density ρ. What is the electric field for $r < R$ and for $r > R$?

8. The field of a cylinder. A cylinder with radius R is uniformly filled with a charge of space density ρ. What is the electric field inside the cylinder? Neglect end effects.

9. Charges affect chemical isomers. 1,2-dibromoethane is a molecule that can be in either the *trans* form or the *gauche* form, as shown in Figure 20.13.

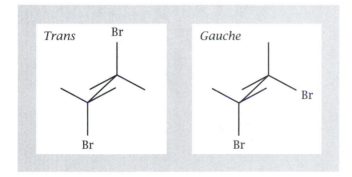

Figure 20.13

Using Table 20.2, which describes the relative populations of *gauche/trans* conformations in solvents having different dielectric constants D, determine whether bromines attract or repel each other.

Table 20.2 Equilibrium constants $N_{\text{gauche}}/N_{\text{trans}}$ for the *trans/gauche* isomerism of 1,2-dibromoethane at 25 °C. Source: JE Leffler and E Grunwald, *Rates and Equilibria of Organic Reactions*, Dover Publications, New York, 1989.

Medium	D	$N_{\text{gauche}}/N_{\text{trans}}$
Gas	1	0.164
n-Hexane	1.9	0.251
Cyclohexane	2.0	0.260
CCl$_4$, dilute	2.2	0.307
CS$_2$	2.7	0.453
1,2-Dibromoethane, pure liquid	4.8	0.554
Methanol	33	0.892

10. The field of a parallel plate capacitor. A parallel-plate capacitor is an arrangement of two parallel conducting plates that carry opposite surface charges, one σ, the other $-\sigma$. Extend Example 20.6 to show that the field outside a parallel plate capacitor vanishes, in accordance with Gauss's law.

11. Small charge cluster. Based on the NaCl crystal lattice of Example 20.2, is a single Na^+ surrounded by its six nearest neighboring Cl^- ions a stable structure?

References

[1] RP Feynman. *The Feynman Lectures on Physics*. Vol II, Mainly Electromagnetism and Matter, Addison-Wesley, Reading, MA, 1964.

Suggested Reading

Excellent introductory texts on electrostatics:

D Halliday and R Resnick, *Fundamentals of Physics*, 3rd edition, Wiley, New York, 1988.

EM Purcell, *Electricity and Magnetism*, McGraw-Hill, New York, 1965.

More advanced treatments:

R Becker, *Electromagnetic Fields and Interactions*, Dover Publications, New York, 1982.

IS Grant and WR Phillips, *Electromagnetism*, 2nd edition, Wiley, New York, 1990.

P Lorrain, DR Corson and F Lorrain, *Fields and Waves*, 3rd edition, WH Freeman, New York, 1988.

21 The Electrostatic Potential

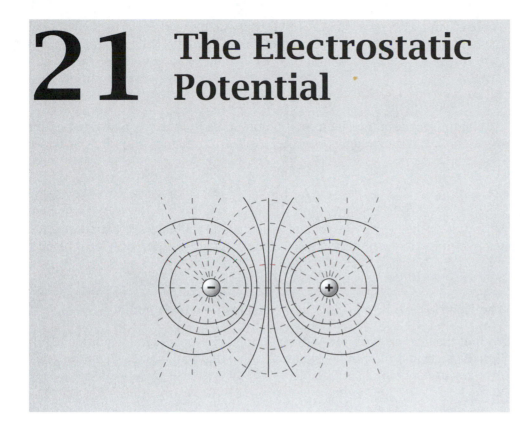

Working with Electrostatic Potentials (Energies) Rather than Fields (Forces)

While the *electrostatic field*, introduced in Chapter 20, describes forces (vector quantities), the *electrostatic potential* in this chapter describes energies, which are scalar quantities. We switch our attention from forces to energies for two reasons. First, scalars are easier to use than vectors. Second, our aim in Chapters 22 and 23 is to predict equilibria in systems in which charges play a role. To predict equilibria, we need energies.

The main tool that we develop here is Poisson's equation, one of Maxwell's four famous equations of electricity and magnetism. Poisson's equation gives a simpler way to treat many electrostatic interactions. We will use it in Chapter 22 for computing the energy for moving a charged particle, such as a salt ion or protein, in the vicinity of other charged objects such as proteins, colloids, DNA, polyelectrolytes, membranes, electrodes, or dielectric boundaries.

What is the Electrostatic Potential?

The work dw you must perform to move a charge q through a small distance $d\ell$ in the presence of a fixed electrostatic field $\mathbf{E}$ is the negative of the dot product of the force $\mathbf{f} = q\mathbf{E}$ and the displacement $d\ell$ (see Equation (3.2)):

$$\delta w = -\mathbf{f} \cdot d\ell = -q\mathbf{E} \cdot d\ell. \tag{21.1}$$

(a) Positive Charge

$\psi(r)$

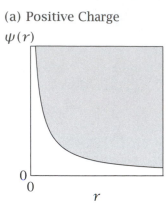

(b) Negative Charge

$\psi(r)$

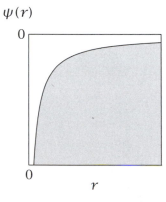

Figure 21.1 The electrostatic potential $\psi(r)$ as a function of the distance r from (a) a positive point charge and (b) a negative point charge.

There is a minus sign in Equation (21.1) because this work is performed *against* the field, not *by* the field as in Equation (3.2). To move a charge from point A to B, the total work w_{AB} is given by the *path integral* (see page 307),

$$w_{AB} = -q \int_A^B \mathbf{E} \cdot d\boldsymbol{\ell}. \tag{21.2}$$

The difference in the **electrostatic potentials** ψ_B and ψ_A is defined as the work w_{AB} of moving a unit test charge q_{test} from point A to point B, divided by the test charge q_{test},

$$\psi_B - \psi_A = \frac{w_{AB}}{q_{\text{test}}} = -\int_A^B \mathbf{E} \cdot d\boldsymbol{\ell}. \tag{21.3}$$

If you know the field $\mathbf{E}$ due to some constellation of charges, Equation (21.3) tells you how to compute the electrostatic potential difference for moving a test charge from point A to B within that field.

The Electric Field is the Gradient of the Electrostatic Potential

To find the electric field, given the potential, put the integral, Equation (21.3), into the form of a differential equation. To do this, first express Equation (21.3) in terms of its vector components,

$$\psi_B - \psi_A = -\int_A^B \mathbf{E} \cdot d\boldsymbol{\ell}$$
$$= -\int_{x_A}^{x_B} E_x \, dx - \int_{y_A}^{y_B} E_y \, dy - \int_{z_A}^{z_B} E_z \, dz. \tag{21.4}$$

Now convert from an integral to a differential equation. Suppose points A and B are very close together: A is at (x, y, z) and B is at $(x + \Delta x, y, z)$, so $dy = dz = 0$. Then Equation (21.4) gives

$$\psi_B - \psi_A = \Delta\psi = -\int_{x_A}^{x_B} E_x \, dx = -E_x \Delta x. \tag{21.5}$$

At the same time, the Taylor series for $\Delta\psi$ gives

$$\Delta\psi = \left(\frac{\partial\psi}{\partial x}\right)\Delta x. \tag{21.6}$$

Comparing Equations (21.5) and (21.6) shows that

$$E_x = -\left(\frac{\partial\psi}{\partial x}\right). \tag{21.7}$$

By the same reasoning, taking B at $(x, y + \Delta y, z)$ and at $(x, y, z + \Delta z)$, you will find that $E_y = -(\partial\psi/\partial y)$ and $E_z = -(\partial\psi/\partial z)$. In more compact notation (see page 304),

$$\mathbf{E} = -\nabla\psi. \tag{21.8}$$

The electric field is the negative of the gradient of the electrostatic potential ψ.

To compute the potential, it is customary to put the starting point A at an infinite distance from the end point B and to set $\psi_\infty = 0$, so the potential difference is $\psi_B - \psi_A = \psi_B$. The *electrostatic potential* is defined as the work *per unit charge*, so it is *not an energy* and does not have units of energy. The electrostatic potential *multiplied by charge* is an energy. Like electrostatic forces and fields, electrostatic potentials are additive. They are the sums of the potentials from the individual fixed charges, no matter how the charges are distributed.

Example 21.1 gives the potential around a point charge.

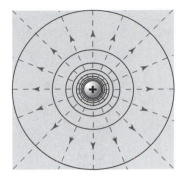

Figure 21.2 The electrostatic potential is constant at distance r from a point charge. Lines of equipotential are circles in two dimensions. Broken lines indicate the direction of force and the field **E**.

EXAMPLE 21.1 The electrostatic potential around a point charge. Consider the simplest possible charge constellation, namely a single fixed point charge q_{fixed}. How much work is required to bring a test charge from far away to a distance r from the fixed point charge? The potential only varies along a single coordinate, the radial direction, so you can dispense with vector notation for this problem. That is, $\mathbf{E} \cdot d\boldsymbol{\ell} = E\,dr$ because the field **E** and the vector $d\boldsymbol{\ell}$ are parallel to each other (both point in the radial direction). You need only to multiply the magnitudes of E and dr, then integrate. The field that emanates radially from the fixed charge has strength $E(r) = \mathcal{C}q_{fixed}/Dr^2$, according to Equation (20.13). The change in electrostatic potential ψ_{test} upon moving a test charge radially inward from $r' = \infty$ to $r' = r$ toward the fixed charge is (see Figure 21.1)

$$\psi_{test} = -\int_\infty^r E\,dr' = -\frac{\mathcal{C}q_{fixed}}{D}\int_\infty^r \frac{1}{(r')^2}\,dr' = \frac{\mathcal{C}q_{fixed}}{Dr} = \frac{q_{fixed}}{4\pi\varepsilon_0 Dr}. \tag{21.9}$$

The prime $'$ is used to indicate that r' is the variable of integration that ranges from r to ∞. If the test particle has charge Q, then the work of bringing it to a distance r from the fixed particle is $w = Q\psi_{test}$ (see Equation (21.3)).

If instead of a single charge q_{fixed}, you have a fixed charge density $\rho_{fixed} = \Sigma q_i/V$ in a volume V around a point 1, the electrostatic potential ψ_2 at point 2 is

$$\psi_{test} = \frac{1}{4\pi\varepsilon_0 D}\int_V \frac{\rho_{fixed}}{r_{12}}\,dV, \tag{21.10}$$

where r_{12} is the distance between all the pairs of fixed and test charges.

What Are Electrostatic Potential Surfaces?

To illustrate the idea of electrostatic potential, consider a distribution of charges in a plane. Positive charges tend to move toward negative regions of electrostatic potential. Negative charges spontaneously move toward positive regions of electrostatic potential.

Think of an electrostatic potential surface as a contour map. No work is required to move charges along contours of constant electrostatic potential, just as no gravitational work is done in walking along level paths on a hillside. For a point charge, a surface of constant electrostatic potential is any sphere of radius r centered on the point. It requires no work to move a test charge anywhere on that sphere (see Figure 21.2). The electric field **E** is always perpendicular to an equipotential surface.

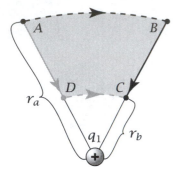

Figure 21.4 The same amount of work is required to move a charge from point A to D to C, as from point A to B to C, because work is done only in the radial steps.

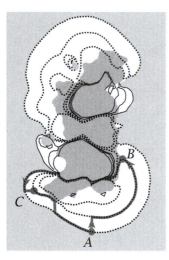

Figure 21.5 Even with an array as complex as the charges in the protein superoxide dismutase (gray), no work is done in moving a charge in a cycle. One such cycle is shown by the heavy line. Dashed lines are equipotential contours. Source: I Klapper, R Hagstrom, R Fine, K Sharp and B Honig, *Proteins: Structure, Function and Genetics* **1**, 47–59 (1986).

The Equipotential Surfaces Around Two Positive Point Charges

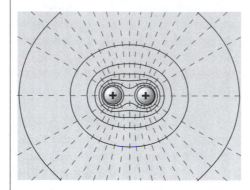

Figure 21.3 Lines of equal electrostatic potential around two positive point charges.

The broken lines in Figure 21.3 show the electric field $\mathbf{E}$ in the xy plane due to two charges of the same sign; q at $x = -l/2$, and q at $x = +l/2$. The x-axis is an axis of symmetry of this electric field and charge constellation. The solid curves indicate the intersection of the plane of the page and the equipotential surfaces. If you rotate the solid curves about the x-axis, out of the plane of the page, you generate the full three-dimensional equipotential surfaces. The equipotential surfaces become more nearly perfect spheres as their distance from the charges increases. Therefore, at a distance far away from the point charges, you can view the electrostatic potential as though it were due to a single point charge equal to $2q$ at $x = 0$.

Electrostatic Interactions are Conservative Forces.

As long as there are no frictional losses, electrostatic work is reversible work, a pathway-independent quantity that sums to zero around a cycle. Figure 21.4 shows two different pathways for moving a charge q_2 from point A to point C in the field of a single point charge q_1. Path 1 is from A to B to C. Path 2 is from A to D to C. Moving a particle along circumference segments AB and DC involves no work because there is no change in radius. Both radial segments, AD and BC, involve the same change in radius, so the work in moving a charge along either pathway ABC or ADC is the same:

$$
\begin{aligned}
w &= -q_2 \frac{\mathcal{C}}{D} \int_{r_a}^{r_b} \frac{q_1}{r^2} \, dr \\
&= \frac{\mathcal{C} q_1 q_2}{D} \left(\frac{1}{r_b} - \frac{1}{r_a} \right).
\end{aligned}
\tag{21.11}
$$

The reversible electrostatic work sums to zero around a cycle, no matter how complex the electrostatic potential surface. Any path between any two points A and C can be approximated as a sequence of radial and equipotential segments. The work along the equipotential curves will be zero, and the work along the radial segments is given by Equation (21.11). Figure 21.5 shows electrostatic equipotential curves around the protein superoxide dismutase. A path is shown for moving a charge in a cycle from point A to B to C and back to A: the work is zero around this cycle.

Now we calculate the electrostatic potential between two planes.

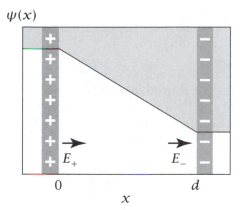

$\psi(x)$

x

Figure 21.6 The electrostatic potential $\psi(x)$ as a function of the distance x between two charged planes at $x = 0$ and $x = d$. The negative slope indicates that a positive charge moves to the right to go energetically downhill. Because a positive charge between the plates feels a repulsive field E_+ to the right and an attractive field E_- to the right, the total field is $E_{\text{inside}} = E_+ + E_-$.

EXAMPLE 21.2 The electrostatic potential in a parallel plate capacitor. A capacitor is a device made of two conductors separated by an insulator. It stores positive charge on one conductor and negative charge on another. Often the conductors are arranged as two parallel charged planes.

We want to compute the work of moving a charge from one plate to the other in the capacitor shown in Figure 21.6. This calculation is useful for treating the energies of ions that approach surfaces and membranes. Suppose that the positive plate has $+\sigma$ charges per unit area, giving an electrostatic field E_+, and the negative plate has $-\sigma$ charges per unit area and electrostatic field E_-. The plates are separated by a distance d. Each square plate has area A. If d is much smaller than $\sqrt{A}$, you can neglect edge effects. To find the work of moving a charge, first determine the total electric field E_{inside} between the plates.

In Example 20.6 we used Gauss's law to find that the field due to a charged plate is equal to $\sigma/2\varepsilon_0 D$, so we have $E_+ = \sigma/2\varepsilon_0 D$ and $E_- = -\sigma/2\varepsilon_0 D$. The total field inside the capacitor in Figure 21.6 acting to drive a positive charge to the right is the sum of the fields from each plate, $E_{\text{inside}} = E_+ + E_- = \sigma/\varepsilon_0 D$. The potential is (see Equation (21.3)):

$$\Delta\psi = -\int E_{\text{inside}}\, dx' = -\int_0^x \left(\frac{\sigma}{\varepsilon_0 D}\right) dx' = \frac{-\sigma x}{\varepsilon_0 D}. \tag{21.12}$$

For moving a positive charge from $x = 0$ to $x = d$, $\Delta\psi = -\sigma d/D\varepsilon_0$. Equation (21.12) implies that it takes work to move a positive charge to the positive plane in a parallel plate capacitor, and the work increases with charge density and plate separation.

The field outside the plates is zero because a positive charge that is located to the left of the positive plate, for example, is pulled to the right by E_- but pushed to the left by E_+. The capacitance C_0 is defined as the total amount of charge separation, $A\sigma$, per unit of potential difference:

$$C_0 = \frac{A\sigma}{|\Delta\psi|} = \frac{A\varepsilon_0 D}{d}. \tag{21.13}$$

Capacitors can be used to measure the dielectric constants D of materials. To do this, you first measure the capacitance of a capacitor of given area A and separation d having only air or vacuum between the plates ($D = 1$). Then fill the space between the plates with the material of interest and measure the capacitance again. Equation (21.13) says that the ratio of the measured capacitances gives the dielectric constant.

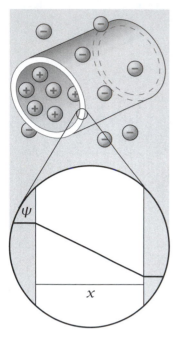

Figure 21.7 The cell membrane of a nerve acts as a capacitor. The membrane is composed of lipids that, like oil, have a low dielectric constant. If the inside and outside solutions have different electrostatic potentials, there is a gradient of electrostatic potential across the membrane.

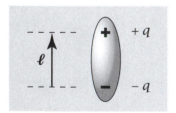

Figure 21.8 A dipole.

Biological membranes are capacitors. Let's calculate the capacitance of the membrane of a nerve cell.

EXAMPLE 21.3 The capacitance of a nerve cell membrane. Assume that a nerve cell is a long cylinder enclosed by a thin planar bilayer membrane of lipids (see Figure 21.7). You can treat the lipid bilayer as a parallel plate capacitor. Lipid bilayers have oil-like interiors, for which we will assume $D = 2$, and thickness approximately $d \approx 20$ Å. Equation (21.13) gives the capacitance per unit area as

$$\frac{C_0}{A} = \frac{\varepsilon_0 D}{d} = \frac{(8.85 \times 10^{-12} \text{ farad m}^{-1})(2)}{20 \text{ Å}} \left(\frac{1}{10^2} \frac{\text{m}}{\text{cm}}\right) \left(\frac{10^8 \text{ Å}}{\text{cm}}\right)$$

$$= 8.85 \times 10^{-7} \text{ farad cm}^{-2} \approx 0.9 \, \mu\text{farad cm}^{-2}. \qquad (21.14)$$

This gives a reasonable model for the capacitance of nonmyelinated nerve cells, which is about $1 \, \mu\text{farad cm}^{-2}$.

Dipoles Are Equal and Opposite Charges Separated by a Distance

A dipole is an arrangement of charges $+q$ and $-q$ separated by a distance ℓ (see Figure 21.8). If there is a charge -3.6 at one end and $+5.2$ at the other end, you need a more complex model, described in Chapter 24. A dipole is defined as a charge $+q$ at one end, and exactly the same amount of charge—but opposite sign—at the other end. Dipoles are oriented in space, indicated by their vector ℓ pointing from $-q$ to $+q$. The **dipole moment** is a vector

$$\boldsymbol{\mu} = q\boldsymbol{\ell}. \qquad (21.15)$$

Now we calculate the work of orienting an electric dipole.

EXAMPLE 21.4 The energy of a dipole in an electric field. Figures 21.9 and 21.10 show a dipole with its center fixed in space. The dipole is subject to an orienting force from an electric field **E**. Compute the work w that the field performs in rotating the dipole from 0 (parallel to the field) to an angle θ,

$$w = \int \mathbf{f} \cdot d\boldsymbol{\ell}. \qquad (21.16)$$

The magnitude of the force on a charge q acting in the direction of the field is qE. The force can be decomposed into two vector components: f_c acting in the circumferential direction to rotate the dipole, and f_s acting to stretch the dipole.

We are interested only in f_c. As Figure 21.9 indicates, $f_c = f \sin \theta = Eq \sin \theta$. What is the charge displacement $d\ell$ in the circumferential direction? If a point at radial distance a from a pivot point in the middle of the dipole is rotated through an angle $d\theta$, the circumferential displacement is $a \, d\theta$ (see Figure 21.10). Because the force acts in the opposite direction of increasing θ, it introduces a minus sign and we have

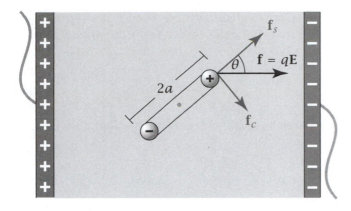

Figure 21.9 A dipole of length $\ell = 2a$ orients in an applied electric field, **E**. The force acting on the dipole is **f**. The component of the force that acts to orient the dipole is $f_c = f \sin\theta$.

$$\mathbf{f} \cdot d\boldsymbol{\ell} = -(f_c)(a\,d\theta) = -Eqa\sin\theta\,d\theta. \qquad (21.17)$$

To get the work of rotating the dipole from angle 0 to θ, integrate Equation (21.17) and multiply by 2 because there is an equal torque acting on each end of the dipole:

$$w = -2Eqa\int_0^\theta \sin\theta'\,d\theta' = Eq\ell(\cos\theta - 1) = E\mu(\cos\theta - 1). \qquad (21.18)$$

Equation (21.18) will be useful for modelling polar molecules orienting in electric fields and gating in ion channels (see page 417).

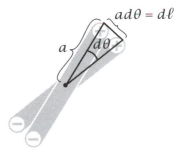

Figure 21.10
Displacement due to dipole orientation. The dipole of length $2a$ in Example 21.4 changes angle by an amount $d\theta$, so $d\ell = a\,d\theta$.

EXAMPLE 21.5 The potential field around a dipole. Consider the electrical dipole shown in Figure 21.11. It is formed by charges $-q$ at $x = -\ell/2$ and $+q$ at $x = +\ell/2$, so the dipole moment vector points in the $+x$ direction. Here we will determine the electrostatic potential at any point in the xy plane. The complete three-dimensional potential field is found by rotating the equipotential lines about the x-axis.

Let's find the potential at any arbitrary point $P(x, y)$ at distance $r = (x^2 + y^2)^{1/2}$ from the origin in Figure 21.11(b). The vector **r** is at an angle θ with respect to the x-axis. Use Equation (21.9) to define the potential at P from each of the two fixed point charges. The total electrostatic potential ψ is the sum of the potentials from each charge of the dipole:

$$\psi = -\frac{\mathcal{C}q}{Dr_-} + \frac{\mathcal{C}q}{Dr_+}. \qquad (21.19)$$

The distances r_- from $-q$ and r_+ from $+q$ are found by considering the right triangles formed by drawing a perpendicular line from point P to the x-axis. You can compute r_-^2 using Figure 21.11(b) and the expression $r^2 = x^2 + y^2$:

$$r_-^2 = \left(x + \frac{\ell}{2}\right)^2 + y^2 = x^2 + 2\left(\frac{\ell x}{2}\right) + \left(\frac{\ell}{2}\right)^2 + y^2$$

$$= r^2 + \ell x + \left(\frac{\ell}{2}\right)^2 \approx r^2\left(1 + \frac{\ell x}{r^2}\right), \qquad (21.20)$$

where the approximation arises from neglecting the second-order term $(\ell/2)^2$ in Equation (21.20). This applies when P is far away from the dipole, $r \gg \ell$. The same logic gives r_+^2 in terms of r:

$$r_+^2 = \left(x - \frac{\ell}{2}\right)^2 + y^2 \approx r^2\left(1 - \frac{\ell x}{r^2}\right). \tag{21.21}$$

Substitute r_+ and r_- from Equations (21.20) and (21.21) into Equation (21.19):

$$\psi \approx \frac{Cq}{Dr}\left(-\left(1 + \frac{\ell x}{r^2}\right)^{-1/2} + \left(1 - \frac{\ell x}{r^2}\right)^{-1/2}\right). \tag{21.22}$$

The terms in the inner parentheses in Equation (21.22) can each be approximated with the first term in the series expansion for $(1+x)^{-1/2} \approx (1-(x/2)\ldots)$ given in Appendix C, Equation (C.6):

$$\psi \approx \frac{Cq}{Dr}\left(-1 + \frac{\ell x}{2r^2} + 1 + \frac{\ell x}{2r^2}\right) = \frac{Cq}{Dr}\left(\frac{\ell x}{r^2}\right). \tag{21.23}$$

Collecting terms and using $x/r = \cos\theta$, you have

$$\psi \approx \frac{Cq\ell}{Dr^2}\cos\theta = \frac{C\mu\cos\theta}{Dr^2}, \tag{21.24}$$

where $\mu = q\ell$ is the magnitude of the dipole moment. When $0 < \theta < \pi/2$, $\cos\theta$ is positive so ψ is positive, indicating that a positive charge at P will be repelled by the dipole, because it is closest to the $+$ end of the dipole. Because $\mu = q\ell$ is the dipole moment, you can write Equation (21.24) in terms of the dot product of the dipole moment vector $\boldsymbol{\mu}$ and the radius vector $\mathbf{r}$:

$$\psi = \frac{C\boldsymbol{\mu} \cdot (\mathbf{r}/r)}{Dr^2}. \tag{21.25}$$

Lines of constant ψ surrounding a dipole are shown in Figure 21.11(a), or surrounding two charges of the same sign in Figure 21.3.

Now we use Equation (21.24) to determine the energy of interaction between a charged ion and a dipolar molecule.

Interactions Between a Charge and a Dipolar Molecule

An ion is located at point P and has a point charge Q. Another molecule has a dipole moment of magnitude $q\ell$. The interaction energy $u(r, \theta)$ is the work of bringing the two molecules from infinite separation to a separation of r, where θ is the angle of the dipole relative to the line between the centers of the dipole and point charge (see Figure 21.11). From Equations (21.3) and (21.24), you get

$$u(r, \theta) = \psi Q = \frac{C\mu Q \cos\theta}{Dr^2}. \tag{21.26}$$

You would get the same result by summing with Coulomb's law for the charge Q and the two dipolar charges.

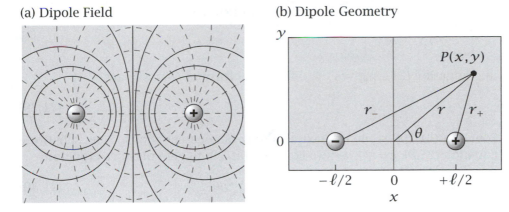

(a) Dipole Field

(b) Dipole Geometry

Figure 21.11 (a) Electric field (– – –) and equipotential lines (———) for a dipole in the xy plane. (b) Definition of quantities r_+ and r_- in Equations (21.20) and (21.21).

The most notable aspect of Equation (21.26) is the dependence on the inverse *square* of the distance, $u \propto 1/r^2$. According to Coulomb's law (Equation (20.1)), two ions interact with a dependence $u \propto 1/r$. Replacing one ion by a dipole causes the interaction to become shorter-ranged. At close range, the single charge 'sees' both charges on the dipole, a sum of two $1/r$ terms. As the dipole moves further away, however, the single charge sees the dipole charges as neutralizing each other because they are close together relative to their distance from the ion. So the interaction weakens even more than the $1/r$ dependence would predict. A similar treatment would show that when two dipoles move apart, the interaction becomes even more short-ranged, $u \propto 1/r^3$ at large separations (see Chapter 24).

The Poisson Equation is Used to Compute the Potential for Any Constellation of Charges

So far, we have computed the electrostatic potentials for very simple constellations of charges. In Chapter 20, we found that we could treat more complex problems of electrostatic forces by using a little vector calculus to find Gauss's law. Now, for energies rather than forces, we resort again to some vector calculus, which will lead us to Gauss's divergence theorem (different from Gauss's law) and to Poisson's equation.

Gauss's theorem, Equation (17.24), says that for any vector field $\mathbf{v}$,

$$\int_{\text{surface}} \mathbf{v} \cdot d\mathbf{s} = \int_{\text{volume}} \nabla \cdot \mathbf{v} \, dV. \tag{21.27}$$

Substituting $\mathbf{v} = D\mathbf{E}$ gives **Gauss's theorem** applied to electric fields,

$$\int_{\text{surface}} D\mathbf{E} \cdot d\mathbf{s} = \int_{\text{volume}} D\nabla \cdot \mathbf{E} \, dV. \tag{21.28}$$

Gauss's theorem equates the *flux* of the electrostatic field *through a closed surface* with the *divergence* of that same field *throughout its volume*.

Substituting Gauss's law (Equation (20.18)),

$$\int_{\text{surface}} D\mathbf{E} \cdot d\mathbf{s} = \int_{\text{volume}} \frac{\rho}{\varepsilon_0} \, dV,$$

into Equation (21.28) gives the **differential form of Gauss's law**:

$$D\nabla \cdot \mathbf{E} = \frac{\rho}{\varepsilon_0}. \tag{21.29}$$

Equation (21.29) is one of Maxwell's four famous equations. Now substitute Equation (21.8), $\mathbf{E} = -\nabla\psi$, into Equation (21.29) to get **Poisson's equation**:

$$\nabla^2\psi = -\frac{\rho}{\varepsilon_0 D}, \qquad \text{where} \qquad \nabla \cdot \mathbf{E} = -\nabla^2\psi. \tag{21.30}$$

You now have two methods for deriving the electrostatic potential from a given distribution of fixed electrical charges. First, you can use Coulomb's law. But this becomes unwieldy for complex constellations of charges or for systems with two or more different dielectric media. Second, you can solve Poisson's equation. The second method is more general, and it can also be used for heterogeneous dielectric media (see page 401).

Examples 21.6 and 21.7 show how to use Poisson's equation to compute the electrostatic potential around a charged sphere and cylinder.

EXAMPLE 21.6 A charged spherical shell: Poisson's equation. Consider a thin uniformly charged spherical shell located at radius $r = a$. The shell has a total charge q. The interior of the sphere is uncharged with dielectric constant D. The shell is immersed in an external medium of the same dielectric constant D. This system might model the charged membrane of a spherical vesicle or a biological cell.

Place a test charge inside or outside the sphere and find the electrostatic potential at that point due to the fixed charges on the spherical shell. Divide space into three regions: inside the charged sphere ($r < a$, where there is no net fixed charge), on the spherical shell ($r = a$, where all the charge is located), and outside ($r > a$, where there is also no net fixed charge). Compute the electrostatic potential everywhere by using Poisson's Equation (21.30).

The problem has spherical symmetry, so the first step is to look up the spherical form of the vector operators (see Table 17.1 on page 312). The potential ψ depends only on r and not on angles θ or ϕ, because the charge is distributed with spherical symmetry. Therefore, $(\partial\psi/\partial\theta) = 0$ and $(\partial^2\psi/\partial\phi^2) = 0$, and $\nabla^2\psi$ in Equation (21.30) becomes

$$\nabla^2\psi = \frac{1}{r^2}\frac{d}{dr}\left(r^2\frac{d\psi}{dr}\right) = \frac{1}{r}\frac{d^2(r\psi)}{dr^2}. \tag{21.31}$$

Our strategy is to solve Poisson's equation inside and outside the sphere. There is no charge in either region so $\nabla^2\psi = 0$ for both. Then we'll use Gauss's law to establish a boundary condition for the charged shell at $r = a$, which will allow us to complete the solution.

For both the inside and outside regions, since $\rho = 0$, Poisson's equation is

$$\frac{1}{r}\frac{d^2(r\psi)}{dr^2} = 0 \qquad \text{for } r > a, \text{ and for } r < a. \tag{21.32}$$

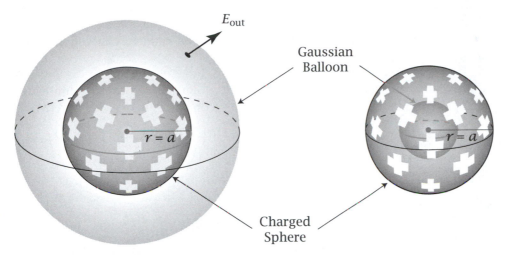

(a) All the Charge is Contained in the Gaussian Balloon

E_{out}

(b) No Charge is Contained in the Gaussian Balloon

Gaussian Balloon

$r = a$

$r = a$

Charged Sphere

Figure 21.12 (a) To find the potential $\psi(r)$ outside a charged sphere, construct a Gaussian balloon that contains the sphere. (b) To find the potential $\psi(r)$ inside, construct the balloon inside the charged sphere.

To solve Equation (21.32), multiply by r and integrate once to get

$$\frac{d(r\psi)}{dr} = A_1, \tag{21.33}$$

where A_1 is the first integration constant. Integrating again yields another integration constant and the general solution,

$$r\psi = A_1 r + A_2 \implies \psi(r) = A_1 + \frac{A_2}{r}. \tag{21.34}$$

You can use boundary conditions to get the constants A_1 and A_2. But here's an easier way, in this case, based on the principle that ψ must be continuous at the boundaries. First, outside the sphere ($r > a$), $E_{out}(r) = \mathcal{C}q/Dr^2$ (see Equation (20.13)). This equation was derived for a point charge q, but it applies to any spherical charge distribution because a Gaussian balloon of radius $r > a$ fully contains the charge (see Figure 21.12(a)). Use Equation (21.3) to get

$$\psi_{out}(r) = -\int_{\infty}^{r} E_{out}(r') \, dr' = \frac{\mathcal{C}q}{Dr}, \qquad \text{for } r > a. \tag{21.35}$$

Now to get $\psi_{in}(r)$, put a spherical Gaussian balloon inside $r < a$ (see Figure 21.12(b)). There is no charge inside this balloon because all the charge is on the shell at $r = a$, so $E_{in}(r) = 0$, and Equation (21.35) gives

$$\psi_{in}(r) = -\int_{\infty}^{r} E_{in}(r') \, dr' = \text{constant}. \tag{21.36}$$

Finally, a condition of *continuity* requires that $\Delta\psi = \psi_{out} - \psi_{in} = 0$ for an infinitesimal step across the boundary, so comparing Equations (21.35) and

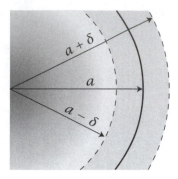

Figure 21.13 The definition of δ for the proof of the continuity principle for a sphere of radius a.

(21.36) at $r = a$ gives

$$\psi_{\text{in}} = \frac{Cq}{Da} \qquad \text{for } r \leq a. \tag{21.37}$$

This is the electrostatic potential everywhere inside a charged spherical shell.

What is continuity? Use Equation (21.4) to find $\Delta\psi = \psi_{\text{out}} - \psi_{\text{in}} = \psi(a+\delta) - \psi(a-\delta)$, the difference in electrostatic potential across an infinitesimal change δ in radius from the inside to the outside of a sphere (see Figure 21.13):

$$\Delta\psi = \psi_{\text{out}} - \psi_{\text{in}} = -\left[\int_a^{a+\delta} E_{\text{out}}(r)\,dr - \int_{a-\delta}^a E_{\text{in}}(r)\,dr \right]. \tag{21.38}$$

Because $E_{\text{in}} = 0$, and $E_{\text{out}} = Cq/Dr^2$, integrating Equation (21.38) gives

$$\Delta\psi = \frac{Cq}{Dr}\Big|_a^{a+\delta} = \frac{Cq}{D}\left(\frac{1}{a+\delta} - \frac{1}{a} \right). \tag{21.39}$$

Equation (21.39) proves continuity because as $\delta \to 0$, it shows that $\Delta\psi \to 0$. Continuity holds across an interface of any shape.

The electrostatic potential is constant everywhere inside the sphere, so the field is zero. This is true inside any closed conducting surface. This is why it is a good idea to be inside a car or other conducting container in a lightning storm—there is no electrostatic force inside. This is also why your car radio loses its signal when you are driving inside a metal bridge or tunnel.

EXAMPLE 21.7 The electrostatic potential around a line charge. Let's compute the electrostatic field around a long charged cylinder of radius a. You can do this in different ways. One way is to use Poisson's equation as in Example 21.6. This problem has cylindrical symmetry, so use the cylindrical form of the vector calculus expressions. For the charge-free region outside the cylinder, Poisson's equation (using Equation (17.30) for a cylinder) is

$$\nabla^2\psi = \frac{1}{r}\frac{d}{dr}\left(r\frac{d\psi}{dr} \right) = 0. \tag{21.40}$$

Multiply by r and integrate Equation (21.40) once over r to get

$$r\frac{d\psi}{dr} = A_1 \quad \Longrightarrow \quad \frac{d\psi}{dr} = \frac{A_1}{r}, \tag{21.41}$$

where A_1 is a constant. Integrate $d\psi = (A_1/r)\,dr$ to get

$$\psi = A_1 \ln r + A_2. \tag{21.42}$$

There's an alternative way to derive Equation (21.42). Equation (20.20) shows that the electrostatic field around a charged cylinder is $E(r) = \lambda/(2\pi\varepsilon_0 Dr)$. Equation (21.3) says that you can integrate $E(r)$ over r from the surface at $r = a$ to get the potential:

$$\Delta\psi(r) = -\int_a^r E(r')\,dr' = -\frac{\lambda}{2\pi\varepsilon_0 D}\ln\left(\frac{r}{a}\right). \tag{21.43}$$

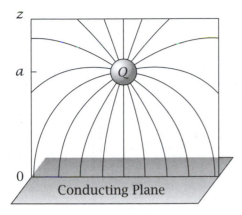

Figure 21.14 The field lines from a point charge Q must intersect a conducting plane at 90°.

The electrostatic potential around an infinitely long line charge is different from the potential around a point charge in one important respect: $\psi \to -\infty$ at large distances r away from a line charge. This means that you cannot define a boundary condition $\psi = 0$ at infinite distance.

The most readily solved electrostatics problems have some intrinsic geometric symmetry—spheres and point charges, cylinders and line charges, or planes and parallel-plate capacitors, for example. Some situations involve two or more different types of symmetry, as when a spherical ion approaches a planar interface such as an oil/water interface. Electrostatic fields and potentials can be found in such cases by the *method of image charges*.

The Attractions and Repulsions of Ions for Conductors or Dielectric Interfaces Can Be Computed by the Method of Image Charges

Ions are attracted to metal or other conductors *that have no charge on them.* The attraction arises from a process of *induction.* An ion in the neighborhood of a conducting surface induces a field. That field attracts the ion to the surface. Induction is readily modelled by recognizing an extraordinary mathematical property, the *uniqueness* of differential equations such as Poisson's equation. According to the uniqueness theorem, if you can find *any* solution to Poisson's equation that satisfies the boundary conditions for the problem of interest, you have found the *only solution,* even if you found it by wild guesses or clever tricks.

Herein lies the strategy of image charges [1, 2]. Rather than trying to solve Poisson's equation for the given charge distribution, look instead for some other imaginary distribution of charges that satisfies the same boundary conditions. The uniqueness theorem says that that imaginary charge distribution will substitute perfectly well to give the correct electrostatic potential.

Let's use this approach to find the electrostatic potential that attracts a point charge Q at a distance a from a conducting planar surface. By definition, charges move freely in a conducting surface. As long as no electric current is flowing, the charges in a conductor experience no electrostatic force within the

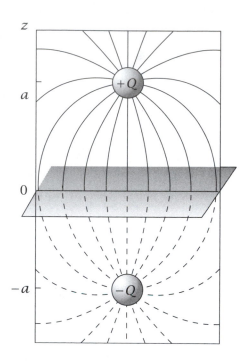

Figure 21.15 The attraction of a charge Q for a conducting plane separated by a distance a can be modelled as though the charge Q were attracted to an 'image charge,' $-Q$ at $z = -a$.

conductor. The field lines cannot have any component that is parallel to the conducting plane, because that would imply forces acting on charges within the plane. So the only field lines near a conductor must be perpendicular to it (see Figure 21.14).

We want to find the electrostatic potential for an ion near a conducting plane. According to the uniqueness theorem, if we can find any other constellation of charges that causes all the field lines to be parallel to the z-axis at a distance a from the point charge, then it will give exactly the right electrostatic potential function. We have already solved this problem in Example 21.5, the potential field around a dipole. Figures 21.11 and 21.15 show that the plane that bisects a dipole has only perpendicular intersections with the field lines from the two dipolar charges, so this must be the solution we are looking for.

The method of image charges says that you can replace the conducting plane in the present problem with an imaginary charge of opposite sign ($-Q$ at $z = -a = -\ell/2$) as though it were the mirror image of the real charge Q, reflected over to the other side of the conducting plane (see Figure 21.15). This pair—the real charge and the imaginary charge—will give exactly the same electrostatic potential as the ion in the presence of the actual conducting plane everywhere on the $z > 0$ side of the conducting plane. With this trick, you now have a very graphic way to view the attraction of the charge Q for the plane. You can think of it as being attracted to its image $-Q$. You save the trouble of computing the electrostatic potential because you have already done it in Equation (21.19). This is shown in Example 21.8.

EXAMPLE 21.8 The attraction of an ion toward a planar conductor. To compute the attraction of an ion Q for a conducting plane, use Figure 21.11(a) on page 395. Now the y-axis coincides with the conducting plane. The ion is indicated by the positive charge at $x = \ell/2$, and the image charge is indicated by the negative charge at $x = -\ell/2$. We are interested in the potential $\psi(\ell/2)$ arising from the image charge $-Q$ at $x = -\ell/2$ that acts on the charge at $x = \ell/2$. You can use Equation (21.19) with $r_- = \ell = 2a$, where a is the distance of the ion from the plane. The second term of Equation (21.19) is not appropriate here because it is not a potential that acts on $+Q$. Equation (21.19) gives

$$\psi = -\frac{\mathcal{C}Q}{2Da} = -\frac{\mathcal{C}Q}{D\ell}. \tag{21.44}$$

The field E is the force per unit charge acting on the particle along the z-axis in Figure 21.14,

$$E = -\frac{d\psi}{d\ell} = -\frac{\mathcal{C}Q}{D\ell^2} = -\frac{\mathcal{C}Q}{4Da^2}. \tag{21.45}$$

The negative sign in Equation (21.45) indicates that a charge Q of either sign is attracted to the conducting plane owing to the induced field. As a gets smaller, the attraction gets stronger.

Now let's apply the same reasoning to a charge near a *dielectric* boundary.

A Charge Near a Dielectric Interface

A charged ion in water will be repelled as it approaches an oil/water interface. In the language of image charges, the charge in water sees its own image across the interface in the oil phase (not a negative image this time) and is repelled by it. Why does it see a positive image? And how strong is the repulsion?

Suppose that you have a charge q at point A in water at a distance a from an oil/water interface (see Figure 21.16). The dielectric constant of water is D_w and the dielectric constant of oil is D_o. In general, we want to know the electrostatic potential at an arbitrary point P in water, at a distance r from A. At P, the electrostatic potential from the charge would be $\psi_P = \mathcal{C}q/D_w r$ if there were no interface. But the presence of the dielectric interface perturbs the field at P.

To find the true electrostatic potential ψ_P, our strategy will be to invent some arrangement of image charges that will satisfy the boundary conditions imposed by the interface. We will make a few arbitrary decisions in doing this. But as long as we preserve the symmetries in the problem so that we satisfy the boundary conditions, the uniqueness theorem says that we'll get the right answer!

We begin by recognizing that there is a fundamental symmetry in this problem. If we violate that symmetry by our invention of image charge positions, we will be unable to satisfy the boundary conditions. The oil/water interface is not a conductor, so the field lines need not be perpendicular to it, as they were in Example 21.8. (Indeed, the field lines look much like the light rays at optical boundaries: Snell's law of optical refraction has its roots in similar physics [2].)

The z-axis is a symmetry axis for the field, so the only reasonable places to put image charges are along the z-axis. Previously, we needed only one image charge. That's because we had help from the condition that the field must be perpendicular to the plane everywhere. Now we need two image charges, q' and q''. We will choose their values later to satisfy the boundary conditions.

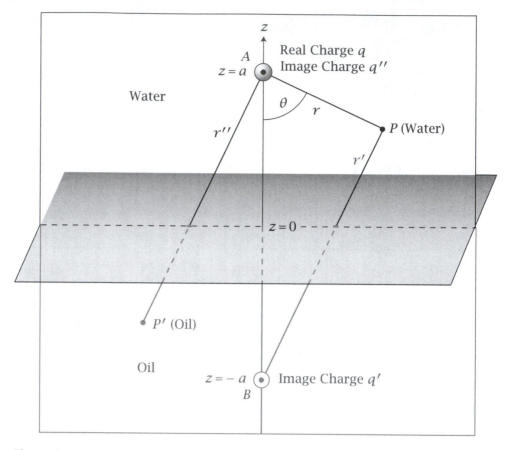

Figure 21.16 The arrangement of the image charges and the electrostatic field for finding the field of a charge near an oil/water interface. The real charge q is at A. It has two images: image charge q'' is also at A (where $z = a$), and the other image q' is at B (where $z = -a$). We are interested in the potential from the charge and its images at P in the water, or P' in the oil.

Begin with the field in the water phase. To describe this, put one image charge q' at point B where $z = -a$. Suppose this image acts on the point P *as though the intervening medium were water*, even though in reality the intervening region includes two media and an interface. The electrostatic potential at point P in water is the sum of the potentials from the real charge at A and the image charge at B:

$$\psi_P = \frac{\mathcal{C}q}{D_w r} + \frac{\mathcal{C}q'}{D_w r'}, \tag{21.46}$$

where r' is the distance from B to P, and P is in the water phase. (The arbitrary decisions so far have been: putting the image at $z = -a$, and assuming that the intervening medium has a single dielectric constant D_w.) If we can ultimately find a value of q' that causes Equation (21.46) to satisfy the boundary conditions, then these arbitrary choices will have been justified, and we will get the true potential at every point, according to the uniqueness theorem.

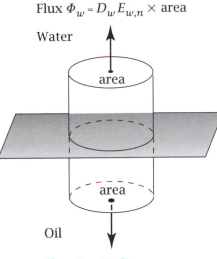

Flux $\Phi_w = D_w E_{w,n} \times$ area

Water

area

area

Oil

Flux $\Phi_o = D_o E_{o,n} \times$ area

Figure 21.17 The Gaussian box at the oil/water interface and the electrostatic fluxes normal to the plane.

How do we determine the field in the oil phase, say at point P'? Again, symmetry can help. Just as we need an image in the oil to account for the forces at P in the water, we also need an image in the water to account for the forces at P' in the oil phase. So let's suppose the field at P' comes from *another image charge* q'' that is located at point A exactly where the real charge q is located. The image at A has a yet unknown charge q'' and we will assume that the intervening medium has the dielectric constant D_o of oil. This gives

$$\psi_{P'} = \frac{\mathcal{C}q''}{D_o r''}. \tag{21.47}$$

If we can choose q' and q'' to satisfy the two appropriate boundary conditions, tangential and normal, then our strategy succeeds.

The boundary condition for the parallel component is the continuity of ψ (see page 398). We must have $\psi_{oil} = \psi_{water}$ at the interface. For points located exactly at the interface, the distances to the charges are $r = r' = r''$, so you can get one boundary condition by combining Equations (21.46) and (21.47):

$$\frac{\mathcal{C}q}{D_w r} + \frac{\mathcal{C}q'}{D_w r} = \frac{\mathcal{C}q''}{D_o r} \quad \Longrightarrow \quad \frac{q}{D_w} + \frac{q'}{D_w} = \frac{q''}{D_o}. \tag{21.48}$$

The second boundary condition comes from the components of the field that are perpendicular to the interface. Make a small Gaussian box that sandwiches the surface (see Figure 21.17). Because the surface is uncharged, the net flux out is zero, so the perpendicular electrostatic flux into the oil phase ($D_o E_{o,n} \times$ area) must equal the perpendicular electrostatic flux into the water phase ($D_w E_{w,n} \times$ area), and

$$D_w E_{w,n} = D_o E_{o,n}, \tag{21.49}$$

where the subscript n indicates the component normal to the interface. Let's evaluate these normal components at the point $z = 0$ along the z-axis. The field components at $z = 0$ from the two charges in the water phase are

$$E_{w,n} = \frac{\mathcal{C}q}{D_w a^2} + \frac{\mathcal{C}q''}{D_w a^2}. \tag{21.50}$$

The oil phase component is

$$E_{o,n} = \frac{\mathcal{C}q'}{D_o a^2}. \tag{21.51}$$

Substituting Equations (21.50) and (21.51) into (21.49) gives

$$q - q' = q''. \tag{21.52}$$

Substitute Equation (21.52) for q'' into Equation (21.48) to get

$$\frac{q}{D_w} + \frac{q'}{D_w} = \frac{q}{D_o} - \frac{q'}{D_o}. \tag{21.53}$$

Rearranging allows you to solve for the image charge q':

$$q' = \frac{D_w - D_o}{D_w + D_o} q. \tag{21.54}$$

Because $D_w > D_o$, the image charge q' has the same sign as the real charge. So the real ion is repelled by the oil interface.

Substitute Equation (21.54) into Equation (21.52), and solve for the image charge q'':

$$q'' = \frac{2D_o}{D_w + D_o} q. \tag{21.55}$$

Because the dielectric constant is much greater for water than for oil, $D_w \gg D_o$, you can approximate Equations (21.54) and (21.55) with $q' \approx q$, and $q'' \approx 0$. You can think of the ion as seeing its own reflection in a mirror and being repelled by it. The ion is driven away from the oil phase by the repulsion. In this limit, the electrostatic potential lines will look like those between two charges of the same sign (see Figure 21.3). If the phase we have labeled 'o' had the *higher* dielectric constant, $D_o > D_w$, then q' would change sign and the interface would become attractive. In the limit $D_o \to \infty$, $q' \to -q$, and the interface attracts as if it were a conducting plane. A charge is attracted to whichever medium has the higher dielectric constant. Example 21.9 illustrates an application.

EXAMPLE 21.9 A sodium ion in water is repelled by an oil surface. Let's compute the work of moving a sodium ion at room temperature from a distance $d_A = 20$ Å to a distance $d_B = 5$ Å from a planar oil interface (see Figure 21.18). Assume that $D_w \gg D_o$. Then Equation (21.54) gives an image charge $q' \approx q = e$ and Equation (21.55) gives $q'' \approx 0$. This means that you can regard all the electrostatic forces in the system as coming from either the real charge q or its first image q'. So, according to Equation (21.46), the electrostatic potential ψ acting on the ion due to the presence of the interface is

$$\psi = \frac{\mathcal{C}q'}{D_w(2a)} = \frac{\mathcal{C}e}{2D_w a}, \tag{21.56}$$

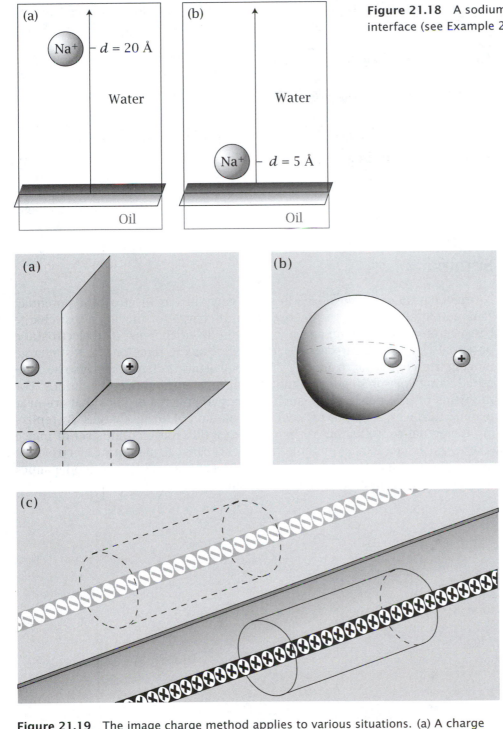

Figure 21.18 A sodium ion in water is repelled by an oil interface (see Example 21.9).

Figure 21.19 The image charge method applies to various situations. (a) A charge (dark) between two conducting planes has three image charges (light). (b) A charge (dark) outside a conducting sphere has an image charge inside (light). (c) A line of charges (dark) next to a conducting plane has an image line of charges behind the plane.

where a is the distance to the interface, and $r' = 2a$ is the distance from the ion to its image. The work of moving the ion from $a_1 = 20$ Å to $a_2 = 5$ Å is

$$w = e\psi(a_2) - e\psi(a_1) = \frac{\mathcal{C}e^2}{D_w}\left(\frac{1}{2a_2} - \frac{1}{2a_1}\right). \tag{21.57}$$

In terms of the dimensionless Bjerrum length ℓ_B defined on page 374,

$$\frac{w}{kT} = \ell_B\left(\frac{1}{2a_2} - \frac{1}{2a_1}\right) = (7.13\text{Å})\left(\frac{1}{10} - \frac{1}{40}\right) = 0.535,$$

$$\implies w = (0.535)(1.987)(300\text{ K}) = 318\text{ cal mol}^{-1}. \tag{21.58}$$

Figure 21.19 shows other symmetries for which there are simple image-charge solutions [2, 3].

Summary

A constellation of charges fixed in space generates an electrostatic potential. This potential describes the energy per unit charge for moving a test charge between any two points in space, due to the fixed charges. The electrostatic potential can be computed as a path integral of the force on the test charge moving from one point to another. The electrostatic potential can also be computed using Poisson's equation and the appropriate boundary conditions. For problems of mixed symmetry, such as charged spheres or lines that are near planar interfaces, an important tool is the method of image charges. By replacing the actual constellation of charges with a different fictitious set of image charges, it is sometimes possible to simplify the calculation of electrostatic fields. In the next chapter we will combine electrostatics with thermodynamics to consider the equilibria of charged systems.

Problems

1. Compute $\Delta\psi$ for a capacitor. You have a parallel-plate capacitor with a distance of 0.20 cm between two 20 cm × 20 cm plates, filled with an oil of dielectric constant $D = 2$. The upper plate has a total charge of 10^{-10} C relative to the lower plate. Neglecting edge effects, what is the potential difference between the upper and lower plates?

2. The potential around an ion. What is the potential at 25 °C at a distance of 10 Å from a monovalent ion in air? And in pure water?

3. A dipole inside a sphere. What is the average potential over a spherical surface that has a dipole at its center?

4. A dipole moment. What is the dipole moment of two charges −e and +e that are 1 Å apart? The unit of dipole moment is 1 Debye = 1 D = 3.33×10^{-30} C m.

5. The charge–dipole interaction of Na^+ with water. What is the maximum attraction in units of kT

(a) between a bare Na^+ ion and a water molecule in air?

(b) between a hydrated Na^+ ion in water and a water molecule?

Take a sphere with radius 0.95 Å for the bare Na^+ ion and a sphere with radius 2.3 Å for the hydrated Na^+ ion. Model a water molecule as a sphere with radius 1.4 Å and a point dipole of moment 1.85 D at its center.

6. Water is a dipole. A water molecule in a vacuum has a dipole moment of 0.62×10^{-29} C m.

(a) Assuming that this dipole moment is due to charges +e and −e at distance ℓ, calculate ℓ in Å.

(b) Consider the water molecule as a sphere with radius 1.5 Å. If the dipole moment in (a) is due to charges $+q$ and $-q$ at the north and south poles of the sphere, how large is q?

7. The potential around a charged sphere. You have an evenly charged spherical surface with radius a and a total charge q in a medium with dielectric constant D. Derive the potential inside and outside the surface by using Coulomb's law and integrating over the charged surface.

8. The potential around a cylindrical shell. You have a hollow inner cylinder with radius a, surrounded by a concentric outer cylinder with radius $a_2 > a_1$. Charges $-\lambda$ and $+\lambda$ per unit length are distributed uniformly over the inner and outer cylinders, respectively. The arrangement is in a medium with dielectric constant D.

(a) What is the potential $\psi(r)$ as a function of the axial distance r?

(b) What is the capacitance per unit length of the arrangement of cylinders?

9. The work of moving a micelle to an oil/water interface. A spherical micelle in water has a charge $q = -60$ e. What is the work required to bring it from deep in a water phase to 100 Å away from an oil/water interface, at $T = 300$ K?

10. The work of bringing an ion near a protein. An ion has negative charge $Q = -2$. A protein has negative charge $q = -10$. What is the work of bringing the ion from infinite distance to within 10 Å of the protein at $T = 300$ K?

References

[1] EM Purcell. *Electricity and Magnetism*. McGraw-Hill, New York, 1965.

[2] IS Grant and WR Phillips. *Electromagnetism*, 2nd edition, Wiley and Sons, New York, 1990.

[3] L Eyges. *The Classical Electromagnetic Field*. Dover Publications, New York, 1972.

Suggested Reading

Clearly written elementary treatments of electrostatics include:

RP Feynman, RB Leighton and M Sands, *The Feynman Lectures on Physics*, volume II, Mainly Electromagnetism and Matter, Addison-Wesley Publishing Company, Reading, Massachusetts, 1964.

EM Purcell, *Electricity and Magnetism*, McGraw-Hill, New York, 1965.

More advanced treatments include:

R Becker, *Electromagnetic Fields and Interactions*, Dover Publications, New York, 1982.

DK Cheng, *Field and Wave Electromagnetics*, 2nd edition, Addison-Wesley, Reading, Massachusetts, 1989.

L Eyges, *The Classical Electromagnetic Field*, Dover Publications, New York, 1972.

IS Grant and WR Phillips, *Electromagnetism*, 2nd edition, Wiley, New York, 1990.

22 Electrochemical Equilibria

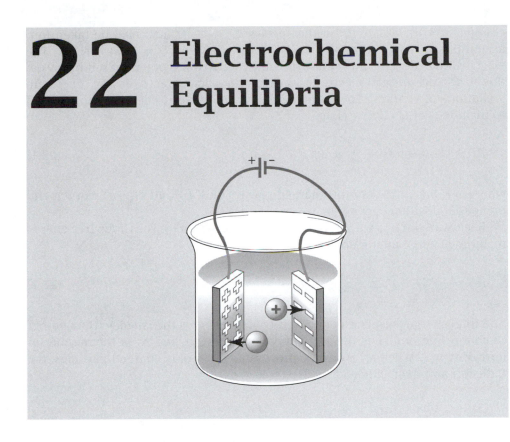

Electrochemical Potentials Describe Equilibria in Solutions of Charged Species

The electrostatic forces and charges on atoms and molecules can affect chemical equilibria. For example, acid–base equilibria are shifted by the presence of charged surfaces or dielectric boundaries. Electric fields can gate the flow of ions through membrane channels. And chemical reactions can cause charge separations, which is the basis for batteries. To explore these processes, we combine the laws of electrostatics, through Coulomb's law and Poisson's equation, with the laws of thermodynamic equilibrium, through the Boltzmann equation. We begin with the partitioning of charged particles, which is governed by a generalization of the chemical potential that is called the *electrochemical potential*.

What Drives the Partitioning of Ions?

If a molecule is charged, it will be driven to partition from one environment to another by both chemical forces—described by chemical potentials—and by electrical forces, if an electrostatic potential is present. To see how these forces combine, we first generalize the thermodynamics of Chapters 7 and 8 to treat electrostatics.

In Chapter 7, we defined *simple systems* for which the internal energy is a function of the extensive variables S, V, and $\mathbf{N}$. If electrical changes occur, the

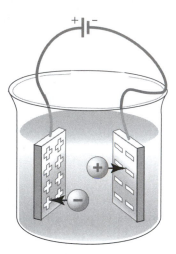

Figure 22.1 A voltage applied to electrodes produces an electrostatic potential $\psi(x)$ that acts on particle i having charge q_i. A negative ion will move toward a positive potential, while a positive ion will move toward a negative potential.

charges $\mathbf{q} = q_1, q_2, \ldots, q_M$ on the species $1, 2, \ldots, M$ must also be taken into account as extensive variables in the energy function $U = U(S, V, \mathbf{N}, \mathbf{q})$. The electrostatic energy is $q\psi$, where ψ is the electrostatic potential felt by an ion due to the presence of electrodes, nearby charged surfaces, or any other constellations of charges. The fundamental equation (see Chapter 8), augmented to include charge effects, is now

$$dU = TdS - pdV + \sum_{j=1}^{t} \mu_j dN_j + \sum_{i=1}^{M} \psi dq_i, \tag{22.1}$$

where $i = 1, 2, \ldots, M$ are the charged species and $\psi = \psi(x, y, z)$ can depend on spatial position in general.

Because $G = U + pV - TS$, the differential equation for the Gibbs free energy including charge interactions is

$$dG = -SdT + Vdp + \sum_{j=1}^{t} \mu_j dN_j + \sum_{i=1}^{M} \psi dq_i. \tag{22.2}$$

The total charge on species i is $q_i = z_i e N_i$, where z_i is the *valency* (the number of charges per ion), e is the unit charge on a proton, and N_i is the number of ions of type i. When an exchange process involves only charged particles, the indices j and i coincide and the free energy is

$$dG = -SdT + Vdp + \sum_{i=1}^{M} (\mu_i + z_i e\psi) dN_i. \tag{22.3}$$

The quantity $\mu_i + z_i e\psi$ defines the *electrochemical potential* μ_i',

$$\mu_i' = \mu_i + z_i e\psi. \tag{22.4}$$

For uncharged species at constant T and p, equilibrium occurs when the chemical potentials are equal. For charged species, equilibrium occurs when the *electrochemical* potentials are equal. Partitioning, solvation, and the transport of charged particles are governed by both chemical and electrostatic forces.

The Nernst Equation: Electrostatic Fields Drive Ion Distributions

Here's how you use Equation (22.4). Suppose that you are interested in the distribution of ions that are subject to a fixed spatial gradient of electrostatic potential. Consider two different locations in space, $\mathbf{r}_1$ and $\mathbf{r}_2$, within a single solution. To eliminate some vector complexity, let's look at a one-dimensional problem. Suppose that $\mathbf{r}_1 = x_1$ and $\mathbf{r}_2 = x_2$. At location x_1, some constellation of fixed charges or electrodes creates the electrostatic potential $\psi(x_1)$, and at x_2 you have $\psi(x_2)$. Now consider a single species of mobile ions (so that you can drop the subscript i) that is free to distribute between locations x_1 and x_2. The mobile ions will be at equilibrium when the electrochemical potentials are equal,

$$\mu'(x_1) = \mu'(x_2). \tag{22.5}$$

Substituting the expression for the chemical potential, $\mu(x) = \mu° + kT \ln c(x)$, into Equation (22.4) gives

$$\mu'(x) = \mu° + kT \ln c(x) + ze\psi(x). \tag{22.6}$$

By regarding $\mu°$ as independent of x in this case, we are restricting our attention to situations where x_1 and x_2 are in the same phase of matter.

In earlier chapters we used mole fraction concentrations. Other concentration units are often useful, and just require a conversion factor that can be absorbed into $\mu°$. Here we express concentrations in a general way as $c(x)$, in whatever units are most useful for the problem at hand. Substituting Equation (22.6) into (22.5) gives the **Nernst equation**, named after WH Nernst (1864–1941), a German chemist who was awarded the 1920 Nobel Prize in Chemistry for his work in chemical thermodynamics:

$$\ln \frac{c(x_2)}{c(x_1)} = \frac{-ze[\psi(x_2) - \psi(x_1)]}{kT}, \tag{22.7}$$

or

$$c(x_2) = c(x_1) \exp\left[\frac{-ze[\psi(x_2) - \psi(x_1)]}{kT} \right]. \tag{22.8}$$

Equation (22.8) is Boltzmann's law for systems involving charges.

According to the Nernst Equations (22.7) and (22.8), positive ions tend to move away from regions of positive potential $\psi(x) > 0$, and toward regions of negative potential. Negative charges move the opposite way (see Figure 22.1). These tendencies are greater, the higher the charge on the ion or the lower the temperature. Figure 22.2 shows an example of a potential that is positive at a plane at $x = 0$, and diminishes with x. While $\mu = \mu° + kT \ln c(x)$ describes a balance between translational entropy and chemical affinity for non-electrolytes, the Nernst Equation (22.8) describes a balance between translational entropy and electrostatic forces for electrolytes. Beginning on page 412 we explore some applications of the Nernst equation.

You should be aware of one important difference between the electrochemical potential and the chemical potential. The chemical potential describes the free energy of inserting one particle into a particular place or phase, subject to any appropriate constraint. Constraints are introduced explicitly. In contrast, the electrochemical potential always carries an *implicit* constraint with it: overall electroneutrality must be obeyed. This is a very strong constraint. You can never insert a single ion in a volume of macroscopic dimensions because that would violate electroneutrality. You can insert only an electroneutral combination of ions.

It is possible to measure the chemical potentials for salts, but not for the individual ions that compose them because electroneutrality must be maintained. So when you are dealing with simple systems, for example systems in which salts are distributed uniformly throughout a macroscopic volume, you can work with the *chemical* potentials of neutral molecules such as NaCl. In contrast, when you are interested in microscopic nonuniformities, such as the counterion distribution near a DNA molecule, or the electrostatic potential microscopically close to an electrode surface, then you can use the electrochemical potential. Charge neutrality need not hold in a microscopic region of

(a)

$\psi(x)$

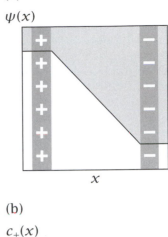

(b)

$c_+(x)$

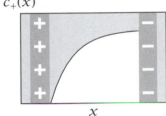

Figure 22.2 If the electrostatic potential $\psi(x)$ (a) is a linear function of distance x, then (b) the concentration $c(x)$ of ions depends exponentially on x, according to Equation (22.8). Positive ions are depleted near the positively charged surface.

interest. But even when charge neutrality does not hold locally, it must hold throughout the larger volume.

The Chemical Potential of Neutral Salts

What is the chemical potential of a neutral salt that ionizes in solution? For salts such as NaCl, add the electrochemical potentials of the separate ions to get the chemical potential of the neutral salt. For NaCl:

$$\mu_{NaCl} = \mu'_{Na^+} + \mu'_{Cl^-}$$

or, with Equation (22.6),

$$\mu_{NaCl} = \mu^\circ_{Na^+} + kT \ln c_{Na^+} + e\psi_{Na^+} + \mu^\circ_{Cl^-} + kT \ln c_{Cl^-} - e\psi_{Cl^-},$$

where ψ_{Na^+} and ψ_{Cl^-} are the electrostatic potentials felt by a Na^+ ion and a Cl^- ion, respectively, due to whatever field is present from external electrodes or nearby charges, for example. To get the standard chemical potential of the salt, μ°_{NaCl}, add the constants $\mu^\circ_{Na^+}$ and $\mu^\circ_{Cl^-}$:

$$\mu_{NaCl} = \mu^\circ_{NaCl} + kT \ln (c_{Na^+} c_{Cl^-}) + e(\psi_{Na^+} - \psi_{Cl^-}). \tag{22.9}$$

In solutions where the electrostatic potential is spatially uniform, $\psi_{solution} = \psi_{Na^+} = \psi_{Cl^-}$, and Equation (22.9) becomes

$$\mu_{NaCl} = \mu^\circ_{NaCl} + kT \ln (c_{Na^+} c_{Cl^-}). \tag{22.10}$$

In terms of the salt concentration $c_{NaCl} = c_{Na^+} = c_{Cl^-}$,

$$\mu_{NaCl} = \mu^\circ_{NaCl} + 2kT \ln c_{NaCl}. \tag{22.11}$$

The factor of 2 in Equation (22.11) comes from assuming the complete ionization of NaCl.

This description is too simple in two situations. First, sometimes salts do not ionize completely. Second, near large charged surfaces such as colloids, large electrostatic potentials can lead to variations in ion concentrations, $c_{Na^+} \neq c_{Cl^-}$. Incomplete dissociation and microscopic nonuniformities are treated in Chapter 23 using the Debye–Hückel and Poisson–Boltzmann theories.

The Nernst equation provides the operating principle for batteries, electrolysis (the production of hydrogen and oxygen gases from water), electroplating, oxidations and reductions, and corrosion. These processes all depend on the equilibrium between a charged solid surface and its corresponding ions in solution.

The Nernst Equation for an Electrode

If you put a silver electrode into a liquid solution that contains silver ions, the silver ions can either deposit onto the electrode or dissolve in the solution. The ions exchange between the electrode and the solution until they reach an equilibrium that is dictated by both the chemical potential and the electrostatic interactions of the silver ions with the charged surface. Such equilibria have important applications. If you apply an electrostatic potential difference (a voltage) to such a solution/electrode interface, it can drive metal ions onto solid surfaces. This is the operating principle for electroplating. Conversely, if the solution is an acid that dissolves the metal electrode, it can cause the flow of ions and electrons, forming a battery. Here's a description of such equilibria.

Suppose that a silver electrode is in equilibrium with a solution of silver nitrate ($AgNO_3$), as shown in Figure 22.3. Silver is called the 'common ion' in this case because it is the exchangeable component that is common to both the liquid and the solid. Silver ions (Ag^+) can plate onto the solid metal or dissolve to increase the $AgNO_3$ concentration in the liquid. Use the concentration c of Ag^+ in solution as the degree of freedom. The condition for equilibrium is that the electrochemical potential of the silver ions in the solid must equal the electrochemical potential of the silver ions in the liquid:

$$\mu'_{Ag^+}(\text{solid}) = \mu'_{Ag^+}(\text{liquid}). \tag{22.12}$$

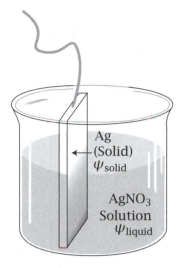

Figure 22.3 A silver electrode in a silver nitrate solution. The silver ions are free to plate onto the metal or dissolve in solution.

According to Equation (22.6) the electrochemical potential for the liquid solution is

$$\mu'_{Ag^+}(\text{liquid}) = \mu^\circ_{Ag^+}(\text{liquid}) + kT \ln c + ze\psi_{\text{liquid}}, \tag{22.13}$$

where ψ_{liquid} is the electrostatic potential in the liquid. ψ_{liquid} depends on voltages applied to the system, or on chemical reactions in the solution. In the pure solid $c = 1$, and the electrochemical potential of the silver on the electrode is

$$\mu'_{Ag^+}(\text{solid}) = \mu^\circ_{Ag^+}(\text{solid}) + ze\psi_{\text{solid}}. \tag{22.14}$$

According to Equation (22.12), the equilibrium condition is

$$\mu^\circ_{Ag^+}(\text{solid}) + ze\psi_{\text{solid}} = \mu^\circ_{Ag^+}(\text{liquid}) + kT \ln c + ze\psi_{\text{liquid}}. \tag{22.15}$$

Equation (22.15) has four quantities that are not independently measurable, two μ°'s and two ψ's. To use Equation (22.15), focus on a process that reduces the number of indeterminate quantities. Compare two measurements: an electrode contacts a liquid solution having Ag^+ concentration c_1 in the first measurement, and c_2 in the second. Because the solid is identical in the two cases, you can set the right-hand side of Equation (22.15) equal for the two different solutions, liquid$_1$ and liquid$_2$, at equilibrium with the solid:

$$\mu^\circ_{Ag^+}(\text{liquid}) + kT \ln c_1 + ze\psi_{\text{liquid}_1}$$
$$= \mu^\circ_{Ag^+}(\text{liquid}) + kT \ln c_2 + ze\psi_{\text{liquid}_2}, \tag{22.16}$$

which rearranges to the Nernst equation:

$$\Delta\psi = \psi_{\text{liquid}_1} - \psi_{\text{liquid}_2} = \left(\frac{kT}{ze}\right) \ln \left(\frac{c_2}{c_1}\right), \tag{22.17}$$

with $z = 1$ for $AgNO_3$. Using $R = \mathcal{N}k$, and the Faraday constant, which is the unit charge per mole, $F = \mathcal{N}e = 23{,}060$ cal mol^{-1} V^{-1} = 96,500 C mol^{-1}, Equation (22.17) can be given in molar units:

$$\Delta\psi = \frac{RT}{zF} \ln \left(\frac{c_2}{c_1}\right). \tag{22.18}$$

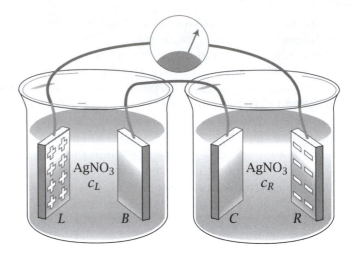

Figure 22.4 Two different salt solutions can create a voltage, hence a battery, as in Example 22.1. Electrical current flows freely between electrodes B and C, holding the two solutions at the same potential.

For $T = 300$ K and univalent charges $z = +1$, you have $RT/zF = 0.0258$ V, and $\ln x = 2.3026 \log_{10} x$, so the Nernst equation can be written as (in volts)

$$\Delta \psi = 0.0594 \log_{10} \left(\frac{c_2}{c_1} \right). \tag{22.19}$$

According to Equation (22.19), for every factor of 10 increase in the concentration of $AgNO_3$ in solution, the electrostatic potential of the liquid increases by 59.4 mV.

It is conventional to define a reference solution of activity $a_0 = 1$ having electrostatic potential ψ_0. So you can express Equation (22.18) as

$$\psi = \psi_0 - \frac{RT}{zF} \ln a, \tag{22.20}$$

where a is the activity of the ion of interest.

ψ_0 is called a *half-cell potential*. A reaction is called a *half*-cell because it takes two such solid/liquid equilibria to form a complete battery, or cell. The reference potential ψ_0 is the following collection of the unmeasurable quantities in Equation (22.15):

$$ze\psi_0 = \mu^\circ(\text{solid}) - \mu^\circ(\text{liquid}) + ze\psi_{\text{solid}}. \tag{22.21}$$

Half-cell potentials are usually measured from some standard level that, in itself, is arbitrary. The accepted standard is $\psi_0 = 0$ for the hydrogen electrode reaction $H^+ + e^- \rightarrow (1/2)H_2(g)$.

EXAMPLE 22.1 Making a battery out of salt solutions. A battery is a device that creates a difference in electrostatic potential (voltage) between two electrodes. One way to create a difference $\Delta \psi$ is to use two different salt solutions. Figure 22.4 shows how to do this in a device called a *concentration cell*. Suppose you choose Ag^+ as the common ion. Electrodes L (left) and R (right) are silver, and the salt in solution is $AgNO_3$. Electrodes B and C are *reference* electrodes (their electrostatic potentials do not depend on the concentration of either Ag^+ or NO_3^-). Ag^+ is the common ion, also called the *potential determining ion*.

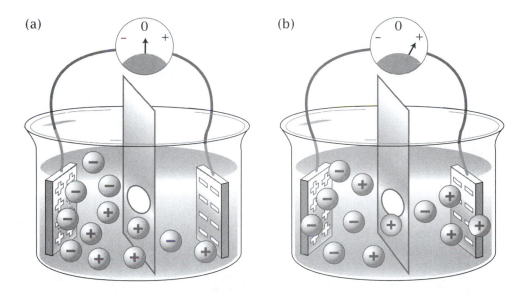

Figure 22.5 Model for an ion channel. (a) The initial concentration of positive ions on the left is high, before the system equilibrates. If positive ions can flow, but negative ions cannot, then positive ions tend to flow down their concentration gradient from left to right. This creates a net positive electrostatic potential on the right that opposes further flow. (b) The final equilibrium state is more positive on the right.

Using Equation (22.15) for electrodes L and R, you have

$$\mu^\circ(\text{solid}, L) + ze\psi_{\text{solid}\, L} = \mu^\circ(\text{liquid}, L) + kT \ln c_L + ze\psi_{\text{liquid}\, L} \qquad (22.22)$$

and

$$\mu^\circ(\text{solid}, R) + ze\psi_{\text{solid}\, R} = \mu^\circ(\text{liquid}, R) + kT \ln c_R + ze\psi_{\text{liquid}\, R}. \qquad (22.23)$$

Electrodes L and R are identical, so $\mu^\circ(\text{solid}, L) = \mu^\circ(\text{solid}, R)$. The standard states are identical, $\mu^\circ(\text{liquid}, L) = \mu^\circ(\text{liquid}, R)$. Because reference electrodes B and C hold the two solutions at the same electrostatic potential, you have $\psi_{\text{liquid}\, L} = \psi_{\text{liquid}\, R}$. Subtracting Equation (22.23) from (22.22) gives

$$\Delta\psi = \psi_{\text{solid}\, L} - \psi_{\text{solid}\, R} = \frac{kT}{ze} \ln\left(\frac{c_L}{c_R}\right). \qquad (22.24)$$

Because a voltage difference is created across the two electrodes by the difference between the salt concentrations c_L and c_R, the system acts as a battery. But since you get only about 60 mV per factor of ten difference in concentration, a one volt battery would require concentration differences greater than 10^{16}! There are better ways to make batteries. This example just shows how differences in salt concentration can create voltages.

It is difficult to measure the potential difference between two phases such as an electrode and a solution. This is because a minimal measuring circuit must involve at least two boundaries. If you have two different electrodes in a liquid, you have two electrode/solution boundaries. Any potential that you measure cannot be unambiguously attributed to one particular electrode/solution

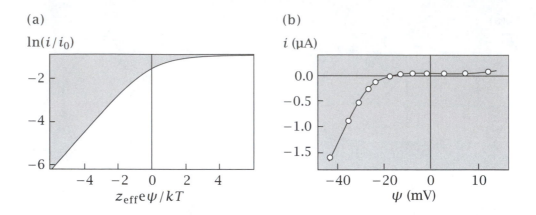

Figure 22.6 (a) Relationship between the current i and the applied voltage ψ from Equation (22.29). (b) Current-voltage measurements for a potassium channel in the egg cell membrane of a starfish. Source: (a) TW Weiss, *Cellular Biophysics, Volume 2: Electrical Properties*, MIT Press, Cambridge, 1996. (b) B Hille, *Ionic Channels of Excitable Membranes*, Sinauer, Sunderland, MA, 1984. Data are from S Hagiwara, S Miyazaki and NP Rosenthal, *J Gen Physiol* **67**, 621–638 (1976).

interface. For example, it is difficult to know the potential difference between a biological membrane and water, because any measurement must also involve electrodes, which have their own potentials.

Selective Ion Flow Through Membranes Can Create a Voltage

Some biological membranes have protein pores or channels that allow certain ions to pass through, but not others. Figure 22.5 shows a hole in a membrane to represent a potassium channel, through which the permeant K^+ ions pass freely but other ions cannot. Suppose you put a high concentration of KCl on the left and a low concentration on the right. At first, each side is electroneutral so there are no electrical driving forces. K^+ flows from left to right down its concentration gradient. There is no counterbalancing flow of Cl^-, because Cl^- cannot permeate the membrane. Therefore, a net positive charge builds up on the right side of the membrane. Equilibrium is achieved when the electrochemical potential for K^+ is the same on both sides,

$$\mu'_{K^+}(L) = \mu'_{K^+}(R)$$

$$\implies \quad \mu^\circ_L + kT\ln[K^+]_L + e\psi_L = \mu^\circ_R + kT\ln[K^+]_R + e\psi_R$$

$$\implies \quad \Delta\psi = \psi_L - \psi_R = \frac{kT}{e}\ln\frac{[K^+]_R}{[K^+]_L}, \tag{22.25}$$

because $\mu^\circ_L = \mu^\circ_R$ and $z = 1$.

Equation (22.25) defines the *potassium potential*. If Na^+ or Ca^{2+} were the permeant ion instead of K^+, there would be a different potential. You can think of this potential in two different ways. First, it is the voltage that arises as a counterbalance when an ion flows down its concentration gradient. Second, if you applied the K^+ potential across a membrane, it would drive K^+ to flow until it reached this particular ratio of concentrations inside and outside.

> **EXAMPLE 22.2 A potassium potential in skeletal muscle.** In mammalian skeletal muscle, the extracellular potassium concentration is $[K^+]_{out} = 4\,mM$ and the intracellular concentration is $[K^+]_{in} = 155\,mM$. The K^+ potential at $37\,°C$ is given by Equation (22.25):
>
> $$\psi_{in} - \psi_{out} = \frac{(8.314\,J\,K^{-1}\,mol^{-1})(310\,K)}{(9.65 \times 10^4\,C\,mol^{-1})} \ln\left(\frac{4\,mM}{155\,mM}\right) = -98\,mV.$$

Voltage-Gated Ion Channels May Be Modeled as Orienting Dipoles

Ions sometimes flow through a biological membrane in one direction but not the other. Figure 22.6(b) shows experimental evidence that applying an external voltage opens a protein channel to allow ions to pass through. Here is a simple model for voltage-gated ion conductance, based on the assumption that the applied field orients a dipole within the protein channel (see Figure 22.7).

The equilibrium constant for channel opening can be expressed in terms of the ratio of the probabilities that the channel is in the open or closed states,

$$\frac{p_{open}}{p_{closed}} = e^{-(G_0-w)/kT}, \tag{22.26}$$

where G_0 is the free energy required to open the channel in the absence of the applied field, and w is the work performed by the applied electrostatic field in orienting the dipole.

Applying an electrostatic field can shift the equilibrium toward the open state. According to Equation (21.18), the work performed by a field in aligning a dipole is

$$w(\theta) = Eq\ell\left(\cos\theta_{open} - \cos\theta_{closed}\right) = -\frac{\psi q\ell}{d}(\Delta\cos\theta). \tag{22.27}$$

We have used $E = -\psi/d$ to convert from the field E to the voltage difference across the membrane (ψ is the potential of the lower solution relative to 0 in the upper solution) because the solutions on the two sides of the membrane act as parallel plate electrodes across a membrane of thickness d. Because the quantities $\Delta\cos\theta = \cos\theta_{open} - \cos\theta_{closed}$ and q and ℓ may not be known independently, we can reduce the number of unknown parameters in the model by defining an 'effective charge' $z_{eff}e$,

$$z_{eff}e = \frac{-q\ell}{d}(\Delta\cos\theta). \tag{22.28}$$

We introduced the minus sign in Equation (22.28) because we are free to define z_{eff} any way we want, and this simplifies the interpretation of the signs. In particular, any favorable arrangement of the dipole (negative end near the positive electrode, for example, as shown in Figure 22.7) leads to a negative value of $q\psi$ or a positive value of $z_{eff}e$.

(a) No Field: Closed Channel

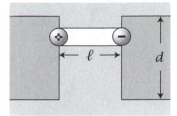

(b) Applied Field: Open Channel

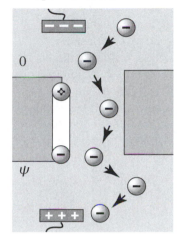

Figure 22.7 A voltage-gated channel. (a) gate closed; (b) applied potential opens gate.

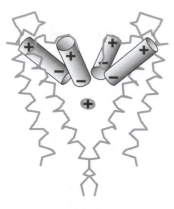

Figure 22.8 The structure of a membrane protein channel composed of helices that have dipole moments. A K^+ ion is shown at the center. Source: B Roux and R MacKinnon, *Science* **285**, 100–102 (1999).

The ion flow through the channel is proportional to the fraction of time that the channel is open, which you can compute by substituting Equation (22.28) into (22.27) and then into Equation (22.26) to get

$$\frac{p_{open}}{p_{open} + p_{closed}} = \frac{1}{1 + (p_{closed}/p_{open})} = \frac{1}{1 + e^{(G_0 - z_{eff}e\psi)/kT}}. \tag{22.29}$$

Figure 22.6(a) shows the prediction of Equation (22.29) that the gate opens for positive applied voltages, and becomes increasingly closed for negative applied voltages. The structure of a bacterial K^+ channel shown in Figure 22.8 illustrates that membrane proteins are sometimes composed of helices. They might serve as dipolar gates [1, 2].

Electric fields can not only move charged molecules, they can also shift chemical equilibria.

Acid–Base Equilibria Are Shifted by Electrostatic Fields

Acid–base equilibria can be shifted in the presence of electrical fields. For example, the pK_a of an acid group changes if the acid is near a charged surface, such as a membrane or a protein. H^+ ions will be repelled near a positively charged surface, causing an acid attached to the surface to give up its protons more easily than in the absence of the surface. This makes the acid appear to be stronger. Let's make this more quantitative.

Consider the dissociation of an acid HA in solution,

$$HA \xrightarrow{K_a} A^- + H^+ \qquad \text{where} \qquad K_a = \frac{[A^-][H^+]}{[HA]}. \tag{22.30}$$

The subscript a on K_a indicates that this is an *acid* dissociation constant (not an *association* constant: K_b is the dissociation constant for bases). Taking the logarithm of both sides of Equation (22.30) for K_a gives the **Henderson–Hasselbalch** equation,

$$\log K_a = \log \frac{[A^-]}{[HA]} + \log[H^+], \tag{22.31}$$

which relates the equilibrium constant K_a to the hydrogen ion concentration. Using the notation $p = -\log_{10}$, you have

$$pK_a = pH - \log \frac{[A^-]}{[HA]}. \tag{22.32}$$

Now consider acid groups A that are rigidly tethered at a fixed distance x from a charged surface (see Figure 22.9). The H^+ ions are distributed throughout the solution, not just at x. For a tether of length x, you can express the equilibrium constant as

$$K_a(x) = \frac{[A^-]_x[H^+]_x}{[HA]_x},$$

where $[H^+]_x$, $[A^-]_x$, and $[HA]_x$ are the concentrations of hydrogen ions and unprotonated and protonated acid groups at a distance x from the surface.

But $K_a(x)$ is not a particularly useful quantity because $[H^+]_x$ is difficult to determine. More useful is the *apparent equilibrium constant*,

$$K_{app}(x) = \frac{[A^-]_x[H^+]_\infty}{[HA]_x},$$

because $-\log[H^+]_\infty$ is the pH of the bulk solution, far away from the surface, which can be readily measured.

$pK_a(x)$ and $pK_{app}(x)$ are related. The Nernst equation (22.8) shows that the electrochemical potentials of the hydrogen ions must be the same everywhere at equilibrium, $\mu'_{H^+}(x) = \mu'_{H^+}(\infty)$ (see also Equation (22.5)), so Equation (22.8) with $\psi(\infty) = 0$ gives

$$[H^+]_x = [H^+]_\infty e^{-e\psi(x)/kT}. \tag{22.33}$$

An additional useful relationship is $K_a(x) = K_a(\infty)$, which is shown as follows. The condition for electrochemical equilibrium is

$$\mu_{HA}(x) = \mu'_{A^-}(x) + \mu'_{H^+}(x)$$

$$\implies \mu^\circ_{HA} + kT\ln[HA]_x = \mu^\circ_{A^-} + kT\ln[A^-]_x - e\psi(x)$$

$$+ \mu^\circ_{H^+} + kT\ln[H^+]_x + e\psi(x). \tag{22.34}$$

Rearranging Equation (22.34) gives

$$K_a(x) = \frac{[A^-]_x[H^+]_x}{[HA]_x} = e^{-\Delta\mu^\circ/kT}, \tag{22.35}$$

where $\Delta\mu^\circ = \mu^\circ_{HA} - \mu^\circ_{A^-} - \mu^\circ_{H^+}$ is independent of x. So $K_a(x) = K_a(\infty)$ is independent of x. The positive surface stabilizes A^- to the same degree that it destabilizes H^+, so the $e\psi(x)$ terms cancel.

Now substitute $K_a(x) = K_a(\infty)$ and Equations (22.33) and (22.35) into the definition of $K_{app}(x)$ to get

$$K_{app}(x) = \frac{[A^-]_x[H^+]_x}{[HA]_x} e^{e\psi/kT} = K_a(\infty)e^{e\psi(x)/kT}. \tag{22.36}$$

Use $p = -\log_{10} = -0.4343\ln$ to simplify Equation (22.36) to

$$pK_{app}(x) = pK_a(\infty) - \frac{0.4343\,e\psi(x)}{kT}. \tag{22.37}$$

A positively charged surface with a potential $\psi > 0$ causes the pK_{app} to be lower than the solution value $pK_a(\infty)$ (that is $K_{app} > K_a(\infty)$), so an acid near a positively charged surface will seem to be a stronger acid than the same group in bulk solution. A positive surface leads to greater dissociation and a higher value of $[A^-]/[HA]$ for a given bulk concentration of H^+ ions. Conversely, an acid near a negatively charged surface will seem to be weaker than the corresponding acid in solution.

EXAMPLE 22.3 Histidine in a spherical protein. Histidine is a basic amino acid with $pK_a(\infty) = 6.0$. Suppose that a histidine is attached to the surface of a spherical protein of radius $a = 30\,\text{Å}$ in water. Suppose that the protein has

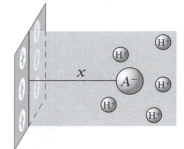

Figure 22.9 An acid group A^- is tethered at a distance x from a charged surface. It undergoes a reaction $HA \rightarrow H^+ + A^-$.

Table 22.1 pK_{app}, the apparent pK_a of an acid group, as a function of distance a from an oil/water interface into the water phase.

a Å	pK_{app}
∞	4.6
5	4.91
4	4.99
3	5.12
2	5.37

a net charge of $q = -10e$. What is the pK_{app} of the histidine? For a sphere in water you have the surface potential from Equation (21.37):

$$\frac{e\psi_a}{kT} = \frac{\mathcal{C}qe}{DkTa} = -10\left(\frac{\ell_B}{a}\right) = -10\left(\frac{7.13}{30}\right) = -2.38.$$

With Equation (22.37), you have

$$pK_{app} = 6.0 + (0.4343 \times 2.38) = 6.0 + 1.03 = 7.03.$$

Acid–base equilibria are affected not only by charged interfaces, but also by dielectric interfaces, as shown in Example 22.4.

EXAMPLE 22.4 Acid dissociation near an oil/water interface. Consider an acid group in water near a dielectric interface. For example, an acidic group may be tethered to an uncharged bilayer membrane in contact with water. Suppose that the intrinsic pK_a of the acid in the bulk solution is p$K_a(\infty) = 4.6$. What is the pK_a of the same acidic group as a function of the tethering distance from the oil interface?

First determine the electrostatic image potential that will be felt by the acid at a distance a from the interface. Assuming that $D_{water} \gg D_{oil}$, Equations (21.54) and (21.55) give $q' = q$ and $q'' = 0$, and Equation (21.56) gives a negative potential,

$$\frac{e\psi}{kT} = -\frac{\mathcal{C}e^2}{2aDkT} = -\frac{\ell_B}{2a} = -\frac{7.13\text{ Å}}{2a}. \tag{22.38}$$

Substituting Equation (22.38) into Equation (22.37) gives Table 22.1, the apparent pK_a of an acid in water as a function of its distance from a planar interface with oil. This treatment neglects atomic detail, which may be important at such short distances, so it is only approximate.

Now we switch from the thermodynamics to the kinetics of moving charges. What are the forces that determine the velocities of ions or electrons in flow?

Gradients of Electrostatic Potential Cause Ion Flows

Concentration gradients can cause particles to flow, whether the particles are neutral or charged. In addition, charged particles flow if they are subject to gradients of *electrostatic* potential. According to Equation (20.12), the force on a charged particle equals (charge ze) $\times$ (electrostatic field E),

$$f = zeE = -ze\left(\frac{\partial\psi}{\partial x}\right), \tag{22.39}$$

where Equation (21.7) relates E to ψ. Here we treat simple instances where the electrostatic potential $\psi(x)$ varies only in the x-direction.

A particle's velocity v is related to the force f acting on it through Equation (18.41), $f = \xi v = (1/u)v$, where $u = 1/\xi$ is the *mobility*, the inverse of the friction coefficient. Substitute this expression into Equation (22.39) and use $J = cv = cuf$ (Equation (18.1)) to get the flux of ions J that results from a

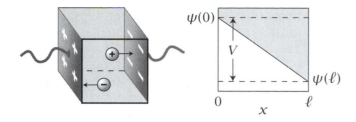

Figure 22.10 Applying a voltage $V = \psi(0) - \psi(\ell)$ to plates separated by distance ℓ leads to ion flow and electrical conductance.

gradient of electrostatic potential,

$$J = -zecu\left(\frac{\partial \psi}{\partial x}\right) = -\kappa\left(\frac{\partial \psi}{\partial x}\right),$$ (22.40)

where κ is a measure of *electrical conductivity*. Like Fick's law and Fourier's law, Equation (22.40) describes a flow that is proportional to the steepness of a gradient. While the diffusion constant describes the flux of particles per unit of concentration gradient, $\kappa = zecu$ describes the flux of charged particles per unit of electrostatic potential gradient.

Equation (22.40) is an expression of **Ohm's law**, which says that electrical current flow is linearly proportional to the difference in applied electrostatic potential, or voltage. When a voltage $V = \psi(0) - \psi(\ell)$ is applied, charge carriers (ions or electrons) will flow, giving an electrical current I,

$$I = JA = \frac{\kappa A}{\ell}V,$$ (22.41)

where A is the cross-sectional area of the conducting medium, and the electrical *resistance* is $R = \ell/\kappa A$. Ion conductivities and mobilities can be measured in the type of device shown in Figure 22.10. A voltage is applied, and the resulting electrical current flow is measured.

The Nernst–Planck Equation Describes Ion Flow due to Both Electrical Gradients and Concentration Gradients

If an ion is driven by both a gradient of its concentration and a gradient of electrostatic potential, then its flux J_p will be the sum of the component fluxes,

$$J_p = -D\left(\frac{\partial c}{\partial x}\right) - zecu\left(\frac{\partial \psi}{\partial x}\right).$$ (22.42)

Equation (22.42) is the **Nernst–Planck equation**. The subscript p on the flux J_p indicates that this is the flux of *particles*. Because each particle carries a charge ze, the flux of *charge* J_c can be expressed as

$$J_c = zeJ_p.$$ (22.43)

Substituting Equation (22.43) and the Einstein relation $D = ukT$ into Equation (22.42) gives

$$J_c = -zecu\left(\frac{kT}{c}\left(\frac{\partial c}{\partial x}\right) + ze\left(\frac{\partial \psi}{\partial x}\right)\right) = -zecu\left(kT\left(\frac{\partial \ln c}{\partial x}\right) + \left(\frac{\partial(ze\psi)}{\partial x}\right)\right)$$

$$= -zecu\left(\frac{\partial \mu'}{\partial x}\right).$$ (22.44)

Table 22.2 Electric mobilities u of ions at 25 °C. Source: B Hille, *Ionic Channels of Excitable Membranes*, Sinauer, Sunderland, 1984.

Ion	u $(cm^2\,V^{-1}\,s^{-1})$
H^+	36.25
Li^+	4.01
Na^+	5.19
K^+	7.62
Cs^+	8.01
NH_4^+	7.52
Mg^{2+}	2.75
Ca^{2+}	3.08
F^-	5.74
Cl^-	7.92
Br^-	8.09
I^-	7.96
NO_3^-	7.41
SO_4^{2-}	4.15

Equation (22.44) is analogous to Equations (18.4) to (18.6), but here the driving force for charged particles is the electrochemical potential, $\mu' = \mu° + kT \ln c(x) + ze\psi(x)$, instead of the chemical potential that drives the flow of neutral particles.

In the next section we use the Nernst–Planck equation to show that the electrostatic potentials across membranes depend not only on the difference in ion concentrations, but also on the *ion mobilities*.

Ion Flows Through Membranes

Suppose that you start with a high concentration c_ℓ of salt on the left side of a membrane and a lower concentration c_r on the right. Suppose that the membrane is permeable to both the cations and anions. The positive and negative ions both tend to flow from left to right, down their concentration gradients. The flux of positive ions is given by Equation (22.44):

$$J_+ = -z_+ecu_+ \left[\frac{kT}{c} \left(\frac{\partial c}{\partial x} \right) + z_+e \left(\frac{\partial \psi}{\partial x} \right) \right], \tag{22.45}$$

and the flux of negative ions is

$$J_- = -z_-ecu_- \left[\frac{kT}{c} \left(\frac{\partial c}{\partial x} \right) + z_-e \left(\frac{\partial \psi}{\partial x} \right) \right], \tag{22.46}$$

where u_+ is the mobility of the positive ions, u_- is the mobility of the negative ions, and z_+ and z_- are the valencies of the positive and negative ions. Electroneutrality requires that $J_+ + J_- = 0$. To make the math simple, consider monovalent ions, so that $z_+ = +1$ and $z_- = -1$. Adding Equation (22.45) to (22.46) gives

$$0 = (u_- - u_+)ekT \left(\frac{\partial c}{\partial x} \right) - (u_+ + u_-)e^2c \left(\frac{\partial \psi}{\partial x} \right)$$

$$\implies \quad \left(\frac{\partial \psi}{\partial x} \right) = \frac{(u_- - u_+)}{(u_+ + u_-)} \frac{kT}{ec} \left(\frac{\partial c}{\partial x} \right). \tag{22.47}$$

Integrating both sides of Equation (22.47) over x, across the membrane gradients both of concentration and electrostatic potential, gives

$$\Delta\psi = \psi_\ell - \psi_r = \frac{kT}{e} \frac{(u_- - u_+)}{(u_+ + u_-)} \ln \frac{c_\ell}{c_r}, \tag{22.48}$$

where ψ_ℓ and ψ_r are the electrostatic potentials on the two sides of the membranes.

Let's consider two different situations. First, the positive ions can cross the membrane but the negative ions cannot ($u_- = 0$), so Equation (22.48) reduces to the result that we found in the Nernst Equation (22.7). Then if $c_\ell = 0.1\,M$ and $c_r = 0.01\,M$, the electrostatic potential difference is $\Delta\psi = -60\,mV$.

But second, the additional implication of Equation (22.48) is that the electrostatic potential across the membrane can depend not only on the salt concentrations on the two sides, but also on the *rates* of ions flowing across the membrane. This is called a *diffusion potential*. Table 22.2 lists ion mobilities u in water at 25 °C.

Creating a Charge Distribution Costs Free Energy

So far in this chapter we have considered only a limited class of electrostatic processes: moving charges from one point to another within a fixed electrostatic field. Now we consider the other main class of electrostatic problems: computing the free energies for creating the electrostatic fields in the first place. You can view this either as a process of bringing a constellation of charges together into the correct configuration, or as a process of 'charging up' an assembly of originally uncharged particles that are already in the correct configuration. Such processes of charging, discharging, and assembly are used to describe how ions partition between different phases, move across membranes, or become buried in proteins, for example.

The Method of Assembling Charges

Imagine assembling a system of charged particles, one at a time. The first particle has charge q_1. The electrostatic work of bringing a second particle, having charge q_2, from infinity into the electrostatic field of the first is

$$w_{\text{el}} = q_2\psi_2, \tag{22.49}$$

where ψ_2 is the electrostatic potential felt by particle 2 due to particle 1. Taking the potential at infinite separation to be zero, the electrostatic potential ψ_2 at the location of charge q_2 is

$$\psi_2 = \frac{\mathcal{C}q_1}{Dr_{12}}, \tag{22.50}$$

where r_{12} is the distance between the charges. When you have more than two charges, the superposition principle says that the electrostatic potential acting on charge i is the sum of the electrostatic potentials from all other charges j:

$$\psi_i = \sum_{j \neq i}\frac{\mathcal{C}q_j}{Dr_{ij}}. \tag{22.51}$$

The total electrostatic free energy ΔG_{el} for assembling the charges is

$$\Delta G_{el} = w_{el} = \frac{1}{2} \sum_i q_i \psi_i = \frac{\mathcal{C}}{2D} \sum_i \sum_{j \neq i} \frac{q_i q_j}{r_{ij}}. \tag{22.52}$$

The factor $1/2$ corrects for the double counting: charge j is included as contributing to the electrostatic potential experienced by i, and i is also counted as contributing to the electrostatic potential experienced by j if the sum in Equation (22.52) is over all charges $i = 1, 2, \ldots, N$.

When a system has many charges, it may be more convenient to use integrals rather than sums, and to approximate the distribution of discrete charges as a continuous function $\rho(x, y, z)$ describing charge densities. For a continuous charge distribution with ρ unit charges per unit volume in a total volume V, the free energy is

$$\Delta G_{el} = \frac{1}{2} \int_V \rho \psi_V \, dV, \tag{22.53}$$

where, according to Equation (21.10), $\psi_V = (\mathcal{C}/D) \int_V \rho(\mathbf{r}) r^{-1} \, dV$.

If you have a continuous *surface* distribution of charge, with σ charges per unit *area* on a surface s, then the electrostatic free energy is given by a similar integral

$$\Delta G_{el} = \frac{1}{2} \int_s \sigma \psi_s \, ds, \tag{22.54}$$

where $\psi_s = (\mathcal{C}/D) \int_s \sigma r^{-1} \, ds$ is the surface equivalent of Equation (21.10).

These continuum equations contain an error, which is often small. While Equation (22.51) correctly sums over all charges *except* i, the conversion to integrals in Equation (22.53) and (22.54) incorrectly includes the interaction of charge i with itself, called the *self-energy*. This error is small when the number of charges is large enough to warrant the conversion from the sum to the integral. Example 22.6 is an application of Equation (22.54).

EXAMPLE 22.6 The free energy of charging a sphere. What is the free energy required to charge a sphere of radius a to a total charge q uniformly distributed over its surface? The total area of the sphere is $4\pi a^2$. The surface charge density, total charge divided by total area, is $\sigma = q/4\pi a^2$. At the surface of a sphere, Equation (21.37) gives $\psi_s = \mathcal{C}q/Da$. Substitute these quantities into Equation (22.54):

$$\Delta G_{el} = \frac{1}{2} \int_s \sigma \psi_s \, ds = \frac{1}{2} \left(\frac{q}{4\pi a^2} \right) \left(\frac{\mathcal{C}q}{Da} \right) 4\pi a^2 = \frac{\mathcal{C}q^2}{2Da}. \tag{22.55}$$

Charging Processes

Another way to derive Equation (22.55) is based on the work done in a reversible charging process. Start with the uncharged sphere and 'charge it up' in small increments $dq = q\,d\lambda$. The parameter λ is a 'progress coordinate' for the charging process. As λ goes from 0 to 1, the surface goes from no charge to full charge. At any intermediate stage λ the total surface charge is λq and

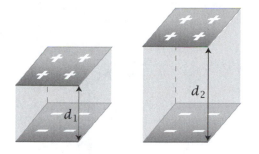

Figure 22.11 Parallel-plate capacitor with two different plate separations, d_1 and d_2.

the surface potential is (see Equation (21.37))

$$\psi_s(\lambda) = \frac{\mathcal{C}\lambda q}{Da}.$$ (22.56)

The work for the complete charging process is

$$\Delta G_{el} = \int_0^1 \psi_s(\lambda) q \, d\lambda = \frac{\mathcal{C}q^2}{Da} \int_0^1 \lambda \, d\lambda = \frac{\mathcal{C}q^2}{2Da},$$ (22.57)

which is identical to the result in Equation (22.55).

Example 22.7 shows how much free energy is involved in charging up two parallel planes. This is useful when considering the electrostatic repulsion between biological membranes or colloids.

EXAMPLE 22.7 The free energy of charging a parallel-plate capacitor. What is the free energy for charging up two parallel plates? Start with two uncharged plates and transfer charge from one to the other. At the end of this process, the upper plate has a charge density (charge per unit area) $+\sigma$, and the lower plate has charge density $-\sigma$. The area of each plate is A. The dielectric constant of the material between the plates is D. Use Equation (22.54), with surface density σ and $\psi_s = (\sigma d)/(\varepsilon_0 D)$ (see Equation (21.12)) to get

$$\Delta G_{el} = \frac{1}{2}\sigma\psi_s A = \frac{\sigma^2 A d}{2\varepsilon_0 D}.$$ (22.58)

Let's use this result to compute the free energy of bringing two charged parallel planes from separation d_1 to d_2 (see Figure 22.11).

EXAMPLE 22.8 Bringing charged planes together. Suppose that two planes with surface charge $\sigma = 10^{-3} \, \text{C m}^{-2}$ are separated from each other by $d_1 = 10$ cm (see Figure 22.11). Inside is oil, with a dielectric constant $D = 2$. Equation (22.58) gives the electrostatic free energy per unit area of charging up the planes at this fixed separation:

$$\frac{\Delta G_{el}}{A} = \frac{\sigma^2 d}{2\varepsilon_0 D} = \frac{(10^{-6} \, \text{C}^2 \, \text{m}^{-4})(0.1 \, \text{m})}{(2)(8.85 \times 10^{-12} \, \text{C}^2 \, \text{J}^{-1} \, \text{m}^{-1})(2)}$$

$$= 2.83 \, \text{kJ m}^{-2}.$$ (22.59)

Bringing two oppositely charged planes together from $d_1 = 10$ cm to $d_2 = 2$ cm can be regarded as a process of discharging the planes when they are separated by distance d_1, moving them together while they are uncharged ($\Delta G_{el} = 0$ for

---

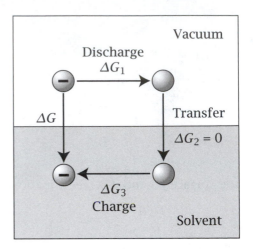

Figure 22.12 Transferring an ion from one medium to another—vacuum to solvent in this case—can be modelled with Born energies in a thermodynamic cycle. We compute here only the electrostatic free energy $\Delta G = \Delta G_{el}$. ΔG_1 is the free energy of discharging the ion in a vacuum, $\Delta G_2 = 0$ is the electrostatic free energy of transferring the neutral particle to the solvent, and ΔG_3 is the free energy of charging up the ion in the solvent.

this step), and charging them up at separation d_2. This process lowers the free energy by

$$\frac{\Delta G_{el}}{A} = \frac{\sigma^2 (d_2 - d_1)}{2\varepsilon_0 D} = \frac{-0.08\sigma^2}{2\varepsilon_0 D} = -2.26 \text{ kJ m}^{-2}. \tag{22.60}$$

The *Born energy* is the energy cost of transferring an ion from a medium having dielectric constant D_1 to a medium having dielectric constant D_2.

Ion Solvation: Born Energy

Although the short-ranged interactions between the ion and the molecules of the medium discussed in earlier chapters contribute to solvation energies, here we consider just the electrostatic contribution. According to the *Born model*, this free energy is found by discharging the sphere in the first medium, transferring the neutral sphere, and then charging up the sphere in the second medium (see Figure 22.12).

Suppose that the ion is a sphere of radius a with a total charge q uniformly distributed on its surface. The free energy ΔG_1 for *discharging* the sphere in the medium with dielectric constant D_1 is given by Equation (22.55),

$$\Delta G_1 = \frac{-\mathcal{C}q^2}{2D_1 a}. \tag{22.61}$$

In step 3, the sphere is charged up in the medium with dielectric constant D_2, so Equation (22.55) gives

$$\Delta G_3 = \frac{\mathcal{C}q^2}{2D_2 a}. \tag{22.62}$$

Step 2, transferring the uncharged sphere to the second medium, involves no change in the electrostatic component of the free energy, $\Delta G_2 = 0$. The electrostatic free energy of transfer ΔG_{el} is the sum of the free energies for steps 1 through 3:

$$\Delta G_{el} = \Delta G_1 + \Delta G_2 + \Delta G_3 = \frac{-\mathcal{C}q^2}{2D_1 a} + 0 + \frac{\mathcal{C}q^2}{2D_2 a} = \frac{\mathcal{C}q^2}{2a}\left(\frac{1}{D_2} - \frac{1}{D_1}\right). \tag{22.63}$$

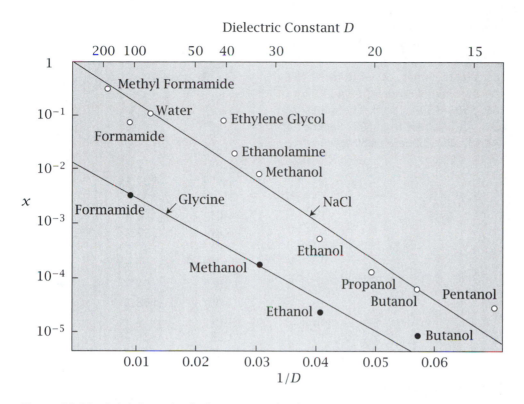

Figure 22.13 Solubilities (mole fraction x) of polar molecules (sodium chloride, $\circ$; and glycine, $\bullet$) in media of different dielectric constants, D. Media of high D are the best solvents, and the lines are predicted by Equation (22.63). Source: JN Israelachvili, *Intermolecular and Surface Forces with Applications to Colloidal and Biological Systems*, Academic Press, New York, 1985. Solubility data are from *Gmelins Handbuch*, Series 21, Volume 7, for NaCl; DR Lide Editor, *CRC Handbook of Chemistry and Physics*, 81st edition, CRC Press, Boca Raton, 2000, for glycine.

For the transfer of ions from water, where the dielectric constant is about $D = 80$, to oil, where the dielectric constant is about $D = 2$, ΔG_{el} is positive, as Example 22.9 shows. This is why ion-forming compounds such as salts are less soluble in oil than water.

EXAMPLE 22.9 The cost of partitioning an ion into oil. Compute the free energy of partitioning an ion of radius $a = 2\,\text{Å}$ and charge $q = e$ from water ($D_w = 80$) into oil ($D_o = 2$). To simplify, use the Bjerrum length in a vacuum $\ell_B = \mathcal{C}e^2/kT = 560\,\text{Å}$. Equation (22.63) becomes

$$\frac{\Delta G_{el}}{kT} = \frac{\ell_B}{2a}\left(\frac{1}{2} - \frac{1}{80}\right) \approx \frac{\ell_B}{4a} = \frac{1}{4}\left(\frac{560\,\text{Å}}{2\,\text{Å}}\right) = 70.$$

At $T = 300\,\text{K}$, $\Delta G_{el} = 70 \times (1.987\,\text{cal mol}^{-1}\,\text{K}^{-1})(300\,\text{K}) \approx 42\,\text{kcal mol}^{-1}$. This is why ions are seldom found inside nonpolar media, such as lipid bilayers or the cores of globular proteins.

Figure 22.13 shows that polar molecules partition most readily into media with high dielectric constants, according to the $1/D$ dependence predicted for

Table 22.3 Experimental and theoretical values of the heats of solvation (in kcal mol^{-1}). Source: The calculated enthalpies are from AA Rashin and B Honig, *J Phys Chem* **89**, 5588–5593 (1985). The values of $\Delta H_{\text{experimental}}$ are from JO'M Bockris and AKN Reddy, *Modern Electrochemistry*, Vol 1, Plenum Press, New York, 1977.

Salt	$\Delta H_{\text{experimental}}$	$\Delta H_{\text{calculated}}$
LiF	−245.2	−243.9
LiCl	−211.2	−212.8
LiBr	−204.7	−206.6
LiI	−194.9	−197.9
NaF	−217.8	−216.4
NaCl	−183.8	−185.4
NaBr	−177.3	−179.2
NaI	−197.5	−170.4
KF	−197.8	−194.0
KCl	−163.8	−162.9
KBr	−157.3	−156.7
KI	−147.5	−147.9
RbF	−192.7	−189.3
RbCl	−158.7	−158.3
RbBr	−152.2	−152.4
RbI	−142.4	−143.3
CsF	−186.9	−183.5
CsCl	−152.9	−152.4
CsBr	−146.4	−146.2
CsI	−136.6	−137.5

ions by Equation (22.63). The Born model also adequately predicts ion solvation enthalpies, provided that a is taken not as the ionic radius but as the somewhat larger radius of the solvent cavity containing the ion [3].

The Enthalpy of Ion Solvation

You can calculate the enthalpies of ion solvation from the vapor phase to water (dielectric constant D) from the Born model by using the Gibbs–Helmholtz Equation (13.41):

$$\Delta H_{\text{el}} = \Delta G_{\text{el}} - T\frac{d\Delta G_{\text{el}}}{dT}. \tag{22.64}$$

Substitute the Born electrostatic free energy ΔG_{el} from Equation (22.63) into Equation (22.64), use $D_{\text{vapor}} = 1$ and recognize that water's dielectric constant depends on temperature,

$$\Delta H_{\text{el}} = -\frac{\mathcal{C}q^2}{2a}\left(1 - \frac{1}{D} - T\frac{d}{dT}\left(1 - \frac{1}{D}\right)\right). \tag{22.65}$$

Use the relations $d\ln x = dx/x$ for the term dT/T, and $d(1/x) = -dx/x^2 = -(d\ln x)/x$ for the term $d(1/D)$, to rewrite Equation (22.65) as

$$\Delta H_{\text{el}} = -\frac{\mathcal{C}q^2}{2a}\left(1 - \frac{1}{D} - \frac{1}{D}\frac{d\ln D}{d\ln T}\right). \tag{22.66}$$

ΔH (kcal mol^{-1})

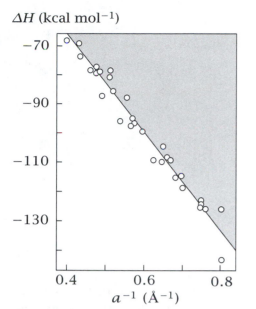

Figure 22.14 Enthalpies of hydration versus ion radius, a. Source: AA Rashin and B Honig, *J Phys Chem* **89**, 5588–5593 (1985). The radii are from JA Dean (ed), *Lange's Handbook of Chemistry*, 11th edition, McGraw-Hill, New York, 1973; and ES Gold, *Inorganic Reactions and Structure*, Holt, Reinhart & Winston, New York, 1960. The experimental enthalpies are from J O'M Brockis and AKN Reddy, *Modern Electrochemistry*, Vol 1, Plenum Press, New York, 1977.

Dielectric Constant

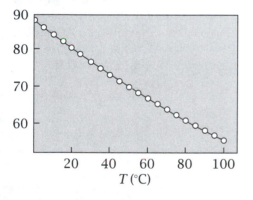

Figure 22.15 Experimental values for the dielectric constant of liquid water as a function of temperature.

Table 22.3 and Figure 22.14 show experimental enthalpies of ion solvation in water. Because D depends on temperature, the solvation of ions also involves entropy, which can arise from the degree of alignment of solvent dipoles. Dielectric constants generally decrease with temperature (see Figure 22.15), indicating that the thermal motion weakens the degree of polarization at higher temperatures.

EXAMPLE 22.10 Calculating the heat of solvation of NaCl. Let's compute the enthalpy of solvation of NaCl by using the Born model. If NaCl is sufficiently dilute, the heats of ionization of Na^+ and Cl^- can be treated as independent. Use Equation (22.66) with $q^2 = e^2$, and the Bjerrum length $\ell_B = 7.13\,\text{Å}$. For water at 298 K, Figure 22.15 shows $d\ln D/d\ln T = -1.357$, and the enthalpy of solvation of NaCl in water is

$$\Delta H_{\mathrm{el}} = \frac{-l_B}{2a}\left[D - 1 - \left(\frac{\partial \ln D}{\partial \ln T}\right)\right]RT$$

$$= \frac{-7.13\,\text{Å}}{2a}(78.54 - 1 + 1.357)(0.592\,\text{kcal mol}^{-1})$$

$$= \left(\frac{-166.5}{a}\right)\text{kcal mol}^{-1}. \qquad (22.67)$$

For Na^+, $a = 1.680\,\text{Å}$, so $\Delta H_{\mathrm{el}} = -166.5/1.680 = -99.1\,\text{kcal mol}^{-1}$. For Cl^-, $a = 1.937\,\text{Å}$, so $\Delta H_{\mathrm{el}} = -166.5/1.937 = -86.0\,\text{kcal mol}^{-1}$. Then $\Delta H_{\mathrm{el}}\,(NaCl) = \Delta H_{\mathrm{el}}\,(Na^+) + \Delta H_{\mathrm{el}}\,(Cl^-) = -185.1\,\text{kcal mol}^{-1}$, which agrees well with the experimental value in Table 22.3.

Summary

The electrochemical potential is a generalization of the chemical potential to situations that involve charged molecules. It is useful for predicting the equilibria of particles that are driven by both chemical interactions and electrical fields. Applications of electrochemical potentials include the partitioning of ions from one medium to another, electroplating, batteries, and the binding of ions to surfaces. Electric fields can perturb chemical equilibria such as acid–base interactions. The process of moving charges from one medium to another can be described by charging and discharging cycles. An example is the Born model for the partitioning of ions between different media.

Problems

1. A charged protein.
Model a protein as a sphere with a radius of 20 Å and a charge of 20e in water at 25 °C and $D_w = 78.54$. Assume the sphere is uniformly charged on its surface.

   (a) In units of kT, what is the potential at a distance 30 Å from the protein surface?

   (b) What is the electrostatic free energy of the charge distribution on the protein in kcal mol^{-1}?

2. pK_a's in a protein.
Suppose that the protein in problem 1 has an aspartic acid residue and a lysine residue at its surface. In bulk water the pK_a of the aspartic acid side group is 3.9, and the pK_a of the lysine NH_2 side group is 10.8. What are the pK_a's of these groups in a protein if it has a net charge of $+20e$?

3. Acid dissociation near a protein.
Now put the protein of problems 1 and 2 in a 0.01 M aqueous solution of acetic acid (HAc). In bulk solution the acetic acid has a $pK_a = 4.0$. It is dissociated to give ion concentrations $[H^+]_{bulk}$ and $[Ac^-]_{bulk}$.

   (a) What are the concentrations of H^+ and Ac^- ions at a location x near the protein where the potential is $\psi(x)$?

   (b) What is the pK_a of the acetic acid near the protein at x? Explain the difference between the situation of this acid and the situation of the aspartic acid in problem 2.

4. Acids and bases near a charged surface.
On a negatively charged surface:

   (a) Are basic groups ionized more or less than in bulk water?

   (b) Are acid groups ionized more or less than in bulk water?

5. Free energy of charging a sphere.
A sphere with radius a in a medium with dielectric constant D is uniformly filled with a charge of volume density ρ. Derive the electrical free energy of this sphere in two different ways:

   (a) by deriving the potential field and then charging the sphere from charge density 0 to ρ;

   (b) by adding fully charged shells of thickness dr from $r = 0$ to $r = a$.

6. Burying a charge in a protein.
As an estimate for the free energy of burying a charged amino acid such as aspartic or glutamic acid in protein folding, compute the free energy of transferring an ion of radius 3 Å and charge +1 from water to oil. Assume that water has a dielectric constant $D_w = 80$, and oil has $D_o = 2$.

7. A solvated charge in a protein.
The Born energy in Equation (22.55) allows you to estimate the free energy cost for a charged group to be deep inside a fully folded protein (instead of at the protein surface in contact with water). Estimate the free energy cost for a charged group to stay inside a *partly* folded protein. Consider the ionic group as a sphere with radius 2 Å and total surface charge e (see Figure 22.16). This group is first surrounded by a shell of bulk water, 4 Å thick. This in turn is surrounded by a shell of protein, 14 Å thick with dielectric constant $D_P = 4$, which is in contact with bulk water at 25 °C. Derive the electrostatic free energy of the ionic group

   (a) in water,

   (b) in the partly folded protein,

   (c) as in (b) but with the water shell replaced by protein, $D_P = 4$, to simulate the completely folded protein.

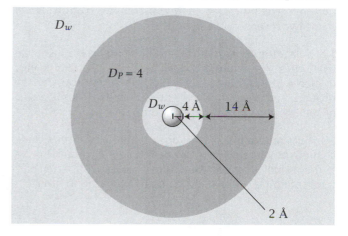

Figure 22.16

8. An interfacial potential.
Consider an uncharged oil/water interface. On the aqueous side, the proximity of the oil phase biases the orientation of the water molecules in an unknown way. Calculate the resulting potential across the boundary layer for the maximum bias, the complete line-up of the first layer of water dipoles perpendicular to the interface. Treat this layer as a parallel plate capacitor, with one water molecule occupying 10 Å^2 of the interfacial area. The dipole moment is $\mu = 1.85$ Debye (1 Debye $= 3.336 \times 10^{-30}$ C m) per water molecule. Take two values of the dielectric constant D between the capacitor plates:

   (a) $D = 2$, as for oil, and

   (b) $D = 80$, as for bulk water.

(If water is perfectly oriented, a situation called dielectric saturation, $D = 2$ is more likely.)

9. Small electrostatic potentials. For a monovalent ion at $T = 300\,K$, what is the value of ψ such that $e\psi = kT$?

10. Sodium potential in a frog muscle. Inside a frog muscle cell, the sodium concentration is $[Na]_{in} = 9.2\,mM$; outside, the sodium concentration is $[Na]_{out} = 120\,mM$.

(a) Compute the sodium potential $\Delta\psi$ across the membrane, at $T = 300\,K$.

(b) Which side of the membrane has the more positive potential, the inside or the outside?

11. Membrane pores. A neutral protein 'carrier' may help an ion to transfer into and cross a lipid membrane.

(a) What is the electrostatic free energy change when a monovalent ion is transferred from water at 25 °C to a hydrocarbon solvent with dielectric constant $D_{hc} = 2$? The radius of the ion is 2 Å.

(b) Now wrap the ion in a neutral protein to produce a spherical complex with radius $b = 15$ Å. What is the electrostatic free energy of transfer from water to hydrocarbon of the ion–protein complex?

12. Solvating a protein. A spherical protein has a valence of $z = -5$ and a radius of $a = 10$ Å. The change in electrostatic free energy, ΔG_{el}, when you transfer the protein from vacuum ($D = 1$) to water ($D \approx 80$) is a part of the 'solvation free energy.' Compute ΔG_{el}.

13. Burying an ion pair in oil. What is the free energy cost of transferring a monovalent anion of radius $a = 2$ Å and a monovalent cation of the same radius a from vacuum into oil ($D = 2$) at an ion-paired separation of $2a$?

References

[1] B Zagrovic and R Aldrich. *Science* **285**, 59–61 (1999).

[2] B Roux and R MacKinnon. *Science* **285**, 100–102 (1999).

[3] AA Rashin and B Honig, *J Phys Chem* **89**, 5588–5593 (1985).

Suggested Reading

Excellent texts covering electrostatic forces and flows:

CMA Brett and AMO Brett, *Electrochemistry: Principles, Methods, and Applications*, Oxford University Press, New York, 1993.

B Hille, *Ionic Channels of Excitable Membranes*, Sinauer, Sunderland, 1984. Excellent discussion of the dipolar gate model and ion potentials.

B Katz, *Nerve, Muscle, and Synapse*, McGraw-Hill, New York, 1966. Excellent overview of the Nernst equation and neural physics.

N Lakshminarayanaiah, *Equations of Membrane Biophysics*, Academic Press, Orlando, 1984.

SG Schultz, *Basic Principles of Membrane Transport*, Cambridge University Press, Cambridge, 1980.

TF Weiss, *Cellular Biophysics*, Volume 1, MIT Press, Cambridge, Massachusetts, 1996.

23 Salt Ions Shield Charged Objects in Solution

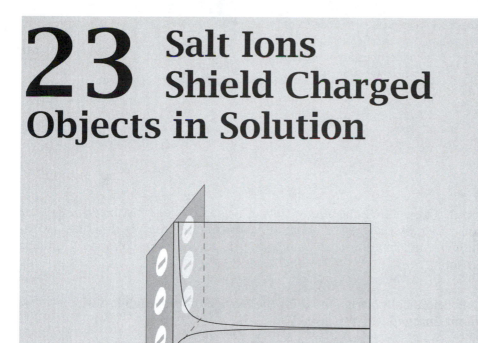

Salts Dissociate into Mobile Ions and Shield Charged Objects in Water

Any molecule that can dissolve in polar liquids, dissociate into ions, and carry an electrical current is an *electrolyte*. Molecules that dissociate completely are called *strong electrolytes*, and those that dissociate only partly are called *weak electrolytes*. Such ions can regulate the chemical and physical equilibria of other charged objects. For example, DNA is a molecular chain of negative phosphate charges. Because its charges repel each other, DNA is relatively expanded in water. When NaCl is added to an aqueous solution of DNA, the salt dissociates into positive ions (Na^+) and negative ions (Cl^-). The positive ions seek the DNA's negative charges, surrounding and shielding them. This shielding weakens the intra-DNA charge repulsions, causing the DNA to become more compact. Two biological cells each having a net negative surface charge will also repel each other. Adding salt weakens the charge repulsions, so the cells can come together and fuse. Adding salt precipitates colloids for the same reason. Salts can also influence the rates of chemical reactions.

In this chapter we explore how ions shield charged objects. First we define what we mean by *added salt*. When you buy a bottle of some charged molecule, such as DNA, it will be electroneutral, so it will already have a stoichiometric complement of counterions. So you may get Na^+-DNA or Mg^{2+}-DNA, for example. With no additional salt, such solutions remain electroneutral. This is called the *no salt* condition. We are interested in conditions when additional salt is

Figure 23.1 Dissociated mobile salt ions move in the presence of a charged surface. The counterions are attracted to the surface and the co-ions are repelled.

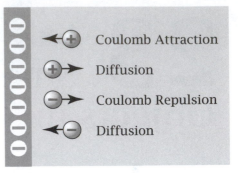

present. The electrostatic interactions of a charged object with added salt can be described by the Poisson equation, and its equilibrium is described by the Boltzmann distribution law. Combining these gives the Poisson–Boltzmann equation.

The Poisson–Boltzmann Model for the Distribution of Mobile Salt Ions Around Charged Objects

A charged molecule or surface P attracts the mobile salt ions that have the opposite charge to P, called the *counterions*. P repels the mobile ions of the same sign, called the *co-ions* (see Figure 23.1). The counterions distribute around P and act as a sort of electrostatic shield, reducing the electrostatic potential that is felt by a particle more distant from P. The interface between P and the neighboring salt solution is called the *electrical double layer*: the first layer is the charge on P, and the second layer is the adjacent diffuse sea of excess counterions.

In the absence of electrolyte, two negatively charged P particles repel each other. But if a salt such as NaCl is added to the solution, the small ions intervene to weaken the charge repulsions between the two negative P particles, often to the point that other weak attractions can prevail and cause the P particles to associate. In this way charged colloids can be induced to aggregate by the addition of salts.

River deltas are interesting cases of electrostatic shielding, for example where the Mississippi River meets the ocean. In the upstream fresh river water, fine silt particles are suspended, owing to soil erosion from the surrounding land. Silt particles are charged colloids. The fresh water carries the silt downstream toward the ocean. Where the fresh water meets the salty seawater, the high salt concentration causes the colloidal silt to aggregate and precipitate to form the macroscopic landforms known as deltas.

The simplest model of charge shielding and colloidal stability against aggregation was developed around 1910 independently by L-G Gouy (1854–1926), a French physicist, and DL Chapman (1869–1958), a British chemist. They combined Poisson's equation of electrostatics with the Boltzmann distribution law.

In the simplest case, the surface of a charged colloidal particle is described as a plane. Suppose that a charged P particle, which is fixed in space, produces an electrostatic potential $\psi(x)$, which you take to be a function of a single spatial coordinate x. P is in a salt solution of dissociated mobile ions. Let

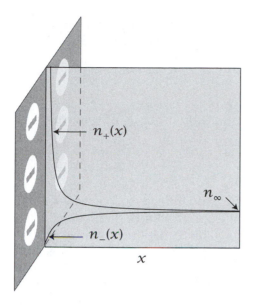

Figure 23.2 The concentration of counterions $n_+(x)$, and concentration of co-ions $n_-(x)$ as a function of distance x away from a negatively charged surface.

$n_+(x)$ represent the concentration of mobile positive ions (the number of ions per unit volume) at a distance x from the plane. This quantity is given by the Nernst or Boltzmann equation (22.8),

$$n_+(x) = n_\infty e^{-ze\psi(x)/kT}, \tag{23.1}$$

where n_∞ is the concentration of positive ions in the bulk solution, distant from the particle P, where the potential is $\psi(\infty) = 0$ (see Figure 23.2). An identical expression applies to the negative ions, except that z is replaced by $-z$, if the electrolyte is symmetrical in its charge. Charge neutrality in the bulk requires that the number density of negative ions in the bulk far away from P is also n_∞, so

$$n_-(x) = n_\infty e^{+ze\psi(x)/kT}. \tag{23.2}$$

The mobile ions not only *experience* the electrostatic field from P. They also *contribute to it*. You can compute the electrostatic potential that arises from both the fixed charge from P and from the mobile charges of the dissociated salt ions. First, compute the total charge density $\rho(x)$ as a function of the number of ions at position x:

$$\rho(x) = \sum_i z_i e n_i(x) = ze\left[n_+(x) - n_-(x)\right], \tag{23.3}$$

where z_i is the valence of an ion of species i, and n_i is its concentration at position x. Now use the Poisson equation (21.30) to relate the charge density ρ to the electrostatic potential ψ:

$$\nabla^2 \psi = -\frac{\rho}{D\varepsilon_0},$$

where D is the dielectric constant of the solution. D is approximately equal to the dielectric constant of the solvent, because the ions themselves (if their concentration is low) contribute little to D.

Table 23.1 The Debye length $1/\kappa$ describes the range of the potential, given here for various concentrations c of aqueous monovalent salt solutions at 25 °C.

c (mol L^{-1})	$1/\kappa$ Å
0.5	4.30
0.2	6.80
0.1	9.62
0.05	13.6
0.02	21.5
0.01	30.4
0.005	43.0
0.002	68.0
0.001	96.2

Now you have all the terms necessary to formulate the **Poisson–Boltzmann equation**. Substitute $n_+(x)$ and $n_-(x)$ from Equations (23.1) and (23.2) into Equation (23.3) for ρ and then into Equation (21.30):

$$\nabla^2 \psi = \frac{zen_\infty}{D\varepsilon_0}(e^{ze\psi/kT} - e^{-ze\psi/kT}). \tag{23.4}$$

Equation (23.4) can be expressed more compactly in terms of the hyperbolic sine function, $\sinh(x) = (e^x - e^{-x})/2$:

$$\nabla^2 \psi = \frac{2zen_\infty}{D\varepsilon_0}\sinh(ze\psi/kT). \tag{23.5}$$

The Poisson–Boltzmann equation (23.5) is a nonlinear second-order differential equation from which you can compute ψ if you know the charge density on P and the bulk salt concentration, n_∞. This equation can be solved numerically by a computer. However, a linear approximation, which is easy to solve without a computer, applies when the electrostatic potential is small. For small potentials, $ze\psi/kT \ll 1$, you can use the approximation $\sinh(x) \approx [(1+x) - (1-x)]/2 = x$ (which is the first term of the Taylor series expansion for the two exponentials in $\sinh(x)$ (see Appendix C, Equation (C.1)). Then Equation (23.5) becomes

$$\nabla^2 \psi = \frac{2zen_\infty}{D\varepsilon_0}(ze\psi/kT) = \kappa^2 \psi, \tag{23.6}$$

where κ^2 is defined by

$$\kappa^2 = \frac{2z^2e^2n_\infty}{D\varepsilon_0 kT}. \tag{23.7}$$

Equation (23.6) is called either the **linearized Poisson–Boltzmann** or the **Debye-Hückel equation**. $1/\kappa$ is called the **Debye length**. The Debye length is a screening, or shielding, distance. A charge that is closer to P than $1/\kappa$ 'sees' the charged plane and interacts with it. A charge that is further than $1/\kappa$ from P is shielded from it by the intervening salt solution, which weakens its attraction or repulsion for the plane. Table 23.1 shows that increasing the concentration of the salt decreases the Debye length. Example 23.1 shows how the Debye length is computed.

EXAMPLE 23.1 Computing Debye lengths. For a monovalent salt, $z = 1$, with concentration 0.1 mol L^{-1}, show that the Debye length is 9.62 Å if $D = 78.54$ (see Table 23.1).

Multiply the numerator and denominator on the right-hand side of Equation (23.7) by Avogadro's number $\mathcal{N}$ to get

$$\kappa^2 = \frac{2(ze)^2 n_\infty \mathcal{N}}{D\varepsilon_0 RT}.$$

From the units given in the box on page 370, $e^2 \mathcal{N}/(4\pi\varepsilon_0) = 1.386 \times 10^{-4}$ J m mol^{-1}, so

$$\kappa^2 = 2(4\pi)\left(1.386 \times 10^{-4}\,\text{J m mol}^{-1}\right)\left(0.1\,\text{mol L}^{-1}\right)\left(10^3\,\text{L m}^{-3}\right)$$

$$\times \frac{\left(6.022 \times 10^{23} \text{ molecule mol}^{-1}\right)}{\left[(78.54)\left(8.314\,\text{J K}^{-1}\,\text{mol}^{-1}\right)(298\,\text{K})\right]}$$

$$= 0.01078 \times 10^{20}\,\text{m}^{-2} \quad \Longrightarrow \quad \frac{1}{\kappa} = 9.62 \times 10^{-10}\,\text{m} = 9.62\,\text{Å}.$$

Equation (23.4) may also be used for *asymmetrical* electrolyte solutions, in which the valency z_- of the anion is not the same as the valency z_+ of the cation. If ionic species i has valency z_i and molar concentration m_i, replace $n_\infty z^2$ in Equation (23.7) with the quantity $I = (1/2)\sum_i m_i z_i^2$, which is the **ionic strength** of the solution.

The Poisson–Boltzmann model is used to compute the electrostatic potential ψ if you know the charge on a surface and the concentration of salt in the solution. In Example 23.2, we compute $\psi(x)$ in the direction normal to a charged plane in a salt solution.

EXAMPLE 23.2 The potential near a uniformly charged plane in a salt solution. Because ψ depends only on x in this case, you need to solve the Poisson–Boltzmann Equation (23.6) in only one dimension to find the potential at a distance x from a plane having surface charge density σ:

$$\frac{d^2\psi}{dx^2} = \kappa^2\psi. \tag{23.8}$$

You can verify that the function

$$\psi(x) = A_1 e^{\kappa x} + A_2 e^{-\kappa x} \tag{23.9}$$

satisfies Equation (23.8) for any constants A_1 and A_2 by substituting Equation (23.9) into Equation (23.8) and carrying out the two differentiations indicated.

To solve Equation (23.8), we need two boundary conditions. We follow the same conventions that we used in solving the Poisson equation. First, the electrostatic potential is defined to be zero $\psi(\infty) = 0$ as $x \to \infty$; this gives $A_1 = 0$. Second, we require that at $x = 0$, $\psi = \psi_0$, where ψ_0 is the potential at the surface. This gives $A_2 = \psi_0$, so the electrostatic potential is

$$\psi(x) = \psi_0 e^{-\kappa x}. \tag{23.10}$$

Equation (23.10) predicts that the electrostatic potential in salt solutions approaches zero *exponentially* as a function of distance from the charged plane. In contrast, Equation (21.12) shows that in the absence of salt, the electrostatic potential varies *linearly* with distance from a charged plane.

To complete Equation (23.10), compute the surface potential ψ_0 from the given surface charge σ, the number of charges per unit area. This can be done in different ways. The surface charge on the plane must be compensated by the net charge (of opposite sign) in the solution, so

$$\sigma = -\int_0^\infty \rho(x)\,dx = D\varepsilon_0 \int_0^\infty \frac{d^2\psi}{dx^2}\,dx \tag{23.11}$$

where the second equality comes from the Poisson Equation (21.30). Integrating Equation (23.11) and substituting Equation (23.10) gives

$$\int_0^\infty \frac{d^2\psi}{dx^2}\, dx = \left[\frac{d\psi}{dx}\right]_\infty - \left[\frac{d\psi}{dx}\right]_0 = -\left[\frac{d\psi}{dx}\right]_0 = \kappa\psi_0. \tag{23.12}$$

Combining Equations (23.11) and (23.12) gives

$$\sigma = D\varepsilon_0\kappa\psi_0. \tag{23.13}$$

Substituting this expression for ψ_0 into Equation (23.10) gives the electrostatic potential near a planar surface of charge density σ,

$$\psi(x) = \frac{\sigma}{\kappa\varepsilon_0 D} e^{-\kappa x}. \tag{23.14}$$

Equation (23.10) shows that the range of the electrostatic potential from plane P shortens as κ increases. Because κ is proportional to $\sqrt{c}$, where $c \propto n_\infty$ is the salt concentration (see Equation (23.7)), the range of the potential $\psi(x)$ is shortened by the added salt. At a distance equal to the Debye length, $x = 1/\kappa$, the potential ψ is decreased by a factor of $1/e$, to $\psi(1/\kappa) = \psi_0/e$. The Debye length shortens as the salt concentration increases. For example, for a z-z valent salt ($+z$ on each cation and $-z$ on each anion) in water at 25 °C, where $D = 78.54$, Equation (23.7) gives $1/\kappa = 3.044/(z\sqrt{c})$ Å.

Table 23.1 shows the dependence of the Debye length on the salt concentration in aqueous monovalent salt solutions. When the radius of curvature of particle P is much larger than $1/\kappa$, the double layer can be regarded as planar. Planar double layers have an important role as models in colloid chemistry, where particles often range in length from 10^2 to 10^5 Å.

Example 23.3 finds the potential near a charged sphere, a model for proteins, small ions, and micelles in salt solutions.

EXAMPLE 23.3 The spherical double layer. Let's compute the electrostatic potential $\psi(r)$ as a function of the radial distance from a charged sphere in a salt solution. The sphere has radius a, net charge Q, and a uniform surface charge density $\sigma = Q/(4\pi a^2)$. The sphere is in a monovalent salt solution that has a Debye length $1/\kappa$. The potential changes only in the radial direction so $(\partial\psi/\partial\theta) = 0$ and $(\partial^2\psi/\partial\Phi^2) = 0$. For spherical coordinates (see Equation (17.33)), the linearized Poisson–Boltzmann Equation (23.6) becomes

$$\nabla^2\psi = \frac{1}{r}\frac{d^2(r\psi)}{dr^2} = \kappa^2\psi. \tag{23.15}$$

Multiply both sides of Equation (23.15) by r to get

$$\frac{d^2(r\psi)}{dr^2} = \kappa^2 r\psi. \tag{23.16}$$

Comparison of Equation (23.16) with Equation (23.8) shows that $r\psi(r)$ has the same role as $\psi(x)$ in the planar problem. So the general solution of Equation (23.16) is

$$r\psi = A_1 e^{\kappa r} + A_2 e^{-\kappa r} \implies \psi = A_1 \frac{e^{\kappa r}}{r} + A_2 \frac{e^{-\kappa r}}{r}. \tag{23.17}$$

Again, you need two boundary conditions to establish A_1 and A_2. It is conventional to choose $\psi(\infty) = 0$, so $A_1 = 0$. The second boundary condition is at $r = a$, where $\psi = \psi_a = A_2 e^{-\kappa a}/a$. Inverting this expression gives $A_2 = a\psi_a e^{\kappa a}$. Then Equation (23.17) becomes

$$\psi = \frac{a\psi_a}{r} e^{-\kappa(r-a)} \qquad \text{for } r > a. \tag{23.18}$$

Now determine the surface potential ψ_a from the total charge Q. You can find this by taking the derivative $d\psi/dr$ of Equation (23.18) at $r = a$:

$$\left(\frac{d\psi}{dr}\right)_{r=a} = -a\psi_a e^{\kappa a}\left[\frac{\kappa}{r}e^{-\kappa r} + \frac{1}{r^2}e^{-\kappa r}\right]_{r=a} = -\psi_a \frac{1 + \kappa a}{a}. \tag{23.19}$$

Now use Equation (21.7) to relate ψ and E, and Gauss's Law for $E(r = a)$ around a sphere (Equation (21.37)) to get

$$\left(\frac{d\psi}{dr}\right)_{r=a} = -E_{r=a} = \frac{-\mathcal{C}Q}{Da^2}. \tag{23.20}$$

Setting Equation (23.20) equal to (23.19) gives ψ_a in terms of Q and κ:

$$\psi_a = \frac{\mathcal{C}Q}{Da(1 + \kappa a)}. \tag{23.21}$$

In the absence of added salt, $\kappa = 0$ and $\psi_a = \mathcal{C}Q/Da$. Equation (23.21) shows that the effect of added salt is to shield the charge on the sphere and reduce the surface potential by the factor $(1 + \kappa a)$. Substituting Equation (23.21) into (23.18) gives the potential we seek:

$$\psi(r) = \frac{\mathcal{C}Q}{Dr(1 + \kappa a)} e^{-\kappa(r-a)}. \tag{23.22}$$

In the absence of added salt, $\kappa = 0$ so $\psi \propto r^{-1}$, but the presence of salt introduces shielding that causes ψ to diminish with an additional exponential dependence on r.

Equation (23.22) gives the electrostatic potential from a point charge Q ($a = 0$) in a salt solution as

$$\psi(r) = \frac{\mathcal{C}Q}{Dr} e^{-\kappa r}. \tag{23.23}$$

The Poisson–Boltzmann equation describes spatial variations in salt concentrations over microscopic length scales near the surfaces of molecules and particles. Equation (23.22) holds for a large spherical particle having a uniform surface charge Q, but also for a small ion, like sodium or chloride, which has a single charge.

Equations (23.22) and (23.23) come from the *linearized* Poisson–Boltzmann Equation (23.6). It is a general property of linear differential equations that you can sum their solutions. For example, if you have two point charges in a uniform salt solution, q_1 at distance r_1 from P, and q_2 at r_2 from P, the

Table 23.2

Freezing-point depression of water by K_2SO_4 salt. At low concentrations, the salt approaches full dissociation, $i \to 3$. Source: CH Langford and RA Beebe, *The Development of Chemical Principles*, Dover Publications, Minnesota, 1969.

Molality	i
0.10	2.32
0.05	2.45
0.01	2.67
0.005	2.77
0.001	2.84

Debye–Hückel potential at P is

$$\psi_P(r_1, r_2) = \frac{\mathcal{C}q_1}{D}\frac{e^{-\kappa r_1}}{r_1} + \frac{\mathcal{C}q_2}{D}\frac{e^{-\kappa r_2}}{r_2}. \tag{23.24}$$

Summing potentials this way is sometimes the easiest way to solve the linear Poisson–Boltzmann equation for an arbitrary constellation of charges.

Before electrolytes can shield charged objects, they must first dissociate. Some electrolytes dissociate fully and others only partly. One of the main uses of the Poisson–Boltzmann equation has been to understand the degree of dissociation in electrolyte solutions.

Electrolytes Are Strong or Weak, Depending on Whether They Fully Dissociate in Water

Salts and acids and bases are *electrolytes*. In aqueous solutions, electrolytes dissolve and dissociate into ions. Molecules that dissociate completely are called *strong electrolytes*, and those that dissociate only partly are called *weak electrolytes*. The degree of dissociation can be represented by the equilibrium dissociation constant K,

$$MX \xrightarrow{K} M^+ + X^-, \quad \text{where } K = \frac{[M^+][X^-]}{[MX]}. \tag{23.25}$$

A strong electrolyte has a large K, typically $K > 0.1$. Using the notation $pK = -\log_{10}$, this means that a strong electrolyte has a $pK < 1$ (the pK can be negative, which occurs if $K > 1$). For a weak electrolyte, $K < 0.1$, so $pK > 1$.

Much of what is known about ion dissociation comes from two types of experiment: measurements of colligative properties and ion conductivities. Chapter 16 shows that the freezing point depression ΔT_f should increase with the concentration of a solute species m according to $\Delta T_f = K_f m$, where K_f is a constant that is independent of the solute (see page 288). But what value of m should you use for a solute, like NaCl, that can dissociate? Each mole of NaCl dissociates into two moles of ions, Na^+ and Cl^-. Should m be the concentration of NaCl or twice the concentration of NaCl? If you let m be the concentration of the neutral species then the observed value i in the colligative expression $\Delta T_f = imK_f$ gives a measure of the degree of dissociation. For example, if you observe that $i = 2$ for NaCl, it implies full dissociation.

The degree of ion dissociation can also be measured by *ion conductivities*. Ion conductivities are found by dissolving salts in water and applying an electric field, which causes the ions to flow, resulting in a measurable electrical current. The conductivity is the observed current flow (number of ions per unit time) divided by the applied voltage. Increasing the electrolyte concentration increases the number of charge carriers and thus the conductivity. For example, at low concentrations, KCl fully dissociates into K^+ and Cl^- ions, so twice the KCl concentration gives the number of ions that carry current.

A strong electrolyte is defined by two observations: (1) the value of i measured by colligative properties equals the number of ion types that would be expected from complete dissociation (for NaCl, $i = 2$; for Na_2SO_4, $i = 3$, etc.), and (2) doubling the electrolyte concentration leads to a doubling of the con-

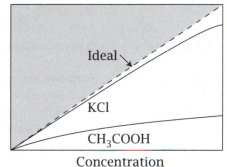

Conductance

Ideal

KCl

CH₃COOH

Concentration

Figure 23.3 For strong electrolytes like KCl, the conductance is proportional to the electrolyte concentration (except at very high concentrations). For weak electrolytes like acetic acid (CH_3COOH), it is not. Source: CH Langford and RA Beebe, *The Development of Chemical Principles*, Dover Publications, Minnesota, 1969.

ductivity of the solution, implying that the ions dissociate completely and that each ion acts as an independent carrier of electrical current. The principle that ions contribute independently to conductivity is called Kohlrausch's law, named after the German chemist F Kohlrausch (1840–1910). NaCl, HCl, KCl, and K_2SO_4 are strong electrolytes when they are at low concentrations.

Figure 23.3 shows that electrolyte dissociation can deviate in two different ways from the ideal behavior predicted by Kohlrausch's law. First, weak electrolytes, like acetic acid and ammonia, do not fully dissociate at any concentration, so their conductivities and colligative i values are smaller than predicted by Kohlrausch's law.

Second, even strong electrolytes that show ideal full dissociation behavior at low electrolyte concentrations may become nonideal at higher concentrations because the positive ions tend to be surrounded by negative ions, and vice versa. Table 23.2 shows that a strong electrolyte, K_2SO_4, appears to dissociate completely into three ions ($i \rightarrow 3$, where i is the dissociation parameter) as the electrolyte concentration approaches zero. Our aim here is to explain why the apparent degree of dissociation decreases with increasing salt concentration. At low concentrations, such nonidealities depend only on the valence and electrolyte concentration, and not on the chemical identity of the ions. Each ion is partly *shielded* by its neighboring ions, leading to an apparently incomplete dissociation. This is described by the Debye–Hückel model.

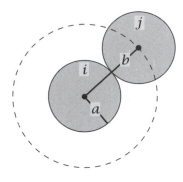

Figure 23.4 In the Debye–Hückel theory, a central ion of type i and radius a is surrounded by counterions of type j. The distance of closest approach is b.

Debye–Hückel Theory for the Nonidealities of Strong Electrolytes

In 1923, E Hückel and P Debye, winner of the 1936 Nobel Prize in Chemistry, adapted the Poisson–Boltzmann theory to explain the nonidealities of dilute solutions of strong electrolytes. To visualize Na^+ ions surrounded by Cl^- ions and Cl^- surrounded by Na^+ at the same time, think of a NaCl crystal that is expanded uniformly. Now add fluctuations. Debye and Hückel focused on one ion as a charged sphere, and used the linear approximation to the Poisson–Boltzmann equation to compute the electrostatic free energy of creating the nonuniform distribution of its surrounding counterions and co-ions.

We focus on one ion, called the central ion (Figure 23.4), in a salt solution. Let b be the distance of closest approach between the centers of two oppositely charged ions. The charge Q on the central ion is given in terms of its valence, $Q = z\mathrm{e}$. The nonideality of the solution can be described in terms of an activity coefficient y in the expression for the chemical potential of the salt,

$$\mu_{NaCl} = \mu^\circ + 2kT \ln c_{NaCl} + 2kT \ln y, \tag{23.26}$$

where the factor of 2 describes the ideal behavior of the complete dissociation of NaCl to Na^+ and Cl^-.

Debye and Hückel regarded all the nonideality as arising from the electrostatics, and none from short-ranged interactions. The nonideality reflected in the activity coefficient is modelled as arising from the nonuniformity in the distribution of counterions and co-ions that the central ion creates in its neighborhood. That is, an excess of counterions and a depletion of co-ions surround the central ion, in a spherically symmetrical way.

The activity coefficient for one of the ions in the pair, say Na^+, can be computed by assuming a reversible process of charging up the central ion from $Q = 0$ to $Q = ze$, using a parameter $0 \leq \lambda \leq 1$ to describe the degree of charging:

$$kT \ln y = \int_0^1 \psi_{dist}(\lambda) ze \, d\lambda. \tag{23.27}$$

The quantity $kT \ln y$ in Equation (23.27) is the electrostatic free energy that accounts for the ion-induced distribution of neighboring ions.

How do we compute ψ_{dist}? Charging up the central ion results in two kinds of work. First, even if there were no added salt present, charging the central ion would give rise to a Born energy for the central ion in the pure solvent. The Born energy Equation (22.62) is not dependent on the salt concentration, so it will be included in μ° in Equation (23.26), rather than in the activity coefficient. Second, charging the central ion also leads to a redistribution of the surrounding salt ions, drawing counterions nearer to the central ion and pushing co-ions further away. We regard ψ_{dist}, the electrostatic potential responsible for the nonideality, as the difference between the total potential ψ_b for charging up the central ion, and the potential $\psi_{no\ salt}$ for charging up the central ion in the pure solvent where $\kappa = 0$. Inside the radius b the total charge is ze. So you can use Equation (23.21) with the radius b, instead of the ion radius a, to get ψ_b. Then

$$\psi_{dist} = \psi_b - \psi_{no\ salt} = \frac{\mathcal{C}ze}{Db(1 + \kappa b)} - \frac{\mathcal{C}ze}{Db} = -\frac{\mathcal{C}ze}{D}\frac{\kappa}{1 + \kappa b}. \tag{23.28}$$

At any intermediate state of charging, $0 \leq \lambda \leq 1$, the potential is

$$\psi_{dist}(\lambda) = -\lambda \frac{\mathcal{C}ze}{D}\frac{\kappa}{1 + \kappa b}. \tag{23.29}$$

Substitute $\psi_{dist}(\lambda)$ from Equation (23.29) into Equation (23.27) and integrate:

$$kT \ln y = -\frac{\mathcal{C}z^2 e^2}{D}\left(\frac{\kappa}{1 + \kappa b}\right)\int_0^1 \lambda \, d\lambda = -\frac{\mathcal{C}z^2 e^2}{2D}\left(\frac{\kappa}{1 + \kappa b}\right). \tag{23.30}$$

If Na^+ is the central ion, Equation (23.30) gives $y = y_+$. Because b measures the shortest distance between ionic centers, the result is the same for Cl^- ions, $y = y_-$. Because only properties of electroneutral solutions can be measured, it is usually more useful to write $y^2 = y_+ y_-$.

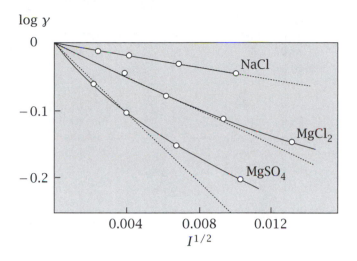

log y

Figure 23.5 Experimental activity coefficients for various electrolytes (——) compared with the Debye–Hückel limiting law (·········). Source: P Atkins, *Physical Chemistry*, Freeman, New York, 1998.

Equation (23.30) predicts that $\ln y \propto \kappa \propto \sqrt{c}$ at low salt concentrations c where $\kappa b \ll 1$. Figure 23.5 shows $\log y$ versus $\sqrt{I}$ for several salts, where I is the ionic strength of the solution. The straight lines are calculated from the Debye–Hückel Equation (23.30) for $b = 0$. This is called a limiting law because all ionic solutions are predicted to obey it in the limit of sufficiently low concentrations. For $b = 0$ the Debye–Hückel theory fits the limiting law slopes very well. Furthermore, with reasonable values of the distance of closest approach b, Equation (23.30) gives excellent agreement for higher concentrations, up to $c = 0.1\,\text{M}$ for monovalent salts at 25 °C.

The Debye–Hückel model is useful for explaining how the rates of chemical reactions depend on salt concentration.

EXAMPLE 23.4 Salts affect the rates of chemical reactions. Consider the chemical reaction in Equation (19.18):

$$A + B \rightarrow (AB)^{\ddagger} \rightarrow P.$$

The rate k_0 of the reaction in the absence of salt (subscript 0) is proportional to the concentration of activated species:

$$k_0 \propto \left[(AB)^{\ddagger}\right]_0 = \text{constant } [A][B]. \tag{23.31}$$

The rate k_s of the reaction in the presence of salt (subscript s) depends, in addition, on electrostatic nonidealities, which are accounted for by activity coefficients:

$$k_s \propto \left[(AB)^{\ddagger}\right]_s = \text{constant } [A][B]\frac{y_A y_B}{y_{AB^{\ddagger}}}. \tag{23.32}$$

Dividing Equation (23.32) by (23.31) gives

$$\ln\left(\frac{k_s}{k_0}\right) = \ln\frac{y_A y_B}{y_{AB^{\ddagger}}}. \tag{23.33}$$

If the total charge is conserved in the transition state, $z_{(AB)^{\ddagger}} = z_A + z_B$, substituting the Debye–Hückel Equation (23.30) for the activity coefficients into Equation (23.33) gives

Figure 23.6 Chemical reaction rates of ions as a function of the square root of the ionic strength $I^{1/2}$, for various reactions. The top curve, for example, is $2[\text{Co(NH}_3)_5\text{Br}]^{2+} + \text{Hg}^{2+} + 2\text{H}_2\text{O} \rightarrow 2[\text{Co(NH}_3)_5\text{H}_2\text{O}]^{3+} + \text{HgBr}_2$. The numbers on the curves are the values of $z_A z_B$, the signed product of the valencies of the reacting ions. The lines show predictions from the Debye-Hückel model. Source: ML Bender, RJ Bergeron and M Komiyama, *The Bioorganic Chemistry of Enzymatic Catalysis*, Wiley, New York, 1984.

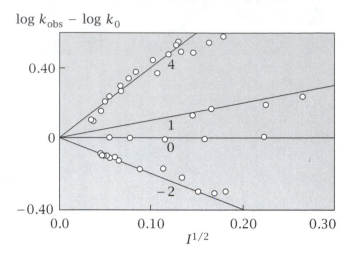

$$\ln\left(\frac{k_s}{k_0}\right) = \frac{-\mathcal{C}e^2}{2DkT}\left(\frac{\kappa}{1+\kappa b}\right)\left(z_A^2 + z_B^2 - (z_A + z_B)^2\right)$$

$$= \frac{\mathcal{C}z_A z_B e^2}{DkT}\left(\frac{\kappa}{1+\kappa b}\right) \approx 2A_0 z_A z_B \sqrt{I}, \qquad (23.34)$$

where A_0 is a constant, I is the ionic strength, and we have assumed that $\kappa b \ll 1$. Figure 23.6 shows experimental evidence that reaction rates depend on the product of the valences of the ions involved and the square root of the ionic strength of the medium.

The Poisson-Boltzmann and Debye-Hückel models are relatively successful predictors of the shielding of charged objects by dissociated salt ions. But these models are approximate, and they have limitations. The Debye-Hückel model shows that when salts dissociate into ions, the ions are not distributed uniformly, but are 'clumpy.' There is an enhanced concentration of negative ions around each mobile positive ion, and vice versa. The main nonideality of salt solutions is attributed to this clumpiness.

But the Poisson-Boltzmann theory treats only the distribution of mobile ions as a function of distance from the macroion P, and does not treat the clumpiness of the counterion/co-ion distribution. It involves an implicit averaging over the distributions of the mobile ions which eliminates the discreteness of the small ions [1]. The Poisson-Boltzmann model works best for low ion concentrations, and for monovalent mobile ions because the clumpiness is greatest for multivalent ions, $z = 2$ or $z = 3$, etc. Better approaches, such as integral equation treatments or Monte Carlo simulations, take into account all the charge interactions, but at the expense of simplicity [2].

Summary

When ions are dissolved in solution, they dissociate. The Poisson-Boltzmann equation predicts that mobile ions form electrostatic shields around charged objects. Sometimes the dissociation of strong electrolytes is complete. In other cases, dissociation is incomplete. The apparent incomplete dissociation of strong electrolytes is described in terms of the long-range electrostatic interactions between the ions. Strong electrolyte behavior can be described by the Debye–Hückel theory, a linear approximation to the Poisson–Boltzmann equation.

Problems

1. The potential around colloidal spheres. What is the dimensionless surface potential, $\Phi = e\psi/kT$, at a distance of 50 Å from

(a) a colloidal sphere with a radius of 20 Å and charge 20e in pure water?

(b) the same sphere in 0.1 M NaCl at 25 °C?

(c) What are the potentials in volts?

2. The potential near a protein in salt solution. Consider a protein sphere with a radius of 18 Å, and charge $Q = -10\,e$, in an aqueous solution of 0.05 M NaCl at 25 °C. Consider the small ions as point charges and use the Debye–Hückel linear approximation of the Poisson–Boltzmann equation.

(a) What is the dimensionless surface potential $e\psi_a/kT$ of the protein?

(b) What is the concentration of Na^+ ions and of Cl^- ions at the surface of the protein?

(c) What is the concentration of Na^+ and Cl^- ions at a distance of 3 Å from the protein surface?

3. Surface potentials and Debye lengths. You have a uniformly charged sphere with radius $a = 50$ Å in a 0.02 M NaCl solution. At a distance of 30 Å from the surface of the sphere the potential $\psi = 20$ mV.

(a) What is the Debye length $1/\kappa$ in the solution?

(b) What is the surface potential ψ_a of the sphere? (Assume that the potential field in the solution around the sphere can be derived from the linear Poisson–Boltzmann equation.)

(c) What is the charge Q on the sphere?

(d) Sketch the potential as a function of distance from the sphere.

4. The potential near a plane. You have a uniformly charged flat plate in contact with a 0.02 M NaCl solution at 25 °C. At a distance of 3 nm from the plate the potential is 30 mV.

(a) What is the Debye length $1/\kappa$ in the solution?

(b) What is the surface potential of the charged plane in mV, and in units of kT/e? Use the linear Poisson–Boltzmann equation.

(c) If you had used the nonlinear Poisson–Boltzmann equation, would you find the surface potential to be larger or smaller than you found under (b)? Why?

(d) Use the surface potential from (b) to find the surface charge density σ of the plane.

5. A charged protein near a chromatographic surface. Consider a spherical protein in water with charge $-q$ ($q > 0$) at a distance R_0 from a planar ion chromatography column surface with net positive charge, shown in Figure 23.7.

(a) Draw the field lines approximately. Include arrows to indicate the field direction.

(b) Draw the equipotential contours.

(c) If the solvent water were replaced by a water/methanol mixture with a lower dielectric constant, would it weaken or strengthen the attraction?

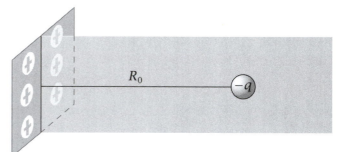

Figure 23.7

6. The potential of a membrane. Consider a phospholipid bilayer membrane consisting of a mixture of 90% uncharged lipid (zwitterionic phosphatidylcholine) and 10% acid lipid (singly charged phosphatidylserine or phosphatidylglycerol). Assume 68 Å² surface area per lipid head group. The membrane is in contact with an aqueous solution of NaCl concentration c_{NaCl} at 25 °C. Calculate the surface potential of the membrane for $c_{NaCl} = 0.05$ and 0.1 mol^{-1}.

7. Binding to a membrane. What is the electrostatic free energy of binding to the membrane description in problem 6 of a trivalent positive ion such as spermidine (a biologically active polyamine) assuming that

(a) binding occurs at the membrane surface and

(b) owing to steric factors the charges of the bound spermidine stay in the water 5 Å distant from the membrane surface?

8. Electrostatic potential near a protein. A protein in aqueous solution with 0.1 M monovalent salt has 9 positive and 22 negative charges at 25 °C. It is modelled as a sphere with radius 20 Å and uniform surface charge. What is the potential in units of kT/e

(a) at the surface of the protein and

(b) at a distance 10 Å from the surface into the solution?

Use (i) Coulomb's equation (no shielding by small ions) and (ii) the linear Poisson–Boltzmann equation.

9. Debye–Hückel model. Apply the Debye–Hückel theory to a 0.01 M monovalent salt solution at 25 °C. Treat the ions as hard spheres with a radius of 2 Å. Assume $D = 80$. What is the change in chemical potential in cal mol^{-1} due to the ion interactions? What is the activity coefficient of the ions?

10. Electrostatic potential around a line charge. A line charge with density λ and with length ℓ is in a salt solution with Debye length $1/\kappa$. Use Equation (23.24) to find an integral expression for the potential in the plane bisecting the line charge.

11. Electrostatic potential near a vesicle. A simple model for a spherical vesicle is a spherical shell with radius a, charge density σ, permeated by a salt solution with Debye length $1/\kappa$. Use Equation (23.24) to derive the potential outside the shell at distance r from the center.

    (a) Derive and discuss the limit of $\psi(r)$ for $\kappa \to 0$.

    (b) For the situation in (a), derive the potential inside the spherical shell. What is the limit for $\kappa \to 0$?

    (c) From the results of (a) and (b), derive the expression for the limit $\kappa a \to \infty$ and compare these with the result of Example 23.4.

12. Debye lengths. In the Debye–Hückel theory of monovalent salt solutions there is a characteristic length quantity κ, defined by $\kappa^2 = (2e^2 n_\infty)/(\varepsilon_0 D k T)$, where n_∞ is the salt concentration.

    (a) Express κ in terms of the Bjerrum length ℓ_B.

    (b) For water at room temperature, $\ell_B = 7.13$ Å. Compute the Debye length $1/\kappa$ in Å for a solution of $n_\infty = 1$ mol L^{-1}.

13. Ion binding to a sphere. Compute the free energy of bringing a divalent ion into contact with a spherical particle of radius $a = 14$ Å, from far away. The ion has valence $z = +2$ and the particle has a valence of $Z = +20$. For water at room temperature, compute the free energy

    (a) in a solution having monovalent salt concentration 0.1 M;

    (b) for a solution having monovalent salt concentration 0.01 M.

References

[1] R Kjellander. *Ber Bunsen Ges Phys Chem* **100**, 894–904 (1996).

[2] V Vlachy. *Annu Rev Phys Chem* **50**, 145–165 (1999).

Suggested Reading

Excellent treatments of the Poisson–Boltzmann and Debye–Hückel theories:

AW Adamson, *Physical Chemistry of Surfaces*, 3rd edition, Wiley, New York, 1976.

CMA Brett and AMO Brett, *Electrochemistry: Principles, Methods and Applictions*, Oxford University Press, New York, 1993.

PC Hiemenz, *Principles of Colloid and Surface Chemistry*, Marcel Dekker, New York, 1977.

RJ Hunter, *Foundations of Colloid Science*, Volume 1, Oxford University Press, New York, 1987.

RJ Hunter, *Introduction to Modern Colloid Science*, Oxford University Press, New York, 1993.

J Lyklema, *Fundamentals of Interface and Colloid Science*, Vol 1, Fundamentals, San Diego, 1991.

WB Russel, DA Saville and WR Schowalter, *Colloidal Dispersions*, Cambridge University Press, New York, 1989.

24 Intermolecular Interactions

Atoms and molecules, even uncharged ones, are attracted to each other. They may form noncovalent, as well as covalent, bonds. We know this because they condense into liquids and solids at low temperatures. Noncovalent interactions between uncharged particles are relatively weak and short-ranged, but they are the fundamental driving forces for much of chemistry, physics, and biology. These intermolecular interactions can be understood through measurements of the pressures of nonideal gases.

The laws of electrostatics explain the attractions between charged atoms, say Na^+ and Cl^-. But what forces bond neutral atoms together? To a first approximation, the bonding between neutral molecules can also be explained by electrostatic interactions. Even when the atoms have no net charge, they have *charge distributions*, and *polarizabilities*, which lead to weak attractions.

Molecules Repel Each Other at Very Short Range and Attract at Longer Distances

A bond between two particles is described by a *pair potential* $u(r)$, the energy as a function of the separation r between the particles.

The force $f(r)$ between two particles is the derivative of the pair potential,

$$f(r) = -\frac{du(r)}{dr}. \tag{24.1}$$

Pair potentials have three main features (see Figure 24.1). First, particles do

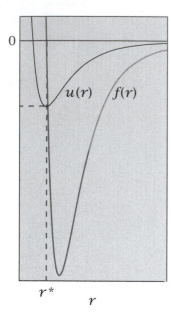

Figure 24.1 The energy of interaction $u(r)$ between two particles as a function of their separation r, and the corresponding force $f = -du/dr$; r^* is the equilibrium bond length.

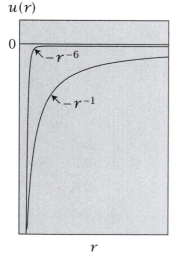

Figure 24.2 The range of an interaction: r^{-1} is long-ranged, and r^{-6} is short-ranged.

not interact if they are far apart ($u \to 0$ and $f \to 0$ as $r \to \infty$). Second, if two particles are close enough, they attract each other ($f < 0$ indicates that a positive force acts along the $-r$ direction to bring the particles together). Third, particles repel if they are too close together ($f > 0$ indicates that a positive force acts along the $+r$ direction to separate the particles).

At $r = r^*$ the attraction balances the repulsion, and the pair potential is at a minimum, so the force between two particles is zero. r^* represents the average length of an intermolecular 'bond.' Although the functional form for $u(r)$ is different for covalent bonds than for weaker interactions, all bonds have these main features: no interaction at very long range, attraction at short range, and repulsion at very short range, leading to an average equilibrium bond length where the net force is zero.

Intermolecular forces can be divided into two classes: those that are *long-ranged* and those that are *short-ranged*. The *range* is defined by the dependence of u on the separation r. Intermolecular interactions are commonly modelled as a power law,

$$u(r) = (\text{constant})r^{-p}, \tag{24.2}$$

where p is a positive integer. Interactions are called short-ranged if $p > 3$ and long-ranged if $p \leq 3$. The coulombic interaction $u(r) \propto \pm 1/r$ is long-ranged, while $u(r) \propto \pm 1/r^6$ is short-ranged, whether the interactions are attractive ($u(r)$ is negative) or repulsive ($u(r)$ is positive) (see Figure 24.2).

When two atoms are very close together in space, they repel, owing to the Pauli Principle that electrons in the same state can't occupy the same space. Quantum mechanical calculations show that these very short-ranged repulsions can best be modelled as exponential functions or power laws, typically with $p = 9, 12$, or 14.

Short-Ranged Attractions Can Be Explained as Electrostatic Interactions

Electrostatic interactions are long-ranged. The exponent in Equation (24.2) is $p = 1$ in Coulomb's law, $u(r) = Cq_1q_2/r$. Yet Coulomb's law can also explain weak intermolecular attractions, which are short-ranged. Even neutral atoms and molecules, which have no net charge, have *charge distributions*, some positive charge here and some negative charge there.

There are three different ways in which a shorter-ranged exponent ($p > 1$) can arise from coulombic interactions between neutral molecules. (1) Atoms and molecules may not be *monopoles*, having a net charge (like Na^+ or Cl^-). Rather, particles may be *multipoles*, such as *dipoles*, *quadrupoles*, etc. In multipoles the charge density varies throughout the molecule, even when the net charge sums to zero. The interactions between multipoles are shorter-ranged than the interactions between monopoles. (2) When dipolar atoms or molecules are free to orient, orientational averaging reduces the range of interaction. (3) Molecules are *polarizable*. They can be induced to have charge distributions by the electric field from nearby atoms or molecules or other sources. The interactions between polarizable molecules are shorter-ranged than the interactions between molecules having fixed dipoles. Here are the details.

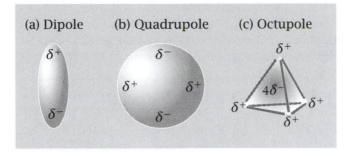

(a) Dipole (b) Quadrupole (c) Octupole

Figure 24.3 The moments of charge distributions are monopoles (not shown), (a) dipoles, (b) quadrupoles, (c) octupoles, etc.; δ^- and δ^+ indicate partial negative and positive charges.

A Charge Distribution Is Characterized by Its Multipole Expansion

Any spatial distribution of charge $\rho(x, y, z) = \rho(\mathbf{r})$ can be described by a *multipole expansion*, a series in which the first term is called the *monopole*, the second is the *dipole*, the third is the *quadrupole*, then the *octupole*, etc. The various terms are *moments* of the distribution in the same way that the mean and standard deviation are related to the first two moments of a probability distribution (see Chapter 1).

The zeroth moment, the monopole, is the total charge of the distribution, $\int \rho(\mathbf{r}) d\mathbf{r}$. The first moment $\int \mathbf{r}\rho(\mathbf{r}) d\mathbf{r}$, the dipole, describes a symmetric arrangement of equal amounts of positive and negative charge separated by a vector $\mathbf{r}$. The dipole is the lowest-order distribution for a molecule that has no net charge (see Figure 24.3). A quadrupole, $\int \mathbf{r}^2 \rho(\mathbf{r}) d\mathbf{r}$, is the next lowest-order distribution for a molecule that has no net charge and no net dipole moment. An octupole has no net charge, or dipole, or quadrupole moment. There is no tripole or pentapole, etc., because these distributions can always be expressed more simply as linear combinations of some amount of net charge plus some amount of dipole moment, plus some amount of quadrupole moment, etc. In this way, any charge distribution can be described by a multipole expansion, a sum of moments. For example, a molecule having a dipole moment in addition to a net charge can be described by a sum of the first two terms in a multipole expansion.

Let's consider dipoles. There are two types: permanent and induced. A *permanent* dipole occurs when a charge separation is always present in a molecule, even in the absence of any external electric fields. An *induced dipole* is a charge separation that arises only in the presence of an applied electric field. Some molecules have both permanent and induced dipole moments.

A permanent dipole is characterized by its dipole moment $\boldsymbol{\mu} = q\boldsymbol{\ell}$ (see Equation (21.15)). Dipole moments are measured in units of Debyes (D). One Debye equals 1×10^{-18} esu-cm $= 3.33564 \times 10^{-30}$ coulomb meter (C m). When one positive charge (such as a proton) is separated from one negative charge (such as an electron) by 1 Å, the magnitude of the dipole moment is $q\ell = (1.6022 \times 10^{-19} \text{ C}) \times (1 \times 10^{-10} \text{ m}) \times (1/(3.33564 \times 10^{-30} \text{ C m})) = 4.8033 \text{ D}$.

Table 24.1 gives some dipole moments. For example, water, which has no net charge, has a permanent dipole moment of 1.85 D, due to the partial negative charge on the oxygen and the partial positive charges on the two hydrogens. Figure 24.4 shows how dipole moments can arise from molecular asymmetry.

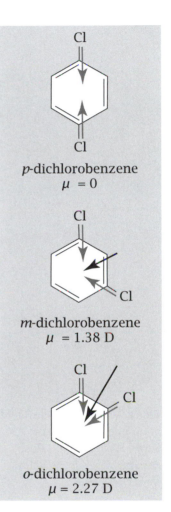

p-dichlorobenzene
$\mu = 0$

m-dichlorobenzene
$\mu = 1.38$ D

o-dichlorobenzene
$\mu = 2.27$ D

Figure 24.4 Permanent dipole moments can depend on molecular symmetry. The dipole moment (◄—) is the vector sum of unit dipoles (◄—). Source: PW Atkins, *Physical Chemistry*, 6th edition, WH Freeman, New York, 1998.

Table 24.1 Dipole moments μ and polarizability volumes, $\alpha' = \alpha/4\pi\varepsilon_0 = \alpha\mathcal{C}$ where α is the polarizability. Source: PW Atkins, *Physical Chemistry*, 6th edition, WH Freeman, New York, 1998; Handbook of Chemistry and Physics, CJF Böttcher and P Bordewijk, *Theory of Electrical Polarization*, Elsevier, Amsterdam, 1978.

Molecule	μ (10^{-30} C m)	μ (D)	α' (10^{-24} cm^3)
H_2	0	0	0.819
N_2	0	0	1.77
CO_2	0	0	2.63
CO	0.390	0.117	1.98
HF	6.37	1.91	0.51
HCl	3.60	1.08	2.63
HBr	2.67	0.80	3.61
HI	1.40	0.42	5.45
H_2O	6.17	1.85	1.48
NH_3	4.90	1.47	2.22
CCl_4	0	0	10.5
$CHCl_3$	3.37	1.01	8.50
CH_2Cl_2	5.24	1.57	6.80
CH_3Cl	6.24	1.87	4.53
CH_4	0	0	2.60
CH_3OH	5.70	1.71	3.23
CH_3CH_2OH	5.64	1.69	
C_6H_6	0	0	10.4
$C_6H_5CH_3$	1.20	0.36	
o-$C_6H_4(CH_3)_2$	2.07	0.62	
He	0	0	0.20
Ar	0	0	1.66

Neutral Molecules Attract Because of Charge Asymmetries

Table 24.2 Some covalent bond energies. Source: RS Berry, SA Rice, and J Ross, *Physical Chemistry*, Wiley, New York, 1980.

Bond	Energy (kcal mol^{-1})
C–C	80.6
C=C	145.2
C≡C	198.1
C–H	98.3
O=O (in O_2)	118.1
F–F (in F_2)	37.0

Interactions become shorter-ranged and weaker as higher multipole moments become involved. When a monopole interacts with a monopole, Coulomb's law says $u(r) \propto r^{-1}$. But when a monopole interacts with a distant dipole, coulombic interactions lead to $u(r) \propto r^{-2}$ (see Equation (21.26)). Continuing up the multipole series, two permanent dipoles that are far apart interact as $u(r) \propto r^{-3}$. Such interactions can be either attractive or repulsive, depending on the orientations of the dipoles. Table 24.2 gives typical energies of some covalent bonds, and Table 24.3 compares covalent to noncovalent bond strengths.

Orientational Averaging Shortens the Range of Interactions

Compare two situations: (1) a charge interacts with a dipole that has a fixed orientation, and (2) a charge interacts with a dipole that orients freely over all possible angles (see Figure 24.5). The energy of a charge interacting with

Type of Interaction	u (kcal mol^{-1})	Distance r Dependence
Ionic	66	$1/r$
Ion/Dipole	4	$1/r^2$
Dipole/Dipole	0.5	$1/r^3$
Dipole/Induced Dipole	0.012	$1/r^6$

Table 24.3 Various types of energy u (at 5 Å) and the dependence on distance r. Source: RS Berry, SA Rice, and J Ross, *Physical Chemistry*, Wiley, New York, 1980.

a rotating dipole is shorter-ranged than the energy of interacting with a fixed dipole. Here is the explanation.

Equation (21.26) shows that when a charge Q is separated by a distance r from a dipole having a dipole moment of magnitude $\mu = q\ell$, constrained to a fixed angle θ, the pair energy is

$$u(r, \theta) = u_0 \cos\theta, \quad \text{where} \quad u_0 = \frac{\mathcal{C}Q\mu}{Dr^2}. \tag{24.3}$$

However, if a dipole is free to rotate over all angles θ, the *average* interaction energy is

$$\langle u(r) \rangle = \frac{\int_0^\pi u(r, \theta) e^{-u(r,\theta)/kT} \sin\theta \, d\theta}{\int_0^\pi e^{-u(r,\theta)/kT} \sin\theta \, d\theta}. \tag{24.4}$$

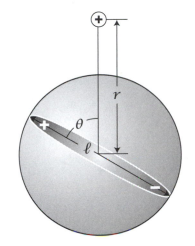

Figure 24.5 A charge is at a distance r from a dipole of length ℓ that is oriented at an angle θ from the axis between them.

The factor of $\sin\theta$ accounts for the different numbers of dipoles that point in the different directions θ (see Example 1.23 and Figure 24.5). Relatively few dipole orientations point in the directions $\theta = 0$ and $\theta = 180°$ where the area elements are small. More dipole orientations point toward $\theta = 90°$.

If the energy is small, $u/kT \ll 1$, you can use a Taylor series expansion for the exponential terms (see Appendix C, Equation (C.1)), $e^{-u/kT} \approx 1 - u/kT \ldots$, and Equation (24.4) becomes

$$\langle u(r) \rangle \approx \frac{\int_0^\pi u_0 \cos\theta \left(1 - \left(\frac{u_0 \cos\theta}{kT}\right)\right) \sin\theta \, d\theta}{\int_0^\pi \left(1 - \left(\frac{u_0 \cos\theta}{kT}\right)\right) \sin\theta \, d\theta}. \tag{24.5}$$

Equation (24.5) has two terms in the numerator and two terms in the denominator. The first term in the denominator is

$$\int_0^\pi \sin\theta \, d\theta = -\cos\theta \Big|_0^\pi = 2.$$

To evaluate the other three terms, let $x = \cos\theta$. Then $dx = -\sin\theta d\theta$. The first term in the numerator of Equation (24.5) and the second term in the denominator are zero because

$$\int_0^\pi \cos\theta \sin\theta \, d\theta = -\int_1^{-1} x \, dx = 0. \tag{24.6}$$

The second term in the numerator of Equation (24.5) becomes

$$-\frac{u_0^2}{kT} \int_0^\pi \cos^2 \theta \sin \theta \, d\theta = \frac{u_0^2}{kT} \int_0^\pi \cos^2 \theta \, d\cos \theta = \frac{u_0^2}{kT} \int_1^{-1} x^2 \, dx$$

$$= \left(\frac{u_0^2}{kT}\right) \frac{x^3}{3}\bigg|_1^{-1} = -\frac{2}{3}\frac{u_0^2}{kT}.$$

Combining all the terms in Equation (24.5) gives

$$\langle u(r) \rangle = -\frac{u_0^2}{3kT}$$

$$= -\frac{1}{3kT}\left(\frac{\mathcal{C}Q\mu}{D}\right)^2 \frac{1}{r^4}. \tag{24.7}$$

Compare Equations (24.7) and (24.3). If the dipole angle is fixed, $u(r) \propto -(1/r^2)$. But if the dipole rotates, the orientational average gives $\langle u(r) \rangle \propto -(1/r^2)^2 = -(1/r^4)$. The range of interaction is shorter when the dipole tumbles freely. This argument readily generalizes: because the energy of interaction of two permanent dipoles with fixed orientations is $u(r) \propto -1/r^3$, the interaction between two tumbling permanent dipoles is $u(r) \propto -(1/r^6)$. The interaction energy between two permanent dipoles that are free to rotate, that have moments μ_A and μ_B, and that have a center-to-center separation of r, is

$$\langle u(r) \rangle = -\frac{2}{3kT}\left(\frac{\mu_A \mu_B}{4\pi\varepsilon_0 D}\right)^2 \frac{1}{r^6}, \tag{24.8}$$

when averaged over all the possible angles of each dipole. Example 24.1 computes the magnitude of a dipole–dipole interaction.

EXAMPLE 24.1 The dipole–dipole interaction of ethanol molecules. If two ethanol molecules are $r = 10\,\text{Å}$ apart in the gas phase, what is their average interaction energy $\langle u(r) \rangle$? Table 24.1 gives the dipole moment of ethanol as $\mu = 5.70 \times 10^{-30}\,\text{C m}$ per molecule. Equation (24.8) gives

$$\langle u(r) \rangle = -\frac{2}{3kT}\left(\frac{\mu^2}{4\pi\varepsilon_0}\right)^2 \frac{1}{r^6}$$

$$= -\frac{2}{3\left(1.38 \times 10^{-23}\,\text{J K}^{-1}\text{per molecule}\right)(300\,\text{K})}$$

$$\times \left(\frac{(5.70 \times 10^{-30}\,\text{C m per molecule})^2}{4\pi \times 8.85 \times 10^{-12}\,\text{C}^2\,\text{J}^{-1}\,\text{m}^{-1}}\right)^2 \left(\frac{1}{10^{-9}\,\text{m}}\right)^6$$

$$\times 6.02 \times 10^{23}\,\text{molecules per mol}$$

$$= -8.27\,\frac{\text{J}}{\text{mol}}.$$

At $r = 10\,\text{Å}$, this attraction is very weak owing to the r^{-6} distance dependence. At $r = 5\,\text{Å}$, the interaction is $2^6 = 64$ times stronger, $-529\,\text{J mol}^{-1}$.

London Dispersion Forces Are Due to the Polarizabilities of Atoms

Attractive interactions are universal. Molecules need not have net charge, or internal charge asymmetry, or even an ability to orient, to experience attractions. Even spherical uncharged inert gas atoms condense into liquids at very low temperatures. Such attractions were first described in 1937 by F London (1900-1957), an American physicist. They are called *London forces*, or *dispersion forces*. Two molecules can induce an attraction in each other because they are polarizable.

A *polarizable* atom or medium is one that responds to an applied electric field by redistributing its internal charge. In the simplest case, when an electrostatic field is applied to a polarizable atom, charge inside the atom redistributes to form a dipole pointing in the direction opposite to the applied field (see Figure 24.6). The dipole moment μ_{ind} that is induced by the field $\mathbf{E}$ is often found to be proportional to the applied field, if the field is sufficiently small,

$$\mu_{ind} = \alpha \mathbf{E}, \tag{24.9}$$

where α is called the *polarizability* of the atom or medium. (If the field is large, the dipole moment will not be linear in E but may depend on higher powers, $\mu = \alpha E + \alpha_2 E^2 + \alpha_3 E^3, \ldots$.) The polarizability is the induced dipole moment per unit of applied electric field. Table 24.1 lists some polarizabilities. Polarizabilities have units of volume, and they are typically about the size of a molecular volume, a few cubic Ångströms.

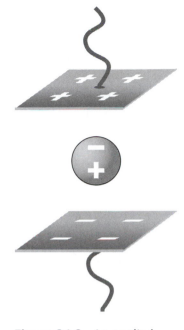

Figure 24.6 An applied field can induce a dipolar charge distribution in a neutral atom.

A Charge Will Polarize a Neutral Atom and Attract It

When a neutral atom is in the electric field of a charge, the atom will be attracted to the charge. Here's a simple model. Suppose a charge Q is at a distance r from the center of a neutral atom, as shown in Figure 24.7. The electric field from Q causes a redistribution of the charge inside the neutral atom, into amounts q and $-q$ separated by a distance Δr, giving an induced dipole moment $\mu_{ind} = q\Delta r$. Induction of the dipole results in an attractive force because Q is closer to the partial charge of opposite sign ($-q$ in this case), which it attracts, than to the partial charge of the same sign ($+q$), which it repels. The net force on the neutral atom is $f = (\text{charge}) \times E$ (see Equation (20.12)). It is the sum of the force $(-q)E[r - (\Delta r/2)]$ of Q on nearby charge $-q$, plus the force $qE[r + (\Delta r/2)]$ of Q on the more distant charge q,

$$f = -q\left[E\left(r - \frac{\Delta r}{2}\right) - E\left(r + \frac{\Delta r}{2}\right)\right] \approx q\Delta r \frac{dE}{dr} = \mu_{ind}\frac{dE}{dr}$$

$$= \alpha E \frac{dE}{dr}. \tag{24.10}$$

The field from a charge Q at a distance r is $E = \mathcal{C}Q/Dr^2$ (see Equation (20.13)). Substituting $E = \mathcal{C}Q/Dr^2$ into Equation (24.10) and taking the derivative gives

$$f = -2\alpha\left(\frac{\mathcal{C}Q}{D}\right)^2 \frac{1}{r^5}. \tag{24.11}$$

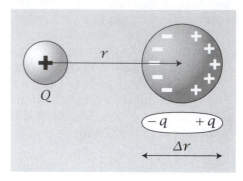

Figure 24.7 When a neutral atom (on the right) is in the electrostatic field of a charge Q, a charge separation is induced in the atom, modelled as a dipole. This attracts the neutral atom to the charge.

Integrating to express Equation (24.11) instead as an intermolecular potential yields

$$u(r) = -\int f(r)\, dr = -\frac{\alpha}{2}\left(\frac{CQ}{D}\right)^2 \frac{1}{r^4}. \tag{24.12}$$

The minus signs in Equations (24.11) and (24.12) show that the interaction of a charge with a neutral atom is attractive, irrespective of the sign of Q. Equations (24.11) and (24.12) show that the interaction is short-ranged ($p > 3$) and that the attraction increases with the polarizability of the neutral atom.

Similarly, when two neutral atoms interact, they induce dipoles in each other, with a pair interaction $u(r) \propto -1/r^6$ that is also proportional to the polarizabilities α of the two dipoles.

All the short-ranged attractions we have described, involving multipoles, induced polarization, and orientational averaging, lead to power laws with $p \geq 3$, and are collectively called *van der Waals* forces.

The fact that dipoles interact with a distance dependence r^{-6} gives a physical basis for the form of the attractive interaction in a widely used energy function called the **Lennard–Jones potential**,

$$u(r) = \frac{a}{r^{12}} - \frac{b}{r^6},$$

where a and b are parameters that depend on the types of interacting atoms [1, 2]. The positive sign in the first term implies a repulsion, and the minus sign in the second term implies an attraction. The repulsive part of this potential, a/r^{12}, was originally chosen because it can be calculated rapidly by computers using the square of r^{-6}. The virtue of this model is that it captures the universal features of a short-ranged attraction and even shorter-ranged repulsion, and the two parameters a and b give enough flexibility for the model to predict experimental data accurately.

Hydrogen Bonds

Hydrogen bonds are weak interactions (typically a few kcal mol^{-1}) that occur when a hydrogen is situated between two other atoms. For example, a hydrogen bond can form between an amide and carbonyl group, N–H$\cdots$O=C. In this case, the N–H group is called the *hydrogen bond donor* and the C=O group is called the *hydrogen bond acceptor*. To a first approximation, a hydrogen bond can be described as an electrostatic interaction between the N–H dipole and the O=C dipole. Hydrogen bonds act to align the N–H and O=C bonds, and to stretch the N–H bonds. To describe hydrogen bonds more accurately requires a quantum mechanical treatment of the charge distribution.

Empirical Energy Functions

A popular approach to modeling the interactions and conformations of large molecules, particularly in solution, is to assume an energy U that is a sum of terms: (1) coulombic interactions between charged atoms, (2) spring forces that stretch and bend bonds, (3) periodic potentials for torsional rotations around bonds, and (4) a Lennard–Jones potential for nonbonded interactions [3, 4, 5]. For example,

$$U = \sum_{\text{bond lengths } b} \frac{K_b}{2}(b - b_0)^2$$

$$+ \sum_{\text{bond angles } \theta} \frac{K_\theta}{2}(\theta - \theta_0)^2$$

$$+ \sum_{\text{dihedral angles } \phi} K_\phi[1 + \cos(n\phi - \delta)]$$

$$+ \sum_{\text{nonbonded pairs } i<j} \left(\frac{a_{ij}}{r_{ij}^{12}} - \frac{b_{ij}}{r_{ij}^6} + \frac{\mathcal{C}q_iq_j}{Dr_{ij}} \right), \tag{24.13}$$

where K_b is the spring constant for stretching bonds, b_0 is the equilibrium bond length, K_θ is the spring constant for bending bonds, and θ_0 is the equilibrium bond angle. K_ϕ and δ are torsional constants, and a_{ij} and b_{ij} are Lennard–Jones constants for atoms i and j separated by a distance r_{ij}. The summation over the index $i < j$ is a convenient way to count every interaction only once. Summing the pair interaction energies u_{ij} over all indices, $\sum_{\text{all } i} \sum_{\text{all } j} u_{ij}$, counts every interaction exactly twice. For example, the interaction of particle 2 with particle 3, u_{23}, is equal to u_{32}. Thus

$$\sum_{i<j} u_{ij} = \frac{1}{2} \sum_{\text{all } i} \sum_{\text{all } j} u_{ij}. \tag{24.14}$$

Much of what is known about intermolecular interactions comes from measuring the pressures of nonideal gases. The simplest model that relates intermolecular interactions to the pressures of gases is the van der Waals model.

The van der Waals Gas Model

Real gases are more complex than ideal gases. The ideal gas model predicts that the pressure depends only on the temperature and the gas density ($\rho = N/V = p/kT$). The pressure of an ideal gas does not depend on the types or atomic structures of the gas molecules. But ideal behavior applies only at low densities where molecules don't interact much with each other. For denser gases, intermolecular interactions affect the pressures, and gases differ from

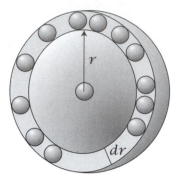

Figure 24.8 To compute the internal energy of a gas, identify one particle in a gas as a test particle. A spherical shell at a distance r from that particle has volume $4\pi r^2 dr$. The number of particles in the shell is $\rho 4\pi r^2 dr$.

each other depending on their atomic structures. The ideal gas law does not predict that gases condense to liquids. The van der Waals gas law does.

The **van der Waals equation of state** for the pressure $p(T, V, N)$ is

$$p = \frac{NkT}{V - Nb} - \frac{aN^2}{V^2} = \frac{\rho RT}{1 - b\rho} - a\rho^2, \tag{24.15}$$

where a and b are parameters (not to be confused with the Lennard–Jones parameters). Equation (24.15) was developed in 1873 in the Ph.D. thesis of the Dutch physicist JD van der Waals (1837–1923), who won the Nobel prize in 1910 for this work.

The van der Waals gas law can be derived in various ways from an underlying model of intermolecular interactions. The next two sections give a simple derivation.

We will find the pressure as a sum of energy and entropy components (combining Equations (8.12) and (8.29)),

$$p = -\left(\frac{\partial F}{\partial V}\right)_{T,N} = -\left(\frac{\partial U}{\partial V}\right)_{T,N} + T\left(\frac{\partial S}{\partial V}\right)_{T,N}. \tag{24.16}$$

The lattice model in Example 7.1 gives the second term, the entropic component, as $T(\partial S/\partial V)_{T,N} = NkT/(V - Nb)$. The next section shows that intermolecular attractions lead to $-(\partial U/\partial V)_{T,N} = -aN^2/V^2$, the energetic component.

The Energetic Component of the van der Waals Gas Pressure

To describe the energetic component of the pressure, you need a model for the energy as a function of volume, $U(V)$. The total interaction energy U of the gas is the sum of all the interparticle interactions. You can treat only the *interactions* between the particles and ignore the contributions from the internal partition functions of the particles because they don't change with V, and won't contribute to the pressure $(\partial U/\partial V)_{T,N}$. (The translational partition function does depend on V, but this contribution is treated in the entropy component.)

Choose one particle, and call it the *test particle*. Divide the space around the test particle into spherical shells. Figure 24.8 shows a shell at radius r from a test particle. What is the energy U' of interaction between the test particle and all the other particles in the system? To compute U', multiply the number of particles in the shell at radius r by the interaction energy $u(r)$ between the test particle and all the particles in that shell, then integrate over all shells.

How many particles are in the shell at radius r? A gas of density $\rho = N/V$ has N particles in a volume V. If there is a uniform spatial distribution, the number of particles in the shell at radius r will be (density) × (volume of the shell) $= \rho \times 4\pi r^2 dr$. So the interaction energy of the test particle with all the other molecules is $U' = \int u(r)\rho 4\pi r^2\,dr$.

To get the total energy U, sum the energy U' over all N particles (each taken once as the test particle), and divide by 2 to correct the double-counting. (If you multiplied U' by N, you would count every interaction exactly twice, once when a particle was the test particle at the center, and once when it was in a shell at radius r from the other test particle.) The total energy is

$$U = \frac{N}{2}U' = \frac{N}{2}\int_0^\infty u(r)\rho 4\pi r^2\, dr. \qquad (24.17)$$

Equation (24.17) holds when particles are distributed uniformly and when the energies are pairwise additive, i.e., when the pair interaction is not influenced by the presence of a third particle.

To compute U from Equation (24.17), you need to know the pair interaction energy function $u(r)$. Because repulsions between atoms and molecules are very steep short-ranged functions for small r, it is simplest to assume 'hard-core' repulsions approximately like those between billiard balls. If you assume that the attractions fall off as r^{-6} (see Figure 24.9), the pair potential is

$$u(r) = \begin{cases} \infty & \text{if } r < r^*, \\ -u_0(r^*/r)^6 & \text{if } r \geq r^*, \end{cases} \qquad (24.18)$$

where r^* is the minimum distance between the pair of particles. Substituting the pair potential Equation (24.18) into Equation (24.17) gives

$$U = \frac{N}{2}\int_{r^*}^\infty \left[-u_0\left(\frac{r^*}{r}\right)^6\right]\left(\frac{N}{V}\right)4\pi r^2\, dr$$

$$= \frac{-u_0 N^2 2\pi (r^*)^6}{V}\int_{r^*}^\infty \frac{dr}{r^4}. \qquad (24.19)$$

In principle, the upper limit of the integration in Equation (24.19) should be some dimension of the gas container, but because the integral converges to its final value for as few as 10–20 shells, you can define the upper limit as $r = \infty$. No particle is located at a distance between $r = 0$ and $r = r^*$ from the center of another particle, so you have $\int_0^\infty = \int_{r^*}^\infty$.

Because the integral $\int_{r^*}^\infty r^{-4}\, dr = -[1/(3r^3)]_{r^*}^\infty$ equals $1/(3r^{*3})$, Equation (24.19) reduces to

$$U = -\frac{aN^2}{V}, \qquad \text{where } a = \frac{2\pi (r^*)^3}{3}u_0. \qquad (24.20)$$

The minus sign in Equation (24.20) indicates that attractive interactions reduce the energy relative to that of an ideal gas.

Now you can complete the free energy by considering the entropic component.

The Entropic Component of the Pressure

The entropy of a van der Waals gas can be derived in different ways. Let's use a simple lattice model. Recall from the lattice model of gases (Equation (7.9)) that the entropy S of distributing N particles onto a lattice of M sites is

$$\frac{S}{k} = -N\ln\left(\frac{N}{M}\right) - (M - N)\ln\left(\frac{M - N}{M}\right).$$

Let V equal the total volume of the system, and let $b_0 = V/M$ be the volume per lattice site. Substituting $M = V/b_0$ into Equation (7.9), and using $F = U - TS$

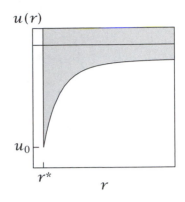

$u(r)$

Figure 24.9 A model for the attractive energy between two particles, used to derive the van der Waals gas law.

with Equation (24.20) for U gives

$$F = -\frac{aN^2}{V} + kT \left[N \ln \left(\frac{Nb_0}{V} \right) - \left(\frac{V}{b_0} - N \right) \ln \left(1 - \frac{Nb_0}{V} \right) \right]. \qquad (24.21)$$

To get the pressure p from the free energy F, use Equation (8.12):

$$p = -\left(\frac{\partial F}{\partial V} \right)_{T,N}$$

$$= -\frac{aN^2}{V^2} - \frac{kT}{b_0} \ln \left(1 - \frac{Nb_0}{V} \right). \qquad (24.22)$$

For low densities, $\ln (1 - x) \approx -x - x^2/2 \dots$ (Appendix C, Equation C.4), so for $x = Nb_0/V$ you have

$$-\frac{kT}{b_0} \ln \left(1 - \frac{Nb_0}{V} \right) \approx \frac{NkT}{V} \left[1 + \frac{1}{2} \left(\frac{Nb_0}{V} \right) + \cdots \right]. \qquad (24.23)$$

The first term on the right-hand side of Equation (24.23) gives the ideal gas law. For densities that are higher than those of ideal gases, include the next higher order term in Equation (24.23). Using the approximation $(1+x/2) \approx 1/(1-x/2)$ (Appendix C, Equation C.6) gives the **van der Waals equation**,

$$p = \frac{NkT}{V(1 - Nb/V)} - \frac{aN^2}{V^2}$$

$$= \frac{NkT}{V - Nb} - \frac{aN^2}{V^2}.$$

The van der Waals constant $b = b_0/2$ represents half the volume of each particle. In practice a and b in the van der Waals equation are taken as adjustable parameters for fitting pressure/volume/temperature data.

The van der Waals equation is just one of many different models for the pressures of real gases. It is perhaps the simplest model that can describe the phase transition between gas and liquid states (boiling) (see Chapter 25). Another model that accounts for phase transitions is the **Redlich–Kwong equation** for the pressure [6, 7],

$$p = \frac{kT}{V - b_1} - \frac{a_1}{T^{1/2}V(V + b_1)}$$

in terms of parameters a_1 and b_1.

One assumption that we made for computing the energy of the van der Waals model is that the gas is distributed uniformly in space. In reality, if particles have attractions, they tend to cluster. The next section shows how you can correct for this.

Radial Distribution Functions

The true density in the shell at radius r from a test particle can be described by $\rho g(r)$, where $\rho = N/V$ is the average density and $g(r)$ is called the *radial distribution function* or sometimes the *pair correlation function*. $g(r)$ is the ratio of the actual density of particles at r to the mean density ρ. The pair correlation function $g(r)$ equals one when the local density in a shell is the

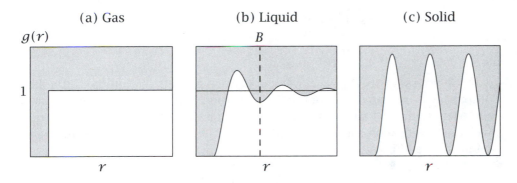

Figure 24.10 Examples of pair correlation functions: (a) the model we have used for the van der Waals gas—hard core repulsions, and uniform distributions otherwise; (b) typical liquids (B is the limit used to define the first shell of neighbors); (c) solids, which have long-range order.

same as the mean density averaged over the whole volume. Where $g(r) > 1$, it means the density of particles in that shell is higher than mean density ρ.

Figure 24.10 shows three examples of correlation functions. Figure 24.10(a) shows the correlation function that we assumed for the van der Waals gas: particles cannot be closer together than a distance r^*, and they are randomly distributed otherwise. Figure 24.10(b) shows a correlation function for a typical liquid: there is a depletion $g(r) = 0$ for small r because particles cannot overlap, then a peak indicates there is an excess of particles that make up a first neighbor shell, then a slight depletion, followed by a second neighbor shell, etc. Beyond the third or fourth shell, there is little correlation and $g(r) = 1$. Figure 24.10(c) illustrates the *long-ranged order* in crystalline solids.

Now we show how the number of neighbors and pairwise energies of interaction are computed from pair correlation functions.

Using Pair Correlation Functions

Pair correlation function $g(r)$ can be found from experiments using the scattering of electromagnetic radiation. From this function, you can get other quantities, such as the average number of nearest neighbors of any molecule. Because the volume of a spherical shell at radius r is $4\pi r^2 dr$, the number of molecules in the first neighbor shell is density × (volume of the shell) = $\rho g(r)4\pi r^2 dr$. The total number of molecules in the first shell of a molecule is the integral

$$\text{number of nearest neighbors} = \int_0^B \rho g(r)4\pi r^2 \, dr,$$

where B is the location shown in Figure 24.10(b). If you integrate over all the shells, you get n, the total number of particles (minus the test particle),

$$\int_0^\infty \rho g(r)4\pi r^2 \, dr = n - 1 \approx n.$$

In a system where particles are not distributed uniformly, such as a liquid, the energy U' of interaction between one particle and all the other particles requires the factor of $g(r)$ in the integral:

$$U' = \int_0^\infty \rho u(r) g(r) 4\pi r^2 \, dr. \tag{24.24}$$

In Chapters 14 to 16, we developed lattice models of liquids and solids composed of uncharged particles. We assumed that each molecule interacts only with its nearest neighbor through a contact energy w. The next section gives the justification for using such contact energies and for neglecting more distant interactions.

The Lattice Model Contact Energy w Approximates Intermolecular Interactions

The lattice model quantity w (for example, in Equations (14.6) and (15.4)) is $u(r^*)$, the pair potential evaluated at the equilibrium bond separation distance $r = r^*$. In the lattice model the sum over all pairs in Equation (24.17) is treated as a sum over only nearest neighbors,

$$U = \frac{N}{2} \sum_{r=0}^\infty u(r) \rho g(r) 4\pi r^2 \approx \frac{N}{2} u(r^*) z = \frac{Nwz}{2}. \tag{24.25}$$

The justification for neglecting all other shells in the lattice model of liquids and solids is that nearest-neighbor pair interactions contribute most of the energy for dense phases of particles having short-ranged interactions. For London interactions, the energy of interaction diminishes as r^{-6}, but the number of molecules increases with distance as $4\pi r^2$. The product of these two factors gives $r^{-6} \times r^2 = r^{-4}$. So, even though the number of particles increases with distance from a test particle, the total energy contribution diminishes with distance.

Let's see what fraction of the total energy comes from the second shell and beyond. On the lattice, you can sum over shells numbered by integers $r = 1, 2, 3, \ldots$. You can look up these sums in math tables [8]:

$$\frac{\displaystyle\sum_{r=2}^\infty (1/r)^4}{\displaystyle\sum_{r=1}^\infty (1/r)^4} = \frac{(\pi^4/90) - 1}{(\pi^4/90)} = 0.076. \tag{24.26}$$

The total energy due to all neighbors beyond the first neighbor shell is only about 7.6% of the total. This is the size of the error that we make in neglecting all except nearest-neighbor interactions.

If you had substituted any attractive short-ranged potential (one that falls off more rapidly than r^{-3}) into Equation (24.17), it would have led to the pressure $\rho \propto V^{-2}$, just as in Equation (24.22). For this reason, many different types of interaction are collectively called van der Waals forces. They all lead to the same contribution to $p(V)$. For longer-ranged forces, such as coulombic interactions, the integral in Equation (24.17) becomes infinite, and Equation (24.20) would not apply. For long-range interactions, you need to resort to the methods of Chapters 20 to 23, such as the Poisson equation.

In Chapters 15 and 16, we noted that two uncharged atoms or molecules A and B generally prefer to self-associate than to form AB interactions, leading normally to a positive value of the exchange energy, $\chi_{AB} > 0$. Example 24.2 shows how this arises from dipole interactions.

EXAMPLE 24.2 Like dissolves like: a consequence of induced dipole interactions. For uncharged molecules having permanent dipoles, the interaction energy is proportional to the product $\mu_A \mu_B$ (see Equation (24.8)). For molecules having induced dipoles, a similar derivation would show that the interaction energy is proportional to the product of polarizabilities $\alpha_A \alpha_B$. Such products imply the following pair interactions

$$w_{AA} = -c\alpha_A^2, \qquad w_{BB} = -c\alpha_B^2, \quad \text{and} \quad w_{AB} = -c\alpha_A \alpha_B, \qquad (24.27)$$

where c is a positive constant that is approximately the same for atoms and for molecules of similar size. Thus χ_{AB}, defined in Equation (15.11), can be given in terms of the dipole moments or polarizabilities:

$$\chi_{AB} = \left(\frac{z}{kT}\right)(-c)\left(\alpha_A \alpha_B - \frac{\alpha_A^2 + \alpha_B^2}{2}\right) = \left(\frac{cz}{2kT}\right)(\alpha_A - \alpha_B)^2 > 0. \qquad (24.28)$$

The quantity χ_{AB} is positive because $(\alpha_A - \alpha_B)^2$ is always positive. This is the basis for the general rule that like dissolves like.

Summary

Intermolecular bonds are the weak interactions that hold liquids and solids together. Atoms and molecules repel each other when they are too close, attract when they are further apart, and are stable at the equilibrium *bond length* in between. Intermolecular attractions make the pressures of van der Waals gases lower than the pressures of ideal gases at low densities. To a first approximation, intermolecular attractions can be explained as electrostatic interactions between charge distributions, due to internal charge asymmetries in molecules, to the freedom of molecules to rotate, or to molecular polarizabilities.

Problems

1. Interpreting Lennard–Jones parameters.
Intermolecular interactions are often described by the Lennard–Jones potential $u(r)$, which gives the internal energy of interaction between two molecules as a function of intermolecular separation:

$$u(r) = 4\varepsilon \left[\left(\frac{\sigma}{r} \right)^{12} - \left(\frac{\sigma}{r} \right)^{6} \right],$$

where ε and σ are characteristic energy and bond length parameters.

(a) At low temperatures, entropy is relatively unimportant, and the free energy is minimized at the intermolecular separation of the molecules at which the potential energy is minimized. At what separation does that occur? What is the energy of that state?

(b) It's often convenient to divide energies ε by Boltzmann's constant k to give energies in units of temperature (K). In these units, the constants in Table 24.4 have been found. What physical properties of these systems can you deduce from this information?

Table 24.4   Source: PW Atkins, *Physical Chemistry*, 6th edition, WH Freeman and Co, New York, 1998.

Molecule	ε/k (K)	σ(Å)
He	10.22	2.58
Ethane	205	4.23
Benzene	440	5.27

2. How the dielectric constant affects electrostatic binding.
An anesthesiologist has isolated two proteins A and B that bind to each other by electrostatic attraction. A has net positive charge and B has net negative charge. In water, which has a dielectric constant of $D = 80$ at room temperature, their binding constant is $K_1 = 1000$. What is their binding constant if the medium is trifluoroethanol with a dielectric constant of $D = 27$? (Assume that the charges on each protein and the distance between them are the same in water and trifluoroethanol.)

3. An electrostatic model for the chemical bond.
The simplest model for a chemical bond is two fixed unit positive charges separated by a distance r_0, and a unit negative charge that is free to move along a dividing line (see Figure 24.11).

(a) For fixed r_0, how does the coulombic energy depend on r_0 and θ?

(b) For what angle θ is the coulombic energy a minimum?

(c) What is the energy at its minimum value?

(d) At what angle θ is the energy equal to zero?

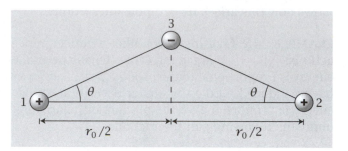

Figure 24.11   Two positive charges (1 and 2) at a fixed separation r_0, with a minus charge (3) that is free to move vertically between them.

4. Direction and distance in a dipolar interaction.

(a) Which dipole pair in Figure 24.12 has the lower energy, the parallel pair in Figure 24.12(a) or the antiparallel pair in Figure 24.12(b)?

(b) If you double the distance between the dipole centers by what factor does each pair interaction change? Does it increase or decrease?

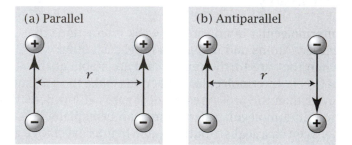

Figure 24.12   Dipoles separated by a distance r, (a) parallel, and (b) antiparallel.

5. The r dependence of pair potential functions.
You have two different pair potential functions:

$$u_1(r) = -\frac{1}{r} \quad \text{and} \quad u_2(r) = -\frac{1}{r^6}.$$

(a) Plot both functions.

(b) At $r = 1$, which pair potential has the stronger attraction?

(c) At $r = 2$, which pair potential has the stronger attraction?

6. Relating force and equilibria to a pair potential.
Consider the interaction potential between two particles shown in Figure 24.13.

(a) Draw the corresponding force curve $f(r)$.

(b) Identify the points of equilibrium.

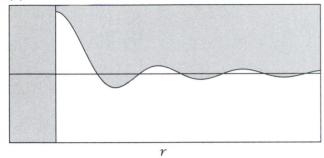

Figure 24.13 Energy $u(r)$ and axes for drawing the force curve $f(r)$.

7. Predicting electrostatic attraction. The system shown in Figure 24.14 has two fixed positive charges separated by a distance $3r$. The negative charge is closer to the left positive charge. Which way will the negative charge move?

Figure 24.14 Two fixed positive charges and a movable negative charge on the line between them.

References

[1] U Burkert and NL Allinger. *Molecular Mechanics*, ACS Monograph 177, American Chemical Society, Washington, 1982.

[2] M Rigby et al. *The Forces between Molecules*, Oxford University Press, New York, 1986.

[3] JA McCammon and SC Harvey. *Dynamics of Proteins and Nucleic Acids*, Cambridge University Press, New York, 1987.

[4] WD Cornell, AE Howard, and P Kollman. *Curr Opin Struct Biol* **1**, 201 (1991).

[5] CL Brooks, M Karplus and BM Pettitt. *Advances in Chemical Physics*, Volume LXXI *Proteins: A Theoretical Perspective of Dynamics, Structure, and Thermodynamics*, Wiley, New York, 1988.

[6] M Modell and RC Reid. *Thermodynamics and Its Applications*, 2nd edition, Prentice Hall, Englewood Cliffs, 1983.

[7] O Redlich and JNS Kwong. *Chem Rev* **44**, 233 (1949).

[8] MR Siegel. *Mathematical Handbook*, Schaum's Ouitline Series, McGraw-Hill, New York, 1968.

Suggested Reading

Detailed discussions of intermolecular interactions and polarizability are found in:

RS Berry, SA Rice and J Ross, *Physical Chemistry* (Intermolecular Forces, Chapter 10), Wiley, New York, 1980.

JN Israelachvili, *Intermolecular and Surface Forces*, 2nd edition, Academic Press, London, 1992.

Equations of state are discussed in:

HT Davis, *Statistical Mechanics of Phases, Interfaces, and Thin Films*, VCH, New York, 1996.

JC Slater, *Introduction to Chemical Physics*, McGraw-Hill, New York, 1963.

25 Phase Transitions

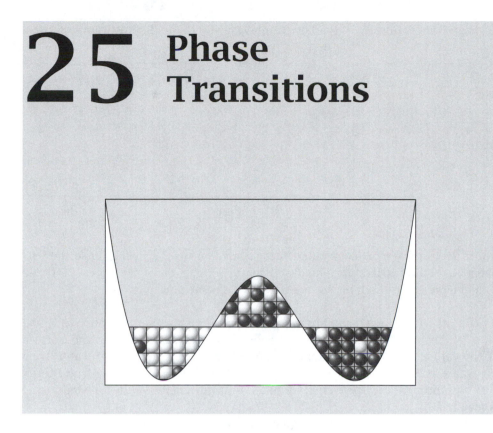

Two States Can Be Stable at the Same Time

Phase transitions are dramatic changes in the collective behavior of a system. When water is heated by 2 °C at 99 °C, it boils. Water transforms from a liquid to a gas, decreasing its density 1600-fold, losing its liquid properties. In contrast, if you heat water by 2 °C at 25 °C, its density decreases by only about 0.2%. The dramatic change in density at 99 °C is a phase transition: the decrease in density at 25 °C is just a gradual change. Other phase transitions include freezing, the magnetization of metals, solubilization and precipitation in liquids, and the sharp increase in the alignment of liquid crystal molecules (used in watches and computer displays) with increasing concentration. While 'phase transition' refers to changes in interatomic and intermolecular behavior in macroscopic systems, 'cooperativity' more broadly also refers to sharp changes of properties of single molecules, for example in protein folding, helix–coil transitions, or ligand binding (see Chapter 26). In this chapter we focus on phase transitions.

At the midpoint of a phase transition, two phases are in equilibrium with each other. In this chapter our first example is liquid–liquid immiscibility. Put oil and water together: at equilibrium, you get two phases having different compositions—one is mostly oil and the other is mostly water. We also consider boiling water: at equilibrium, you get two phases having different densities— one is steam and the other is liquid water.

To see how two different phases can be stable at the same time, recall that if a system has a degree of freedom x, the stable state is identified as the value

467

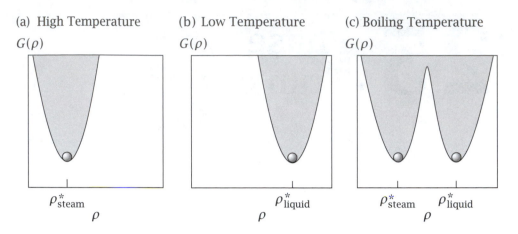

(a) High Temperature

$G(\rho)$

ρ^*_{steam}

ρ

(b) Low Temperature

$G(\rho)$

ρ^*_{liquid}

ρ

(c) Boiling Temperature

$G(\rho)$

ρ^*_{steam} ρ^*_{liquid}

ρ

Figure 25.1 (a) At high temperature, steam with density ρ^*_{steam} is the stable state of water. (b) At low temperature, liquid with density ρ^*_{liquid} is the stable state. (c) At the boiling temperature, both phases are stable at the same time.

T (K)

390

370

350

330

0   0.5   1

x

Single Phase

Two Liquid Phases

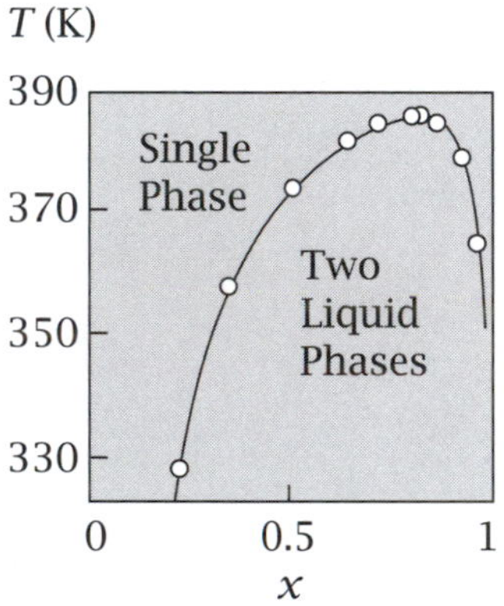

Figure 25.2 A phase diagram for liquid benzene (mole fraction x) in perfluoroheptane. Source: JN Murrell and AD Jenkins, *Properties of Liquids and Solutions*, John Wiley & Sons, New York, 1982. Data are from JH Hildebrand, BB Fisher, and HA Benesi, *J Am Chem Soc* **72**, 4348 (1950).

$x = x^*$ at which the free energy is a minimum. Using the metaphor of a ball on a landscape (Chapter 2), the ball rolls along x until it reaches the bottom of the valley, $x = x^*$. If the degree of freedom is the density ρ of water, then for water at $T = 25$ °C, $\rho = \rho^* = 1\,\text{g cm}^{-3}$. But at $T = 150$ °C the free energy is a minimum where the density equals an appropriate value for steam (see Figure 25.1).

What is ρ^* at the boiling temperature, $T = 100$ °C? At the boiling point, there are two equal minima in the free energy function. Both the liquid and the vapor states of water are stable at the same time. There is more than one valley on the energy landscape, and balls roll with equal tendency into either of them. Changing the temperature changes the relative depths of the two minima.

Let's look first at liquid–liquid immiscibility before returning to boiling.

Liquids or Solids May Mix at High Temperatures but Not at Low Temperatures

If you put equal amounts of oil and water into a test tube, you find two phases in equilibrium: an oil phase on top of a water phase. A *phase diagram* like the one in Figure 25.2 is a sort of a road map to the phases. The lines on phase diagrams indicate which conditions result in a single phase, and which conditions result in multiple phases.

Let's follow the type of experiment that creates the phase diagram in Figure 25.2. Make six different solutions, numbered 1 to 6, having different compositions of the two components. All of the solutions are kept at the same temperature T_0. The y-axes of Figures 25.2 and 25.3 represent the temperatures of the solutions, and the x-axes represent the compositions x_B of those solutions. If B in Figure 25.3 is oil and A is water, x_B represents the mole fraction of the oil in the water. The left end of the diagram ($x_B = 0$) represents pure water, and the right end ($x_B = 1$) represents pure oil.

1.   **A SMALL AMOUNT OF OIL DISSOLVED IN WATER.**   Test tube 1 has a very small amount of oil in a large volume of water. The oil fully dissolves in the

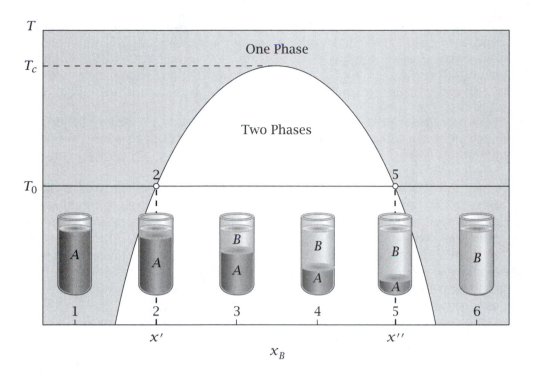

Figure 25.3 Six different solutions help to map out the phase diagram. T_0 is the temperature of the solution, x' is the concentration of B that saturates liquid A, and x'' is the concentration of B at which A saturates B. T_c is the critical temperature.

water, meaning that most of the individual oil molecules are dispersed and surrounded by water molecules. This is called a *single-phase* solution.

2. THE SOLUBILITY LIMIT OF OIL IN WATER. The oil concentration in test tube 2 is greater than in tube 1. This is the lowest oil concentration at which you can detect a second phase, a thin layer of oil on the water. Tube 2 has enough oil to *saturate* the capacity of the water to hold oil. At the temperature and composition where you first see a new phase, you mark a *phase boundary* point to begin mapping out the curve in Figure 25.3.

3. AN OIL PHASE ON TOP OF A WATER PHASE. Test tube 3 has a higher oil concentration than tube 2. As a result, test tube 3 has a larger volume of the oil phase and a smaller volume of the water phase. This is a *two-phase solution*. The top phase is not pure oil and the bottom phase is not pure water. Rather, the water-rich phase on the bottom is saturated with oil, having oil concentration x'. The oil-rich phase on top is saturated with water, having oil concentration x''.

4. THE OIL PHASE ON TOP OF THE WATER PHASE, BUT MORE OF IT. Test tube 4 continues the trend of higher oil/water ratios. Tube 4 has a larger volume of the oil-rich phase than tube 3, and a smaller volume of the water-rich phase. But the *compositions* of the oil-rich and water-rich phases remain the same as they are in test tube 3: the bottom phase still has oil concentration x', and the top phase still has oil concentration x''.

5. THE SOLUBILITY LIMIT OF WATER IN OIL. Now the oil concentration reaches the point at which you first detect the disappearance of the water phase. The system becomes a single phase again. You now have an oil-rich phase saturated with water. Mark a second point on the phase diagram to indicate the changeover from a two-phase to a one-phase system.

6. A SMALL AMOUNT OF WATER DISSOLVED IN OIL. Test tube 6 contains a single phase, in which a very small amount of water is fully dissolved in a large amount of oil.

Solutions 2 and 5 each define a point on the *phase boundary*, also called the *solubility curve* or *coexistence curve*. A phase is a region in which the composition and properties are uniform. Inside the phase boundary is the *coexistence region* where two phases are in equilibrium: an oil-rich phase sits on a water-rich phase (as in test tubes 3 and 4). Outside the phase boundary, you have a one-phase solution, either oil dissolved in water or water dissolved in oil (test tubes 1 and 6). From the six experiments at temperature $T = T_0$, two compositions, x' and x'', define the phase boundary.

Now perform the same kind of experiment at a higher temperature $T = T_1$. Again, mark the two phase boundary points, $x'(T_1)$ and $x''(T_1)$. At higher temperatures, the phase boundaries are closer together: more oil dissolves in water and more water dissolves in oil. Continue increasing the temperature and finding the two phase boundary points for each temperature. This is how you map out liquid–liquid phase diagrams such as the one shown in Figure 25.2. For any temperature higher than the *critical temperature* T_c (see Figure 25.3) the two components are *miscible in all proportions*. That is, above T_c, there are not two distinct phases. Any proportion of the two components form a single mixed system.

Once you know the phase diagram you can reverse the process to find the compositions x' and x'' of the two phases in equilibrium. For example, focus on solution 3 at temperature T_0. Follow the isotherm (that is, the horizontal line of constant temperature, called the *tie line*), to the left, where it intersects the phase boundary at composition x' at point 2. Also, follow it to the right where it intersects the phase boundary at composition x'' (point 5). Points 2 and 5 define the two compositions of the coexisting phases. For the oil phase, the mole fraction of oil is x'' and the mole fraction of water $1 - x''$. For the water phase, the mole fraction of oil is x' and of water is $1 - x'$.

In this series of test tubes, we fixed the temperature and varied the composition. Now instead, fix the composition and change the temperature, say of solution 4. Read the results from the same phase diagram. At low temperatures, the solution has two phases. At a higher temperature, above the phase boundary, the two components dissolve in each other. This is the standard lore of bench chemistry: heating helps liquids to mix and dissolve.

If you are not skilled with a pipette, phase equilibria can help. Suppose you need a highly reproducible concentration of an oil–water mixture. Your mixing skills need only be good enough to produce any composition between solutions 2 and 5. The phase equilibrium will regulate the concentrations to be precisely x' or x''.

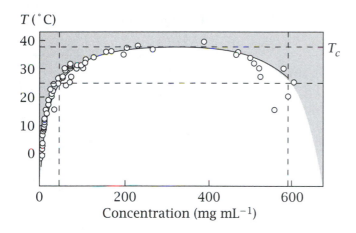

Figure 25.4 The phase diagram for γ-crystallin in water. The solubility limits at 25 °C are the points where the horizontal dashed line intersects the experimental curve. The higher horizontal line shows that the critical solution temperature is 37 °C, which coincides with body temperature. Source: ML Broide, CR Berland, J Pande, OO Ogan and GB Benedek, *Proc Natl Acad Sci USA* **88**, 5660–5664 (1991).

EXAMPLE 25.1 The phase diagram for the eye lens protein crystallin in water. Figure 25.4 shows the phase diagram for mixtures of water and the lens protein crystallin.

What is the solubility limit for the protein in water at 25 °C? The horizontal line at 25 °C intersects the left side of the coexistence curve at about 50 mg mL^{-1}. This is the solubility limit, the most crystallin that can dissolve in water (at 25 °C).

What is the concentration of protein in the 'mostly protein' phase at 25 °C? The same horizontal line intersects the right side of the coexistence curve at about 600 mg mL^{-1}.

What is the solubility limit at body temperature ($T = 37$ °C)? This temperature is near the critical point, so the solubility limit varies strongly with small changes in temperature. Slightly below 37 °C, the solubility limit is 100–200 mg mL^{-1}. Slightly above 37 °C, the protein is miscible in water at all concentrations. This phase behavior may be responsible for the formation of some cataracts in the eye.

Phase Separations Are Driven by the Tendency to Lower the Free Energy

Our aim now is to predict phase diagrams from a model of the underlying intermolecular interactions. We'll use the lattice model of solutions. The lattice model predicts that at high temperatures, where the free energy of mixing is dominated by the entropic component, solutions mix to gain translational entropy. At low temperatures, where the free energy is dominated by the energy (enthalpy), solutions form separate phases because AA and BB attractions are stronger than AB attractions. Here is more quantitative detail.

Begin with two liquid phases, one of pure A and one of pure B. Combine them to reach a mole fraction x of B in A. The free energy of mixing ΔF_{mix} for that process is given in the lattice model by Equation (15.14),

$$\frac{\Delta F_{mix}}{NkT} = x \ln x + (1-x) \ln(1-x) + \chi_{AB} x (1-x).$$

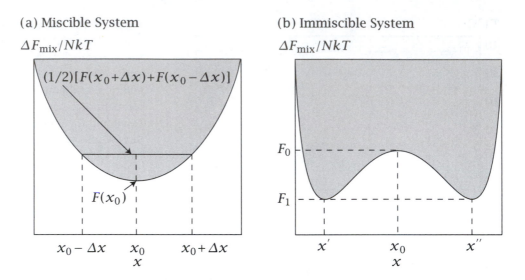

(a) Miscible System $\Delta F_{\text{mix}}/NkT$

$(1/2)[F(x_0+\Delta x)+F(x_0-\Delta x)]$

$F(x_0)$

$x_0-\Delta x$ x_0 $x_0+\Delta x$

x

(b) Immiscible System $\Delta F_{\text{mix}}/NkT$

F_0

F_1

x' x_0 x''

x

Figure 25.5 (a) If a system's free energy $\Delta F_{\text{mix}}(x)$ is concave upwards, the system is miscible. (b) For compositions (near $x = x_0$, in this case) where $\Delta F_{\text{mix}}(x)$ is concave downwards, systems are immiscible.

Varying the temperature changes the balance of mixing forces. To explore the effects of temperature, recall that $\chi_{AB} = c_1/T$ is a quantity that is inversely proportional to temperature (see Equation (15.11)), where

$$c_1 = \frac{z}{k}\left(w_{AB} - \frac{w_{AA} + w_{BB}}{2}\right) \tag{25.1}$$

is a constant, independent of temperature, that describes the sum of pairwise energies. Higher temperatures correspond to smaller values of χ_{AB} in Equation (15.14).

The stability of the system to phase separation is determined by the curvature of $\Delta F_{\text{mix}}/NkT$ versus x. Figure 25.5 shows $\Delta F_{\text{mix}}/NkT$ versus x from Equation (15.14) for two different temperatures (different values of χ_{AB}). This function is concave upward when the temperature T is high ($\chi_{AB} \approx 0$) (see Figure 25.5(a)). The free energy function is concave downward when T is low ($\chi_{AB} \gg 0$) (see Figure 25.5(b)). Where $\Delta F_{\text{mix}}(x)/NkT$ is concave upward, systems are miscible. Where it is concave downward, systems are immiscible. Here's why the curvature of this function determines the stability.

Look at Figure 25.5(b). Imagine putting two balls at the top of the hill at $x = x_0$. One rolls left to x' and the other rolls right to x''. That is, if the system were to mix as a single phase at $x = x_0$, it would have a high free energy F_0. By splitting into two equal phases, it lowers its free energy to F_1. Let's now make the condition for phase stability more general and quantitative.

When a free energy $F(x)$ is concave upward as a function of composition x (see Figure 25.5(a)), you have

$$(1/2)F(x_0 + \Delta x) + (1/2)F(x_0 - \Delta x) > F(x_0), \tag{25.2}$$

and the system is more stable as a single phase ($F = F(x_0)$) than as two slightly different phases of compositions $x_0 + \Delta x$ and $x_0 - \Delta x$. Rearrange

Equation (25.2) to get the condition for stability against small fluctuations:

$$F(x_0 + \Delta x) - F(x_0) - [F(x_0) - F(x_0 - \Delta x)] > 0. \tag{25.3}$$

Equation (25.3) is a difference of a difference,

$$\Delta \Delta F > 0.$$

As $\Delta \Delta x \to 0$, this double difference describing the condition for stability can be expressed as a second derivative:

$$\left(\frac{\partial^2 F}{\partial x^2}\right)_{x_0} > 0. \tag{25.4}$$

When the second derivative is positive, the mixed system is stable and does not form separate phases. When the second derivative is negative, the system does form separate phases. To decide whether the system is stable, look at whether the free energy function is concave upward $((\partial^2 F)/(\partial x^2) > 0)$ or concave downward $((\partial^2 F)/(\partial x^2) < 0)$. This stability criterion can be applied to models for $G(x)$ (when pressure is constant) or $F(x)$ (when volume is constant) to predict phase boundaries. We return to questions of *local stabilities* and small fluctuations on page 477. First, we consider *global stabilities*.

The Common Tangent Predicts the Compositions of Phases

How can you predict the compositions of the two phases if you know the function $F(x)$? Consider the general situation, including free energy functions that are not symmetrical around $x = 0.5$ (see Figure 25.6). You can find the compositions x' and x'' by drawing a line that is tangent to $F(x)$ at two points. The two points of tangency predict the compositions of the two phases. Here is why the common tangent line identifies the compositions of the phases in equilibrium.

To get the free energy $F(x)$ at fixed T and V, start with Equation (8.25) or Equation (8.10). For a system having N_A particles of type A with chemical potential μ_A, and N_B particles of type B with chemical potential μ_B,

$$F = \mu_A N_A + \mu_B N_B$$

$$= N(\mu_A x_A + \mu_B x_B). \tag{25.5}$$

Use $x_A = 1 - x_B$, and take the derivative $(\partial F / \partial x_B)$ to get

$$\left(\frac{\partial F}{\partial x_B}\right) = N\left(\mu_B - \mu_A\right). \tag{25.6}$$

In a two-phase equilibrium, the chemical potential of each component, A or B, must be the same in both phases, so $\mu_A' = \mu_A''$ and $\mu_B' = \mu_B''$. Subtract $\mu_A' = \mu_A''$ from $\mu_B' = \mu_B''$, and multiply by N to get

$$N(\mu_B' - \mu_A') = N(\mu_B'' - \mu_A''). \tag{25.7}$$

Substitute Equation (25.6) into Equation (25.7) for each phase. This leads to the common tangency condition that the slope of the line must be the same at

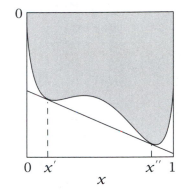

Figure 25.6 The common tangent to $F(x)$, the free energy as a function of composition, identifies the stable phases having compositions x' and x''.

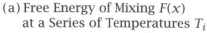

(a) Free Energy of Mixing $F(x)$ at a Series of Temperatures T_i

$\Delta F_{\text{mix}}/NkT$

(b) Free Energies from (a) Combined

$\Delta F_{\text{mix}}/NkT$

(c) Invert Dashed Curve from (b)

T_i

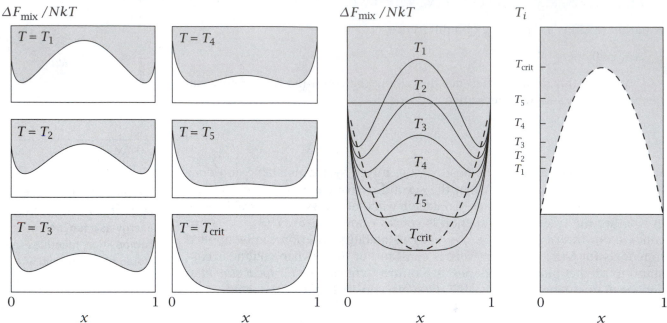

Figure 25.7 Use the lattice model to compute the mixing phase diagram.
(a) Compute $\Delta F_{\text{mix}}(x)/NkT$ for a series of temperatures given c_1, using $\chi_{AB} = c_1/T$ in Equation (25.2). (b) Put those curves on a single figure. The dashed curve represents the two minima in $\Delta F_{\text{mix}}(x)/NkT$ against x for each temperature. Inside the dashed curve is the two-phase region. (c) Replot the dashed curve of (b) so that the temperatures of the free energy minima are on the y-axis.

x' and x'',

$$\left(\frac{\partial F}{\partial x}\right)_{x'} = \left(\frac{\partial F}{\partial x}\right)_{x''}. \tag{25.8}$$

If the function $F(x)$ has minima of two different depths, the common tangency condition means that the points x' and x'' are not exactly the points at which the free energy is a local minimum. However, when $F(x)$ has two minima of equal depths, as in the binary solution lattice model, then the common tangent line has a slope of zero, $(\partial F/\partial x) = 0$.

Now we use the lattice model to compute the coexistence curve, the points x' and x'' where $(\partial F/\partial x) = 0$.

The Coexistence Curve: Computing the Compositions of Two Phases in Equilibrium

Here's a graphical way to determine the coexistence curve if you have a microscopic model for the free energy. For any two materials A and B, suppose the energy constant c_1 is known. Pick a temperature T_1. Compute $\chi_{AB} = c_1/T_1$. For this value of χ_{AB}, plot $\Delta F_{\text{mix}}/NkT$ as a function of x, using Equation (15.14). If

$\chi_{AB} > 2$, the free energy function has two minima. One minimum defines the point x' and the other defines x'' on the coexistence curve. On a plot of $T(x)$ against x, mark these two compositions x' and x'' for temperature $T = T_1$. This gives two points on the phase diagram. Choose another temperature T_2 and find its corresponding coexistence points in the same way, using the same value of c_1 as for T_1.

Doing this for a series of temperatures maps out a full coexistence curve, as indicated in Figure 25.7. The coexistence curve obtained from this simple lattice theory captures the main feature of typical experiments: the phase diagram is concave downward, with a two-phase region inside, a one-phase region outside, and a critical point at the top. The theory predicts a phase diagram that is symmetrical around $x = 0.5$. However, experiments are often not so symmetrical (see Figure 25.2). Example 25.2 shows how to compute the coexistence curve for the lattice model.

EXAMPLE 25.2 The phase diagram for the lattice model of solutions. We want to locate the minima in the free energy function. We seek the points where the derivatives are zero. That is, the coexistence curve is the set of values x' and x'' that satisfy the equation

$$\left[\frac{\partial F}{\partial x} \right]_{x'} = \left[\frac{\partial F}{\partial x} \right]_{x''} = 0. \tag{25.9}$$

Substitute the lattice model free energy, Equation (15.14), into Equation (25.9):

$$\left[\frac{\partial \Delta F}{\partial x} \right]_{x' \text{ or } x''} = NkT \left[\ln \left(\frac{x}{1 - x} \right) + \chi_{AB}(1 - 2x) \right]_{x' \text{ or } x''}$$
$$= 0.$$

You can compute x' and x'' for a series of different values of χ_{AB} by rearranging and solving the equation

$$\ln \left[\frac{x'}{1 - x'} \right] = -\chi_{AB}(T)(1 - 2x'). \tag{25.10}$$

This is a *transcendental* equation. You can solve for $x'(T)$ by iterative methods, such as those described in Chapter 4. Because the lattice model free energy is symmetrical around $x = 1/2$, $x'' = 1 - x'$. If the solute is only sparingly soluble (oil in water, for example), then $x' \ll 1$. In that case, Equation (25.10) simplifies, and predicts that the solubility is

$$\ln x' \approx -\chi_{AB} \quad \text{or} \quad x' = \exp[-\chi_{AB}]. \tag{25.11}$$

If χ_{AB} is large (A has very little affinity for B), then the solubility is small, according to Equation (25.10).

In summary, when χ_{AB} is small (less than 2), corresponding to high temperature, the disaffinity of A for B is small, and the entropic tendency to mix is greater than the energetic tendency to separate. The free energy function is concave upward for all compositions. However, if the disaffinity of A for B is strong ($\chi_{AB} \gg 0$, or $T \to 0$), AA and BB attractions are stronger than AB attractions. The energetic affinities overwhelm the mixing entropy so there is a phase separation. Very dilute solutions are miscible, despite the strong disaffinities, because there is a large mixing entropy at those compositions. Phases in equilibrium are seldom perfectly pure because perfect purity is strongly opposed by the mixing entropy.

So far, we have found *compositions* x' and x''. Now we want to know the *amounts*: how much of the total volume of solution is in the top and how much is in the bottom phase?

The Lever Rule:
Computing the *Amounts* of the Stable Phases

You can determine the amounts (the volumes) of the phases from the phase diagram by using the *lever rule*. What are the amounts in tube 3? The 'lever' is the tie-line from point 2 to point 5 on Figure 25.3. Imagine that it 'pivots' at point 3. The relative length of the line from point 3 to point 2 compared with the whole length, from 5 to 2, gives the fraction of the volume of solution that is in the $''$ phase. Similarly the fractional distance from point 3 to point 5 gives the fraction if the volume that is in the $'$ phase.

The lever rule has a simple basis. The amounts of B in each phase must sum to the total amount originally put into the solution. Let the relative amount of material in the A-rich phase be represented by the fraction

$$f = \frac{\text{number of molecules in } A\text{-rich phase}}{\text{total number of molecules in both phases}}.$$

You also have

$$x' = \frac{\text{number of } B \text{ molecules in } A\text{-rich phase}}{\text{number of molecules in } A\text{-rich phase}},$$

so

$$fx' = \frac{\text{number of } B \text{ molecules in } A\text{-rich phase}}{\text{total number of molecules in both phases}}.$$

In addition

$$(1-f)x'' = \frac{\text{number of } B \text{ molecules in } B\text{-rich phase}}{\text{total number of molecules in both phases}}.$$

Combining the last two expressions gives

$$fx' + (1-f)x'' = x_0$$
$$= \frac{\text{number of } B \text{ molecules in both phases}}{\text{total number of molecules in both phases}}. \qquad (25.12)$$

Rearranging Equation (25.12) leads to the **lever rule**,

$$f = \frac{x_0 - x''}{x' - x''}. \qquad (25.13)$$

Phase diagrams give both the compositions and the amounts of the coexisting phases.

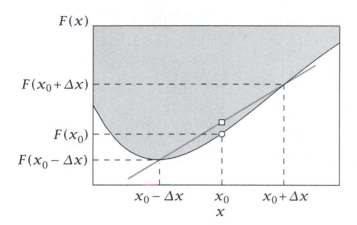

Figure 25.8 The free energy $F(x)$ is concave upward so the system is stable to small fluctuations near the composition $x = x_0$. This is because $F(x_0)$ (○) is lower than the free energy $(1/2)[F(x + \Delta x) + F(x - \Delta x)]$ (□) of a locally phase-separated system.

The Spinodal Curve
Describes the Limit of Metastability

So far, we have described systems in equilibrium. Phase equilibrium is achieved if you shake or stir the solutions well, and wait long enough. But you can also make *metastable* solutions if you are careful not to shake or stir them. For example, a solution can be *supersaturated*. It can have slightly more oil dissolved in the water than is predicted by the coexistence curve: x can be larger than x' for a given temperature. Or, a solution can be *supercooled*: T can be lower than the equilibrium value, for a given x. 'Global' stability (equilibrium) results when a system reaches the lowest free energy that is accessible to it through shaking, stirring, or other large perturbations, for example from composition x_0 to the two states x' and x''. 'Local' stability (metastability) results when a system reaches the lowest free energy that is accessible to it when it is subjected to only small perturbations, for example from composition x_0 to the two states $x_0 + \Delta x$ and $x_0 - \Delta x$.

The coexistence curve for thermodynamic stability is called the *binodal* curve. It is given by a condition on the first derivative of the free energy function, the common tangent rule Equation (25.9). We now aim to find a second type of curve, called the *spinodal*, which identifies the limit of metastability. The spinodal curve is given by a condition on the second derivative. Figure 25.8 shows that where the free energy is concave upward, a system is locally stable because a small change in composition, from x_0 to $x_0 + \Delta x$ and $x_0 - \Delta x$, would lead to an increase in free energy. The spinodal curve is defined by the transition between local stability and local instability, which occurs where the free energy function is no longer convex upward (see Equation (25.4)),

$$\left(\frac{\partial^2 F}{\partial x^2}\right) = 0. \tag{25.14}$$

EXAMPLE 25.3 The spinodal (metastability) curve for the lattice model. You can compute the spinodal curve for the lattice model by taking the second derivative of Equation (15.14) and setting it equal to zero:

Figure 25.9 In region 1P the system is always in one phase. M is the metastable region: the system is a single phase if perturbations are small, or two phases if perturbations are large enough to cause demixing. In region 2P the system is always in two phases. (——) binodal curve; (– – –) spinodal curve. The spinodal and binodal curves meet at the critical point (○).

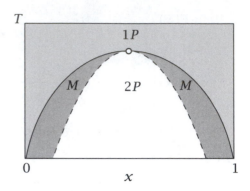

Figure 25.10 Binodal (–●–) and spinodal (– ○ -) curves for a binary liquid of n-hexane in n-tetradecafluorohexane. Source: GI Pozharskaya, NL Kasapova, VP Skripov and YD Kolpakov, *J Chem Thermodyn* **16**, 267 (1984); WJ Gaw and RL Scott, *J Chem Thermodyn* **3**, 335 (1971).

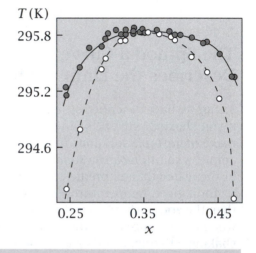

$$\left(\frac{\partial^2 F}{\partial x^2}\right) = NkT\left[\frac{1}{x} + \frac{1}{1-x} - 2\chi_{AB}\right] = 0. \tag{25.15}$$

Figure 25.9 shows that the spinodal curve falls inside the binodal curve. For systems that are sparingly soluble ($x \approx 0$), the spinodal composition x'_s is given by $x'_s = 1/(2\chi_{AB} - 1)$.

Here's how to interpret Figure 25.9. The metastable region M is between the one-phase region 1P and the two-phase region 2P. In the metastable region, the system remains a single-phase mixture if the external perturbations are small, but becomes unstable (it demixes) if the fluctuations are large. In region 2P, the system is unstable against all fluctuations, large or small, and therefore always separates into two phases. Careful experiments are required if you want to find both binodal and spinodal curves, because it is often hard to tell whether your perturbations are small enough to define the true limits of metastability. Binodal and spinodal curves are shown for hexane–tetradecafluorohexane mixtures in Figure 25.10.

The Critical Point is Where Coexisting Phases Merge

The *critical point* is the intersection of the binodal and spinodal curves. At the critical point, there is no longer a distinction between the A-rich and the B-rich phases. Above the critical temperature T_c, the two components are miscible in all proportions (see Figure 25.9). The binodal curve is the set of compositions at which the first derivatives of the free energy function are zero, and the spinodal curve is where the second derivatives are zero. The critical composition $x = x_c$ is where both the first and second derivatives of the free energy are zero. The merging of the phases x' and x'' implies that the third derivative is also zero:

$$\lim_{x' \to x''} \left[\left(\frac{\partial^2 F}{\partial x^2} \right)_{x'} - \left(\frac{\partial^2 F}{\partial x^2} \right)_{x''} \right] = 0 \implies \left(\frac{\partial^3 F}{\partial x^3} \right) = 0. \quad (25.16)$$

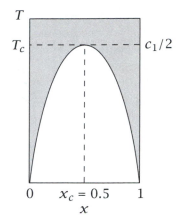

Figure 25.11 The lattice model phase diagram for mixing.

EXAMPLE 25.4 Lattice model critical point. To find the critical point (x_c, χ_c, T_c) for the lattice mixture model, determine the point where both the second and third derivatives of the free energy (given by Equation (15.14)) equal zero:

$$\left(\frac{\partial^3 F}{\partial x^3} \right)_{x_c} = NkT \left[\frac{1}{(1 - x_c)^2} - \frac{1}{x_c^2} \right] = 0 \implies x_c = \frac{1}{2}. \quad (25.17)$$

Now use $x_c = 1/2$ and Equation (25.15) to get the second derivative,

$$\left(\frac{\partial^2 F}{\partial x^2} \right)_{x_c} = \frac{1}{(1/2)} + \frac{1}{(1 - \frac{1}{2})} - 2\chi_c$$

$$= 4 - 2\chi_c = 0 \implies \chi_c = 2, \quad (25.18)$$

where χ_c is the value of χ at the critical point.

From Equations (25.17) and (25.18), you can compute the critical point x_c and T_c for the lattice model. Because $\chi_c = c_1/T_c = 2$, you have $T_c = c_1/2$, where c_1 is the quantity that describes the strengths of interactions between the A's and B's. If $c_1 = 0$, then the system is fully miscible at all temperatures, and there is no two-phase region at all. Figure 25.11 shows the lattice model phase diagram.

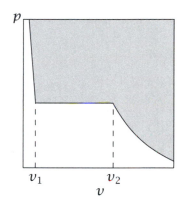

Figure 25.12 A pressure–volume isotherm for a gas: v is the volume per molecule and v_1 and v_2 are the volumes of the coexistent vapor and liquid phases.

The Principles of Boiling Are Related to the Principles of Immiscibility

The boiling of water would seem to be a very different type of phase equilibrium from the mixing of oil and water. But in both cases, phase boundaries are concave downward with a two-phase region inside and a critical point at the top. Remarkably, both can be described, to a first approximation, by the same physical model.

To create the phase diagram for boiling, put N particles in a container of volume V at temperature T. Follow the system pressure $p(T, V, N)$ as you decrease the volume, starting with low densities ρ, that is large volumes per molecule, $v = V/N = \rho^{-1}$.

At low densities (right side of Figure 25.12), the system is a single phase: a gas. As you compress it, decreasing the volume per molecule to v_2, the pressure increases. At volumes between v_1 and v_2, a second phase begins to emerge in the container: some liquid forms in equilibrium with the gas phase. Compress it more, and you get more liquid and less gas. When the volume per molecule becomes smaller than v_1, the system becomes a single phase again, but now it

Figure 25.13 The isotherms of pressure p versus molar volume v for carbon dioxide. The broken line is the phase diagram. Like a miscibility phase diagram, this is concave downward, and has a critical point at the top. Inside the dashed region, the system is boiling. Source: AJ Walton, *Three Phases of Matter*, Oxford University Press, New York, 1983.

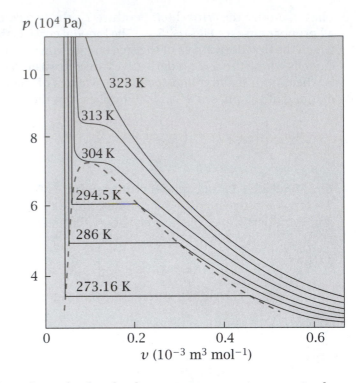

p (10^4 Pa)

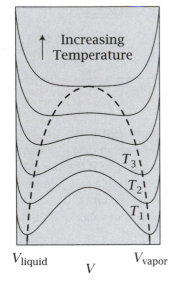

Figure 25.14 The free energy $G(V)$ for a gas–liquid equilibrium. The dashed line encloses the two-phase (boiling) region.

is a liquid. Reducing the volume further leads to a very steep increase in the pressure because liquids are not very compressible.

Volumes v_1 and v_2 define two phase boundaries. To the right of v_2, the system is a single phase (gas). To the left of v_1, the system is a single phase (liquid). For any intermediate volume per molecule between v_1 and v_2, the system is two phases in equilibrium: it is boiling. In the boiling region, the gas has volume per molecule v_2 and the liquid has volume per molecule v_1. For constant temperature, the pressure–volume curve in Figure 25.12 is called an *isotherm*. The total amounts of liquid and gas phases can be found by using a lever rule (see page 476).

Compress again, but now at higher temperature. You'll get two more points on the phase boundary. Just as with the liquid solubilization transition, the phase boundaries are closer together at higher temperatures. Another similarity between boiling and liquid immiscibility is the existence of a critical point. Above the critical temperature, you can continuously compress the gas to make it denser, ultimately reaching liquid densities, without ever passing through a two-phase region: the fluid never boils. The dashed line in Figure 25.13 shows the phase diagram for CO_2. The critical point for CO_2 is 304 K. Above the critical temperature, CO_2 is called a *supercritical fluid*, rather than a supercritical gas or liquid. The supercritical fluid does not have two free energy minima that would distinguish liquid and gas states. Supercritical fluids have practical applications, such as decaffeinating coffee (see Figure 25.16).

Because the phase diagram for boiling is concave downward with a two-phase region inside and a critical point at the top, it must arise from a free energy function with two minima. However, for boiling, the degree of freedom is the volume per molecule (or its inverse, the density), rather than the composition x_B of a liquid mixture. In general, the degree of freedom that changes

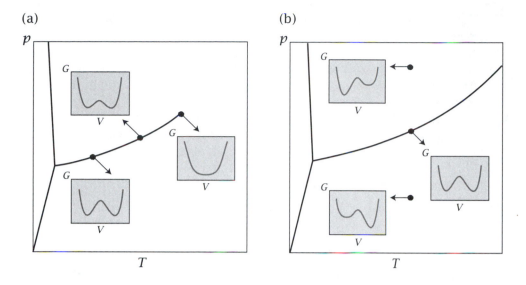

Figure 25.15 (a) $G(V)$ has two minima that become one minimum as the critical point is approached. (b) On the liquid–gas coexistence curve, the free energy has two minima of equal depth. Above it, the liquid phase is more stable. Below it, the gas phase is more stable. Source: HB Callen, *Thermodynamics and an Introduction to Thermostatistics*, Wiley, New York, 1985.

dramatically in a phase transition is called the *order parameter*. Different types of phase transition have different order parameters, but the common thread among different phase transitions is the shape of the free energy function versus its order parameter. For both boiling and binary mixture equilibria, the plot of free energy versus the order parameter has two minima that merge together as the temperature increases through the critical point (see Figure 25.14). A phase transition is called *first-order* when the free energy has two minima separated by a hill (as at temperatures $T = T_1, T_2, T_3$ in Figure 25.14), and is called *higher-order* when the free energy has no hill, as at the critical temperature.

The similarities between boiling and immiscibility transitions can be traced to similar underlying principles. Remarkably, the same model can be used for both. For liquid mixtures, we used a lattice to count arrangements of A's and B's to get the entropy, and we summed the AA, BB, and AB interactions to get the energy. For gases, the same lattice-counting procedure gives the number of arrangements of filled sites (particles) and empty sites (free space). For lattice gases, the energy is just the sum of particle pairwise interactions. For the liquid–vapor equilibrium, the lattice model gives the free energy per unit volume (see Equation (24.21)). You can express this free energy in terms of the density $\rho = N/V$, where N is the number of particles and V is the volume of the system,

$$\frac{F}{V} = -a\rho^2 + kT\left[\rho \ln(b_0\rho) - \left(\frac{1}{b_0} - \rho\right)\ln(1 - b_0\rho)\right], \tag{25.19}$$

where a represents the model attractive energy and b_0 represents the excluded volume between two particles. The free energy function versus ρ (Equation (25.19)) has double minima of the type shown in Figure 25.14.

Liquid–gas coexistence (boiling) is usually studied by controlling the temperature and pressure, so the relevant free energy is $G(T, p)$, rather than $F(T, V)$. Figures 25.15(a) and 25.15(b) show the double minima in $G(V)$ at constant T and p that define the equilibrium molar volumes of the gas and liquid phases. Figure 25.15(a) shows how the free energy minima merge as the temperature and pressure increase toward the critical point.

As with liquid–liquid immiscibility, the coexistence curve for boiling a simple lattice model liquid is the set of points for which $(\partial F / \partial V) = 0$. The metastability curve is where $(\partial^2 F / \partial V^2) = 0$ and the critical point is where you also have $(\partial^3 F / \partial V^3) = 0$. Because the pressure is defined as $p = -(\partial F / \partial V)$, it is often more convenient to express these derivatives for metastability and the critical point in terms of p:

$$\left(\frac{\partial p}{\partial V}\right) = 0 \quad \text{and} \quad \left(\frac{\partial^2 p}{\partial V^2}\right) = 0. \tag{25.20}$$

Example 25.5 shows how to compute the critical properties for a van der Waals gas.

EXAMPLE 25.5 The critical point of the van der Waals gas. The van der Waals Equation (24.15) is

$$p = \frac{NRT}{V - Nb} - \frac{aN^2}{V^2}.$$

To compute the critical temperature, pressure, and volume, substitute Equation (24.15) into Equations (25.20) taking $N = 1$,

$$\left(\frac{\partial p}{\partial V}\right) = \frac{-RT_c}{(V_c - b)^2} + \frac{2a}{V_c^3} = 0, \tag{25.21}$$

and

$$\left(\frac{\partial^2 p}{\partial V^2}\right) = \frac{2RT_c}{(V_c - b)^3} - \frac{6a}{V_c^4} = 0. \tag{25.22}$$

Rearrange Equation (25.21) to find RT_c,

$$RT_c = \frac{2a}{V_c^3}(V_c - b)^2, \tag{25.23}$$

and substitute this expression for RT_c into Equation (25.22) to get

$$V_c = 3b. \tag{25.24}$$

Substitute Equation (25.24) for V_c into Equation (25.23) to get

$$T_c = 8a/27Rb. \tag{25.25}$$

Substitute Equations (25.24) and (25.25) into Equation (24.15) to get

$$p_c = a/27b^2. \tag{25.26}$$

This gives the critical temperature, pressure, and volume of the van der Waals gas as a function of the energy parameters a and b.

Supercritical Fluids Are Used to Decaffeinate Coffee

Figure 25.16 shows how caffeine is removed from coffee. Coffee is mixed with CO_2 at a temperature and pressure above the critical point of the CO_2. The supercritical CO_2 dissolves the caffeine (small black dots), decaffeinating the coffee beans. The fluid mixture of CO_2 with caffeine then flows into a chamber where the pressure is lowered below the critical point so caffeine partitions into water. The carrier CO_2 is recaptured and the caffeine is dumped in the aqueous phase.

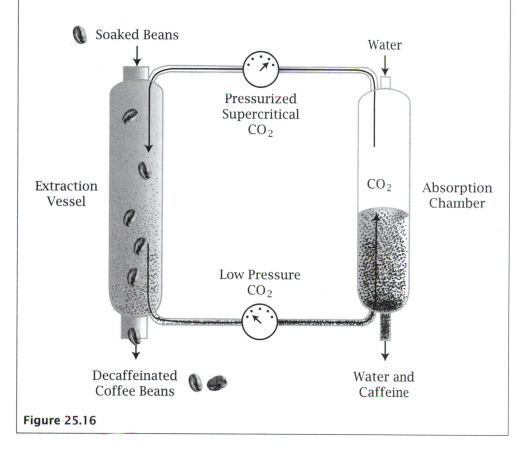

Figure 25.16

The Law of Corresponding States: a Universal Description of Gases and Boiling

A revolution in understanding phase transitions and critical phenomena has occurred since the 1940s. A remarkable underlying unity was found among different types of transition, and for different types of material undergoing the same type of transition. An important step in this unification was the *law of corresponding states*. Different fluids have different critical temperatures and pressures, and different parameters a and b when modelled by the van der Waals equation. However, when equations of state are expressed in terms of dimensionless variables that are normalized by their values at the critical point (critical temperature T_c, pressure p_c, and volume V_c), universal

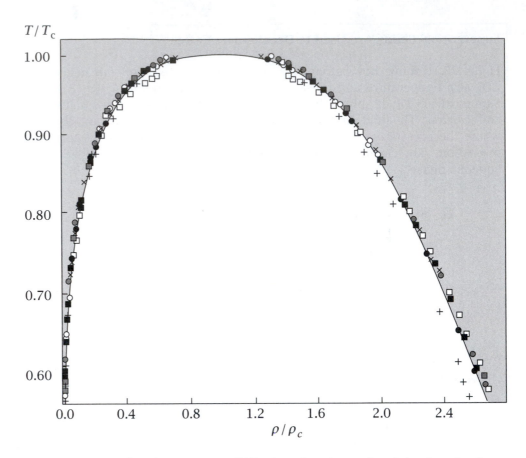

Figure 25.17 Reduced temperature T/T_c plotted against reduced density ρ/ρ_c for eight different gases, illustrating corresponding states: (+) Ne; (■) Kr; (×) Xe; (◉) N_2; (▦) O_2; (□) CO; (○) CH_4; (●) Ar. Source: EA Guggenheim, *J Chem Phys* **13**, 253 (1945).

behavior is often observed. To see this, express the van der Waals equation in terms of reduced variables, $\pi = p/p_c$, $\theta = T/T_c$, and $\phi = V/V_c$. For $N = 1$, rearrange Equation (24.15) to

$$\left(p + \frac{a}{V^2}\right)(V - b) = RT. \tag{25.27}$$

Divide both sides of Equation (25.27) by p_c and use Equation (25.26) to get

$$\left[\frac{p}{p_c} + \frac{a}{V^2}\left(\frac{27b^2}{a}\right)\right](V - b) = \left(\frac{27b^2}{a}\right)RT. \tag{25.28}$$

On the left side of Equation (25.28), replace each factor of b with $V_c/3$ from Equation (25.24) and then replace each factor of V/V_c with ϕ. On the right side of Equation (25.28), replace T with θT_c and use Equation (25.25) to get

$$\left(\pi + \frac{3}{\phi^2}\right)\left(\phi - \frac{1}{3}\right) = \frac{8}{3}\theta. \tag{25.29}$$

Figure 25.17 shows the advantage of using reduced variables: it illuminates the underlying similarities between different materials. By plotting the reduced

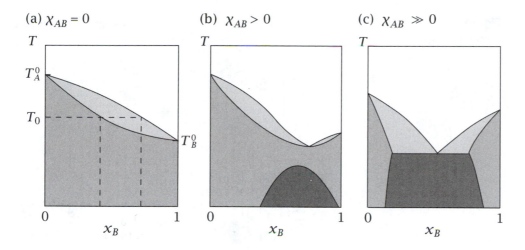

Figure 25.18 Phase diagrams for boiling liquid mixtures: (▢) both components are in the vapor phase; (▣) a vapor phase enriched in one component is in equilibrium with a liquid phase enriched in the other component; (▣) two liquid phases in equilibrium; (▣) a single liquid-phase solution. Temperature T_0 is between the boiling point of pure A, T_A°, and the boiling temperature of pure B, T_B°. Source: JH Brophy, RM Rose and J Wulff, *The Structure and Properties of Materials, Vol II: Thermodynamics of Structure*, Wiley, New York, 1964.

density ρ/ρ_c against the reduced temperature T/T_c, it is found that the data for eight different gases fall on the same line. This is called the law of corresponding states. The importance of this universality is discussed in Chapter 26. Now we consider multicomponent boiling.

Boiling a Liquid Mixture Involves Two Types of Phase Transition

Consider a liquid having two components A and B. The system may have a miscibility phase transition. But it will also boil at a sufficiently high temperature. Each component might boil at a different temperature. Boiling a mixture, say of alcohol and water, is the basis for *fractional distillation* (see page 486). Figure 25.18 shows three typical temperature–composition phase diagrams for boiling a mixture. Figure 25.18(a) shows an ideal mixture ($\chi_{AB} = 0$), having no liquid–liquid two-phase region. The progression from Figure 25.18(a) to (c) represents decreased affinity of A for B (increasing χ_{AB}, in the terminology of the lattice model): the two-phase region grows and the one-phase liquid region shrinks.

Figure 25.18(a) is a phase diagram for the boiling equilibrium of an ideal two-component liquid. At low temperatures, below the canoe-shaped region, both components are liquids. Their liquids are miscible in all proportions. At high temperatures, above the canoe, both components are gases. Inside the canoe is a two-phase region: a liquid and a gas. Pure B boils at temperature T_B^0. Pure A boils at temperature T_A^0. If B is more *volatile* than A, then $T_B^0 < T_A^0$. For example, alcohol is more volatile than water (see Figure 25.27).

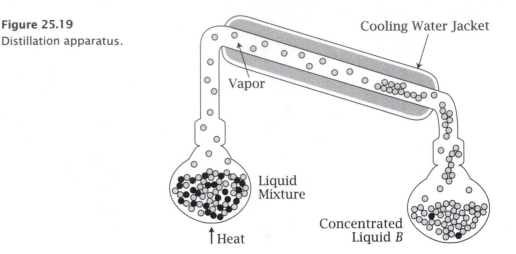

Figure 25.19
Distillation apparatus.

Cooling Water Jacket

Vapor

Liquid
Mixture

Concentrated
Liquid B

↑ Heat

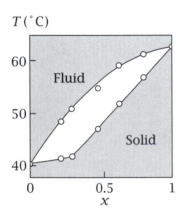

T (˚C)

Fluid

Solid

60

50

40

0   0.5   1

x

Figure 25.20 A fluid–solid phase diagram for lipid bilayer membranes containing two different phospholipids (DPPC–DPPE), showing that at some compositions x and temperatures T, a bilayer is disordered (fluid), at others it is ordered (solid), and at intermediate compositions and temperatures a bilayer has coexisting solid and liquid phases. Source: EJ Shimshick and HM McConnell, *Biochemistry* **12**, 2351–2360 (1973).

Heat the mixture in Figure 25.18(a) to a temperature T_0 above the boiling point T_B^0, but below the boiling point T_A^0, $T_B^0 < T_0 < T_A^0$. Follow the tie-line to the right phase boundary to get the concentration of B in the vapor phase, and to the left to get the concentration of B in the liquid. Heating concentrates the B in the vapor and concentrates the A in the liquid.

Distillation Can Result from Boiling a Liquid Mixture

Figure 25.19 shows how this principle is used to distill liquids. A mixture of A (●) and B (○) is heated in the left flask to a temperature above the boiling temperature of the more volatile component B. The volatile component B concentrates in the vapor phase. For example in alcohol–water mixtures, alcohol is more volatile (○). The vapor is cooled and condenses into the right flask. B is more concentrated in the right flask than in the left flask. By repeating the process, heating the liquid in the right flask at a lower temperature than before (but above the boiling temperature of pure B), B can be concentrated further.

Phase diagrams for solid mixtures and their melting behavior often resemble those of liquid mixtures and their boiling behavior. In the same way that repeated boiling can purify a liquid mixture, repeated melting can purify a solid mixture. This is called *zone refining*: it is used to purify metals.

Figure 25.20 shows an example of phase equilibria of 'solid-like' and 'liquid-like' lipid bilayer membranes containing two kinds of phospholipid. In this case, solid-like refers to a high degree of orientational order in the hydrocarbon chains of the lipids, and liquid-like refers to a more disordered state of the chains. A sharp transition between the solid-like and liquid-like phases happens at certain temperatures and compositions.

Phase transitions are not limited to boiling, freezing, and mixing. Several different forms of molecular organization of surfactant molecules as a function of their concentration in water are shown in Figure 25.21. At very low temperatures, they crystallize. At higher temperatures, low concentrations dissolve in water. Increasing their concentration to beyond the critical micelle concentration leads to spherical micelles, then cylindrical micelles, then lamellar phases such as bilayers.

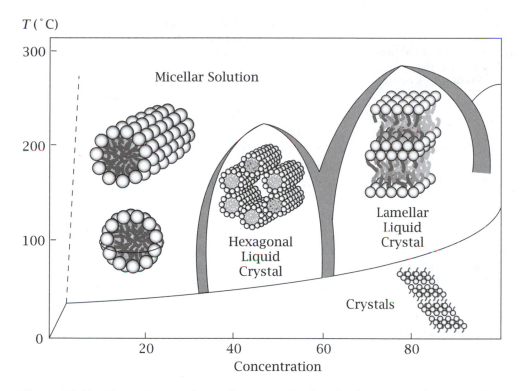

Figure 25.21 Phase diagram for surfactant molecules. Surfactants can form micelles that are spheres, cylinders, or planar bilayers, with increasing surfactant concentration. Critical micelle concentration (– – –). Source: HT Davis, *Statistical Mechanics of Phases, Interfaces, and Thin Films*, VCH, New York, 1996.

Summary

Phase transitions occur at the point where two or more states are stable at the same time. For example, when liquids *A* and *B* are mixed, an *A*-rich phase can be in equilibrium with a *B*-rich phase. Boiling is also an equilibrium between two phases: a vapor phase and a liquid phase. Changing the temperature or other variables can shift the equilibrium from the point of transition toward one phase or the other. The common feature is that the free energy as a function of a degree of freedom (or order parameter) has two minima at the phase transition points. Heating causes the two phases to become more similar until they ultimately merge at the *critical point*. The compositions of the stable phases are found by identifying the points of common tangency on the free energy surface—the ends of tie-lines on phase diagrams. The volumes of each phase are determined by using the lever rule. The common physical basis for such transitions is the gain in translational or mixing entropy at high temperatures, compared with the attractions that favor self-association at low temperatures. A virtue of models that are as simplified as the lattice model is that they show the underlying unity of such processes by leaving out other details.

Problems

1. Protein aggregation. You have a solution with mole fraction x of proteins. The proteins can aggregate and thus change their local concentration. The entropy of mixing is

$$\Delta S = -k[x \ln x + (1-x) \ln(1-x)],$$

and the internal energy is $\Delta U = \chi x(1-x)$. Draw curves for the free energy of the system as a function of x (between 0 and 1), for $\chi = 0, 1, 2$, and $4 kT$. Identify the stable states of the system. For what values of χ does the system prefer to phase separate?

2. Closed-loop phase diagram. Figure 25.22 is a *closed-loop* phase diagram for mixing molecules A and B as a function of temperature T, and mole fraction composition of B, x. Suppose you mix the two liquids at temperature T_1 and composition x_B.

(a) By drawing lines on the figure, determine the compositions x' and x'' of the two phases that will be in equilibrium under those conditions.

(b) If there is 10 mL of solution, estimate the amounts of A-rich and B-rich components.

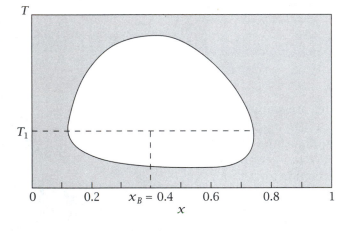

Figure 25.22 Closed-loop phase diagram. White region: two phases. Gray region: one phase.

3. Inverted mixing phase diagrams. In the free energy of mixing of a regular solution, the pair potential energies w are assumed to be constants independent of temperature. Therefore the exchange energy χ_{AB} depends inversely on temperature. Plot the miscibility phase diagram of temperature against mole fraction:

(a) When χ_{AB} is a constant independent of temperature. (This occurs when local orientational entropy of at least one component is important.)

(b) When χ_{AB} increases linearly with temperature. (This occurs for solutions with large heat capacities: nicotine in water shows this behavior.)

4. Phase diagram for uric acid in gout. Old Ma Kettle has gout. Gout is a disease of crystallization of uric acid in the blood. Figure 25.23 is the phase diagram.

(a) What is the solubility limit according to this diagram for uric acid at 10 °C? At 37 °C?

(b) Ma Kettle claims she can always tell when winter's here. How does she do it?

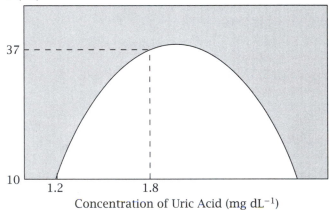

Figure 25.23 Phase diagram of uric acid. Gray region: one phase. White region: two phases.

5. Underlying free energy functions. Draw a diagram of free energy versus composition that would give rise to the phase diagram in Figure 25.24.

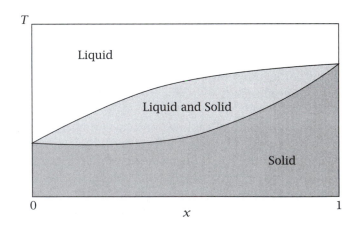

Figure 25.24 General phase diagram of a two-component freezing process.

6. Phase diagram for solder. The phase diagram in Figure 25.25 describes an alloy of tin (Sn) and lead (Pb) that can be melted at relatively low temperature.

(a) If you wish to design solder to have the lowest possible melting temperature, what composition x_{Sn} would you choose?

(b) Using graphical methods for the solid mixture (x_0, T_0), determine the compositions of the two phases x' and x'' into which the solid will partition.

(c) Compute the relative weight percents of the two phases.

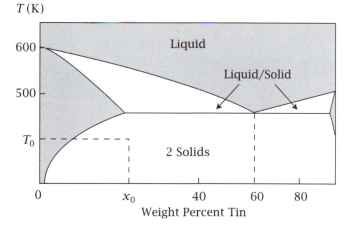

Figure 25.25 Phase diagram for tin and lead (solder).

7. Lattice model mixture. You have a solution with molecules of type A mixed with molecules of type B at $T = 300$ K. Suppose that every two AB molecular contacts are less stable than one AA plus one BB contact, by an energy kT, and that every molecule has an average of six neighbor contacts. For $x_A = x_B = 1/2$, to what temperature do you have to heat the system to get A to dissolve fully in B?

8. Phase diagram for lattice model mixing. You have a lattice solution mixture at a temperature for which $\chi_{AB} = 5.0$.

(a) At what concentrations $x_b = x'$ and $x_t = x''$ are the binodal points?

(b) At what concentrations are the spinodal points?

(c) Over what concentration range of x_B is the solution *meta*stable?

(d) If $\chi_{AB} = 5$ at temperature $T = 300$ K, then what is the critical temperature T_c of the solution?

9. Boiling acetone–cyclohexane. Using the liquid-vapor binary phase diagram in Figure 25.26, describe what is qualitatively different about the boiling of a liquid that has $x = 0.4$ acetone in comparison to a liquid that has $x = 0.9$ acetone.

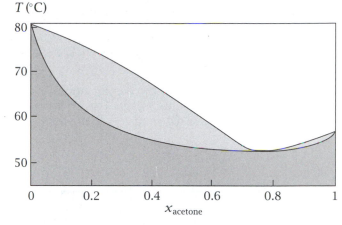

Figure 25.26 Liquid–vapor equilibria for mixtures of acetone in cyclohexane. Source: WE Acree Jr, *Thermodynamic Properties of Nonelectrolyte Solutions*, Academic Press, Orlando, 1984. From: KVK Rao and CV Rao, *Chem Eng Sci* **7**, 97 (1957).

10. Azeotrope for alcohol/water mixtures. The azeotrope is the point on Figure 25.27 where $x = 0.105$ and $T = 78.2$ °C. Describe what the azeotrope point represents and what it means for distilling alcohol.

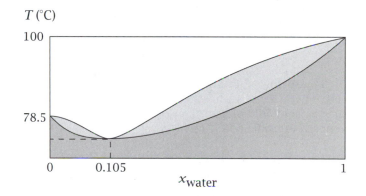

Figure 25.27 Phase diagram for boiling alcohol water mixtures. Source: J deHeer, *Phenomenological Thermodynamics*, Prentice Hall, Englewood Cliffs, 1986.

11. Boiling of a van der Waals fluid. The first region of the pressure-volume isotherm of a given van der Waals fluid extends from $V' = 38 \, cm^3 \, mol^{-1}$ to $V'' = 58 \, cm^3 \, mol^{-1}$. If this fluid is compressed to $41 \, cm^3 \, mol^{-1}$, is it a gas, or liquid, or mixture of both? If it is a mixture, what is its composition?

12. Pressure-induced protein denaturation. Figure 25.28 shows a (p, T) phase diagram for the denaturation of proteins (1 bar = 1 atm).

(a) What is the denaturation temperature at 1 atm applied pressure?

(b) What pressure is needed to denature a protein at 25 °C?

(c) With $(\partial \ln K / \partial p) = (-\Delta v / RT)$ (see Equation 13.48), does denaturation involve an increase or decrease of volume? Explain what volume is changing.

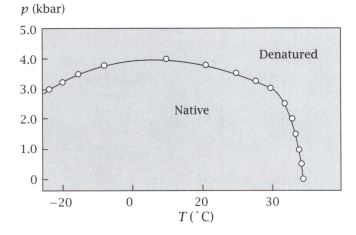

Figure 25.28 Pressure can denature proteins, ribonuclease A in this case. Source: J Zhang, X Peng, A Jones and J Jones, *Biochemistry* **34**, 8631–8641 (1995).

13. Measuring solubilities to determine χ_{AB}. The solubility of ethylbenzene in water is 0.00162 mol L^{-1} at 25 °C, and the solubility of propylbenzene in water at the same temperature is 0.000392 mol L^{-1}.

(a) Compute the χ interaction parameter for ethylbenzene in water.

(b) Compute the χ parameter for propylbenzene in water.

(c) Because propylbenzene has one more CH$_2$ group than to ethylbenzene, compute the free energy of transferring the CH$_2$ group from hydrocarbon to water at 25 °C by using the data given here.

14. Lower and upper critical solution temperatures. Figure 25.29 shows three different liquid–liquid phase diagrams. Explain what they tell us about intermolecular interactions in these systems.

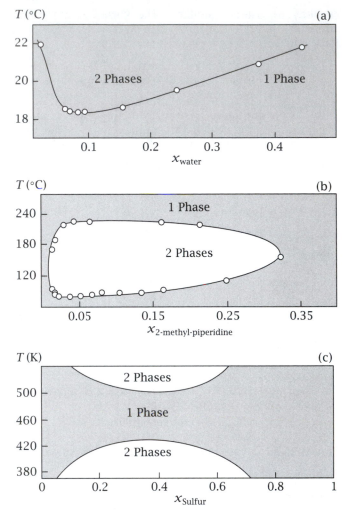

Figure 25.29 Phase diagram for mixtures: (a) Water + triethylamine, (b) water + 2-methyl-piperidine, (c) sulfur + benzene. Source: WE Acree Jr, *Thermodynamic Properties of Nonelectrolyte Solutions*, Academic Press, Orlando, 1984. From: (a) F Kohler and OK Rice, *J Chem Phys* **26**, 1614 (1957); (b) O Flaschner and B Marewen, *J Chem Soc* **93**, 1000 (1908).

15. Critical points of water and CO$_2$. The van der Waals gas constants are $a = 3.592$ L^2 atm mol^{-2} and $b = 0.043$ L mol^{-1} for CO$_2$; and $a = 5.464$ L^2 atm mol^{-2} and $b = 0.0305$ L mol^{-1} for H$_2$O. Compute the critical temperature T_c and pressure p_c for both CO$_2$ and water, and suggest why CO$_2$ is a more practical working fluid for supercritical fluid chromatography.

16. Solubilities of organic liquids. From Table 25.1

(a) Compute $\chi_{\text{benzene/water}}$.

(b) Describe the relative affinities for water of aliphatic hydrocarbons, aromatic hydrocarbons (benzenes), unsaturated hydrocarbons, and alcohols.

Table 25.1 Mole fraction of some common nonpolar organic liquids saturated with water. Source: RP Schwartzenbach, PM Gschwend and DN Imoboden, *Environmental Organic Chemistry: Illustrative Examples, Problems, and Case Studies*, Wiley, New York, 1995.

Organic Liquid	x_0
Heptane[a]	0.99916
Octane[a]	0.99911
Benzene[a]	0.9977
Chlorobenzene[b]	0.9975
Trichloroethylene[b]	0.9977
Tetrachloroethylene[b]	0.99913
Chloroform[b]	0.9946
1,1,1-Trichloroethane[b]	0.9974
Diethyl ether[c]	0.942
Methyl acetate[c]	0.89
2-Butanone[c]	0.69
Pentanol[d]	0.64
Octanol[d]	0.79

[a]W Gerrard, *Gas Solubilities, Widespread Applications*, Pergamon, New York, 1980. [b]AL Horvath, *Halogenated Hydrocarbons. Solubility–Miscibility with water*, Marcel Dekker, New York, 1982. [c] JA Riddick and WB Bunger, *Organic Solvents*, Wiley, New York, 1970. [d] R Stephenson, J Stuart, and M Tabak, *J Chem Eng Data* **29**, 287–290 (1984).

17. Lattice model spinodal curve. Using the lattice binary solution theory,

(a) Express the *spinodal* curve concentration x' as a function of temperature T, where $\chi_{AB} = \theta/T$, and θ is a constant.

(b) Plot this spinodal function $x'(T)$.

(c) What is x' when $T = \theta/4$?

Suggested Reading

There are several outstanding general texts on phase equilibria:

RS Berry, SA Rice and J Ross, *Physical Chemistry*, Wiley, New York, 1980.

HB Callen, *Thermodynamics and an Introduction to Thermostatistics*, 2nd edition, Wiley, New York, 1985.

G Careri, *Order and Disorder in Matter.* English language translation by K Jarrat, Benjamin Cummings, Reading, 1984.

PG Debenedetti, *Metastable Liquids: Concepts and Principles*, Princeton University Press, Princeton, 1966.

JH Hildebrand and RL Scott, *The Solubility of Nonelectrolytes*, 3rd edition, Reinhold, New York, 1950.

R Kubo, in cooperation with H Ichimura, T Usui and N Itashitsumo, *Statistical Mechanics*, Elsevier, New York, 1965.

JS Rowlinson and FL Swinton, *Liquids and Liquid Mixtures*, 3rd edition, Butterworth's Scientific, Boston, 1982.

EKH Salje, *Phase Transitions in Ferroelastic and Co-elastic Crystals: An Introduction for Minerologists, Material Scientists, and Physicists*, Cambridge University Press, New York, 1990.

HE Stanley, *Introduction to Phase Transitions and Critical Phenomena*, Oxford University Press, New York, 1971.

JM Yeomans, *Statistical Mechanics of Phase Transitions*, Oxford University Press, New York, 1992.

26 Cooperativity: the Helix–Coil, Ising & Landau Models

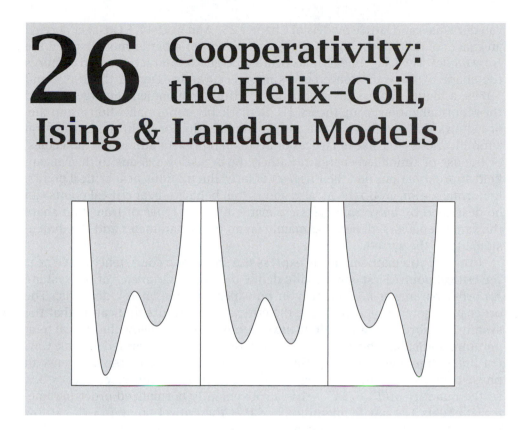

Abrupt Transitions Occur in Many Different Systems

The principles that govern phase transitions in macroscopic systems also underlie the dramatic transformations that take place within individual molecules. Small changes in the temperature or the solvent can cause a polymer molecule to undergo a helix–coil transition, cause the two strands of DNA to zip together, or cause a protein or RNA molecule to fold. A ligand molecule can induce a conformational transition, as when hemoglobin picks up or releases oxygen in the blood. When sickle-cell hemoglobin polymerizes into rods, it disrupts blood cells and causes anemia. In all cases—macroscopic phase transitions or single-molecule conformational changes—the cooperativity is predicted from the shape of a free energy surface.

Transitions and Critical Points Are Universal

In Chapter 25 we noted that the shapes of the coexistence curves for liquid–liquid immiscibility are about the same as for boiling. Both display critical points, and both are described approximately by the same simple model. Advances in theory and experiments since the 1940s have led to a revolution in understanding critical phenomena; it goes beyond those models.

   This revolution started with two important developments. In 1945, EA Guggenheim demonstrated that the 'corresponding states' liquid–gas coexistence curve (Figure 25.17) was slightly flatter than you would expect from the

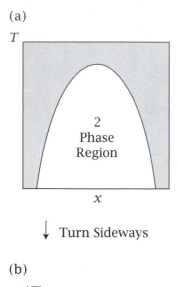

(a)

T

2 Phase Region

x

↓ Turn Sideways

(b)

$m(T)$

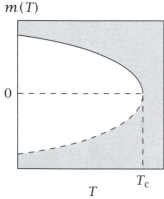

0

T

T_c

Figure 26.1 (a) A coexistence curve for solubilization is temperature versus composition, $T(x)$. (b) Turn it sideways to get $x(T)$. The order parameter in this case is $m(T) = 2x(T) - 1$.

van der Waals and lattice models of Chapter 25. And in 1944, L Onsager derived an exact solution to the two-dimensional version of a better model, called the *Ising model*. The Ising model treats nearest-neighbor interactions, and captures the shape of the coexistence curves more accurately. During the 1960s and 1970s, a highly successful new theory of critical phenomena emerged, called the renormalization group theory [1]. Remarkably, renormalization group theory shows that the flaws in the simpler models do not lie in their structural simplifications—the use of lattices or beads to represent atoms and molecules, or the use of simplified energies. Rather, the breakdown is due to their mean-field approximations and their neglect of large fluctuations near critical points. Renormalization group theory also shows that behavior near critical points can be described by *universality classes*: many different types of transition share the same behaviors, depending mainly on an order parameter and the dimensionality of the system.

How can you mathematically express the shape of a coexistence curve near the critical point? First, you need to define the *order parameter* of the system. An order parameter is a quantity m on which the free energy depends. The order parameter m is zero above the critical temperature, indicating that the system is disordered or randomly mixed. m is non-zero below the critical temperature, indicating that the system is ordered or phase-separated in some way. For liquid–liquid immiscibility, the order parameter could be the difference in phase compositions $m = x'' - x'$. At the critical temperature, $x' = x'' = 0.5$, so the quantity $m(T) = 2x'' - 1$ is a conveniently normalized order parameter: it equals 1 at low temperature, and 0 at the critical point $T = T_c$. At low temperatures, $m \to 1$ indicates that the two coexisting phases are very different. As $T \to T_c$, $m \to 0$ indicates that the two states of the system become indistinguishable from each other.

To see the function $m(T)$ for liquid immiscibility, turn the phase diagram (shown in Figure 26.1(a)) on its side to get $x(T)$ rather than $T(x)$, then multiply by 2 and subtract 1: $m(T) = 2x(T) - 1$ (see Figure 26.1(b)). For boiling, an appropriate order parameter is the difference in density between the liquid and the gas phases (see Figure 26.2). A function with a similar shape near the critical point is the enthalpy of vaporization as a function of the temperature (see Figure 26.3).

Metal alloys provide another example. Brass is an alloy made from copper and zinc. Some brass has two interpenetrating body-centered cubic lattices, one lattice containing the copper atoms, and the other the zinc atoms. Figure 26.4(a) shows a two-dimensional representation of the low-temperature structure, where the system is perfectly ordered. Figure 26.4(b) shows a higher temperature structure, where the alloy is more disordered.

The disordering with temperature is called an order–disorder transition, and the process is called *symmetry-breaking*. The order parameter is the fraction of atoms located on correct lattice sites minus the fraction located on incorrect sites. The order parameter ranges from zero in the fully disordered state to one in the fully ordered state. The mixing of metals in alloys can be treated with the lattice model of previous chapters. Brass with A's and B's alternating throughout the structure is a stable state of a system that has a negative value of the mixing parameter, $\chi_{AB} < 0$, indicating that AB interactions are more favorable than AA and BB. This contrasts with liquid solutions,

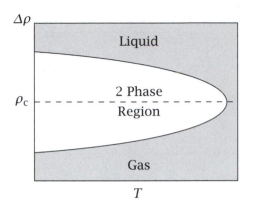

$\Delta\rho$

Liquid

2 Phase
Region

ρ_c ----

Gas

T

Figure 26.2 For boiling, a useful order parameter is the difference between liquid and gas densities $\Delta\rho = \rho_{\text{liquid}} - \rho_{\text{gas}}$. ρ_c is the density of the fluid at the critical point.

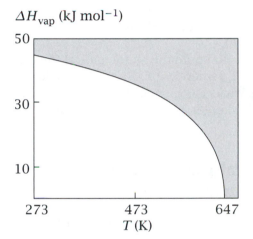

ΔH_{vap} (kJ mol^{-1})

50

30

10

273   473   647

T (K)

Figure 26.3 The enthalpy of vaporization of water as a function of temperature has a functional form similar to the order parameter, becoming zero at the critical point. Source: JN Murrell and AD Jenkins, *Properties of Liquids and Solutions*, Wiley, New York (1982).

(a) Ordered   (b) Disordered

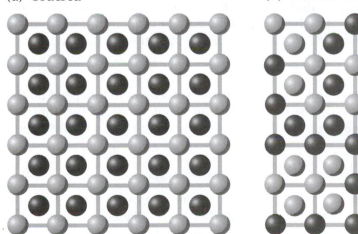

Figure 26.4 Ordering in metal alloys. Two different atom types are interspersed: (a) ordered at low temperature, and (b) disordered at high temperature. Source: G Careri, *Order and Disorder in Matter*, English language translation by K Jarrett, Benjamin Cummings, Reading, 1984.

Figure 26.5 The dependence of the order parameter of beta-brass on temperature T. The points are neutron-scattering results, and the line is the theoretical result for a compressible Ising model. T_c is the critical temperature. Source: J Als-Nielsen, *Phase Transitions and Critical Phenomena*, Volume 5a, C Domb and MS Green, eds, Academic Press, New York, 1976.

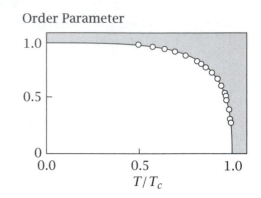

for which χ_{AB} is usually positive. Figure 26.5 shows how the order parameter changes with temperature for brass; $m(T)$ for this solid resembles $m(T)$ for liquid–liquid and boiling equilibria. Many types of system have transitions and critical points with similar shapes for $m(T)$.

To predict the experiments, we want to compute the mathematical form of $m(T)$ near the critical temperature $T \approx T_c$, where $m(T) \to 0$. We could get this function from the lattice or van der Waals gas models, but instead we will find the same result from a model that is simpler and more general, called the *Landau model*.

The Landau Model Is a General Description of Phase Transitions and Critical Exponents

We now describe the Landau model, named for the Russian physicist LD Landau (1908–1968), who won the 1962 Nobel prize in Physics for his work on condensed phases of matter. The Landau model is a generic treatment of phase transitions and critical points. It is based on the idea that coexistence curves are mathematical functions that have two minima at low temperatures, merging into a single minimum at higher temperatures. You can capture that mathematically by expressing the free energy $F(m)$ as a polynomial function,

$$F(m) = F_0 + a_1 m + a_2 m^2 + a_3 m^3 + a_4 m^4 + \dots, \tag{26.1}$$

of an order parameter $0 \leq m \leq 1$. The a_i's are coefficients that depend on the physical problem at hand. For the simplest problems, we want this function to be symmetrical around $m = 0$. So we can eliminate the terms involving odd powers, leaving

$$F(m) = F_0 + a_2 m^2 + a_4 m^4. \tag{26.2}$$

Figure 26.6 shows that this function gives the shapes we want. Suppose $a_4 > 0$. Then $F(m)$ has a single minimum when $a_2 > 0$ (Figure 26.6(a)). $F(m)$ has a broad flat minimum implying large fluctuations in m when $a_2 = 0$ (Figure 26.6(b)). $F(m)$ has two minima, indicating two stable states, when $a_2 < 0$ (Figure 26.6(c)). Lowering a_2 affects the polynomial function $F(m)$ in the same way that reducing the temperature affects phase equilibria free-energy functions. At high temperatures, $T > T_c$, the free energy has a single minimum; at the critical temperature $T = T_c$, the free energy minimum is broad; and below the critical temperature, $T < T_c$, the free energy function has two minima.

A phase transition is called *first-order* when the free energy function has two minima separated by a hill (Figure 25.5), and is called *higher-order* when the transition involves a broad flat minimum without a hill, as in Figure 26.6(b).

To capture the correspondence between a_2 and temperature, let t represent the fractional deviation of the temperature T away from the critical temperature T_c,

$$t = \frac{T - T_c}{T_c}. \tag{26.3}$$

At $T = T_c$, this dimensionless temperature is $t = 0$. If you take a_2 to be proportional to t,

$$a_2 = at, \tag{26.4}$$

and if you choose a to be a positive constant, then the function $F(m)$ will have one narrow minimum when $t > 0$, a broad minimum when $t = 0$ (the critical point), and two minima when $t < 0$. Equation (26.2) becomes

$$F(m) = F_0 + atm^2 + a_4m^4. \tag{26.5}$$

To determine how the equilibrium value of the order parameter m^* depends on T near T_c, find the value $m = m^*$ that minimizes $F(m)$,

$$\left(\frac{dF}{dm}\right)_{m^*} = 2at(m^*) + 4a_4(m^*)^3 = 0. \tag{26.6}$$

Solving Equation (26.6) for m^* gives

$$(m^*)^2 = -\frac{at}{2a_4}. \tag{26.7}$$

The main result is the dependence of the equilibrium value of the order parameter m^* on temperature t. You can write Equation (26.7) as a power law,

$$m^* = \text{constant}(t)^\beta, \tag{26.8}$$

where β is called the *critical exponent*. Equation (26.7) shows that the critical exponent in the Landau model is $\beta = 1/2$.

In general, independently of the model, the critical exponent λ for a function $g(x)$ is

$$\lambda = \lim_{x \to 0} \frac{\ln|g(x)|}{\ln|x|}. \tag{26.9}$$

EXAMPLE 26.1 Finding a critical exponent. What is the critical exponent of the free energy function $g(t) = t^2 + t^{1/2}$? Equation (26.9) gives

$$\lambda = \lim_{t \to 0} \left(\frac{\ln(t^2 + t^{1/2})}{\ln t}\right) \approx \frac{\ln(t^{1/2})}{\ln t} = \frac{1}{2}.$$

This approximation applies because $t^{1/2}$ becomes much larger than t^2 as $t \to 0$.

(a) $a_2 > 0$

$F(m)$

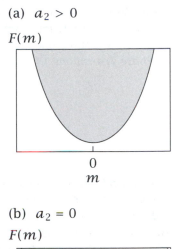

(b) $a_2 = 0$

$F(m)$

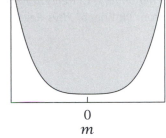

(c) $a_2 < 0$

$F(m)$

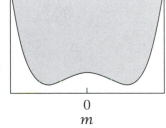

Figure 26.6 (a) The Landau model of one stable state. (b) The Landau model of a critical point. The free energy is flat and broad at the minimum. This implies large fluctuations in m at equilibrium. (c) The Landau model of two stable states.

Table 26.1 Critical exponents α, β, γ, ν, and η for boiling. The correlation length ξ gives roughly the sizes of droplets. Source: JM Yeomans, *Statistical Mechanics of Phase Transitions*, Clarendon Press, New York (1992).

Property	Critical Exponent
Specific heat at constant volume	$c_V \propto |t|^{-\alpha}$
Liquid–gas density difference	$(\rho_\ell - \rho_g) \propto (-t)^\beta$
Isothermal compressibility	$\kappa \propto |t|^{-\gamma}$
Correlation length	$\xi \propto |t|^{-\nu}$
Pair correlation function at T_c	$g(r) \propto 1/r^{1+\eta}$

Figure 26.7 Measurements of the helium coexistence curve near the critical point, on a log-log scale, indicating that the critical exponent is $\beta = 0.354$. Source: HE Stanley, *Introduction to Phase Transitions and Critical Phenomena*, Clarendon Press, Oxford (1971). Data are from PR Roach, *Phys Rev* **170**, 287 (1968).

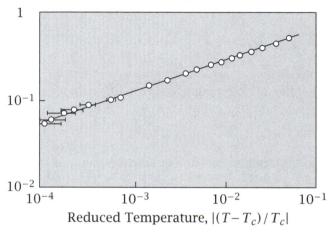

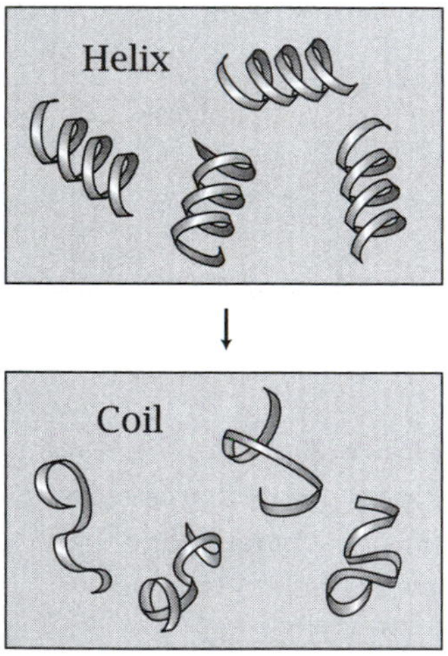

Reduced Density, $|(\rho - \rho_c)/\rho_c|$

Reduced Temperature, $|(T - T_c)/T_c|$

Different physical properties of a system have various critical exponents (see Table 26.1). Because physical properties are often related through thermodynamic relationships, there are algebraic relationships among critical exponents [1, 2, 3].

The importance of critical exponents is their *universality*. While the *critical temperature* depends on the details of the interatomic interactions, the *critical exponent* does not. For example, $\beta = 1/2$ arises in the Landau model just from the general notion that two minima in the free energy function merge to a single minimum as the temperature approaches the critical point, irrespective of whether the transition involves boiling, mixing, metal alloy order–disorder, or magnetisation, and irrespective of microscopic parameters such as χ_{AB}. It can be shown that the van der Waals and lattice mean-field models also give a critical exponent of $\beta = 1/2$ [4]. Guggenheim's data in Figure 25.17 show that eight different types of atom have identical coexistence curves. The shapes of these coexistence curves do not depend on the types of atoms that are interacting or on their interaction energies.

The revolution in understanding critical phenomena began with Guggenheim's experiments. The critical exponent for those data and others is found to be $\beta \approx 1/3$ (see Figure 26.7), in disagreement with the value $\beta = 1/2$ predicted by the Landau model [4]. The discrepancy is evidently not in the neglect of atomic detail, because the coexistence curves for different types of atom superimpose upon each other. Rather, the problem is that very near the critical

Figure 26.8 Some polymers undergo a helix–coil transition. In the single helical conformation, the polymer molecule is ordered. The coil state is a collection of disordered conformations.

Helix

Coil

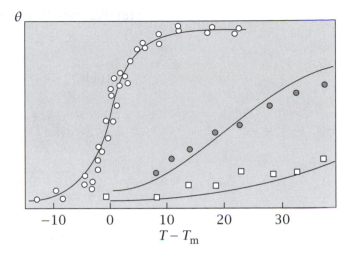

Figure 26.9 The helix–coil transition of poly-γ-benzyl-l glutamate chains: ($\square$) 26-mers, ($\bullet$) 46-mers, ($\circ$) 1500-mers. Fractional helicity θ versus temperature T, relative to the midpoint T_m for the long chains, from optical-rotation data. Curves are calculated from Zimm–Bragg theory by using $\sigma = 2 \times 10^{-4}$ and $\Delta h = 890$ cal mol^{-1}, where $s = \exp[-\Delta h / kT]$. Source: CR Cantor and PR Schimmel, *Biophysical Chemistry*, WH Freeman, San Francisco (1980). After BH Zimm, P Doty and K Iso, *Proc Natl Acad Sci* USA **45**, 1601 (1959).

point, higher-order terms beyond m^4 are needed. The true free energy surfaces are flatter near critical points—so the fluctuations are broader—than would be predicted by mean-field or Landau models. A type of model in which $F(m)$ can be expressed exactly, without mean-field approximations or low-order expansions in m, is the *Ising model*, also called the *nearest-neighbor model*. It handles critical points more accurately. We illustrate this type of model for a one-dimensional system, the helix–coil transition. It can be solved exactly.

Polymers Undergo Helix–Coil Transitions

Figure 26.8 illustrates a transformation that occurs in some polymers, including RNA, DNA, and polypeptides. In aqueous solution, at high temperatures, the chain has a large ensemble of disordered conformations. This ensemble is collectively called the *coil state*. When the temperature is lowered, each molecule undergoes a transition to a helix (see Figure 26.9). These transitions are called *cooperative*, which is sometimes taken to mean the following. (1) Some property (the amount of helix, in this case) changes in a sigmoidal way as a function of an external variable (such as temperature or solvent). (2) The sigmoidal curve steepens with increasing size of the system (the chain length, in this case). Interestingly, some of the most important helix–coil cooperativity data (see Figure 26.9) were taken in a solvent (80% dichloroacetic acid, 20% ethylene dichloride) that inverts the transition: increasing the temperature increases the helix content. Cooperativity in either type of solvent is treated by the same models.

The simplest models of this cooperative behavior suppose that each monomeric unit along the chain has one of two possible states, H (helix) or C (coil). The chain conformations are represented as one-dimensional sequences of H and C units. An example conformation is $HHCCCCCHCHHHHC$.

Our aim is to compute the partition function, based on the combinatorics for counting the numbers of arrangements of H and C units in the chain. We will consider three different models for the partition function. To show the nature of cooperativity, it is useful to start with a model that has no cooperativity.

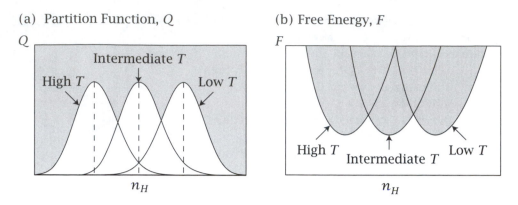

(a) Partition Function, Q

(b) Free Energy, F

Figure 26.10   For the noncooperative model, lowering the temperature increases the helicity gradually, not sharply. n_H is the number of helical units.

Noncooperative Model: Neighboring Units Are Independent of Each Other

Suppose a chain has N monomer units. Assume that the state (H or C) of each monomer $j = 1, 2, 3, \ldots, N$ in the chain is independent of the state of every other monomer. Let q_C be the partition function for each C unit. Let q_H be the partition function for each H unit. Because any conformational change can always be expressed as the ratio of these quantities, you are free to choose $q_C = 1$, which is equivalent to setting the free energy of a coil unit equal to zero. Let $q_H = \exp(-\beta\varepsilon)$, where $\varepsilon < 0$ represents the energy of forming a helical unit relative to a coil unit, and $\beta = 1/kT$. The ratio

$$s = \frac{q_H}{q_C} = e^{-\beta\varepsilon} \tag{26.10}$$

defines the relative probabilities of H and C states for a unit of the chain and represents its helix-coil equilibrium constant. The partition function is the sum over all the possible sequences of H and C, which is given by the binomial (coin flip) distribution:

$$Q = \sum_{n_H=0}^{N} q_C^{n_C} q_H^{n_H} \frac{N!}{n_C! n_H!} = (q_C + q_H)^N = q_C^N (1+s)^N, \tag{26.11}$$

where N is the total number of units, n_H is the number of helical units, n_C is the number of coil units in the chain, and s is the helix-coil equilibrium constant given by Equation (26.10).

We now compute $f_H = n_H^*/N$, the equilibrium fraction of the N monomers that are H's, as a function of the chain length N and the temperature T. We want to know the value $n_H = n_H^*$ that causes the free energy F to be a minimum. Figure 26.10 shows the result of substituting Equation (26.11) into the free energy expression $F = -kT \ln Q$. The noncooperative model predicts that lowering the temperature increases the helicity gradually, as indicated by a shift in the minimum of the free energy curve with temperature.

To calculate $f_H(T) = n_H^*/N$, you could compute the free energy for each T, find the value of $n = n_H^*$ that causes F to be at a minimum, then plot n_H^* as a function of T. An easier way is to use Equations (26.11) and (26.15). Because

$$\left(\frac{\partial Q}{\partial s}\right) = q_C^N N (1 + s)^{N-1},$$

you have

$$\langle n \rangle = \frac{s}{Q}\left(\frac{\partial Q}{\partial s}\right) = \frac{Ns(1+s)^{N-1}}{(1+s)^N} = \frac{Ns}{1+s}.$$

So the average helicity is

$$f_H = \frac{\langle n \rangle}{N} = \frac{s}{1+s}, \tag{26.12}$$

which depends on temperature through the definition of s in Equation (26.10).

The noncooperative model predicts that the helicity changes gradually, not cooperatively, with temperature. It proves that independent units are not cooperative. Cooperativity requires some interdependence. Equation (26.12) shows that the noncooperative model fails to predict the dependence of f_H on chain length that is indicated by the experiments (see Figure 26.9).

Gradual noncooperative changes are characterized by free energy functions in which a single minimum $m^*(T)$ shifts with T or with some other external variable (see Figure 26.10). Here's a different model that defines the opposite extreme, maximum cooperativity.

The Two-State Model: Maximum Cooperativity

Suppose the helix–coil system has only two possible states: $CCCC\ldots C$ or $HHH\ldots H$. Assume that all other states have zero probability. Maximum cooperativity means that if one monomer is H, all are H; if one monomer is C, all are C. Now the partition function for each of the two states is q_C^N or q_H^N, so the partition function for the system is

$$Q = q_C^N + q_H^N = q_C^N(1 + s^N), \qquad \text{and}$$

$$F = -kT \ln Q = -kT \ln(1 + s^N) + \text{constant}.$$

Figure 26.11(b) shows that the probability density function for this model has two peaks, separated by a trough. At low temperatures, all molecules are fully helical. At high temperatures, the molecules in coil conformations substantially outnumber the molecules in helical states. At the midpoint of the transition, the helicity is $1/2$, not because the individual molecules are half helix and coil but because half of the molecules are all-helix, and half of them are all-coil. In the two-state model no molecule is in an *intermediate* state. In this regard, the two-state model differs markedly from the noncooperative model.

For the two-state model,

$$\langle n \rangle = \frac{s}{Q}\left(\frac{\partial Q}{\partial s}\right) = \frac{sNs^{N-1}}{1+s^N} = N\frac{s^N}{1+s^N}, \qquad \text{so} \qquad f_H = \frac{s^N}{1+s^N}.$$

Comparing Figure 26.12 with Figure 26.9 shows that the two-state model for the helix coil transition is too cooperative to explain the experimental data.

Population

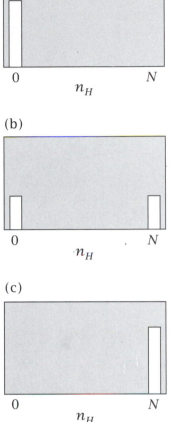

Figure 26.11 The two-state model: populations are helix ($n_H = N$) and coil ($n_H = 0$), and there are no 'intermediate' states. (a) $s < 1$ (high temperature), (b) $s = 1$, (c) $s > 1$ (low temperature).

An Expression for Average Helix Content

Suppose you have a partition function with the general form

$$Q = 1 + a_1 s + a_2 s^2 + a_3 s^3 + \cdots + a_N s^N, \tag{26.13}$$

where the a_i's are constants relevant for the particular problem. The equilibrium constant $s = e^{-\beta\varepsilon}$ represents a Boltzmann factor, so $s^2 = e^{-\beta(2\varepsilon)}$, $s^3 = e^{-\beta(3\varepsilon)}$, etc. This polynomial has appeared before in dice problems (Equation (6.20)) and vibrational partition functions (Equation (11.25)). Now we'll use it for helix–coil and ligand-binding processes.

The probability that the system is in state n is

$$p_n = \frac{a_n s^n}{Q}.$$

We want to compute the average 'state number' $\langle n \rangle$ because this defines the average helicity. In later chapters it will define the average numbers of ligands bound to a molecule or surface. We follow the logic described in Equations (4.10) to (4.13) to arrive at the average,

$$\langle n \rangle = \sum_{n=1}^{N} n p_n = \frac{a_1 s + 2a_2 s^2 + 3a_3 s^3 + \cdots + N a_N s^N}{Q}. \tag{26.14}$$

To get this expression, take the derivative of Equation (26.13),

$$\left(\frac{\partial Q}{\partial s} \right) = a_1 + 2a_2 s + 3a_s^2 + \cdots + N a_N s^{N-1}.$$

Multiply by s and divide by Q to get

$$\frac{s}{Q} \left(\frac{\partial Q}{\partial s} \right) = \left(\frac{\partial \ln Q}{\partial \ln s} \right) = \frac{a_1 s + 2a_2 s^2 + 3a_3 s^3 + \cdots + N a_N s^N}{Q},$$

which is identical to Equation (26.14). The average state number is

$$\langle n \rangle = \left(\frac{\partial \ln Q}{\partial \ln s} \right). \tag{26.15}$$

Equation (26.15) gives the helix content for the models in this chapter.

In summary, one-state and two-state transformations are distinguished by the shapes of their free energy functions, $F(m^*)$. At the midpoint of a two-state transition, the free energy function has two minima (see Figure 26.13). Under those conditions, half the molecules are in state m_1^* and half the molecules are in state m_2^*. There is little or no population of 'intermediate states' having intermediate values of the order parameter. In contrast, at the midpoint of a one-state transition, the free energy function has a single minimum, and the molecules in those systems have a distribution of order parameter values, $m_1^* \le m^* \le m_2^*$.

How can experiments determine whether a transition is one-state or two-state? Both models can give sigmoidal curves of $f_H(T)$, or of an order parameter $m(x)$ as a function of an externally controlled variable x. So a sigmoidal

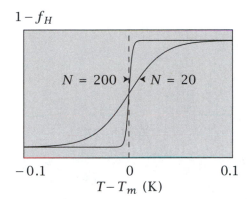

$1 - f_H$

$N = 200$ ▸ ◂ $N = 20$

-0.1 0 0.1

$T - T_m$ (K)

Figure 26.12 The two-state model predicts more cooperativity than is observed in experiments (compare with Figure 26.9).

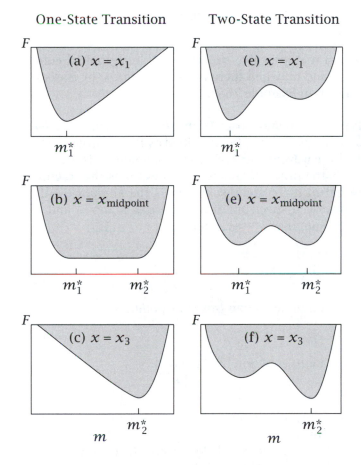

One-State Transition

F

(a) $x = x_1$

m_1^*

F

(b) $x = x_{\text{midpoint}}$

m_1^* m_2^*

F

(c) $x = x_3$

m m_2^*

Two-State Transition

F

(e) $x = x_1$

m_1^*

F

(e) $x = x_{\text{midpoint}}$

m_1^* m_2^*

F

(f) $x = x_3$

m m_2^*

Figure 26.13 Transitions are *two-state* if there are two minima in the free energy, and *one-state* if there is one minimum. In both types of system, the system changes from state m_1^* when $x = x_1$ to m_2^* when $x = x_2$. At the midpoint $x = x_{\text{midpoint}}$ of a two-state transition, there are two stable states, m_1^* and m_2^*. At the midpoint of a one-state transition, there is a broad distribution of stable states from m_1^* to m_2^*.

shape alone is not sufficient to establish whether the free energy has one or two minima. The strongest proof of a two-state transition would be direct evidence of two populations. Figure 26.14 shows an example of direct evidence of a two-state process, the unfolding of a protein with increasing concentrations of urea, a denaturing agent. Three other methods are commonly accepted evidence for two-state transitions: (1) when the enthalpy measured by calorimetry equals the enthalpy based on a two-state model (see Problem 11, page 513), (2)

Figure 26.14 This chromatography experiment resolves the two states of a protein. The native state (N) is compact and has a long elution time. The unfolded state (U) is expanded and has a short elution time. Urea is a denaturing agent. It shifts the equilibrium from N to U. The peak areas represent the relative fractions of the two species. Source: RJ Corbett and RS Roche, *Biochemistry* **23**, 1888–1894 (1984).

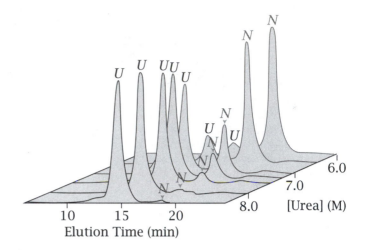

Fluorescence Intensity

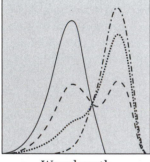

Wavelength

Figure 26.15 Fluorescence spectra of napthol at various pH values. The common intersection point is an isosbestic point, and indicates a two-state transition, in this case $A + H^+ \rightarrow AH^+$. Source: W Hoppe, W Lohmann, H Markl, eds, *Biophysics*, Springer-Verlag, New York (1983).

when different experimental methods give superimposable sigmoidal curves (see Problem 10, page 512), and (3) when there is an *isodichroic* or *isosbestic point* in the spectra (as shown in Example 26.2).

> **EXAMPLE 26.2 Identifying a two-state transition by a spectroscopic isosbestic point.** Suppose that a state A can be identified by its absorbance spectrum $f(\lambda)$ as a function of the wavelength λ. Suppose that state B is identified by its spectrum $g(\lambda)$. m is the order parameter that describes the fractional amount of B that is present in the system, so $1 - m$ is the amount of A. In a two-state process, the observed spectrum of a mixture of states A and B is $h(\lambda)$, which is the weighted sum over states A and B,
>
> $$h(\lambda) = mg(\lambda) + (1 - m)f(\lambda)$$
>
> $$= f(\lambda) + m(g(\lambda) - f(\lambda)). \qquad (26.16)$$
>
> Suppose that the two spectra intersect at an *isosbestic point* λ_0,
>
> $$g(\lambda_0) = f(\lambda_0). \qquad (26.17)$$
>
> Substituting Equation (26.17) into Equation (26.16) gives
>
> $$h(\lambda_0) = f(\lambda_0),$$
>
> which is independent of m. This means that if you vary a parameter, such as the temperature or the concentration of a chemical agent that changes m, and changes the relative amounts of A and B, all the spectra from those experiments intersect at the wavelength $\lambda = \lambda_0$ (if the physical process is adequately modelled with two states). So if you observe a single point where all the spectra for different stages of the transition superimpose, you have evidence for a two-state transition (see Figure 26.15).

Now we explore a third model, in which the monomers are neither fully independent of each other nor totally cooperative.

| Monomer State | | Zimm–Bragg Statistical |
j − 1	j	Weight for j
C	C	$q(C \mid C) = 1$
H	C	$q(C \mid H) = 1$
C or ∅	H	$q(H \mid C) = \sigma s$
H	H	$q(H \mid H) = s$

Table 26.2 Statistical weights for the Zimm–Bragg model. The statistical weight q for having a monomer type H or C at position j in the chain is *conditional* upon what monomer type is at position j.

The Zimm–Bragg Model of the Helix–Coil Transition: Nearest-Neighbor Cooperativity

The central idea of this model, developed by BH Zimm and JK Bragg [5], is that the probability that monomer j is in state C or H depends on whether its neighboring monomer $j − 1$ is in state C or H. The statistical weight q for monomer j is *conditional* upon the state of its neighboring monomer. There are four different *monomer conditional statistical weights*: $q(C \mid C)$, $q(C \mid H)$, $q(H \mid C)$, and $q(H \mid H)$. Statistical weights are quantities that describe relative populations or probabilities. For example $q(C \mid H)$ is the relative amount of monomer j that is in state C if monomer $j − 1$ is in state H. You can multiply together appropriate monomer statistical weights to get the *sequence statistical weight* for a chain of N H and C monomers of length N. Then the partition function is the sum over the statistical weights of all the possible sequences. Table 26.2 lists all four possible monomer statistical weights, in the Zimm–Bragg notation.

Every coil unit C is assigned a statistical weight of 1, by convention, whether the preceding monomer is an H or a C. Any helical unit H is assigned a statistical weight s. You can think of s as an equilibrium constant for converting a C to an H,

$$s = \frac{[H]}{[C]}.$$

Assign a statistical weight σs if an H follows a C, or if the H is the first monomer in the chain. Think of σ as a *nucleation parameter* for initiating a helix. For most amino acids in proteins, s is slightly greater than 1 and $\sigma \approx 10^{-3}$ to 10^{-4} [6]. Propagation is easy but nucleation is difficult.

Using the rules of conditional probability, Equation (1.11), you can compute the statistical weight for a sequence, say CHH, as the product $q(CHH) = q(H \mid H)q(H \mid C)q(C \text{ first}) = s \times \sigma s \times 1 = \sigma s^2$. Table 26.3 lists the statistical weights of all sequences of C's and H's that have three monomers. Each is a product of a factor of 1 for every C, a factor of s for every H, and a factor of σ for each beginning of a helix. For the first monomer of a chain, you have $q(C) = 1$ or $q(H) = \sigma s$.

Because each chain of length N can have any possible configuration—any sequence of H's and C's—the partition function is the sum over all the possible chain configurations. You can compute the partition function using a simple matrix method. First, express the statistical weights for the two possible states of the first monomer as a statistical weight vector $\mathbf{q_1}$:

$$\mathbf{q_1} = [q(C),\ q(H)].$$

Table 26.3 The Zimm–Bragg partition functions for all eight possible trimer sequences of H's and C's.

Trimer Sequence	Statistical Weight
CCC	1
CCH 1·1·σs	σs
CHC 1·σs·1	σs
CHH 1·σs·s	σs^2
HCC	σs
HCH	$\sigma^2 s^2$
HHC	σs^2
HHH	σs^3

Refresher on Matrix and Vector Multiplication

When a vector $[a, b]$ multiplies a matrix $\begin{bmatrix} x & y \\ z & w \end{bmatrix}$, the result is

$$[a, b]\begin{bmatrix} x & y \\ z & w \end{bmatrix} = [ax + bz, ay + bw].$$

When two matrices are multiplied, the result is

$$\begin{bmatrix} a & b \\ c & d \end{bmatrix}\begin{bmatrix} x & y \\ z & w \end{bmatrix} = \begin{bmatrix} ax + bz & ay + bw \\ cx + dz & cy + dw \end{bmatrix}.$$

Note that the order of matrix multiplication matters. In general for two matrices $\mathbf{A}$ and $\mathbf{B}$, $\mathbf{AB} \neq \mathbf{BA}$.

Now express the monomer conditional statistical weights in terms of matrix $\mathbf{G}$:

$$\mathbf{G} = \begin{bmatrix} q(C \mid C) & q(H \mid C) \\ q(C \mid H) & q(H \mid H) \end{bmatrix} = \begin{bmatrix} 1 & \sigma s \\ 1 & s \end{bmatrix}. \tag{26.18}$$

To get the statistical weight vector $\mathbf{q_2}$ for all the two-monomer sequences, multiply $\mathbf{q_1}$ by $\mathbf{G}$:

$$\begin{aligned}
\mathbf{q_2} &= \mathbf{q_1 G} \\
&= [q(C)q(C \mid C) + q(H)q(C \mid H), \\
&\qquad q(C)q(H \mid C) + q(H)q(H \mid H)] \\
&= [q(CC) + q(CH), \quad q(HC) + q(HH)] \\
&= \left[1 + \sigma s, \quad \sigma s + \sigma s^2\right].
\end{aligned} \tag{26.19}$$

The leftmost term in the vector $\mathbf{q_2}$ is the sum of the statistical weights of all of the sequences that end in C. The rightmost term counts all of the sequences that end in H. To get the partition function Q_2, which is the sum over the statistical weights of all sequences for a two-monomer chain, multiply $\mathbf{q_2}$ by a column vector of 1's,

$$Q_2 = \mathbf{q_2}\begin{bmatrix} 1 \\ 1 \end{bmatrix} = q(CC) + q(CH) + q(HC) + q(HH)$$

$$= 1 + 2\sigma s + \sigma s^2. \tag{26.20}$$

This procedure readily generalizes for a chain of any length N. Every multiplication by the matrix $\mathbf{G}$ 'grows' the chain by one unit. For example,

$$\mathbf{q_3} = \mathbf{q_2 G} = \mathbf{q_1 G}^2.$$

The partition function Q_N for a chain of N monomers is

$$Q_N = [1, \ \sigma s]\mathbf{G}^{N-1}\begin{bmatrix} 1 \\ 1 \end{bmatrix}. \tag{26.21}$$

Equation (26.21) gives the sum of the statistical weights for all of the possible sequences of H's and C's.

EXAMPLE 26.3 Compute the partition function for a trimer. Substitute $N = 3$ into Equation (26.21):

$$Q_3 = [1, \sigma s]\mathbf{G}^2 \begin{bmatrix} 1 \\ 1 \end{bmatrix}$$

$$= [1, \sigma s]\begin{bmatrix} 1 & \sigma s \\ 1 & s \end{bmatrix}\begin{bmatrix} 1 & \sigma s \\ 1 & s \end{bmatrix}\begin{bmatrix} 1 \\ 1 \end{bmatrix}$$

$$= [1 + \sigma s, \ \sigma s + \sigma s^2]\begin{bmatrix} 1 & \sigma s \\ 1 & s \end{bmatrix}\begin{bmatrix} 1 \\ 1 \end{bmatrix}$$

$$= [1 + \sigma s + \sigma s + \sigma s^2, \ \sigma s + \sigma s^2 + \sigma^2 s^2 + \sigma s^3]\begin{bmatrix} 1 \\ 1 \end{bmatrix}$$

$$= 1 + 3\sigma s + 2\sigma s^2 + \sigma^2 s^2 + \sigma s^3. \tag{26.22}$$

You can check Equation (26.22) by noting that it gives the same sum as adding all the statistical weights in Table 26.3. From the partition function, Equation (26.21), you can compute the properties of the model, such as the fractional helicity.

EXAMPLE 26.4 Compute the fractional helicity of the trimer. Following Equation (26.15), take the derivative of Equation (26.22) to get the fractional helicity:

$$f_H = \frac{1}{N}\left(\frac{\partial \ln Q}{\partial \ln s}\right) = \left(\frac{1}{N}\right)\left(\frac{3\sigma s + 4\sigma s^2 + 2\sigma^2 s^2 + 3\sigma s^3}{1 + 3\sigma s + 2\sigma s^2 + \sigma^2 s^2 + \sigma s^3}\right). \tag{26.23}$$

In the Zimm–Bragg model, the cooperativity is controlled by the parameter σ. Helix formation is cooperative when σ is small. When $\sigma = 1$ in Equation (26.23), the Zimm–Bragg model reduces to the noncooperative independent units model,

$$\theta = \frac{s}{1 + s}.$$

As σ decreases, nucleation becomes more difficult, and helix formation becomes more cooperative. That is, it takes a long stretch of helix (j large) if σ is small, to achieve an equilibrium constant $\sigma s^j > 1$ that is favorable for helix formation. Short helices are not stable unless the temperature is low. If $\sigma \to 0$, this model becomes identical to the two-state model.

Figure 26.9 shows that the Zimm–Bragg model captures the chain-length dependence of the cooperativity in the helix–coil transitions of polypeptides. This model simplifies the problem of the many different three-dimensional polymer chain configurations into a one-dimensional problem of enumerating the strings of arrangements of nearest neighbors.

Transitions in other kinds of system can also be treated by the nearest-neighbor model. One example is the magnetization of metals.

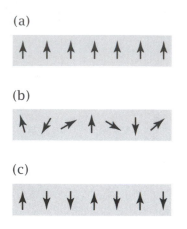

(a)

(b)

(c)

Figure 26.16 Magnetic spins in a one-dimensional system. (a) Fully aligned, for example at low temperatures, (b) disordered, at high temperatures, (c) a simple 'binary' model of the disordered states in which spins are either up or down.

The Ising Model Describes Magnetization

You can magnetize certain materials, such as iron, by putting them into a magnetic field. Below a certain critical temperature, such materials have a net magnetic moment. They are magnetized. But if you heat them above that temperature, they lose magnetization. What is the microscopic basis for this behavior?

Atomic spins act like small magnets: they are magnetic dipoles with north and south poles that become aligned in applied magnetic fields. At high temperatures the spins are randomly oriented (see Figure 26.16(b) and (c)), but below a critical temperature $T = T_c$, the spins align with each other (Figure 26.16(a)).

The net magnetic moment in the up direction is represented by the number of upward arrows ($\uparrow$) minus the number of downward arrows ($\downarrow$).

Magnetic alignment is cooperative because of *nearest-neighbor* interactions. At low temperatures the spins in ferromagnets line up to satisfy the favorable nearest neighbor interactions, but at high temperatures the arrow directions randomize to maximize the entropy. In a *ferromagnet*, such as iron, certain iron oxides, or cerium antimonide, each spin prefers to orient in the same direction as its neighbors. In *antiferromagnets*, like MnO, each spin prefers to orient in the direction opposite to its neighbors.

In the *Ising model*, each atomic magnet occupies one lattice site with a spin that is either up or down. This model is named for physicist E Ising, who solved it in his Ph.D. thesis work with W Lenz in 1925. If you model each spin as an $\uparrow$ or $\downarrow$ on a one-dimensional lattice, the partition function can be computed in the same way as the partition function for a sequence of H or C units in the helix–coil model. The difference between the one-dimensional Ising model and the helix–coil model is just the statistical weights. For the magnet model, the weights are: $q(\uparrow\uparrow) = q(\downarrow\downarrow) = e^J$ and $q(\uparrow\downarrow) = q(\downarrow\uparrow) = e^{-J}$, where $J = J_0/kT$ is a dimensionless energy. Ferromagnets are modelled by $J > 0$, and antiferromagnets by $J < 0$.

The partition function of this magnet model is found by growing the linear sequence of spins in the same way as you did for the helix–coil model. For a single site, the partition function is $Q_1 = 2$, because there are two states: $\uparrow$ or $\downarrow$. To get the two-site partition function, you multiply Q_1 by $(e^J + e^{-J})$, because the second arrow can point either in the same direction as the first or in the opposite direction. In terms of the hyperbolic cosine function, $\cosh(x) = (e^x + e^{-x})/2$, you have

$$Q_2 = Q_1(e^J + e^{-J}) = Q_1(2\cosh J). \tag{26.24}$$

Every site that you add to the one-dimensional array of spins multiplies the partition function by a factor of $(e^J + e^{-J})$, so the partition function for a linear lattice of N magnetic spins is

$$Q_N = Q_1(2\cosh J)^{N-1} = 2^N \cosh^{N-1} J. \tag{26.25}$$

The thermodynamic properties of the one-dimensional Ising model are computed from the partition function in the standard way (see Table 10.1).

Real magnets are three-dimensional. The importance of the one-dimensional Ising model is that it illustrates the principles of the nearest-neighbor model and that it can be solved exactly. For the three-dimensional Ising model, no

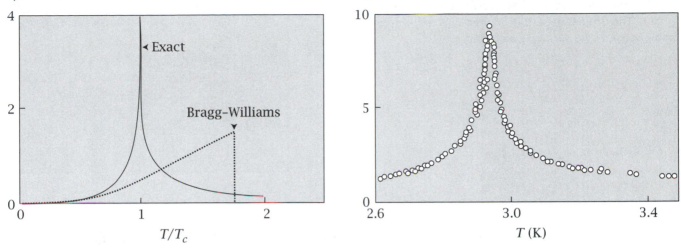

(a) Heat Capacity Predicted from Theory

C_V/Nk

◄ Exact

Bragg–Williams

T/T_c

(b) Observed Heat Capacity

C_V/Nk

T (K)

Figure 26.17 (a) Theoretical predictions for the heat capacity C_V near the critical temperature, for two-dimensional systems. The Bragg–Williams mean-field lattice model of Chapter 25 leads to a triangular function, while the exact solution of the two-dimensional Ising model shows a sharp peak. Source: R Kubo, in cooperation with H Ichimura, T Usui and N Hashitsome, *Statistical Mechanics*, Elsevier Pub. Co., New York (1965). (b) Experimental data for helium on graphite closely resembles the Ising model prediction. Source: RE Ecke and JG Dash, *Phys Rev B* **28**, 3738 (1983).

analytical solution has yet been found, despite extensive efforts. The two-dimensional Ising model was solved by the Norwegian–American physicist L Onsager (1903–1976; Nobel prize in Chemistry, 1968) through a mathematical *tour de force* in 1944. It serves as an exact standard with which other approximations are often compared. Figure 26.17 shows a plot of the predicted order parameter for the two-dimensional Ising model versus temperature, compared with the Bragg–Williams approximation, Equation (15.9), that we used in the lattice model in Chapter 25. Near the critical point, these two models give results that are quite different.

Now we consider one of the simplest models for the kinetics of phase transitions.

The Kinetics of Phase Transitions Can Be Controlled by Nucleation

How fast is a phase transition? If you cool a binary liquid mixture, what limits the rate of phase separation? How fast do crystals form in a liquid? How fast do micelles form in a surfactant solution? For such processes, there are two different mechanisms. First, if you supercool a single-phase solution into the always-unstable part of the two-phase region (see Figure 25.9), the process is spontaneous. It is called *spinodal decomposition*. But a different mechanism applies if you supercool a single-phase solution into the metastable region

Figure 26.18 The free energy Δg of forming a spherical droplet that nucleates a phase transition. A droplet must reach a minimum radius r^* to grow into a macroscopic phase. The droplet growth is favored by a 'bulk' term ε_0 and opposed by a surface-area term γ.

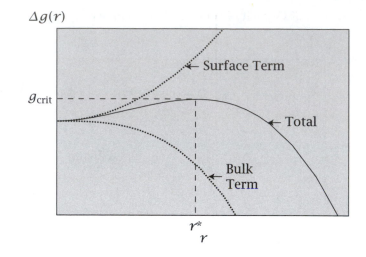

between the binodal and spinodal curves. Then phase separation can be slow. To form a B-rich phase in a metastable solution of components A and B requires that enough B molecules come together to form a nucleus that is so large that further growth of the B-rich phase will be spontaneous. The process of *nucleation* is the formation of this *critical nucleus*, which requires that the system reach the top of an energy barrier. Here's a simple model of the activation energy for this process.

What is the free energy Δg of forming a spherical nucleus of radius r for a B-rich phase in a background sea of mostly A molecules? There are two contributions to the free energy of droplet formation (see Figure 26.18):

$$\Delta g = 4\pi r^2 \gamma - \frac{4}{3}\pi r^3 \varepsilon_0. \tag{26.26}$$

Growth is driven by ε_0, the energy per unit volume for forming BB contacts inside the droplet, multiplied by the droplet volume $(4/3)\pi r^3$. Growth is opposed by the surface tension γ at the interface between the B-rich droplet and the A-rich solution, multiplied by the surface area $4\pi r^2$. The radius r^* of the critical nucleus is the point at which $\Delta g(r^*) = \Delta g_{\text{crit}}$ is a maximum,

$$\left.\frac{d\Delta g}{dr}\right|_{r^*} = 0 \quad \Longrightarrow \quad r^* = \frac{2\gamma}{\varepsilon_0}. \tag{26.27}$$

A large surface tension γ implies that large droplets are required to nucleate the phase separation. Substitute Equation (26.27) into Equation (26.26) to compute the free energy,

$$\Delta g_{\text{crit}} = 4\pi (r^*)^2 \left[\gamma - \frac{r^* \varepsilon_0}{3}\right]$$

$$= \frac{4\pi (r^*)^2 \gamma}{3} = \frac{16\pi \gamma^2}{3\varepsilon_0^2}. \tag{26.28}$$

Droplet formation is an activated process. The rate of forming a droplet is proportional to $\exp(-\Delta g_{\text{crit}}/kT)$.

Summary

Macroscopic matter or individual molecules can undergo dramatic transitions. Changing an external variable x, such as the temperature, can cause a change in an order parameter $m(x)$, such as the helix content of a polymer, the density of a liquid, the composition of a binary mixture or alloy, or the magnetization of a metal. Widely different physical processes can often be described by the same mathematical models. If the underlying free energy surface has two minima at the transition point, it is a *two-state* or *first-order* transition. If there is only a single minimum, it is a *one-state* or *higher-order* transition. Such transitions can be modelled with the nearest-neighbor model, which takes correlations between neighboring molecules into account. This model has played an important role in understanding critical exponents. Very close to critical points, fluctuations can occur over large spatial scales, and models based on low-order expansions predict incorrect critical exponents.

Problems

1. A micellization model.
You have developed a model for the formation of micelles, based on expressions for the chemical potential $\mu(mono)$ of the free monomeric molecules in solution and the chemical potential $\mu(mic)$ of the aggregated state of the monomers in micelles as a function of the mole fraction x of monomers in solution,

$$\mu(mono) = 0.5 + 0.1x^2, \quad \text{and}$$

$$\mu(mic) = 1.3 - 25x^2.$$

(a) What is the state of the system at low concentration?

(b) What is the state of the system at high concentration?

(c) What is the concentration at which the micelle transition occurs?

2. Stabilities of droplets.
Certain molecular aggregates (like surfactant micelles and oil droplets) shrink owing to surface tension, and expand owing to volumetric forces. Suppose the free energy $g(r)$ as a function of the radius r of the aggregate is $g(r) = 2r^2 - r^3$. Such a system has two equilibrium radii.

(a) Calculate the two radii.

(b) Identify each radius as stable, unstable, neutral, or metastable.

3. The energy of the Ising model.
Derive an expression for the energy of the Ising model from the partition function.

4. The Landau model for the critical exponent of $C_V(T)$.
In the Landau model, show that $C_V \propto T$ as T increases toward the critical temperature T_c.

5. Nucleation droplet size.
What is the number of particles n^* in a critical nucleus for phase separation?

6. Broad energy wells imply large fluctuations.
A stable state of a thermodynamic system can be described by the free energy $G(x)$ as a function of the degree of freedom x. Suppose G obeys a square law with spring constant k_s, $G(x)/kT = k_s x^2$. A small spring constant k_s implies a broad shallow well (see Figure 26.19).

(a) Show that the fluctuations $\langle x^2 \rangle$ are larger when k_s is small.

(b) Suppose two phases in equilibrium have narrow energy wells, each with spring constant k_1, and that the critical point of the system has a broad well with spring constant $k_2 = 1/4k_1$. What is the ratio of the fluctuations, $\langle x_2^2 \rangle / \langle x_1^2 \rangle$?

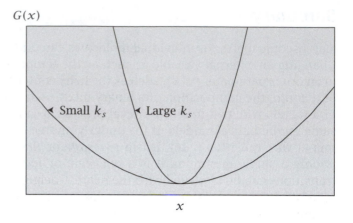
$G(x)$

Figure 26.19 Two different square-law free energies for problem 6: one minimum is broad and the other is narrow.

7. Critical exponents.
What are the critical exponents for the following free energy functions? A is a constant, and t is the reduced temperature $t = |(T - T_c)/T_c|$.

(a) $g(t) = At^2 e^{-t}$.

(b) $g(t) = At^{5/3}$.

(c) $g(t) = A$.

8. Zimm–Bragg helix–coil theory for $N = 4$ chain units.

(a) Write the Zimm–Bragg partition function Q_4 in terms of σ and s for a four-unit chain, where $HHHH$ is the helical state.

(b) Write an expression for $f_H(s)$ for this transition.

9. The Schellman helix–coil model.
A helix–coil model developed by JA Schellman [7] is simpler than the Zimm–Bragg model, and works well for short chains. Consider a chain having N units.

(a) Write an expression for Ω_k, the number of configurations of a chain that has all its H units in a single helix k units long, as a function of N and k.

(b) If σ is the parameter for nucleating a helix, and s is the propagation parameter, write an expression for the partition function Q_N over all possible helix lengths k.

(c) Write an expression for $p_k(N)$, the probability of finding a k-unit helix in the N-mer.

10. Overlapping transition curves as evidence for a two-state transition.
Consider a two-state transition $A \rightarrow B$. m is an order parameter: when $m = 0$, the system is fully in state A; when $m = 1$, the system is in state B. Suppose some agent x (which could be the temperature or the concentration of ligands or salts) shifts the equilibrium as indicated in Figure 26.20. An experiment measures a property (which might be fluorescence,

light scattering, circular dichroism, heat, etc.) measures a property

$$g(m) = g_B m + g_A(1 - m),$$

where g_B and g_A are the measured values for pure B and A, respectively. A different type of experiment measures another property h,

$$h(m) = h_B m + h_A(1 - m),$$

where h_B and h_A are also the measured values for the pure states. Show that if the baselines of the two types of experiment are superimposed ($g_A = h_A$), and the amplitudes are scaled to be the same ($h_B - h_A = g_B - g_A$), then the curve $h(x)$ must superimpose on $g(x)$.

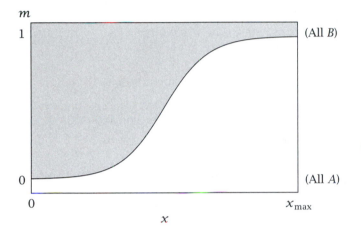

Figure 26.20 A transition curve, for folding, binding, or conformational change, in which the order parameter m is a function of some agent x that shifts the equilibrium.

11. Calorimetric evidence for a two-state transition: the relationship between ΔH_{cal} and $\Delta H_{van\,'t\,Hoff}$. A calorimeter measures the heat capacity $C_p(T)$ as a function of temperature T (see Figure 26.21). The area under the full curve describes an absorption of heat

$$q_0 = \Delta H_{cal} = \int_{T_A}^{T_B} C_p \, dT.$$

Suppose the absorption of heat results from a two-state process $A \xrightarrow{K} B$, where $K = [B]/[A]$.

(a) Express the fraction $f = [B]/([A]+[B])$ in terms of K.

(b) Express $C_p(T)$ as a function of the heat q_0 and the equilibrium constant $K(T)$.

(c) Assume a two-state model (Equation (13.37)),

$$\left(\frac{\partial \ln K}{\partial T}\right) = \frac{\Delta H_{van\,'t\,Hoff}}{kT^2},$$

and $K = 1$ at $T = T_{max}$, the point at which C_p is a maximum.

Derive a relationship between $\Delta H_{van\,'t\,Hoff}$, ΔH_{cal}, and $C_p(T_{max})$.

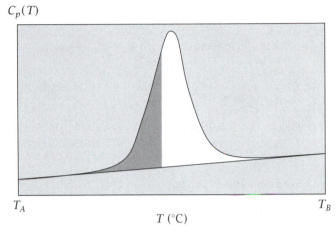

Figure 26.21 The heat capacity change in a two-state transition.

12. **Cooperativity in a three-state system.** Perhaps the simplest statistical mechanical system having cooperativity is the three-level system shown in Figure 26.22.

(a) Write an expression for the partition function q as a function of energy ε, degeneracy y, and temperature T.

(b) Write an expression for the average energy $\langle \varepsilon \rangle$ versus T.

(c) For $\varepsilon/kT = 1$ and $y = 1$, compute the populations, or probabilities p_1, p_2, and p_3, of the three energy levels.

(d) Now if $\varepsilon = 2$ kcal mol^{-1} and $y = 1000$, find the temperature T_0 at which $p_1 = p_3$.

(e) Under the condition of part (d), compute p_1, p_2, and p_3 at temperature T_0. In what sense is this system cooperative?

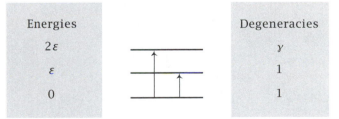

Figure 26.22 Diagram of three energy levels and degeneracies.

References

[1] JJ Binney, NJ Dowrick, AF Fisher and MEJ Newman. *The Theory of Critical Phenomena: an Introduction to Renormalization Group Theory.* Oxford University Press, New York, 1992.

[2] HE Stanley. *Introduction to Phase Transitions and Critical Phenomena.* Oxford University Press, New York, 1971.

[3] JM Yeomans. *Statistical Mechanics of Phase Transitions.* Clarendon Press, New York, 1992.

[4] R Kubo, with H Ichimura, T Usui and N Hashitsome. *Statistical Mechanics.* Elsevier Publishing Company, New York, 1965.

[5] BH Zimm and JK Bragg. *J Chem Phys* **31**, 526 (1959).

[6] BH Zimm, P Doty and K Iso. *Proc Natl Acad Sci USA* **45**, 1601 (1959).

[7] JA Schellman. *J Phys Chem* **62**, 1485 (1958).

Suggested Reading

Detailed treatments of Zimm–Bragg helix–coil theory:

CR Cantor and PR Schimmel, *Biophysical Chemistry.* WH Freeman, San Francisco (1980).

BH Zimm and JK Bragg, *J Chem Phys* **31**, 526 (1959).

Excellent treatments of phase transitions, critical phenomena, and Ising models:

JJ Binney, NJ Dowrick, AF Fisher and MEJ Newman, *The Theory of Critical Phenomena: an Introduction to Renormalization Group Theory*, Oxford University Press, New York, 1992.

N Goldenfeld, *Lectures on Phase Transitions and the Renormalization Group*, Addison-Wesley, Reading, 1992.

R Kubo, with H Ichimura, T Usui and N Hashitsome, *Statistical Mechanics.* Elsevier , New York, 1965.

HE Stanley, *Introduction to Phase Transitions and Critical Phenomena*, Oxford University Press, New York, 1971.

JM Yeomans, *Statistical Mechanics of Phase Transitions.* Clarendon Press, New York, 1992.

Excellent treatments of nucleation theory:

PG Debenedetti, *Metastable Liquids: Concepts and Principles*, Princeton University Press, Princeton (1996).

DV Ragone, *Thermodynamics of Materials*, Vol 2, Wiley, New York, 1995.

JW Mullin, *Crystallization*, 3rd edition, Butterworth–Heineman, Boston, 1993.

G. Wilemski, *J Phys Chem* **91**, 2492 (1987).

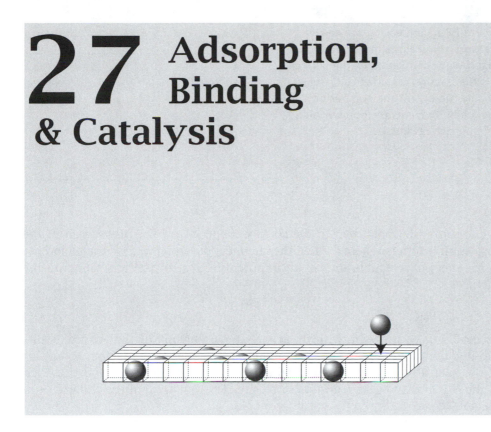

27 Adsorption, Binding & Catalysis

515-520

Binding and Adsorption Processes Are Saturable

Atoms and molecules bind, or adsorb, to surfaces. Adsorption has a key role in catalysis, filtration, oxidation and corrosion, chromatographic separation, and in the pharmacology of some drugs and metabolites. Atoms or molecules that bind to surfaces are called *adsorbates* or *ligands*. Increasing the concentration of an adsorbate will ultimately *saturate* a surface, filling all the available space.

The simplest model of binding and saturation is the Langmuir model. It applies not only to adsorbates on macroscopic solid surfaces but also to pH titration, the kinetics of enzyme reactions, and the transport of particles through biological membranes. The Langmuir model describes the balance between the energetic tendency of the particles to stick to the surfaces and the entropic tendency of the particles to gain translational freedom by escaping from them.

The Langmuir Model Describes Adsorption of Gas Molecules on a Surface

If a solid surface is available to a gas, some of the gas atoms or molecules may adsorb to it. Increasing the pressure of the gas will increase the number of gas molecules bound to the surface (see Figure 27.1). The simplest treatment of adsorption and binding is the Langmuir model, named after I Langmuir (1881–1957), an American chemist who won the 1932 Nobel prize in Chemistry for his work in surface science. Our aim is to compute how N, the number of

Figure 27.1 Increasing the gas pressure p increases the amount of gas adsorbed until it saturates the surface. θ is the fraction of surface sites filled by adsorbate; N is the number of ligands bound; A is the number of binding sites available.

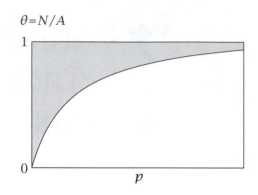

ligand molecules adsorbed to the surface, depends on the pressure p of the gas. Equilibrium is reached when the chemical potential μ_{gas} of the adsorbate in the gas phase equals the chemical potential μ_{surf} of the adsorbate on the surface,

$$\mu_{surf} = \mu_{gas}. \tag{27.1}$$

To model μ_{surf}, we begin with the translational entropy of the adsorbate distributed on the surface. A lattice approach is simplest. We consider the surface of the solid to be a two-dimensional lattice having A sites (Figure 27.2). Each of the N ligand molecules occupies one site. The density θ of the adsorbate on the surface is

$$\theta = \frac{N}{A}, \tag{27.2}$$

the fraction of surface sites that are filled with ligand. The number of arrangements of the particles on the surface lattice is given by Equation (1.19),

$$W = \frac{A!}{N!(A-N)!}. \tag{27.3}$$

To get the translational entropy, apply the Boltzmann relation and Stirling's approximation, Equation (4.26), to Equation (27.3),

$$\frac{S}{k} = \ln W = -N \ln \left(\frac{N}{A}\right) - (A-N) \ln \left(\frac{A-N}{A}\right)$$

$$\implies \frac{S}{Ak} = -\theta \ln \theta - (1-\theta) \ln(1-\theta). \tag{27.4}$$

Equation (27.4) has appeared before as the translational entropy of a lattice gas (Equation (7.9)) and as the entropy of mixing in a three-dimensional system (Equation (15.2)). Remarkably, this distributional entropy does not depend on whether a system is one-, two-, or three-dimensional. Nor does it depend on the arrangement of the binding sites. They could form a contiguous two-dimensional plane, or each binding site could be found on a different protein or polymer molecule.

Now we compute the adsorption free energy. If N particles stick to the surface, each with energy $w < 0$, then binding contributes an amount

$$U = Nw \tag{27.5}$$

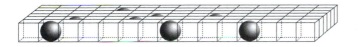

Figure 27.2 Two-dimensional lattice for the Langmuir model. There are N particles. The lattice has A sites.

to the total energy U of the system. If the internal degrees of freedom of the particle also change because particle orientations become restricted upon binding, or because new modes of vibrations are created in the surface, there is an additional contribution to the free energy, $-NkT \ln q_{surf}$. The free energy F of the adsorbed gas is

$$\frac{F}{AkT} = \frac{U - TS}{AkT} = \theta \ln \theta + (1 - \theta) \ln(1 - \theta) + \left(\frac{w}{kT}\right) \theta - \theta \ln q_{surf}. \quad (27.6)$$

The degrees of freedom are (N, A, T). They correspond to (N, V, T) in a single-component three-dimensional system of size V. To get the chemical potential μ_{surf} of the adsorbate at the surface, take the derivative with respect to N, holding A and T constant:

$$\frac{\mu_{surf}}{kT} = \left(\frac{\partial(F/kT)}{\partial N}\right)_{A,T}$$

$$= \ln\left(\frac{N}{A - N}\right) + \frac{w}{kT} - \ln q_{surf} = \ln\left(\frac{\theta}{1 - \theta}\right) + \frac{w}{kT} - \ln q_{surf}. \quad (27.7)$$

For the adsorbate molecules in the gas phase, the chemical potential is given by Equation (14.5),

$$\mu_{gas} = kT \ln\left(\frac{p}{p_{int}^\circ}\right), \quad (27.8)$$

where $p_{int}^\circ = q'_{gas}kT$, and q'_{gas} is the partition function of the gas molecules with the volume V factored out ($q_{gas} = q'_{gas}V$; see Equation (11.48)).

Substitute Equation (27.7) for μ_{surf}, and Equation (14.5) for μ_{gas}, into the condition for equilibrium given in Equation (27.1):

$$p = \left(\frac{q'_{gas}kT}{q_{surf}}\right)\left(\frac{\theta}{1 - \theta}\right) e^{w/kT}. \quad (27.9)$$

Collect terms together to form a *binding constant*,

$$K = \frac{q_{surf}}{q'_{gas}kT} e^{-w/kT} \quad (27.10)$$

(see Equation (13.10), page 237). Substituting K into Equation (27.9) gives

$$Kp = \frac{\theta}{1 - \theta}, \quad (27.11)$$

which rearranges to the **Langmuir adsorption equation,**

$$\theta = \frac{Kp}{1 + Kp}. \quad (27.12)$$

Figure 27.3 shows this function. For low pressures, Equation (27.12) predicts that the ligand coverage of the surface is linearly proportional to the gas pressure ($\theta = Kp$). This is called the Henry's law region, by analogy with the vapor

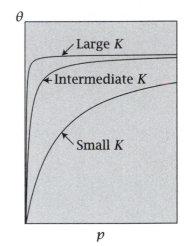

Figure 27.3 Increasing the binding affinity of the adsorbate for the surface increases the surface coverage, at fixed pressure.

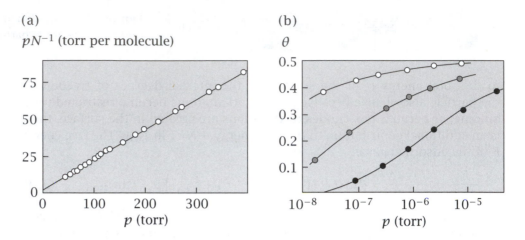

(a) pN^{-1} (torr per molecule)

(b) θ

Figure 27.4 (a) A plot of p/N $(= 1/(KA) + p/A)$ versus p for propane in zeolite pores at 273 K is linear, following the Langmuir model (see Equation (27.13)). Source: DM Ruthven, *Principles of Adsorption and Adsorption Processes*, Wiley, New York, 1984. Data are from DM Ruthven and KF Loughlin, *J Chem Soc Faraday Trans* **68**, 696 (1972). (b) Increasing temperature desorbs CO from palladium surfaces: (○) 383 K; (◉) 425 K; (●) 493 K. Note the logarithmic x-axis. Source: A Zangwill, *Physics at Surfaces*, Cambridge University Press, New York, 1988. Data are from G Ertl and J Koch, *Z Naturforsch* **25A**, 1906 (1970).

pressure of a mixture (see page 281). For adsorbates that have a strong affinity for the surface ($K \gg 1$), little pressure is required to saturate the surface (see Figure 27.3). The Langmuir equation predicts that the surface will saturate ($\theta \to 1$) at high pressures ($Kp \gg 1$) as all the sites become filled.

If you want to test whether your data follow the Langmuir model, there is a better functional form than Equation (27.12). Equation (27.12) requires data over a wide range of pressures to ensure that you reach saturation, and this can often be difficult to obtain. Instead, you can rearrange Equation (27.12) into a linearized form:

$$\frac{1}{\theta} = \frac{1 + Kp}{Kp} \implies \frac{p}{\theta} = \frac{1}{K} + p. \tag{27.13}$$

A plot of p/θ against p will yield a straight line with intercept $1/K$ when the Langmuir model applies. Figure 27.4(a) shows that the Langmuir model predicts the adsorption of propane in zeolite pores.

To determine how binding depends on temperature, take the derivative of Equation (27.10) (neglecting the small dependence $q(T)$):

$$\left(\frac{\partial \ln K}{\partial T} \right) = \frac{1}{T} \left(\frac{w}{kT} - 1 \right). \tag{27.14}$$

Or the temperature dependence can be expressed approximately with the van 't Hoff form, Equation (13.37),

$$\left(\frac{\partial \ln K}{\partial T} \right) = \left(-\frac{\Delta h}{kT^2} \right). \tag{27.15}$$

The energetic tendency of the particles to stick to the surface dominates at low temperatures, while the entropic tendency of the ligands to escape dominates at high temperatures (see Figure 27.4(b)).

The Langmuir Model Also Treats Binding and Saturation in Solution

Put ligand molecules of type X into a solution in equilibrium with some type of particle, polymer, or surface, P. Ligands can bind the particles to form a 'bound complex' PX,

$$X + P \overset{K}{\to} PX. \tag{27.16}$$

$[X]$ is the concentration of free ligand molecules, $[P]$ is the concentration of free particles (each of which has a single binding site), and $[PX]$ is the concentration of the bound complex. The binding, or association, equilibrium constant is

$$K = \frac{[PX]}{[P][X]}. \tag{27.17}$$

We want to know how θ, the fraction of binding sites on P that are filled by ligand, depends on the solution concentration of the ligand $[X]$. θ is the ratio (filled sites on P) / (empty + filled sites on P):

$$\begin{aligned} \theta &= \frac{[PX]}{[P] + [PX]} \\ &= \frac{K[P][X]}{[P] + K[P][X]} = \frac{Kx}{1 + Kx}. \end{aligned} \tag{27.18}$$

To simplify the notation slightly, we replaced $[X]$ by x. Equation (27.18) resembles the Langmuir Equation (27.12), but with the pressure replaced by the ligand concentration. Equation (27.18) is general, and applies whether the concentration x refers to mole fractions, molarities, molalities, or any other units. The units that you use for x determine the units of K.

For binding in solution, the equilibrium constant K in Equation (27.18) results from both the direct ligand–surface interactions and the interactions of the ligand and surface with the solvent.

A simple way to determine K from a binding curve $\theta(x)$ is to find the point $\theta = 1/2$. At that point, $x = 1/K$, as shown in Examples 27.1 and 27.2.

EXAMPLE 27.1 Finding a binding constant. Figure 27.5 shows the curve θ for binding *E. coli* cyclic AMP receptor protein to a 32-base-pair DNA molecule as a function of DNA concentration. If you assume Langmuir binding, the dashed lines give $K = 1/(16\,\text{nM}) \approx 6 \times 10^7 \,\text{M}^{-1}$. Note that in general the true Langmuir binding does not reach a plateau until very large values of x (compare Figure 27.6, in which the concentration of ligand is hundreds of μM to ensure saturation). The preferred method for determining a binding constant is to linearize the plot with Equations (27.13) or (28.20).

Figure 27.5 DNA molecule ligands bind to the *E. coli* cyclic AMP receptor protein. The fraction of filled sites θ is shown as a function of the DNA ligand concentration c_s. The dashed line indicates the ligand concentration at half-saturation of the binding sites (see Example 27.1). Source: T Heyduk and JC Lee, *Proc Natl Acad Sci USA* **87**, 1744 (1980).

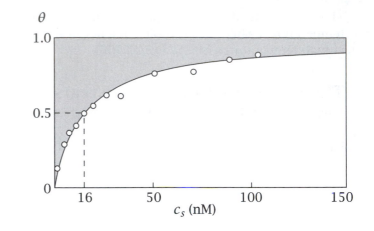

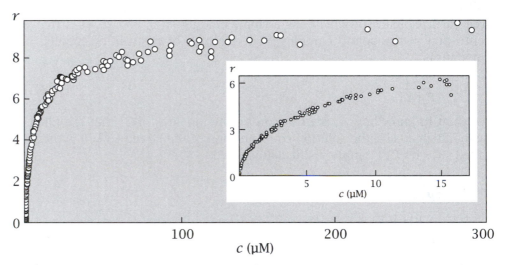

Figure 27.6 Laurate ion binding r to human serum albumin protein follows the Langmuir model. c is the laurate ion concentration. Note that to reach saturation requires a wide range of ligand concentrations. Source: AO Pedersen, B Hust, S Andersen, F Nielsen and R Brodersen, *Eur J Biochem* **154**, 545–552 (1986).

EXAMPLE 27.2 pH titration is an example of Langmuir binding. The Langmuir model is useful for treating pH titration (see Figure 27.7). In this case, protons are the ligands and the binding sites are acidic or basic groups on molecules. For an acid A^-, the equilibrium is $HA \rightarrow H^+ + A^-$, and the acid dissociation constant is $K_a = [H^+][A^-]/[HA]$. Varying the hydrogen ion concentration (pH) in the solution can change the protonation state of acids and bases according to $\theta = [H^+]K_a^{-1}/(1 + [H^+]K_a^{-1})$. If instead you use a *logarithmic* x-axis, the Langmuir function has a sigmoidal shape. This is the standard x-axis for *a pH titration curve*: $-\log x = -\log[H^+] = pH$. At the midpoint of titration, $\theta = 1/2$ and $Kx = 1$: half the acid or base groups have a proton on them, and the other half do not.

The Langmuir model also applies to chemical separation processes such as chromatography.

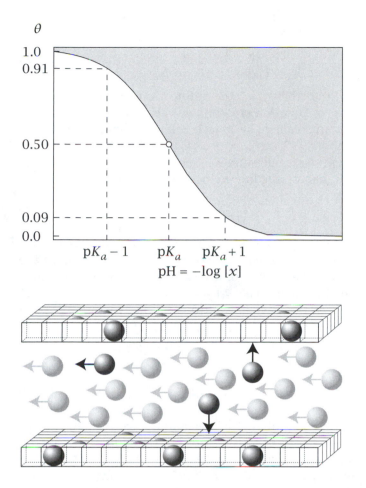

θ

1.0
0.91

0.50

0.09
0.0

$pK_a - 1$ pK_a $pK_a + 1$

pH = $-\log[x]$

Figure 27.7 A pH titration curve is an example of Langmuir binding. The fractional saturation curve θ is sigmoidal because the pH on the x-axis is proportional to the *logarithm* of the ligand concentration. When the pH is low, the concentration of hydrogen ions (the ligand) in solution is high, so the fractional binding θ to acid groups is high. As the available ligands (protons) decrease (pH increases), fewer protons bind to acid groups.

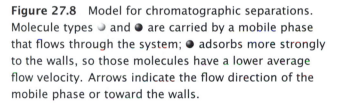

Figure 27.8 Model for chromatographic separations. Molecule types ⊙ and ● are carried by a mobile phase that flows through the system; ● adsorbs more strongly to the walls, so those molecules have a lower average flow velocity. Arrows indicate the flow direction of the mobile phase or toward the walls.

The Independent Site Model Describes the Principle of Adsorption Chromatography

Gas and liquid chromatographies are based on the principle of selective adsorption (see Figure 27.8). Solutes are carried along in the flow of a *mobile phase*, which can be a gas or a liquid. Assume that the flow is slow enough for the solute to be in exchange equilibrium between the mobile phase and adsorption sites on a *stationary phase*. We approximate the fluid flow profile by using two velocities: v_m is the velocity of the mobile phase fluid down the center of the channel, and v_s is the velocity of flow very close to the wall surface. To a first approximation, the particles that are close to the surface are stuck to the wall and don't move very quickly, so $v_s \approx 0$. The average flow velocity $\langle v \rangle$ for the solute passing through the chromatographic column is

$$\langle v \rangle = f_m v_m + f_s v_s \approx f_m v_m, \tag{27.19}$$

where f_m and $f_s = 1 - f_m$ are the fractions of the solute molecules that are in the mobile and stationary phases, respectively. The fraction f_m is given in terms of the concentration of solute c_m in the mobile phase, the concentration of solute c_s in the stationary phase, the volume V_m of the mobile phase, and

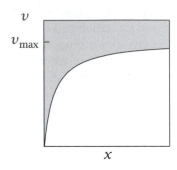

Figure 27.9 The rate v of enzyme catalysis saturates ($v = v_{max}$) at high substrate concentrations x.

the volume V_s of the stationary phase,

$$f_m = \frac{c_m V_m}{c_m V_m + c_s V_s} = \frac{1}{1 + (c_s V_s)/(c_m V_m)} = \frac{1}{1 + K\phi}. \tag{27.20}$$

$K = c_s/c_m$ is the partition coefficient of the solute from the mobile phase into the stationary phase (see Equation (16.39)). $\phi = V_s/V_m$ is a property of the chromatography apparatus called the '*phase ratio.*' Langmuir binding is reflected in the expression $f_s = 1 - f_m = K\phi/(1 + K\phi)$. One purpose of chromatography experiments is to measure K. Substituting Equation (27.20) into Equation (27.19) gives the average flow velocity of the solute,

$$\langle v \rangle = \frac{v_m}{1 + K\phi}. \tag{27.21}$$

Solute molecules that have a greater affinity for the stationary phase have larger K, so they have a smaller average flow velocity $\langle v \rangle$. Molecules that have no affinity for the stationary phase, $K = 0$, do not adsorb, and so they flow faster, at the rate of the mobile phase, $\langle v \rangle = v_m$.

Rearranging Equation (27.21) shows how chromatography can be used to measure partition coefficients. If the phase ratio is known, the partition coefficient is found from measuring the flow delay of a solute relative to the velocity of the carrier fluid,

$$K\phi = \frac{v_m - \langle v \rangle}{\langle v \rangle}. \tag{27.22}$$

So far, we have applied the Langmuir model to equilibria. Now we switch our attention to kinetics. Some *rate* processes saturate as a function of ligand concentration, resembling Langmuir behavior. An example is Michaelis–Menten kinetics in which the rate of enzymatic catalysis of biochemical reactions reaches a maximum speed as the substrate concentration reaches levels that saturate the available enzyme molecules (see Figure 27.9).

The Michaelis–Menten Model Describes Saturation in Rate Processes

Consider the reaction

$$E + S \xrightarrow{K} ES \xrightarrow{k_2} E + P, \tag{27.23}$$

where E is the enzyme, S is the substrate, ES is the enzyme with substrate bound, and P is the product. Assume that the rate indicated by k_2 is small enough for the binding of substrate to enzyme to be considered to be at equilibrium. The equilibrium constant is

$$K = \frac{[ES]}{[E]x}, \tag{27.24}$$

where $x = [S]$ is the substrate concentration, $[ES]$ is the concentration of the enzyme-substrate complex ES, and $[E]$ is the concentration of free enzyme. The velocity v of the reaction is defined in terms of the rate constant k_2 for

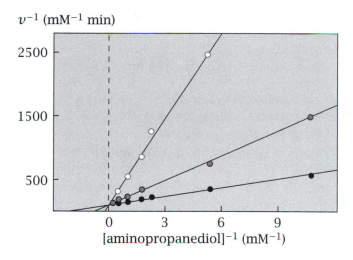

v^{-1} (mM^{-1} min)

[aminopropanediol]$^{-1}$ (mM^{-1})

Figure 27.10 A double-reciprocal plot for the initial phosphorylation rate v of glycerol kinase against the concentration of the substrate aminopropanediol, at various concentrations of MgATP. (○) 0.126 mM; (●) 0.504 mM; (●) 2.52 mM. Source: J Kyte, *Mechanism in Protein Chemistry*, Garland, New York, 1995. Adapted from WB Knight and WW Cleland, *Biochemistry* **28**, 5728 (1989).

the reaction step shown in Equation (27.23),

$$v = \frac{dP}{dt} = k_2[ES] = k_2K[E]x. \tag{27.25}$$

Because the total enzyme concentration is

$$E_T = [E] + [ES] = [E](1 + Kx), \tag{27.26}$$

you can get the velocity per enzyme molecule by dividing Equation (27.25) by Equation (27.26):

$$\frac{v}{E_T} = \frac{k_2K[E]x}{[E](1 + Kx)} = \frac{k_2Kx}{1 + Kx}. \tag{27.27}$$

Product formation is fastest when the enzyme is fully saturated by substrate, $Kx/(1 + Kx) = 1$. Then the maximum rate is $v_{\max} = k_2E_T$. Expressed in terms of this maximum rate, the velocity is

$$\frac{v}{v_{\max}} = \frac{Kx}{1 + Kx}. \tag{27.28}$$

The Langmuir equation appears in this rate process because of an underlying saturable binding step—the substrate must bind the enzyme before the reaction can be catalyzed.

The **Michaelis–Menten** Equation (27.28) is often expressed in a slightly different way, in terms of the *dissociation constant* or *Michaelis constant* $K_m = 1/K$. K_m is the inverse of the binding constant when the rate constant k_2 is small enough for the binding to be in equilibrium, as we have assumed. In those terms, you have

$$\frac{v}{v_{\max}} = \frac{x/K_m}{1 + x/K_m} = \frac{x}{x + K_m}. \tag{27.29}$$

You can linearize Equation (27.29) in the same way as the Langmuir equation (Equation (27.13)),

$$\frac{v_{\max}}{v} = \frac{K_m + x}{x} \implies \frac{1}{v} = \frac{K_m}{v_{\max}}\frac{1}{x} + \frac{1}{v_{\max}}. \tag{27.30}$$

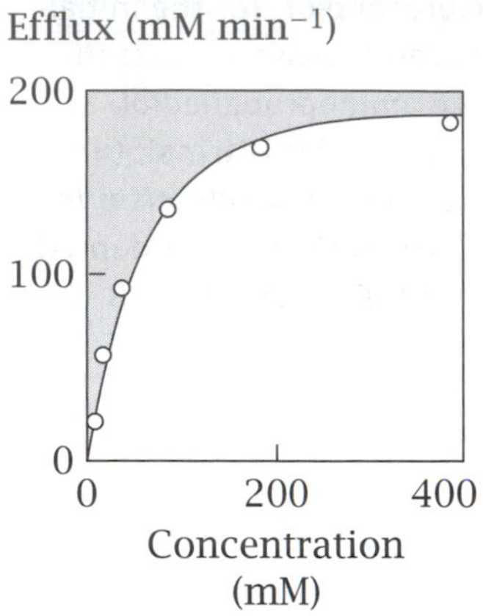

Efflux (mM min^{-1})

Figure 27.11 The rate of flow of galactose out of erythrocyte membranes as a function of the intracellular galactose concentration. Source: TF Weiss, *Cellular Biophysics*, Vol 1, MIT Press, Cambridge, 1996. From: H Ginsburg, *Biochim Biophys Acta* **506**, 119 (1978).

If a plot of $1/v$ against $1/x$ is linear, it supports the use of the Michaelis–Menten model. Then the slope will be K_m/v_{max} and the intercept will be $1/v_{max}$. Such plots are called *Lineweaver–Burk* or *double-reciprocal* plots. Figure 27.10 shows this type of plot for the phosphorylation of (S)-1-amino-2,3-propanediol by glycerol kinase.

Another example of saturable kinetics is the carrier-mediated transport of sugars and other molecules across cell membranes (see Figure 27.11).

Carrier Proteins Transport Solutes Across Biological Membranes

Ligands bind to receptor proteins in membranes, which can be carriers that ferry the ligands across from one side of a membrane to the other. Such processes are saturable. The transport of each solute molecule across the membrane can be described by a four-step cycle (see Figure 27.12).

1. THE CARRIER BINDS THE SOLUTE. A carrier protein binds to a ligand that is at concentration c_ℓ in the solution compartment on the left side of the membrane.

2. THE CARRIER HAULS THE SOLUTE ACROSS. The carrier protein transports the solute from the left to the right side of the membrane.

3. THE CARRIER UNLOADS THE SOLUTE. The carrier protein releases the solute into the compartment on the right at solute concentration c_r.

4. THE CARRIER RESETS. The carrier protein returns to its unbound state, ready to bind another solute molecule on the left.

We want to compute the flux of the solute across the membrane as a function of binding constants, rate constants, and ligand concentrations c_ℓ and c_r.

In step 2, the rate of solute transport from left to right is

$$J_{solute} = k_r[ES]_\ell - k_\ell[ES]_r, \tag{27.31}$$

where k_r and k_ℓ are the rate constants for the carrier-mediated transport of the solute to the right and left, respectively.

Because the total concentration E_T of the carrier proteins in the membrane is constant,

$$E_T = [ES]_\ell + [E]_\ell + [ES]_r + [E]_r. \tag{27.32}$$

At steady state, the rate J_{solute} of solute transport in step 2 must equal the return rate of the carrier $J_{carrier\ return}$ in step 4,

$$J_{solute} = J_{carrier\ return}. \tag{27.33}$$

Substitute Equation (27.31) for the solute, and a corresponding equation for the (empty) carrier return into Equation (27.33), to get

$$k_r[ES]_\ell - k_\ell[ES]_r = k_\ell[E]_r - k_r[E]_\ell. \tag{27.34}$$

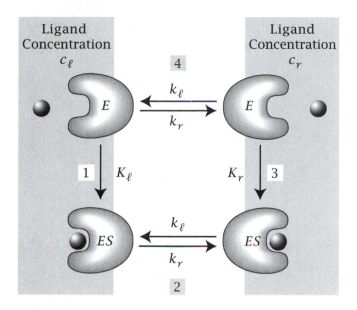

Figure 27.12 A model of the carrier-mediated transport of solutes across membranes. (1) The protein takes up solute on the left side of the membrane, (2) transports the ligand across to the right side, (3) releases the ligand, then (4) resets. Arrows (1) and (3) show the direction that defines the binding constant. Source: TF Weiss, *Cellular Biophysics*, Vol 1, MIT Press, Cambridge, 1996.

Collecting terms in Equation (27.34) and substituting Equation (27.32) gives

$$k_r([ES]_\ell + [E]_\ell) = k_\ell([ES]_r + [E]_r) = k_\ell(E_T - [ES]_\ell - [E]_\ell). \qquad (27.35)$$

Divide both sides of Equation (27.35) by k_ℓ and rearrange to get

$$E_T = \left(\frac{k_r}{k_\ell} + 1\right)([ES]_\ell + [E]_\ell). \qquad (27.36)$$

Now express E_T in terms of binding constants. To keep the discussion general for now, the binding affinity K_ℓ on the left side can differ from the binding affinity K_r on the right,

$$K_\ell = \frac{[ES]_\ell}{[E]_\ell c_\ell}, \quad \text{and} \quad K_r = \frac{[ES]_r}{[E]_r c_r}. \qquad (27.37)$$

Substitute Equation (27.37) for K_ℓ into Equation (27.36) to get

$$E_T = \left(\frac{k_r}{k_\ell} + 1\right)[ES]_\ell \left(1 + \frac{1}{K_\ell c_\ell}\right), \qquad (27.38)$$

and rearrange to

$$[ES]_\ell = E_T \left(\frac{k_\ell}{k_r + k_\ell}\right)\left(\frac{K_\ell c_\ell}{1 + K_\ell c_\ell}\right). \qquad (27.39)$$

Because the system is symmetric, you can interchange r's and ℓ's to get an additional expression,

$$[ES]_r = E_T \left(\frac{k_r}{k_r + k_\ell}\right)\left(\frac{K_r c_r}{1 + K_r c_r}\right). \qquad (27.40)$$

Finally, substitute Equations (27.39) and (27.40) into Equation (27.31) to get the solute flux in terms of the rate and binding constants,

Figure 27.13 Some mechanisms of surface catalysis: (a) a surface might help to distort a reactant into a product-like form; (b) a charged surface can help to pull apart a charged molecule; (c) two reactants can be drawn together into a confined space to react.

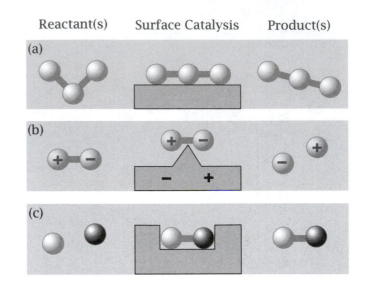

Reactant(s)   Surface Catalysis   Product(s)

(a)

(b)

(c)

$$J_{\text{solute}} = E_T \left(\frac{k_\ell k_r}{k_r + k_\ell} \right) \left(\frac{K_\ell c_\ell}{1 + K_\ell c_\ell} - \frac{K_r c_r}{1 + K_r c_r} \right). \tag{27.41}$$

Equation (27.41) illustrates how the flux reaches a plateau when the solute concentration is high enough to keep the carrier proteins shuttling at their maximum rate. The maximum flux is achieved when the quantity in the second parentheses of Equation (27.41) approaches 1,

$$J_{\text{max}} = E_T \left(\frac{k_\ell k_r}{k_r + k_\ell} \right). \tag{27.42}$$

In terms of J_{max}, the rate equation becomes

$$\frac{J_{\text{solute}}}{J_{\text{max}}} = \left(\frac{K_\ell c_\ell}{1 + K_\ell c_\ell} - \frac{K_r c_r}{1 + K_r c_r} \right). \tag{27.43}$$

Equation (27.43) gives a model for two different transport mechanisms: *passive transport* and *active transport*. In passive transport, ligands are moved downhill along their concentration gradients, while in active transport energy is used to carry ligands uphill against a concentration gradient. In the simplest model of passive transport, the binding affinity of the protein for the ligand is the same on both sides of the membrane, $K = K_\ell = K_r$. In passive transport, the flux J_{solute} is positive toward the right if $c_\ell > c_r$, and negative if $c_\ell < c_r$. The solute is always carried down its concentration gradient. For passive transport, when the ligand concentrations are small, $Kc_\ell \ll 1, Kc_r \ll 1$, you recover an expression similar to Fick's law (Equation (18.17)), in which the flux is linear in the concentration difference,

$$J_{\text{solute}} = J_{\text{max}} K (c_\ell - c_r). \tag{27.44}$$

For Fick's law processes, the quantity J_{max} in Equation (27.44) is replaced by the diffusion constant divided by the membrane thickness. However, for carrier transport, the permeability is defined as $P = KJ_{\text{max}}$ (compare with Equation (18.18)).

Equation (27.43) can also describe active transport. If the two binding affinities K_ℓ and K_r differ, then the ligands can be pumped uphill against a concentration gradient. The sign of J_{solute} is no longer determined by the sign of $c_\ell - c_r$ but by the sign of $K_\ell c_\ell - K_r c_r$. If the ligand concentration gradient is downhill toward the right ($c_\ell > c_r$), and if the left-side affinity is lower than the right-side affinity, $K_\ell c_\ell - K_r c_r < 0$, the flux is toward the left. This model is also useful for describing processes in which binding and release events are coupled to pumping, motor, or catalysis actions.

The Langmuir model also explains how surfaces that bind reactants can catalyze reactions. A catalyst binds a reactant and activates it in some way for conversion to product. Figure 27.13 illustrates some of the possible mechanisms in which binding to a catalyst increases a reaction rate. When an appropriate surface binds a reactant, surface forces can help to pull the reactant into a structure resembling the product. This mechanism is called *strain*. Or, for catalyzing a bond-breaking reaction, surface forces can help to pull the atoms apart. For catalyzing bond-forming reactions, $A + B \rightarrow AB$, the surface can provide a good binding site for both A and B, bringing them into positions and/or orientations that are favorable for reaction.

Sabatier's Principle Describes How Surfaces Can Stabilize Transition States

We now explore *Sabatier's principle*, named for the French chemist P Sabatier (1854–1941), who won the 1912 Nobel Prize in Chemistry for his work on catalysis. This principle explains why the best catalysts are surfaces that bind reactants neither too weakly nor too strongly.

Consider the reaction

$$A \rightarrow B.$$

Suppose this reaction is catalyzed by a surface S to which reactant A adsorbs. The surface 'activates' the A molecules in some way, making them more 'B-like.' How does the strength of binding A to the surface affect the rate of the reaction? Divide the reaction into two steps:

$$A + S \overset{K}{\rightarrow} AS \overset{k_2}{\rightarrow} B + S$$

Step 1 is the adsorption of A to the surface to reach an activated state AS. We use the Langmuir model for this step. Step 2 involves the conversion of A to B and the desorption of B from the catalyst. This second step involves crossing an energy barrier (see Figure 27.15), which occurs with rate constant k_2. At low affinities, changing the surface S to increase its affinity for A accelerates the reaction because more AS molecules are formed. But if the surface binds too tightly, it won't allow the product to escape, so further increases in affinity slow the reaction. Here's a simple model, due to AA Balandin [1, 2, 3].

Following Equations (27.25) and (19.14), the overall rate r of the reaction will be the product (the rate k_2 of conversion plus desorption) × (the number of molecules in the state AS), so

(a) Step 1

ln (amount bound)

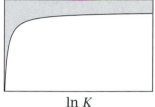

ln K

(b) Step 2

ln (rate)

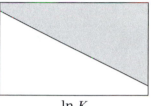

ln K

(c) Overall Reaction

ln (rate)

ln K

Figure 27.14 In a transformation of A to B catalyzed by surface S: (a) the amount of A bound to S increases with binding affinity K; (b) the rate of release of B decreases with binding affinity K (Equation (27.47)); (c) the net result of both steps, binding and release, is an increase then decrease of overall rate with K (Equation (27.48)).

Figure 27.15 Definitions of the energies used in the model of Sabatier's principle. ΔG_1 is the free energy of binding A to the surface; ΔG_2 is the free energy of converting the bound complex AS to B, and releasing it. $\Delta G = \Delta G_1 + \Delta G_2$ is the free energy of converting reactant A to product B. E_a is the activation energy of conversion.

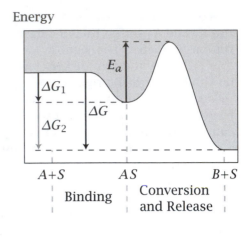

Energy

$A+S$ AS $B+S$

Binding   Conversion and Release

$$\frac{r}{A_S} = k_2\theta = \frac{k_2 Kp}{1 + Kp}, \tag{27.45}$$

where K is the affinity of ligand A for the surface, p is the gas pressure of A, A_s is the number of surface sites, and θ is the fraction of the surface that is covered by A molecules. Now use the Arrhenius rate law, Equation (19.16). Express the reaction rate k_2 in terms of the activation energy E_a for the conversion of A to B and the desorption of B,

$$k_2 = c_1 e^{-E_a/kT}, \tag{27.46}$$

where c_1 is a constant. If the Brønsted Equation (19.45) holds, relating rates to equilibria, you have $E_a = a\Delta G_2 + b$ (see Figure 19.16), where $a > 0$ and $b > 0$ are constants. ΔG_2 is the free energy change for converting AS to B and releasing B from the surface (see Figure 27.15).

Figure 27.15 also defines ΔG_1 for the binding of A to S. Because the free energy difference ΔG for the conversion of A to B is independent of the catalyst, the free energies are related by $\Delta G_2 = \Delta G - \Delta G_1$. When this is expressed in terms of the binding free energy ΔG_1, you have $E_a = a(\Delta G - \Delta G_1) + b = (\text{constant} - a\Delta G_1)$, and Equation (27.46) becomes

$$k_2 = c_2 e^{a\Delta G_1/kT} = c_2 K^{-a}, \tag{27.47}$$

where $K = e^{-\Delta G_1/kT}$ and $c_2 = c_1 e^{-(a\Delta G + b)/kT}$ is a constant, independent of affinity K.

Equation (27.47) shows that the desorption rate k_2 decreases as the binding affinity of A for the surface increases (K increases), because $a > 0$. Substituting Equation (27.47) for k_2 into Equation (27.45) gives the overall rate of reaction in the presence of the surface,

$$\frac{r}{A_S} = \frac{c_2 K^{1-a} p}{1 + Kp}. \tag{27.48}$$

Figure 27.14 shows how the binding increases with the affinity K, and the release rate decreases, leading to the overall result (c) that there is an optimal binding affinity for catalytic effectiveness. Figure 27.16 shows a plot of the rate of formic acid dehydrogenation as a function of the type of metal surface that is used as a catalyst. Such plots are sometimes called *volcano curves*

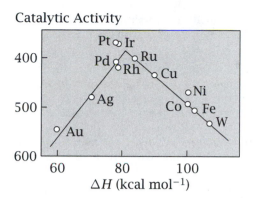

Figure 27.16 A volcano curve shows the catalytic activities of metal surfaces in formic acid dehydrogenation, as a function of the enthalpy ΔH of formation of the metal formates. The enthalpies represent the strength of the substrate–surface interaction. Source: BC Gates, *Catalytic Chemistry*, Wiley, New York, 1992. Data are from WJM Rootsaert and WMH Sachtler, *Z Phys Chem NF* **26**, 16 (1960).

because of their inverted 'V' shapes. The main idea expressed by this model is that the best catalysts bind neither too weakly nor too strongly. Similarly, the most effective enzymes have a relatively high affinity for the transition states of the reaction, but their binding affinities for the reactants should be neither too weak nor too strong [4].

Summary

Ligands bind to surfaces and to sites on molecules. Increasing the concentration of a ligand increases the coverage of the available binding sites and leads to saturation. The Langmuir model of independent site binding is the simplest treatment of binding and saturation. It accounts for gases sticking to metals, pH titration, rates of catalysis, and the saturable rates of transport through cell membranes. A catalyst should bind strongly enough to capture the reactant, but weakly enough to release the product. In the next chapter we go beyond the model of binding to independent sites, and consider cooperative binding processes.

Problems

1. Adsorption in gas masks.
You have a gas mask that absorbs a certain toxic compound with absorption constant $K = 1000 \, \text{atm}^{-1}$ at $T = 25 \, °C$, and binding energy $w = -5 \, \text{kcal mol}^{-1}$. In the desert, where the temperature is $T = 35 \, °C$, what is the adsorption constant?

2. Binding in solution.
A ligand A in solvent S binds to a cavity in protein P, as shown in Figure 27.17.

(a) Write an expression for the log of the association constant, $\ln K_{assoc}$, for the case in which the binding pocket is originally occupied by a solvent molecule which is then displaced by the ligand.

(b) Write an expression for the association constant, $\ln K_{assoc}$, for the case in which the binding pocket is originally an empty cavity, not occupied by a solvent molecule, before occupation by the ligand.

(c) If the magnitude of the solvent–solvent attraction dominates all the other interactions, and is equal to $w_{ss} = -4 \, \text{kcal mol}^{-1}$ at room temperature, then which model (a or b), leads to the stronger ligand binding, and by how much?

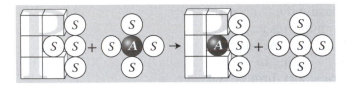

Figure 27.17 Ligand A displaces solvent S when it binds to a cavity in protein P.

3. Polymer adsorption.
You are designing a new drug delivery polymer to which you plan to attach an active compound covalently. You have three polymers, A, B, and C, with different properties, but they all have the drawback that they adsorb to some unidentified site in the body. Adsorption is of the Langmuir type. At a concentration of 1 mM, half of the body's binding sites are filled with polymer A. At a concentration of 5 mM, half of the sites are filled with B. At 10 mM, half of the sites are filled with C. You have the further problem that you can attach only one-tenth as much drug to polymer A as to either B or C.

(a) What is the desorption (dissociation) constant for removing polymer B from the body's binding sites?

(b) What is the best approach for delivering the maximum amount of drug with the minimum amount of adsorption?

4. Chromatographic separations.
A typical chromatography column is 15 cm long, has a phase ratio $\phi = 0.2$, and has a mobile phase velocity of 2 mm s^{-1}.

(a) What is the velocity of a solute that has a partition coefficient $K = 20$ for binding to the stationary phase?

(b) Suppose you can distinguish two peaks as close as 2 mm together when they reach the end of the column. If one peak corresponds to $K = 20$, what is the partition coefficient for the other peak?

5. How binding affects catalysis.
When a catalyst binds a substrate with binding constant K, the reaction rate r is given by

$$r = \frac{k_o K^{1-a} p}{1 + K p},$$

where p is the gas pressure of the substrate, and k_o and a are constants. Plot $r(K)$ for $a < 0$ and explain what it means.

6. The binding affinity of a ligand to a protein.
Figure 27.18 shows the binding constant K for binding cytidine monophosphate to ribonuclease A, as a function of $1/T$ where T is temperature. For $T = 300 \, \text{K}$, compute

(a) the enthalpy,

(b) the Gibbs free energy, and

(c) the entropy of binding.

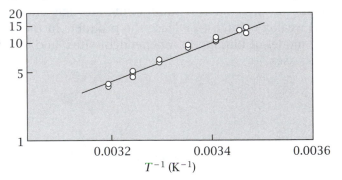

Figure 27.18 Source: IM Klotz, *Ligand–Receptor Energetics*, Wiley, New York, 1997. Data are from H Naghibi, A Tamura and JM Sturtevant, *Proc Natl Acad Sci USA* **92**, 5597 (1995).

7. Ligand–protein binding thermodynamics.
A drug D binds to a protein P with binding constant $K_1 = 10^6$ at $T_1 = 300 \, \text{K}$ and with $K_2 = 1.58 \times 10^6$ at $T_2 = 310 \, \text{K}$.

(a) Calculate $\Delta G°$ for binding at $T = T_1 = 300 \, \text{K}$.

(b) Assuming that the binding enthalpy $\Delta H°$ is independent of temperature, calculate $\Delta H°$.

(c) Assuming that the binding entropy $\Delta S°$ is independent of temperature, calculate $\Delta S°$ at $T_1 = 300$ K.

8. Binding of insulin to its receptor. Figure 27.19 shows the binding of insulin to the insulin receptor in a detergent solution. Use the Langmuir model to estimate the binding constant.

Bound Insulin (counts per minute)

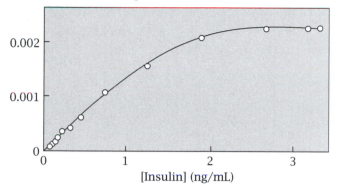

[Insulin] (ng/mL)

Figure 27.19 Source: J Kyte, *Mechanism in Protein Chemistry*, Garland, New York, 1995. Adapted from WB Knight and WW Cleland, *Biochemistry* **28**, 5728–5734 (1989).

9. Ion flow through membrane pores. Figure 27.20 shows the passive flow of sodium ion current through a membrane channel.

(a) Make a model for the ion current $i(x)$ as a function of sodium ion concentration $x = [Na^+]$.

(b) Write an expression that might linearize the data.

i (pA)

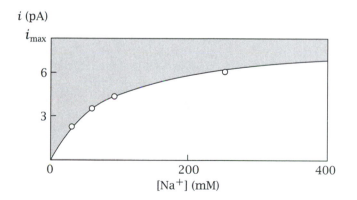

$[Na^+]$ (mM)

Figure 27.20 Source: B Hille, *Ionic Channels of Excitable Membranes*, Sinauer Associates, Sunderland, 1984. Data are from R Coronado, RL Rosenberg and C Miller, *J Genl Physiol* **76**, 425 (1980).

References

[1] RI Masel. *Principles of Adsorption and Reaction on Surfaces*, Wiley, New York, 1996.

[2] AA Balandin. *Adv Catal*, **10**, 96 (1958).

[3] AA Balandin. *Adv Catal*, **19**, 1 (1969).

[4] A Fersht. *Enzyme Structure and Mechanism*, 2nd edition, WH Freeman, New York, 1985.

Suggested Reading

Excellent treatments of surfaces, adsorption, and catalysis are:

A Adamson, *Physical Chemistry of Surfaces*, 5th edition, Wiley, New York, 1990.

RPH Gasser, *An Introduction to Chemisorption and Catalysis by Metals*, Oxford University Press, New York (1985).

MJ Jaycock and GD Parfitt, *Chemistry of Interfaces*, Halstead Press, New York, 1981.

RI Masel, *Principles of Adsorption and Reaction on Surfaces*, Wiley, New York, 1996.

H Maskill, *The Physical Basis of Organic Chemistry*, Oxford University Press, New York, 1985.

GD Parfitt and CH Rochester, *Adsorption from Solution at the Solid/Liquid Interface*, Academic Press, New York, 1983.

DM Ruthven, *Principles of Adsorption and Adsorption Processes*, Wiley, New York, 1984.

RA van Santen and JW Niemantsverdriet, *Chemical Kinetics and Catalysis*, Plenum Press, New York, 1995.

GA Somorjai, *Introduction to Surface Chemistry and Catalysis*, Wiley, New York, 1994.

A Zangwill, *Physics at Surfaces*, Cambridge University Press, New York, 1988.

An exceptionally detailed and careful analysis of membrane transport is:

TF Weiss, *Cellular Biophysics*, Vol 1, MIT Press, Cambridge, 1996.

Concise introductions to ligand–protein binding include:

IM Klotz, *Ligand–Receptor Energetics: A Guide for the Perplexed*, Wiley, New York, 1997.

DJ Winzor and WH Sawyer, *Quantitative Characterization of Ligand Binding*, Wiley–Liss, New York, 1995.

J Wyman and SJ Gill, *Binding and Linkage: Functional Chemistry of Biological Macromolecules*, University Science Books, Mill Valley, 1990.

Lattice model of adsorption:

TL Hill, *Introduction to Statistical Thermodynamics*, Addison-Wesley, Reading, 1960.

For catalysis in biology see:

A Fersht, *Enzyme Structure and Mechanism*, 2nd edition, WH Freeman, New York, 1985.

WP Jencks, *Catalysis in Chemistry and Enzymology Publications*, Dover, New York, 1987.

J Kyte, *Mechanism in Protein Chemistry*, Garland, New York, 1995.

28 Multi-site & Cooperative Ligand Binding

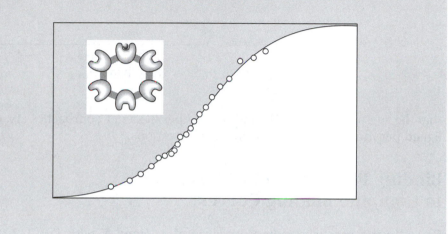

Some binding and association processes are too complex for the independent-site model of Chapter 27. For example, saturation behavior is not observed for gases that form multilayers on a surface. And sometimes binding at different sites may be *cooperative*, rather than independent. That is, the binding at one site may influence the binding at another. An example is oxygen binding to hemoglobin. Hemoglobin can hold up to four oxygen molecules. The second O_2 molecule has more affinity for the protein than the first. Binding, enzyme catalysis, and transport processes can also be inhibited or regulated by different molecules. For example, changing pH changes oxygen binding to hemoglobin.

In this chapter we devise models for interpreting binding experiments using *binding polynomials*. Experiments give you the average number of ligands bound, for example to a protein, as a function of the solution concentration of the ligand. To interpret binding experiments, you make a model and derive its *binding polynomial*. You start with the simplest models (models with the fewest parameters) that are consistent with the data and any known symmetries. From the binding polynomial you calculate the number of bound ligand molecules as a function of the bulk ligand concentration, adjusting the model parameters to best fit the data. If that model does not fit the data to within acceptable errors, you try a different model.

Figure 28.1 shows data for the binding of a ligand (TNP-ATP, trinitrophenyl adenosine triphosphate) to DnaB helicase protein, a hexamer of six identical subunits. This protein is involved in the replication of DNA. The figure shows

Figure 28.1 A binding curve. Nucleotide ligands bind to DnaB, a hexameric protein. The average number ν of ligands bound per hexamer protein is a function of the ligand concentration in solution. Source: W Bujalowski and MM Klonowska, *Biochemistry* **32**, 5888–5900 (1993).

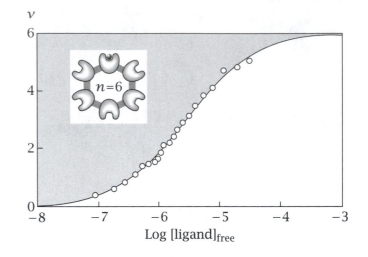

that the average number of ligands bound per hexamer approaches six, or one ligand per monomer, at high ligand concentrations.

Binding Polynomials Are Used to Compute Binding Curves

Consider the *multiple binding equilibria* of a ligand X with $i = 1, 2, 3, \ldots, t$ different binding sites on a protein or polymer P:

$$P + X \xrightarrow{K_1} PX_1,$$

$$P + 2X \xrightarrow{K_2} PX_2,$$

$$\vdots$$

$$P + iX \xrightarrow{K_i} PX_i,$$

$$\vdots$$

$$P + tX \xrightarrow{K_t} PX_t.$$

Each equilibrium constant K_i is defined by

$$K_i = \frac{[PX_i]}{[P]x^i},$$

where $x = [X]$ is the concentration of free ligand, $[P]$ is the concentration of free P molecules, and $[PX_i]$ is the concentration of those P molecules that have i ligands bound. There can be many different ligation states of P molecules in solution at the same time. Some P molecules contain no ligand; their concentration is $[P]$. Some P molecules contain 1 ligand molecule; their concentration is $[PX_1]$. Some P molecules contain i ligands; their concentration is $[PX_i]$.

What are the fractions of all the P molecules in solution that are i-liganded? To get these quantities, you first need the sum of the concentrations of all of the species of P, which is expressed in terms of *the binding polynomial Q*,

$$Q[P] = [P] + [PX_1] + [PX_2] + \ldots + [PX_t]$$

$$= [P](1 + K_1x + K_2x^2 + K_3x^3 + \ldots + K_tx^t)$$

$$\implies Q = 1 + K_1x + K_2x^2 + K_3x^3 + \ldots + K_tx^t$$

$$= \sum_{i=0}^{t} K_ix^i. \tag{28.1}$$

The fraction of P molecules that are in the i-liganded state is $[PX_i]/(Q[P])$.

A key quantity for comparing models with experiments is the average number $v = \langle i \rangle$ of ligands bound per P molecule, as a function of the ligand concentration x. To compute this average, use $\langle i \rangle = \sum i p(i)$, where $p(i) = [PX_i]/(Q[P])$ is the probability that a P molecule is in state i, to get

Average # ligands/mole

$$v = \langle i \rangle = \frac{[PX_1] + 2[PX_2] + 3[PX_3] + \ldots + t[PX_t]}{[P] + [PX_1] + [PX_2] + [PX_3] + \ldots + [PX_t]}$$

$$= \frac{[P](K_1x + 2K_2x^2 + 3K_3x^3 + \ldots + tK_tx^t)}{[P](1 + K_1x + K_2x^2 + K_3x^3 + \ldots + K_tx^t)}$$

$$= Q^{-1} \sum_{i=0}^{t} iK_ix^i$$

$$= \frac{x}{Q}\frac{dQ}{dx} = \frac{d\ln Q}{d\ln x} \tag{28.2}$$

(compare Equations (4.13) and (4.15) and the box on page 502).

The binding polynomial Q is the sum over all possible ligation states. It resembles a partition function, which is a sum over all possible energy levels. Equation (28.2) shows that the derivative of Q with respect to x gives an average number of ligands bound per P molecule. This corresponds to taking the derivative of a partition function with respect to temperature to get the average energy per molecule, Equation (10.31).

Other properties can also be computed from the binding polynomial. For example, you can find the *fluctuations* in the number of ligands bound from the second moment $\langle i^2 \rangle$. Take the second derivative of the binding polynomial,

$$x\frac{d}{dx}\left(x\frac{dQ}{dx}\right) = \sum_{i=0}^{t} i^2K_ix^i. \tag{28.3}$$

Then the second moment, which is used to determine the variance in the number of ligands bound, is

$$\langle i^2 \rangle = \frac{x}{Q}\frac{d}{dx}\left(x\frac{dQ}{dx}\right) = \frac{\displaystyle\sum_{i=0}^{t} i^2K_ix^i}{\displaystyle\sum_{i=0}^{t} K_ix^i}. \tag{28.4}$$

Binding polynomials describe ligand binding cooperativity. The simplest case of cooperative binding is a molecule P that has two binding sites that are not independent of each other.

The Simplest Model of Binding Cooperativity Involves Two Binding Sites

Suppose that a ligand molecule of type X can bind to P at either site a with an affinity described by the equilibrium binding constant K_a, or at site b with affinity K_b, or that two ligand molecules X can bind at the two sites at the same time with affinity K_c:

$$P + X \xrightarrow{K_a} P_a X, \qquad P + X \xrightarrow{K_b} P_b X, \qquad \text{and} \qquad P + 2X \xrightarrow{K_c} P_{ab} X_2.$$

The equilibrium constants are

$$K_a = \frac{[P_a X]}{[P]x}, \qquad K_b = \frac{[P_b X]}{[P]x}, \qquad \text{and} \qquad K_c = \frac{[P_{ab} X_2]}{[P]x^2},$$

where $x = [X]$ is the concentration of free ligand, $[P]$ is the concentration of P molecules that have no ligands bound, $[P_a X]$ and $[P_b X]$ are the concentrations of the two types of singly liganded P molecules, and $[P_{ab} X_2]$ is the concentration of doubly liganded P.

We want to find v, the fraction of binding sites on P that are filled, as a function of the ligand concentration x. The sum over all the ligation states of P is

$$[P]Q = [P] + [P_a X] + [P_b X] + [P_{ab} X_2]$$
$$= [P](1 + K_a x + K_b x + K_c x^2), \tag{28.5}$$

where
$$Q = 1 + K_a x + K_b x + K_c x^2 \tag{28.6}$$

is the binding polynomial for this process. The fraction of P molecules that have a single ligand at site a is $[P_a X]/Q[P]$, and the total fraction of P molecules that have a single ligand bound is $([P_a X] + [P_b X])/Q[P]$. The fraction of P molecules that have two ligands bound is $[P_{ab} X_2]/Q[P]$.

The average number of ligands bound per P molecule, $0 \leq v \leq 2$, is

$$v(x) = \frac{[P_a X] + [P_b X] + 2[P_{ab} X_2]}{[P]Q}$$
$$= \frac{K_a x + K_b x + 2K_c x^2}{1 + K_a x + K_b x + K_c x^2} = \frac{d \ln Q}{d \ln x}, \tag{28.7}$$

because $d \ln Q/d \ln x = (x/Q)(dQ/dx)$ and $dQ/dx = K_a + K_b + 2K_c x$.

When the two sites are independent, $K_c = K_a K_b$. Then there is no cooperativity so Equation (28.6) reduces to

$$Q = (1 + K_a x + K_b x + K_a K_b x^2) = (1 + K_a x)(1 + K_b x), \tag{28.8}$$

and the average number of ligands bound is given by

$$\nu = \frac{x(K_a + K_b + 2K_aK_bx)}{(1 + K_ax)(1 + K_bx)} = \frac{K_ax}{1 + K_ax} + \frac{K_bx}{1 + K_bx}, \tag{28.9}$$

which corresponds to two Langmuir binding events, one at each site.

Binding *cooperativity* occurs when $K_c \neq K_aK_b$. *Positive cooperativity* occurs when $K_c > K_aK_b$, that is, when two ligands bind with higher affinity than would be expected from the two individual binding affinities alone. *Negative cooperativity* or *anti-cooperativity* occurs when $K_c < K_aK_b$. Then the two ligands bind with lower affinity than would be expected from the individual affinities. A *cooperativity free energy* is $\Delta G = -kT \ln[K_c/(K_aK_b)]$. It arises from the interdependence of the two binding events.

The treatment above is *site-based*. It is useful when experiments provide all three quantities K_a, K_b, and K_c. But this amount of information is often not available. Notice that the terms K_a and K_b always appear as a sum $K_a + K_b$ in Equation (28.7). If you use the binding constants as parameters to fit experimental data of ν versus x, and if you have no additional information, you won't be able to choose K_a and K_b independently. You can only determine their sum, so you are only justified in using two independent parameters to fit the data, not three. The *stoichiometry-based* model in the next section uses only two parameters to describe the same two-site binding process.

The Stoichiometric Approach to Two Binding Sites

Suppose that one ligand X can bind to P with equilibrium constant K_1, and the second ligand X can bind with equilibrium constant K_2:

$$P + X \xrightarrow{K_1} PX_1, \quad \text{and} \quad PX_1 + X \xrightarrow{K_2} PX_2.$$

We are no longer assuming that we can distinguish sites a and b. When the reactions are written in this form, the equilibrium constants are

$$K_1 = \frac{[PX_1]}{[P]x} \quad \text{and} \quad K_2 = \frac{[PX_2]}{[PX_1]x} = \frac{[PX_2]}{K_1[P]x^2},$$

where $x = [X]$ is the concentration of free ligand, $[P]$ is the concentration of free P molecules, $[PX_1]$ is the concentration of all P molecules with one ligand bound and $[PX_2]$ is the concentration of P molecules with two ligands bound. Now the binding polynomial is

$$Q = 1 + K_1x + K_1K_2x^2. \tag{28.10}$$

In the stoichiometry-based approach, the binding isotherm is

$$\nu(x) = \frac{d\ln Q}{d\ln x} = \frac{K_1x + 2K_1K_2x^2}{1 + K_1x + K_1K_2x^2}. \tag{28.11}$$

When this stoichiometry-based model fits your data, it is preferable to the site-based model because it requires one less parameter. Example 28.1 demonstrates both approaches.

Figure 28.2 Glycine has two titratable binding sites for protons. When the concentration of ligand [H$^+$] is low (high pH), the number of protons bound is small ($v \to 0$). As the ligand concentration increases (toward low pH), the glycines become saturated and have $v \to 2$ protons each. Source: J Wyman and SJ Gill, *Binding and Linkage*, University Science Books, Mill Valley, CA (1990).

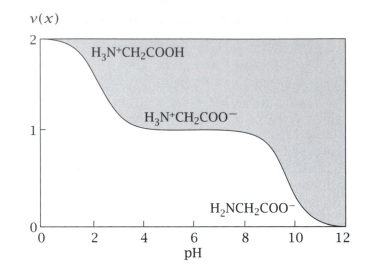

EXAMPLE 28.1 Two binding sites for titratable protons on a glycine. The amino acid glycine has two sites for binding a proton. At the amino end

$$R\text{—}NH_2 + H^+ \xrightarrow{K_a} R\text{—}NH_3^+,$$

and at the carboxyl end

$$R\text{—}COO^- + H^+ \xrightarrow{K_b} R\text{—}COOH.$$

Figure 28.2 shows the binding curve, $v(x)$, the average number of protons bound, versus x. (In the figure, $x = $ [H$^+$] is plotted as pH $= -\log$[H$^+$].) At high pH (low [H$^+$]), no ligands are bound. At intermediate pH, one ligand is bound. And at low pH (high [H$^+$]), both binding sites are filled with ligand.

The two separate transitions in Figure 28.2 show that the two sites have very different affinities, $K_1 \gg K_2$. In that case, for low to intermediate ligand concentrations ($K_2 x \ll 1$), Equation (28.11) gives $v(x) \approx K_1 x/(1 + K_1 x)$. You can find K_1 from the midpoint ($v = 1/2$) of the transition on the right side of Figure 28.2, where $x = K_1^{-1} = 2 \times 10^{-10}$ M. So $K_1 = 5 \times 10^9$ M^{-1}. To get K_2, focus on the second ligand (proton), from $v = 1$ to $v = 2$. For this ligand, it is most convenient to look at the quantity $v - 1$, which ranges from 0 to 1. Equations (28.10) and (28.11) give

$$v - 1 = \frac{K_1 x + 2K_1 K_2 x^2 - Q}{Q}. \tag{28.12}$$

Because the second titration takes place for intermediate ligand concentrations x, where $K_1 x \gg 1$, $K_2 x \approx 1$, and $Q \approx K_1 x + K_1 K_2 x^2$, Equation (28.12) gives

$$v - 1 \approx \frac{K_1 K_2 x^2}{K_1 x + K_1 K_2 x^2}$$

$$= \frac{K_2 x}{1 + K_2 x}.$$

The midpoint of this second titration is given from Figure 28.2 by $K_2^{-1} = 5 \times 10^{-3}$ M. So $K_2 = 2 \times 10^2$ M^{-1}. K_1 and K_2 are the two affinities from the stoichiometric model. To obtain the same binding isotherm from the site-based model, equate the coefficients of the binding polynomials,

$$Q_{site} = 1 + (K_a + K_b)x + (K_a K_b c)x^2$$
$$= Q_{stoichiometric} = 1 + K_1 x + K_1 K_2 x^2, \qquad (28.13)$$

where $c = K_c/(K_a K_b)$ defines the degree of cooperativity. If a is the high-affinity site, comparison of the first terms of Equation (28.13) gives

$$K_a \approx K_1 = 5 \times 10^9 \text{ M}^{-1}.$$

Comparing the second terms of Equation (28.13) gives

$$K_b c = K_2 = 2 \times 10^2 \text{ M}^{-1}.$$

Independent experiments show that $K_b = 2 \times 10^4$ M^{-1} [1], leading to the conclusion that $c = 10^{-2}$. This negative cooperativity has been attributed to electrostatic repulsion between the two sites [1].

'Cooperativity' Describes the Depletion of Intermediate States

In Chapter 26, we distinguished between two types of transition (see Figure 28.3), two-state and one-state. At the midpoint of a two-state transition, two states are populated, and the states between them have small populations. At the midpoint of a one-state transition, there is a single broad population distribution, so intermediates are highly populated.

You can apply these same classifications to binding cooperativity. Using the binding polynomial Equation (28.10), you can express the relative populations of the three states as: no ligands, $1/Q$; one ligand, $K_1 x/Q$; and two ligands, $2K_1 K_2 x^2/Q$. Figure 28.4(a) shows a one-state transition as a function of ligand concentration. At the left (small x) most P molecules are unliganded. In the middle (intermediate x) three states are equally populated: zero-liganded, one-liganded, and two-liganded. At the right (high x) most P molecules have two ligands. Figure 28.4(b) shows two-state cooperativity: at the midpoint concentration in x, most P molecules are either zero-liganded or two-liganded. There is little population of the one-liganded intermediate state. More quantitative details are given in Example 28.2.

EXAMPLE 28.2 Binding intermediate states and cooperativity. Figure 28.5 shows three different cases of systems that may have zero, one or two ligands bound. In each case, suppose that the midpoint of the binding isotherm is at $x_{mid} = 0.01$ (see Figure 28.3).

In case (a), $K_1 = 1$ and $K_2 = 10^4$, so Equation (28.10) gives $Q = 2.01$. This is an example of two-state binding. At low ligand concentrations, no ligand is bound. At high ligand concentrations, two ligands are bound. At the midpoint of the transition, singly liganded 'intermediate-state' P molecules are much rarer than unliganded or doubly liganded P molecules. The relative populations are 1 : 0.01 : 1 for unbound : singly liganded : doubly liganded species at the midpoint of the binding transition. The midpoint is

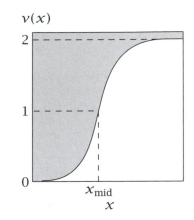

$v(x)$

Figure 28.3 A binding curve. To determine the type of cooperativity, look at the ligation distribution at the transition midpoint, x_{mid} where $v = 1$. Examples of distributions are given in Figures 28.4 and 28.5.

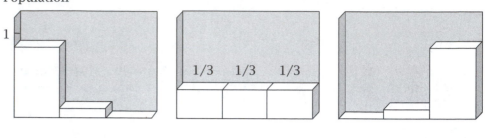

(a) One-State Transition

Population

1/3 1/3 1/3

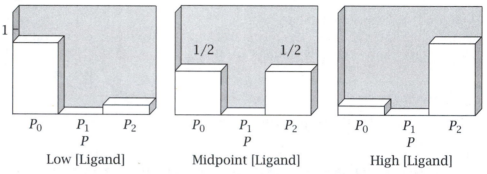

(b) Two-State Transition

Population

1/2   1/2

P_0 P_1 P_2 P_0 P_1 P_2 P_0 P_1 P_2

P P P

Low [Ligand]   Midpoint [Ligand]   High [Ligand]

Figure 28.4 Ligation states for P molecules with two binding sites: P_0 is the fraction of P molecules with no ligands, P_1 is the fraction with one ligand bound, and P_2 is the fraction having two ligands bound. (a) One-state transition. (b) Two-state transition. From left to right: at low ligand concentrations, P molecules are mostly unliganded. At the midpoint of binding, a one-state transition has a substantial intermediate population while a two-state transition involves no intermediates. At high ligand, most P molecules are doubly ligated.

$$v = \frac{K_1 x_{\mathrm{mid}} + 2K_1 K_2 x_{\mathrm{mid}}^2}{1 + K_1 x_{\mathrm{mid}} + K_1 K_2 x_{\mathrm{mid}}^2} = \frac{0.01 + 2}{1 + 0.01 + 1} = 1, \qquad (28.14)$$

where x_{mid} is the concentration of ligand at half saturation of the binding curve.

In case (b), $K_1 = 1000$ and $K_2 = 10$, so $Q = 12$ at $x_{\mathrm{mid}} = 0.01$. This is an example of one-state binding. Starting at low ligand concentrations $x \ll 1$, the P molecules have no ligands. Increasing the ligand concentration leads to predominantly singly ligated P molecules, then ultimately to doubly ligated P molecules at high ligand concentrations. At the midpoint of the transition, intermediate states are highly populated. The relative populations are $1 : 10 : 1$.

In case (c), $K_1 = 100$ and $K_2 = 100$, so $Q = 3$ at $x_{\mathrm{mid}} = 0.01$. This system has equal populations of the three states, $1 : 1 : 1$, at the midpoint of the binding isotherm. It involves independent binding.

The next section shows how to construct binding polynomials for more complicated systems.

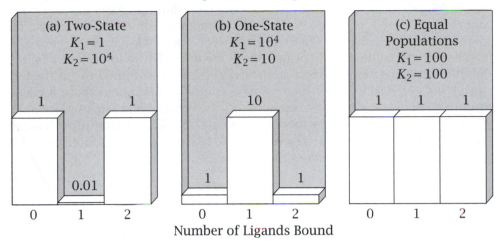

Figure 28.5 Relative populations of P molecules having zero, one, or two ligands bound at the midpoint of the binding transition described in Example 28.2. (a) Two-state cooperativity; (b) one-state cooperativity; (c) independent, equal populations.

Binding Polynomials Can Be Constructed by Using the Addition and Multiplication Rules of Probability

To predict binding equilibria, the first step is to devise an appropriate model and compute the binding polynomial. In this section we show a short-cut for writing the binding polynomial directly from the model, skipping the step of writing out the equilibria. We illustrate it first on simple cases, but the main power of the method is for describing more complex cases. The key to this approach is to use the addition and multiplication rules of probability (see Chapter 1).

EXAMPLE 28.3 The Langmuir model illustrates the binding polynomial. The binding polynomial is the sum of concentrations over all the ligation states of P. For the Langmuir model, this is

$$[P]Q = [P] + [PX] = [P](1 + Kx) \implies Q = 1 + Kx. \tag{28.15}$$

Using $dQ/dx = K$ and Equation (28.2), you can compute the fraction θ of sites filled as

$$\theta = \frac{d\ln Q}{d\ln x} = \frac{x}{Q}\frac{dQ}{dx} = \frac{Kx}{1 + Kx}. \tag{28.16}$$

You can also get the binding polynomial in Equation (28.15) by using the addition rule of probabilities described in Chapter 1. According to the addition rule, if two states are mutually exclusive (bound and unbound, for example), then you can sum their statistical weights, in the same way that terms are summed in partition functions. Use 1 as the statistical weight for the empty site and Kx as the statistical weight for the filled site, and add them to get Q.

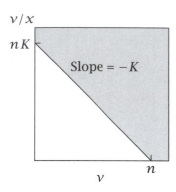

Figure 28.6 Scatchard plot of $v(x)/x$ versus v, where v is the average number of ligands bound per P molecule, and x is the ligand concentration. When there are n independent binding sites with affinity K, the slope is $-K$ and the intercepts are nK and n.

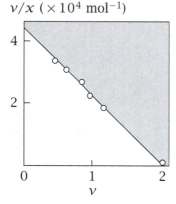

Figure 28.7 Scatchard plot for binding tryptophan to Trp aporepressor. Source: DN Arvidson, C Bruce and RP Gunsalus, *J Biol Chem* **261**, 238 (1986).

Similarly, according to the multiplication rule (also described in Chapter 1) if two ligands bind independently, then the contribution to the binding polynomial is the product of the two statistical weights. For example, for two independent sites, Equation (28.8) gives $Q = (1 + K_a x)(1 + K_b x)$. The a site can be empty OR filled, and because these are mutually exclusive options, the addition rule dictates that their statistical weights are summed to get the factor $(1 + K_a x)$. Likewise the b site can be empty OR filled, and these are also mutually exclusive, giving the factor $(1 + K_b x)$. Because a and b sites are independent of each other, the factors $(1 + K_a x)$ and $(1 + K_b x)$ are multiplied to get the full binding polynomial. The main value of this logic is for more complex equilibria, such as those in the next section.

n Independent Sites with Identical Affinities

Suppose there are n independent binding sites on P, all having the same affinity, K. Using the addition rule, the statistical weight for each site is $(1 + Kx)$. Because there are n independent sites, the binding polynomial is the product

$$Q = (1 + Kx)^n. \tag{28.17}$$

Taking the derivative according to Equation (28.2) (see also Equation (4.13)) gives v, the average number of ligands bound per P molecule,

$$v(x) = \frac{x}{Q}\left(\frac{dQ}{dx}\right) = \frac{nxK(1 + Kx)^{n-1}}{(1 + Kx)^n} = \frac{nKx}{1 + Kx}. \tag{28.18}$$

Equation (28.18) can also be derived from the Langmuir adsorption equation, because θ is the fraction of sites filled,

$$\theta = \frac{v}{n} = \frac{Kx}{1 + Kx}$$

$$= \frac{\text{average number of ligands bound per } P \text{ molecule}}{\text{total number of sites on each } P \text{ molecule}}. \tag{28.19}$$

The Scatchard Plot for n Independent Binding Sites

In the same way that a linear formulation of the Langmuir and Michaelis-Menten equations has practical advantages for fitting experimental data, the *Scatchard plot* (Figure 28.6) is a useful linearized form of Equation (28.18). Rearranging Equation (28.18) gives

$$\frac{v}{x} = \frac{nK}{1 + Kx} \implies \frac{v}{x} + vK = nK \implies \frac{v}{x} = nK - vK. \tag{28.20}$$

A plot of v/x versus v will be linear with a slope of $-K$, and it will have an intercept (at $v = 0$) of nK, when a binding process involves n independent sites of binding constant K. The limiting values at $v = 0$ and $v/x = 0$ are found by extrapolation, because they cannot be measured. When such a plot is not linear, however, you should not try to fit it with multiple straight lines: instead you should try a different model [2].

EXAMPLE 28.4 A Scatchard plot. Figure 28.7 shows a Scatchard plot for the binding of tryptophan to the *E. coli* tryptophan aporepressor protein. The x-intercept indicates that the number of sites is $n = 2$. The binding constant is found to be about $K = 2.0 \times 10^4 \text{ M}^{-1}$, taken either from the slope of this figure, or from the y-axis intercept, which is nK.

The Hill Plot for Cooperative Binding

If a P molecule binds to exactly n ligand molecules at a time, then the Hill model is useful:

$$nX + P \xrightarrow{K} PX_n, \quad \text{where} \quad K = \frac{[PX_n]}{[P]x^n}. \tag{28.21}$$

The sum over the states of P is given by

$$[P]Q = [P] + [PX_n] = [P](1 + Kx^n), \tag{28.22}$$

so the binding polynomial is $Q = 1 + Kx^n$. Then $dQ/dx = nKx^{n-1}$, and the average number of ligands bound per P molecule is

$$v(x) = \left(\frac{d \ln Q}{d \ln x}\right) = \left(\frac{x}{Q}\right)\left(\frac{dQ}{dx}\right) = \frac{nKx^n}{1 + Kx^n}, \tag{28.23}$$

according to Equation (28.2).

For a Hill process, a plot of v versus x will be sigmoidal (see Figure 28.8(a)). The steepness of the transition region increases with n. To put Equation (28.23) into a linearized form, use the average number of ligands bound per site $\theta = v/n$,

$$\theta = \frac{Kx^n}{1 + Kx^n}. \tag{28.24}$$

Because $1 - \theta = 1/(1 + Kx^n)$, you have

$$\frac{\theta}{1 - \theta} = Kx^n. \tag{28.25}$$

A *Hill plot* is $\ln(\theta/(1 - \theta)) = \ln K + n \ln x$ versus $\ln x$. The slope gives the Hill exponent (or Hill coefficient) n, and the intercept is $\ln K$.

EXAMPLE 28.5 Biochemical switching is a cooperative process. Some biological processes are all-or-none, like electrical switches. For example a dose of the hormone progesterone that is above a threshold concentration can trigger a frog's eggs to mature. Below the threshold concentration, the progesterone has no effect. Such behavior can be expressed in terms of large Hill coefficients, and may result from cooperative binding processes. Figure 28.9 shows a step in the progesterone process in which the concentration of a protein called MOS leads to a sharp increase in the phosphorylation of a protein called MAPK, with a Hill exponent $n \geq 5.1$. This is the Hill coefficient observed for groups of egg cells; within individual cells, the Hill coefficient is found to be much larger, $n \approx 35$ [3]. Such ultrasensitivity to ligand concentrations is often important in biological regulation.

(a) Cooperative Binding

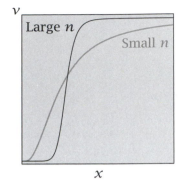

(b) Hill Plot

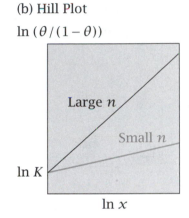

Figure 28.8 Cooperativity described by the Hill model. (a) When the fraction v of sites filled is plotted versus ligand concentration x, the steepness of the curve increases with n, the Hill coefficient. (b) The Hill plot linearizes the data. The slope gives the Hill coefficient n and the intercept gives the binding constant K.

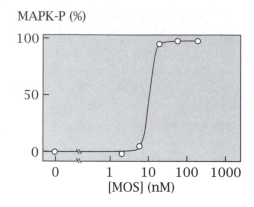

Figure 28.9 For the maturation of frog eggs, the production of phosphorylated mitogen-activated protein kinase (MAPK-P) is governed by small changes in the concentration of the MOS protein. This relationship can be described by the Hill model. Source: JE Ferrel and EM Machleder, *Science* **280**, 895–898 (1998).

The cooperative process of micellization is described by a similar model in the next section.

Aggregation and Micellization Processes

Surfactant molecules can aggregate into clusters called *micelles* (see Figure 25.21). A typical spherical micelle may contain 60 surfactant molecules. Micelles usually have narrow size distributions, say from 55 to 65 surfactant molecules. To make the mathematics simple, let's suppose that every micelle is an *n*-mer that has exactly *n* surfactant molecules. The concentration of micelles is $[A_n]$. The rest of the surfactant molecules are isolated as individual molecules. Their concentration is $[A_1]$. The equilibrium is

$$nA_1 \xrightarrow{K} A_n,$$

where the equilibrium constant is

$$K = \frac{[A_n]}{[A_1]^n}.$$

The total number of *objects* (monomers plus micelles) in the solution (per unit volume) is $[A_1] + [A_n]$. The total number of *surfactant molecules* is $[A_1] + n[A_n]$. So the number of molecules per object is

$$\nu = \frac{[A_1] + n[A_n]}{[A_1] + [A_n]}. \tag{28.26}$$

Let $x = [A_1]$, and use the definition of K to express Equation (28.26) as

$$\nu(x) = \frac{[A_1](1 + nKx^{n-1})}{[A_1](1 + Kx^{n-1})} = \frac{1 + nKx^{n-1}}{1 + Kx^{n-1}}. \tag{28.27}$$

This model predicts a sharp transition at *the critical micelle concentration* (cmc), approximately the concentration at which $x^{n-1} = 1/K$. At surfactant concentrations below the cmc, the solution contains mostly monomers. For small x, $x^{n-1} \ll 1/K$, so the number of surfactant molecules per object is $\nu \to 1$. At low concentrations, the high cost in translational entropy of recruiting the molecules and assembling them into a cluster outweighs the energetic advantage of sequestering their hydrocarbon tails from water. But at surfac-

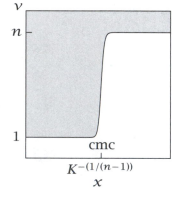

Figure 28.10 Micelle formation is cooperative. When the surfactant concentration is below the cmc, most surfactant molecules are monomers. When the surfactant concentration is above the cmc, most molecules are in *n*-mer micelles .

tant concentrations above the cmc, the solution contains mostly micelles. For large x, you have $x^{n-1} \gg 1/K$, so the number of surfactant molecules per object is $\nu \to (nKx^{n-1})/(Kx^{n-1}) = n$. At high concentrations, there is less translational entropy to overcome, because the average distance between the surfactant molecules in solution is small. Then the sticking energy dominates, and the molecules form micelles. The sharpness of the transition increases with n.

What determines the size of a micelle? Why does each micelle have about n molecules, rather than being smaller or growing into a macroscopic surfactant-rich phase and a water-rich phase? Micelles are driven to be as large as possible by the incompatibility of oil and water, which causes micelles to have hydrophobic cores. But typical surfactants have charged head groups at the micellar surface. Micelles that are too large would bury the charged groups in hydrophobic cores, or would squeeze charged groups too close together, so further growth is disfavored by electrostatic forces.

Example 28.6 describes the cooperative formation of multilayers of molecules on surfaces.

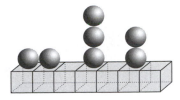

Figure 28.11 Ligands can bind on top of each other on surfaces in the BET model.

EXAMPLE 28.6 The BET adsorption model of layers of ligands. Sometimes ligands can 'pile up' on a site, layer upon layer, so a surface does not become saturated. This type of adsorption behavior can often be described by a model developed by S Brunauer, PH Emmett, and E Teller in 1938, called the BET model [4, 5]. The first molecule that adsorbs directly to a site on the surface binds with affinity K_1. Second and subsequent ligands need not bind directly to the surface; they can bind on top of a molecule that has already bound (see Figure 28.11). Suppose that such 'stacked' ligands bind with affinity K_2. The binding polynomial is

$$Q = 1 + K_1 x + K_1 K_2 x^2 + K_1 K_2^2 x^3 + K_1 K_2^3 x^4 + \dots, \tag{28.28}$$

because a site can be unbound OR have one ligand stuck to the surface (with statistical weight $K_1 x$) OR have two ligands piled up (with statistical weight $K_1 K_2 x^2$), etc. You can simplify Equation (28.28) to

$$Q = 1 + K_1 x \left[1 + K_2 x + (K_2 x)^2 + (K_2 x)^3 + \dots \right]$$

$$= 1 + \frac{K_1 x}{1 - K_2 x} = \frac{1 - (K_2 - K_1)x}{1 - K_2 x}, \tag{28.29}$$

by using Appendix C, Equation (C.6): $1 + K_2 x + (K_2 x)^2 + (K_2 x)^3 + \dots = 1/(1 - K_2 x)$, if $K_2 x < 1$. To find ν, the average number of ligands bound per surface site, as a function of the ligand concentration, take the derivative of Equation (28.29),

$$\frac{dQ}{dx} = \frac{K_1}{(1 - K_2 x)^2}, \tag{28.30}$$

to get, according to Equation (4.13),

$$\nu(x) = \frac{x}{Q} \left(\frac{dQ}{dx} \right) = \frac{K_1 x}{(1 - K_2 x)(1 - (K_2 - K_1)x)}. \tag{28.31}$$

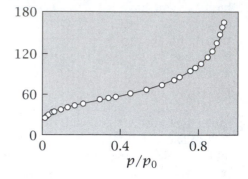

Figure 28.12 Adsorption of nitrogen on silica as a function of nitrogen pressure (relative to the nitrogen saturation pressure p_0) at 77 K follows the BET model. Source: PC Heimenz, *Principles of Colloid and Surface Chemistry*, M Dekker, New York (1977). Data are from DH Everett, GD Parfitt, KSW Sing and R Wilson, *J Appl Chem Biotechnol* **24**, 199–217 (1974).

In this model, the first layer of ligands seeds the formation of subsequent layers if $K_2 > K_1$. The amount of ligand adsorbed to the surface increases sharply with increasing ligand concentration as $(K_2 - K_1)x \to 1$. An example of the BET isotherm is shown in Figure 28.12.

Oxygen Binding to Hemoglobin Is a Cooperative Process

Hemoglobin is a protein that binds four oxygen molecules. Probably more is known about this process than about any other protein–ligand interaction. Hemoglobin has four subunits. So the first model that you might consider would involve four identical independent binding sites, each with affinity K, giving a binding polynomial

$$Q = (1 + Kx)^4 = 1 + 4Kx + 6(Kx)^2 + 4(Kx)^3 + (Kx)^4,$$

where x represents the oxygen concentration (pressure). The binomial coefficients in this expression reflect the numbers of ways of arranging a given number of ligands on the four sites: $4!/4!0! = 1$ is the number of arrangements in which a hemoglobin tetramer has either no ligands or four ligands bound. There are $4!/3!1! = 4$ different ways to arrange one or three ligands on the four sites. There are $4!/2!2! = 6$ ways two ligands can bind to four sites. This is just the Langmuir model for four independent identical sites, $\nu = 4Kx/(1 + Kx)$. But even the earliest experiments showed that oxygen binding to hemoglobin is cooperative and does not fit the independent site model. For example, the fourth oxygen molecule binds with 500-fold more affinity than the first oxygen molecule.

One of the earliest models of hemoglobin cooperativity was *the Adair equation*, developed by G Adair in 1925 [6], which can be expressed as

$$Q = 1 + 4K_1x + 6K_2x^2 + 4K_3x^3 + K_4x^4. \tag{28.32}$$

The binding parameters K_1, K_2, K_3, and K_4 represent statistical weights for the states of one, two, three, and four oxygen molecules bound. An extensive literature has developed since 1925 to improve upon the Adair model: to reduce the number of parameters to fewer than four, to interpret the parameters

Table 28.1 In the Pauling model, the four subunits of hemoglobin are assumed to be arranged in a tetrahedron. Each sphere represents one bound ligand molecule. Nearest-neighbor ligand interactions are indicated by continuous lines. The table shows the count of these interactions.

Number of Ligands Bound	0	1	2	3	4
Number of Ligand Pairs	0	0	1	3	6
Number of Pairwise Interactions	0	0	1	3	6

in terms of the protein structure, and to account for how ligands other than oxygen influence hemoglobin–oxygen binding. Among the first and simplest improvements was the Pauling model.

The Pauling Model

In 1935, Linus Pauling began with the model that each subunit of hemoglobin is identical and binds one O_2 molecule with affinity K [7]. He then proposed that there is an additional interaction free energy δ whenever two oxygen molecules are bound to adjacent subunits. He described the additional energy in terms of a Boltzmann weight f,

$$f = e^{-\delta/kT}.$$

Pauling postulated that the four subunits of hemoglobin could have various possible geometric relationships to each other. Two geometries that he considered were the square and the tetrahedron. Let's consider the tetrahedron (see Table 28.1). Every vertex of a tetrahedron is adjacent to every other. For every pair of oxygen molecules that are bound at adjacent sites, Pauling introduced the interaction factor f. Oxygen bound to one subunit might propagate some conformational change to the adjacent subunit that would affect its affinity for its oxygen ligand. When only one oxygen binds, there is no interaction energy so f is not included. For any of the four possible arrangements in which two oxygens are adjacent to each other, the factor f appears once. When three ligands are bound, there will be three pairwise nearest neighbor interactions, so the interaction factor is included as f^3. When four ligands are bound, there will be six pairwise nearest-neighbor interactions, so the interaction factor appears

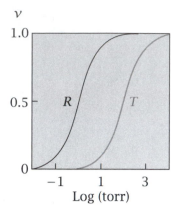

v

1.0

0.5 R T

0

−1   1   3

Log (torr)

Figure 28.13 In the MWC model of hemoglobin cooperativity, the ligand oxygen has greater affinity for the R conformation than for T. Less ligand is required to cause the R transition than the T transition. Source: J Wyman and SJ Gill, *Binding and Linkage*, University Science Books, Mill Valley 1990.

as f^6. Combining the intrinsic binding energies with these extra interaction energies gives the Pauling model for the binding polynomial,

$$Q = 1 + 4Kx + 6(Kx)^2 f + 4(Kx)^3 f^3 + (Kx)^4 f^6.$$

The Pauling model treats oxygen binding to hemoglobin using only two parameters, and attributes the binding cooperativity to pairwise interactions between adjacent subunits. This model, and a more recent extension called the KNF model (named after D Koshland, G Nemethy, and D Filmer), interpret binding in terms of ligand-induced conformational changes or 'induced fit' [8].

A different model of oxygen–hemoglobin cooperativity arose after M Perutz et al. showed in the 1960s that hemoglobin has two different structures [9, 10]. This model was due to J Monod, J Wyman, and P Changeux [11], and the mechanism is called *allostery*. We describe a simplified version of their model in the next section.

The Monod–Wyman–Changeux (MWC) Allosteric Model

Monod, Wyman, and Changeux proposed that hemoglobin has two different conformations, T and R (originally *tense* and *relaxed*, but these are just labels). T and R are in equilibrium with equilibrium constant L

$$R \xrightarrow{L} T, \qquad \text{where} \qquad L = \frac{[T]}{[R]}.$$

The ligand has different affinities for the two protein conformations (see Figure 28.13):

$$R + X \xrightarrow{K_R} RX \qquad \text{and} \qquad T + X \xrightarrow{K_T} TX.$$

In the absence of ligand, T is more stable than R. That is $L > 1$. Cooperativity results if the ligand binds to R more tightly than to T. In that case, adding ligand shifts the system from T states (liganded plus unliganded) toward the R states. To illustrate the principle, consider the simplest case, the binding of one ligand. There are four possible states of the system: R, T, RX, and TX, where R and T have no ligand and RX and TX each have one ligand bound. The binding polynomial Q is the sum over all ligation states of both T and R species:

$$([R] + [T]) Q = [R] + [RX] + [T] + [TX]. \tag{28.33}$$

In terms of the equilibrium constants, $[T] = L[R]$, $[RX] = K_R x [R]$, and $[TX] = K_T x [T] = K_T x L[R]$, Equation (28.33) becomes

$$Q = \left(\frac{1}{1+L}\right) [(1 + K_R x) + L(1 + K_T x)]. \tag{28.34}$$

The fraction f_R of protein molecules that are in the R state is

$$f_R = \frac{[R] + [RX]}{[R] + [RX] + [T] + [TX]} = \frac{1 + K_R x}{1 + K_R x + L(1 + K_T x)}, \tag{28.35}$$

Fraction Populated

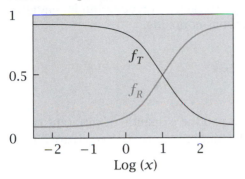

Figure 28.14 Increasing ligand concentration x in the MWC model of allostery causes a conformational change from T to R (Equation (28.35) and Equation (28.36)). The fraction f_T of molecules in the T state falls off as the fraction of molecules in the R state increases to 1.

and the fraction f_T of protein molecules in the T state is

$$f_T = \frac{[T] + [TX]}{[R] + [RX] + [T] + [TX]} = \frac{L(1 + K_T x)}{1 + K_R x + L(1 + K_T x)}. \tag{28.36}$$

Dividing Equation (28.36) by (28.35) gives

$$\frac{f_T}{f_R} = \frac{L(1 + K_T x)}{(1 + K_R x)}.$$

Consider the two limits:

$$x \to 0 \quad \Longrightarrow \quad \frac{f_T}{f_R} \to L,$$

$$x \to \infty \quad \Longrightarrow \quad \frac{f_T}{f_R} \to \frac{L K_T}{K_R}.$$

If $L > 1$ and $K_T/K_R < 1$, then increasing the ligand concentration shifts the equilibrium from T to R. Figure 28.14 shows how increasing the ligand concentration shifts the equilibrium from T to R, for $K_R = 1$, $K_T = 0.01$, and $L = 10$.

Generalized to handle four binding sites, the MWC binding polynomial for hemoglobin is $Q = (1 + L)^{-1}[(1 + K_R x)^4 + L(1 + K_T x)^4)]$. In that case, the average number ν of ligands bound is

$$\nu(x) = \frac{d \ln Q}{d \ln x} = Q^{-1}[4 K_R x (1 + K_R x)^3 + 4 L K_T x (1 + K_T x)^3]. \tag{28.37}$$

The MWC model describes oxygen binding to hemoglobin in terms of three parameters, K_R, K_T, and L. The key assumption is that the R and T structures of hemoglobin have different affinities for the ligand.

While the Pauling model supposes that the binding of one ligand can influence the next, the MWC model assumes that the affinity of oxygen for either the T or R state does not depend on how many other oxygens are already bound. These models were among the first and simplest to capture the principles of hemoglobin-binding cooperativity. Better models and more extensive data are now available, and give a more refined perspective on hemoglobin–oxygen interactions [1, 9, 10, 12, 13]. Equation (28.37) is also useful for understanding how ligands can drive the opening and closing of membrane channels [14].

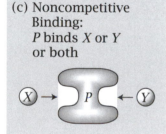

(a) Competitive Binding:
P Binds X or Y

(b) Prebinding:
P Binds X first, then Y

(c) Noncompetitive Binding:
P binds X or Y or both

Figure 28.15 Models of inhibitors. (a) Competitive inhibitor Y competes for the site where X binds to P.
(b) Uncompetitive inhibitor Y does not bind unless a ligand X also binds.
(c) Noncompetitive inhibitor Y has no effect on the binding of X to P, for example, because it binds at an independent site. Y affects P in some other way than through the process of binding X.

For many systems, including hemoglobin, more than one type of ligand can bind at a time. For example, the binding of oxygen to hemoglobin is regulated by pH, and it can also be inhibited by carbon monoxide.

Binding Polynomials Can Treat Multiple Types of Ligand

Binding polynomials can be applied when you have ligands of different types. Some molecules can inhibit or regulate the binding of others (see Figure 28.15).

Competitive Inhibition

Suppose a ligand of type X binds with affinity K_X to a site on molecule P. A different ligand molecule, of type Y, can bind to the same site with affinity K_Y,

$$P + X \xrightarrow{K_X} PX \quad \text{and} \quad P + Y \xrightarrow{K_Y} PY. \quad (28.38)$$

The concentrations of each type of ligand in solution are $x = [X]$ and $y = [Y]$. The binding polynomial is

$$Q = 1 + K_X x + K_Y y, \quad (28.39)$$

because each site can be free (with a statistical weight of 1), OR it can bind X (weighted by $K_X x$) OR it can bind Y (with weight $K_Y y$). The average fraction of P sites that are filled by X is $v_X = (\partial \ln Q / \partial \ln x)$,

$$v_X = \frac{K_X x}{1 + K_X x + K_Y y}. \quad (28.40)$$

Similarly, the fraction of P sites filled by Y is $v_Y = (\partial \ln Q / \partial \ln y)$. Y is called a *competitive inhibitor*, because Y competes with X for the binding site. The amount of X bound is reduced as Y increases in concentration or affinity (v_X decreases as $K_Y y$ increases). Equation (28.40) also characterizes impurities that can 'poison' a surface or catalyst by displacing the ligand. Figure 28.16(a) shows how the binding of X is affected by Y.

Prebinding: Ligand Y Binds Only if X Also Binds

Suppose that molecule Y can bind to P only if X also binds,

$$P + X \xrightarrow{K_X} PX + Y \xrightarrow{K_Y} PXY. \quad (28.41)$$

The binding polynomial is

$$Q = 1 + K_X x + K_X K_Y x y, \quad (28.42)$$

because each site can be free (statistical weight of 1) OR bind X (weighted $K_X x$) OR bind X AND Y (weighted $K_X K_Y x y$). The average number of X molecules bound per P is

$$v_X = \frac{K_X x + K_X K_Y x y}{1 + K_X x + K_X K_Y x y}. \quad (28.43)$$

Figure 28.16(b) illustrates prebinding behavior. The BET adsorption in Example 28.6 is a type of prebinding.

Related models describe *noncompetitive* and *uncompetitive* inhibitors.

Noncompetitive Inhibition

A noncompetitive inhibitor has the binding polynomial $Q = 1 + K_X x + K_Y y + K_X K_Y x y = (1 + K_x X)(1 + K_Y y)$ and can be described by the process

$$P + X \xrightarrow{K_X} PX$$
$$P + Y \xrightarrow{K_Y} PY$$
$$P + X + Y \xrightarrow{K_X K_Y} PXY.$$

In the noncompetitive inhibitor model, the binding of X to P is independent of the binding of Y to P; $v(x)$ is independent of $K_Y y$. This can result from an independence of the two sites. Y regulates, controls, or shuts down an effect or activity of P in some way that does not involve the binding of X to P.

Uncompetitive Inhibition and Activators

An uncompetitive inhibitor has the binding polynomial $Q = 1 + K_X x + K_X K_Y x y$ and corresponds to the process

$$P + X \xrightarrow{K_X} PX$$
$$P + X + Y \xrightarrow{K_X K_Y} PXY.$$

Uncompetitive inhibition is the same as prebinding and can be interpreted in terms of a site that binds either X alone or X and Y. According to Equation (28.43), increasing $K_Y y$ will *increase* the amount of X that binds to P. But this model can also describe *inhibition* if only the state PX is active (and PXY is inactive). Then the relevant measure of activity is the fraction of P sites that have only X bound

$$v = \frac{K_X x}{1 + K_X x + K_X K_Y x y}. \tag{28.44}$$

Wyman's Linkage Theory

Linkage is a term that describes how the binding of one ligand affects the binding of another [1]. Linkage can be expressed by $(\partial v_X / \partial y)$, the change in the degree of binding of X as the concentration of Y is changed,

$$\left(\frac{\partial v_X}{\partial y} \right) = \left(\frac{\partial}{\partial y} \right) \left(\frac{\partial \ln Q}{\partial \ln x} \right) = \left(\frac{\partial^2 \ln Q}{\partial y \partial \ln x} \right), \tag{28.45}$$

where v_X is given by Equation (28.2).

Also, using Maxwell's relations (Chapter 8), you have

$$\left(\frac{\partial v_X}{\partial y} \right) = \left(\frac{\partial v_Y}{\partial x} \right).$$

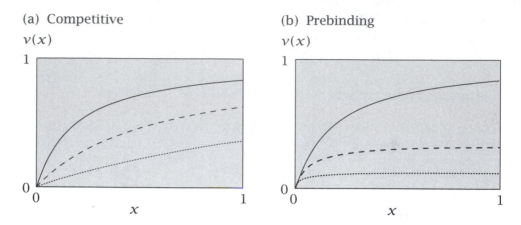

(a) Competitive

$\nu(x)$

(b) Prebinding

$\nu(x)$

Figure 28.16 Binding isotherms for a ligand X (a) in the presence of a competitive inhibitor Y, and (b) in the presence of a molecule Y that can bind only if X also binds (Equation (28.44)) (———, $K_Y y = 0$); (– – –, $K_Y y = 2$); (·········, $K_Y y = 8$).

Relationships of this type have been used extensively for understanding the regulation of oxygen binding to hemoglobin by salts, pH, carbon monoxide, and other ligands. Such linked processes are often the basis for the regulation of DNA, RNA, and protein molecules by hormones, metabolites, and drugs.

The Model of McGhee and von Hippel Treats Ligands that Crowd Out Other Ligands

Figure 28.17(b) shows how one ligand molecule X can crowd out another ligand from binding to a molecule P. The ligand X occupies more than a single binding site on P. At low ligand concentrations, each ligand molecule can readily find space available on P to bind (Figure 28.17(a)). But high concentrations of ligand lead to a 'parking problem' or 'excluded-volume problem.' Figure 28.17(c) shows P saturated by X, the true binding equilibrium that will be reached at very high concentrations of X, given enough time. But Figure 28.17(b) shows the problem at intermediate concentrations: P is not yet fully saturated by X, but nevertheless is unable to accept any new ligands. Sites remain available but they are *not distributed appropriately* to allow additional ligands to bind. We have not faced this excluded-volume problem before because we have considered only ligands that occupy a single site. However, with multi-site ligands, a ligand that binds at one site can prevent another ligand from binding at a different site.

An apparent negative cooperativity can arise from *ligand shape and packing*, rather than from the energetic factors that we have considered so far. Also, the *kinetics* of multisite binding can be very slow, because new ligands can bind only if ligands that are already bound dissociate and rebind. JD McGhee and PH von Hippel developed a model to predict the binding isotherms for multisite ligands [15].

In the model of McGhee and von Hippel, each P molecule is a linear lattice of N sites. Each bound ligand occupies n sites on P. The binding affinity is

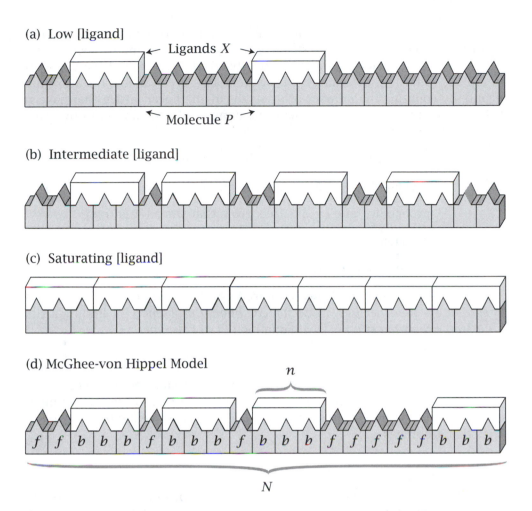

(a) Low [ligand]

Ligands X

Molecule P

(b) Intermediate [ligand]

(c) Saturating [ligand]

(d) McGhee-von Hippel Model

n

| f | f | b | b | b | f | b | b | b | f | b | b | b | f | f | f | f | f | b | b | b |

N

Figure 28.17 Multi-site ligands can bind to a linear molecule such as DNA, excluding other ligands from binding. Molecule P is represented as a linear lattice having N sites. Each ligand occupies n sites ($n = 3$ shown here). (a) Low ligand concentrations. (b) At ligand concentrations that are less than full coverage, further binding is impossible because the remaining free sites are not distributed in stretches long enough for additional ligands to bind. (c) Full saturation of P by ligand molecules. (d) In the model of McGhee and von Hippel, lattice sites are either blocked (b) by a bound ligand, or free (f).

$$K = \frac{[PX]}{[P]x},$$
(28.46)

where x is the concentration of free ligand (the number of molecules per unit volume) and $[PX]$ is the number of ligand molecules that are bound (per unit volume). $[P]$ is the number of ligand-sized stretches of sites on P molecules that are empty, per unit volume. In our previous modelling, $[P]$ would have been the concentration of 'receptor' molecules, the number of free P molecules per unit volume, because this gives the number of binding sites that are accessible to a ligand. But for multi-site binding, determining $[P]$, the number of binding sites that are accessible to a ligand, is a little more challenging.

Figure 28.17(d) shows one particular distribution of ligands bound to P, expressed as a one-dimensional string of two characters, f and b. A site is free (f) if it is not occupied by a ligand. A site is blocked (b) if a bound ligand covers it. You can find $[P]$, the number of locations that are available for an n-mer ligand, by counting the number of stretches of the linear lattice that have n free sites in a row. You can estimate the number of such stretches by using conditional probabilities:

$$[P] = \text{(number of individual sites (i) that are free)}$$
$$\times \text{ probability (site $i + 1$ is free, given that site i is free)}$$
$$\times \text{ probability (site $i + 2$ is free, given that site $i + 1$ is free)...}$$
$$\times \text{ probability (site $n + i$ is free,}$$
$$\text{given that site $n + i - 1$ is free).} \tag{28.47}$$

If $p(f)$ is the *fraction* of sites that are free, then $Np(f)$ is the *number* of sites that are free. Let $p(f|f)$ be the *conditional probability* of finding a site $i + 1$ free, given that site i is free. Equation (28.47) becomes

$$[P] = Np(f)p(f|f)^{n-1}. \tag{28.48}$$

First we compute $p(f)$. The number of ligands bound (per unit volume) is $B = [PX]$. Each ligand molecule blocks n sites at a time, so the total number of blocked sites is nB. The rest of the $N - nB$ sites are free. So the fraction of sites that are free is

$$p(f) = \frac{N - nB}{N}. \tag{28.49}$$

To determine the conditional probability $p(f|f)$, first determine how to satisfy the condition. The horse race problem, Example 1.10, shows how to compute conditional probabilities by deleting the options that do not satisfy the given conditions. In this case, the condition that must be satisfied to identify an appropriate site is that an f must occupy the site immediately to its left. The b's always come in runs: $b_1 b_2 b_3 \ldots b_n$, where b_1 is the first site blocked by a ligand, b_2 is the second site blocked, etc. The only b sites that have an f to their left are b_1 positions, because positions $b_2, b_3, \ldots, b_n$ all have b sites to their left. The number of sites that do not satisfy our condition is $(n - 1)B$ because this is the number of $b_2, b_3, \ldots, b_n$ sites. All the other

$$N - (n - 1)B \tag{28.50}$$

sites satisfy the condition of having an f to the left.

The number of f sites is $N - nB$. So the conditional probability that a site is f, given that the site to the left of it is f, is

$$p(f|f) = \frac{\text{number of (f) sites}}{\text{number of (ff) pairs} + \text{number of (fb_1) pairs}}$$
$$= \frac{N - nB}{N - (n - 1)B}. \tag{28.51}$$

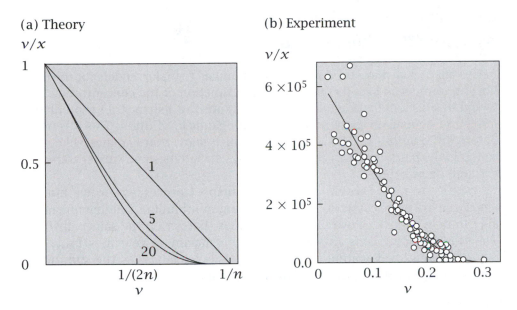

(a) Theory

(b) Experiment

Figure 28.18 (a) Predicted binding isotherms from the McGhee–von Hippel model for block sizes $n = 1, 5, 20$. Source: JD McGhee and PH von Hippel, *J Mol Biol* **86**, 469 (1974). (b) Experiments on daunomycin binding to calf thymus DNA. The data fit the McGhee-von Hippel model, with parameters $n = 3$ and $K = 5.9 \times 10^5 \, \text{M}^{-1}$. Source: JB Chaires, *Methods in Enzymology* **340**, 3–22 (2001).

We have made an approximation by neglecting cases of two ligands with no f sites separating them, i.e., cases of a b_1 following a b_n. Substituting Equation (28.51) into Equation (28.48) gives

$$[P] = (N - nB) \left(\frac{N - nB}{N - (n-1)B} \right)^{n-1}. \tag{28.52}$$

To put the binding equilibrium in terms of a Scatchard plot, let $v = B/N$ represent the number of ligands bound per P site. This gives $[P]$ in terms of v

$$[P] = N(1 - nv) \left(\frac{1 - nv}{1 - (n-1)v} \right)^{n-1}. \tag{28.53}$$

Using $[PX] = B = vN$ in Equation (28.46) gives $K = vN/([P]x)$. Rearranging this expression and substituting Equation (28.53) gives the binding isotherm

$$\frac{v}{x} = K(1 - nv) \left(\frac{1 - nv}{1 - (n-1)v} \right)^{n-1}. \tag{28.54}$$

Figure 28.18(a) shows the prediction of Equation (28.54) that excluded volume causes curvature in Scatchard plots. Figure 28.18(b) shows experimental data that fit the model. McGhee and von Hippel also generalized this model to allow for next-neighbor interactions among ligands to explain, for example, how gene-32 protein from bacteriophage T4 can line up within milliseconds and saturate single-strand DNA molecules to assist in DNA replication.

The next section shows that binding polynomials are just as applicable to rate processes as to binding thermodynamics.

Rates Can Often Be Treated by Using Binding Polynomials

In this chapter we have used binding polynomials to make models for the equilibrium number of ligands bound to P as a function of the concentration x of ligand in solution. Such equilibrium expressions are also useful for predicting how *rates* depend on the concentrations of ligands in solution. This was noted before in Equations (27.27) and (27.43), which show that Langmuir-like saturation is the basis for Michaelis–Menten enzyme kinetics and carrier transport through membranes.

When enzymes, carriers, and other transport processes involve binding steps that are more complex than simple Langmuir binding, binding polynomials can help. Consider two different enzymes that act on the same substrate. They have binding constants K_1 and K_2 and maximum velocities v_{max1} and v_{max2}. Generalize Equation (27.27) to give the velocity v as the sum of the velocities from the two different enzyme reactions,

$$v = \frac{K_1 x v_{max1}}{1 + K_1 x} + \frac{K_2 x v_{max2}}{1 + K_2 x}. \tag{28.55}$$

This corresponds to a two-site binding process.

Inhibitors can also slow down enzyme reactions in ways that are predicted by binding polynomials. For example, if ligand Y is a competitive inhibitor of substrate X, then Equation (28.40) can be used to predict the rate

$$\frac{v}{v_{max}} = \frac{K_x x}{1 + K_x x + K_y y}. \tag{28.56}$$

If an enzyme requires n identical substrate molecules to bind at the same time before it can catalyze a reaction, and if each molecule binds with affinity K, the Hill model can be used to predict the rate v of product formation:

$$\frac{v}{v_{max}} = \frac{K x^n}{1 + K x^n}, \tag{28.57}$$

where n is the Hill coefficient.

In the next section we give the statistical mechanical basis for binding polynomials.

The Grand Canonical Ensemble Gives the Basis for Binding Polynomials

The canonical ensemble (T, V, N) that we used in earlier chapters applies when the number N of molecules in the system is fixed. This applies to a beaker containing DNA and ligand molecules, for example. The N ligand molecules are conserved; they are either bound to the DNA or in solution. But now we focus on the DNA molecule itself, not the beaker. On the DNA molecules the number N of bound ligands is not fixed: it fluctuates from one DNA molecule to the next. The fixed quantity is the chemical potential μ (the concentration, or activity) of the ligand in the 'bath' surrounding the DNA. For this system, the ensemble of interest is the *Grand Canonical* (T, V, μ) ensemble. Binding polynomials in powers of the ligand concentration derive from this ensemble.

First, we need to identify the extremum function that applies to the (T, V, μ) ensemble. We consider here only a single chemical component, a ligand that has chemical potential μ. The Legendre transforms of Chapter 8 show that the appropriate function is

$$F - \mu N = U - TS - \mu N. \tag{28.58}$$

At equilibrium, this function is at a minimum, and the differential expression is $-SdT - pdV - Nd\mu = 0$. Because T and μ are fixed, the energy and particle number are the quantities that fluctuate and exchange with the bath that surrounds the system.

The macroscopic energy $U = \langle E \rangle$ is the ensemble average over the microscopic energy levels E_j, $j = 1, 2, \ldots, t$. Similarly, the macroscopically observable particle number N is the ensemble average over the microscopic numbers of particles in the system $M = 0, 1, 2, \ldots, M_0$, so $N = \langle M \rangle$. (For ligands binding to a system, such as a molecule of DNA, the number N of particles in the system is the average number of ligand molecules bound to a DNA molecule.) In the canonical ensemble, the only fluctuating quantity is the energy, which can go up a little or down a little, but its average value is fixed by holding the temperature constant. In the Grand Canonical ensemble, the number M of particles that are bound or associated with the system can go up a little or down a little, but the average particle number is equal to the macroscopic quantity N. This is fixed by holding the chemical potential (bulk concentration) constant. In terms of these microscopic quantities, Equation (28.58) becomes $F - \mu N = \langle E \rangle - TS - \mu \langle M \rangle$, or in differential form:

$$d(F - \mu N) = d\langle E \rangle - TdS - \mu d\langle M \rangle \tag{28.59}$$

because $d\mu$ and dT are zero when (T, V, μ) are constant.

The constraints are: that the probabilities sum to one; that the energies, summed over energy levels and particle numbers, must give the appropriate average; and that the number of particles, summed over all the states of the system, must equal the appropriate average particle number. Let p_{jM} represent the probability that the system is in energy level j and has M particles. Then the constraints are

$$\sum_{j=1}^{t} \sum_{M=0}^{M_0} p_{jM} = 1, \tag{28.60}$$

$$\sum_{j=1}^{t} \sum_{M=0}^{M_0} E_{jM} p_{jM} = \langle E \rangle, \qquad \text{and} \tag{28.61}$$

$$\sum_{j=1}^{t} \sum_{M=0}^{M_0} M p_{jM} = \langle M \rangle. \tag{28.62}$$

To minimize $F - \mu N$, set $d(F - \mu N) = 0$ (Equation (28.59)), subject to constraint Equations (28.60) to (28.62). The method of Lagrange multipliers (page 71) leads to

$$\sum_j \sum_M \left[E_{jM} + kT(1 + \ln p_{jM}^*) + \alpha - \mu M \right] dp_{jM}^* = 0, \tag{28.63}$$

where α is the Lagrange multiplier for constraint Equation (28.60). Because the term in the brackets in Equation (28.63) equals zero according to the Lagrange method, you have

$$\ln p_{jM}^* = -\frac{E_{jM}}{kT} - \frac{\alpha}{kT} - 1 + \frac{M\mu}{kT}$$

$$\Rightarrow p_{jM}^* = \frac{e^{-E_{jM}/kT} e^{M\mu/kT}}{\Xi}, \tag{28.64}$$

where Ξ is the grand partition function. It is a sum over both energy levels and particle numbers,

$$\Xi(T, V, \mu) = \sum_{j=1}^{t} \sum_{M=0}^{M_0} e^{-E_{jM}/kT} e^{M\mu/kT}$$

$$= \sum_{M=0}^{M_0} e^{M\mu/kT} \sum_{j=1}^{t} e^{-E_{jM}/kT}$$

$$= \sum_{M=0}^{M_0} Q(T, V, M) e^{M\mu/kT}, \tag{28.65}$$

where $Q(T, V, M)$ is the canonical partition function of Equation (10.10).

Substituting Equation (28.64) into Equation (28.62) gives an expression for the average particle number,

$$\langle M \rangle = \frac{\sum_M \sum_j M e^{-E_{jM}/kT} e^{M\mu/kT}}{\Xi} = kT \left(\frac{\partial \ln \Xi}{\partial \mu} \right)_{V,T}. \tag{28.66}$$

Finally, to make the connection with binding polynomials, note that the concentration of ligand in the bath is $x = e^{(\mu/kT)}$ (since $\mu = kT \ln x$) and that $K_M = \sum_{j=1}^{t} e^{-E_{jM}/kT}$ represents a partition-function-like sum of Boltzmann weights for all the arrangements of M ligands with the system, bound (the particles that are contained within the system) and unbound (the particles that are outside the system, in the bath).

Summary

Binding processes can be described by binding polynomials. A binding polynomial is a sum over all possible ligation states, similar to a partition function, which is a sum of Boltzmann factors over energy levels. A binding polynomial embodies a model that can be used to calculate quantities such as the binding isotherm, the average number of ligands bound versus ligand concentration. This function can be used to interpret experiments. In particular, some binding processes are cooperative, meaning that when a ligand binds to one site, the affinity of another site may change.

Problems

1. A molecular machine. You have a complex binding process, a 'molecular machine,' in which three ligands, X, Y, and Z bind to a molecule P. The free concentrations of the ligands in the solution are x, y, and z, respectively. Molecule X can bind with either 'foot' on P, with binding constant K_1, or with both feet on P, with binding constant K_1^2. Molecule Y can bind only if X is bound with both feet on P. Molecule Z can bind to P only if X is doubly bound and Y is bound. The binding equilibria are shown below:

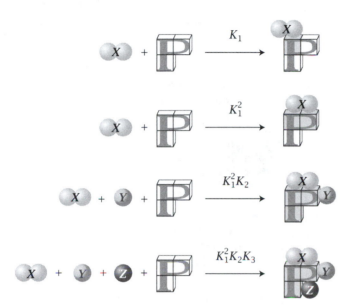

(a) Write the binding polynomial for this molecular machine.

(b) What is the fraction of P molecules that have Y and Z and both feet of X bound?

(c) What species dominate(s) at small x?

2. Saturation of myoglobin. Suppose that O_2 molecules bind to myoglobin with association constant $K = 2 \text{ torr}^{-1}$ at 25 °C and pH 7.4.

(a) Show a table of the fractional saturation of myoglobin for pressures of 1, 2, 4, 8, and 16 torr O_2.

(b) Does the fractional saturation double for each doubling of the pressure?

3. The entropy in the grand ensemble. Write an expression for the entropy in the Grand Canonical ensemble (T, V, μ).

4. Three-site binding. A ligand X can bind to a macromolecule P at three different binding sites with the binding constants K_1, K_2, and K_3:

$$X + P \xrightarrow{K_1} PX, \quad X + PX \xrightarrow{K_2} PX_2,$$

$$\text{and} \quad X + PX_2 \xrightarrow{K_3} PX_3.$$

(a) Write the binding polynomial, Q.

(b) Write an expression for the number of ligands ν bound per P molecule.

(c) Compute ν for $x = [X] = 0.05$, assuming $K_1 = 1$, $K_2 = 1$, and $K_3 = 1000$.

(d) Assume the same K values as in (c). Below ligand concentration $x = x_0$ most of the macromolecular P molecules have 0 ligands bound. Above $x = x_0$ most of the P molecules have three ligands bound. Compute x_0.

(e) For $x = x_0$ in part (d), show the relative populations of the ligation states with zero, one, two, and three ligands bound.

5. Drug binding to a protein. A drug D binds to a protein P with two different equilibrium binding constants, K_1 and K_2,

$$D + P \xrightarrow{K_1} PD_1, \quad \text{and} \quad D + PD_1 \xrightarrow{K_2} PD_2,$$

where $K_1 = 1000 \text{ M}^{-1}$, and $K_2 = 2000 \text{ M}^{-1}$ at $T = 300$ K.

(a) Write an algebraic expression for the fraction of protein molecules that have two drug molecules bound as a function of drug concentration $x = [D]$.

(b) If the drug concentration is $x = 10^{-3}$ M, then what is the fraction of protein molecules that have one drug molecule bound?

(c) If the drug concentration is $x = 10^{-3}$ M, then what is the fraction of protein molecules that have zero drug molecules bound?

(d) If the drug concentration is $x = 10^{-3}$ M, then what is the fraction of protein molecules that have two drug molecules bound?

6. Polymerization equilibrium. Consider the process of polymerization and assume the principle of equal reactivity, that each monomer adds with the same equilibrium constant as the previous one:

$$2X_1 \xrightarrow{K} X_2,$$

$$3X_1 \xrightarrow{K^2} X_3,$$

$$\vdots$$

$$nX_1 \xrightarrow{K^{n-1}} X_n.$$

(a) Write the binding polynomial Q.

(b) Write the average chain length $\langle n \rangle$ in terms of Q.

(c) Plot $\langle n \rangle$ versus x for $K = 1$, and $Kx < 1$.

7. Oxygen shifts the dimer/tetramer equilibrium in hemoglobin. Oxygen binding alters the equilibrium between hemoglobin dimers and tetramers. The free energies shown in Figure 28.19 have been measured.

(a) What is ΔG_4?

(b) Does oxygen binding shift the equilibrium toward the dimers or toward the tetramer?

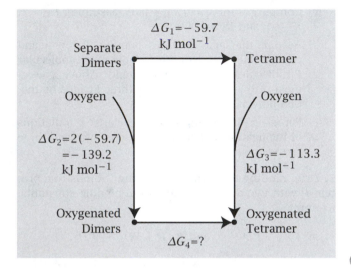

Figure 28.19 Thermodynamic cycle for O_2 binding and tetramer formation of hemoglobin. Source: FC Mills and GK Ackers, *J Biol Chem* **259**, 2881–2887 (1979); GK Ackers, *Biophys J* **32**, 331–346 (1980).

8. Helicases unwind DNA. Suppose DNA has two states: wound (helical) and unwound. In chromosomal DNA, the wound state is more stable than the unwound form of DNA by -4.1 kcal mol^{-1} (per unit length).

(a) Compute the equilibrium constant K at $T = 300\,\mathrm{K}$ for unwound DNA $\xrightarrow{K}$ wound DNA.

(b) Helicase is a protein that binds with equilibrium binding constant $B_h = 10$ to wound DNA (see Figure 28.20). Using the thermodynamic cycle in Figure 28.20, compute the binding constant B_u for binding helicase to unwound DNA that would be required to unwind the DNA (half the DNA is unwound and half is helical when helicase is unbound and $K_0 = 1$).

(c) Write an expression for the fraction of DNA molecules f_u that are unwound (including both u and ux) as a function of the concentration x of helicase in solution and of K, B_h, and B_u.

(d) For the values of K and B_u that you found, and for $B_h = 0.1$, find the value of x that gives $f_u \approx 1/2$.

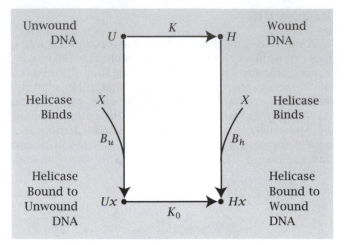

Figure 28.20 Thermodynamic cycle for binding helicase to DNA. B_u and B_h are the equilibrium binding constants for binding helicase protein to unwound and wound DNA, respectively. $K = $ [helical DNA]/[unwound DNA] is the equilibrium constant for helix formation in DNA in the absence of helicase, and K_0 is that equilibrium constant in the presence of ligand.

9. Scatchard plot for an antibody. The plot in Figure 28.21 shows the result of binding dinitrophenol (DNP) ligand to an antibody (Ab). Compute the binding constant K, and the number of binding sites per Ab molecule n.

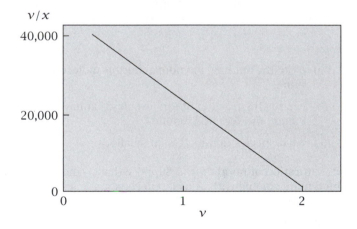

Figure 28.21 v is the number of moles of DNP bound per mole of Ab, and x is the bulk concentration of DNP in solution in moles per liter. Source: J Darnell, H Lodish, and D Baltimore, *Molecular Cell Biology*, 2nd edition, WH Freeman, New York, 1990.

10. Relating stoichiometric to site constants. Derive the relationship between stoichiometric-model binding constants K_1 and K_2 with site-model binding constants K_a

and K_b. Suppose the sites are independent, $K_c = K_a K_b$.

11. Scatchard plots don't apply to multiple types of binding site. If a binding process involves n_1 sites with affinity K_1, a Scatchard plot of v/x versus v gives a straight line with K_1 as slope and n_1 as x intercept (see Figure 28.6). Show that if you have two types of sites having (number, affinity) = (n_1, K_1) and (n_2, K_2), a Scatchard plot doesn't give two straight lines from which you can get (n_2, K_2). To fit multiple types of site you need a different model.

12. Oxygen binding to hemoglobin and myoglobin. Figure 28.22 compares the single site binding of oxygen to myoglobin to the four-site cooperative binding to hemoglobin. What biological advantage of hemoglobin is evident from these curves?

Fraction of O_2 Binding Sites Filled

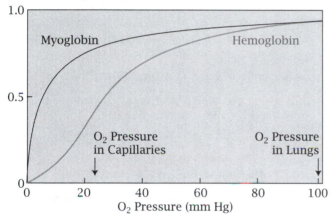

Figure 28.22 The binding of O_2 to hemoglobin and myoglobin. Source: J Darnell, H Lodish and D Baltimore, *Molecular Cell Biology*, 2nd edition, WH Freeman, 1990.

13. Inhibitors of enzyme kinetics. The Michaelis–Menten model of enzyme kinetics gives the reaction velocity v in terms of a maximum rate v_{max} as

$$v = v_{max}\left(\frac{Kx}{1 + Kx}\right), \qquad (28.67)$$

where K is the binding constant ($K = 1/K_M$ where K_M is the Michaelis constant) and x is the substrate concentration. A Lineweaver–Burk plot is the linearized version of Equation (28.67), a plot of $1/v$ versus $1/x$.

(a) Give the quantities (i), (ii), and (iii) shown in Figure 28.23 in terms of K and v_{max}.

(b) Write the linearized form, $1/v$ versus $1/x$.
Competitive inhibitors obey the expression

$$v = v_{max}\left(\frac{K_x x}{1 + K_x x + K_y y}\right).$$

Noncompetitive inhibitors obey the expression

$$v = v_{max}\left(\frac{K_x x}{1 + K_x x + K_y y + K_x K_y xy}\right).$$

Uncompetitive inhibitors obey the expression

$$v = v_{max}\left(\frac{K_x x}{1 + K_x x + K_x K_y xy}\right).$$

(c) Sketch the plots of $1/v$ versus $1/x$ for competitive, noncompetitive, and uncompetitive inhibition. In each case include the curve for the uninhibited rates.

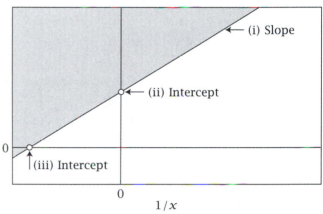

Figure 28.23 A Lineweaver–Burk plot has a slope and two intercepts that can help determine the nature of an inhibitor.

14. DNA-binding protein cooperativity. Single-strand binding protein (SSB) is a tetrameric protein that binds single-stranded DNA oligomers. One DNA oligomer can bind to each protein monomer. A maximum of four oligomers can bind one tetrameric protein. The binding constant for one oligonucleotide is K. x is the concentration of oligonucleotide. Binding is observed to be cooperative. Bujalowski and Lohman [16] have proposed a Pauling-like model of cooperativity. Think of the four protein subunits as being arranged in a square. A multiplicative 'cooperativity' factor f applies whenever two ligands occupy *adjacent* sites on the protein square, but not when two ligands are *diagonally across* from each other.

(a) Write the binding partition function in terms of K, f, and x.

(b) Compute $v(x)$, the average number of DNA ligands bound per protein tetramer.

15. Helicase nucleotide binding site. DNA B helicase is a hexameric protein that binds nucleotide ligands (see

Figure 28.1). The binding fits a Pauling-like model involving six sites with binding constant K, arranged in a hexamer, and with interaction parameter f whenever two ligands are on adjacent sites.

(a) Write the binding polynomial for this model.

(b) Write an expression for the average number of sites filled as a function of ligand concentration x.

References

[1] J Wyman and SJ Gill. *Binding and Linkage*, University Science Books, Mill Valley, 1990.

[2] IM Klotz. *Ligand-Receptor Energetics*, Wiley, New York, 1997.

[3] JE Ferrell Jr and EM Machleder. *Science* **280**, 895–898 (1998).

[4] DM Ruthven. *Principles of Adsorption and Adsorption Processes*, Wiley, New York, 1984.

[5] PC Hiemenz. *Principles of Colloid and Surface Chemistry*, Marcel Dekker, New York, 1977.

[6] GS Adair. *J Biol Chem* **63**, 529 (1925).

[7] L Pauling. *Proc Natl Acad Sci USA* **21**, 186 (1935).

[8] D Koshland, G Nemethy and D Filmer. *Biochem* **5**, 364 (1966).

[9] MF Perutz, AJ Wilkinson, M Paoli and GG Dodson. *Ann Rev Biophys Biomol Struc* **27**, 1 (1998).

[10] WA Eaton. *Dahlem Workshop on Simplicity and Complexity in Proteins and Nucleic Acids*, Berlin, 1998.

[11] J Monod, J Wyman and JP Changeaux *J Mol Biol* **12**, 881 (1965).

[12] A Szabo and M Karplus. *J Mol Biol* **72**, 163–197 (1972).

[13] GK Ackers. *Adv Prot Chem* **51**, 185 (1998).

[14] J Li, WN Zagotta and HA Lester. *Quart Rev Biophys* **30**, 177–193 (1997).

[15] JD McGhee and PH von Hippel. *J Mol Biol* **86**, 469 (1974).

[16] W Bujalowski and T Lohman. *J Mol Biol* **207**, 249 (1989).

Suggested Reading

Treatments of the principles of binding polynomials:

CR Cantor and PR Schimmel, *Biophysical Chemistry*, Vol 3, WH Freeman, San Francisco, 1980.

E Di Cera, *Thermodynamic Theory of Site-Specific Binding Processes in Biological Macromolecules.* Cambridge University Press, New York, 1995.

JT Edsall and H Gutfreund, *Biothermodynamics: the Study of Biochemical Processes at Equilibrium*, Wiley, New York, 1983.

IM Klotz, *Ligand-Receptor Energetics.* Wiley, New York, 1997.

JA Schellman, *Biopolymers* **14**, 999–1008 (1975).

DJ Winzor and WH Sawyer, *Quantitative Characterization of Ligand Binding*, Wiley-Liss, New York, 1995.

J Wyman and SJ Gill, *Binding and Linkage: Functional Chemistry of Biological Macromolecules*, University Science Books, Mill Valley, 1990.

Treatments of the Grand Canonical Ensemble:

TL Hill, *An Introduction to Statistical Thermodynamics.* Addison-Wesley Pub Co, Reading, 1960.

DA McQuarrie, *Statistical Mechanics*, Harper & Row, New York, 1976.

The BET isotherm is treated in:

PC Hiemenz, *Principles of Colloid and Surface Chemistry*, Marcel Dekker, New York, 1977.

DM Ruthven, *Principles of Adsorption and Adsorption Processes*, Wiley, New York, 1984.

Excellent reviews of hemoglobin cooperativity:

GK Ackers, *Adv Prot Chem* **51**, 185–253 (1998).

WA Eaton, *Dahlem Workshop on Simplicity and Complexity in Proteins and Nucleic Acids*, Berlin, 1998.

MF Perutz, AJ Wilkinson, M Paoli and GG Dodson, *Annu Rev Biophys Biomol Struc* **27**, 1 (1998).

A Szabo and M Karplus, *J Mol Biol* **72**, 163–197 (1972).

A detailed discussion of inhibitors is given in:

IH Segal, *Biochemical Calculations: How to Solve Mathematical Problems in General Biochemistry*, 2nd edition, Wiley, New York, 1976.

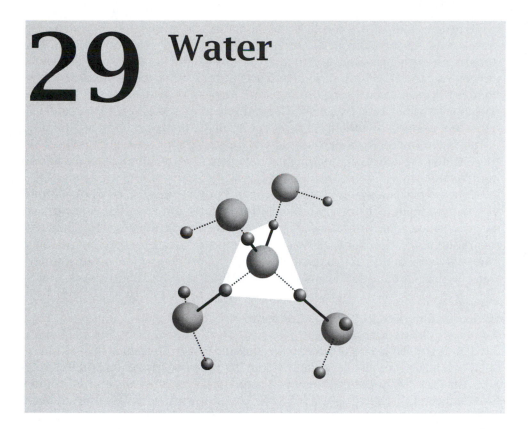

29 Water

Water Has Anomalous Properties

Water is the most abundant liquid on Earth. Yet it is often considered *anomalous* because it behaves somewhat differently from simpler and better understood liquids such as argon. Argon atoms can be modeled as Lennard–Jones spheres. Water molecules attract and repel each other like argon atoms do, but water can also form hydrogen bonds. Hydrogen-bonding introduces orientation dependence into the intermolecular interactions. So water has some of the properties of a normal liquid and some of the properties of a fluctuating tetrahedral network of hydrogen bonds. In this chapter we look at the properties of pure water. In Chapter 30 we look at water as a solvent for polar and nonpolar molecules.

Hydrogen Bonds Are a Defining Feature of Water Molecules

Figure 29.1 shows the atomic geometry of a water molecule. Composed of only three atoms, water is small and compact, so to a first approximation it is spherical. Two hydrogen atoms and two lone pairs of electrons are arranged in near-tetrahedral symmetry around an oxygen atom (see Figure 29.2). Water has no net charge but it has a permanent dipole moment, which can be described as a partial negative charge on the oxygen and partial positive charges near the hydrogens. Water molecules can form hydrogen bonds with each other.

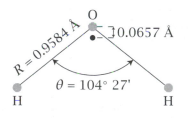

Figure 29.1 Geometry of a water molecule; (R) OH bond distance; (●) center of gravity.

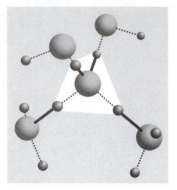

Figure 29.2 Water forms a tetrahedral bonding arrangement with neighboring water molecules.

Energy (kcal mol^{-1})

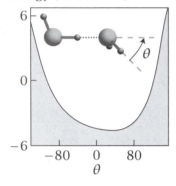

Figure 29.3 Energy versus angle θ of a hydrogen bond between two water molecules in the gas phase, from molecular simulations. Source: FH Stillinger, *Science* **209**, 451–457 (1980).

Hydrogen bonds are formed when a hydrogen *donor* atom is near an *acceptor* atom in the appropriate orientation. Some types of atom are good at donating hydrogens, some are good at accepting, and some do neither. Water molecules can act as both donors and acceptors for hydrogen bonds. The OH bond is a donor: it can stretch to share its hydrogen with an acceptor. Each lone pair of electrons on the oxygen of a water molecule can be an acceptor, and can bond to a donor's hydrogen. A water molecule can participate in any number of hydrogen bonds from zero to four. Figure 29.2 shows a water molecule in the tetrahedral hydrogen-bonding arrangement that is common in ice and in liquid water.

How strong is a water–water hydrogen bond? Figure 29.3 shows a plot of energy versus hydrogen-bond angle for a water dimer in the vapor phase, based on molecular orbital calculations reviewed by Stillinger [1]. The most favorable orientation between two water molecules occurs when the hydrogen bond between them is approximately straight. The energy of an optimal water–water hydrogen bond in the gas phase is estimated to be about 5.5 kcal mol^{-1}.

Hydrogen Bonding in Water Is Cooperative

Small polygonal clusters of water molecules have been studied in the gas phase. Such clusters serve as useful models for the types of hydrogen-bonded polygons that are components of bulk water and of solvation shells around solutes (see Chapter 30). The structures of water clusters are governed by the tendency to maximize the number of hydrogen bonds. The hydrogen bonds in water are called *cooperative* because the strength of one hydrogen bond is not independent of neighboring hydrogen bonds. For example, Figure 29.4 shows how the oxygen–oxygen distance between water molecules and the average dipole moment per water molecule change with the number of water molecules in the cluster. The dipole moment is 1.855 Debyes for a single isolated water molecule; 2.7 Debyes for some of the polyhedral clusters having five to eight water molecules; and 2.4–2.6 Debyes for the average water molecule in bulk liquid water at 0 °C.

Hydrogen-bonding cooperativity arises from the shifting of electron density. Suppose water molecule *A* donates a hydrogen from its OH group to the lone pair of electrons on the acceptor water molecule *B*. When a hydrogen bond forms, the hydrogen atom shifts toward molecule *B*, while electron density shifts toward water molecule *A*. The loss of electron density on molecule *B* results in more positive charge on its hydrogens. So molecule *B* can more readily donate its hydrogens to water molecule *C*. The shifting of electron density accounts for the higher dipole moments of water molecules in cluster and bulk states than in the gas phase. Similar cooperativities appear in other highly polar hydrogen-bonded liquids, such as HF, H_2O_2, and HCN.

Water Has Tetrahedral Coordination

The tetrahedral symmetry of a water molecule defines the structural framework for the solid, ice. Crystal structures show that the water molecules in ice I$_h$ (the form that is stable at 0 °C at 1 atm pressure) have tetrahedral symmetry (see

(a)

Monomer Dipole Moment (D)

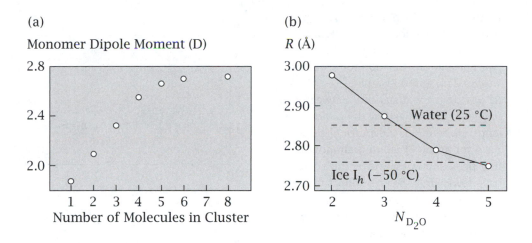

(b)

R (Å)

Figure 29.4 Two measures of hydrogen bond cooperativity.
(a) Quantum-mechanical calculations show that the dipole moment per water
increases with the size of a water cluster; (b) the average oxygen–oxygen distance R
decreases with increasing cluster size N_{D_2O}. Source: JK Gregory, DC Clary, K Liu,
MG Brown and RJ Saykally, *Science* **275**, 814–817 (1997); and JD Corzan, LB Braly,
K Liu, MG Brown, JG Loeser and RJ Saykally, *Science* **271**, 59–62 (1996).

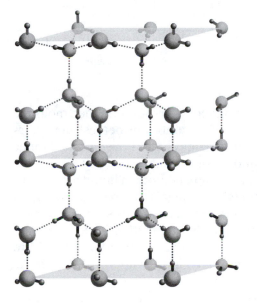

Figure 29.5 The structure of ice I_h.

Figure 29.5). Hydrogen-bonding appears to be a dominant component of the
energetics of water in its solid phase.

Liquid water also has tetrahedral symmetry because the melting process
is not disruptive enough to disorder the water molecules fully. This is known
from the radial distribution functions for liquid water that are observed in x-ray
and neutron-diffraction experiments. Figure 29.6 shows the radial distribution
functions of two liquids, water, and argon.

Figures 29.6 and 29.7 indicate that liquid water is tetrahedrally coordinated.
Integrating under the first peak of the radial distribution function in Figure

Figure 29.6 The oxygen–oxygen pair correlation function $g(R)$ for liquid water at 4 °C (– – –) compared with the pair correlations function for argon at $T = 84.25$ K and 0.71 atm (———). $R^* = r/\sigma$ is the distance r between two oxygens, normalized by the bond length, where σ is 2.82 Å for water and 3.4 Å for argon. The integral under the first peak shows that water molecules have fewer first-shell neighbors (about 4.4) than argon (about 10). Source: F Franks, *Water*, The Royal Society of Chemistry, London, 1983.

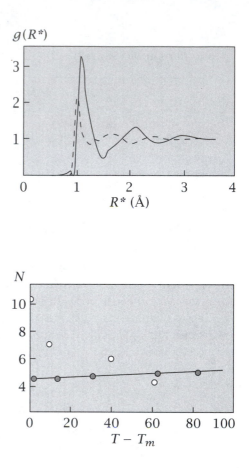

Figure 29.7 The number N of closest neighbors around water (●), and around argon (○) as a function of the temperature difference from the melting point, $T - T_m$. Water has 4 to 5 nearest neighbors through the full liquid range. Argon has 10 neighbors in its cold liquid state. Source: F Franks, *Water*, The Royal Society of Chemistry, London, 1983.

29.6 gives the average number of nearest neighbors surrounding each molecule (see discussion of Figure 24.10, page 461). The number of nearest neighbors around a water molecule is 4.4 at 4 °C, and remains about 4 over a broad range of temperatures from 4 °C to 200 °C (under high pressures at 200 °C, to keep water liquid). In contrast, the number of nearest neighbors in liquid argon is 10 at low temperatures. That number diminishes at higher temperatures. Argon packing is comparable to the closest packing of hard spheres, in which each sphere has 12 nearest neighbors. The peak of the radial distribution function shows that liquid water has oxygen–oxygen distances from 2.82 Å at 4 °C to 2.92 Å at 200 °C. Water dimers have longer bonds in the vapor phase (2.98 Å), and shorter ones in ice (2.74 Å) near 0 K.

Liquid water is more ordered than argon. This is indicated in Figure 29.6. The width of the first peak of the correlation function provides a measure of disorder in the first neighbor shell. It indicates the amount of variation in the numbers of first neighbors. Also, the entropy of vaporization is larger for water (109 J K^{-1} mol) than for other simple liquids (70-90 J K^{-1} mol), indicating that more structure is broken by boiling water.

The oscillations in the oxygen–oxygen pair correlation function become negligible after about the third neighbor shell in water, as in argon. This indicates that the structuring influence of each water molecule on other waters is relatively localized.

Intensity

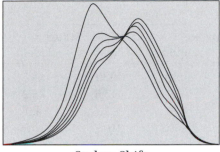

Stokes Shift

Figure 29.8 Raman spectra of water at different temperatures. The isosbestic point supports a model involving two states. Source: G D'Arrigo, G Maisana, F Mallamace, P Migliardo and F Wanderlingh, *J Chem Phys* **75**, 4264–4270 (1981).

Hydrogen Bonding Weakens with Temperature in Liquid Water

What is the strength of a hydrogen bond in liquid water, and how does it change with temperature? Estimates of the numbers, strengths, and angles of the hydrogen bonds in liquid water come from vibrational spectroscopy. Raman spectroscopy has been used to study the stretch vibration of OD bonds when a small amount of deuterated water, D_2O, is dissolved in a solvent of H_2O. The spectral shift from the D_2O peak at $2500 \, cm^{-1}$ (at 20 °C) toward $2650 \, cm^{-1}$ (at 400 °C, liquid at high pressure) is generally taken to indicate that hydrogen bonds bend or break with increasing temperature.

To say that a hydrogen bond has a 'strength' requires that there be two distinguishable states, made and broken, for which a free energy difference can be determined. For liquid water, it has long been debated whether hydrogen bonds fall into two such classes, or whether they are stretched or bent through a continuum of states. The first 'two-state' model was due to WK Röntgen in 1892 [2], who was better known for his discovery of x-rays, for which he won the first Nobel prize in Physics in 1901. He proposed that liquid water has two components: bulk and dense. A model by Nemethy and Scheraga [3] postulates five species—waters having zero to four hydrogen bonds. While much evidence supports the view that there is a continuum of hydrogen bonding in water, the two-state model is often found to be a useful approximation. Support for the two-state model comes partly from Raman experiments that show an *isosbestic point* (see Example 26.2, page 504), in which spectra at a series of different temperatures all intersect at a single point (see Figure 29.8).

Computer simulations of liquid water at room temperature also show that the continuum of hydrogen bonding in liquid water falls into two classes. They show that breaking a hydrogen bond increases the free energy by $\Delta G = 480 \, cal \, mol^{-1}$, with corresponding enthalpy and entropy increases of $\Delta H = 1.9 \, kcal \, mol^{-1}$ and $\Delta S = 4.77 \, cal \, mol^{-1} K^{-1}$ [4]. Figure 29.9 shows the prediction for how the amount of hydrogen bonding in liquid water decreases from the melting point to the boiling temperature. Most models predict that not all the hydrogen bonds are made at the melting point and not all the hydrogen bonds are broken at the boiling point.

Computer simulations suggest that liquid water has the structure of a fluctuating network of hydrogen bonds. Figure 29.10 shows a model distribution

Fraction Hydrogen-bonded

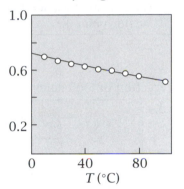

Figure 29.9 Fraction of hydrogen bonds made in liquid water over the liquid temperature range. Sources: DE Hare and CM Sorensen, *J Chem Phys* **93**, 6954–60 (1990); KAT Silverstein, ADJ Haymet, KA Dill, *J Am Chem* **122**, 8037–8041 (2000).

Table 29.1 Physical properties of some typical liquids. Source: F Franks, *Water*, The Royal Society of Chemistry, London (1983).

	Argon Solid	Argon Liquid	Benzene Solid	Benzene Liquid	Water Solid	Water Liquid
Density (kg m^{-3})	1636	1407	1000	899	920	997
Heat capacity (J (mol K)$^{-1}$)	30.9	41.9	11.3	132	37.6	75.2
Isothermal compressibility ((N m^2)$^{-1}$)	1	20	3	8.7	2	4.9
Self-diffusion coefficient (m^2 s^{-1})	10^{-13}	1.6 × 10^{-9}	10^{-13}	1.7 × 10^{-9}	10^{-14}	2.2 × 10^{-9}
Thermal conductivity (J (s m K)$^{-1}$)	0.3	0.12	0.27	0.15	2.1	0.58
Liquid range (K)		3.5		75		100
Surface tension (mJ m^{-2})		13		28.9		72
Viscosity (Poise)		0.003		0.009		0.01
Latent heat of fusion (kJ mol^{-1})		1.18		10		5.98
Latent heat of evaporation (kJ mol^{-1})		6.69		35		40.5
Melting point (K)		84.1		278.8		273.2

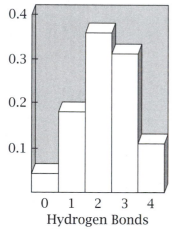

Figure 29.10 Simulated distribution of the number of hydrogen bonds per molecule in liquid water at 25 °C. Source: FH Stillinger, *Science* **209**, 451–457 (1980).

of hydrogen bonds in cold water, around room temperature. The distribution shifts with increasing temperature toward fewer average hydrogen bonds per water molecule. *Random network models* and *continuum models* predict that hydrogen bonds are stretched and bent through a continuum of angles. Figure 29.11 shows how hydrogen bonds loosen (stretch and bend) with temperature and how they compress and bend with pressure, according to a model of Henn and Kauzmann [5] that is based partly on measured spectral changes with temperature.

Pure Water Has Anomalous Properties

Increasing the temperature of water causes the solid phase (ice) to melt to a liquid, and then causes the liquid phase to boil to become a gas. In this regard, water is normal. But water is unusual in more subtle ways. The physical properties of water are compared with those of other liquids and solids in Table 29.1. Water has a high dielectric constant, in part because hydrogen bonds are polarizable. So water is a better solvent for ions than many other liquids. Water is more cohesive than liquids of comparable molecular size because it can form hydrogen bonds. This cohesion is indicated in Figure 29.12, which shows the high boiling and melting temperatures of water, and in Table 29.1, which shows water's relatively high enthalpy of vaporization. Water also has high melting and critical temperatures, and a high surface tension (about 70 dyn cm^{-1}) compared with the surface tensions of alkanes (about 30 dyn cm^{-1}).

Liquid water has a relatively large heat capacity for its size (see Table 29.1 and Figure 29.13(a)). The heat capacity $C_p = (\partial H/\partial T)$ describes the storage of energy (or enthalpy) in bonds that break or weaken with increasing temperature. A statistical mechanical model of Dahl and Andersen [6] shows that water is able to store energy in hydrogen bonds that can weaken, as well as van der Waals interactions that can break or weaken.

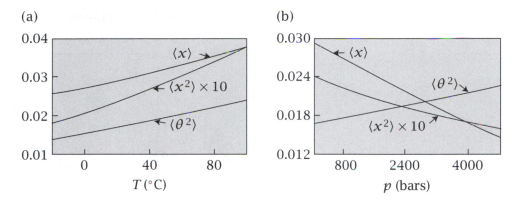

(a)

(b)

Figure 29.11 The model of Henn and Kauzmann predicts that (a) increasing the temperature T stretches and bends the water–water hydrogen bonds, while (b) increasing the pressure p shortens and bends those bonds. x is the hydrogen-bond length and θ is the hydrogen-bond angle. Source: AR Henn and W Kauzmann, *J Phys Chem* **93**, 3776–3783 (1989).

Another interesting property of water is that changes of pH move through the liquid unusually fast, faster than predicted by the diffusion constant for the hydrogen ions. Why? A model of Stanley and Teixeira [7] shows that liquid water is above a 'critical percolation threshold,' which means that uninterrupted pathways in hydrogen-bonding networks extend over macroscopic distances in water. The H^+ and OH^- ions can be transported rapidly via a 'bucket brigade' mechanism: one water donates a hydrogen to the next water, which passes a different hydrogen to water number three, and so on.

Another view of the cohesion of water comes from comparing H_2O with its deuterated and tritiated isotopes D_2O and T_2O. Figure 29.14 shows the entropy of transferring a single water molecule from the vapor phase into the pure liquid of each of these materials. The heavier isotopes have stronger bonds, presumably due to lower zero-point energies. The entropies measured in these experiments show that water composed of the heavier isotopes is more ordered than H_2O.

The Volumetric Anomalies of Water

Perhaps the most recognizable anomaly of water is its volume as a function of temperature. Figure 29.13(b) shows that the volume per molecule of water *decreases* upon melting, the reverse of simpler systems. For most materials, the solid is denser than the liquid: atoms are usually more closely packed together in solids than in liquids. The reverse is true for water. Ice floats. Ice is less dense than liquid water because the regular tetrahedra of ice are open, and the broken tetrahedra in liquid water are more dense.

This volumetric anomaly is also manifested in the $p(T)$ phase diagram (see Figure 29.15). For simple materials, the slope of the $p(T)$ phase boundary between the liquid and the solid is positive. Applying pressure to simple liquids squeezes the molecules together and drives them to freeze. But for water, the opposite happens: pressure can melt ice. For water, the slope of $p(T)$, of the liquid–solid phase boundary, is negative.

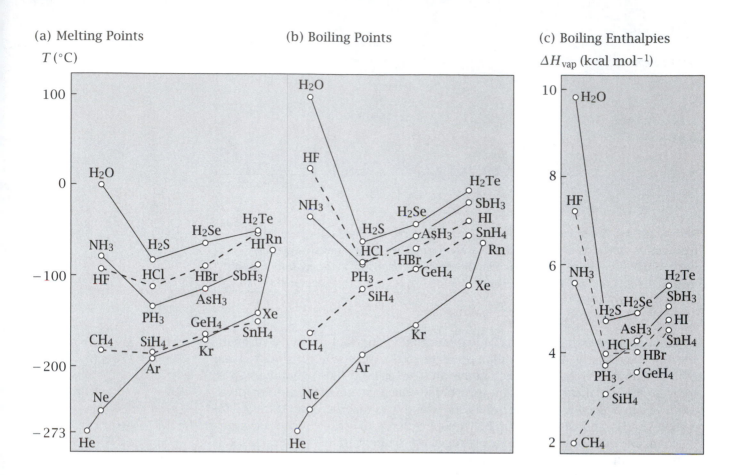

(a) Melting Points

$T\,(°C)$

(b) Boiling Points

(c) Boiling Enthalpies

ΔH_{vap} (kcal mol^{-1})

Figure 29.12 Water has higher (a) melting and (b) boiling temperatures and (c) higher enthalpy of vaporization than comparable molecules. Each connected line represents a series of isoelectronic molecules, stepping downward in the periodic table. Alternate lines are broken to make them easier to follow. Source: L Pauling, *General Chemistry*, Dover Publications, New York, 1970.

The slope of the phase boundary gives useful information about structural changes through the Clapeyron Equation (14.19), $dp/dT = \Delta s/\Delta v$, where Δs and Δv are the changes in entropy and volume between the two phases. Fundamentally, the entropy of a liquid is always higher than its corresponding solid, $\Delta s = s_{liquid} - s_{solid} > 0$ (provided that the liquid is the stable phase at the higher temperature). If dp/dT is positive, as it is for simple materials, the solid must be the denser phase, $\Delta v = v_{liquid} - v_{solid} > 0$. But if dp/dT is negative, as it is for water, the liquid must be denser, $v_{liquid} < v_{solid}$. Example 29.1 shows how Δs is computed from the slope of the phase boundary.

EXAMPLE 29.1 Computing Δs from the phase diagram. The liquid–solid boundary for the water-to-ice transition in Figure 29.15(a) has a slope

$$\frac{dp}{dT} = -133 \text{ atm K}^{-1}.$$

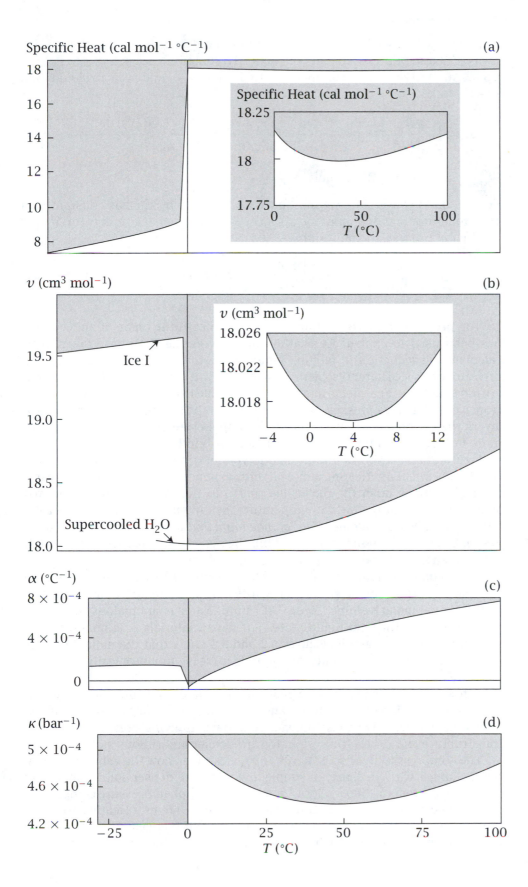

Figure 29.13 (a) The heat capacity of water, versus temperature T, from ice at below $-25\,°C$ to the boiling point. Comparison with Figure 8.11 shows that melting ice leads to a bigger increase in heat capacity, $C_p = dH/dT$, than in simpler liquids. An expanded scale of the specific heat of water versus temperature is shown over the liquid range. Source: D Eisenberg and W Kauzmann, *The Structure and Properties of Water*, Oxford University Press, New York, 1969. (b) The partial molar volume v of water versus temperature T. Ice is less dense than liquid water. The inset shows an expanded scale around the temperature of maximum density, $3.984\,°C$. (c) The thermal expansion coefficient $\alpha = (1/V)(\partial V/\partial T)_p$ of water versus temperature T. (d) The isothermal compressibility $\kappa = (1/V)(\partial V/\partial p)_T$ of water. The minimum is around $46\,°C$. Source of (b), (c), and (d): GS Kell, *J Chem Eng Data* **12**, 66 (1967).

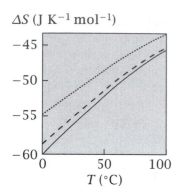

ΔS (J K^{-1} mol^{-1})

T (°C)

Figure 29.14 Entropies of transferring ($\cdots\cdots$) H_2O, ($-\,-\,-$) D_2O, or (———) T_2O from vapor to liquid water, versus temperature. The heavier isotopes have more ordered liquid phases. Source: A Ben-Naim, *Hydrophobic Interactions*, Plenum Press, New York, 1980.

This means that you can lower the melting temperature of water by 1 K for every 133 atm of pressure that you apply. To compute Δs, use $v_{\text{liquid}} - v_{\text{solid}} = 18.02\,\text{cm}^3\,\text{mol}^{-1} - 19.66\,\text{cm}^3\,\text{mol}^{-1} = 1.64\,\text{cm}^3\,\text{mol}^{-1}$, and rearrange Equation (14.19) to get

$$\Delta s = \Delta v \frac{dp}{dT} = \frac{\left(1.64\,\text{cm}^3\,\text{mol}^{-1}\right)\left(-133\,\text{atm}\,\text{K}^{-1}\right)\left(1.987\,\text{cal}\,\text{mol}^{-1}\text{K}^{-1}\right)}{\left(82.058\,\text{cm}^3\,\text{atm}\,\text{K}^{-1}\,\text{mol}^{-1}\right)}$$

$$= 5.28\,\text{cal}\,\text{mol}^{-1}\text{K}^{-1}.$$

For the units conversion, we multiplied and divided by the gas constant R in the numerator and denominator. As a check, you can also get Δs from Δh, for the fusion process at equilibrium

$$\Delta s_{\text{fusion}} = \frac{\Delta h_{\text{fusion}}}{T_{\text{fusion}}} = \frac{1.44\,\text{cal}\,\text{mol}^{-1}}{273.15\,\text{K}} = 5.28\,\text{cal}\,\text{mol}^{-1}\text{K}^{-1}.$$

The pressure-induced melting of the ice beneath the blade of an ice skate has been used to explain ice skating. But if you use the phase diagram to calculate the reduction in melting temperature that would be caused by the pressure of an ice skater (Problem 1, page 576), you will find that the pressures are too small. Surface effects might also have a role in pressure-induced melting from ice skates. Less pressure is required to melt the first few molecular surface layers than is predicted by the phase diagram, because surface molecules are more fluid, owing to their fewer intermolecular bonds.

Figure 29.13(b) shows a more subtle volumetric property of water called the *density anomaly*. Heating simple liquids expands them because the greater thermal motions push the molecules apart. In this regard, water is normal above 3.984 °C. But below that temperature, heating *increases* water's density. 3.984 °C is called the *temperature of maximum density*. The maximum density is subtle (see the inset in Figure 29.13(b)): the density increase from 0 °C to 4 °C is less than 1% of the density increase upon melting.

The volume anomalies are also reflected in Figure 29.13(c), which shows the thermal expansion coefficient $\alpha = (1/V)(\partial V/\partial T)_p$, the temperature derivative of the partial molar volume (Figure 29.13(b)). Water is anomalous for $\alpha < 0$, which occurs in the cold liquid state. For most materials α is positive at all temperatures. For example, Figures 9.2 and 9.5 show that the molar volumes of polyethylene, benzene, and other liquids increase monotonically with temperature.

A few other systems, including silicon and zirconium tungstate, ZrW_2O_8, are also orientationally structured and, like water, have negative thermal expansion coefficient. For these unusual materials, applying pressure melts out the structure and drives the system toward denser amorphous states.

The compressibility $\kappa = (1/V)(\partial V/\partial p)_T$ describes how the volume of a material changes with pressure. In simple materials, the denser solid state is less compressible than the looser liquid state. A given applied pressure compresses looser systems more than it compresses denser systems. Heating weakens the noncovalent bonds, decreasing the density and increasing the compressibility. This is observed for hot water (above 46 °C), but not for cold water.

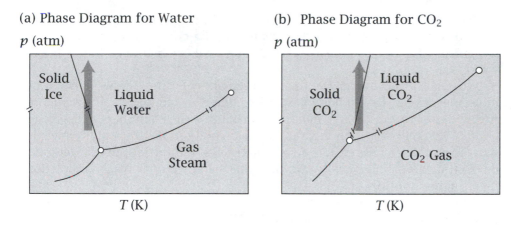

(a) Phase Diagram for Water

p (atm)

Solid Ice | Liquid Water | Gas Steam

T (K)

(b) Phase Diagram for CO_2

p (atm)

Solid CO_2 | Liquid CO_2 | CO_2 Gas

T (K)

Figure 29.15 Phase diagrams, pressure p versus temperature T for (a) water and (b) CO_2 show that $dp/dT < 0$ for water and $dp/dT > 0$ for CO_2. Applied pressure melts ice, but freezes CO_2. Source: P Atkins, *Physical Chemistry*, 6th edition, WH Freeman, New York, 1977.

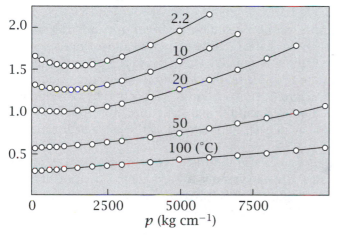

Viscosity (centipoise)

2.0

1.5

1.0

0.5

2.2

10

20

50

100 (°C)

0   2500   5000   7500

p (kg cm^{-1})

Figure 29.16 The viscosity of hot water increases with pressure, resembling simple liquids. But at low pressures, the viscosity of cold water (below about 20 °C) decreases with applied pressure. Curves are for different temperatures, in °C. Source: T DeFries and J Jones, *J Chem Phys* **66**, 896–901 (1977).

Figure 29.13(d) shows that water has an anomalous minimum in its isothermal compressibility. Heating cold water (below 46 °C) makes it less compressible. Interestingly, the minimum in the compressibility is at 46 °C, which is not the same as the temperature of maximum density. Figures 9.5 and 9.6 show that the thermal expansion coefficient and compressibility are both substantially smaller for water than for small organic molecules, indicating that water's underlying framework is more solid-like and less deformable than the frameworks of other liquids.

Another anomalous property of water is the pressure dependence of its viscosity. Because the viscosity of a liquid is a measure of the friction that hinders the flow of molecules past each other, strengthening the intermolecular attractions increases the viscosity. So normally, heating a liquid loosens the noncovalent bonds and lowers the viscosity. Applying pressure normally increases the viscosity. Water follows this behavior with temperature, and with

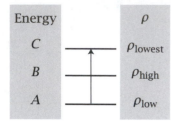

Figure 29.17 A qualitative energy diagram for three states of liquid water molecules. In state A tetrahedral ice-like molecules have low density. In state B higher-energy structures have weaker hydrogen bonding, and higher density than A. In C, molecules have weaker interactions and looser structures than B. Increasing temperature leads from mainly A molecules through B to C, increasing—then decreasing—the density.

pressure for high pressures. But Figure 29.16 shows that applying pressure to cold water (at pressures of less than 1000 kg cm^{-1}) decreases its viscosity, presumably because pressure bends and weakens hydrogen bonds in cold water.

What causes the unusual volumetric behaviors of water? Between two water molecules, the hydrogen bond is stronger than the van der Waals interaction. At low temperatures, water molecules maximize their numbers of hydrogen bonds. This makes ice tetrahedral. To form good tetrahedra, water molecules must form relatively open, low-density networks. But the tetrahedral geometry doesn't maximize the density. On the other hand, van der Waals interactions favor higher densities.

The general features of water's behavior can be described in terms of at least three general states (see Figure 29.17): A is the lowest energy state of liquid water molecules. Molecules in state A are at low density and have the structure of tetrahedral ice I_h. B is the next higher energy state of molecules in liquid water. Molecules in state B have some disruption of the tetrahedral structure, some broken or weakened bonding. They have higher density than A and perhaps better van der Waals bonding. A key issue in modelling water is to understand the structural basis for state B. Molecules in state C are at an even higher energy than in state B. In state C the structure is further loosened and weakened. State C has the lowest density.

At the lowest temperatures, water molecules are predominantly in state A. Increasing the temperature (through the melting point and above) increases the population of state B, increasing the density. Further heating of hot water increases the population of state C, lowering the density. While water undergoes thermal transitions from A to B to C, simpler liquids can generally be modelled by just the transition from B to C. The additional structural state in water gives a qualitative explanation for the density anomalies and the extra heat capacity of water.

This qualitative model also gives a guide to the effects of pressure. By Le Chatelier's principle, pressure drives systems toward their states of highest density. In hot water, as in simpler liquids, pressure drives the system from C to B, increasing structure, density, and viscosity. But in cold water, pressure drives the system from A to B, because B has higher density. B also has decreased structure and viscosity. It remains a challenge to account for these properties more quantitatively.

Supercooled and Stretched Water

Water can be supercooled to $-39.5\,°C$ at a pressure of one atmosphere, if ice nucleation and impurities are prevented. Water can also remain liquid if its pressure is lowered below its freezing point, at fixed temperature: this is called *stretched water*. Compared with other supercooled and stretched materials, water has two anomalous properties: the heat capacity and the isothermal compressibility grow large as the temperature approaches $-45\,°C$ (see Figure 29.18). Such divergences often indicate a phase transition. The microscopic origins of this behavior are not yet understood [8].

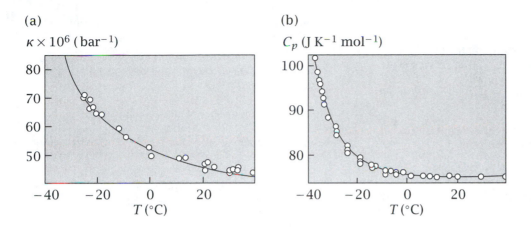

(a) $\kappa \times 10^6 \, (\text{bar}^{-1})$

(b) $C_p \, (\text{J K}^{-1} \, \text{mol}^{-1})$

Figure 29.18 (a) The isothermal compressibility κ and (b) the heat capacity C_p of water diverge upon supercooling to $-45\,°C$. Source: PG DeBenedetti, *Metastable Liquids, Concepts and Principles*, Princeton University Press, Princeton, 1966.

Summary

Water molecules have tetrahedral symmetry and the ability to form strong orientation-dependent hydrogen bonds. The tetrahedral symmetry appears in the structure of ice and in the underlying fluctuating hydrogen-bonded network structure of liquid water. Hot water acts like a normal liquid. It expands when heated. However, cold water has anomalous volumetric properties because of the competition between the hydrogen bonds that favor open tetrahedral structures, and the van der Waals interactions that favor denser disordered structures.

Problems

1. Does pressure-induced melting explain ice skating?

Suppose that an ice skater weighs $m = 70$ kg. An ice skate exerts a pressure $\Delta p = mg/A$ on ice, where $A = 1.0 \text{ cm}^2$ is the surface area of a skate, and g is the gravitational constant.

(a) Compute the decrease in freezing temperature of water caused by this applied pressure.

(b) Is this pressure sufficient to melt ice?

2. Comparing water with other small molecules.

Table 29.2 shows the melting and boiling temperatures and enthalpies of several substances.

(a) Compute the melting and boiling entropies for these substances.

(b) Plot the enthalpies, entropies, boiling and freezing temperatures versus molecular size, and explain the anomalies.

Table 29.2 Enthalpies of fusion, ΔH_{fus} at various pressures and enthalpies of evaporation, ΔH_{vap} at 1 atm at the transition temperatures, T_{fus} and T_{vap}. Source: P Atkins, *Physical Chemistry*, 6th edition, Freeman, New York (1997).

Substance	T_{fus} (K)	ΔH_{fus} (kJ mol^{-1})	T_{vap} (K)	ΔH_{vap} (kJ mol^{-1})
Ar	83.81	1.188	87.29	6.506
H_2	13.96	0.117	20.38	0.9163
N_2	63.15	0.719	77.35	5.586
O_2	54.36	0.444	90.18	6.820
Cl_2	172.12	6.406	239.05	29.410
Br_2	265.90	10.573	332.35	29.45
I_2	386.75	15.52	458.39	41.80
CO_2	217.0	8.33	194.64	25.23^a
H_2O	273.15	6.008	373.15	40.656
NH_3	195.40	5.652	239.73	23.351
H_2S	187.61	2.377	212.80	18.673
CH_4	90.68	0.941	111.66	8.18
C_2H_6	89.85	2.86	184.55	14.7
C_6H_6	278.65	10.59	353.25	30.8
CH_3OH	175.25	3.159	337.22	35.27

^{a} sublimation

3. The residual entropy of ice.

The entropy of ice at $T = 0$ K is -3.43 J K^{-1} mol^{-1}. This can be explained in two steps. If there are N water molecules in an ice crystal, there are $2N$ hydrogens. Each hydrogen is situated between two oxygens (see Figure 29.2). Imagine that the oxygens define a three-dimensional grid, and that the hydrogens are located between oxygens.

(a) How many configurations of the hydrogens would there be in the absence of any other constraints?

(b) How many configurations of the hydrogens would there be if every oxygen must have exactly 2 hydrogens assigned to it?

4. Global warming of the oceans.

The oceans on Earth weigh 1.37×10^{21} kg [9]. If global warming caused the average ocean temperature to increase by 1 °C, what would be the enthalpy increase ΔH?

5. The tip of the iceberg.

When ice floats on water, only a fraction h of the ice is above the water. Compute h.

References

[1] FH Stillinger. *Science* **209**, 451–457 (1980).

[2] WK Röntgen. *Ann Phys Chem* **45**, 91 (1892).

[3] G Nemethy and HA Scheraga. *J Chem Phys* **36**, 3382 (1962).

[4] KAT Silverstein, ADJ Haymet and KA Dill. *J Am Chem Soc* **122**, 8037–8041 (2000).

[5] AR Henn and W Kauzmann. *J Phys Chem* **93**, 3770 (1989).

[6] LW Dahl and HC Andersen. *J Chem Phys* **78**, 1980–1993 (1983).

[7] HE Stanley and J Teixeira. *J Chem Phys* **73**, 3404 (1980).

[8] PG DeBenedetti. *Metastable Liquids: Concepts and Principles*, Princeton University Press, Princeton, 1996.

[9] W Stumm and JJ Morgan. *Aquatic Chemistry: Chemical Equilibria and Rates in Natural Waters*, 3rd edition, Wiley, New York, 1996.

Suggested Reading

Four of the main references on the properties of water:

PG Debenedetti, *Metastable Liquids: Concepts and Principles*, Princeton University Press, Princeton, 1996.

D Eisenberg and W Kauzmann, *The Structure and Properties of Water*, Oxford University Press, New York, 1969.

F Franks, *Water*. The Royal Society of Chemistry, London, 1983.

F Franks (ed), *Water, A Comprehensive Treatise*. Plenum Press, New York, 1972–1982.

30 Water as a Solvent

Water is called a *universal solvent*. Because it is polar and can donate and accept hydrogen bonds, it can solvate a wide range of polar and hydrogen bonding molecules and ions. But water has little ability to dissolve inert gases or hydrocarbons. The low solubility of oil in water is sometimes called *the hydrophobic effect*. However, the inability to mix is not remarkable *per se*. Many compounds are insoluble in others. Benzene won't dissolve in water, but it also won't dissolve in formamide or glycerol. What is so special about the disaffinity of oil and water?

Oil and Water Don't Mix: The Hydrophobic Effect

Three properties set the mixing of oil and water apart from other solvation processes. The mixing of oil and water is opposed by a much larger free energy. The opposition to mixing is mainly entropic at 25 °C. And the transfer of oil into water is accompanied by a large positive change in the heat capacity.

Two processes are of interest: transferring a solute from the vapor phase into water (see Figure 16.3, page 282), and transferring a solute from a pure liquid phase into water (see Figure 16.11, page 293). For comparison, let's use the lattice model of solutions, described in Chapters 14 to 16, to define 'normal' solvation. According to Equation (16.4), transferring a lattice model solute B from the vapor phase into solvent A involves opening a cavity in the solvent,

(a) Aqueous Solution

kJ mol^{-1}

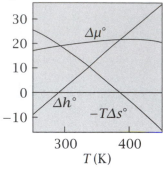

(b) Simple Solution

kJ mol^{-1}

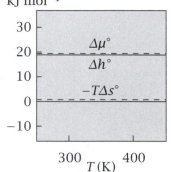

Figure 30.1 Comparison of the molar enthalpy Δh°, entropy Δs°, and free energy $\Delta\mu^\circ$ for transferring (a) benzene from its pure liquid into water, and (b) a solute from its pure liquid into a simple solution. The entropy at high temperature is extrapolated assuming constant heat capacity. Sources: (a) PL Privalov and SJ Gill, *Adv Prot Chem* **39**, 191–234 (1988); (b) KA Dill, *Biochemistry* **29**, 7133–7155 (1990).

and A interacts with B, giving a vapor pressure:

$$\frac{p_B}{p^\circ_{B,\text{int}}} = x_B e^{\Delta\mu^\circ/kT} \quad \text{where} \quad \Delta\mu^\circ = z\left[w_{AB} - \frac{w_{AA}}{2}\right]. \tag{30.1}$$

Transferring B from the pure *liquid* state, rather than from the vapor phase, differs in two respects, as shown in Figure 16.11. Liquid BB bonds must be broken, and the cavity in the liquid B must be closed after the transfer. Equations (16.39) and (16.40) give the partition coefficient:

$$K^A_B = e^{-\Delta\mu^\circ/kT}, \tag{30.2}$$

where

$$\Delta\mu^\circ = z\left[w_{SB} - w_{SA} + \frac{w_{AA}}{2} - \frac{w_{BB}}{2}\right]. \tag{30.3}$$

According to the lattice model of simple solutions, the pair attractions described by the energies w_{AA}, w_{BB}, and w_{AB} are independent of temperature, so $\Delta\mu^\circ$ is independent of temperature. In the lattice model, the disaffinity of A for B has no entropic component, and arises because the AA and BB attractive energies are stronger than the AB attractions. The next section describes how oil–water immiscibility differs from this.

The Signature of Hydrophobicity Is Its Temperature Dependence

To determine $\Delta\mu^\circ$ from experiments, you measure the concentration c_{SW} of the solute in the water phase and the concentration c_{SO} of the solute in the oil phase, at equilibrium. The ratio is the partition coefficient $K(T) = c_{SW}/c_{SO}$, which can change with temperature T. Such measurements give $\Delta\mu^\circ(T) = -kT\ln K(T)$. (This works only if A and B are approximately the same size and nearly spherical. Chapter 31 shows that a different form of $\Delta\mu^\circ$ is needed if one species is a polymer and is much larger than the other.)

You can express $\Delta\mu^\circ(T)$ in terms of its enthalpic and entropic components (see Equation (9.25)),

$$\Delta\mu^\circ = \Delta h^\circ - T\Delta s^\circ, \tag{30.4}$$

where Δh° is the molar enthalpy and Δs° is the molar entropy. Because Δh° and Δs° in Equation (30.4) may themselves depend on temperature, you cannot determine these quantities uniquely unless you have an independent way to measure them. You can make such measurements by calorimetry. Calorimetry gives the molar heat capacity change upon transfer, Δc_p. According to Equations (8.34) and (8.35), Δc_p can be used to uniquely determine both the enthalpy and entropy,

$$\Delta h^\circ(T) = \Delta h^\circ(T_1) + \int_{T_1}^{T} \Delta c_p \, dT' \quad \text{and} \tag{30.5}$$

$$\Delta s^\circ = \Delta s^\circ(T_1) + \int_{T_1}^{T} \frac{\Delta c_p}{T'} \, dT'. \tag{30.6}$$

Equations (30.4) to (30.6) lead to a recipe for handling experimental data.

Table 30.1 Thermodynamic properties for the transfer of small hydrocarbons from their pure liquid state to water, based on solubilities at $T = 298$ K. Source: SJ Gill and I Wadsö, *Proc Natl Acad Sci USA* **73**, 2955–2958 (1976).

Compound		$\Delta\mu°$ (kJ mol^{-1})	$\Delta h°$ (kJ mol^{-1})	$\Delta s°$ (J deg^{-1} mol^{-1})	$\Delta C_p°$ (J deg^{-1} mol^{-1})
C_6H_6	Benzene	19.33	2.08	−57.8	225
C_7H_8	Toluene	22.82	1.73	−70.7	263
C_8H_{10}	Ethyl benzene	26.19	2.02	−81.1	318
C_9H_{12}	Propyl benzene	28.8	2.3	−88.9	391
C_5H_{12}	Pentane	28.62	−2.0	−102.7	400
C_6H_{12}	Cyclohexane	28.13	−0.1	−94.7	360
C_6H_{14}	Hexane	32.54	−0.0	−109.1	440

Using a calorimeter, measure Δc_p for solute transfer. For *simple solutions*, Δc_p is small, so $\Delta h°$ and $\Delta s°$ are constants, independent of temperature. Then the quantities $\Delta h°$ and $\Delta s°$ can be determined from $-kT \ln K(T)$. $\Delta\mu°$ is either independent of temperature (if $\Delta s° \approx 0$), or it is a linear function of temperature. As we noted, $\Delta s° = 0$ for simple systems, so $\Delta\mu° \approx \Delta h°$ is independent of temperature (see Figure 30.1(b)).

However, when oils are dissolved in water, experiments show that $\Delta c_p \gg 0$. If Δc_p is independent of temperature, Equations (30.5) and (30.6) become

$$\Delta h°(T) = \Delta h°(T_1) + \int_{T_1}^{T} \Delta c_p dT' = \Delta h°(T_1) + \Delta c_p(T - T_1) \tag{30.7}$$

and

$$\Delta s°(T) = \Delta s°(T_2) + \int_{T_2}^{T} \frac{\Delta c_p}{T} dT' = \Delta s°(T_2) + \Delta c_p \ln\left(\frac{T}{T_2}\right), \tag{30.8}$$

where T_1 and T_2 are temperatures at which $\Delta h°$ and $\Delta s°$ are known. So Equation (30.4) becomes

$$\Delta\mu°(T) = \Delta h°(T_1) - T\Delta s°(T_2) + \Delta c_p \left[(T - T_1) - T\ln\left(\frac{T}{T_2}\right)\right]. \tag{30.9}$$

Here's a convenient way to determine the quantities in Equation (30.9). Find the temperature T_h at which the enthalpy is zero (you can show that $\Delta h° = 0$ occurs where $\Delta\mu°/kT$ has a maximum or minimum). Also find the temperature T_s at which the entropy is zero ($\Delta s° = 0$ where $\Delta\mu°$ has a maximum or minimum). Then find the value of Δc_p that causes Equation (30.9),

$$\Delta\mu°(T) = \Delta c_p \left[(T - T_h) - T\ln\left(\frac{T}{T_s}\right)\right], \tag{30.10}$$

to best fit the data.

Experimental results for $\Delta\mu°$ and its components $\Delta h°$, $T\Delta s°$, and Δc_p for transferring nonpolar solutes from their pure liquid phase into liquid water are given in Table 30.1. The table shows that $\Delta\mu°$ is positive, $\Delta c_p \gg 0$, $\Delta h° \approx 0$, and

Figure 30.2 Solubility of benzene in water versus temperature T. Benzene is least soluble around 18 °C. Source: C Tanford, *The Hydrophobic Effect: Formation of Micelles and Biological Membranes*, 2nd edition, Wiley, New York (1980). Data are from SJ Gill, NF Nichols, I Wadsö, *J Chem Thermo* **8**, 445 (1976).

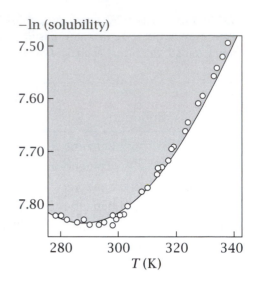

$\Delta s° \ll 0$ around room temperature. The aversion of oil for water is entropic at 25 °C. It becomes enthalpic at higher temperatures. The large heat capacity implies curvature in the free energy function (see Figure 30.1(a)).

Figure 30.2 shows this curvature in $\ln(\text{solubility}) = \Delta\mu°/RT$ for benzene in water. There is a minimum in the solubility as a function of temperature, around 18 °C. The same general behavior is observed with alkanes, alcohols, or inert gases dissolved in water.

What is the microscopic basis for the hydrophobic effect? Because the quantities $\Delta\mu°$, $\Delta h°$, $\Delta s°$, and Δc_p are proportional to the solute surface area, hydrophobicity is proportional to the numbers of water molecules in the first solvation shell (see, for example, Jorgensen et al. [1]).

The Hydrophobic Entropy
Is Due to the Ordering of Waters Around the Solute

What is the basis for the excess entropy $\Delta s° \ll 0$ that opposes the mixing of oil and water? It cannot be due to solute–solute interactions because the experiments are performed at very low solute concentrations. Rather, it seems that water is ordered by the presence of solutes. For example, water forms polygonal cages around nonpolar molecules in *natural gas clathrate hydrates* [2]. A clathrate is a polyhedral cage of water molecules that contains a single hydrocarbon molecule such as CH_4. Because of their commercial importance, much is known about clathrate structures. When natural gas flows through pipelines from offshore drilling rigs, it collects water. They combine to form solid clathrates of water and hydrocarbons that clog up the pipelines.

Computer simulations indicate that a solute that is introduced into water at 25 °C orders the first shell of neighboring water molecules, although to a smaller degree than the solutes in clathrates. The first-shell water molecules orient to avoid 'wasting' hydrogen bonds (see Figure 30.3). Independent evidence that the hydrophobic entropy is due to water–water hydrogen bonding comes from comparing water with hydrazine ($H_2N = NH_2$) [3]. Hydrazine resembles water

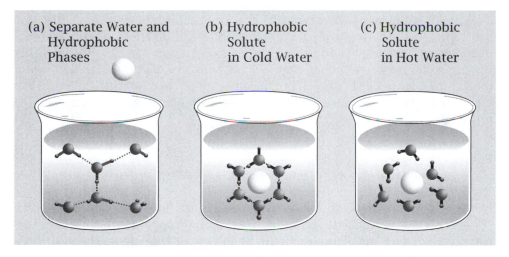

(a) Separate Water and Hydrophobic Phases

(b) Hydrophobic Solute in Cold Water

(c) Hydrophobic Solute in Hot Water

Figure 30.3 The 'iceberg' model of the heat capacity of (a) the transfer of nonpolar solutes into water [4, 5]. (b) In cold water (near room temperature), the water molecules surrounding the nonpolar solute form good hydrogen bonds (low energy), in structured cages (low entropy) that avoid 'wasting' hydrogen bonds. (c) In hot water, more first-shell water conformations are accessible (higher entropy), but some of them have weaker or unformed hydrogen bonds and/or van der Waals interactions (higher energy). This contributes to the heat capacity because the system enthalpy increases with temperature.

in many of its normal properties, but it cannot form hydrogen-bonded networks and does not have water's volumetric anomalies. Oil is no more soluble in hydrazine than in water, but the insolubility in hydrazine is enthalpic.

Why is the hydrophobic entropy so large? Here's a qualitative argument. Suppose that each first-shell water molecule could form good water–water hydrogen bonds in half of all its possible orientations around the solute. Table 30.2 shows that small solutes are surrounded by about 15–30 first-shell water molecules. If each of the 15 water molecules in the shell were restricted independently, then the hydrophobic entropy would be

$$\Delta S = k \ln W = k \ln(1/2)^{15} = -86 \, \text{J K}^{-1}\text{mol}^{-1}.$$

The factor of $1/2$ is a reasonable estimate [6, 7]. A water molecule W on a tetrahedral lattice has six orientations of its two hydrogen bonds and two lone pairs in which it is fully hydrogen bonded (see Figure 30.4(a)). Remove a water from a site adjacent to W (the gray dot at the back of the tetrahedron in Figure 30.4(a)), and replace it with a nonpolar sphere. The number of hydrogen-bonding orientations of W is now 3. The entropy of transfer (per solute molecule) into water is large, not because the solute molecule is ordered, but because the *many neighboring water molecules* surrounding the solute molecule are ordered.

Why Is $\Delta h° \approx 0$ at 25 °C?

For simple solutions, the transfer of solute A from its pure liquid into solvent B typically has an unfavorable enthalpy, $\Delta h° > 0$. However, the interaction of

Table 30.2 Surface areas A of small nonpolar solutes, and numbers N_w of first-neighbor waters. Source: SJ Gill, SF Dec, G Oloffson and I Wadsö, *J Phys Chem* **89**, 3758–3761 (1985).

Molecule	$A(\text{Å}^2)$	N_w
CH_4	152.36	17
C_2H_6	191.52	21
C_3H_8	223.35	25
N_2	142.41	16
CO	145.31	16
O_2	135.74	15
He	105.68	12
Ne	116.13	13
Ar	143.56	16
Kr	155.70	17
Xe	168.33	19

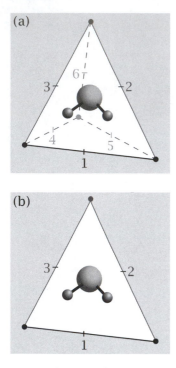

Figure 30.4 (a) A water molecule has six hydrogen-bonding configurations when it is surrounded by four neighboring waters. Neighboring water molecules are shown as vertices of a tetrahedron. The six edges of the tetrahedron indicate the pairs of acceptors for the two hydrogen-bond donors of the central water. (b) When a nonpolar solute replaces the water at the back vertex, the central water now has only three fully hydrogen-bonded configurations with neighboring water molecules. Source: G Ravishankar and DL Beveridge in *Theorteical Chemistry of Biological Systems*, DL Beveridge and WL Jorgensen, eds, Elsevier, Amsterdam, 1986.

a nonpolar solute with water typically has a more favorable enthalpy, $\Delta h° \approx 0$; see Table 30.1. The relatively favorable enthalpy of inserting oil into water is observed in experiments and in computer simulations [8, 15]. It seems that inserting oil into water creates cage-like ordering in which the first-shell water molecules have better hydrogen bonds, on average, than tetrahedral (pure) water. Water favors polygons and clathrate cages that are slightly larger than the tetrahedra that water forms in the pure liquid.

What Is the Basis for the Large Hydrophobic Heat Capacity?

For transferring nonpolar solutes from their pure liquid to water, there is a change in the heat capacity $\Delta c_p \gg 0$. This implies that a nonpolar solute surrounded by water has a greater ability to absorb enthalpy from the surroundings than the corresponding pure components alone (because $\Delta c_p = \partial h/\partial T$; see Chapter 12). Figure 30.5 shows that the hydrophobic heat capacity change Δc_p grows with the solute size, in proportion to the number of water molecules in the first shell, implying that the mechanism of enthalpy absorption may be localized in the first shell.

Figure 30.3 shows the 'iceberg' model of heat capacity, from Frank and Evans [5]. At low temperatures (25 °C), first-shell water molecules are ordered (low entropy) and form good hydrogen bonds with other water molecules (low enthalpy). Heating melts these icebergs. Any melting process involves an increase of entropy and enthalpy. Warming up cold water increases the enthalpy by bending, breaking, or loosening water–water hydrogen bonds in the first solvation shell around the solute. Water gains entropy from this bending and loosening. But the term 'iceberg' should not be taken to imply crystal-like ordering in cold liquid water. The order of liquid water around nonpolar solutes is substantially less than the order in ice.

Hydrophobic Solvation Near the Critical Temperature of Water

Figure 30.6 shows how the volume and enthalpy in a system of neon dissolved in water changes near the critical temperature of water, 647.3 K (at $p = 218$ atm). The hydrophobic enthalpy and the partial molar volume of a nonpolar solute in water grow large and diverge as the critical temperature is approached. Why? Fluctuations become large near critical points (see Chapters 25 and 26). In some regions of space, water is clustered as in the liquid, while in other regions, there are large holes, corresponding to many broken intermolecular interactions, a state of high enthalpy. Near the critical point, nonpolar solutes create large holes.

Water Is Structured Near Cavities and Planar Surfaces

Computer simulations by Postma et al. [10] show that water molecules around small empty cavities behave much like water molecules around nonpolar solutes: the first-shell water molecules orient to avoid pointing their hydrogen-bonding groups toward the cavity and wasting hydrogen bonds. In addition,

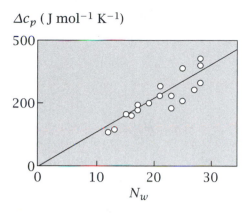

Δc_p (J mol^{-1} K^{-1})

N_w

Figure 30.5 Experimental heat capacity of transfer Δc_p versus the number N_w of water molecules in the first solvation shell for hydrocarbon gases. Source: SJ Gill, SP Dec, G Oloffson and I Wadsö, *J Phys Chem* **89**, 3758–3761 (1985). .

the energy of opening a cavity increases substantially with radius d (see Figure 30.7). So most cavities in water are relatively small.

Water molecules adjacent to planar nonpolar surfaces behave differently from those adjacent to small spherical solutes [11, 12]. In both cases, cold water molecules (room temperature or below) are driven to be maximally hydrogen bonded. But water molecules cannot straddle a planar interface. A planar geometry forces each first-shell water molecule to waste one hydrogen bond, which it points directly at the surface. Sacrificing one hydrogen bond is the best the water can do near a plane. In contrast, water near small spheres can form all four possible hydrogen bonds. Inserting large planar interfaces into water costs enthalpy, rather than entropy. Because of this high energetic cost, when two planar surfaces are brought close together, the intervening water molecules evacuate the intervening space [13].

So far we have considered the insertion of a single nonpolar molecule into water. This is sometimes called *hydrophobic hydration* or *hydrophobic solvation*. Now consider bringing two nonpolar solutes together in water. This is called *hydrophobic interaction*.

Potentials of Mean Force Describe Interactions in Fluids

Imagine putting two solute molecules in water. Brownian motion separates them by different distances r at different times. Suppose you plot how often the solute molecules are found at separation r, for all possible values $r = 0$ to $r \to \infty$. This is the radial distribution function $g(r)$ (see page 461). $w(r)$, the *potential of mean force* (pmf), is defined as the corresponding free energy,

$$w(r) = -kT \ln g(r).$$

The potential of mean force $w(r)$ applies to any system having a pair distribution function $g(r)$, including pure liquids and solids. Figure 30.8 shows $g(r)$ and $w(r)$ for two spherical solutes in a solvent.

If the same two molecules were brought together in the gas phase, $w(r)$ would simply be the pair potential $u(r)$ (see Figure 24.1), which has only a single minimum. But the potential of mean force between two particles in liquids—even simple liquids—oscillates with maxima and minima. For a given

(a) Partial Molar Volume of Solvation

v (cm^3 mol^{-1})

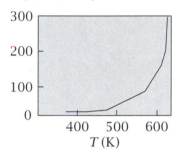

T (K)

(b) Enthalpy of Solvation

$\Delta h°$ (kJ mol^{-1})

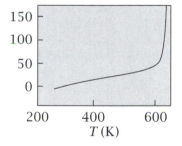

T (K)

Figure 30.6 (a) The partial molar volume v and (b) enthalpy $\Delta h°$ of solvation of neon in water diverge as the temperature approaches the critical temperature of water. Source: R Crovetto, R Fernandez-Prini, and ML Japas, *J Chem Phys* **76**, 1077–1086 (1982).

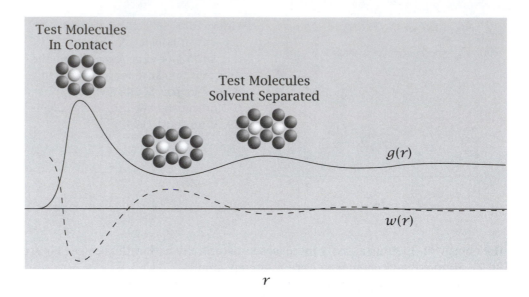

Figure 30.8 Two white spheres represent solute test particles in a solvent of darker colored spheres. The x axis shows the separation r between the test particles. $g(r)$ (——) is a typical distribution of the different separations of the test particles. In this case the particles prefer to be in contact. Slightly greater separation is unfavorable. A second small peak indicates a small preference of the solutes to be separated by a single layer of solvent. The potential of mean force $w(r)$ (– – –) is the corresponding free energy.

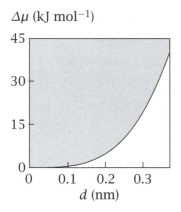

Figure 30.7 The free energy of creating a cavity of radius d in water. Source: G Hummer, S Garde, AE Garcia, A Pohorille and LR Pratt, *Proc Natl Acad Sci USA* **93**, 8951 (1996).

separation r between the two molecules of interest, the pmf describes an average over all the conformations of the surrounding solvent molecules. It shows the apparent attractions and repulsions between solute molecules that arise from the surrounding solvent.

Figure 30.8 illustrates the main features of the pmf. First, the test molecules may come into contact if they are attracted to each other. This separation can be the position of the deepest well in the pmf. The second well in the pmf is called the *solvent-separated* minimum. This is where the two test particles are separated by a single layer of solvent. Each test particle protects the near side of the other particle from collisions with the solvent that would otherwise drive two solute molecules apart. So it is more probable that solvent collisions will drive two nearby solute particles together than that they will drive them apart. This means that there is an apparent solvent-driven attraction of the test particles even when they are *solvent-separated* by one solvation shell. The system is unstable when two particles that are too close together for a solvent molecule to fit between them.

In water, nonpolar solute molecules associate with each other. According to the simplest argument, pairing up offers the molecules the advantage of reducing the net surface area of contact with water, reducing the number of first-shell water molecules, and thus reducing the unfavorable entropy at 25 °C. Figure 30.9 illustrates that the solute–solvent contact surface area is reduced when two lattice solutes come into contact.

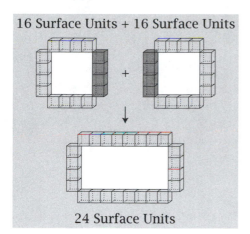

16 Surface Units + 16 Surface Units

24 Surface Units

Figure 30.9 Why do nonpolar solutes associate in water? Association reduces the total surface area of contact between solute molecules and the surrounding water. This lattice picture shows that when two solutes are isolated, they have 32 units of surface contact with the solvent, but when the solutes contact each other, they have only 24 units of surface contact with water.

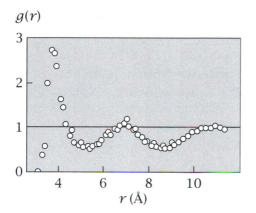

$g(r)$

r (Å)

Figure 30.10 Computed krypton–krypton pair correlation function in water. $g(r)$ describes the enhancement (or deficit) of krypton atoms at a given distance from another krypton, relative to the value at large separations, $g(r) = 1$ as $r \to \infty$. The second peak indicates that some solute pairs are separated by water. Source: K Watanabe and HC Andersen, *J Phys Chem* **90**, 795 (1986).

Modeling by Pratt and Chandler [14], Zichi and Rossky [9, 15], Watanabe and Andersen [16], and Ravishanker et al. [17, 18] shows that the solvent-separated minimum can sometimes be most stable. For example, Figure 30.10 shows the krypton–krypton pair correlation function in water computed by Watanabe and Andersen [16]; the second small peak represents a population of solvent-separated pairs. This state is stable because each nonpolar molecule is surrounded by a clathrate cage. Solvent-separated states are most likely to occur when the solutes are small [19].

Much information about water as a solvent comes from mixtures with alcohols, which are often highly soluble.

Alcohols Constrict the Volumes in Mixtures with Water

At high concentrations, solutes can disrupt the structure of water. Figure 30.11 shows the partial molar volumes of alcohols in mixtures of alcohol and water, as a function of alcohol concentration. Adding a small amount of alcohol to water leads to a constriction. The volume of the mixture is smaller than the sum of volumes of the pure alcohol and pure water components. The water

Figure 30.11 Partial molar volumes v of alcohol water mixtures, versus alcohol concentration, for methanol (MeOH), ethanol (EtOH), n-propanol (n-PrOH), and t-butanol (t-BuOH). Source: F Franks *Water*, The Royal Society of Chemistry, London, 1983.

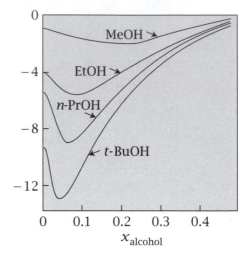

molecules pack better around the alcohol molecules than around other water molecules. However, at higher concentrations, alcohol molecules pervade the system so extensively that they disrupt the fluctuating network of hydrogen-bonded water polygons.

Ions Can Make or Break Water Structure

Ions Order Water Molecules by an Electrostatic Mechanism

Like nonpolar solutes, ions can cause the structuring of water molecules. But the ordering of water molecules that is induced by ions is different from the ordering induced by oils. While structure around nonpolar solutes in cold water is driven by the maximization of water–water hydrogen bonds, the structure around ions is driven by the electrostatic interaction of the ion with the nearby water molecules. Water's oxygen is the negative end of a dipole, so it is attracted to positive ions. And water's hydrogens are the positive end of a dipole, so they are attracted to negative ions. Inserting nonpolar solutes can *strengthen* water–water hydrogen bonds. Inserting ions can either strengthen *or weaken* them. Both mechanisms involve changes in entropy, according to computer simulations [20].

The electrostatic mechanism of water ordering depends on not just the charge of an ion, but on its charge density [21]. High charge density results either from a high charge on an ion or a small ionic radius. The electrostatic potential at the surface of a spherical ion is proportional to the ion's charge and inversely proportional to its radius (see Equation (21.37)). Water molecules bind to small or multivalent ions very tightly. This is reflected in the enthalpies of solvation (see Figure 22.14): water molecules are bonded more tightly to small ions such as Li^+ than to large ions such as K^+. These strong ion–water attractions also cause *electrostriction*, the negative partial molar volume that results when ions are inserted into water.

	Cl⁻	Br⁻	I⁻	NO₃⁻
Li⁺	0.139	0.106	0.081	0.101
Na⁺	0.07866			
K⁺	−0.0140	−0.0480		−0.0531
Rb⁺	−0.037	−0.061	−0.11	
Cs⁺	−0.050		−0.1184	−0.092

Table 30.3 Jones–Dole B coefficients at $T = 25°$ C (see Equation (30.11)). Source: RW Gurney, *Ionic Processes in Solution*, Dover Publications, New York, 1962. Data are from G Jones and SK Talley, *J Am Chem Soc*, **55**, 624 (1933); G Jones and RE Stauffer, *J Am Chem Soc*, **62**, 336 (1940); and VD Laurence and JH Wolfenden, *J Chem Soc*, 1144 (1934).

Figure 30.12 illustrates the continuum of mechanisms of water ordering. Small highly charged ions strongly orient water's dipole and may cause water molecules to break hydrogen bonds. Large weakly charged ions cause less electrostatic ordering and allow more hydrogen bonding between first-shell water molecules. Neutral solutes induce a high degree of first-shell water–water hydrogen bonding. This spectrum of mechanisms is reflected in heat capacities: inserting nonpolar solutes into water increases the heat capacity, but inserting ions can decrease it. We describe how ions affect water structure in the next section.

Electrostatic Ordering by Kosmotropes and Disordering by Chaotropes

Ions are sometimes divided into two classes, called *kosmotropes* (*structure makers*) and *chaotropes* (*structure breakers*) (see Figure 30.12). The distinction between making and breaking structure is not always well defined and can depend on what experimental method is used to monitor it.

Among the earliest studies were the experiments of Jones and Dole in 1929 on the viscosities of ions in dilute aqueous solutions [22]. Jones and Dole found that the viscosity η of an aqueous solution depends on the ion concentration c according to the relation

$$\eta = \eta_0(1 + A\sqrt{c} + Bc), \tag{30.11}$$

where η_0 is the viscosity of pure water and A and B are constants determined by the ionic solute. The term $A\sqrt{c}$ dominates at low concentrations (below about 0.01 M). It is positive for all ions, and originates in the same type of electrostatic interactions that give rise to the $\sqrt{c}$ term in the Debye–Hückel theory of activity coefficients (see page 441). Positive and negative ions are attracted to one another, contributing extra cohesion to the liquid, making it more viscous.

The Jones–Dole B term becomes important at higher ion concentrations and defines so-called *specific ion* effects. The B term can be either negative or positive. It carries information about how much structure the ions impose on water. Ions with $B > 0$ are considered structure makers because increasing the concentrations of those ions increases the viscosity of an aqueous solution. Each flowing ion drags along some water. Ions with $B < 0$ are considered to be structure breakers because increasing their concentrations decreases the viscosity.

Table 30.3 lists some B coefficients. Comparison with Figure 30.13 shows that structure-making cations ($B > 0$) have high charge density and structure

(a) Hydrophobic Hydration

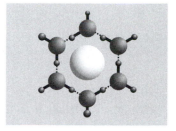

(b) Structure-breaking Ionic Hydration

(c) Structure-making Ionic Hydration

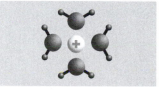

Figure 30.12 A continuum of water ordering mechanisms, from (a) zero charge density on the solute, where first-shell water molecules form hydrogen bonds, to (c) high charge density on an ion, where water's dipole moment orients in the field of the charge.

Figure 30.13 Relative radii of ions. Source: *CRC Handbook of Chemistry and Physics*, 81st edition, DR Lide, editor, CRC Press, Boca Raton, 2000.

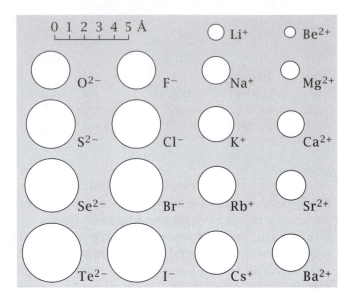

Table 30.4 Molar solubilities of salts. Source: KD Collins, *Biophys J* **72**, 65–76 (1997).

	Solubility (Molarity)			
	F^-	Cl^-	Br^-	I^-
Li^-	0.1	19.6	20.4	8.8
Na^-	1.0	6.2	8.8	11.9
K^-	15.9	4.8	7.6	8.7
Rb^-	12.5	7.5	6.7	7.2
Cs^-	24.2	11.0	5.1	3.0

breakers ($B < 0$) have low charge density. So viscosity is another property that reflects how tightly water binds to ions. As temperature increases, B values grow more positive.

Another property that seems to depend on charge density is ionic solubility in water.

Ion Pairing Preferences in Water Depend on Charge Densities

Table 30.4 shows the solubilities of various salts in water. The process in Figure 30.14 describes a model for ion solubilities due to Collins [21]. In this model, water solubility results from a balance of three forces: the anion–cation attraction, the water–water attraction, and the ion–water interaction. Ions are water-soluble if ion–water attractions are stronger than the other interactions. When the anion and cation are both small, they are strongly attracted to each other, resulting in low water solubility. When the anion and cation are both large, the water–water attraction is dominant, and again the water solubility is low. But when one ion is large and the other small, the ion–water attractions dominate, and there is high solubility in water.

Figure 30.14 Ion association in water involves both direct interactions and desolvation.

The Hofmeister Series: Nonpolar Solvation in Salt Solutions

Nonpolar compounds dissolve in salt water to different degrees, depending on the type of salt. In 1888, F Hofmeister [23] discovered that different salts have different propensities to precipitate or dissolve proteins in solution. This series is of practical importance for 'salting-in' (dissolving) or 'salting-out' (precipitating) proteins. The effect of salt on proteins is usually regarded as an effect of salt on the hydrophobic effect. Salts that strengthen hydrophobic interactions cause proteins to associate with each other, leading to protein precipitation. The *Hofmeister series* is a rank ordering of salt types that affect the partitioning of oils into aqueous salt solutions. For example, the solubilities of benzene and acetyltetraglycyl ethyl ester increase in the following order: for anions, SO_4^{2-}, CH_3COO^-, Cl^-, Br^-, ClO_4^-, CNS^-; and for cations, NH_4^+, K^+, Na^+, Li^+, Ca^{2+}. Hofmeister effects are subtle because they describe the small variations in the hydrophobic effect due to salt effects on the ordering of water.

Summary

The hydrophobic effect describes the unusual thermodynamics of the disaffinity of oil for water. In cold water, nonpolar solutes induce structuring in first-shell waters. Increasing the temperature melts out this structure. Ions, too, can cause the ordering of waters, by an electrostatic mechanism. The charge densities on ions have a role in determining their heats of dissolution and the viscosities of aqueous solutions, and can modulate the solubilities of oil in salt water.

Problems

1. **Compute the extrema of $\Delta\mu°(T)$ and $\Delta\mu°(T)/kT$.**

   (a) Use the Gibbs–Helmholz Equations (13.42) and (13.43) to show that $\Delta s° = 0$ at the temperature at which $\mu°(T)$ is a maximum or minimum.

   (b) Show that $\Delta h° = 0$ at the temperature at which $\Delta\mu°/kT$ is a maximum or minimum.

2. **The heat capacity of protein unfolding.** Figure 30.15 shows the enthalpy of folding a protein.

   (a) Determine ΔC_p for folding from this graph.

   (b) If $\Delta S_{\text{fold}} \approx 0$ at $T = 28\,°C$, compute ΔS_{fold} at $T = 100\,°C$.

   (c) Compute ΔG_{fold} at $T = 25\,°C$.

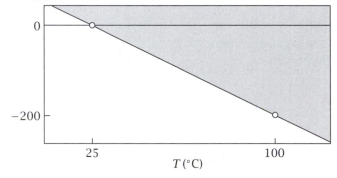

ΔH_{fold} (kcal mol^{-1})

Figure 30.15

3. **The forces of micelle formation.** A micelle contains 80 sodium dodecyl sulfate molecules, each of which has a headgroup with one negative charge at the surface. The micellar radius is 15 Å.

   (a) Compute the electrostatic energy required to charge up the head groups at the surface of the spherical micelle in water at $T = 300\,K$.

   (b) The free energy of transfer of a CH$_2$ group from water to oil is $-0.8\,$kcal mol^{-1}. Dodecyl chains have 12 CH$_2$ groups. Is hydrophobic association sufficient to overcome the electrostatic repulsions to form micelles?

4. **'Icebergs' are less ordered than ice.** Compare the entropy per molecule of inserting a nonpolar solute into water with the entropy of freezing water.

5. **The electrostriction of ions in water.** Formic acid (HCOOH) at infinite dilution in water at $T = 298.15\,K$ has partial molar volume $v = 34.69\,$cm^3 mol^{-1}. The change in volume upon ionization is $\Delta v = -8.44\,$cm^3 mol^{-1}. If this

Δv is distributed uniformly among the first-shell water molecules, how much volume reduction can be attributed to each first-shell water molecule?

6. **Ion solvation enthalpies scale inversely with ion radius.** Ionic radii increase from Li$^+$ to Na$^+$ to K$^+$, yet the enthalpies of solution decrease (see Figure 22.14). Write an electrostatic expression to rationalize this trend.

7. **Estimating the cavity size distribution in water.** Suppose that the energy cost of creating a spherical cavity of radius r in water is

$$\varepsilon(r) = 4\pi r^2 \gamma,$$

where $\gamma = 7.2\times10^{-2}\,$N m^{-1} is the surface tension of water at $T = 300\,K$.

   (a) Write an expression for the size distribution $p(r)$ of cavities in water.

   (b) Compute the average radius $\langle r\rangle$ of a cavity at $T = 300\,K$.

References

[1] WL Jorgensen, J Gao and C Ravimohan. *J Phys Chem* **89**, 3470–3473 (1985).

[2] AM Buswell and WH Rodebush. *Sci American* **194**, 76 (1959).

[3] MS Ramadan, DF Evans and R Lumry. *J Phys Chem* **87**, 4538 (1983).

[4] KA Dill. *Biochemistry* **29**, 7133–7155 (1990).

[5] HS Frank and MW Evans. *J Chem Phys* **13**, 507 (1945).

[6] SJ Gill, SF Dec, G Oloffson and I Wadsö. *J Phys Chem* **89**, 3758–3761 (1985).

[7] G Ravishankar and DL Beveridge. *Theoretical Chemistry of Biological Systems*, DL Beveridge and WL Jorgensen, eds, Elsevier, Amsterdam, 1986.

[8] A Geiger, A Rahman and FH Stillinger. *J Chem Phys* **70**, 263–276 (1979).

[9] DA Zichi and PJ Rossky. *J Chem Phys* **84**, 2814–2822 (1986).

[10] JPM Postma, HJC Berendsen and JR Haak. *Faraday Symp Chem Soc* **17**, 55–67 (1982).

[11] CY Lee, JA McCammon and PJ Rossky. *J Chem Phys* **80**, 4448–4455 (1984).

[12] NT Southall and KA Dill. *J Phys Chem B* **104**, 1326 (2000).

[13] K Lum, D Chandler and JD Weeks. *J Phys Chem B* **103**, 4570–4577 (1999).

[14] LR Pratt and D Chandler. *J Chem Phys* **67**, 3683–3704 (1977).

[15] DA Zichi and PJ Rossky. *J Chem Phys*, **83**, 797–808 (1985).

[16] K Watanabe and HC Andersen. *J Phys Chem* **90**, 795 (1986).

[17] G Ravishanker, M Mezei and DL Beveridge. *Faraday Symp, Chem Soc* **17**, 79–91 (1982).

[18] WH New and BJ Berne. *J Am Chem Soc* **117**, 7172–7179 (1995).

[19] RH Wood and PT Thompson. *Proc Natl Acad Sci USA* **87**, 946 (1990).

[20] BM Pettitt and PJ Rossky. *J Chem Phys* **84**, 5836–5844 (1986).

[21] KD Collins. *Biophys J* **72**, 65 (1997).

[22] RW Gurney. *Ionic Processes in Solution* Dover Publications, New York, 1953.

[23] PH von Hippel and T Schleich. *Acc Chem Res* **2**, 257 (1960).

Suggested Reading

Solvation thermodynamics is treated extensively in:

A Ben-Naim, *Hydrophobic Interactions*, Plenum Press, New York (1980).

A Ben-Naim, *Solvation Thermodynamics*, Plenum Press, New York (1987).

PL Privalov and SJ Gill, *Adv Protein Chem* **39**, 191–234 (1988).

Biological applications of hydrophobicity are discussed in:

NT Southall, KA Dill, and ADJ Haymet, *J Phys Chem B* **106**, 521–533 (2002).

C Tanford, *The Hydrophobic Effect: Formation of Micelles and Biological Membranes*, 2nd edition, Wiley, New York (1980).

Potentials of mean force are treated in statistical mechanics texts such as:

HL Friedman, *A Course in Statistical Mechanics*, Prentice-Hall, Englewood Cliffs (1985).

DA McQuarrie, *Statistical Mechanics*, Harper & Row, New York (1976).

31 Polymer Solutions

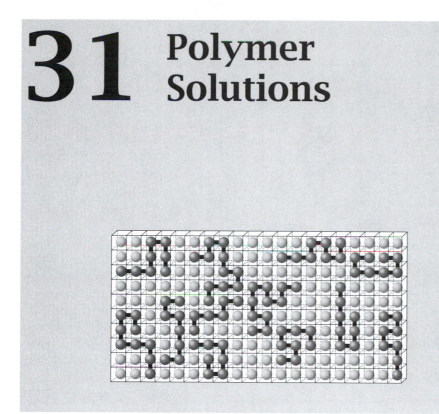

Polymer Properties Are Governed by Distribution Functions

Polymer molecules are chains of covalently bonded monomer units. Some polymers are *linear*, like beads on a necklace; others are *branched* (see Figure 31.1). Statistical mechanics has a prominent role in predicting the properties of polymers because two types of distribution function are intrinsic to them. First, polymer chains have *distributions of conformations*. Second, synthetic polymers have *distributions of chain lengths*. Many of the unique properties of polymeric solutions and solids, such as the elasticity of rubber and the viscoelasticity of slimy liquids, are consequences of the entropies that arise from the conformational freedom of chain molecules. We begin with the properties of polymer mixtures.

Polymers Have Distributions of Conformations

Chain molecules have conformational degrees of freedom. Different conformations arise when bonds in the chains have rotational isomers of nearly equal energies. Multiple conformers of polymers are often populated at room temperature.

The energies of rotation around the C–C single bonds in butane and ethane are shown in Figure 31.2. At low temperatures, the molecules are found mainly in the *trans* conformation, following the Boltzmann distribution. Molecular

(a) Linear Polymer

(b) Branched Polymer

Figure 31.1 Polymers may be (a) linear or (b) branched chains of monomer units.

(a) Ethane

$U(\phi)$ (kJ mol^{-1})

(b) *n*-Butane

$U(\phi)$ (kJ mol^{-1})

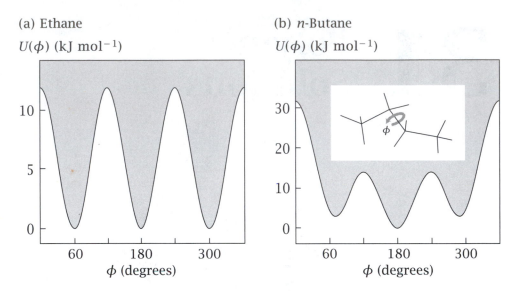

ϕ (degrees)

ϕ (degrees)

Figure 31.2 Bonds in polymers have torsional freedom. In saturated hydrocarbons, three favored torsional angles ϕ, called *trans*, *gauche*$^+$, and *gauche*$^-$, are favored by the minima in $U(\phi)$. (a) For ethane, the three *rotational isomers* have equal energy; (b) for butane, one minimum has lower energy than the other two. Source: WL Mattice and UW Suter, *Conformational Theory of Large Molecules*, Wiley, New York, 1994.

motions at low temperature are confined to vibrations within the lowest potential well. When the temperature is high enough, however, some molecules escape over the potential barrier (about 2.9 kcal mol^{-1} in butane) to populate two states called *gauche*$^+$ and *gauche*$^-$. Polyethylene, a chain of CH$_2$ groups, is one of the most common synthetic polymers. At high temperatures a polyethylene chain of N bonds has approximately 3^N conformations, a large number if N is large. Such configurational freedom underlies the properties of polymers.

Polymer Solutions Differ from Small Molecule Solutions

Solutions of polymer molecules differ from solutions of small molecules in several ways. First, the colligative properties differ from the simple mixtures that we considered in Chapters 15 and 16 because polymer molecules are typically much larger than solvent molecules. Second, small molecule solutions have a strong entropic tendency to mix, but polymer solutions do not. Polymers rarely mix with other polymers. This is unfortunate because it would be useful to blend polymers in the same way that metals are alloyed, to gain the advantageous properties from each of the components. Third, polymeric liquids, solids, and solutions are rubbery owing to chain conformational freedom.

In this chapter we describe the *Flory–Huggins theory*, the simplest model that explains the first two of these thermodynamic properties of polymer solutions. We consider the elasticity and viscoelasticity of polymeric materials in Chapters 32 and 33.

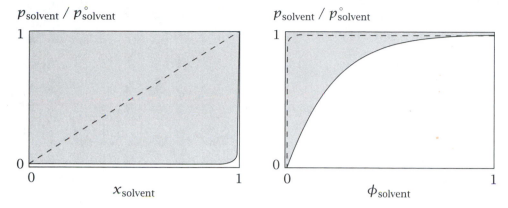

Figure 31.3 (a) The vapor pressure over an ideal solution (– – –) and the vapor pressure of a small-molecule solvent over a polymer solution (——) plotted as a function of mole fraction x. (b) The vapor pressure of a small molecule solvent over a polymer solution (——) may have only a small deviation from the diagonal, if plotted versus *volume fraction* ϕ rather than against *mole fraction*. Source: PJ Flory, *Principles of Polymer Chemistry*, Cornell University Press, Ithaca, 1953.

A Polymer Chain Is Typically Much Larger Than a Solvent Molecule

A polymer molecule can be composed of up to tens of thousands of monomers. Because each monomer is about the size of a solvent molecule, a polymer molecule occupies much more volume than a solvent molecule. That makes the volume fraction of a polymer in solution very different from its mole fraction. If a polymer chain has 1000 monomers, and if each monomer is the size of a solvent molecule, and if there is one polymer chain in the solution for every 1000 solvent molecules, then the volume fraction concentration of the polymer in the solution is 0.5, while the mole fraction is only 0.001 (see problem 1). The solvent volume fraction is 0.5, and its mole fraction is 0.999.

Colligative properties reflect this difference. Figure 31.3 shows the vapor pressure of the solvent benzene over a solution containing rubber, which is a polymer: (a) as a function of mole fraction, and (b) as a function of volume fraction. The vapor pressure of a small-molecule solvent over a polymer solution shows nonideal behavior when plotted versus mole fraction x. The Flory–Huggins theory described in the next section shows that a better measure of concentration in polymer solutions is the volume fraction ϕ.

Figure 31.3(a) shows that, for a given mole fraction of solvent, say $x = 0.5$, the solvent has a much smaller vapor pressure over a polymer solution than it would have over the corresponding small-molecule solution at the same mole fraction concentration. Polymer molecules have low vapor pressures because each polymer molecule is so large that it has many points of attraction to the neighboring solvent molecules. It cannot readily escape. But why should the *solvent molecules* be so strongly attracted to the polymers in the solution? Some nonidealities are not due to interaction energies. Independent work by PJ Flory and ML Huggins in the early 1940s showed how the nonidealities of polymer solutions can arise from the very different sizes of a polymer chain

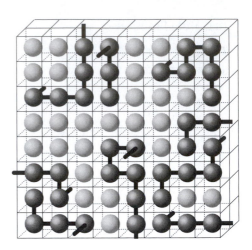

Figure 31.4 Polymer lattice model used for the Flory–Huggins theory. The dark beads show the polymer chain(s). The light beads represent the solvent molecules.

and its small-molecule solvent. In short, to compute the entropies of polymer solutions, you don't treat each *polymer molecule* on an equal footing with each solvent molecule. Rather, you treat each *solvent-sized chain segment* of the polymer molecule as equivalent to a solvent molecule. Here's the idea in more detail.

The Flory–Huggins Model Describes Polymer Solution Thermodynamics

To compute the thermodynamics of polymer solutions, consider a lattice of M sites (see Figure 31.4). Each site has z nearest neighbors. Suppose that there are n_s solvent molecules, each of which occupies a single lattice site, and that there are n_p polymer molecules. Each polymer molecule has N *chain segments*, representing the monomers. Each chain segment occupies one lattice site. If the polymer and solvent molecules completely fill the lattice, then

$$M = Nn_p + n_s. \tag{31.1}$$

The mole fractions of polymer and solvent are $n_p/(n_p + n_s)$ and $n_s/(n_p + n_s)$, respectively. The volume fractions, ϕ_s for the solvent and ϕ_p for the polymer, are

$$\phi_s = \frac{n_s}{M} \qquad \text{and} \qquad \phi_p = \frac{Nn_p}{M}. \tag{31.2}$$

The Entropy of Mixing

To compute the entropy of mixing, you can count the number of arrangements of the n_p polymers and the n_s solvent molecules on the lattice. Let's first count the number of configurations of a single chain molecule by using a 'chain growth' method. The first monomer can be located on any of the M lattice sites. If you were counting the arrangements of the molecules of a gas or a simple lattice solution, the second monomer could be located on any of the remaining $M - 1$ sites. But since the second monomer is in a chain and is connected to the first monomer, there are only z sites available to monomer 2 that are adjacent to monomer 1. Likewise the third monomer must be connected to the second, so it has $(z - 1)$ options (since it cannot be in the same site as monomer 1). The fourth connects to the third, etc. The total number ν_1 of conformations of one chain on the lattice is

$$\nu_1 = Mz(z - 1)^{N-2} \approx M(z - 1)^{N-1}, \tag{31.3}$$

which we have simplified by approximating the lone factor of z by $z - 1$.

Equation (31.3) is incomplete because it does not account for *excluded volume* among distant segments, the possibility that one monomer may land on a lattice site that has already been occupied by a more distant monomer. Equation (31.3) overestimates the number of conformations ν_1. To account for excluded volume, PJ Flory (1910-1985), who won the 1974 Nobel prize in Chemistry for his contributions to polymer science, made the approximation that the volume excluded to each segment is proportional to the amount of space filled by the chain segments that have already been placed on the lattice, as if they were randomly dispersed in space. According to this approximation, $(M-1)/M$ is the fraction of sites available for the second monomer, $(M-2)/M$ is the fraction available for the third monomer, and so on. Therefore a better estimate of the number of conformations available for one chain is

$$\nu_1 = M \left[(z) \left(\frac{M - 1}{M} \right) \right] \left[(z - 1) \left(\frac{M - 2}{M} \right) \right] \dots \left[(z - 1) \left(\frac{M - N + 1}{M} \right) \right]$$

$$\approx \left(\frac{z - 1}{M} \right)^{N-1} \frac{M!}{(M - N)!}. \tag{31.4}$$

Now we use the same logic for putting all of the n_p chains onto the lattice. Each of the n_p chains has a first monomer unit. The number of arrangements ν_{first} for placing the first monomer segments for all n_p chains is

$$\nu_{\text{first}} = M(M - 1)(M - 2) \dots (M - n_p + 1) = \frac{M!}{(M - n_p)!}. \tag{31.5}$$

Next, we count the number of arrangements of the $n_p(N - 1)$ remaining segments of the n_p chains. So far, n_p monomers (the first segment of each chain) have been placed on the lattice. This means that $M - n_p$ sites are available for the first of the subsequent segments. And $M - Nn_p - 1$ sites are available for the last segment. So, following the logic of Equation (31.4), the number of arrangements of all the subsequent segments, $\nu_{\text{subsequent}}$, is

$$\nu_{\text{subsequent}} = \left(\frac{z - 1}{M} \right)^{n_p(N-1)} \frac{\left(M - n_p \right)!}{\left(M - Nn_p \right)!}. \tag{31.6}$$

To compute the total number of arrangements $W(n_p, n_s)$ of the system of all n_p polymer molecules and n_s solvent molecules, multiply the factors together:

$$W(n_p, n_s) = \frac{\nu_{\text{first}} \nu_{\text{subsequent}}}{n_p!}$$

$$= \left(\frac{z - 1}{M} \right)^{n_p(N-1)} \frac{M!}{(M - Nn_p)! n_p!}. \tag{31.7}$$

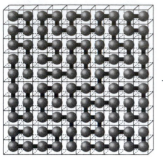 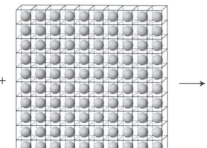 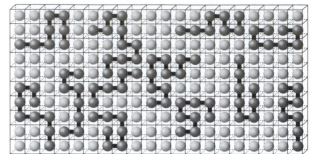

Figure 31.5 A lattice model for mixing n_p polymer molecules, and n_s solvent molecules, to get a solution of n_p polymer molecules and n_s solvent molecules.

The factor $n_p!$ in the denominator on the right-hand side of Equation (31.7) accounts for the indistinguishability of one polymer chain from another. The factor $(M - Nn_p)!$ accounts for the indistinguishability of the solvent molecules.

To mix the polymer with the solvent, start with a container of pure solvent (n_s solvent molecules and no polymer) and a container of pure polymer (n_p polymer molecules and no solvent), and combine them to get a solution of n_s solvent molecules and n_p polymer molecules (see Figure 31.5). The Boltzmann expression gives the entropy change upon mixing in terms of the multiplicities W:

$$\frac{\Delta S_{\mathrm{mix}}}{k} = \ln\left(\frac{W(n_p, n_s)}{W(0, n_s)W(n_p, 0)}\right). \tag{31.8}$$

For a lattice of pure solvent molecules, $W(0, n_s) = 1$. You can compute $W(n_p, 0)$ for the pure polymer by substituting Nn_p for M into Equation (31.7) to get

$$W(n_p, 0) = \left(\frac{z-1}{Nn_p}\right)^{n_p(N-1)} \frac{(Nn_p)!}{n_p!}. \tag{31.9}$$

Use Equation (31.9) and substitute $n_s = M - Nn_p$ into Equation (31.7) for the mixture $W(n_p, n_s)$ to get

$$\frac{W(n_p, n_s)}{W(0, n_s)W(n_p, 0)} = \frac{\left(\dfrac{M!}{n_p! n_s!}\right)\left(\dfrac{z-1}{M}\right)^{n_p(N-1)}}{\left(\dfrac{(Nn_p)!}{n_p!}\right)\left(\dfrac{z-1}{Nn_p}\right)^{n_p(N-1)}}$$

$$= \left(\frac{M!}{(Nn_p)! n_s!}\right)\left(\frac{Nn_p}{M}\right)^{n_p(N-1)}. \tag{31.10}$$

Substituting Equation (31.10) into Equation (31.8) and using Stirling's approximation (Equation (4.26)) gives

$$\frac{\Delta S_{\mathrm{mix}}}{k} = M \ln M - (Nn_p)\ln(Nn_p) - n_s \ln n_s$$

$$+ n_p(N-1)\ln\left(\frac{Nn_p}{M}\right). \tag{31.11}$$

Replace $M = n_s + Nn_p$ in the first term, and use the definitions of volume fractions, Equations (31.2), to get

$$\frac{\Delta S_{mix}}{k} = (n_s + Nn_p)\ln M - n_s \ln n_s - Nn_p \ln(Nn_p)$$

$$+ Nn_p \ln\left(\frac{Nn_p}{M}\right) - n_p \ln\left(\frac{Nn_p}{M}\right)$$

$$= -n_s \ln\left(\frac{n_s}{M}\right) - Nn_p \ln\left(\frac{Nn_p}{M}\right) + Nn_p \ln\left(\frac{Nn_p}{M}\right)$$

$$- n_p \ln\left(\frac{Nn_p}{M}\right)$$

$$= -n_s \ln \phi_s - n_p \ln \phi_p, \tag{31.12}$$

or, per lattice site,

$$\frac{\Delta S_{mix}}{Mk} = -\phi_s \ln \phi_s - \left(\frac{\phi_p}{N}\right)\ln \phi_p. \tag{31.13}$$

When $N = 1$, the Flory–Huggins mixing entropy reduces to that of the simple solution lattice model, Equation (15.3).

Now we compute the energy of mixing.

The Energy of Mixing

We follow the same procedure that we used for simple lattice solutions (see Equation (15.4)). If m_{ss}, m_{pp}, and m_{sp} are the numbers of contacts between solvent molecules (s) and monomeric segments (p) of the polymer chain, and if w_{ss}, w_{pp}, and w_{sp} are the corresponding contact energies for each type of pair, then the total contact energy is

$$U = m_{ss}w_{ss} + m_{pp}w_{pp} + m_{sp}w_{sp}. \tag{31.14}$$

The conservation relations between the numbers of contacts and the numbers of lattice sites are

$$zn_s = 2m_{ss} + m_{sp} \quad \text{and}$$

$$zNn_p = 2m_{pp} + m_{sp}. \tag{31.15}$$

We have made a simplification here. The quantity zNn_p in the second Equation (31.15) could better be approximated as $(z-2)Nn_p$. The factor of $z-2$ would be more appropriate than z because each p monomer is not fully accessible for making noncovalent contacts, because it is covalently connected to two other p monomers in the chain (or one if it is at the end of the chain). But at the level of simplicity of the model, this refinement is neglected. Substitute Equations (31.15) into Equation (31.14) to replace the unknown m's by the known n's. Now use the Bragg–Williams approximation (Equation (15.9)), which estimates the number of sp contacts as the product (n_s, the number of s monomers) $\times$ (the number z of neighboring sites per s monomer) $\times$ (the fraction Nn_p/M of such lattice sites that are occupied by p monomers). That is, $m_{sp} \approx (z n_s N n_p)/M$, so

$$\frac{U}{kT} = \left(\frac{z w_{ss}}{2kT}\right) n_s + \left(\frac{z w_{pp}}{2kT}\right) N n_p + \chi_{sp} \frac{n_s n_p N}{M}, \tag{31.16}$$

where

$$\chi_{sp} = \frac{z}{kT}\left(w_{sp} - \frac{w_{ss} + w_{pp}}{2}\right). \tag{31.17}$$

The Free Energy and Chemical Potential

To get the free energy, combine the entropy Equation (31.12) with the energy Equation (31.16),

$$\frac{F}{kT} = \frac{U}{kT} - \frac{S}{k} = n_s \ln \phi_s + n_p \ln \phi_p + \left(\frac{z w_{ss}}{2kT}\right) n_s$$

$$+ \left(\frac{z w_{pp}}{2kT}\right) N n_p + \chi_{sp} \frac{n_s n_p N}{M}. \tag{31.18}$$

To generalize, suppose that you have a mixture of n_A molecules of a polymer A containing N_A monomer units, and n_B molecules of a polymer B having N_B monomers per chain. Then Equation (31.18) gives

$$\frac{F_{\text{mix}}}{kT} = n_A \ln \phi_A + n_B \ln \phi_B + \left(\frac{z w_{AA}}{2kT}\right) N_A n_A$$

$$+ \left(\frac{z w_{BB}}{2kT}\right) N_B n_B + \chi_{AB} \frac{n_A n_B N_A N_B}{M}. \tag{31.19}$$

The volume fractions are $\phi_A = (n_A N_A)/M$ and $\phi_B = (n_B N_B)/M$, where $M = n_A N_A + n_B N_B$. As a check, notice that if $N_B = 1$, Equation (31.19) reduces to Equation (31.18).

Now we compute the chemical potential of the B molecules (which can be either a polymer or small molecule, depending on N_B). Hold n_A (not M) constant and take the derivative of Equation (31.19) (see Equations (15.15) and (15.16)):

$$\frac{\mu_B}{kT} = \left(\frac{\partial}{\partial n_B}\left(\frac{F_{\text{mix}}}{kT}\right)\right)_{n_A, T}$$

$$= \ln \phi_B + 1 - \frac{n_B N_B}{M} - \frac{n_A N_B}{M}$$

$$+ \chi_{AB} N_B (1 - \phi_B)^2 + \left(\frac{z w_{BB}}{2kT}\right) N_B. \tag{31.20}$$

Use the following expression

$$\frac{n_A N_B}{M} = \frac{n_A N_A}{M}\left(\frac{N_B}{N_A}\right) = \phi_A \left(\frac{N_B}{N_A}\right) = (1 - \phi_B)\left(\frac{N_B}{N_A}\right)$$

and the definition $\phi_B = n_B N_B/M$ to reduce Equation (31.20) to

$$\frac{\mu_B}{kT} = \ln \phi_B + (1 - \phi_B)\left(1 - \frac{N_B}{N_A}\right)$$

$$+ \chi_{AB} N_B (1 - \phi_B)^2 + \left(\frac{z w_{BB}}{2kT}\right) N_B. \tag{31.21}$$

Now that you have the chemical potential, you can follow the same procedures you use for the lattice model of simple solutions to predict the colligative properties and phase separations of polymer solutions.

Flory–Huggins Theory Predicts Nonideal Colligative Properties for Polymer Solutions

Let's compute the vapor pressure of a solvent over a polymer solution by using the Flory-Huggins theory. Let component A represent the polymer and B represent the small molecule. To describe the equilibrium, follow the strategy of Equation (16.2). Use Equation (31.20), and set the chemical potential of B in the vapor phase equal to the chemical potential of B in the polymer solution. The vapor pressure of B over a polymer solution is

$$\frac{p_B}{p_{B,\text{ref}}} = \phi_B \exp\left[(1 - \phi_B)\left(1 - \frac{N_B}{N_A}\right) + N_B \chi_{AB}(1 - \phi_B)^2 \right.$$
$$\left. + \left(\frac{zw_{BB}}{2kT}\right)N_B\right]. \tag{31.22}$$

Equation (31.22) resembles Equation (16.2) for small molecule solutions, except that the volume fraction ϕ_B replaces the mole fraction x_B, and a term $(1 - \phi_B)(1 - N_B/N_A)$ accounts for nonideality due to the difference in molecular sizes. The Flory-Huggins theory predicts that it is the number of solvent-sized monomer units in polymer chains that determine the colligative properties, not the number of polymer chains. This is the reason that volume fraction concentrations rather than mole fractions are more useful for polymer solutions.

The Phase Behavior of Polymers Differs from that of Small Molecules

To explore the phase behavior of polymer solutions, we want the free energy as a function of volume fraction ϕ. Compute ΔF_{mix} as the difference in free energy resulting from mixing the two pure phases together. Using Equation (31.19) divided by M and taking this difference gives

$$\frac{\Delta F_{\text{mix}}}{MkT} = \left(\frac{\phi_A}{N_A}\right)\ln \phi_A + \left(\frac{\phi_B}{N_B}\right)\ln \phi_B + \chi_{AB}\phi_A\phi_B,$$

or, in terms of the volume fraction of A, $\phi = \phi_A$,

$$\frac{\Delta F_{\text{mix}}}{MkT} = \left(\frac{\phi}{N_A}\right)\ln \phi + \left(\frac{1 - \phi}{N_B}\right)\ln(1 - \phi) + \chi_{AB}\phi(1 - \phi). \tag{31.23}$$

When the two types of molecule in a solution have different sizes, the mixing free energy will be an asymmetrical function of ϕ as in Figure 31.6. Systems are predicted to be miscible for compositions in which the free energy is concave upward, and immiscible where that function is concave downward (see Chapter 25, Figure 25.5). In contrast, Figures 25.7 and 25.11 show that lattice-model phase diagrams for small-molecule solutions are symmetrical around $\phi_A = \phi_B = 1/2$.

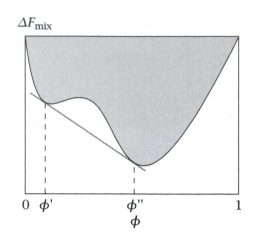

Figure 31.6 The free energy of mixing a polymer solution versus volume fraction ϕ. Find the phase equilibria by using the common tangent method (Figure 25.6). The two phases in equilibrium have volume fractions ϕ' and ϕ''.

We now show the basis for the asymmetry in polymer solutions. The coexistence curve is defined by the common tangent,

$$\left(\frac{\partial \Delta F_{\text{mix}}}{\partial \phi}\right)' = \left(\frac{\partial \Delta F_{\text{mix}}}{\partial \phi}\right)'', \tag{31.24}$$

where the $'$ and $''$ indicate the two phases that are in equilibrium. In general, calculation of the coexistence curve involves the solution of two simultaneous nonlinear equations. You can compute the coexistence curve by substituting Equation (31.18) or (31.19) into Equation (31.24) and finding the points of common tangency. We won't do this here. Our goal here is just to show the asymmetry in polymer phase diagrams and polymer–polymer immiscibility. You can do this by focusing on the critical point.

The spinodal decomposition curve is given by the values of ϕ that cause the second derivative of Equation (31.21) to equal zero,

$$\frac{\partial^2}{\partial \phi^2}\left(\frac{\Delta F_{\text{mix}}}{MkT}\right) = \frac{1}{N_A \phi} + \frac{1}{N_B(1 - \phi)} - 2\chi_{AB} = 0. \tag{31.25}$$

The critical point ϕ_c is where the spinodal curve and its derivative are zero (see Chapter 25, page 478),

$$\frac{\partial^2}{\partial \phi^2}\left(\frac{\Delta F_{\text{mix}}}{MkT}\right) = \frac{\partial^3}{\partial \phi^3}\left(\frac{\Delta F_{\text{mix}}}{MkT}\right) = 0. \tag{31.26}$$

Taking the derivative of Equation (31.25) gives

$$\left(\frac{\partial^3}{\partial \phi^3}\left(\frac{\Delta F_{\text{mix}}}{MkT}\right)\right) = -\frac{1}{N_A \phi_c^2} + \frac{1}{N_B(1 - \phi_c)^2} = 0$$

$$\Longrightarrow \frac{\phi_c}{1 - \phi_c} = \left(\frac{N_B}{N_A}\right)^{1/2}$$

$$\Longrightarrow \phi_c = \frac{N_B^{1/2}}{N_A^{1/2} + N_B^{1/2}}. \tag{31.27}$$

Now let's use this expression to interpret phase equilibria in polymer solutions.

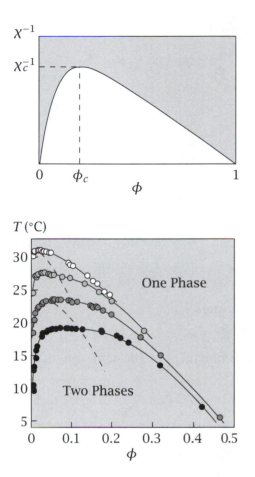

Figure 31.7 A polymer phase diagram. χ_c^{-1} on the y-axis is proportional to temperature. Mixtures of polymers and small molecules have asymmetric phase diagrams, in which the critical volume fraction of polymer $\phi_c < 1/2$ (see Equation (31.28)). Equation (31.29) shows that as $N \to \infty$, the critical exchange parameter $\chi_c \to 1/2$.

Figure 31.8 Experimental phase diagrams of polystyrene in cyclohexane, showing the strong asymmetry indicated in Figure 31.7. Molecular weights (○) 1.27×10^6, (◐) 2.5×10^5, (◑) 8.9×10^4, (●) 4.3×10^4. The dashed lines indicate the Flory–Huggins theory predictions for the first and third curves from the top. Source: LH Sperling, *Introduction to Physical Polymer Science*, 2nd edition, Wiley, New York, 1992.

Polymer Phase Diagrams Are Asymmetrical Owing to Size Differences between Polymer and Solvent Molecules

Two cases are of special interest. First, suppose that component A is a long-chain polymer and B is a small molecule, $N = N_A \gg 1$ and $N_B = 1$. Then Equation (31.27) gives the critical volume fraction of the polymer (A) as

$$\phi_c = \frac{1}{1 + N^{1/2}}. \tag{31.28}$$

Substituting Equation (31.28) into Equation (31.25) gives the value of the interaction parameter χ_c at the critical temperature,

$$\chi_c = \frac{1 + N^{1/2}}{2N^{1/2}}. \tag{31.29}$$

As $N \to \infty$, $\phi_c \to 0$, which means that the peak of the mixing phase diagram shifts strongly to the left in Figure 31.7 as the chain length increases. The phase diagram becomes highly asymmetrical because of the difference in size between the polymer and the solvent molecules. Figure 31.8 shows the asymmetry of the liquid–liquid phase diagram of polystyrene in cyclohexane. In the next section we treat a second case of interest, in which both the solute and the solvent are polymers.

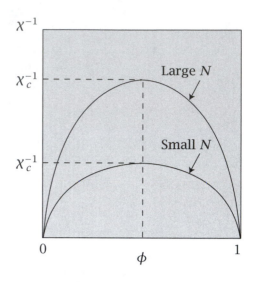

Figure 31.9 Miscibility phase diagram for two polymers having the same length N. Long polymers don't readily mix with each other: the two-phase region in the middle becomes large as $N \rightarrow \infty$.

Polymers Are Immiscible with Other Polymers Because the Mixing Entropy Is so Small

Now consider the case of two long-chain polymers of the same length $N = N_A = N_B$. Equations (31.27) and (31.25) give

$$\phi_c = \frac{1}{2} \quad \text{and} \quad \chi_c = \frac{2}{N}. \tag{31.30}$$

As $N \rightarrow \infty$, $\chi_c \rightarrow 0$ and the chains become immiscible at all temperatures (see Figure 31.9). The entropy of mixing polymers is so small that even an extremely small unfavorable enthalpy of mixing can prevent mixing. For example, even deuterated polybutadiene will not mix with protonated polybutadiene below a critical temperature of 61.5 °C, if $N = 2300$. Rather than mix, polymer blends often separate into domains of each of the component polymers.

EXAMPLE 31.1 Polymer mixing entropy. The mixing entropy for small molecules having mole fractions $x_A = x_B = 0.5$ is

$$\frac{\Delta S_{\text{mix}}}{Mk} = -0.5 \ln 0.5 - 0.5 \ln 0.5 = 0.69,$$

for a lattice having M sites. The mixing entropy for two polymers $N_A = N_B = 10,000$ having mole fractions $x_A = x_B = \phi_A = \phi_B = 0.5$ is

$$\frac{\Delta S_{\text{mix}}}{Mk} = -\left(\frac{\phi_A}{N_A}\right) \ln \phi_A - \left(\frac{\phi_B}{N_B}\right) \ln \phi_B$$

$$= -\left(\frac{0.5}{10,000}\right) \ln 0.5 - \left(\frac{0.5}{10,000}\right) \ln 0.5$$

$$= 6.9 \times 10^{-5}.$$

(Take the entropic part of Equation (31.19).)

Because this entropy is so small, polymers have little tendency to mix.

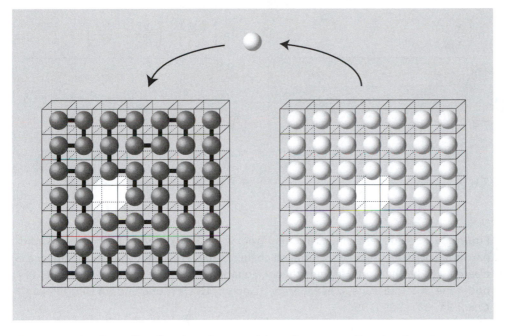

Figure 31.10 A small-molecule solute partitions from a small-molecule solvent phase to a polymeric solvent phase. Solutes are attracted to dense polymer phases, because diluting these polymer phases increases the chain conformational entropy.

Dilution Entropy Drives Solutes to Partition into Polymeric Phases

The partitioning of a solute molecule between solvents A and B was treated in Chapter 16. Solutes tend to concentrate in the phases for which they have the greatest chemical affinity, but concentration gradients are opposed by the translational entropy. Now suppose that one of the solvents is a polymer, such as octanol (see Figure 31.10). Then there is an additional driving force: solutes are attracted to the polymeric phase. This attraction results because diluting a polymeric system with small molecules causes the polymers to gain conformational entropy. In such cases, you can compute the partition coefficient by using Flory–Huggins theory.

Suppose a solute s partitions between a polymer liquid A having chain length N_A and a polymer liquid B having length N_B. If the solute is dilute in both phases, then its concentrations in the two phases are $\phi_{sA}, \phi_{sB} \ll 1$. At equilibrium, the chemical potential of the solute must be equal in the two phases: $\mu_s(A) = \mu_s(B)$. Flory–Huggins theory Equation (31.20) gives the two chemical potentials as

$$\frac{\mu_s(A)}{kT} = \ln \phi_{sA} + \left(1 - \frac{N_s}{N_A}\right) + N_s \chi_{sA} + \left(\frac{z w_{ss}}{2kT}\right) N_s \quad \text{and}$$

$$\frac{\mu_s(B)}{kT} = \ln \phi_{sB} + \left(1 - \frac{N_s}{N_B}\right) + N_s \chi_{sB} + \left(\frac{z w_{ss}}{2kT}\right) N_s,$$

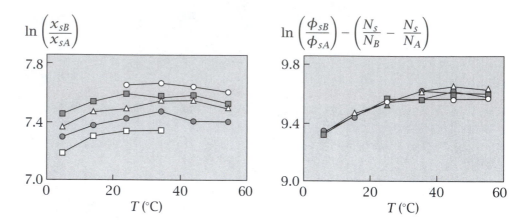

Figure 31.11 For benzene partitioning from water into different hydrocarbons, the partition coefficient depends on the oil chain length. The chain-length dependence is treated well by the Flory–Huggins theory, Equation (31.31). (□ octane, ● decane, △ dodecane, ■ tetradecane, ○ hexadecane.) Source: LR DeYoung and KA Dill, *J Phys Chem* **94**, 801–809 (1990).

where N_s is the number of s molecules. Setting μ_A and μ_B equal gives

$$\ln\left(\frac{\phi_{sB}}{\phi_{sA}}\right) = N_s(\chi_{sA} - \chi_{sB}) + \left(\frac{N_s}{N_B} - \frac{N_s}{N_A}\right). \qquad (31.31)$$

Compare Equation (31.31) for partitioning a solute in polymeric solvents with Equation (16.39) for partitioning in small-molecule solvents. Again, the volume fraction is a more natural measure of concentration than the mole fraction. Also note that the term $(\chi_{sA} - \chi_{sB})$, which describes the interaction energy per chain segment, is multiplied by N_s: the free energy of partitioning is proportional to the size of the solute molecule. Finally, there is an additional term that depends on the relative sizes of the solute and the polymer molecules, $(N_s/N_B) - (N_s/N_A)$.

Figure 31.11 shows that Equation (31.31) accounts for the chain-length dependence of partitioning benzene into hydrocarbon phases of different chain lengths. Longer chains gain more conformational entropy as solutes dilute them out, releasing steric constraints that are imposed by the surrounding chains. So small molecules are drawn more strongly into longer-chain solvents than into shorter-chain solvents.

Now we look at a factorization assumption of the Flory–Huggins model, and its implication for how polymer concentration affects chain conformations.

The Flory Theorem Says that the Intrinsic Conformations of a Chain Molecule Are Not Perturbed by Surrounding Chains

A fundamental assumption that underlies the polymer solution theory in this chapter is the factorization of the chain partition function in Equations (31.4) and (31.7) into a product of two terms. One term, which depends on factors

$(z - 1)$, describes *local* interactions that determine the number of bond rotational isomers between connected neighbors. The other term, which depends on factors $(M - i)/M$, accounts approximately for *nonlocal* interactions. Nonlocal interactions are due to excluded volume among monomers of different chains or among monomers that are not neighbors within a single chain. The assumption that local factors are independent of nonlocal factors is such an important premise that it has been given a name, the *Flory theorem*. According to this premise, increasing the concentration of chain molecules (or the compactness of a single-chain molecule) should diminish the partition function uniformly, not biasing the distribution of conformations in any one particular way or another, since the factors $(M - i)/M$ do not affect the factors $(z - 1)$. This implies that the *distribution of the* $(z - 1)^{N-1}$ *conformations* of a chain are exactly the same whether the chain is in a concentrated polymer solution or not (provided that it is homogeneous). The remarkable prediction is that chains will have exactly the same conformations when they are in highly tangled bulk polymer solutions as they would have if they were free in a simple solvent. What does change strongly with polymer concentration, however, is the excluded-volume entropy, which disfavors the dense states. In short, increasing the chain density diminishes the total number of conformations, without affecting their distribution.

High polymer concentrations are predicted to decrease the populations of all the conformations uniformly. Is this remarkable theorem correct? To a very good approximation, neutron-scattering experiments first performed in the 1970s showed that it is [1]. A labeled chain in a bulk polymer medium has the same average radius as that expected for an ideal chain, which we describe in Chapter 32.

Summary

A polymer molecule is much larger than a typical small solvent molecule, often by more than a thousandfold. This difference in size has important consequences for the thermodynamics of polymer solutions. First, the colligative attraction of a polymer solution for solvent small-molecule molecules is much higher than would be expected on the basis of the mole fractions. Rather the attraction is proportional to the volume fraction. Second, polymers are typically not miscible with other polymers because of small mixing entropies. Third, the size asymmetry between a polymer and a small-molecule solvent translates into an asymmetry in miscibility phase diagrams.

Problems

1. Concentrations: mole fractions x versus volume fractions ϕ. Consider a solution of polymers P and small molecules s, with $\phi_s = \phi_P = 0.5$. The polymer chain length is $N = 1000$.

What are the mole fractions x_s and x_P?

2. Excluded volume in protein folding.

(a) Use the Flory–Huggins theory to estimate the number $v_1(c)$ of conformations of a polymer chain that are confined to be maximally compact, $M = N$. Simplify the expression by using Stirling's approximation.

(b) If the number of conformations of the unfolded chain is $v_1(u) = M(z-1)^{N-1}$ (since $M \gg N$), then compute the entropy of folding,

$$\Delta S_{\text{fold}} = k \ln\left[\frac{v_1(c)}{v_1(u)}\right].$$

3. Flory–Huggins mixing free energy. A polymer A with chain length $N_A = 1000$ is mixed with a small molecule B. The volume fractions are $\phi_A = 0.3$ and $\phi_B = 0.7$, and $\chi_{AB} = 0.1$. The temperature is 300 K. Using the Flory–Huggins theory,

(a) Compute $\Delta S_{\text{mix}}/Mk$.

(b) Compute $\Delta U_{\text{mix}}/MkT$.

(c) Compute $\Delta F_{\text{mix}}/M$.

4. The critical point of a Flory–Huggins solution. For the mixing of a polymer A ($N_A = 10^4$) with a small molecule B ($N_B = 1$), compute the critical volume fraction ϕ_c and the critical exchange parameter χ_c.

5. Volume fractions versus mole fractions. For what class of model mixing process are volume fractions identical to mole fractions?

6. Solvent chain lengths affect partition coefficients. What is the maximum effect of the solvent chain length on the partition coefficient according to the Flory–Huggins model?

7. The colligative properties of polymers. Show how a small molecule diluent can reduce the freezing point of a polymer solution.

References

[1] PJ Flory. *Faraday Discussions Chem Soc* **68**, 14-25 (1979).

Suggested Reading

Excellent texts on Flory–Huggins theory are:

PG deGennes, *Scaling Concepts in Polymer Physics*, Cornell University Press, Ithaca, 1979.

M Doi, *Introduction to Polymer Physics*, Clarendon Press, Oxford, 1996.

PJ Flory, *Principles of Polymer Chemistry*, Series title: George Fisher Baker nonresident lectureship in chemistry at Cornell University, Cornell University Press, Ithaca, 1953.

UW Gedde, *Polymer Physics*, Chapman and Hall, London, 1955.

TL Hill, *An Introduction to Statistical Thermodynamics*, Series title: Addison-Wesley series in chemistry, Addison-Wesley, Reading, MA, 1969.

SF Sun, *Physical Chemistry of Macromolecules: Basic Principles and Issues*, Wiley, New York, 1994.

A discussion of the Flory–Huggins model for partition coefficients:

HS Chan and KA Dill, *Annu Rev Biophys Biomol Struct* **26**, 425–459 (1997).

32 Polymer Elasticity & Collapse

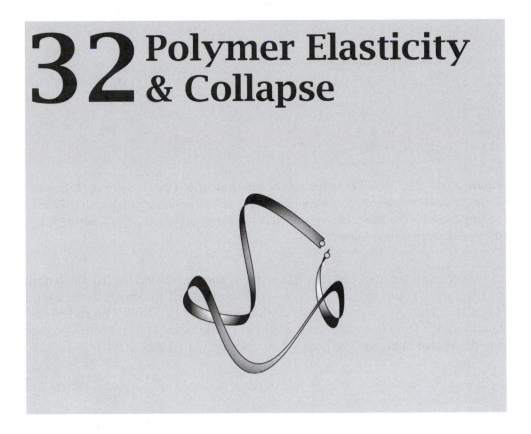

Polymeric Materials Are Often Elastic

Rubber is elastic, and gooey liquids such as raw eggs are viscoelastic, because polymer molecules have many conformations of nearly equal energy. Stretching the polymer chains lowers their conformational entropy. The chains retract to regain entropy. The entropy is also lowered when polymers become compact, as when proteins fold or when DNA becomes encapsulated within virus heads. The simplest molecular description of polymer elasticity is the *random-flight model*.

Polymer Chains Can Be Modelled as Random Flights

We consider a polymer chain to be a connected sequence of N rigid vectors, each of length b. For now, each vector represents a chemical bond, but starting on page 612 we consider other situations in which each vector can represent a *virtual bond*, a collection of more than one chemical bond.

There are different ways to characterize the size of a polymer chain. The *contour length L* is the total stretched-out length of the chain,

$$L = Nb \tag{32.1}$$

(see Figure 32.1(a)). The contour length has a fixed value, no matter what the chain conformation.

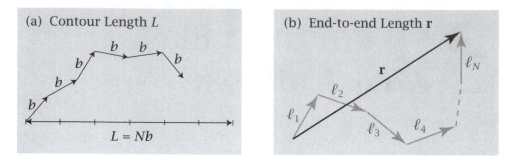

(a) Contour Length L

(b) End-to-end Length $\mathbf{r}$

$L = Nb$

Figure 32.1 Two measures of the size of a polymer molecule. If there are N bonds and each bond has length b, the *contour length* (a) is the sum of the lengths of all bonds, $L = Nb$. The *end-to-end length* (b) is the *vector sum*, $\mathbf{r} = \sum_{i=1}^{N} \boldsymbol{\ell}_i$, where $\boldsymbol{\ell}_i$ is the vector representing bond i.

Another measure of polymer chain size is the *end-to-end* length, the length of the vector pointing through space from one end of the chain to the other, when the chain is in a given conformation (see Figure 32.1(b)). The end-to-end length varies from one chain conformation to the next. The end-to-end vector $\mathbf{r}$ is the vector sum over the bonds $\boldsymbol{\ell}_i$ for bonds $i = 1, 2, 3, \ldots, N$:

$$\mathbf{r} = \sum_{i=1}^{N} \boldsymbol{\ell}_i. \tag{32.2}$$

The end-to-end length cannot be greater than the contour length. The end-to-end length changes when a polymer is subjected to applied forces. It is related to the radius of the molecule, and can be used to interpret the viscosities of polymer solutions, the scattering of light, and some of the dynamic properties of polymer solutions.

For any vector, you can always consider the x, y, and z components individually. The x component of the end-to-end vector is

$$r_x = \sum_{i=1}^{N} x_i, \tag{32.3}$$

where $x_i = b \cos \theta_i$ is the projection on the x-axis of the ith vector and θ_i is the angle of bond i relative to the x-axis. If all the bond vectors have the same length b, you get

$$r_x = b \sum_{i=1}^{N} \cos \theta_i. \tag{32.4}$$

In solution, there are many molecules, each with a different conformation. The physical properties that can be measured by experiments are averages over all the possible conformations. To get these averages, you can use a simple model, called the *freely jointed chain*, or *random-flight* model ('flight' in three dimensions, or 'walk' in two dimensions). In this model the angle of each bond is independent of every other, including the nearest neighboring bonds. Assuming that all conformations have equal probabilities, the average x-component of the end-to-end vector is

$$\langle r_x \rangle = \left\langle b \sum_{i=1}^{N} \cos \theta_i \right\rangle = b \sum_{i=1}^{N} \langle \cos \theta_i \rangle = 0, \tag{32.5}$$

because the average over randomly oriented vectors is $\langle \cos \theta_i \rangle = 0$ (see Example 1.23). This means that monomer N of a random flight chain is in about the same location as monomer 1, because there are as many positive steps as negative steps, on average. Because the x, y, and z components are independent of each other, the y and z components can be treated identically and $\langle r_y \rangle = \langle r_z \rangle = 0$.

Because the mean value of the end-to-end vector is zero ($\langle r_x \rangle = 0$), $\langle r_x \rangle$ doesn't contain useful information about the size of the molecule. The *mean square* end-to-end length $\langle r^2 \rangle$ or $\langle r_x^2 \rangle$ is a more useful measure of molecular size. It is related to the *radius of gyration* R_g by $R_g^2 = \langle r^2 \rangle / 6$ for a linear chain [1]. We won't prove that relationship here. The radius of gyration, which is always positive, is a measure of the radial dispersion of the monomers. We use $\langle r^2 \rangle$ instead of R_g because the math is simpler.

The square of the magnitude of the end-to-end vector can be expressed as a matrix of terms:

$$\mathbf{r} \cdot \mathbf{r} = \left(\sum_{i=1}^{N} \boldsymbol{\ell}_i \right)^2 = \begin{matrix} \boldsymbol{\ell}_1 \boldsymbol{\ell}_1 + \boldsymbol{\ell}_1 \boldsymbol{\ell}_2 + \boldsymbol{\ell}_1 \boldsymbol{\ell}_3 + \cdots + \boldsymbol{\ell}_1 \boldsymbol{\ell}_N + \\ \boldsymbol{\ell}_2 \boldsymbol{\ell}_1 + \boldsymbol{\ell}_2 \boldsymbol{\ell}_2 + \boldsymbol{\ell}_2 \boldsymbol{\ell}_3 + \cdots + \boldsymbol{\ell}_2 \boldsymbol{\ell}_N + \\ \vdots \qquad\qquad\qquad\qquad \vdots \\ \boldsymbol{\ell}_N \boldsymbol{\ell}_1 + \boldsymbol{\ell}_N \boldsymbol{\ell}_2 + \boldsymbol{\ell}_N \boldsymbol{\ell}_3 + \cdots + \boldsymbol{\ell}_N \boldsymbol{\ell}_N. \end{matrix} \tag{32.6}$$

This sum involves two kinds of terms, 'self terms' $\boldsymbol{\ell}_i \boldsymbol{\ell}_i$ (along the main diagonal) and 'cross terms' $\boldsymbol{\ell}_i \boldsymbol{\ell}_j, i \neq j$. The self terms give $\langle \boldsymbol{\ell}_i \cdot \boldsymbol{\ell}_i \rangle = b^2$, because the product of a vector with itself is the square of its length. The cross terms give $\langle \boldsymbol{\ell}_i \cdot \boldsymbol{\ell}_j \rangle = b^2 \langle \cos \theta \rangle = 0$ because there is no correlation between the angles of any two bond vectors in the random-flight model. Therefore the only terms that contribute to the sum are the self terms along the diagonal, and there are N of those, so

$$\langle r^2 \rangle = Nb^2. \tag{32.7}$$

The *root-mean-square* (rms) end-to-end distance is $\langle r^2 \rangle^{1/2} = N^{1/2} b$. Because the molecular weight of a polymer is proportional to the number of monomers it contains, one important prediction of the random-flight theory is that the mean radius (and the rms end-to-end length) of a polymer chain increases in proportion to the square root of the molecular weight. This result is the centerpiece of much of polymer theory.

How do real chains differ from random flights? An important difference is chain stiffness.

Chain Stiffness

Contrast the mathematical description of a random flight with that of a rod, for which all vectors point in the same direction. For a rod, *all* the terms in Equation (32.6) contribute b^2 to the sum, and there are N^2 terms in the matrix, so the end-to-end length equals the contour length, and $\langle r^2 \rangle = (Nb)^2 = L^2$.

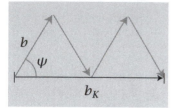

Figure 32.2 The Kuhn model of a polymer. b is the length of the chemical bond, b_K is the length of a virtual, or Kuhn, bond directed along the chain axis, and ψ is the angle between the chain axis and the chemical bond.

Because $\langle r^2 \rangle \sim N$ when bonds are uncorrelated, and $\langle r^2 \rangle \sim N^2$ when bonds are perfectly correlated, you might expect that partial correlation between bonds, which is a more realistic model for polymers, would lead to a dependence of $\langle r^2 \rangle$ on N with some exponent between 1 and 2. But this is not the case.

To illustrate, consider a model in which bonds have a weak angular correlation between first neighbors, $\langle \boldsymbol{\ell}_i \cdot \boldsymbol{\ell}_{i+1} \rangle = b^2 \gamma$, where $\gamma < 1$ is a positive constant. There are no correlations beyond first neighbors. If N is large, Equation (32.6) contains only terms along the main diagonal and the two adjacent diagonals. So for large N, you have $\langle r^2 \rangle \approx Nb^2 + 2Nb^2\gamma = Nb^2(1 + 2\gamma)$.

Angular correlations between neighboring bonds do not change the scaling, $\langle r^2 \rangle \sim N$. Neighbor correlations only change a multiplicative constant. This conclusion also holds if you include second neighbors and third neighbors, etc., provided only that the correlations decay to zero over a number of bonds $|i - j| \ll N$ that is much smaller than the chain length. So the random-flight model is useful even for real chains, for which bonds are not fully independent of their neighbors. *Chain stiffness* is a term sometimes used to describe the degree of angular correlation between neighboring bonds.

In general, the effects of bond correlations can be described by

$$\langle r^2 \rangle = C_N N b^2, \tag{32.8}$$

where the constant C_N is called the *characteristic ratio*. The subscript indicates that C_N can depend on the chain length N. The characteristic ratios of polymers are typically greater than one, indicating that there are correlations of orientations between near-neighbor bonds along the chain.

These arguments suggest a strategy for treating real chains. The idea is to represent a real chain by an equivalent freely jointed chain. The equivalent chain has *virtual bonds*, or *Kuhn segments*, each of which represents more than one real chemical bond. The number of virtual bonds N_K and the length of each bond b_K are determined by two requirements. First, the Kuhn model chain must have the same value of $\langle r^2 \rangle$ as the real chain, but it is freely jointed so its characteristic ratio equals one,

$$\langle r^2 \rangle = C_N N b^2 = N_K b_K^2. \tag{32.9}$$

Second, the Kuhn chain has the same contour length as the real chain,

$$L = Nb \cos \psi = N_K b_K, \tag{32.10}$$

where ψ is the angle between the real bond and the long axis of the extended real chain (see Figure 32.2), since that axis defines the direction of the virtual bonds. Dividing Equation (32.9) by (32.10) gives

$$\frac{b_K}{b} = \frac{C_N}{\cos \psi} \tag{32.11}$$

so

$$\frac{N_K}{N} = \frac{\cos^2 \psi}{C_N}. \tag{32.12}$$

Example 32.1 gives the Kuhn length for polyethylene.

EXAMPLE 32.1 The Kuhn model of polyethylene. Polyethylene has a measured characteristic ratio of 6.7. For tetrahedral valence angles, $\psi = 70.5°/2 = 35.25°$ [1]. Equations (32.11) and (32.12) give the Kuhn length as $b_K/b = 6.7/\cos 35.25° \approx 8$ times the chemical bond length, and there are $N/N_K = 6.7/\cos^2(35.25°) \approx 10$ chemical bonds per virtual bond.

Figure 32.3 Example of a random-walk chain conformation. Source: LRG Treloar, *Physics of Rubber Elasticity*, 2nd edition, Oxford University Press, New York, 1958.

Another model is the *wormlike chain* model [2]. Its measure of stiffness is the *persistence length*, which is half the Kuhn length. The persistence and Kuhn lengths characterize the number of bonds over which orientational correlations decay to zero along the chain. These are simplified models of bond correlations. A better treatment, which we won't explore here, is the *rotational isomeric state model* [1, 3], which accounts not only for the angles and lengths of chemical bonds, but also for the different statistical weights of the various possible bond conformations.

To understand why polymeric materials are elastic, you need to know more than just the first and second moments, $\langle r \rangle = 0$ and $\langle r^2 \rangle = Nb^2$; you need the full distribution function of the end-to-end lengths.

Random-flight Chain Conformations Are Described by the Gaussian Distribution Function

To get the distribution of end-to-end lengths of a polymer chain, we use the random-flight model that predicted the diffusion of a particle in Chapter 4 (pages 57 to 59). In diffusion, a particle moves a distance b in a random direction at each time step. For a polymer, imagine laying down a bond at a time, from bond $i = 1$ to N, so the growing end of a random-flight chain moves a distance b in a random direction as each bond vector is added to the chain. Adding each randomly oriented bond vector to the growing chain corresponds to a particle moving one time step in the diffusion model. Figure 32.3 shows how the conformation of a random-flight polymer chain looks like the path of a diffusing particle.

To be more quantitative, we first consider a one-dimensional random walk, then we generalize to three dimensions. Each bond vector, because it is oriented randomly, can have a different projection onto the x-axis. We first compute the distribution of the number of forward and reverse steps, then we compute the average x-axis distance travelled per step.

We want the probability, $P(m, N)$ that the chain takes m steps in the $+x$ direction out of N total steps, giving $N - m$ steps in the $-x$ direction. Just like the probability of getting m heads out of N coin flips, this probability is given approximately by the Gaussian distribution function, Equation (4.34),

$$P(m, N) = P^* e^{-2(m-m^*)^2/N},$$

where m^* is the most probable number of steps in the $+x$ direction.

Now we convert from the *number of steps m* to *coordinate position x*. The net forward progress x is a product of two factors: (1) the number of forward minus reverse steps, which equals $m - (N - m) = 2m - N$; (2) the average x-axis distance travelled per step. The average x-axis projection of the bond vectors is $\langle r_x \rangle = 0$ (see Equation (32.5)) because of the symmetry between forward and reverse steps. However, because we want the average step *length*, without its sign, we use the rms x-axis projection: $\langle b^2 \cos^2 \theta \rangle^{1/2} = b\langle \cos^2 \theta \rangle^{1/2} = b/\sqrt{3}$ (see Equation (1.45)).

The product of these two factors is $x = (2m - N)b/\sqrt{3}$. Since the average number of forward steps is $m^* = N/2$, you have $x = 2(m-m^*)b/\sqrt{3}$. Squaring both sides and rearranging gives

$$(m - m^*)^2 = 3x^2/4b^2.$$

Substituting this expression into Equation (4.34) gives

$$P(x, N) = P^* e^{-3x^2/2Nb^2}. \tag{32.13}$$

You can find the normalization constant P^* by integrating:

$$\int_{-\infty}^{\infty} P^* e^{-3x^2/2Nb^2} \, dx = 1.$$

This is solved by using the integral (Appendix D, Equation (D.1))

$$\int_{-\infty}^{\infty} e^{-\beta x^2} \, dx = \sqrt{\frac{\pi}{\beta}}, \qquad \text{where } \beta = 3/(2Nb^2). \tag{32.14}$$

Equation (32.13) becomes

$$P(x, N) = \left[\frac{\beta}{\pi}\right]^{1/2} e^{-\beta x^2} = \left[\frac{3}{2\pi Nb^2}\right]^{1/2} e^{-3x^2/2Nb^2}. \tag{32.15}$$

Equation (32.15) gives you the relative numbers of all N-mer chain conformations that begin at the origin of the x-axis and end at x. The Gaussian distribution has a peak at $x = 0$, implying that most of the chains have about as many + steps as − steps, as we noted in Chapter 4.

Relatively few chains are highly stretched. If you want to convert from the *fractions* of chain conformations to *total numbers* of chain conformations, you can multiply by the approximate total number of chain conformations z^N, where z is the number of rotational isomers per bond, to get $z^N P(x, N)$. Now we show that the Gaussian model predicts that rubber and other polymeric materials are highly deformable and elastic.

Polymer Elasticity Follows Hooke's Law

When a polymer is stretched and then released, it retracts like a Hooke's law spring. The retractive force is proportional to the extension. The retractive force is entropic.

Suppose you pull the two ends of a polymer chain apart along the x-direction. Because all the conformations of random-flight chains have the same energy, the free energy is purely entropic, $F = -TS$. Substituting Equation (32.15) into the expression $S/k = \ln P(x, N)$ gives, for the entropy and free energy of stretching,

$$\frac{F}{kT} = -\frac{S}{k} = -\ln P(x, N) = \beta x^2 + \text{constant} = \frac{3x^2}{2Nb^2} + \text{constant}. \tag{32.16}$$

Force (pN)

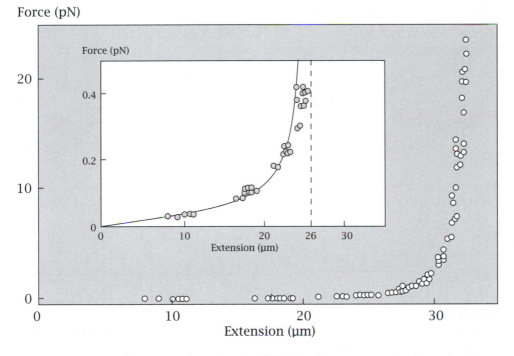

Force (pN)

0.4

0.2

0

0   10   20   26   30

Extension (μm)

0   10   20   30

Extension (μm)

Figure 32.4 Stretching a single molecule of DNA leads to a retractive force that is linear in the extension at low extensions, but steeper at higher extensions. The inset is an enlargement of the y-axis, for small deformations. Source: SB Smith, L Finzi and C Bustamante, *Science* **258**, 1122–1126 (1992).

The retractive force f_{elastic} is defined as the derivative of the free energy,

$$f_{\text{elastic}} = -\frac{dF}{dx} = -2kT\beta x = -\frac{3kTx}{Nb^2}. \tag{32.17}$$

Equation (32.17) shows that the retractive force f_{elastic} is linear in the displacement x, like a Hooke's law spring. Polymeric materials can be stretched to many times their undeformed size. (Metal can be stretched too, but only to a much smaller extent, and by a different, energetic, mechanism.) Fully stretched, the size of a chain is determined by its contour length, $L = Nb$. In its undeformed state, its average size is $\langle r^2 \rangle^{1/2} = N^{1/2}b$, so a polymer chain can be stretched by nearly a factor of $\sqrt{N}$.

Figure 32.4 shows how the retractive force depends on the stretched lengths of single molecules of DNA. You can see the linearity between force and extension for small stretching. At large deformations, the retractive forces become much stronger and the Gaussian distribution no longer applies.

Now we generalize to three dimensions to describe elasticity as a *vector* force that drives the chain ends together (see Figure 32.5).

Elasticity in Two and Three Dimensions

Let **r** represent the vector from the origin, where the chain begins, to (x, y, z), where the chain ends (see Figure 32.6(a)). The length of the end-to-end vector

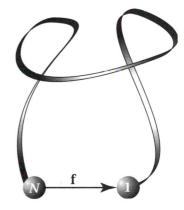

Figure 32.5 Chain conformational elasticity is a vector force tending to cause the chain ends (labeled N and 1) to be near each other, on average.

(a)

$$P(\mathbf{r}, N) = P(x, y, z, N)$$

(b)

$$P(r, N) = 4\pi r P(\mathbf{r}, N)$$

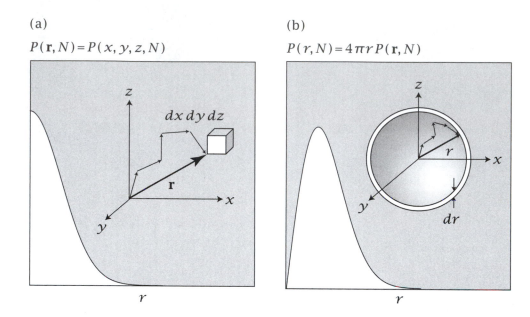

Figure 32.6 Two different probability densities for the termini of chains that begin at the origin $(0, 0, 0)$. (a) The probability density $P(\mathbf{r}, N) = P(x, y, z, N)$ that the chain end is between (x, y, z) and $(x + dx, \ y + dy, \ z + dz)$. The most probable termination point is the origin. (b) The probability density $P(r, N) = 4\pi r^2 P(\mathbf{r}, N)$ that the chain end is anywhere in a radial shell between r and $r + dr$. The peak of this distribution is predicted by Equation (32.23). Source: CR Cantor and PR Schimmel, *Biophysical Chemistry, Part II*, WH Freeman, San Francisco, 1980.

is r, where $r^2 = x^2 + y^2 + z^2$. Because each bond is oriented randomly, the x, y, and z projections of a random-flight chain are independent of each other. So the probability density of finding the chain end in a small volume element between (x, y, z) and $(x + dx, y + dy, z + dz)$ is the product of independent factors:

$$P(x, y, z, N) \, dx \, dy \, dz = P(x, N)P(y, N)P(z, N) \, dx \, dy \, dz$$

$$= \left[\frac{\beta}{\pi} \right]^{3/2} e^{-\beta(x^2 + y^2 + z^2)} \, dx \, dy \, dz. \tag{32.18}$$

In terms of the vector $\mathbf{r}$, you have

$$P(\mathbf{r}, N) = P(x, y, z, N) = \left[\frac{\beta}{\pi} \right]^{3/2} e^{-\beta r^2}. \tag{32.19}$$

Equations (32.18) and (32.19) give the probability that the chain terminus is located at a specific location $\mathbf{r}$ between (x, y, z) and $(x + dx, y + dy, z + dz)$, if the chain originates at $(0, 0, 0)$ (see Figure 32.6(a)). However, sometimes this is not exactly the quantity you want. Instead you may want to know the probability that the chain terminates *anywhere in space at a distance r from the origin* (see Figure 32.6(b)). We denote this quantity $P(r, N)$, without the bold $\mathbf{r}$ notation. Because the number of elements having a volume $dx \, dy \, dz$ grows with distance r from the origin as $4\pi r^2$, the probability of finding the

Number of Molecules

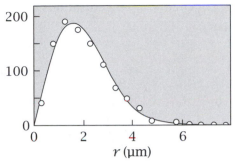

Figure 32.7 Distribution of end-to-end distances r of T3 DNA adsorbed on cytochrome c film (in microns, μm). The line represents the theoretical distribution for two-dimensional random walks. Source: D Lang, H Bujard, B Wolff, and D Russell, *J Mol Biol* **23**, 163–181 (1967).

two ends separated by a distance r in any direction is $P(r, N) = 4\pi r^2 P(\mathbf{r}, N)$. The normalization is given by

$$\int_{-\infty}^{\infty} \int_{-\infty}^{\infty} \int_{-\infty}^{\infty} P(x, y, z, N)\, dx\, dy\, dz = \int_0^{\infty} 4\pi r^2 P(\mathbf{r}, N)\, dr = 1. \qquad (32.20)$$

The function $P(r, N)$ has a peak because it is a product of two quantities: the probability $P(\mathbf{r}, N)$ diminishes monotonically with distance from the origin, but the number of volume elements, $4\pi r^2$, increases with distance from the origin. Figure 32.7 shows that the two-dimensional end-to-end distance distribution of T3 DNA molecules adsorbed on surfaces and counted in electron micrographs is well predicted by two-dimensional random-walk theory.

Example 32.2 shows how the Gaussian distribution function is used for predicting polymer cyclization equilibria and kinetics.

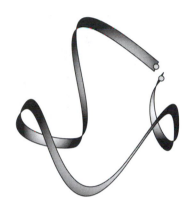

Figure 32.8 For polymer cyclization, the two chain ends must be close together.

EXAMPLE 32.2 Polymer cyclization (Jacobson–Stockmayer theory). What is the probability that the two ends of a polymer chain come close enough together for them to react with each other? This probability is useful for calculating the rate and equilibrium constants for cyclization processes in which linear chains form circles (see Figure 32.8). Here we follow the random-flight model of H Jacobson and WH Stockmayer [4].

Suppose that the polymer has N bonds. To determine the probability that the chain end is within a bond distance b of the chain beginning, integrate $P(r, N)$, the probability of finding the ends a distance r apart, from 0 to b:

$$P_{\text{cyclization}} = \int_0^b P(r, N)\, dr$$

$$= \left[\frac{3}{2\pi N b^2}\right]^{3/2} \int_0^b e^{-3r^2/2Nb^2} 4\pi r^2\, dr. \qquad (32.21)$$

When the ends are together, $r = b$ is small ($r^2/Nb^2 \ll 1$), so $e^{-3r^2/2Nb^2} \approx 1$. Then the integral in Equation (32.21) is $\int_0^b 4\pi r^2\, dr = 4\pi b^3/3$, and

$$P_{\text{cyclization}} = \left[\frac{3}{2\pi N b^2}\right]^{3/2} \left(\frac{4}{3}\pi b^3\right) = \left(\frac{6}{\pi}\right)^{1/2} N^{-3/2}.$$

The main prediction of the Jacobson–Stockmayer theory is that the cyclization probability diminishes with chain length as $N^{-3/2}$. The longer the chain, the smaller is the probability that the two ends are close enough to react. This

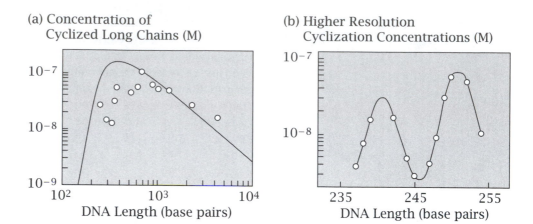

(a) Concentration of
Cyclized Long Chains (M)

(b) Higher Resolution
Cyclization Concentrations (M)

DNA Length (base pairs)

DNA Length (base pairs)

Figure 32.9 (a) For long chains the DNA cyclization probability diminishes as predicted by the Jacobson–Stockmayer theory. Shortening the chain stiffens it, reducing the cyclization probability below the Jacobson–Stockmayer value. (b) At higher resolution the cyclization probability depends on periodicities in the polymer, neglected in the random-flight treatment. Source: D Shore and RL Baldwin, *J Mol Biol* **170**, 957–981 (1983).

Figure 32.10 Experimental molar cyclization equilibrium constants K (in mol dm^{-3}) for cyclic $[O(CH_2)_{10}OCO(CH_2)_4CO]_n$ in poly(decamethylene adipate) melts at 423 K (○) versus chain length n are compared with values calculated (●) from the Jacobson–Stockmayer theory. Source: JA Semylen, *Cyclic Polymers*, Elsevier, London, 1986.

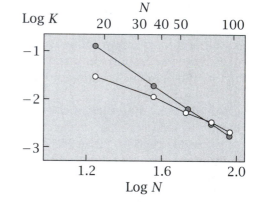

prediction is confirmed by experiments shown in Figures 32.9 and 32.10. Figure 32.9 shows that the probability of DNA cyclization diminishes with chain length for long chains, as predicted by the theory, but you can also see the effect of chain stiffness, which is not treated by the theory. If chains are too short, their stiffness prevents them from bending enough to cyclize, so their cyclization probability is smaller than is predicted by the random-flight model. Also, chemical details can matter for short chains: the ends must be oriented correctly with respect to each other to cyclize.

Example 32.3 shows how to find the most probable radius R_0.

EXAMPLE 32.3 Most probable radius R_0. What is the most probable end-to-end distance $R = R_0$ of a Gaussian chain? Because this is related to the size of the molecule, this quantity is often called the *most probable radius*. Begin with the entropy, $S(r) = k \ln P(r, N)$. Because $F = -TS$, using Equation (32.19) gives

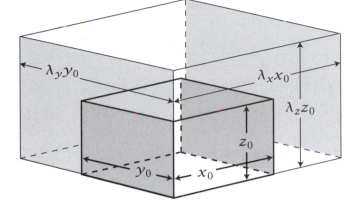

$$\frac{F}{kT} = -\frac{S}{k} = -\ln\left(4\pi r^2 P(\mathbf{r}, N)\right) = \beta r^2 - 2\ln r + \text{constant}. \tag{32.22}$$

To get the equilibrium value $r = R_0$, find the minimum free energy:

$$\frac{d}{dr}\left(\frac{F}{kT}\right)_{r=R_0} = 2\beta R_0 - \frac{2}{R_0} = 0 \tag{32.23}$$

$$\Rightarrow R_0^2 = \frac{1}{\beta} = \frac{2Nb^2}{3}.$$

The quantity $R_0^2 = 2Nb^2/3 = (2/3)\langle r^2 \rangle$ is the square of the most probable radius of the Gaussian chain. We use it starting on page 621 for treating polymer collapse and folding processes.

Now we use the theory of polymer chains to describe elastic materials.

The Elasticity of Rubbery Materials Results from the Sum of Chain Entropies

Rubber and other polymeric materials are elastic. Polymeric elastomers are covalently cross-linked networks of polymer chains. Here we describe one of the simplest and earliest models for the retractive forces of polymeric materials, the *affine network model*.

Suppose you have m chains, all of the same length N. The bond length is b. The chain ends are covalently cross-linked at junction points. Assume there are no intermolecular interactions, and that the total elastic free energy of the material is the sum of the elastic free energies of each of the chains. For the undeformed material, indicated by subscript 0, the end of a chain is at (x_0, y_0, z_0) and

$$r_0^2 = x_0^2 + y_0^2 + z_0^2 = Nb^2. \tag{32.24}$$

If the material is deformed by a factor λ_x in the x-direction, λ_y in the y-direction, and λ_z in the z-direction (see Figure 32.11), so that $x = \lambda_x x_0$, $y = \lambda_y y_0$, and $z = \lambda_z z_0$, the chain end is moved to (x, y, z),

$$r^2 = x^2 + y^2 + z^2 = \lambda_x^2 x_0^2 + \lambda_y^2 y_0^2 + \lambda_z^2 z_0^2. \tag{32.25}$$

Using $F = -kT \ln P(x, y, z, N)$, $\beta = 3/(2Nb^2)$, and Equation (32.19), you have the free energy for deforming a single chain, F_1:

$$\Delta F_1 = F_{\text{deformed}} - F_{\text{undeformed}} = -kT \ln \frac{P(x, y, z, N)}{P(x_0, y_0, z_0, N)} = kT\beta(r^2 - r_0^2).$$

Summing over m independent chains gives the total free energy F_m:

$$\Delta F_m = kT\beta \sum_m (r^2 - r_0^2). \tag{32.26}$$

Assuming that all chains are equivalent, you have $\sum_m (r^2 - r_0^2) = m[\langle r^2 \rangle - \langle r_0^2 \rangle]$, and Equation (32.26) becomes

$$\begin{aligned} \Delta F_m &= kT\beta m(\langle r^2 \rangle - \langle r_0^2 \rangle) \\ &= kT\beta m \left[(\lambda_x^2 - 1)\langle x_0^2 \rangle + (\lambda_y^2 - 1)\langle y_0^2 \rangle + (\lambda_z^2 - 1)\langle z_0^2 \rangle \right]. \end{aligned} \tag{32.27}$$

If the undeformed chains are isotropic (all the directions are equivalent), then $\langle x_0^2 \rangle = \langle y_0^2 \rangle = \langle z_0^2 \rangle = (Nb^2)/3$, and Equation (32.27) becomes

$$\frac{\Delta F_m}{kT} = \frac{m}{2}(\lambda_x^2 + \lambda_y^2 + \lambda_z^2 - 3). \tag{32.28}$$

In a macroscopic material, applying a force in one direction can cause forces and deformations in other directions. The forces per unit area in various directions are called *stresses* and the relative displacements are called *strains*. Stresses are derivatives of the free energy with respect to strains. If you deform a material by a factor α, where the x-axis, y-axis, and z-axis deformations depend on α, the free energy is a function $F(\lambda_x(\alpha), \lambda_y(\alpha), \lambda_z(\alpha))$. The derivative can be written in terms of λ_x^2, λ_y^2, and λ_z^2 as

$$\left(\frac{\partial F}{\partial \alpha} \right) = \left(\frac{\partial F}{\partial \lambda_x^2} \right) \frac{d\lambda_x^2}{d\alpha} + \left(\frac{\partial F}{\partial \lambda_y^2} \right) \frac{d\lambda_y^2}{d\alpha} + \left(\frac{\partial F}{\partial \lambda_z^2} \right) \frac{d\lambda_z^2}{d\alpha}. \tag{32.29}$$

Suppose that you stretch an elastomer along the x-direction by a factor $\alpha = x/x_0$. Then the x-direction force is

$$f_x = - \left(\frac{\partial \Delta F_m}{\partial x} \right) = \frac{1}{x_0} \left(\frac{\partial \Delta F_m}{\partial \alpha} \right) = -\frac{\alpha}{x} \left(\frac{\partial \Delta F_m}{\partial \alpha} \right). \tag{32.30}$$

The stress τ acting in the opposite direction equals the force given by Equation (32.30) divided by the deformed cross-sectional area yz,

$$\tau = -\frac{f_x}{yz} = \frac{\alpha}{V} \left(\frac{\partial \Delta F_m}{\partial \alpha} \right), \tag{32.31}$$

where $V = xyz$ is the final volume of the material [5, 6]. Examples 32.4 and 32.5 are applications of this elasticity model.

EXAMPLE 32.4 Stretch a rubber band along the x-axis. If you stretch rubber, its volume remains approximately constant. Therefore, stretching along the x-direction by $\lambda_x = \alpha$ leads to

$$\lambda_y = \lambda_z = \frac{1}{\sqrt{\alpha}}. \tag{32.32}$$

Then Equation (32.28) becomes

$$\frac{\Delta F_m}{kT} = \frac{m}{2}\left(\alpha^2 + \frac{2}{\alpha} - 3\right).$$

Because $\partial \Delta F_m / \partial \lambda_x^2 = \partial \Delta F_m / \partial \lambda_y^2 = \partial \Delta F_m / \partial \lambda_z^2 = mkT/2$, Equation (32.28) gives

$$\frac{\partial F_m}{\partial \alpha} = mkT\left(\alpha - \frac{1}{\alpha^2}\right). \tag{32.33}$$

Substituting Equation (32.33) into Equation (32.31) gives

$$\tau = \frac{mkT}{V}\left(\alpha^2 - \frac{1}{\alpha}\right). \tag{32.34}$$

Figure 32.12 shows that the model predicts experimental data adequately at extensions below about $\alpha = 3$–5, but that the chains become harder to stretch than the model predicts at higher extensions.

EXAMPLE 32.5 Stretch a rubber sheet biaxially.   Stretch a rubber sheet along the x-axis by an amount $\lambda_x = \alpha_1$ and along the y-axis by an amount $\lambda_y = \alpha_2$, where α_1 is independent of α_2. If the volume is constant, $\lambda_z = 1/(\alpha_1\alpha_2)$. Then the stress along the x-axis is

$$\tau_x = \frac{mkT}{2V}\alpha_1\left(\frac{d\lambda_x^2}{d\alpha_1} + \frac{d\lambda_z^2}{d\alpha_1}\right) = \frac{mkT}{V}\left(\alpha_1^2 - \frac{1}{\alpha_1^2\alpha_2^2}\right). \tag{32.35}$$

The retractive stress in elastomers depends not only on the deformation but also on the cross-link density, through m/V, the density of chains. In a unit volume there are $2m$ total chain ends. If each junction is an intersection of j chain ends, then there will be (1 junction/j chain ends) × ($2m$ chain ends) = ($2m/j$) junctions, so the number of junctions is proportional to m. Therefore the retractive force increases linearly with the cross-link density of the network. Bowling balls are made of a type of rubber that has a much higher cross-link density than rubber bands.

The main advance embodied in theories of polymer elasticity was the recognition that the origin of the force is due mainly to the conformational freedom of the chains, and is mainly entropic, not energetic. The conformational entropies of polymers are important not only for stretching processes. They also oppose the contraction of a polymer chain to a radius smaller than its equilibrium value.

Polymers Expand in Good Solvents, Are Random Flights in Theta Solvents, and Collapse In Poor Solvents

According to the random-flight theory, the 'size' of a molecule increases in proportion to $N^{1/2}$, where size means either the average end-to-end distance $\langle r^2 \rangle = Nb^2$ or the most probable radius $R_0^2 = 2Nb^2/3$. We showed on pages

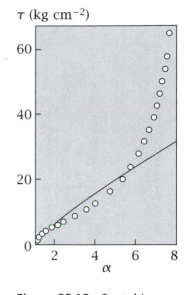

τ (kg cm^{-2})

Figure 32.12   Stretching a rubber string gives stress τ versus elongation α, ($\circ$) experimental data; (——) curve predicted by Equation (32.34). Source: P Munk, *Introduction to Macromolecular Science*, Wiley, New York, 1989. Data are from LRG Treloar, *Trans Faraday Soc* **40**, 59 (1944).

611–613 that near-neighbor bond angle correlations (called *local* interactions) do not change this scaling relationship. But now we show that solvent (*nonlocal*) interactions *can* change this relationship because different solvents or temperatures cause the chain to swell or contract.

Random-flight behavior, which is dominated by local interactions, applies only to a limited class of solvent and temperature conditions, called θ-*conditions*, θ-*solvents*, or θ-*temperatures*. In contrast, in a *good solvent*, monomer–solvent interactions are more favorable than are monomer–monomer interactions. Chains expand in good solvents. The radius of a polymer molecule grows more steeply with N than is predicted by the random-flight model, mainly because of the self-avoidance of the segments of the chain. In a third class of conditions, called *poor solvents*, the chain monomers are attracted to each other more strongly than they are attracted to the solvent. Poor solvents cause an isolated chain to collapse into a compact globule, with radius $\sim N^{1/3}$. Poor solvents can also cause multiple chains to aggregate with each other. For example, oil chains such as polymethylene phase-separate from water because water is a poor solvent. Here is the simplest model of the expansion and collapse of a single chain, due to PJ Flory [7, 8].

How does the conformational free energy depend on the radius? Consider a chain with N monomers. As in Equation (32.23), our strategy is to find the radius $r = R$ that maximizes the entropy, or minimizes the free energy. However, now we include an additional contribution to the free energy. In addition to the elastic free energy F_{elastic}, we now also account for the solvation free energy $F_{\text{solvation}}$, using the Flory–Huggins theory of Chapter 31,

$$\frac{d}{dr}\left(F_{\text{elastic}}(r) + F_{\text{solvation}}(r)\right)_{r=R} = 0. \tag{32.36}$$

The solvation free energy depends on the radius through the mean chain segment density, $\rho = N/M$. M is the number of sites of a lattice that contains the chain. We define M in terms of the chain radius below. Low density, $\rho \to 0$, means the chain is expanded, while high density, $\rho \to 1$, means the chain is compact.

Chain collapse is opposed by the conformational entropy but is driven by the gain in monomer–monomer contacts, which are favorable under poor-solvent conditions. To get the conformational entropy for $F_{\text{solvation}}$, you can begin with Equation (31.4), which gives ν_1, the number of conformations of a single chain. But we make two changes. First, we leave out the factor $(z-1)^{N-1}$ because it is a constant that doesn't change with the density ρ, and leaving it out simplifies the math. (The independence of the conformational entropy from the factor $(z-1)^{N-1}$ in the Flory model has the important implication that chain stiffness and local interactions do not contribute to entropies of collapse or expansion.) Second, our focus on a single chain means that its center-of-mass position in space is irrelevant. This is equivalent to neglecting the placement of the first monomer, so we replace the factor $M^{-(N-1)}$ with M^{-N} and the conformational entropy of the chain is

$$\frac{S}{k} = \ln \nu_1 = \ln \frac{M!}{(M-N)!M^N}. \tag{32.37}$$

Multipy the numerator and denominator of Equation (32.37) by $N!$ to put part

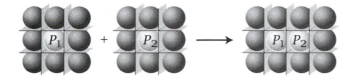

Figure 32.13 The contact free energy Δg describes desolvating two polymer chain segments P_1 and P_2 and bringing them into contact. The interaction parameter χ is defined for the process with the opposite sign (see Figure 15.6).

of this expression into a more familiar form,

$$
\begin{aligned}
\frac{S}{k} &= \ln\left(\frac{M!}{(M-N)!N!}\right) + \ln\left(\frac{N!}{M^N}\right) \\
&= -N\ln\left(\frac{N}{M}\right) - (M-N)\ln\left(1 - \frac{N}{M}\right) + N\ln\left(\frac{N}{M}\right) - N\ln e \\
&= -(M-N)\ln\left(1 - \frac{N}{M}\right) - N,
\end{aligned}
$$
(32.38)

by using Stirling's approximation $N!/M^N \approx (N/eM)^N = (N/M)^N e^{-N}$. Now divide by N to get the entropy per monomer, and express the result in terms of the average density $\rho = N/M$,

$$
\frac{S}{Nk} = -\left(\frac{1-\rho}{\rho}\right)\ln(1-\rho) - 1.
$$
(32.39)

You can check that $S \to 0$ as $\rho \to 0$ and $S \to -Nk$ as $\rho \to 1$.

EXAMPLE 32.6 Polymer collapse entropy. When a protein folds, it collapses to a nearly maximally compact state. What is the entropic component of the free energy opposing collapse at $T = 300$ K? If the chain length is $N = 100$ monomers, the equations below Equation (32.39) give

$$
\begin{aligned}
\Delta F_{\text{collapse}} &= -T(S_{\text{compact}} - S_{\text{open}}) = NkT \\
&\approx (100)(1.987 \text{ cal mol}^{-1}\text{ K}^{-1})(300\text{ K}) \approx 60 \text{ kcal mol}^{-1}
\end{aligned}
$$

Next, we determine how the contact energy depends on the chain radius, or density. Let w_{pp}, w_{ss}, and w_{sp} represent the energies of a contact between two polymer segments, between two solvent molecules, and between a solvent molecule and a polymer segment, respectively. Figure 32.13 and Equation (15.11) indicate that the free energy Δg for desolvating two polymer segments and forming a contact is

$$
\Delta g = -2\left(w_{sp} - \frac{w_{ss} + w_{pp}}{2}\right) = -\frac{2}{z}\chi kT.
$$
(32.40)

The minus sign indicates that the process defining Δg is the reverse of the process defining χ (see Figure 15.6). The mean-field approximation gives an estimate of the number of contacts among chain monomers. The probability that a site adjacent to a polymer segment contains another polymer segment is $\rho = N/M$, and there are z sites that are neighbors of each of the N monomers, so the contact energy U relative to the fully solvated chain is

$$
U = \frac{N\rho z}{2}\Delta g = -NkT\rho\chi.
$$
(32.41)

The factor of $1/2$ corrects for the double counting of interactions. Combining the entropy Equation (32.39) with the energy Equation (32.41) gives the conformational free energy of the chain as a function of its compactness,

$$\frac{F_{\text{solvation}}}{NkT} = \frac{U}{NkT} - \frac{S}{Nk} = \left(\frac{1-\rho}{\rho}\right)\ln(1-\rho) + 1 - \rho\chi. \tag{32.42}$$

When the chain is relatively open and solvated, the density is small, $\rho \ll 1$. Then you can use the approximation $\ln(1-\rho) \approx -\rho - (1/2)\rho^2 - \ldots$ to get

$$\left(\frac{1-\rho}{\rho}\right)\ln(1-\rho) + 1 \approx \left(\frac{1-\rho}{\rho}\right)\left(-\rho - \frac{\rho^2}{2} - \ldots\right) + 1$$

$$\approx \frac{1}{2}\left(\rho + \rho^2\right). \tag{32.43}$$

Substituting Equation (32.43) into Equation (32.42), and keeping only the first-order approximation, gives the solvation free energy as a function of the average segment density,

$$\frac{F_{\text{solvation}}}{kT} = N\rho\left(\frac{1}{2} - \chi\right). \tag{32.44}$$

In Flory theory, a simple approximation relates the density ρ to the size r

$$\rho = \frac{N}{M} = \frac{Nv}{r^3}, \tag{32.45}$$

where v is the volume per chain segment. This relationship defines the value of M, the number of sites on the lattice that contains the polymer chain. Combining Equation (32.45) with Equations (32.22) and (32.44) gives

$$\frac{F_{\text{elastic}}}{kT} + \frac{F_{\text{solvation}}}{kT} = \beta r^2 - 2\ln r + \frac{N^2 v}{r^3}\left(\frac{1}{2} - \chi\right) + \text{constant.} \tag{32.46}$$

Taking the derivative of Equation (32.46) and finding the value $r = R$ that causes the derivative to be zero (the most probable value of r; Equation (32.36)) gives

$$2\beta R - \frac{2}{R} - \frac{3N^2 v}{R^4}\left(\frac{1}{2} - \chi\right) = 0. \tag{32.47}$$

You can express this in terms of $R_0^2 = 2Nb^2/3 = \beta^{-1}$, the most probable radius of the unperturbed chain (see Equation (32.23)). Multiplying both sides of Equation (32.47) by $R^4/(2R_0^3)$, and rearranging gives

$$\left(\frac{R}{R_0}\right)^5 - \left(\frac{R}{R_0}\right)^3 = \left(\frac{3N^2 v}{2R_0^3}\right)\left(\frac{1}{2} - \chi\right)$$

$$= \left(\frac{3}{2}\right)^{5/2}\frac{v}{b^3}\left(\frac{1}{2} - \chi\right)\sqrt{N}. \tag{32.48}$$

Theta Solvents Give Random-flight Behavior

For solvent and temperature conditions that cause $\chi = 0.5$, the right-hand side of Equation (32.48) equals zero. In that case, multiplying Equation (32.48) by

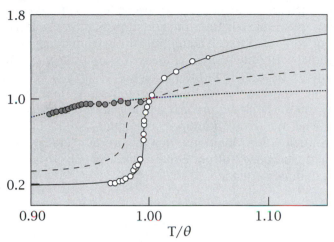

Expansion Factor

Figure 32.14 Homopolymers collapse when the temperature T is less than the temperature θ at which $\chi = 0.5$. Chains expand at higher temperatures ($\chi < 0.5$, good solvents). The transition steepens with increasing chain lengths. (($\bullet$) molecular weight $M = 2.9 \times 10^3$, (- - -) $M = 1 \times 10^5$, ($\circ$) $M = 2.6 \times 10^7$.) Source: ST Sun, I Hishio, G Swislow and T Tanaka, *J Chem Phys* **73**, 5971–5975 (1980).

$(R_0/R)^3$ gives the random-flight prediction,

$$\left(\frac{R}{R_0}\right)^2 = 1 \quad \Rightarrow \quad R^2 = R_0^2 = \frac{2Nb^2}{3}, \tag{32.49}$$

given by Equation (32.23). That is, when the solvation free energy is zero (the monomer–monomer attraction just balances the excluded volume), then the most probable radius of the chain is given simply by the elastic free energy alone.

Good Solvents Expand Polymers

For solvent and temperature conditions that cause $\chi < 0.5$ (called good solvents), the right-hand side of Equation (32.48) is positive. In this regime, excluded volume causes chain expansion. For large N, the fifth-power term is much larger than the third-power term, so

$$\left(\frac{R}{R_0}\right)^5 \approx \left(\frac{3}{2}\right)^{5/2} \frac{v}{b^3} \left(\frac{1}{2} - \chi\right)\sqrt{N} \quad \Rightarrow \quad R^5 \propto R_0^5 N^{1/2} \propto N^3$$

$$\Rightarrow \quad R \propto N^{0.6}. \tag{32.50}$$

In good solvents, the chain radius grows more steeply with chain length ($R \propto N^{0.6}$) than in θ-solvents ($R \propto N^{0.5}$).

Polymers Collapse in Poor Solvents

Solvent and temperature conditions that cause $\chi > 0.5$ are called poor solvents. For a poor solvent, the right-hand side of Equation (32.48) is negative, so $R/R_0 < 1$, and the polymer collapses into a compact conformation. For poor solvent conditions, Equation (32.48) is no longer sufficient because of the low-density approximation we used to derive it. A collapsed chain has a high segment density. If you were to keep the next higher term in the density expansion for $\ln(1 - \rho)$, however, you would find that the model predicts $R \propto N^{1/3}$, as expected for a compact chain.

Polymers can undergo very sharp transitions from the coil to compact states as the solvent and temperature are changed. These are called *coil-to-globule transitions*. The reason they are so sharp is found in Equation (32.48). For large N, the right-hand side of Equation (32.48) can change abruptly from being very positive to being very negative with only a small change in χ at about $\chi = 0.5$. Figure 32.14 shows the collapse process in homopolymers.

Two natural collapse processes are the folding of proteins into their compact native states in water, and the compaction of DNA molecules for insertion into virus heads and cell nuclei. While this homopolymer collapse model illustrates the principle of coil-to-globule transitions, neither protein folding nor DNA collapse follow it exactly, because both polymers also have electrostatic interactions and specific monomer sequences.

Summary

Polymers have many conformations of nearly equal energy, so they have broad distributions of conformations. Stretching or squeezing or otherwise perturbing polymers away from their equilibrium conformations leads to entropic forces that oppose the perturbations. This is the basis for rubber elasticity. One of the most important models is the random-flight theory, in which the distribution of chain end distances is Gaussian. The rms distance between the two ends of the chain increases as the square root of the chain length N, $\langle r^2 \rangle^{1/2} = (Nb^2)^{1/2}$. The random-flight theory applies when the chain is in a θ-solvent. In that case, the steric tendency to expand is just balanced by the self-attraction energy causing the chain to contract. When the solvent is poor, the self-attractions between the chain monomers dominate and chains collapse to compact configurations. When the solvent is good, the self-attractions are weak and chains expand more than would be predicted by the random-flight theory.

Problems

1. Stretching a rubber sheet. What is the free energy for stretching an elastomeric material uniformly along the x and y directions, at constant volume?

2. Stretching DNA. Figure 32.4 shows that it takes about 0.1 pN of force to stretch a DNA molecule to an extension of 20 μm. Use the chain elasticity theory to estimate the undeformed size of of the molecule, $\langle r^2 \rangle^{1/2}$.

3. Stretched polymers have negative thermal expansion coefficients. The thermal expansion coefficient of a material is $\alpha = (1/V)(\partial V/\partial T)_p$. Consider the corresponding one-dimensional quantity $\alpha_p = (1/x)(\partial x/\partial T)_f$ for a single polymer molecule stretched to an end-to-end length x by a stretching force f.

   (a) Compute α_p for the polymer chain.

   (b) What are the similarities and differences between a polymer and an ideal gas?

4. An 'ideal' solvent expands a polymer chain. If a polymer chain is composed of the same monomer units as the solvent around it, the system will be ideal in the sense that the polymer–polymer interactions will be identical to polymer–solvent interactions, so $\chi = 0$.

   (a) Write an expression for the most probable radius R for a chain in an ideal solvent.

   (b) Show that such a chain is expanded relative to a random-flight chain.

   (c) Describe the difference between an ideal solvent and a θ-solvent.

5. Computing conformational averages. Using the expression for the distribution $P(r, N)$ for the end-to-end separation of a polymer chain of length N, compute $\langle r^2 \rangle$ and $\langle r^4 \rangle$.

6. Contour length of DNA. The double-stranded DNA from bacteriophage λ has a contour length $L = 17 \times 10^{-6}$ m. Each base pair has bond length $b = 3.5$ Å.

   (a) Compute the number of base pairs in the molecule.

   (b) Compute the molecular weight of the DNA.

7. Using elasticity to compute chain concentrations. Figure 32.12 shows the stress–strain properties of a rubber band. Use the figure and chain elasticity theory to:

   (a) Estimate the number of polymer chains in a cubic volume 100 Å on each side.

   (b) If each monomer occupies 100 Å^3, what is the length of each chain between junction points?

References

[1] PJ Flory. *Statistical Mechanics of Chain Molecules*, Wiley, New York, 1969.

[2] JE Mark ed. *Physical Properties of Polymers Handbook*, Chapter 5, American Institute of Physics, Woodbury, 1996.

[3] WL Mattice and UW Suter. *Conformational Theory of Large Molecules: the Rotational Isomeric State Model in Macromolecular Systems*, Wiley, New York, 1994.

[4] H Jacobson and WH Stockmayer. *J Chem Phys* **18**, 1600 (1950).

[5] B Erman and JE Mark. *Ann Rev Phys Chem* **40**, 351–374 (1989).

[6] JE Mark and B Erman. *Rubberlike Elasticity: a Molecular Primer*, Wiley, New York, 1988.

[7] PJ Flory. *Principles of Polymer Chemistry*, Series title: George Fisher Baker Nonresident Lectureship in Chemistry at Cornell University, Cornell University Press, Ithaca, 1953.

[8] HS Chan and KA Dill. *Ann Rev Biophys and Biophys Chem* **20**, 447–490 (1991).

Suggested Reading

The classic texts on rotational isomeric state and random-flight models:

CR Cantor and PR Schimmel, *Biophysical Chemistry*, Vol III, WH Freeman, San Francisco, 1980.

M Doi, *Introduction to Polymer Physics*, trans by H See, Oxford University Press, New York, 1996.

PJ Flory, *Principles of Polymer Chemistry*, series title: George Fisher Baker Nonresident Lectureship in Chemistry at Cornell University. Cornell University Press, Ithaca, 1953.

WL Mattice and UW Suter, *Conformational Theory of Large Molecules: the Rotational Isomeric State Model in Macromolecular Systems*, Wiley, New York, 1994.

Excellent summaries of rubber elasticity:

B Erman and JE Mark, *Ann Rev Phys Chem* **40**, 351–374 (1989).

JE Mark and B Erman, *Rubberlike Elasticity: a Molecular Primer*, Wiley, New York, 1988.

One of the first models of a polymer collapse, applied to DNA:

CB Post and BH Zimm, *Biopolymers* **18**, 1487–1501 (1979).

33 Polymers Resist Confinement & Deformation

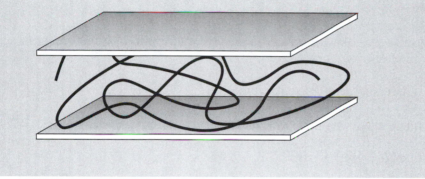

Random-flight polymer chains repel each other. They also repel inert surfaces, and they avoid entering confined spaces, such as the small pores in chromatography columns. These behaviors arise from polymer elasticities and confinement entropies. The same forces also give rise to viscoelasticity, the 'rubbery' property of polymeric liquids. The relaxation mechanism becomes more complex and slows down in concentrated polymeric liquids because the molecules must slither in snakelike motions, called *reptation*, through tubes created by neighboring chain molecules. To begin our discussion of how confinement and deformation affect polymers, we first consider excluded volume.

'Excluded Volume' Describes the Large Volume Inside a Polymer Conformation that Is Inaccessible to Other Chains

Two chain molecules rarely interpenetrate each other, even if each chain is highly expanded and solvated. The chains repel each other because of *excluded volume* (see Figure 33.1).

Suppose that chains A and B each have N monomers. Because the radius of a random-flight chain molecule is proportional to $(Nb^2)^{1/2}$ (Equation (32.7)), the volume circumscribed by molecule A is proportional to $(Nb^2)^{3/2}$. Almost all of this volume is solvent if N is large. The A chain segments themselves occupy very little of that volume. You can see this by computing the concentration of

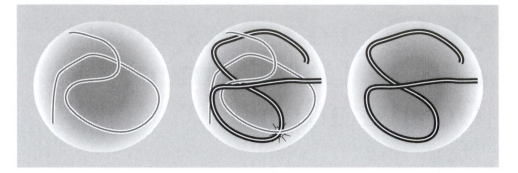

Figure 33.1 Two random-flight polymers (left and right) do not interpenetrate (center) because of *excluded volume*, even though most of the internal volume is solvent. This leads to a repulsion.

chain segments within the volume of the molecule A (the number of segments N divided by volume $(Nb^2)^{3/2}$),

$$\text{chain segment concentration} \approx \frac{N}{(Nb^2)^{3/2}} \approx N^{-1/2}. \tag{33.1}$$

Inside the volume $(Nb^2)^{3/2}$, the concentration of *solvent* is

$$\text{solvent concentration} \approx \left(1 - N^{-1/2}\right). \tag{33.2}$$

So, for typical polymers, $N = 10^3$–10^7, the interior of a random-flight chain is nearly pure solvent.

If the volume within chain A is mostly 'empty,' why won't chain B enter that space? Suppose chain A is in a particular configuration. Think about placing all N chain segments of the B chain one by one in a particular conformation inside the volume circumscribed by molecule A. Any one potential site for a B monomer is not likely to be already occupied by an A segment. But the probability is high that *at least one* of the N segments of a randomly configured polymer B will collide with an A segment. The probability, P_{success}, that all N monomers of B will avoid an A site, is

$$P_{\text{success}} = \left(1 - N^{-1/2}\right)^N. \tag{33.3}$$

Using the approximation $e^{-x} \approx 1 - x \ldots$ (see Appendix C, Equation (C.1)), you can replace $1 - N^{-1/2}$ by $e^{-N^{-1/2}}$ to get

$$P_{\text{success}} = \left(e^{-N^{-1/2}}\right)^N = e^{-N^{1/2}}. \tag{33.4}$$

If N is large, $P_{\text{success}} \approx 0$. Even though the interior of a random-flight polymer chain is mostly solvent, two polymer chains with $N \gg 1$ have a very strong steric tendency not to interpenetrate each other. This is called excluded volume.

EXAMPLE 33.1 Excluded volume. For $N = 10^4$, Equation (33.4) gives

$$P_{\text{success}} = e^{-100}.$$

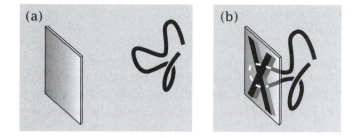

Figure 33.2 (a) Far away from a surface, chain conformations are unperturbed. (b) Close to a surface, a chain loses conformations (those that would otherwise have passed through the surface), leading to an increased free energy, and repulsion from the surface.

Excluded volume can be regarded as a repulsive force that pushes two non-interacting random-flight polymer molecules apart. This repulsion is the basis for the *steric stabilization* of colloids. If two compact colloidal particles are covered with random-flight polymer chains, the particles repel each other to avoid intermingling their polymer chains. The attached chains prevent particle aggregation and *stabilize* the colloidal solution. Some inks are colloidal suspensions that are stabilized in this way. And the glycosylation of proteins may help to prevent proteins from aggregating.

Excluded volume is different in different dimensionalities. To see this, look at Equation (33.1). Let d represent the dimensionality: $d = 3$ for polymers in three-dimensional space, or $d = 2$ for polymers on two-dimensional surfaces, etc. You can express the concentration (N/volume) of polymer segments in any dimensionality d as N/R^d, where $R = (Nb^2)^{1/2}$. Generalizing Equation (33.4) in this way gives

$$P_{\text{success}} \approx e^{-N^{(2-d/2)}}. \tag{33.5}$$

For $d = 2$, $P_{\text{success}} \approx e^{-N}$. Comparing to Equation (33.4) shows that excluded volume is more severe in two dimensions than in three dimensions. Equation (33.5) also shows that for dimensionality $d > 4$, $P_{\text{success}} \to 1$ with increasing chain length. So in high-dimensional spaces, long-chain polymers can readily interpenetrate. For this reason, $d = 4$ is called an 'ideal dimensionality.' However, $d = 4$ is of interest only in principle because real systems live in three dimensions or less.

Steric repulsions also cause random-flight polymers to repel rigid impenetrable surfaces.

Chain Conformations Are Perturbed Near Surfaces

Consider a polymer chain that can be positioned at various distances, z, away from an impenetrable surface. As the chain's center of mass approaches the surface, some of the chain's extended conformations are not physically viable because those conformations would pass through the surface (see Figure 33.2). Because the conformational entropy decreases as the chain approaches the surface, the conformational free energy F increases, and this can be described as a repulsive force, $f = -dF/dz$, between the surface and the chain.

There are various ways to model the distributions of chain conformations near surfaces. Here we describe an approach based on a *reflectance principle*. It resembles the image charge approach in the theory of electrostatic interactions (see Chapter 21, page 399).

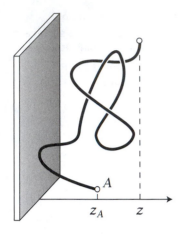

Figure 33.3 A polymer chain begins at z_A, ends at z, and does not pass through the plane.

The DiMarzio–McCrackin Reflectance Principle Predicts Chain Distributions Near Surfaces

To describe how the distribution of chain conformations is perturbed near an impenetrable surface, we need a way to count the conformations that do not cross the plane. Consider an impenetrable plane that extends in the x and y directions, located at $z = 0$ (see Figure 33.3). The chain is on the side where $z > 0$. Consider first the z distribution of the chain. For random flights, the x, y, and z distributions are independent of each other, so we can append multiplicative factors for the x and y dimensions later. The beginning of the chain is at $(0, 0, z_A)$. What is the fraction $w_{\text{viable}}(z_A, z)$ of conformations that originate at z_A, end at z, and do not cross the plane?

If there were no surface at all, the distribution of all conformations of a random-flight chain would be given by the Gaussian function, Equation (32.15),

$$
\begin{aligned}
w_{\text{total}}(z_A, z) &= \left(\frac{3}{2\pi N b^2}\right)^{1/2} \exp\left(-\frac{3(z - z_A)^2}{2N b^2}\right) \\
&= \left(\frac{\beta}{\pi}\right)^{1/2} \exp\left(-\beta(z - z_A)^2\right),
\end{aligned}
\tag{33.6}
$$

where $\beta = 3/(2N b^2)$ simplifies the notation.

But the impenetrable plane eliminates some of these conformations. We want to count only the conformations that do not pass through the plane. The total number of conformations is the sum of *viable* conformations (those that don't cross the plane) and *nonviable* conformations (those that cross the plane):

$$
w_{\text{total}} = w_{\text{viable}} + w_{\text{nonviable}}.
\tag{33.7}
$$

Equation (33.7) says that you can get the quantity you want, w_{viable}, if you know $w_{\text{nonviable}}$, because you have w_{total} from Equation (33.6).

The *reflectance principle*, developed by polymer physicists EA DiMarzio and FL McCrackin [1], gives a simple way to compute $w_{\text{nonviable}}$, the fraction of conformations that begin at A, end at B, and cross the plane at least once (see Figure 33.4). You make a mirror image of the first part of the chain conformation, from A to its first crossing point, C on the surface, by reflecting the chain origin from $+z_A$ to $-z_A$ through the surface. Now count the number of these image chains $(A'CB)$ that begin at $-z_A$ and end at z, and call this quantity $w_{\text{image}}(-z_A, z)$. The reflectance principle says that

$$
w_{\text{nonviable}}(z_A, z) = w_{\text{image}}(-z_A, z).
\tag{33.8}
$$

This is useful because $w_{\text{image}}(-z_A, z)$ is very simple to compute. It is just the count of all the Gaussian random flights that begin at $-z_A$ and end at z. For this, use Equation (33.6) with z_A replaced by by $-z_A$:

$$
w_{\text{image}}(-z_A, z) = \left(\frac{\beta}{\pi}\right)^{1/2} \exp\left(-\beta(z + z_A)^2\right).
\tag{33.9}
$$

What is the justification for Equations (33.8) and (33.9)? The quantity $w_{\text{image}}(-z_A, z)$ has two remarkable properties. (1) For every chain from A to B that crosses the plane at least once, its mirror image is counted exactly once in $w_{\text{image}}(-z_A, z)$. From the crossing point C to B the chain and its image have

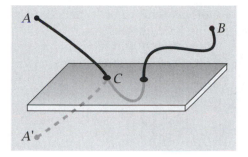

Figure 33.4 A model for computing the loss of entropy that causes polymer repulsions from surfaces. Conformation ACB is not viable because it passes through the plane. It starts at z_A and ends at z_B. To count all the nonviable conformations, create an image of each one, $A'CB$, starting at $-z_A$ and ending at z_B.

exactly the same trajectory, and from A to C the reflected conformation is the perfect image of the crossing conformation, so $w_{\text{image}}(-z_A, z)$ counts each crossing conformation once. (2) $w_{\text{image}}(-z_A, z)$ does not count any extraneous (i.e., noncrossing) conformations because any chain from A' to B *must* cross the plane.

So the reflectance principle gives you a simple way to compute the fraction of all chain conformations that cross the plane. Substituting Equations (33.6), (33.8), and (33.9) into Equation (33.7) gives

$$w_{\text{viable}}(z_A, z) = w_{\text{total}}(z_A, z) - w_{\text{image}}(-z_A, z)$$

$$= \left(\frac{\beta}{\pi}\right)^{1/2} \left[\exp\left(-\beta(z - z_A)^2\right) - \exp\left(-\beta(z + z_A)^2\right)\right]. \quad (33.10)$$

This expression can be simplified further for chains that are tethered to the surface, where z_A is small enough that it represents only a single bond to the surface, $z_A = (b/2) \ll Nb^2$. Then the difference in exponential functions in Equation (33.10) can be approximated as a derivative,

$$\exp\left(-\beta(z - (b/2))^2\right) - \exp\left(-\beta(z + (b/2))^2\right)$$
$$= f(z - (b/2)) - f(z + (b/2))$$
$$= -b\frac{df}{dz} \approx 2\beta b z \exp\left(-\beta z^2\right), \quad (33.11)$$

where f is the function $f(x) = \exp(-\beta x^2)$. Substituting Equation (33.11) into Equation (33.10) gives the distribution of the ends of chains that are tethered to an impenetrable surface:

$$w_{\text{viable, tethered}}(z) = \frac{3z}{Nb}\left(\frac{\beta}{\pi}\right)^{3/2} \exp(-\beta z^2). \quad (33.12)$$

The full three-dimensional distribution function for chains that are tethered to an impenetrable surface and that have their free ends at (x, y, z) can be obtained by multiplying by the distributions for the x and y components,

$$w_{\text{viable}}(x, y, z) = \frac{3z}{Nb}\left(\frac{\beta}{\pi}\right)^{3/2} \exp\left(-\beta(x^2 + y^2 + z^2)\right). \quad (33.13)$$

Tethered chains behave very differently than free chains: compare the front factor of z in Equation (33.13) with the front factor in the free-chain Equations (32.15) and (32.18). Free chains end approximately where they begin,

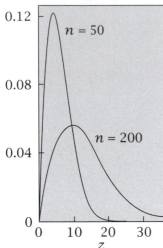

Fraction of Segments at z

Figure 33.5 Computed distribution of the free ends of chains that are tethered by one end to a surface. The free ends tend to be distributed away from the surface. When the chain length increases four-fold, the peak (most probable location of the termini) increases two-fold, supporting the predicted $N^{1/2}$ relationship. Source: EA DiMarzio, Chapter 4, in IC Sanchez, *Physics of Polymer Surfaces and Interfaces*, Butterworth-Heinemann, Stoneham, 1992.

$\langle z \rangle = 0$. In contrast, the termini of tethered chains tend to be 'repelled' from the surface (see Figure 33.5), $\langle z \rangle \neq 0$, as Example 33.2 shows.

EXAMPLE 33.2 The thickness of a layer of tethered polymer chains. You can compute the average thickness $\langle z \rangle$ of a tethered polymer layer by using the distribution function Equation (33.12),

$$\langle z \rangle = \frac{\int_0^\infty z w_{\text{viable}}(z)\,dz}{\int_0^\infty w_{\text{viable}}(z)\,dz} = \frac{\int_0^\infty z^2 \exp(-\beta z^2)\,dz}{\int_0^\infty z \exp(-\beta z^2)\,dz}$$

$$= \left(\frac{1}{4\beta} \left(\frac{\pi}{\beta} \right)^{1/2} \right) \left(\frac{1}{2\beta} \right)^{-1} = \frac{1}{2} \left(\frac{\pi}{\beta} \right)^{1/2} = b \left(\frac{\pi}{6} \right)^{1/2} N^{1/2}. \quad (33.14)$$

The mean thickness of the tethered polymer layer increases with the square root of the chain length (see Figure 33.5).

Excluded volume also has a role when polymers are confined within the pores of chromatography columns, when oil is trapped underground in the microscopic cavities in rocks, and when biopolymers are confined within small spaces. The next section describes a general way to compute chain conformations that are subject to a variety of different constraints.

Polymer Conformations Can Be Described by the Diffusion Equation

A powerful insight into polymer chain conformations is an analogy with the diffusion equation. When polymers are subject to complex geometric constraints, the conformational free energy can sometimes be found as a solution of a diffusion equation.

Imagine growing a chain on a simple cubic lattice, one bond at a time. The chain has its first monomer at $(0,0,0)$. The distance between lattice site centers is b, the bond length. Let $p(N+1,x,y,z)$ represent the probability that the chain has its monomer $N+1$ at lattice site (x,y,z). The probability distribution for monomer $N+1$ is related to the probability that a chain that has monomer N at a neighboring site:

$$p(N+1,x,y,z) = \frac{1}{6}\Big[p(N,x-b,y,z) + p(N,x+b,y,z)$$
$$+ p(N,x,y-b,z) + p(N,x,y+b,z)$$
$$+ p(N,x,y,z-b) + p(N,x,y,z+b) \Big], \quad (33.15)$$

where the factor of $1/6$ represents the probability that the chain steps in the one right direction (out of six, on a cubic lattice) from the site at which monomer N is located.

Now express the difference Equation (33.15) as a differential equation. You have two kinds of partial derivatives here: one is with respect to N, for the addition of each monomer, and the other is with respect to the spatial coordinates

x, y, and z. For the first type of derivative, the Taylor series gives

$$p(N+1,x,y,z) = p(N,x,y,z) + \left(\frac{\partial p}{\partial N}\right)_{x,y,z} \Delta N + \cdots$$

$$\approx p(N,x,y,z) + \left(\frac{\partial p}{\partial N}\right)_{x,y,z}, \tag{33.16}$$

since $\Delta N = 1$. For the second type of derivative, the two x-coordinate Taylor-series terms are:

$$p(N,x-b,y,z) = p(N,x,y,z)$$

$$- \left(\frac{\partial p}{\partial x}\right)b + \frac{1}{2}\left(\frac{\partial^2 p}{\partial x^2}\right)b^2 + \cdots \qquad \text{and} \tag{33.17}$$

$$p(N,x+b,y,z) = p(N,x,y,z)$$

$$+ \left(\frac{\partial p}{\partial x}\right)b + \frac{1}{2}\left(\frac{\partial^2 p}{\partial x^2}\right)b^2 + \cdots. \tag{33.18}$$

Adding Equations (33.17) and (33.18) gives

$$p(N,x-b,y,z) + p(N,x+b,y,z)$$

$$= 2p(N,x,y,z) + \left(\frac{\partial^2 p}{\partial x^2}\right)b^2 + \cdots. \tag{33.19}$$

To express Equation (33.19) in terms of a single differential equation, compute the corresponding terms in y and z, and insert all six terms into the right-hand side of Equation (33.15). Then insert Equation (33.16) into the left-hand side of Equation (33.15) to get

$$\left(\frac{\partial p}{\partial N}\right) = \frac{b^2}{6}\left(\frac{\partial^2 p}{\partial x^2} + \frac{\partial^2 p}{\partial y^2} + \frac{\partial^2 p}{\partial z^2}\right)$$

$$= \frac{b^2}{6}\nabla^2 p. \tag{33.20}$$

Equation (33.20) is identical to the diffusion Equation (18.10) but the polymer chain length N replaces the diffusion time t, and the site probability p for monomer N replaces the concentration c of diffusing particles at time t. Example 18.2 shows that the solution to the diffusion equation from a point source is a Gaussian distribution function. Similarly, the Gaussian function in Equation (32.19) is a solution to Equation (33.20) for unconstrained chains that begin at the origin $(0,0,0)$. For the diffusion equation, the width of the concentration distribution is given by $\langle x^2 \rangle = 2Dt$, while for polymers, the width in one dimension is given by $\langle x^2 \rangle = Nb^2/3$. For diffusion from a point source at $(0,0,0)$, the width of the spatial distribution of the particles increases with the square root of the time that the particles diffuse. For polymers that begin at $(0,0,0)$, the width of the spatial distribution of the chain ends increases with the square root of the chain length.

A polymer chain tends to avoid partitioning into small confined spaces, because doing so would diminish its conformational entropy and increase its free energy. For example, chain molecules avoid entering pores in chromatog-

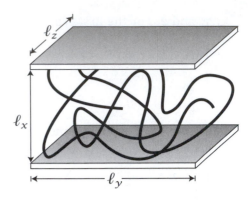

Figure 33.6 A polymer is confined in a box of dimensions ℓ_x, ℓ_y, and ℓ_z. If the box dimensions are small relative to the chain radius, there is a cost in confinement entropy owing to the loss of conformations that the chain would have had if it were in a larger space.

raphy columns, even when the pores are several fold bigger than the radius of gyration of the chain. You can compute this confinement free energy from the diffusion equation analogy.

Polymers Tend to Avoid Confined Spaces

Consider a chain molecule confined within a box of dimensions $\ell_x = \ell$, ℓ_y, and ℓ_z (see Figure 33.6). To count the conformations that are fully contained within the box, you can solve Equation (33.20), subject to the boundary conditions. At impenetrable walls, the concentration of chain ends is zero. This gives six boundary conditions for a box, $p(N, x = 0) = 0$, $p(N, x = \ell_x) = 0$, $p(N, y = 0) = 0$, $p(N, y = \ell_y) = 0$, $p(N, z = 0) = 0$, $p(N, z = \ell_z) = 0$. These are called *absorbing conditions*. Solving Equation (33.20) subject to absorbing boundary conditions [2, 3, 4] gives $w(x_1, x_2, N)$, the x-axis distribution of chains that begin at point x_1 and end at x_2 after N steps,

$$w(x_1, x_2, N) = \frac{2}{\ell} \sum_{j=1}^{\infty} \sin\left(\frac{j\pi x_1}{\ell}\right) \sin\left(\frac{j\pi x_2}{\ell}\right) e^{-(j^2\pi^2 N b^2)/(6\ell^2)}. \qquad (33.21)$$

The same distribution function applies to y and z, because the dimensionalities are independent in the random-flight model.

To get the full x-direction partition function q_x, you must integrate the distribution over all possible beginnings and endings of the chain,

$$q_x = \int_0^\ell dx_1 \int_0^\ell dx_2 \, w(x_1, x_2, N). \qquad (33.22)$$

For each sine function, integration gives

$$\int_0^\ell \sin\left(\frac{j\pi x}{\ell}\right) dx = \begin{cases} \dfrac{2\ell}{j\pi} & \text{for } j = 1, 3, 5, \ldots \\ 0 & \text{for } j = 2, 4, 6, \ldots \end{cases}.$$

Therefore

$$q_x = \frac{8\ell}{\pi^2} \sum_{j=1,3,5,\ldots}^{\infty} \left(\frac{1}{j^2}\right) e^{-(\pi^2 N b^2 j^2)/(6\ell^2)},$$

which is a sum of Gaussian functions. When the box is smaller than the polymer, $\ell^2 \ll Nb^2$, the first term ($j = 1$) dominates this sum and

$$q_x = \frac{8\ell}{\pi^2} e^{-(\pi^2 Nb^2)/(6\ell^2)}. \tag{33.23}$$

If the box is small in all directions, then q_y and q_z are identical to Equation (33.23) but with ℓ_y and ℓ_z replacing $\ell = \ell_x$.

Consider a simple case: a chain confined between infinite plates, where $\ell_x = \ell$ is small but $\ell_y, \ell_z \gg Nb^2$. Then $q_y \approx \ell_y$ and $q_z \approx \ell_z$. In this case, the full partition function is

$$
\begin{aligned}
q &= q_x q_y q_z \\
&= \frac{8V}{\pi^2} e^{-(\pi^2 Nb^2)/(6\ell^2)},
\end{aligned} \tag{33.24}
$$

since $V = \ell_x \ell_y \ell_z$ is the volume of the box. The confinement free energy is

$$\frac{F}{kT} = -\frac{S}{k} = -\ln q = -\ln\left(\frac{8V}{\pi^2}\right) + \frac{\pi^2 Nb^2}{6\ell^2}. \tag{33.25}$$

As the confining box shrinks (ℓ becomes smaller), the conformational free energy increases. The logarithm of the partition coefficient for a polymer entering a confined space is given in terms of the dominant term, $\ln K = -\Delta F/kT \propto -Nb^2/\ell^2$ (see Figure 33.7). Equation (33.25) shows that polymers tend not to partition into small spaces.

Now let's consider a polymer molecule adsorbed on a surface. Adsorption is a form of confinement.

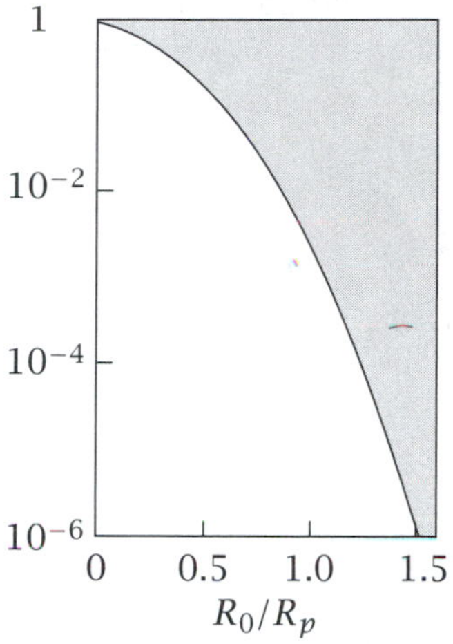

The Thickness of a Polymer Adsorbed on a Surface Is a Balance Between Adsorption Energy and Confinement Entropy

Let's compute the thickness D of a polymer molecule that is adsorbed on a surface (see Figure 33.8). A simple dimensional analysis gives the dependences on the variables that characterize a chain molecule. The thickness of the polymer is determined by a balance between an adsorption energy, which tends to flatten the chain onto the surface, decreasing D, and an 'elastic' entropy that tends to expand the chain, increasing D. The confinement contribution to the free energy is derived from Equation (33.25):

$$\frac{F_{\text{elastic}}}{kT} \propto \frac{Nb^2}{D^2} \tag{33.26}$$

(leaving out constants). The adsorption free energy is the sticking energy per monomer, $\varepsilon > 0$, multiplied by the total number of monomers in the chain N, multiplied by the fraction b/D of the chain's monomers that are in a layer within one bond distance b from the surface, assuming that the monomers are uniformly distributed:

$$\frac{F_{\text{ads}}}{kT} \propto \left(\frac{-N\varepsilon b}{D}\right). \tag{33.27}$$

Partition Coefficient

Figure 33.7 The partitioning of a chain molecule decreases as the size of the box diminishes. R_0 is the radius of gyration and R_p is the pore radius. Source: MG Davidson, VW Suter and WM Deen, *Macromolecules*, **20**, 1141 (1987).

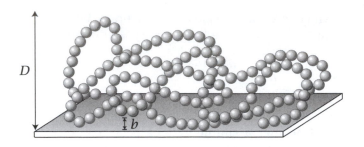

Figure 33.8 The average 'thickness' of a polymer molecule adsorbed on a surface is D. The monomer diameter is b, which defines the thickness of the first layer.

The total free energy F_{total} is the sum of the elastic and adsorption free energies from Equations (33.26) and (33.27):

$$\frac{F_{total}}{kT} = \frac{F_{elastic}}{kT} + \frac{F_{ads}}{kT} = \frac{Nb^2}{D^2} - \frac{N\varepsilon b}{DkT}.$$

The equilibrium thickness D_0 is the value of D that minimizes the total free energy:

$$\frac{\partial}{\partial D}\left(\frac{F_{total}}{kT}\right)_{D_0} = 0$$

$$\implies -\frac{2Nb^2}{D_0^3} + \frac{N\varepsilon b}{D_0^2 kT} = 0$$

$$\implies D_0 = \frac{2bkT}{\varepsilon}.$$

This simple dimensional analysis predicts that the equilibrium thickness gets smaller as the sticking energy increases, does not depend on the chain length, and increases with temperature.

Confinement and elastic entropies are also important for the dynamical relaxation processes of polymeric liquids and solids. For processing polymers and plastics it is often useful to know their relaxation times.

The Rouse–Zimm Model Describes the Dynamics of Viscoelastic Fluids

Some liquids, such as raw eggs, are stretchy and elastic. If you apply a shear force to a *viscoelastic fluid*, the fluid dissipates energy through viscous forces, as simpler liquids do, but it can also store energy through elastic forces. Liquids are often elastic because of dissolved polymers.

To explore viscoelasticity in liquids, consider the simple shear flow experiment shown in Figure 33.9. Begin the experiment by applying a shear force to the solution until it reaches steady state. At that time, the total force f_m that is being applied to the top plate (divided by the number of molecules) equals the sum of the viscous and elastic forces exerted by the molecule,

$$f_{viscous} + f_{elastic} = f_m. \tag{33.28}$$

After the applied force has been turned off, you have

$$f_{viscous} + f_{elastic} = 0. \tag{33.29}$$

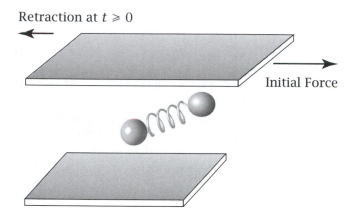

Retraction at $t \geqslant 0$

Initial Force

Figure 33.9 The *Rouse–Zimm model* for viscoelasticity. Applying a shear flow to a polymer solution stretches out the chains, indicated here in terms of a beads-and-spring model, leading to an elastic entropic retractive force.

After the applied force has been turned off, the elastic energy stored in the polymeric fluid pulls the top plate backward (see Figure 33.9) to relax toward equilibrium.

The balance of these forces for dilute solutions of polymer molecules in a small-molecule solvent was first described by PE Rouse [5] and BH Zimm [6]. Here is a simple version of their models. Assume that the polymer molecule has been deformed by the applied shear to have an end-to-end distance $x = x_0$. The elastic retractive force per molecule in a θ-solvent is given by Equation (32.17), $f_{\text{elastic}} = -3kTx/R_e^2$, in terms of the mean-square end-to-end length $R_e^2 = Nb^2 = 3R_0^2/2$. The viscous force exerted per polymer chain on the surrounding medium is $f_{\text{viscous}} = \xi v$, where ξ is the friction coefficient (see Equation (18.41)) and $v = -dx/dt$ is the velocity of retraction in the direction that opposes the original applied force. You can express Equation (33.29) as a differential equation:

$$\xi \frac{dx}{dt} + \frac{3kT}{R_e^2} x = 0 \quad \Longrightarrow \quad \frac{dx}{dt} + \frac{1}{\tau} x = 0, \tag{33.30}$$

where the relaxation time τ is

$$\tau = \frac{\xi R_e^2}{kT}. \tag{33.31}$$

We leave out small numerical factors such as 3 for this dimensional analysis. As another simplification, here we consider only the slowest relaxation time (the Rouse and Zimm theories treat all the relaxation modes of polymers). The solution to Equation (33.30) is

$$x(t) = x_0 \exp(-t/\tau). \tag{33.32}$$

The displacement of the top plate changes exponentially with time as $x \to 0$ after the applied force is turned off.

How does the relaxation time τ depend on the chain length? In the *Rouse model*, the friction coefficient ξ is assumed to be the sum of frictional contributions, ξ_1, from each monomer unit

$$\xi = N\xi_1, \tag{33.33}$$

where ξ_1 is a constant. Substituting $R_e^2 = Nb^2$ and Equation (33.33) into Equa-

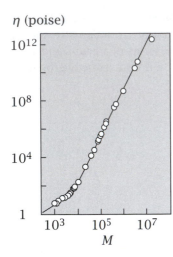

η (poise)

Figure 33.10 The dependence of the viscosity η on the molecular weights M of polybutadiene in the melt: $\propto N$ for small molecular weights, $\propto N^{3.4}$ for large molecular weights. The viscosity and relaxation time are linearly related. Source: RH Colby, LJ Fetters, and WW Graessley, *Macromolecules*, **20**, 2226–2237 (1987).

Figure 33.11 A polymer reptates by moving inside a 'tube,' defined by its neighboring chains.

tion (33.31) gives the Rouse-model relaxation time τ,

$$\tau \propto \frac{\xi_1 b^2}{kT} N^2. \tag{33.34}$$

The Rouse model predicts that longer polymer chains have slower relaxations: the relaxation time increases as N^2.

The *Zimm model* treats the friction differently. In the Zimm model, the interior monomers are shielded (or *screened*) from the flow, so not all the monomers contribute to the friction. This is called *hydrodynamic screening*. In this case, the friction is treated as if the whole polymer molecule were a Stokes–Einstein sphere,

$$\xi = 6\pi\eta_0 R_e \propto \eta_0 b N^{1/2}. \tag{33.35}$$

The Zimm model, Equation (33.35), predicts that the overall friction only increases as the square root of the molecular weight. To get the relaxation time, substitute $R_e^2 = Nb^2$ and Equation (33.35) into Equation (33.31) to get

$$\tau \propto \frac{\eta_0 b^3}{kT} N^{3/2}. \tag{33.36}$$

The Zimm model predicts $\tau \propto N^{3/2}$, in contrast to the Rouse model prediction, $\tau \propto N^2$. Experiments have confirmed the Zimm-model prediction for dilute polymers in small-molecule solvents. These models also predict that the dynamics is slower in more viscous solvents or at lower temperatures.

Of great industrial importance for the processing of plastics, textile fibers, and paints are *highly concentrated* polymer solutions, where there is little or no solvent, in which the molecular organization resembles tangled spaghetti. Such concentrated solutions are called *melts*. For short-chain melts, Rouse dynamics often applies. For long-chain melts, however, the chain dynamics follows a different physical process, called *reptation*.

Reptation Describes the Snakelike Motions of Chains in Melts and Gels

Experiments show that chain dynamics is much slower in polymer melts ($\tau \propto N^{3.4}$) than would be predicted from the Rouse or Zimm theories for dilute solutions (see Figure 33.10). The *reptation model* of PG de Gennes, a French physicist who won the 1991 Nobel prize in Physics for his work on polymers and other complex liquids, explains the dominant dynamics in polymer melts in terms of chains snaking through a medium of surrounding chains. Think about one chain molecule A in a melt. Its surrounding chains define a sort of *tube* within which A can move (see Figure 33.11). In the reptation model, the tube remains in place during the time τ that it takes for the molecule A to slither out of the end of the tube.

The reptation model, like the Rouse model, supposes that the friction involved in dragging the chain through its tube is proportional to the chain length, $\xi = N\xi_1$, Equation (33.33). The diffusion constant D_{tube} for the chain moving through the tube is given by the Einstein–Smoluchowski relation, Equa-

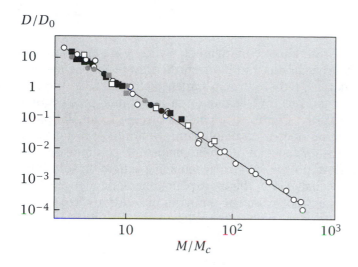

D/D_0

Figure 33.12 Self-diffusion coefficient D of seven different types of polymers (polystyrene, poly(ethyleneoxide), polyisoprene, polybutadiene, poly(dimethylsiloxane), poly(ethylene oxide), poly(methylstyrene)) versus molecular weight shows that $D \propto N^{-2.28 \pm 0.05}$, which differs somewhat from the N^{-2} dependence expected from simple reptation theory. D_0 is the diffusion coefficient for $M/M_c = 10$. Source: H Tao, TP Lodge and ED von Meerwaal, *Macromolecules*, **33**, 1747–1758 (2000).

tion (18.49):

$$D_{\text{tube}} = \frac{kT}{\xi} = \frac{kT}{N\xi_1} = \frac{D_1}{N}, \tag{33.37}$$

where D_1 is a constant with units of a diffusion coefficient. To compute the time required for the chain to escape from the tube, note that the length of the confining tube is approximately the contour length of the chain, $L = Nb$, and combine Equation (33.37) with diffusion Equation (18.53), to get

$$\tau \propto \frac{\langle x^2 \rangle}{D_{\text{tube}}} \propto \frac{L^2}{D_{\text{tube}}} \propto \frac{NL^2}{D_1} \propto N^3. \tag{33.38}$$

Experiments give $\tau \propto N^{3.4}$ (see Figure 33.10), which is slightly steeper than this prediction $\tau \propto N^3$ of the reptation model. There is some evidence that this discrepency is due to the finite lengths of reptation tubes. This model accounts in a simple way for the steep chain-length dependence of the relaxation times of polymers in bulk melts.

The reptation model also predicts a second type of diffusion constant. The diffusion constant D_{tube} cannot be readily measured because experiments cannot track how the chain moves along its tube axis. But you can measure a diffusion constant D that describes how the center of mass of the polymer chain moves in space over time. In this case, the average distance moved is $\langle x^2 \rangle \propto R_e^2$, where $R_e^2 = Nb^2$. The reptation model Equation (33.38) predicts that chain diffusion slows as the square of the chain length,

$$D \propto \frac{R_e^2}{\tau} \propto \frac{N}{N^3} \propto N^{-2}. \tag{33.39}$$

Experiments give $D \propto N^{-2.28}$, which differs somewhat from this reptation model prediction of $D \propto N^{-2}$ (see Figure 33.12). In contrast, the self-diffusion constant for a Rouse chain, which is based on $\tau \propto N^2$ (Equation (33.34)), is

$$D \propto \frac{R_e^2}{\tau} \propto \frac{N}{N^2} \propto N^{-1}.$$

Summary

Chain molecules resist deformation or confinement, owing to elasticity and excluded volume. Chain conformations can be perturbed by the presence of a surface. For polymers that are tethered to an impenetrable surface, the free ends tend not to be at the surface but terminate at a distance proportional to $N^{1/2}$ away from the surface. Confining a chain molecule to small spaces leads to a reduction in the number of available conformations. So there is a free energy that opposes the entrance of polymer chains into confined spaces. After stretching in shear flow, polymers retract, causing elasticity in solutions, called viscoelasticity. The Rouse–Zimm theory explains the dependence of the relaxation time on the polymer molecular weight in dilute solutions. For longer chains at higher concentrations, or in gels, the relaxation and diffusional motions have been modelled in terms of snakelike reptation processes.

Problems

1. Polymer adsorption on a surface.

(a) Write an expression to show how many monomer-sized layers M there are if a polymer of length N sticks to a surface with sticking energy ε_0 per monomer.

(b) Compute M when $\varepsilon_0 = 1/5kT$.

2. Rouse–Zimm dynamics in a good solvent.
Derive the dependence of the relaxation time τ on the chain length N, for a dilute polymer solution in a good solvent.

3. A tethered chain returns to the surface.
A random-flight chain of length N is tethered at one end to an impenetrable surface. Write an expression for the probability $P(N)$ that the free end also happens to land on the surface.

4. c^*, the polymer overlap concentration.
Consider a solution of polymer molecules of length N. At concentrations c below the *overlap concentration* c^* a polymer molecule seldom collides with other polymers. $c = c^*$ is the point at which the concentration of monomer units throughout the solution equals the concentration of monomers inside the circumscribed volume of the polymer chain.

(a) Compute $c^*(N)$, i.e., the dependence of c^* on the chain length N, for random-flight chains (that is, for polymers in a θ-solvent).

(b) Compute $c^*(N)$ for chains in a good solvent.

5. Relaxation times for DNA molecules.
The relaxation time is $\tau \approx 1$ sec for a dilute solution of T_2 DNA molecules having a molecular weight of 10^8 daltons. What is the relaxation time for *Drosophila* (fruit fly) chromosomal DNA having a molecular weight of 20×10^9 daltons?

6. The free energy of localizing DNA in a cell nucleus.
Human chromosomal DNA molecules have about 3×10^9 nucleotides, each separated from its neighbor by a bond distance of about $b = 3.4$ Å.

(a) Compute the contour length, L, if this were all a single DNA molecule.

(b) Compute its end-to-end length $R_e = (Nb^2)^{1/2}$ if this DNA were a single random-flight chain.

(c) If the cell's nucleus has radius $r_c = 4\,\mu$m, compute R_e/r_c from (b).

(d) Compute the excluded-volume free-energy cost of localizing this 'virtual' single molecule into the human cell nucleus at $T = 300$ K.

7. Distribution of the ends of tethered chains.
Derive $\langle z^2 \rangle$ for the termini of chains that are tethered to an impenetrable surface.

References

[1] EA DiMarzio and FL McCrackin. *J Chem Phys* **43**, 539 (1965).

[2] EF Cassasa. *J Polymer Sci* **B5**, 773 (1967).

[3] M Doi and SF Edwards. *The Theory of Polymer Dynamics*, Oxford University Press, New York, 1986.

[4] MG Davidson, UW Suter and WM Deen. *Macromolecules* **20**, 1141 (1987).

[5] PE Rouse. *J Chem Phys* **21**, 1272 (1953).

[6] BH Zimm. *J Chem Phys* **24**, 269 (1956).

Suggested Reading

Excellent fundamental texts on excluded volume and conformational changes in polymers:

PG de Gennes, *Scaling Concepts in Polymer Physics*, Cornell University Press, Ithaca, 1979.

M Doi, *Introduction to Polymer Physics*, Oxford University Press, New York, 1996.

Good general texts on polymer theory are:

RH Boyd and PJ Phillips, *The Science of Polymer Molecules: An Introduction Concerning the Synthesis and Properties of the Individual Molecules that Constitute Polymeric Materials*, Cambridge University Press, New York, 1993.

EG Richards, *An Introduction to the Physical Properties of Large Molecules in Solution*, Cambridge University Press, New York, 1980.

SF Sun, *Physical Chemistry of Macromolecules: Basic Principles and Issues*, Wiley, New York, 1994.

A Yu Grosberg and AR Khokhlov, *Giant Molecules*, Academic Press, San Diego, 1997.

E F Cassasa, *J Polymer Sci*, **B5**, 773 (1967).

General overview of polymer dynamics in dilute solutions:

J Ferry, *Viscoelastic Properties of Polymers*, Wiley, New York, 1961.

Appendix A   Table of Constants

Quantity	Symbol and equivalences	Value
Gas constant	R	$8.31447 \text{ J K}^{-1} \text{ mol}^{-1}$
		$8.31447 \times 10^7 \text{ erg K}^{-1} \text{ mol}^{-1}$
		$1.98717 \text{ cal K}^{-1} \text{ mol}^{-1}$
		$8.20575 \times 10^{-5} \text{ m}^3 \text{ atm K}^{-1} \text{ mol}^{-1}$
		$8.20575 \times 10^{-2} \ \ell \text{ atm K}^{-1} \text{ mol}^{-1}$
		$82.0575 \text{ cm}^3 \text{ atm K}^{-1} \text{ mol}^{-1}$
Boltzmann's constant	$k = R/\mathcal{N}$	$1.380650 \times 10^{-23} \text{ JK}^{-1}$
Avogadro's constant	$\mathcal{N}$	$6.0221367 \times 10^{23} \text{ mol}^{-1}$
Planck's constant	h	$6.626176 \times 10^{-34} \text{ J s}$
Speed of light (vacuum)	c	$2.99792458 \times 10^8 \text{ m s}^{-1}$ (exact)
Standard acceleration of free fall	g	9.80665 m s^{-2} (exact)
Permittivity of vacuum	ε_0	$8.85418782 \times 10^{-12} \text{ F m}^{-1}$
Elementary unit charge	e	$1.6021892 \times 10^{-19} \text{ C}$
Atomic mass unit	u	$1.6605387 \times 10^{-27} \text{ kg}$
Mass of proton	m_p	$1.67262158 \times 10^{-27} \text{ kg}$
Rest mass of electron	m_e	$9.109534 \times 10^{-31} \text{ kg}$
Faraday's constant	$F = \mathcal{N}e$	$9.648534 \times 10^4 \text{ C mol}^{-1}$

Appendix B Table of Units

Physical quantity	Unit name	Unit symbol	Definition
force	newton	N	$J\,m^{-1} = kg\,m\,s^{-2} = 10^5\,dyn$
	dyne	dyn	$g\,cm\,s^{-2}$
pressure	pascal	Pa	$kg\,m^{-1}\,s^{-2} = N\,m^{-2}$ $= 9.87 \times 10^{-6}\,atm$
	atmosphere	atm	$1.01325 \times 10^6\,g\,cm^{-1}\,s^{-2}$ $= 1.01325 \times 10^5\,Pa$
energy	joule	J	$kg\,m^2\,s^{-2} = Nm = 10^7\,erg$
	erg	erg	$g\,cm^2\,s^{-2} = 10^{-7}\,J$
	electron volt	eV	$1.602177 \times 10^{-19}\,J$ $= 23.0605\,kcal/mol$
	calorie	cal	$4.184\,J = 4.184 \times 10^7\,erg$
electric charge	coulomb	C	$s\,A = J\,V^{-1}$
	esu	esu	$3.00 \times 10^9\,C$
electric potential	volt	V	$kg\,m^2\,s^{-3}\,A^{-1} = J\,A^{-1}\,s^{-1}$ $= J\,C^{-1}$
electric current	ampere	A	$C\,s^{-1}$
capacitance	farad	F	$m^{-2}\,kg^{-1}\,s^4\,A^2 = A\,s\,V^{-1}$ $= C\,V^{-1}$
frequency	hertz	Hz	s^{-1}
length	angstrom	Å	$10^{-10}\,m = 10^{-8}\,cm$
viscosity	poise	p	$g\,cm^{-1}\,s^{-1} = dyn\,s\,cm^{-2}$
power	watt	w	$J\,s^{-1} = kg\,m^2\,s^{-3} = 10^7\,erg\,s^{-1}$
		Useful quantities	
length	Bjerrum	l_B	560 Å in vacuum
	length		7.13 Å in water at 25 °C

m = meter
s = sec
g = gram

Appendix C   Useful Taylor Series Expansions

$$e^{ax} = 1 + ax + \frac{(ax)^2}{2!} + \frac{(ax)^3}{3!} + \cdots$$

$$= \sum_{k=0}^{\infty} \frac{(ax)^k}{k!} \qquad \text{for all } x. \tag{C.1}$$

$$\sin ax = ax - \frac{(ax)^3}{3!} + \frac{(ax)^5}{5!} - \frac{(ax)^7}{7!} \cdots$$

$$= \sum_{k=0}^{\infty} \frac{(-1)^k (ax)^{2k+1}}{(2k+1)!} \qquad \text{for all } x. \tag{C.2}$$

$$\cos ax = 1 - \frac{(ax)^2}{2!} + \frac{(ax)^4}{4!} - \frac{(ax)^6}{6!} \cdots$$

$$= \sum_{k=0}^{\infty} \frac{(-1)^k (ax)^{2k}}{(2k)!} \qquad \text{for all } x. \tag{C.3}$$

$$\ln(a + x) = \ln a \left(1 + \frac{x}{a}\right) = \ln(a) + \frac{x}{a} - \frac{x^2}{2a^2} + \frac{x^3}{3a^3} - \cdots + \cdots$$

$$= \ln(a) + \sum_{k=1}^{\infty} \frac{(-1)^{k-1} x^k}{k a^k} \qquad \text{for all } a > 0$$

$$\text{and } -a < x \le a. \tag{C.4}$$

$$\tan^{-1} x = x - \frac{x^3}{3} + \frac{x^5}{5} - \frac{x^7}{7} + \cdots$$

$$= \sum_{k=0}^{\infty} \frac{(-1)^k x^{2k+1}}{2k+1} \qquad \text{for } |x| \le 1. \tag{C.5}$$

$$(1 + x)^p = 1 + px + \frac{p(p-1)}{2!} x^2 + \frac{p(p-1)(p-2)}{3!} x^3 + \cdots$$

$$\text{for all } p \text{ and } |x| < 1. \tag{C.6}$$

Equation (C.6) is the binomial series. When p is a positive integer all the terms after the $(p + 1)$ equal zero, and the series becomes finite:

$$(1 + x)^p = 1 + px + \frac{p(p-1)}{2!} x^2 + \frac{p(p-1)(p-2)}{3!} x^3 + \cdots + x^p$$

$$= \sum_{k=0}^{p} \binom{p}{k} x^k \qquad \text{for all } x.$$

Appendix D   Useful Integrals

$$\int_0^\infty e^{-ax^2}\, dx = \frac{1}{2}\sqrt{\frac{\pi}{a}} \tag{D.1}$$

$$\int_0^\infty xe^{-ax^2}\, dx = \frac{1}{2a} \tag{D.2}$$

$$\int_0^\infty x^2 e^{-ax^2}\, dx = \frac{1}{4a}\sqrt{\frac{\pi}{a}} \tag{D.3}$$

$$\int \sin ax \cos ax\, dx = \frac{\sin^2 ax}{2a} \tag{D.4}$$

$$\int \sin^n ax \cos ax\, dx = \frac{\sin^{n+1} ax}{(n+1)a} \qquad n \neq -1 \tag{D.5}$$

$$\int \cos^n ax \sin ax\, dx = -\frac{\cos^{n+1} ax}{(n+1)a} \qquad n \neq -1 \tag{D.6}$$

$$\int \sin^2 ax\, dx = \frac{x}{2} - \frac{\sin 2ax}{4a} \tag{D.7}$$

$$\int \cos^2 ax\, dx = \frac{x}{2} + \frac{\sin 2ax}{4a} \tag{D.8}$$

$$\int_0^\infty x^n e^{-ax}\, dx = \frac{\Gamma(n+1)}{a^{n+1}} \tag{D.9}$$

$$\int_0^\infty x^m e^{-ax^2}\, dx = \frac{\Gamma[(m+1)/2]}{2a^{(m+1)/2}} \tag{D.10}$$

$$\int e^{ax}\, dx = \frac{1}{a}e^{ax} \tag{D.11}$$

Appendix E   Multiples of Units, Their Names, and Symbols

Multiple	Prefix	Symbol
10^{18}	exa	E
10^{15}	peta	P
10^{12}	tera	T
10^{9}	giga	G
10^{6}	mega	M
10^{3}	kilo	k
10^{2}	hecto	h
10	deca	da
10^{-1}	deci	d
10^{-2}	centi	c
10^{-3}	milli	m
10^{-6}	micro	μ
10^{-9}	nano	n
10^{-12}	pico	p
10^{-15}	femto	f
10^{-18}	atto	a

Index